Stars as Laboratories
for Fundamental Physics

Georg G. Raffelt

Stars as Laboratories for Fundamental Physics

The Astrophysics of Neutrinos, Axions, and Other Weakly Interacting Particles

The University of Chicago Press
Chicago & London

Georg G. Raffelt is a staff researcher at the Max-Planck-Institut für Physik in Munich and a member of the Sonderforschungsbereich Astroteilchenphysik at the Technische Universität München. He is coeditor of the journal *Astroparticle Physics*. He received his doctorate from the University of California, Berkeley.

The University of Chicago Press, Chicago 60637
The University of Chicago Press, Ltd., London
© 1996 by The University of Chicago
All rights reserved. Published 1996
Printed in the United States of America
05 04 03 02 01 00 99 98 97 96   1 2 3 4 5

ISBN: 0-226-70271-5 (cloth)
       0-226-70272-3 (paper)

Library of Congress Cataloging-in-Publication Data

Raffelt, Georg G.
    Stars as laboratories for fundamental physics : the astrophysics of neutrinos, axions, and other weakly interacting particles / Georg G. Raffelt.
        p.   cm. — (Theoretical astrophysics)
    Includes bibliographical references and index.
    1. Neutrino astrophysics.   2. Particles (Nuclear physics)   I. Title.   II. Series.
QB464.2.R34   1996
523.8—dc20   95-39684
                                                                    CIP

Contents

3. Particles Interacting with Electrons and Baryons

4. Processes in a Nuclear Medium

Preface

Ever since Newton proposed that the moon on its orbit follows the same laws of motion as an apple falling from a tree, the heavens have been a favorite laboratory to test the fundamental laws of physics, notably classical mechanics and Newton's and Einstein's theories of gravity. This tradition carries on—the 1993 physics Nobel prize was awarded to R. A. Hulse and J. H. Taylor for their 1974 discovery of the binary pulsar PSR 1913+16 whose measured orbital decay they later used to identify gravitational wave emission. However, the scope of physical laws necessary to understand the phenomena observed in the super-lunar sphere has expanded far beyond these traditional fields. Today, astrophysics has become a vast playing ground for applications of the laws of *microscopic* physics, in particular the properties of elementary particles and their interactions.

This book is about how stars can be used as laboratories to probe fundamental interactions. Apart from a few arguments relating to gravitational physics and the nature of space and time (Is Newton's constant constant? Do all relativistic particles move with the same limiting velocity? Are there novel long-range interactions?), most of the discussion focusses on the properties and nongravitational interactions of elementary particles.

There are three predominant methods for the use of stars as particle-physics laboratories. First, stars are natural sources for photons and neutrinos which can be detected on Earth. Neutrinos are now routinely measured from the Sun, and have been measured once from a collapsing star (SN 1987A). Because these particles literally travel over astronomical distances before reaching the detector one can study modifications of the measured signal which can be attributed to propagation and dispersion effects, including neutrino flavor oscillations or axion-photon oscillations in intervening magnetic fields. It is well known that the discrepancy between the calculated and measured solar neutrino spectra is the most robust, yet preliminary current indication for neutrino oscillations and thus for nonvanishing neutrino masses.

Second, particles from distant sources may decay, and there may be photons or even measurable neutrinos among the decay products. The absence of solar x- and γ-rays yields a limit on neutrino radiative decays which is as "safe" as a laboratory limit, yet nine orders of magnitude more restrictive. An even more restrictive limit obtains from the absence of γ-rays in conjunction with the SN 1987A neutrinos which allows one to conclude, for example, that even ν_τ must obey the cosmological limit of $m_\nu \lesssim 30\,\text{eV}$ unless one invents new invisible decay channels.

Third, the emission of weakly interacting particles causes a direct energy-loss channel from the interior of stars. For neutrinos, this effect has been routinely included in stellar evolution calculations. If new low-mass elementary particles were to exist such as axions or other Nambu-Goldstone bosons, or if neutrinos had novel interactions with the stellar medium such as one mediated by a putative neutrino magnetic dipole moment, then stars might lose energy too fast. A comparison with the observed stellar properties allows one to derive restrictive limits on the operation of a new energy-loss or energy-transfer mechanism and thus to constrain the proposed novel particle interactions.

While these and related arguments as well their application and results are extensively covered here, I have not written on several topics that might be expected to be represented in a book on the connection between particle physics and stars. Neutron stars have been speculated to consist of quark matter so that in principle they are a laboratory to study a quark-gluon plasma. As I am not familiar enough with the literature on this interesting topic I refer the reader to the review by Alcock and Olinto (1988) as well as to the more recent proceedings of two topical conferences (Madsen and Haensel 1992; Vassiliadis et al. 1995).

I have also dodged some important issues in the three-way relationship between cosmology, stars, and particle physics. If axions are not the dark matter of the universe, it is likely filled with a "background sea" of hypothetical weakly interacting massive particles (WIMPs) such as the lightest supersymmetric particles. Moreover, there may be exotic particles left over from the hot early universe such as magnetic monopoles which are predicted to exist in the framework of typical grand unified theories (GUTs). Some of the monopoles or WIMPs would be captured and accumulate in the interior of stars. GUT monopoles are predicted to catalyze nucleon decay (Rubakov-Callan-effect), providing stars with a novel energy source. This possibility can be constrained by analogous methods to those presented here which limit anomalous energy losses. The resulting constraints on the presence of GUT monopoles in the universe have been reviewed, for example, in the cosmology book of Kolb and Turner (1990). Because nothing of substance has changed, a new review did not seem warranted.

WIMPs trapped in stars would contribute to the heat transfer because their mean free path can be so large that they may be orbiting almost freely in the star's gravitational potential well, with only occasional collisions with the background medium. Originally it was thought that this effect could reduce the central solar temperature enough to solve the solar neutrino problem, and to better an alleged discrepancy between observed and predicted solar p-mode frequencies. With the new solar neutrino data it has become clear, however, that a reduction of the central temperature alone cannot solve the problem. Worse, solving the "old solar neutrino problem" by the WIMP mechanism now seems to cause a discrepancy with the observed solar p-mode frequencies (Christensen-Dalsgaard 1992). In addition, a significant effect requires relatively large scattering cross sections and thus rather contrived particle-physics models. Very restrictive direct laboratory constraints exist for the presence of these "cosmions" in the galaxy. Given this status I was not motivated to review the topic in detail.

The annihilation of dark-matter WIMPs captured in the Sun or Earth produces high-energy neutrinos which are measurable in terrestrial detectors such as Kamiokande or the future Superkamiokande, NESTOR, DUMAND, and AMANDA Cherenkov detectors. This indirect approach to search for dark matter may well turn into a serious competitor for the new generation of direct laboratory search experiments that are currently being mounted. This material is extensively covered in a forthcoming review *Supersymmetric Dark Matter* by Jungman, Kamionkowski, and Griest (1995); there is no need for me to duplicate the effort of these experts.

The topics covered in my book revolve around the impact of low-mass or massless particles on stars or the direct detection of this radiation. The highest energies encountered are a few $100\,\mathrm{MeV}$ (in the interior of a SN core) which is extremely small on the high-energy scales of typical particle accelerator experiments. Therefore, stars as laboratories for fundamental physics help to push the low-energy frontier of particle physics and as such complement the efforts of nonaccelerator particle experiments. Their main thrust is directed at the search for nonstandard neutrino properties, but there are other fascinating topics which include the measurement of parity-violating phenomena in atoms, the search for neutron or electron electric dipole moments which would violate CP, the search for neutron-antineutron oscillations, the search for proton decay, or the search for particle dark matter in the galaxy by direct and indirect detection experiments (Rich, Lloyd Owen, and Spiro 1987). It is fascinating that the IMB and Kamiokande water Cherenkov detectors which had been built to search for proton decay ended up seeing supernova (SN) neutrinos instead. The Fréjus detector, instead of seeing proton decay, has set important limits on the oscillation of atmospheric neutrinos. Kamiokande has turned into a major solar neutrino observatory and dark-matter search experiment. The forthcoming Superkamiokande and SNO detectors will continue and expand these missions, may detect a future galactic SN, and may still find proton decay.

◇

All currently known phenomena of elementary particle physics are either perfectly well accounted for by its standard model, or are not explained at all. The former category relates to the electroweak and strong gauge interactions which have been spectacularly successful at describing microscopic processes up to the energies currently available at accelerators. On the other side are, for example, the mass spectrum of the fundamental fermions (quarks and leptons), the source for CP violation, or the relationship between the three families (why three?). There must be "physics beyond the standard model!"

In the standard model, neutrinos have been assigned the most minimal properties compatible with experimental data: zero mass, zero charge, zero dipole moments, zero decay rate, zero almost everything. Any deviation from this simple picture is a sensitive probe for physics beyond the standard model—thus the enthusiasm to search for neutrino masses and mixings, notably in oscillation experiments, but also for neutrino electromagnetic properties, decays, and other effects. In astrophysics, even "minimal neutrinos" play a major role for the energy loss of stars as they can escape unscathed from the interior once produced. Moreover, in spite of their weak interaction there are two astrophysical sites where they actually reach thermal equilibrium: the early universe up to about the nucleosynthesis epoch and in a SN core for a few seconds after collapse. Neutrinos thus play a dominant role in the cosmic and SN dynamical and thermal evolution—little wonder that these environments are important neutrino laboratories.

Nonstandard neutrino properties such as small Majorana masses or magnetic dipole moments would be low-energy manifestations of novel physics at short distances. Another spectacular interloper of high-energy physics in the low-energy world would be a Nambu-Goldstone boson of a new symmetry broken at some large energy scale. Such a particle would be massless or nearly massless. The most widely discussed example is the "invisible axion" which has been postulated as an explanation for the observed absence of a neutron electric dipole moment which in QCD ought to be about as large as its magnetic one. The axion also doubles as a particle candidate for the dark matter of the universe; two beautiful experiments to search for galactic axions are about to go on line in Livermore (California) and Kyoto (Japan). Because of their weak interactions, the role of axions and similar particles in stellar evolution is closely related to that of neutrinos.

◇

The complete or near masslessness of the particles studied here (neutrinos, photons, axions, etc.) opens up a rich phenomenology in its own right. One intriguing issue is the production and propagation of these objects in a hot and dense medium which modifies their dispersion relations in subtle but significant ways. One of the most important neutrino production processes in stars is the "photon decay" $\gamma \rightarrow \bar{\nu}\nu$ which is enabled by the modified photon dispersion relation in a medium, and by an effective neutrino-photon coupling mediated by the ambient electrons. Another example is the process of resonant neutrino oscillations (MSW effect) which is instrumental at explaining the measured solar neutrino spectrum, and may imply vast modifications of SN physics, notably the occurrence of r-process nucleosynthesis. The MSW effect depends on a flavor-birefringent term of the neutrino dispersion relation in a medium. In a SN, neutrinos themselves make an important contribution to the "background medium" so that their oscillations become a nonlinear phenomenon. In the deep interior of a SN core where neutrinos are trapped, a kinetic treatment of the interplay between oscillations and collisions requires a fascinating "nonabelian Boltzmann collision equation."

One may think that low-energy particle physics in a dense and hot medium requires the tools of field theory at finite temperature and density (FTD) which has taken a stormy development over the past fifteen years. In practice, its contribution to particle astrophysics has been minor. The derivation of dispersion relations in a medium can be understood in kinetic theory from forward scattering on the medium constituents—dispersion is a lowest-order phenomenon. The imaginary part of a particle self-energy in the medium is physically related to its emission and absorption rate which in FTD field theory is given by cuts of higher-order graphs. These methods have been applied only in a few cases of direct interest to particle astrophysics where the results seem to agree with those from kinetic theory in the limits which have been relevant in practice. A systematic formulation of astrophysically relevant particle dispersion, emission, and absorption processes in the framework of FTD field theory is a project for another author. Tanguy Altherr had begun to take up this challenge with his collaborators in a series of papers. However, his premature death in a tragic climbing accident on 14 July 1994 has put an abrupt end to this line of research.

My presentation of dispersion, emission, and absorption processes is based entirely on the old-fashioned tools of kinetic theory where, say, an axion emission rate is given by an integral of a squared matrix element over the thermally occupied phase space of the reaction partners. Usually this approach is not problematic, but it does require some tinkering when it comes to the problem of electromagnetic screening or other collective effects. I would not be surprised if subtle but important collective effects had been overlooked in some cases.

A simple kinetic approach is not adequate in the hot nuclear medium characteristic for a young SN core. However, in practice quantities like the neutrino opacities and axion emissivities have been calculated as if the medium constituents were freely propagating particles. At least for the dominant axial-vector current interactions this approach is not consistent as one needs to assume that the spin-fluctuation rate is small compared with typical thermal energies while a naive calculation yields a result much larger than T. Realistically, it probably saturates at $\mathcal{O}(T)$, independently of details of the assumed interaction potential. This conjecture appears to be supported by our recent "calibration" of SN opacities from the SN 1987A neutrino signal. However, a calculation of either neutrino opacities or axion emissivities on the basis of first principles is not available at the present time because FTD effects as well as nuclear-physics complications dominate the problem.

◇

It has been challenging to hammer the multifarious and intertwined aspects of my topic into the linear shape required by the nature of a book. Chapters 1–6 are mainly devoted to the stellar energy-loss argument. It is introduced in Chapter 1 where its general aspects are developed on the basis of the stellar structure equations. Chapter 2 establishes the observational limits on those stellar evolutionary time scales that have been used for the purposes of particle astrophysics. For each case the salient applications are summarized. Chapters 3−6 deal with the interaction of "radiation" (neutrinos, axions, other low-mass bosons) with the main constituents of stellar plasmas (photons, electrons, nucleons). In these chapters all information is pulled together that pertains to the given interaction channel, even if it is not directly related to the energy-loss argument. For example, in Chapter 5 the limits on the electromagnetic coupling of pseudoscalars with photons are summarized; the stellar energy-loss ones are the most restrictive which justifies this arrangement.

Chapter 6 develops the topic of particle dispersion in media. The medium-induced photon dispersion relation allows for the plasma process $\gamma \to \bar{\nu}\nu$ which yields the best limit on neutrino dipole moments by virtue of the energy-loss argument applied to globular-cluster stars. The neutrino dispersion relation is needed for the following discussion of neutrino oscillations, establishing a link between the energy-loss argument and the dispersion arguments of the following chapters.

Dispersion and propagation effects are particularly important for massive neutrinos with flavor mixing. To this end the phenomenology of massive, mixed neutrinos is introduced in Chapter 7. Vacuum and matter-induced flavor oscillations as well as magnetically induced spin oscillations are taken up in Chapter 8. If neutrinos are in thermal equilibrium as in a young SN core or the early universe, neutrino oscillations require a different theoretical treatment (Chapter 9).

Chapters 10−13 are devoted to astrophysical sources where neutrinos have been measured, i.e. the Sun and supernovae (for the latter only the SN 1987A signal exists). Neutrino oscillations, notably of the matter-induced variety, play a prominent role in Chapter 10 (solar neutrinos) and Chapter 11 (SN neutrinos). Radiative particle decays, especially of neutrinos, are studied in Chapter 12 where the Sun and SN 1987A figure prominently as sources. In Chapter 13 the particle-physics results from SN 1987A are summarized, including the ones related to the energy-loss and other arguments.

Chapters 14−16 give particle-specific summaries. While axions play a big role throughout this text, only the structure of the interaction Hamiltonian with photons, electrons, and nucleons is needed. Thus everything said about axions applies to any pseudoscalar low-mass boson for which they serve as a generic example. In Chapter 14 these results are interpreted in terms of axion-specific models which relate their properties to those of the neutral pion and thus establish a nearly unique relationship between their mass and interaction strength. Often-quoted "astrophysical bounds on the axion mass" are really transformed bounds on their interaction strength. This chapter also summarizes recent developments of the putative cosmological role of axions. Chapter 15 takes up a variety of hypotheses which can be tested by the methods developed in this book. Finally, Chapter 16 is an attempt at a bottom line of what we have learned about neutrino properties in the astrophysical laboratory.

◇

Parts of my presentation are devoted to theoretical and calculational fine points of particle dispersion and emission effects in media. I find some of these issues quite intriguing in their own right. Still, the main goal has been to provide an up-to-date overview of what we know about elementary particles and their interactions on the basis of established stellar properties and on the basis of measured or experimentally constrained stellar particle fluxes.

> All those whose lives are spent searching for truth are well aware that the glimpses they catch of it are necessarily fleeting, glittering for an instant only to make way for new and still more dazzling insights. The scholar's work, in marked contrast to that of the artist, is inevitably provisional. He knows this and rejoices in it, for the rapid obsolescence of his books is the very proof of the progress of scholarship. (Henri Pirenne, 1862–1935)

In spite of this bittersweet insight, and in spite of some inevitable errors of omission and commission, I hope that my book will be of some use to researchers, scholars, and students interested in the connection between fundamental physics and stars.

Munich, May 1995.

Acknowledgments

While writing this book I have benefitted from the help and encouragement of many friends and colleagues. In particular, I need to mention Hans-Thomas Janka, Lothar Luh, Günter Sigl, Pierre Sikivie, Thomas Strobel, and Achim Weiss who read various parts of the manuscript, and the referees Josh Frieman and J. Craig Wheeler who read all or most of it. Their comments helped in no small measure to improve the manuscript and to eliminate some errors. Hans-Thomas, in particular, has spared no effort at educating me on the latest developments in the area of supernova physics. Several chapters were written during a visit at the Center for Particle Astrophysics in Berkeley—I gratefully acknowledge the fine hospitality of Bernard Sadoulet and his staff. Staying for that period as a guest in Edward Janelli's house made a huge difference. During the entire writing process Greg Castillo provided encouragement and a large supply of Cuban and Puerto Rican music CDs which have helped to keep my spirits up. At the University of Chicago Press, I am indebted to Vicki Jennings, Penelope Kaiserlian, Stacia Kozlowski, and Eleanore Law for their expert handling of all editorial and practical matters that had to be taken care of to transform my manuscript into a book. David Schramm as the series editor originally solicited this *opus* ("just expand your Physics Report a little bit"). It hasn't quite worked that way, but now that I'm finished I am grateful that David persuaded me to take up this project.

Chapter 1

The Energy-Loss Argument

Weakly interacting, low-mass particles such as neutrinos or axions contribute to the energy loss or energy transfer in stars. The impact of an anomalous energy-loss mechanism is discussed qualitatively and in terms of homology relations between standard and perturbed stellar models. The example of massive pseudoscalar particles is used to illustrate the impact of a new energy-loss and a new radiative-transfer mechanism on the Sun.

1.1   Introduction

More than half a century ago, Gamow and Schoenberg (1940, 1941) ushered in the advent of particle astrophysics when they speculated that neutrinos may play an important role in stellar evolution, particularly in the collapse of evolved stars. Such a hypothesis was quite bold for the time because neutrinos, which had been proposed by Pauli in 1930, were not directly detected until 1954. That their existence was far from being an established belief when Gamow and Schoenberg wrote their papers is illustrated by Bethe's (1939) complete silence about them in his seminal paper on the solar nuclear fusion chains.

Even after the existence of neutrinos had been established they seemed to interact only by β reactions of the sort $e^- + (A, Z) \rightarrow (A, Z-1) + \nu_e$ or $(A, Z-1) \rightarrow (A, Z) + e^- + \bar{\nu}_e$, the so-called URCA reactions which Gamow and Schoenberg had in mind, or by fusion processes like $pp \rightarrow de^+\nu_e$. The URCA reactions and related processes become important only at very high temperatures or densities because of their energy threshold. While the Sun emits two neutrinos for every helium nucleus fused from hydrogen, the energy loss in neutrinos is only a few percent of the total luminosity and thus plays a minor role.

1

Still, neutrinos can be important in normal stars. This became clear when in 1958 Feynman and Gell-Mann as well as Sudarshan and Marshak proposed the universal $V-A$ interaction law which implied a direct neutrino-electron interaction with the strength of the Fermi constant. Pontecorvo (1959) realized almost immediately that this interaction would allow for the bremsstrahlung radiation of neutrino pairs by electrons, and that the absence of a threshold renders this process an important energy loss mechanism for stars. Of course, their typical energies will correspond to the temperature of the plasma (about 1 keV in the Sun) while neutrinos from nuclear reactions have MeV energies.

On the basis of the bremsstrahlung process, Gandel'man and Pinaev (1959) calculated the approximate conditions for which neutrino losses would "outshine" the photon luminosity of stars. For the Sun, thermal neutrino emission is found to be irrelevant.

Subsequently, the neutrino emissivity was calculated by many authors. It was quickly realized that the dominant emission processes from a normal stellar plasma are the photoneutrino process $\gamma e^- \rightarrow e^- \bar{\nu}\nu$, the bremsstrahlung process $e^- + (A, Z) \rightarrow (A, Z) + e^- + \bar{\nu}\nu$, and the plasma process ("photon decay") $\gamma \rightarrow \bar{\nu}\nu$ which is possible because photons have an effective mass in the medium. These and related reactions will be discussed at length in Chapters 3–6.

In the late 1960s Stothers and his collaborators[1] established that the observed paucity of red supergiants is best explained by the fast rate with which these carbon-burning stars spend their nuclear fuel due to neutrino emission. Therefore, the existence and approximate magnitude of the direct electron-neutrino coupling was at least tentatively established by astrophysical methods several years before its experimental measurement in 1976.

While standard neutrino physics today is an integral part of stellar evolution and supernova theory, they could have novel couplings to the plasma, for example by a magnetic dipole moment. Then they could be emitted more efficiently than is possible with standard interactions. Moreover, new concepts of particle physics have emerged that could be equally important despite the relatively low energies available in stellar interiors. In various extensions of the standard model, the spontaneous breakdown of a symmetry of the Lagrangian of the fundamental interactions by some large vacuum expectation value of a new field leads to the prediction of massless or nearly massless particles, the Nambu-Goldstone bosons of the broken symmetry. The most widely discussed

---

[1]For a summary see Stothers (1970, 1972).

example is the axion (Chapter 14) which arises as the Nambu-Goldstone boson of the Peccei-Quinn symmetry which explains the puzzling absence of a neutron electric dipole moment, i.e. it explains CP conservation in strong interactions. The production of axions in stars, like that of neutrinos, is not impeded by threshold effects.

Clearly, the emission of novel weakly interacting particles or the emission of neutrinos with novel properties would have a strong impact on the evolution and properties of stars. Sato and Sato (1975) were the first to use this "energy-loss argument" to derive bounds on the coupling strength of a putative low-mass Higgs particle. Following this lead, the argument has been applied to a great variety of particle-physics hypotheses, and to a great variety of stars.

In the remainder of this chapter the impact of a novel energy-loss mechanism on stars will be discussed in simple terms. The main message will be that the emission of weakly interacting particles usually leads to a modification of evolutionary time scales. By losing energy in a new channel, the star effectively burns or cools faster and thus shines for a shorter time. In the case of low-mass red giants, however, particle emission leads to a delay of helium ignition and thus to an extension of the red-giant phase. Either way, what needs to be observationally established is the duration of those phases of stellar evolution which are most sensitive to a novel energy-loss mechanism. In Chapter 2 such evolutionary phases will be identified, and the observational evidence for their duration will be discussed. In the end one will be able to state simple criteria for the allowed rate of energy loss from plasmas at certain temperatures and densities.

One may be tempted to think that one should consider the hottest and densest possible stars because no doubt the emission of weakly interacting particles is most efficient there. However, this emission competes with standard neutrinos whose production is also more efficient in hotter and denser objects. Because neutrinos are thermally emitted in pairs their emission rates involve favorable phase-space factors which lead to a temperature dependence which is steeper than that for the emission of, say, axions. Thus, for a given axion coupling strength the *relative* importance of axion emission is greater for lower temperatures. Of course, the temperatures must not be so low that neither neutrino nor axion emission is important at all relative to the photon luminosity. Consequently, the best objects to use are those where neutrinos just begin to have an observational impact on stellar observables. Examples are low-mass red giants, horizontal-branch stars, white dwarfs, and old neutron stars. They all have masses of around $1\,\mathcal{M}_\odot$ (solar mass).

For low-mass stars plenty of detailed observational data exist, allowing one to establish significant limits on possible deviations from standard evolutionary time scales.

From the astrophysical perspective, the energy-loss argument requires establishing evolutionary time scales and other observables that are sensitive to a novel "energy sink." This problem will be taken up in Chapter 2. From the particle-physics perspective, one needs to calculate the energy-loss rate of a plasma at a given temperature, density, and chemical composition into a proposed channel. For this purpose it is not necessary to know any of the underlying physics that leads to the hypothesis of nonstandard neutrino properties or the prediction of new particles such as axions. All one needs is the structure of the interaction Hamiltonian with the constituents of stellar plasmas (electrons, nucleons, nuclei, and photons). For example, it is enough to know that an axion is a low-mass particle with a pseudoscalar interaction with electrons to calculate the axion emission rate. As such "axion" stands for any particle with a similar interaction structure. The motivation for "real axions" and their detailed model-dependent properties are not discussed until Chapter 14. In Chapters 3−5 and partly in Chapter 6 various interaction structures will be explored and constraints on the overall coupling strengths will be derived or summarized. Also, they will be put into the context of evidence from sources other than the stellar energy-loss argument.

Putative novel particles almost inevitably must be very weakly interacting or else they would have been seen in laboratory experiments. Still, one is sometimes motivated to speculate about particles which may be so strongly interacting that they cannot freely escape from stars; their mean free path may be shorter than the stellar radius. It is often incorrectly stated that such particles would be astrophysically allowed. Nothing could be further from the truth. Particles which are "trapped," i.e. which cannot freely stream out once produced, contribute to the radiative transfer of energy in competition with photons. Radiative energy transfer occurs because particles produced in a hot region get absorbed in a neighboring somewhat cooler region. This mechanism is more efficient if the mean free path is larger because then ever more distant regions with larger temperature differences are thermally coupled. Therefore, particles which interact more weakly than photons would dominate the radiative energy transfer and thus have a tremendous impact on the structure of stars.

Very roughly speaking, then, novel particles must be either more strongly interacting than photons, or more weakly interacting than neutrinos, to be harmless in stars. Of course, depending on whether the particles are bosons or fermions, and depending on the details of their interaction structure, this statement must be refined. Still, *the impact of novel particles on stellar structure and evolution is maximized when their mean free path is of order the geometric dimension of the system.* The energy-transfer argument is equally powerful as the energy-loss argument. The only reason why it has not been elaborated much in the literature is because there is usually little motivation for considering "strongly" interacting novel particles.

When it comes to the evolution of a supernova (SN) core after collapse even neutrinos are trapped. Such a newborn neutron star is so hot (T of order $30\,\mathrm{MeV}$) and dense (ρ exceeding nuclear density of $3\times10^{14}\,\mathrm{g\,cm^{-3}}$) that neutrinos take several seconds to diffuse to the surface. Particles like axions can then compete in spite of the extreme conditions because they freely stream out if their coupling is weak enough. The energy-loss argument can be applied because the neutrino cooling time scale has been established by the SN 1987A neutrino observations. While the energy-loss argument in this case is fundamentally no different from, say, white-dwarf cooling, the detailed reasoning is closely intertwined with the issue of neutrino physics in supernovae, and with the details of the SN 1987A neutrino observations. Therefore, it is taken up only in Chapter 13. The groundwork concerning the interactions of neutrinos and axions with nucleons, however, is laid in Chapter 4 within the series of chapters devoted to various modes of particle interactions with the constituents of stellar plasmas.

1.2 Equations of Stellar Structure

1.2.1 Hydrostatic Equilibrium

To understand the impact of a novel energy-loss mechanism on the evolution of stars one must understand the basic physical principles that govern stellar structure. While a number of simplifying assumptions need to be made, the theory of stellar structure and evolution has been extremely successful at modelling stars with a vast range of properties. For a more detailed account than is possible here the reader is referred to the textbook literature, e.g. Kippenhahn and Weigert (1990).

One usually assumes spherical symmetry and thus excludes the effects of rotation, magnetic fields, tidal effects from a binary companion, and large-scale convective currents. While any of those effects can be important in special cases, none of them appears to have a noticeable impact on the overall picture of stellar structure and evolution, with the possible exception of supernova physics where large-scale convective overturns may be crucial for the explosion mechanism (Chapter 11).

Second, one usually assumes hydrostatic equilibrium, i.e. one ignores the macroscopic kinetic energy of the stellar medium. This approximation is inadequate for a study of stellar pulsation where the inertia of the material is obviously important, and also inadequate for "hydrodynamic events" such as a supernova explosion (Sect. 2.1.8), and perhaps the helium flash (Sect. 2.1.3). For most purposes, however, the changes of the stellar structure are so slow that neglecting the kinetic energy is an excellent approximation.

Therefore, as a first equation one uses the condition of hydrostatic equilibrium that a spherical shell of the stellar material is held in place by the opposing forces of gravity and pressure,

$$\frac{dp}{dr} = -\frac{G_N \mathcal{M}_r \rho}{r^2}. \tag{1.1}$$

Here, G_N is Newton's constant, p and ρ are the pressure and mass density at the radial position r, and $\mathcal{M}_r = 4\pi \int_0^r dr' \rho\, r'^2$ is the integrated mass up to the radius r.

With apologies to astrophysicists I will usually employ natural units where $\hbar = c = k_B = 1$. In Appendix A conversion factors are given between various units of mass, energy, inverse length and time, temperature, and so forth. Newton's constant is then $G_N = m_{Pl}^{-2}$ with the Planck mass $m_{Pl} = 1.221 \times 10^{19}$ GeV $= 2.177 \times 10^{-5}$ g. Stellar masses are always denoted with the letter $\mathcal{M}$ to avoid confusion with an absolute bolometric brightness which is traditionally denoted by M.

In general, the pressure is given in terms of the density, temperature, and chemical composition by virtue of an equation of state. For a classical monatomic gas $p = \frac{2}{3} u$ with u the density of internal energy. One may multiply Eq. (1.1) on both sides with $4\pi r^3$ and integrate from the center ($r = 0$) to the surface ($r = R$). The r.h.s. gives the total gravitational energy while the l.h.s. yields $-12\pi \int_0^R dr\, p\, r^2$ after a partial integration with the boundary condition $p = 0$ at the surface. With $p = \frac{2}{3} u$ this is $-2U$ with U the total internal energy of the star. Because for a monatomic gas U is the sum of the kinetic energies of the atoms one finds that on average for every atom

$$\langle E_{kin} \rangle = -\tfrac{1}{2} \langle E_{grav} \rangle. \tag{1.2}$$

This is the virial theorem which is the most important tool to understand the behavior of self-gravitating systems.

As a simple example for the beauty and power of the virial theorem one may estimate the solar central temperature from its mass and radius. The material is dominated by protons which have a gravitational potential energy of order $-G_N \mathcal{M}_\odot m_p / R_\odot = -2.14 \text{ keV}$ where $\mathcal{M}_\odot = 1.99 \times 10^{33} \text{ g}$ is the solar mass, $R_\odot = 6.96 \times 10^{10} \text{ cm}$ the solar radius, and m_p the proton mass. The average kinetic energy of a proton is equal to $\frac{3}{2} T$ (remember, k_B has been set equal to unity), yielding an approximate value for the solar internal temperature of $T = \frac{1}{3} 2.14 \text{ keV} = 0.8 \times 10^7 \text{ K}$. This is to be compared with $1.56 \times 10^7 \text{ K}$ found for the central temperature of a typical solar model. This example illustrates that the basic properties of stars can be understood from simple physical principles.

1.2.2 Generic Cases of Stellar Structure

a) Normal Stars

There are two main sources of pressure relevant in stars, thermal pressure and degeneracy pressure. The third possibility, radiation pressure, never dominates except perhaps in the most massive stars. The pressure provided by a species of particles is proportional to their density, to their momentum which is reflected on an imagined piston and thus exerts a force, and to their velocity which tells us the number of hits on the piston per unit time. In a nondegenerate nonrelativistic medium a typical particle velocity and momentum is proportional to $T^{1/2}$ so that $p \propto (\rho/\mu) T$ with ρ the mass density and μ the mean molecular weight of the medium constituents. For nonrelativistic degenerate electrons the density is $n_e = p_F^3 / 3\pi^2$ (Fermi momentum p_F), a typical momentum is p_F, and the velocity is p_F/m_e, yielding a pressure which is proportional to p_F^5 or to $n_e^{5/3}$ and thus to $\rho^{5/3}$.

The two main pressure sources determine two generic forms of behavior of overall stellar models, namely normal stars such as our Sun which is dominated by thermal pressure, and degenerate stars such as white dwarfs which are dominated by degeneracy pressure. These two cases follow a very different logic.

A normal star is understood most easily if one imagines how it initially forms from a dispersed but gravitationally bound gas cloud. It continuously loses energy because photons are produced in collisions between, say, electrons and protons. The radiation carries away energy which must go at the expense of the total energy of the system. If it is roughly in an equilibrium configuration, the virial theorem Eq. (1.2) in-

forms us that a decrease of $\langle E_{\text{kin}} + E_{\text{grav}} \rangle$ causes the gravitational energy
to become more negative, corresponding to a more tightly bound and
thus more compact system. At the same time the average kinetic energy
goes up which corresponds to an increased temperature if the system
can be considered to be locally in thermal equilibrium. Therefore, as
the system loses energy it contracts and heats up. Self-gravitating systems have a negative specific heat!

As a protostar contracts it becomes more opaque and soon the energy loss is limited by the speed of energy transfer from the inner parts
to the surface, i.e. essentially by the photon diffusion speed. Before an
internal energy source for stars was known it was thought that gravitational energy provided for their luminosity. Stellar lifetimes seemed to
be given by the "Kelvin-Helmholtz time scale" for thermal relaxation
which is fixed by the speed of energy transfer. The total reservoir of
gravitational energy of the Sun is estimated by $G_{\text{N}} \mathcal{M}_{\odot}^2 / 2 R_{\odot}$ and its
luminosity is $L_{\odot} = 3.85 \times 10^{26}\,\text{W} = 3.85 \times 10^{33}\,\text{erg/s}$ so that the thermal
relaxation scale is about $\tau_{\text{KH}} \approx \frac{1}{2} G_{\text{N}} \mathcal{M}_{\odot}^2 R_{\odot}^{-1} L_{\odot}^{-1} = 1.6 \times 10^7\,\text{yr}$. Pressed
by thermodynamic theory and the authority of Lord Kelvin, geologists
tried to adjust the age of the Earth to this short time scale against
sound evidence to the contrary. At the beginning of our century the
discovery of radioactivity and thus of nuclear processes revealed that
stars had another source of energy and consequently could live much
longer than indicated by τ_{KH}.

A further contraction is thus intercepted by the onset of hydrogen
burning which commences when the temperature is high enough for
protons to penetrate each other's electrostatic repulsive potential. The
hydrogen-burning reactions are more fully discussed in Chapter 10; the
bottom line is that four protons and two electrons combine to form
a helium nucleus (α particle), releasing 26.73 MeV of energy. A few
percent are immediately lost in the form of two neutrinos which must
emerge to balance the electron lepton number, but most of the energy is available as heat. Because the nuclear reaction rates have a
steep temperature dependence, a further contraction and heating of
the star leads to much more nuclear energy generation, which quickly
increases the average E_{kin} of the nuclei and thus leads to an expansion
and cooling by the same virial-theorem logic that led to contraction and
heating when energy was lost. Thus a stable configuration of "thermal
equilibrium" is reached where the energy lost is exactly balanced by
that produced from nuclear reactions. Stars as fusion reactors are perfectly regulated by the "negative specific heat" of a self-gravitating
system!

The most salient feature of a normal stellar configuration is the interplay of its negative specific heat and nuclear energy generation. Conversely, if the pressure were dominated by electron degeneracy it would be nearly independent of the temperature. Then this self-regulation would not function because heating would not lead to expansion. Thus, stable nuclear burning and the dominance of thermal pressure go inseparably hand in hand. Another salient feature of such a configuration is the inevitability of its final demise because it lives on a finite supply of nuclear fuel.

b) Degenerate Stars

Everything is different for a configuration dominated by degeneracy pressure. Above all, it has a positive heat capacity so that a loss of energy no longer implies contraction and heating. The star actually *cools*. This is what happens to a brown dwarf which is a star so small ($\mathcal{M} < 0.08\,\mathcal{M}_\odot$) that it did not reach the critical conditions to ignite hydrogen: it becomes a degenerate gas ball which slowly "browns out."

The relationship between radius and mass is inverted. A normal star is geometrically larger if it has a larger mass; very crudely $R \propto \mathcal{M}$. When mass is added to a degenerate configuration it becomes geometrically smaller as the reduced size squeezes the electron Fermi sea into higher momentum states, providing for increased pressure to balance the increased gravitational force. The l.h.s. of Eq. (1.1) can be approximated as p/R where $p \propto \rho^{5/3}$ is some average pressure. Because $\rho \approx \mathcal{M}/R^3$ one finds $p/R \propto \mathcal{M}^{5/3}/R^6$ while the r.h.s. of Eq. (1.1) is proportional to $\mathcal{M}\rho/R^2$ and thus to $\mathcal{M}^2/R^5$. Therefore, a degenerate configuration is characterized by $R \propto \mathcal{M}^{-1/3}$.

Increasing the mass beyond a certain limit causes the radius to shrink so much that the electrons become relativistic. Then they move with a velocity fixed at c (or 1 in natural units), causing the pressure to vary only as p_{F}^4 or $\rho^{4/3}$. In this case adding mass no longer leads to a sufficient pressure increase to balance for the extra weight. Beyond this "Chandrasekhar limit," which is about $1.4\,\mathcal{M}_\odot$ for a chemical composition with $Y_e = \frac{1}{2}$ (number of electrons per baryon), no stable degenerate configuration exists.

In summary, the salient features of a degenerate configuration are the inverse mass-radius relationship $R \propto \mathcal{M}^{-1/3}$, the Chandrasekhar limit, the absence of nuclear burning, and the positive specific heat which allows the configuration to cool when it loses energy.

c) Giant Stars

A real star can be a hybrid configuration with a degenerate core and
a nondegenerate envelope with nuclear burning at the bottom of the
envelope, i.e. the surface of the core. The core then follows the logic
of a degenerate configuration. The envelope follows the self-regulating
logic of a normal star except that it is no longer dominated by its self-
gravity but rather by the gravitational force exerted by the compact
core. Amazingly, the envelopes of such stars tend to expand to huge
dimensions. It does not seem possible to explain in a straightforward
way why stars become giants[2] except that "the equations say so." Low-
mass red giants will play a major role in Chapter 2—a further discussion
of their fascinating story is deferred until then.

1.2.3   Energy Conservation

Returning to the basic principles that govern stellar structure, energy
conservation yields another of the stellar structure equations. If the local sources and sinks of energy balance against the energy flow through
the surface of a spherical mass shell one finds

$$dL_r/dr = 4\pi r^2 \epsilon \rho, \tag{1.3}$$

where L_r is the net flux of energy through a spherical shell of radius r
while ϵ is the effective rate of local energy production (units $\mathrm{erg\,g^{-1}\,s^{-1}}$).
It is a sum

$$\epsilon = \epsilon_{\mathrm{nuc}} + \epsilon_{\mathrm{grav}} - \epsilon_\nu - \epsilon_x, \tag{1.4}$$

where ϵ_{nuc} is the rate by which nuclear energy is liberated, ϵ_ν is the
energy-loss rate by standard neutrino production, and ϵ_x is for novel
particles or nonstandard neutrinos with, say, large magnetic dipole moments. Further,

$$\epsilon_{\mathrm{grav}} = c_p T \left(\nabla_{\mathrm{ad}} \dot{p}/p - \dot{T}/T \right) \tag{1.5}$$

is the local energy gain when T and p change because of expansion or
contraction of the star. This term is the dominant heat source in a
red-giant core which contracts because of the mass gained by hydrogen
shell burning. Conversely, it is a sink which absorbs energy when helium
ignites in such a star: the would-be explosion is dissipated by expansion

---

[2]For recent attempts see Faulkner and Swenson (1988), Eggleton and Cannon
(1991), and Renzini et al. (1992).

against the force of gravity. In Eq. (1.5) c_p is the heat capacity at constant pressure. Further, the quantity $\nabla_{\text{ad}} \equiv (\partial \ln T / \partial \ln p)_s$, taken at constant entropy density s, is the "adiabatic temperature gradient." It is not really a gradient. It is a thermodynamic quantity characteristic of the medium at the local conditions of ρ, p, T, and the chemical composition.

The calculation of ϵ_x for a number of hypotheses concerning the existence of novel particles or novel properties of neutrinos will be a major aspect of this book. Even for a given interaction law between the particles and the medium constituents this is not always a straightforward exercise because the presence of the ambient medium can have a significant impact on the microscopic reactions.

This is equally true for the nuclear energy generation rates. Nuclear reactions are slow in stars because the low temperature allows only few nuclei to penetrate each other's Coulomb barriers. Therefore, screening effects are important, and the nuclear cross sections need to be known at energies so low that they cannot be measured directly in the laboratory. Much of the debate concerning the solar neutrino problem revolves around the proper extrapolation of certain nuclear cross sections to solar thermal energies.

1.2.4 Energy Transfer

The transfer of energy is driven by the radial temperature gradient. In the absence of convection heat is carried by photons and electrons moving between regions of different temperature, i.e. by radiative transfer and by conduction. In this case the relationship between the energy flux and the temperature gradient is

$$L_r = -\frac{4\pi r^2}{3\kappa\rho}\frac{d(aT^4)}{dr},$$ (1.6)

where aT^4 is the energy stored in the radiation field ($a = \pi^2/15$ in natural units) and κ is the opacity (units cm^2/g). It is given by a sum

$$\kappa^{-1} = \kappa_\gamma^{-1} + \kappa_c^{-1} + \kappa_x^{-1},$$ (1.7)

where κ_γ is the radiative opacity, κ_c the contribution from conduction by electrons, and κ_x was included for a possible contribution from novel particles. The quantity $(\kappa_\gamma\rho)^{-1} = \langle\lambda_\gamma\rangle_\text{R}$ is the "Rosseland average" of the photon mean free path—its precise definition will be given in Sect. 1.3.4.

One of the main difficulties at calculating the opacity is that heavier elements, notably iron, are only partially ionized for typical conditions. Resonant transitions of electrons between different bound states are very important, an effect which causes stellar models to be rather sensitive to the amount of "metals" (elements heavier than helium). The construction of an opacity table is a major effort as it requires including huge numbers of electronic energy levels. Widely used were the Los Alamos and the Livermore Laboratory opacity tables.

Recently, the Livermore tables were systematically overhauled (Iglesias, Rogers, and Wilson 1990; Iglesias and Rogers 1991a,b), resulting in the new OPAL tables which since have become the standard in stellar evolution calculations. The main differences to the previous tables are at moderate temperatures so that no substantial changes in the deep interior of stellar structures have occurred. However, envelope phenomena are affected, notably convection near the surface and stellar pulsations. A number of previous discrepancies between theory and observations in this area have now disappeared.

If the equation of state, the energy generation rate, and the opacity are known one can construct a stellar model for an assumed composition profile by solving the stellar structure equations with suitable boundary conditions. (It is not entirely trivial to define surface boundary conditions because the star, strictly speaking, extends to infinity. A crude approach is to take $T = 0$ and $p = 0$ at the photosphere.) It may turn out, however, that in some locations this procedure yields a temperature gradient which is so steep that the material becomes unstable to convection—it "boils."

An adequate treatment of convection is one of the main problems of stellar evolution theory. A simplification occurs because convection is extremely effective at transporting energy and so the temperature gradient will adjust itself to a value very close to the "adiabatic gradient" which marks the onset of the instability. At this almost fixed temperature gradient a nearly arbitrary energy flux can be carried by the medium. This approximation tends to be justified for regions in the deep interior of stars while the "superadiabatic convection" found near the surface requires a substantial refinement. One usually applies the "mixing length theory" which contains one free parameter, the ratio between the convective mixing length and the pressure scale height. This parameter is empirically fixed by adjusting the radius of a calculated solar model to the observed value.

Main-sequence stars like our Sun with $M \lesssim M_\odot$ have a radiative interior with a convective surface which penetrates deeper with decreasing $\mathcal{M}$; stars with $\mathcal{M} \lesssim 0.25\,\mathcal{M}_\odot$ are fully convective. For $\mathcal{M} \gtrsim \mathcal{M}_\odot$ the outer regions are radiative while the core is convective out to an ever increasing mass fraction of the star with increasing $\mathcal{M}$. A star with $\mathcal{M}$ near $1\,\mathcal{M}_\odot$ is very special in that it is radiative almost throughout; the Sun is thought to have only a relatively minor convective surface layer.

Besides transporting energy, convection also moves matter and thus affects the composition profile of a star. This is seen, for example, in the upper panels of Fig. 2.4 where the hydrogen depletion of a solar model (which is radiative) is a function of the local nuclear burning rates while for the convective helium core of a horizontal-branch (HB) star the helium depletion reaches to much larger radii than nuclear burning. The long lifetimes of HB stars cannot be understood without the convective supply of fuel to the nuclear furnace at the center. The Sun, on the other hand, will complete its main-sequence evolution when hydrogen is depleted at the center, corresponding to about a 10% global depletion only.

The extent of convective regions can change during the course of stellar evolution. They can leave behind composition discontinuities which are a memory of a previous configuration. For example, on the lower red-giant branch (RGB) the convective envelope reaches so deep that it penetrates into the region of variable hydrogen content caused by nuclear burning. Later, the convective envelope retreats from the advancing hydrogen-burning shell which encounters a discontinuity in the hydrogen profile. This causes a brief "hesitation" on the RGB ascent and thus a "bump" in the distribution of stars in the color-magnitude diagram of globular clusters on the lower RGB. This bump has been identified in several clusters (Figs. 2.18 and 2.19); its location is in good agreement with theoretical expectations (Fusi Pecci et al. 1990).

1.2.5 Gravitational Settling

The composition profile of a star can also change by diffusion, and notably by gravitational settling of the heavier elements. This effect was ignored in most evolution calculations because the time scales are very large. Still, the settling of helium will displace hydrogen from the center of a hydrogen-burning star and thus accelerate the depletion and main-sequence turnoff. The gravitational settling of metals will lead to an opacity increase in the central regions. Helium settling reduces the inferred globular cluster ages by 1 Gyr or more which is about a 10% effect (Proffitt and Michaud 1991; Chaboyer et al. 1992). Because the inferred globular cluster ages are larger than the expansion age of the universe, this "cosmic age problem" is slightly alleviated by gravitational settling.

Helium and metal settling was recently included in solar models (Bahcall and Pinsonneault 1992, 1995; Kovetz and Shaviv 1994; Proffitt 1994). It increased the predicted neutrino fluxes on the $10-30\%$ level, depending on the specific treatment of gravitational settling and on the neutrino source reaction. These modifications are not huge, but surely not entirely negligible. They go in the direction of aggravating the solar neutrino problem.

1.3 Impact of Novel Particles

1.3.1 Energy Loss

One of the most interesting possibilities to use stars as particle-physics laboratories is to study the backreaction of the novel energy-loss rate ϵ_x implied by the existence of new low-mass particles such as axions, or by nonstandard neutrino properties such as magnetic dipole moments. The impact on degenerate stars such as white dwarfs is rather obvious: the new energy-loss rate accelerates the cooling. Therefore, the observationally established cooling speed allows one to constrain this process or to detect evidence for it.

The impact of a novel energy-loss mechanism on a nondegenerate star like the Sun is less obvious. According to the virial theorem one expects that the loss of energy leads to heating and contraction up to a point where the temperature has risen enough that increased nuclear burning provides for the extra energy loss. Because of the steep temperature dependence of ϵ_{nuc} one expects that the overall stellar structure changes very little in response to ϵ_x—the main impact is to accelerate the consumption of nuclear fuel and thus the completion of hydrogen burning.

This argument was cast into a quantitative form by Frieman, Dimopoulos, and Turner (1987). They asked how a given equilibrium structure of a star would change in response to turning on a new energy-loss rate ϵ_x. The main simplifying assumption is that the perturbed configuration is obtained by a homology transformation, so that "the distance between any two points is altered in the same way as the radius of the configuration." Thus, if the new radius of the star is given by $R' = yR$ with a dimensionless scaling factor y, then every point in the star is mapped to a new position $r' = yr$. The mass interior to the new radius is identical with that interior to the old location,

$\mathcal{M}'(r') = \mathcal{M}(r)$, and the chemical composition at r' is the same as that at r. The density is transformed by $\rho'(r') = y^{-3}\rho(r)$, and from Eq. (1.1) one finds that the pressure scales as $p'(r') = y^{-4}p(r)$. The equation of state for a nondegenerate, low-mass star is approximately given by the ideal-gas law where $p \propto \rho T/\mu$, where μ is the average molecular weight of the electrons and nuclei. Since $\mu'(r') = \mu(r)$ by assumption, the temperature is found to scale as $T'(r') = y^{-1}T(r)$, and the temperature gradient as $dT'(r')/dr' = y^{-2}dT(r)/dr$.

The assumption that the star reacts to new particle emission by a homologous contraction imposes restrictions on the constitutive relations for the effective energy generation rate and the opacity. In particular, for a chemically homogeneous star one needs to assume that

$$\epsilon \propto \rho^n T^\nu \quad \text{and} \quad \kappa \propto \rho^s T^p. \tag{1.8}$$

For the opacity, Frieman et al. took the Kramers law with $s = 1$ and $p = -3.5$ which is found to be a reasonable interpolation formula throughout most lower main-sequence interiors. Hence, the local energy flux scales as

$$L'(r') = y^{-1/2} L(r). \tag{1.9}$$

The hydrogen-burning rate $\epsilon_{\rm nuc}$ also has the required form with $n = 1$, and for the pp chain $\nu = 4$–6; it dominates in the Sun and in stars with lower mass. It is assumed that the new energy-loss rate ϵ_x follows the same proportionality; the standard neutrino losses ϵ_ν are ignored because they are small on the lower main sequence. If the star is not in a phase of major structural readjustment one may also ignore $\epsilon_{\rm grav}$ in Eq. (1.4) so that in Eq. (1.3)

$$\epsilon = (1 - \delta_x) \epsilon_{\rm nuc}, \tag{1.10}$$

where $\delta_x < 1$ is a number which depends on the interaction strength of the new particles. From Eq. (1.3) one concludes that

$$L'(r') = y^{-(3+\nu)} (1 - \delta_x) L(r), \tag{1.11}$$

leading to

$$y = (1 - \delta_x)^{2/(2\nu+5)}. \tag{1.12}$$

Assuming $\delta_x \ll 1$, Frieman et al. then found for the fractional changes of the stellar radius, luminosity, and interior temperature,

$$\frac{\delta R}{R} = \frac{-2\delta_x}{2\nu + 5}, \quad \frac{\delta L}{L} = \frac{\delta_x}{2\nu + 5}, \quad \frac{\delta T}{T} = \frac{2\delta_x}{2\nu + 5}. \tag{1.13}$$

Therefore, the star contracts, becomes hotter, and the surface photon luminosity increases—it overcompensates for the new losses. Moreover,

even if the luminosity L_x in "exotics" is as large as the photon luminosity ($\delta_x = \frac{1}{2}$) the overall changes in the stellar structure remain moderate. The predominant effect is an increased consumption of nuclear
fuel at an almost unchanged stellar structure, leading to a decreased
duration of the hydrogen-burning phase of

$$\delta\tau/\tau \approx -\delta_x. \tag{1.14}$$

The standard Sun is halfway through its main-sequence evolution so
that a conservative constraint is $\delta_x < \frac{1}{2}$.

In general, the exotic losses do not have the same temperature and
density dependence as the nuclear burning rate, implying a breakdown
of the homology condition. However, to lowest order these results will
remain valid if one interprets δ_x as a suitable average over the entire star,

$$\delta_x = L_x/(L_x + L_\gamma), \tag{1.15}$$

with the photon luminosity L_γ and that in exotics L_x. To lowest order
L_x can be computed from an unperturbed stellar model.

For a convective structure ($\mathcal{M} \lesssim 0.25\,\mathcal{M}_\odot$ main-sequence stars)
Frieman et al. found by a similar treatment

$$\frac{\delta R}{R} = \frac{-2\delta_x}{2\nu + 11}, \qquad \frac{\delta L}{L} = \frac{-5\delta_x}{2\nu + 11}, \qquad \frac{\delta T}{T} = \frac{2\delta_x}{2\nu + 11}. \tag{1.16}$$

These stars also contract, and the internal temperature increases, but
the surface luminosity decreases.

1.3.2   Application to the Sun

For the Sun, the radius and luminosity are very well measured and
so one may think that small deviations δR and δL from a standard
model were detectable. This is not so, however, because a solar model
is *defined* to produce the observed radius and luminosity at an age
of 4.5 Gyr. The unknown presolar helium abundance Y_{initial} is chosen
to reproduce the present-day luminosity, and the one free parameter
of the mixing-length theory relevant for superadiabatic convection is
calibrated by the solar radius.

In a numerical study Raffelt and Dearborn (1987) implemented ax-
ion losses by the Primakoff process in a $1\,\mathcal{M}_\odot$ stellar model, metallicity $Z = 0.02$, which was evolved to 4.5 Gyr with different amounts
of initial helium and different axion coupling strengths. Details of the
emission rate as a function of temperature and density are studied in

Table 1.1. Initial helium abundance for solar models with axion losses.

g_{10}	Y_{initial}	δ_x	X_c
0	0.274	0.00	0.362
10	0.266	0.16	0.307
15	0.256	0.32	0.292
20	0.241	0.51	0.245
25	0.224	0.65	0.151

Sect. 5.2. For the present discussion the axion losses represent some generic energy-loss mechanism with a rate proportional to the square of the axion-photon coupling strength $g_{a\gamma}$.

Without exotic losses a presolar helium abundance of $Y_{\text{initial}} = 0.274$ was needed to reproduce the present-day Sun. For several values of $g_{10} \equiv g_{a\gamma}/10^{-10}\,\text{GeV}^{-1}$ Raffelt and Dearborn found the initial helium values given in Tab. 1.1 necessary to produce the present-day luminosity. The values for δ_x in Tab. 1.1 are defined as in Eq. (1.15) with L_x the axion luminosity of the (perturbed) present-day solar model which has $L_\gamma = L_\odot$. Also, the central hydrogen abundance X_c of the present-day model is given. For $g_{10} = 30$, corresponding to $\delta_x \approx 0.75$, no present-day Sun could be constructed for any value of Y_{initial}.

The primordial helium abundance is thought to be about 23%, and the presolar abundance is certainly larger. Still, a value of δ_x less than about 0.5 is hard to exclude on the basis of this calculation. Therefore, the approximate solar constraint remains $\delta_x \lesssim \frac{1}{2}$ or $L_x \lesssim L_\odot$ as found from the analytic treatment in the previous section.

One may be able to obtain an interesting limit by considering the oscillation frequencies of the solar pressure modes. Because of the excellent agreement between standard solar models and the observed p-mode frequencies there is little leeway for a modified solar structure and composition. This method has been used to constrain a hypothetical time variation of Newton's constant (Sect. 15.2.3).

1.3.3   Radiative Energy Transfer

If novel particles are so weakly interacting that they escape freely from the star once produced their role is that of a local energy sink. Neutrinos are of that nature, except in supernova cores where they are "trapped" for several seconds. One could imagine new particles with such large interactions that they are even trapped, say, in the Sun. Because their mean free path (mfp) is now less than the geometric dimension of the star, they remove energy from one region and deposit it at an approximate distance of one mfp. This is precisely the mechanism of radiative energy transfer: the particles now contribute to the opacity. In a transition region where the mfp is on the order of the stellar radius this mode of energy transfer couples distant regions and thus cannot be described in the form of the differential equation (1.6). In this case the difference between an energy-loss and an energy-transfer mechanism is blurred.

Equation (1.6) is justified when the mfp is less than the temperature scale height $(d\ln T/dr)^{-1}$. For radiative transfer $(\kappa\rho)^{-1}$ is an average mfp. Therefore, Eq. (1.6) informs us that the energy flux carried through a sphere of radius r is a product of a numerical factor, the area, the photon mfp, the number density of photons, and the temperature gradient. Radiative transfer is more efficient for a larger mfp, i.e. for a more weakly interacting particle!

If a second photon existed with a coupling strength α' instead of $\alpha = \frac{1}{137}$, it would contribute *more* to the energy transfer for $\alpha' < \alpha$. The observed properties of the Sun and other stars confirm that the standard opacities certainly cannot be wrong by more than a factor of a few. Therefore, the new photon must interact about as strongly as the standard one to be in agreement with the properties of stars, or it must interact so weakly that it freely escapes and the integrated volume emission $L_{\gamma'}$ is less than the photon luminosity L_γ.

1.3.4   Opacity Contribution of Arbitrary Bosons

In order to implement the energy-transfer argument one needs a properly defined expression for the Rosseland opacity contribution of arbitrary bosons. Following the textbook derivation of the photon radiative opacity such a definition was provided by Carlson and Salati (1989) as well as Raffelt and Starkman (1989).

For a sufficiently short mean free path ℓ the radiation field of the new bosons is taken to be locally isotropic. The local energy flux is then found to be $F_\omega = -\frac{1}{3}\beta_\omega \ell_\omega \nabla B_\omega$ where the index ω indicates that a quantity refers to the boson energy ω. Moreover, B_ω and F_ω are understood to be "specific," i.e. an energy density and flux per unit energy interval. The velocity is $\beta_\omega = [1 - (m/\omega)^2]^{1/2}$ for a boson with mass m. (Recall that natural units with $\hbar = c = k_B = 1$ are always used.) In local thermal equilibrium the energy density for massive bosons corresponding to the phase-space element $d^3\mathbf{p}$ is $g\,\omega\,(e^{\omega/T} - 1)^{-1}d^3\mathbf{p}/(2\pi)^3$ where g is the number of polarization degrees of freedom and $(e^{\omega/T} - 1)^{-1}$ is the thermal boson occupation number. With an angular integration $d^3\mathbf{p}$ becomes $4\pi p^2 dp$ where $p = |\mathbf{p}|$. Using $p = (\omega^2 - m^2)^{1/2}$ one finds $p\,dp = \omega\,d\omega$ so that

$$B_\omega = \frac{g}{2\pi^2}\frac{\omega^2(\omega^2 - m^2)^{1/2}}{e^{\omega/T} - 1}. \tag{1.17}$$

The total energy flux is found by integrating over all frequencies,

$$F = -\tfrac{1}{3}\,\nabla T \int_m^\infty d\omega\,\beta_\omega \ell_\omega \partial_T B_\omega, \tag{1.18}$$

where $\nabla B_\omega = \partial_T B_\omega \nabla T$ was used with $\partial_T = \partial/\partial T$.

For photons, one usually writes $F = -(3\kappa_\gamma\rho)^{-1}\nabla aT^4$ where aT^4 is the total energy density in photons ($a = \pi^2/15$). Together with Eq. (1.18) this defines the photon Rosseland mean opacity κ_γ. For other bosons one defines a corresponding quantity,

$$\frac{1}{\kappa_x\rho} \equiv \frac{1}{4aT^3}\int_m^\infty d\omega\,\ell_\omega\beta_\omega\partial_T B_\omega. \tag{1.19}$$

The "exotic" opacity thus defined appears in the stellar structure equation in the way indicated by Eq. (1.7).

The production and absorption of bosons involves a Bose stimulation factor. This effect is taken account of by including a factor $(1 - e^{-\omega/T})$ under the integral in Eq. (1.19). The "absorptive" opacity thus derived is usually referred to as the *reduced opacity* κ^* which is the quantity relevant for energy transfer. In practice, it is not very different from κ as ω is typically $3T$ for massless bosons.

In the large-mass limit ($m \gg T$) the reduction factor may be ignored entirely and one finds to lowest order,

$$\frac{1}{\kappa_x\rho} = g\,\frac{15}{4\pi^4}\left(\frac{m}{T}\right)^3 e^{-m/T}\int_0^\infty dy\,\ell_\omega(y)\,y\,e^{-y}, \tag{1.20}$$

where $y \equiv \beta_\omega^2 m/2T$ was used so that the energy of a nonrelativistic boson is given by $\omega = m + yT$.

1.3.5 Solar Bound on Massive Pseudoscalars

In order to illustrate the energy-loss and energy-transfer argument for a boson with arbitrary mass I use the example of pseudoscalars χ which couple to electrons by a "fine-structure constant" α_x. This is the only case in the literature where both arguments have been applied without assuming that the particle mass m_x is small relative to T.

The new bosons can be produced and absorbed by a variety of reactions. For purposes of illustration I focus on the Compton-type process $\gamma e \leftrightarrow e\chi$. For $m_x \ll T$ the energy-loss rate per unit mass will be given in Eq. (3.23) in a more general context. For arbitrary masses it was derived by Raffelt and Starkman (1989). The result is

$$\epsilon_x = \frac{160\alpha\alpha_x}{\pi}\, \frac{Y_e T^6}{m_u m_e^4}\, F(m_x/T), \tag{1.21}$$

where $F(z) = 1$ for $m_x \ll T$ and $F(z) = (20\sqrt{2})^{-1}\, z^{9/2}\, e^{-z}$ for $T \ll m_x \ll m_e$. Further, Y_e is the number of electrons per baryon and m_u is the atomic mass unit. (It was used that approximately $n_e/\rho = Y_e/m_u$ with the electron density n_e.)

For $T \ll m_x \ll m_e$ the Compton absorption rate is found to be $\Gamma_x = 4\pi\alpha\alpha_x m_x^2 m_e^{-4} n_e$, leading to an opacity contribution of the pseudo-

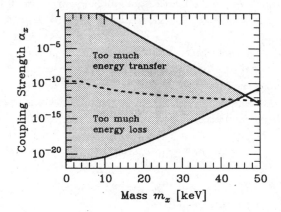

Fig. 1.1. Effects of massive pseudoscalar particles on the Sun (interior temperature about 1 keV). Above the dashed line they contribute to the radiative energy transfer, below they escape freely and drain the Sun of energy. The shaded area is excluded by this simple argument. (Adapted from Raffelt and Starkman 1989.)

scalars of

$$\kappa_x = \frac{2(2\pi)^{9/2}\alpha\alpha_x}{45} \frac{Y_e\, T^{5/2}\, e^{m_x/T}}{m_x^{1/2}\, m_e^4}$$

$$= 4.4\times10^{-3}\,\mathrm{cm}^2\,\mathrm{g}^{-1} \times \alpha_x\, Y_e\, m_{\mathrm{keV}}^{-0.5}\, T_{\mathrm{keV}}^{2.5}\, e^{m_x/T}, \qquad (1.22)$$

where $m_{\mathrm{keV}} = m_x/\mathrm{keV}$ and $T_{\mathrm{keV}} = T/\mathrm{keV}$.

The observed properties of the Sun then allow one to exclude a large range of parameters in the m_x-α_x-plane. The energy-loss rate integrated over the entire Sun must not exceed $L_\odot$. Moreover, κ_x^{-1} must not exceed the standard photon contribution $\kappa_\gamma^{-1} \approx 1\,\mathrm{g/cm}^2$, apart from perhaps a factor of order unity. Taking a typical solar interior temperature of $1\,\mathrm{keV}$, these requirements exclude the shaded region in Fig. 1.1. The dashed line marks the parameters where the mfp is of order the solar radius.

1.4 General Lesson

What have we learned? Weakly interacting particles, if they are not trapped in stars, carry away energy. For a degenerate object this energy loss leads to additional cooling, for a burning star it leads to an accelerated consumption of nuclear fuel. New particles would cause significant effects only if they could compete with neutrinos which already carry away energy directly from the interior.

If new particles are trapped because of a short mfp, they contribute to the energy transfer. They dominate unless their mfp is shorter than that of photons. Such particles probably do not exist or else they would have been found in laboratory experiments. This justifies the usual focus on the energy-loss argument or "free streaming" limit.

Still, the impact of new low-mass particles is maximized for an mfp of order the stellar radius, a fact which is often not appreciated in the literature.

If the particles were so heavy that they could not be produced by thermal processes in a stellar plasma they would be allowed for any coupling strength. How heavy is heavy? The average energy of blackbody photons is about $3T$. Therefore, if $m_x \lesssim 3T$ the particle production will not be significantly suppressed. However, the solar example of Fig. 1.1 illustrates that for a sufficiently large coupling strength even a particle with a mass of $30\,T$ could have a significant impact. While it can be produced only by plasma constituents high up in the tails of the thermal distributions, the Boltzmann suppression factor $e^{-m_x/T}$

can be compensated by a strong coupling. For particles with a different interaction structure the equivalent of Fig. 1.1 would look qualitatively similar, albeit different in detail.

Particles are not harmless in stars just because they are trapped, or just because their mass exceeds a typical temperature. However, in practice new particles are usually thought to be very weakly interacting and either essentially massless on the scales of stellar temperatures, or else very massive on those scales. In this book I will focus on low-mass particles with very weak interactions so that the energy-loss argument will be in the foreground of the discussion.

The opposite case of weakly interacting massive particles (WIMPs) is of great interest in the framework of a hypothesis which holds that such objects are the dark matter of the universe. (Another dark matter candidate are axions which fall into the low-mass category covered in this book.) WIMP masses would be in the 10 GeV range and above so that they cannot be produced in stars, but they can be captured from the galactic background. They would contribute to the energy transfer by conduction. Hence their role in stars is similar to that of electrons, except for their large mfp which allows them to contribute significantly even if their concentration is low. I do not treat this subject because it requires entirely different formal tools from those relevant for low-mass particles. The fascinating story of WIMPs in stars will be told comprehensively within a forthcoming review paper *Supersymmetric Dark Matter* by Jungman, Kamionkowski, and Griest (1995).

Chapter 2

Anomalous Stellar Energy Losses Bounded by Observations

After a description of the main phases of stellar evolution, a review is given of the observations that have been used to constrain novel stellar energy losses. The main arguments involve the cooling speed of white dwarfs and old neutron stars, the delay of helium ignition in low-mass red giants, and the helium-burning lifetime of horizontal-branch stars. The latter two arguments, which are based on observations of globular-cluster stars, are cast into the form of an easy-to-use analytic criterion that allows for a straightforward application in many different cases. The cooling speed of nascent neutron stars is discussed in Chapter 11 in the context of supernova neutrinos.

2.1 Stages of Stellar Evolution

2.1.1 The Main Sequence

To identify those observables of stellar structure and evolution that can be used to constrain or perhaps discover a novel energy-loss mechanism it is necessary to understand how stars live and die. In Chapter 1 the basics of the theory of stellar structure have already been discussed. It is now time to give an overview of the stages of stellar evolution, and how they manifest themselves in the observable properties of stars. Because stellar evolution is an old subject, the literature is vast. I will give reference only to a few recent original papers that I need to support specific points. Otherwise, detailed quantitative accounts of

23

stellar structure and evolution and their observational consequences are found in the textbook and review literature.[3]

How do stars form in the first place? While a detailed understanding of this process is still elusive, for the present discussion it is enough to know that gravitationally bound clouds of gas ultimately fragment and condense because of their continuous energy loss by electromagnetic radiation. As discussed in Sect. 1.2.2, the negative specific heat of a self-gravitating system enforces its contraction and heating. For the fragmentation process a variety of parameters may be important such as overall pressure, angular momentum, or magnetic fields.

Because the formation process of stars is poorly understood the initial mass function (IMF) is not accounted for theoretically. Typically, the number of stars per mass interval may be given by Salpeter's IMF, $dN/d\mathcal{M} \propto (\mathcal{M}/\mathcal{M}_\odot)^{-1.35}$, which means that there are a lot more small than large stars. However, the overall range of stellar masses is quite limited. The largest stars have masses of up to $100\,\mathcal{M}_\odot$; a value beyond this limit does not seem to allow for a stable configuration. At the small mass end, stars with $\mathcal{M} < 0.08\,\mathcal{M}_\odot$ never become hot enough to ignite hydrogen; such "brown dwarfs" have never been unambiguously detected. It has been speculated that they could make up the dark matter of spiral galaxies. A search in our galaxy by the gravitational microlensing technique has yielded first candidates (Alcock et al. 1993, 1995; Aubourg et al. 1993, 1995).

The first stars consist of a mixture of $X \approx 0.75$ (mass fraction of hydrogen) and $Y \approx 0.25$ (mass fraction of helium), the material left over from the big bang of the universe. Subsequent generations also contain a small amount of "metals" which in astronomers' language is anything heavier than helium; our Sun has a metallicity $Z \approx 0.02$ (mass fraction of metals). These heavier elements were bred by nuclear fusion in earlier generations of stars which returned some of their mass to the interstellar

---

[3]An excellent starting point for the nonexpert is Shu (1982). The classics are Chandrasekhar (1939), Schwarzschild (1958), Clayton (1968), and Cox and Giuli (1968). A recent general textbook is Kippenhahn and Weigert (1990). Recent monographs specializing on the Sun in general are Stix (1989), and on solar neutrinos Bahcall (1989). The theory of stellar pulsation is covered in Cox (1980). A classic on the physics of compact stars is Shapiro and Teukolsky (1983). A recent monograph on neutron stars is Lipunov (1992) and on pulsars Mészáros (1992). Recent collections of papers or conference proceedings are available on the physics of red giants (Iben and Renzini 1981), white dwarfs (Barstow 1993), pulsating stars (Schmidt 1989), neutron stars (Pines, Tamagaki, and Tsuruta 1992), and supernovae (Brown 1988; Petschek 1990). Many excellent reviews are found in the *Annual Review of Astronomy and Astrophysics*.

medium. Mass loss in advanced stages of stellar evolution can take on the benign form of a "stellar wind," or it can occur in gigantic "supernova explosions." The end state of stellar evolution must be complete disruption, or a compact object (white dwarf, neutron star, or black hole), because a normal star, supported by thermal pressure, can exist only as long as its nuclear fuel lasts (Sect. 1.2.2). The masses of degenerate stars are constrained by their Chandrasekhar limit of about $1.4\,\mathcal{M}_\odot$ for white dwarfs, and a similar value for neutron stars. Most massive stars seem to return enough mass to the interstellar medium that their typical end states are degenerate stars, not black holes. The primordial material evolves chemically because it is cycled through one generation of stars after another.

Today, the continuing birth and death of stars takes place mostly in the disks of spiral galaxies. Such galaxies also have a population of old halo stars and of globular clusters, about 150 in our Milky Way, each of which is a gravitationally bound system of about 10^6 stars (Fig. 2.1). Most of the globular clusters are in the galactic halo, far away from the disk. The gravitational escape velocity for stars or gas from a cluster is rather small, about $10\,\mathrm{km\,s}^{-1}$. A supernova explosion, on the other

Fig. 2.1. Globular cluster M3. (Image courtesy of Palomar/Caltech.)

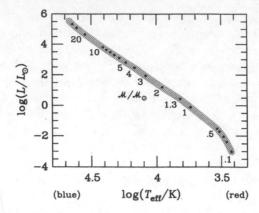

Fig. 2.2. Zero-age main sequence in the Hertzsprung-Russell diagram, with a composition of 68.5% hydrogen, 29.4% helium, and 2.1% metals. (Adapted from Kippenhahn and Weigert 1990.)

hand, ejects its material with typical velocities of several 10^3 km s^{-1}. A single supernova is enough to sweep a globular cluster clean of all gas; then no further star formation is possible.[4] Therefore, these systems are very clean laboratories for studies of stellar evolution as they contain an early generation of stars (typical metallicities $10^{-3}-10^{-4}$) which in a given cluster are all coeval and have almost equal chemical compositions. To a first approximation the stars in a globular cluster differ only in one parameter—their initial mass.

The virial theorem explains (Sect. 1.2.2) that the negative specific heat of a protostellar cloud leads to contraction and heating until it has reached a thermal equilibrium configuration where the nuclear burning rate exactly balances its overall luminosity. At this point the configuration is determined entirely by its mass, apart from a small influence of the metal content. Therefore, after the initial contraction hydrogen-burning stars of different mass form the "zero-age main sequence" in a Hertzsprung-Russell diagram where the effective surface temperature is plotted on the horizontal axis and the stellar luminosity on the vertical axis (Fig. 2.2). As hydrogen is consumed small adjustments of the configuration occur. However, essentially a star remains at its initial main-sequence location for most of its life.

---

[4]The escape velocity from our galaxy is about 500 km s^{-1}. However, in a galactic spiral arm the interstellar medium is dense enough to dissipate a SN explosion which is only able to blow a hole into the interstellar material.

A $1\,\mathcal{M}_\odot$ star lives about $10\,\mathrm{Gyr}$ ($1\,\mathrm{Gyr} = 10^9\,\mathrm{yr}$) on the main sequence; our Sun is thought to have completed about half of this episode. Heavier stars burn brighter (crudely $L \propto \mathcal{M}^3$) and thus live shorter lives. Because the universe is $10{-}20\,\mathrm{Gyr}$ old, stars with $\mathcal{M} \lesssim 0.7{-}0.9\,\mathcal{M}_\odot$ have not yet completed their hydrogen-burning phase, even if they formed shortly after the big bang. Because globular clusters formed very early one expects a main-sequence turnoff

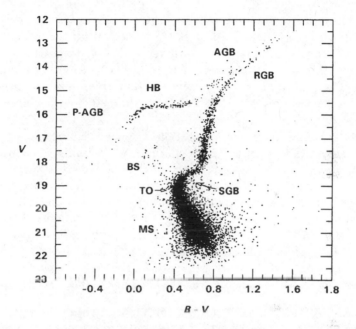

Fig. 2.3. Color magnitude diagram for the globular cluster M3 according to Buonanno et al. (1986), based on the photometric data of 10,637 stars. Following Renzini and Fusi Pecci (1988) the following classification has been adopted for the evolutionary phases. MS (main sequence): core hydrogen burning. BS (blue stragglers). TO (main-sequence turnoff): central hydrogen is exhausted. SGB (subgiant branch): hydrogen burning in a thick shell. RGB (red-giant branch): hydrogen burning in a thin shell with a growing core until helium ignites. HB (horizontal branch): helium burning in the core and hydrogen burning in a shell. AGB (asymptotic giant branch): helium and hydrogen shell burning. P-AGB (post-asymptotic giant branch): final evolution from the AGB to the white-dwarf stage. (Original of the figure courtesy of A. Renzini.)

near this mass, depending somewhat on their metallicity. Indeed, the color-magnitude diagram[5] of the cluster M3 (Fig. 2.3) clearly shows this effect. The location of the turnoff (TO) point in such a diagram can be converted to the mass and thus to the age of a star which is just about to complete hydrogen burning. Globular clusters, therefore, allow one to establish a significant lower limit for the age of the universe.

2.1.2 Becoming a Red Giant

A main-sequence star naturally is hottest and densest at the center where hydrogen is depleted fastest. (See Fig. 2.4 for the overall structure of a typical solar model which represents a low-mass star about halfway through its main-sequence evolution.) After all of the hydrogen is consumed at the center, the further evolution depends on the total mass—I begin with the fascinating story of low-mass stars ($\mathcal{M} \lesssim 2\,\mathcal{M}_\odot$).

After central hydrogen exhaustion a new configuration establishes itself where the "helium ashes" which have accumulated at the center form an ever denser core surrounded by a hydrogen burning shell (Fig. 2.5). At the same time the material above the core begins to *expand* which implies that the stellar surface becomes larger and thus the surface temperature lower according to the Stefan-Boltzmann law. Put another way, the star becomes redder, and ultimately a "red giant." In the color-magnitude diagram of the globular cluster M3 (Fig. 2.3), the collection of stars which begins to form a contracting core and an expanding envelope is marked as the subgiant branch (SGB).

---

[5]The brightness of a star is photometrically measured on the basis of the luminosity in a wavelength band defined by a filter with a well-defined spectral response. There exist many different color systems, corresponding to many different filters. In Fig. 2.3 the visual brightness is defined by $V = -2.5\log L_V + \text{const.}$ with L_V the luminosity measured with the "visual" filter, centered in the yellow-green waveband, and a similar definition applies to the brightness B in the blue. The "color" $B - V$ is a measure of the surface temperature with lower temperatures (redder color) to the right. For technical definitions see Allen (1963). Note that the downward turn of the horizontal branch in the blue is determined by the V filter; bolometrically the HB is truly horizontal. An absolute bolometric brightness is given by

$$M_{\text{bol}} = 4.74 - 2.5\log(L/L_\odot).$$

It is a dimensionless number, but often the unit mag (magnitude) is appended for clarity. The visual brightness in Fig. 2.3 is given in apparent magnitudes which depend on the distance of the cluster. However, because of the logarithmic brightness definition a different distance leads only to a vertical shift of the entire picture; the color and relative brightnesses remain unchanged.

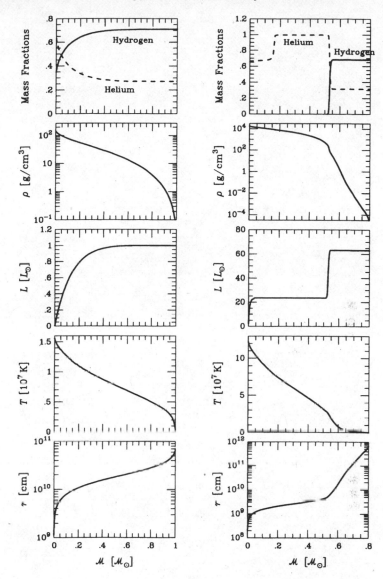

Fig. 2.4. *Left panels:* A typical solar model (Bahcall 1989) as an example for a low-mass star which is halfway through its hydrogen-burning (main-squence) phase. *Right panels:* Horizontal-branch star (central helium burning, shell hydrogen burning) with a metallicity $Z = 0.004$ after about 2.5×10^7 yr which is about a quarter of the HB lifetime. (Model from the calculations of Dearborn et al. 1990.)

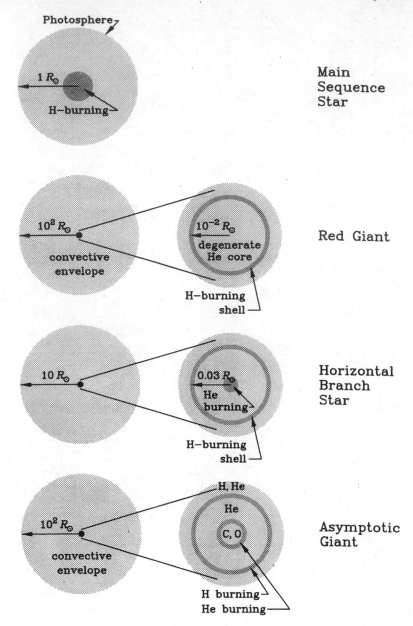

Fig. 2.5. Main evolutionary phases of low-mass stars. The envelope and core dimensions depend on the location on the RGB, HB, or AGB, respectively. The given radii are only meant to give a crude orientation.

As hydrogen burning continues it dumps more and more helium on the core which at first supports itself by thermal pressure. Soon it becomes so dense, however, that the electrons become degenerate, inverting the mass-radius relationship to $R \propto \mathcal{M}^{-1/3}$ (Sect. 1.2.2). Thus, as the helium core mass $\mathcal{M}_c$ grows, the core radius R_c shrinks. The gravitational potential Φ_c at the edge of the core is determined entirely by the core because the envelope contributes little due to its large extension so that $\Phi_c \approx -G_N \mathcal{M}_c/R_c \propto \mathcal{M}_c^{4/3}$. Because the hydrogen-burning shell above the core still supports itself by thermal pressure, the temperature near the core edge is determined by $\Phi_c \propto \mathcal{M}_c^{4/3}$. The growing core mass causes the hydrogen burning shell to become ever hotter! Because of the steep T dependence of the hydrogen burning rates, the growing core causes the star to become ever brighter (Fig. 2.6). *In this*

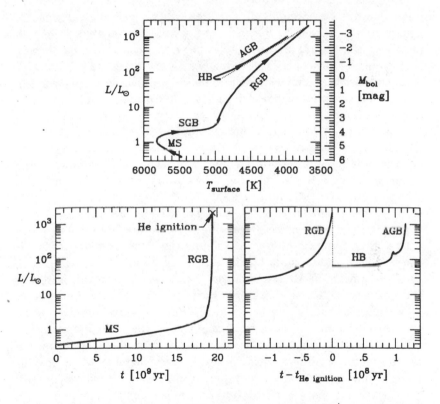

Fig. 2.6. Evolutionary track of a $0.8\,\mathcal{M}_\odot$ star ($Z = 0.004$) from zero age to the asymptotic giant branch. The evolutionary phases are as in Fig. 2.3. (Calculated with Dearborn's evolution code.)

shell-burning phase the luminosity is determined almost entirely by the core mass. On the MS it was determined by the total mass, a parameter which now hardly matters.

The red-giant branch (RGB) ascension is a fast process compared with the main-sequence (MS) evolution (Fig. 2.6). Therefore, the red giants in the observational color-magnitude diagram of Fig. 2.3 differ only by very small amounts of their initial mass. The stars beyond the MS have nearly identical properties while those on the MS differ by their total mass. Therefore, the evolved stars in Fig. 2.3 essentially trace out the evolutionary path of a single star with an approximate mass corresponding to the MS turnoff (TO). For example, the stars on the RGB in Fig. 2.3 essentially constitute snapshots of the single-star evolutionary track shown in the upper panel of Fig. 2.6.

A red giant is a star with a compact energy source at the center and a large convective envelope. In the present case we have a degenerate helium core with shell hydrogen burning, but other configurations are possible. The core is a star unto itself with near to negligible feedback from the envelope. The envelope, on the other side, is strongly influenced by the nontrivial central boundary conditions provided by the core. Convective configurations with a fixed mass and a prescribed luminosity from a central point source occupy the "Hayashi line" in the Hertzsprung-Russell diagram. A low-mass red giant ascends the Hayashi line corresponding to its envelope mass.

The inflation of a star to red-giant dimensions is a remarkable phenomenon which defies an intuitive explanation, except that the stellar structure equations allow for such solutions. Another conceivable structure is a much smaller envelope with radiative rather than convective energy transfer—upper MS stars are of that nature. Prescribing a core with a given luminosity one can imagine these two extreme configurations. Indeed, bright stars ($L \gtrsim 10^2 L_\odot$) are usually found either near the MS or near the RGB which together form a V-shaped pattern in the Hertzsprung-Russell diagram (Fig. 2.9). The empty space in the V is known as the "Hertzsprung gap"—the few stars found there are thought to be on the move from one arm of the V to the other, i.e. from a radiative to a convective envelope structure. Massive stars after completing hydrogen burning move almost horizontally across the Hertzsprung gap and inflate to red-giant dimensions. Remarkably, they can deflate and move horizontally back, executing a "blue loop."

2.1.3   Helium Ignition

The core of a red giant reaches its limiting mass when it has become so hot and dense that helium ignites. Because the nucleus ^{8}Be which consist of two α particles (He nuclei) is not stable, He burning proceeds directly to carbon, $3\alpha \to {}^{12}$C ("triple-α reaction"), via an intermediate ^{8}Be state. Because it is essentially a three-body reaction its rate depends sensitively on ρ and T. In a red-giant core helium ignites when $\mathcal{M}_c \approx 0.5\,\mathcal{M}_\odot$ with central conditions of $\rho \approx 10^6\,\mathrm{g\,cm}^{-3}$ and $T \approx 10^8\,$K where the triple-α energy generation rate per unit mass, $\epsilon_{3\alpha}$, varies approximately as $\rho^2 T^{40}$. This steep temperature dependence allows one to speak of a sharp ignition point even though there is *some* helium burning at any temperature and density.

The helium core of a red giant is like a powder keg waiting for a spark. When the critical temperature is reached where $\epsilon_{3\alpha}$ exceeds the neutrino losses a nuclear runaway occurs. Because the pressure is mainly due to degenerate electrons the energy production at first does not lead to structural changes. Therefore, the rise in temperature is unchecked and feeds positively on the energy generation rate. As this process continues the core expands nearly explosively to a point where it becomes nondegenerate and the familiar self-regulation by the gravitational negative specific heat kicks in (Chapter 1). The "explosion energy" is absorbed by the work necessary to expand the core from about $10^6\,\mathrm{g\,cm}^{-3}$ to about $10^4\,\mathrm{g\,cm}^{-3}$. The core temperature of the final configuration remains at about $10^8\,$K because of the steep temperature dependence of $\epsilon_{3\alpha}$ which allows only for a narrow range of stationary burning conditions.

The final configuration with a helium-burning core and a hydrogen-burning shell (Fig. 2.5) is known as a horizontal-branch (HB) star, a term which is justified by the location of these objects in the color-magnitude diagram Fig. 2.3. Note that the overall luminosity has *decreased* by the process of helium ignition (Fig. 2.6) because of the core expansion which lowers the gravitational potential at the core edge and thus the temperature in the hydrogen-burning shell which continues to be regulated by the core mass and radius. The total luminosity of an HB star is given to about 1/3 by He burning and 2/3 by hydrogen shell burning (Fig. 2.4, right panels).

Because helium ignition is an almost explosive process on dynamical time scales it is known as the "helium flash." For the same reason, a realistic numerical treatment does not seem to exist (for a review, see Iben and Renzini 1984). There are two main problems. First, normal stellar-evolution codes use hydrostatic stellar structure equations which ignore kinetic energies which are small in stationary phases or phases of slow expansion or contraction (Sect. 1.2.1). However, in a phase of fast expansion kinetic energies can be large. It is not known whether the helium flash is a hydrostatic or a hydrodynamic event. Secondly, convection undoubtedly plays a large role at transferring energy from the ignition point. Note that helium ignites off-center because neutrino losses cause a temperature dip at the center. A fundamental theory of convection does not exist; it is likely that one would have to perform a three-dimensional hydrodynamic calculation to develop a full understanding of the helium flash.

2.1.4   The Horizontal Branch

After helium ignition the overall luminosity has dropped as explained above, and the surface has shrunk considerably, leading to a substantially hotter (bluer) configuration. Precisely how blue an HB star becomes depends on the size of its envelope and thus on its total mass. A substantial amount of mass loss occurs on the RGB where the photosphere of the star is so inflated that it is only weakly gravitationally bound; a red giant may typically lose about $0.2 \, \mathcal{M}_\odot$ of its envelope before helium ignites. Presumably, the amount of mass loss is not exactly the same from star to star; a small spread of order $0.03 \, \mathcal{M}_\odot$ is enough to explain the wide range of surface radii and thus colors found for these objects which form a horizontal branch in the color-magnitude diagram.[6] The downward turn of the HB in Fig. 2.3 is an artifact of the filter which determines the visual brightness—the bolometric brightness is the same. It is fixed by the core properties.

In practice, the HB morphology found in different clusters is complicated. Some clusters have widely spread HBs such as that shown in Fig. 2.3 while others have stars only near the red or blue end, and gaps and blue tails occur. One important parameter is the metallicity of the envelope which determines the opacity to which the envelope structure is very sensitive. However, one finds clusters with the same chemical composition but different HB morphologies, a phenomenon which has prompted a search for the "second parameter." Many suggestions have been made by some and refuted by others. One candidate second parameter which appears to remain viable is the cluster age, or rather, the MS mass of the stars observed today on the HB. The envelope mass

---

[6]For recent synthetic HBs see Lee, Demarque, and Zinn (1990, 1994).

remaining for an HB star after mass loss on the RGB will still depend
on its initial value, rendering this a rather natural possibility. However, it implies that the globular clusters of our galaxy did not form
at the same time—an age spread of several Gyr is required with the
older clusters predominantly at smaller galactocentric distances (Lee,
Demarque, and Zinn 1994 and references therein). Apart from some
anomalous cases such an age spread appears to be enough to explain
the second parameter phenomenon.

Returning to the inner structure of an HB star, it evolves quietly
at an almost fixed total luminosity (Fig. 2.6). The core constitutes
essentially a "helium main-sequence star." Because its inner core is
convective it dredges helium into the nuclear furnace at the very center,
leaving a sharp composition discontinuity at the edge of the convective
region (Fig. 2.4). After some ^{12}C has been built up, ^{16}O also forms. At
the end of the HB phase, the helium core has developed an inner core
consisting of carbon and oxygen.

2.1.5   From Asymptotic Giants to White Dwarfs

After the exhaustion of helium at the center a degenerate carbon-
oxygen (CO) core forms with helium shell burning and continuing hydrogen shell burning (Fig. 2.5). Again, the star grows progressively
brighter and inflates. Put another way, it becomes very similar to a
star which first ascended the RGB: it ascends its Hayashi line for a
second time. The track in the Hertzsprung-Russell diagram asymptotically approaches that from the first ascent ("asymptotic giants").
The upper RGB can be observationally difficult to distinguish from the
asymptotic giant branch (AGB) even though they are reasonably well
separated in Fig. 2.3.

In low-mass stars carbon and oxygen never ignite. The shell sources
extinguish when most of the helium and hydrogen has been consumed
so that the star has lost its entire envelope either by hydrogen burning
or by further mass loss on the AGB. The remaining degenerate CO
star continues to radiate the heat stored in its interior. At first these
stars are rather hot, but geometrically very small with a typical radius of 10^4 km. Their small surface area restricts their luminosity in
spite of the high temperature and so they are referred to as "white
dwarfs." Because they are supported by electron degeneracy pressure,
their remaining evolution is cooling by neutrino emission from the interior and by photon emission from the surface until they disappear
from visibility.

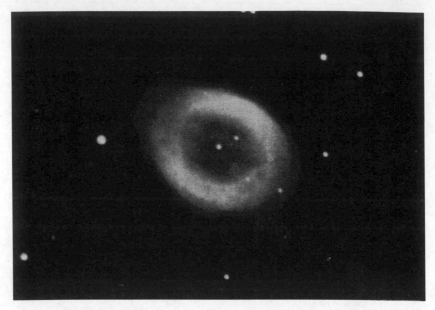

Fig. 2.7. Ring Nebula in Lyra (M57), a planetary nebula. (Image courtesy of Palomar/Caltech.)

For stars which start out sufficiently large the mass loss on the AGB can be so dramatic that one may speak of the ejection of the entire envelope. It forms a large shell of gas which is illuminated by its central star, the newborn white dwarf. Such systems are known as "planetary nebulae" (Fig. 2.7).

2.1.6   Type I Supernovae

A white dwarf can make a spectacular reappearance if it is in a binary system. If the other member expands because it is in an earlier evolutionary phase it can transfer mass, allowing for renewed nuclear burning on the white-dwarf surface. This adds mass to the CO configuration which ultimately becomes so hot and dense that these elements ignite, leading to a (subsonic) deflagration or (supersonic) detonation front which sweeps through the star and disrupts it entirely. This course of events is the standard scenario for type I supernova explosions which are among the most energetic events known in the universe. They must be carefully distinguished from type II supernovae which are related to the collapse of evolved, massive stars.

2.1.7 Intermediate-Mass Stars

The evolutionary scenario described so far applies to stars with a mass below $2-3\,\mathcal{M}_\odot$. For larger masses the core conditions evolve continuously to the ignition of helium which is then a quiet process. For a given chemical composition the transition between the two scenarios is a sharp function of the total mass. Because red giants are very bright they dominate the total luminosity of an old stellar population so that this transition affects its integrated brightness in a discontinuous way. Some authors have used the concept of an "RGB phase transition" to describe this phenomenon (Sweigart, Greggio, and Renzini 1990, and references therein; Bica et al. 1991).

Stars with masses of up to $6-8\,\mathcal{M}_\odot$ are expected to end up as CO white dwarfs. Their evolution on the AGB can involve many interesting phenomena such as "thermal pulses"—for a key to the literature see Iben and Renzini (1984). Intermediate-mass stars have thus far played little role for the purposes of particle astrophysics, and so their evolution does not warrant further elaboration in the present context.

2.1.8 Massive Stars and Type II Supernovae

The course of evolution for massive stars ($\mathcal{M} > 6-8\,\mathcal{M}_\odot$) is qualitatively different because they ignite carbon and oxygen in their core, allowing them to evolve further. This is possible because even after mass loss they are left with enough mass that their CO core grows toward the Chandrasekhar limit. Near that point the density is high enough to ignite carbon which causes heating and thus temporarily relieves the electron degeneracy. Next, the ashes of carbon burning (Ne, Mg, O, Si) form a degenerate core which ultimately ignites Ne burning, and so forth. Ultimately, the star has produced a degenerate iron core, surrounded by half a dozen "onion rings" of different burning shells.

The game is over when the iron core reaches its Chandrasekhar limit because no more nuclear energy can be released by fusion. The temperature is at 0.8×10^{10} K = 0.7 MeV, the density at 3×10^9 g cm^{-3}, and there are about $Y_e = 0.42$ electrons per baryon. Further contraction leads to a negative feedback on the pressure as photons begin to dissociate iron, a process which consumes energy. Electrons are absorbed and converted to neutrinos which escape, lowering the electron Fermi momentum and thus the pressure. Therefore, the core becomes unstable and collapses, a process which is intercepted only when nuclear density (3×10^{14} g cm^{-3}) is reached where the equation of state stiffens.

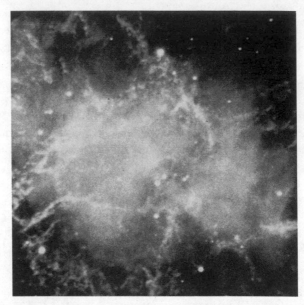

Fig. 2.8. The Crab nebula, remnant of the supernova of A.D. 1054. (Image courtesy of the European Southern Observatory.)

At this point a shock wave forms at the edge of the core and moves outward. The implosion can be said to be reflected and thus turned into an explosion. In practice, it is difficult to account for the subsequent evolution as the shock wave tends to dissipate its energy by dissociating iron. Currently it is thought that a revival of the stalled shock is needed and occurs by neutrinos depositing their energy in the "hot bubble" below the shock. This region has a low density yet high temperature, and thus a high entropy per baryon by common astrophysical standards. Within about 0.3 s after collapse the shock has moved outward and ejects the entire overburden of the mantle and envelope. This course of events is the scenario of a type II supernova (SN) explosion.

What remains is an expanding nebula such as the Crab (Fig. 2.8) which is the remnant of the SN of A.D. 1054, and a central neutron star (radius about 10 km, mass about $1\,\mathcal{M}_\odot$) which often appears in the form of a pulsar, a pulsating source of radiation in some or all electromagnetic wave bands. The pulsed emission is explained by a complicated interplay between the fast rotation and strong magnetic fields (up to $10^{12}-10^{13}$ G) of these objects.

Returning to the moment after collapse of the iron core, it is so dense (nuclear density and above) and hot (temperature of several 10 MeV), that even neutrinos are trapped. Therefore, energy and lepton number are lost approximately on a neutrino diffusion time scale of several seconds. The neutrinos from stellar collapse were observed for the first and only time when the star Sanduleak −69 202 in the Large Magellanic Cloud (a small satellite galaxy of the Milky Way) collapsed. The subsequent explosion was the legendary SN 1987A.

After the exhaustion of hydrogen, massive stars move almost horizontally across the Hertzsprung-Russell diagram until they reach their Hayashi line, i.e. until they have become red supergiants. However, subsequently they can loop horizontally back into the blue; the progenitor of SN 1987A was such a blue supergiant.

The SN rate in a spiral galaxy like our own is thought to be about one in a few decades, pessimistically one in a century. Because many of the ones occurring far away in our galaxy will be obscured by the dust and gas in the disk, one has to be extremely lucky to witness such an event in one's lifetime. The visible galactic SNe previous to 1987A were Tycho's and Kepler's in close succession about 400 years ago. Both of them may have been of type I—no pulsar has been found in their remnants.[7] Of course, in the future it may become possible to detect optically invisible galactic SNe by means of neutrino detectors like the ones which registered the neutrinos from SN 1987A.

2.1.9 Variable Stars

Stars are held in equilibrium by the pull of gravity which is opposed by the pressure of the stellar matter; its inertia allows the system to oscillate around this equilibrium position. If the adiabatic relationship between pressure and density variations is written in the form $\delta P/P = \gamma \, \delta\rho/\rho$, the fundamental oscillation period P of a self-gravitating homogeneous sphere (density $\bar{\rho}$) is found to be[8] $P^{-1} = [(\gamma - \frac{4}{3})G_N\bar{\rho}/\pi]^{1/2}$. This yields the period-mean density relationship $P(\bar{\rho})^{1/2} \approx$ const. which is often written in the form $P(\bar{\rho}/\bar{\rho}_\odot)^{1/2} = Q$ with the average

---

[7]Observationally, type I and II SNe are distinguished by the absence of hydrogen spectral lines in the former which is explained by their progenitor being an accreting white dwarf which explodes after carbon ignition. Therefore, it is difficult to establish the type of a historical SN unless a pulsar is detected as in the Crab.

[8]For $\gamma < 4/3$ the star is dynamically unstable. We have already encountered this magic number for the relationship between pressure and density of a degenerate relativistic electron gas where it led to Chandrasekhar's limiting white-dwarf mass.

solar density $\bar{\rho}_\odot = 1.41\,\mathrm{g\,cm}^{-3}$ and the pulsation "constant" Q. It is in the range $0.5-3\,\mathrm{h}$, depending on the adiabiatic coefficient γ and the detailed structure of the star.

When these oscillations are damped, stars are found in their equilibrium configurations with constant color and brightness. However, it is possible that a small deviation from equilibrium is amplified, leading to a growing oscillation amplitude. Throughout a star there can be regions which try to excite oscillations, and others which damp them. If the driving mechanism is strong enough the star is found in a continuing oscillation which manifests itself in an oscillating lightcurve.

Of particular interest are the Cepheid-type variables. Their oscillations are excited by the "κ mechanism" where the driving force is the heat valve provided by the opacity of the stellar matter near the surface. There, hydrogen and helium are only partially ionized so that a temperature increase leads to increased ionization which increases the opacity because the Thomson cross section on free electrons is much larger than the Rayleigh one on neutral atoms. Therefore, the temperature increase caused by compression makes it more difficult for energy to escape: the valve shuts. In most regions of the star the opposite happens. Increased temperature makes it easier for energy to leak out from the compressed region, the valve opens, and oscillations are damped.

Because the operation of the κ mechanism is determined by conditions near the stellar surface which is characterized by its temperature and luminosity, Cepheid-type variables are found on a certain locus of color and brightness. This "instability strip" extends throughout the Hertzsprung-Russell diagram (hatched band in Fig. 2.9); it applies to stars of radically different internal structure. Variable stars are found wherever the instability strip overlaps with an actual stellar population.

At the bright end of the strip which is close and nearly parallel to the red-giant Hayashi line one finds the classical Cepheids (δ Cepheids) with luminosities $300-30,000\,L_\odot$ and periods $1-50\,\mathrm{days}$. The brightness variation can be up to $1\,\mathrm{mag}$ (visual), or a luminosity which changes by up to a factor of 3. Classical Cepheids are massive stars which cross the otherwise unpopulated Hertzsprung gap from the main-sequence to the red-giant region, and red giants and supergiants which execute blue loops and so temporarily move to the blue of the Hayashi line. The linear relationship between period and brightness of Cepheids as well as their large intrinsic luminosities are the key for their prominent role as standard candles and thus as astronomical distance indicators.

The crossing of the instability strip with the HB is the domain of the RR Lyrae stars. Their luminosities are around $50-100\,L_\odot$, their

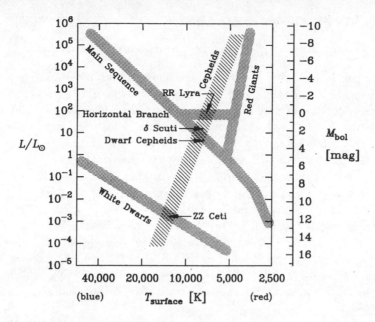

Fig. 2.9. Schematic Hertzsprung-Russell diagram of the main types of stars. The hatched band is the instability strip, the locus of Cepheid-type variables. The "Hertzsprung gap" between the main sequence and the red-giant region is populated with stars crossing between these branches, and by red giants and supergiants which execute blue loops.

periods 1.5 − 24 h. Their color coincides with the MS turnoff in globular clusters (Fig. 2.3). RR Lyrae stars play an important role in the ongoing debate about the age of globular clusters.

Toward fainter luminosities the instability strip crosses the MS, a region where the δ Scuti stars and dwarf Cepheids (periods of 1−3 h) are located.

Fainter still, the only remaining population of stars are white dwarfs. At the crossover of their locus with the instability strip the ZZ Ceti variables are found; they have periods of a few minutes. Because the oscillation period depends on the luminosity, an observed secular change of the period of a ZZ Ceti star can serve as a sensitive diagnostic for a decrease of its luminosity and thus for its cooling speed.

2.2   White-Dwarf Cooling

2.2.1   Theoretical and Observed White-Dwarf Properties

After this general survey of how stars evolve it is time to study individual aspects in more detail, and notably, how stellar evolution is affected by the emission of weakly interacting particles. I begin with the conceptually most transparent case of stars for which the loss of energy simply accelerates their cooling, i.e. white dwarfs and neutron stars.

The former represent the final state of the evolution of stars with initial masses of up to several $\mathcal{M}_\odot$, perhaps up to $8\,\mathcal{M}_\odot$ (Sect. 2.1). For reviews of the theory and observed properties see Hubbard (1978), Liebert (1980), Shapiro and Teukolsky (1983), Weidemann (1990), and D'Antona and Mazzitelli (1990). Because white dwarfs (WDs) are supported by electron degeneracy pressure the hydrostatic and thermal properties are largely decoupled. In Sect. 1.2.2 we had encountered their inverted mass-radius relationship; in a polytropic approximation of the WD structure one finds quantitatively (Shapiro and Teukolsky 1983)

$$R = 10{,}500\,\mathrm{km}\,(0.6\mathcal{M}_\odot/\mathcal{M})^{1/3}\,(2/\mu_e)^{5/3}. \tag{2.1}$$

Here, $\mathcal{M}$ is the stellar mass and $\mu_e = \rho\,m_u^{-1}n_e^{-1} = Y_e^{-1}$ the "mean molecular weight of the electrons" with ρ the mass density, m_u the atomic mass unit, n_e the electron density, and Y_e the number of electrons per baryon. WDs do not contain any hydrogen in their interior—it would immediately ignite—so that $\mu_e = 2$. Typically they consist of carbon and oxygen, the end products of helium burning in the core of the progenitor star. The central density of a polytropic model is

$$\rho_c = 1.46\times10^6\,\mathrm{g\,cm^{-3}}\,(\mathcal{M}/0.6\,\mathcal{M}_\odot)^2\,(\mu_e/2)^5, \tag{2.2}$$

assuming nonrelativistic electrons.

If the mass is so large (the radius so small) that the electrons become relativistic there exists no stable configuration, i.e. the masses of WDs must lie below the Chandrasekhar limit (Shapiro and Teukolsky 1983)

$$\mathcal{M}_{\mathrm{Ch}} = 1.457\,\mathcal{M}_\odot\,(2/\mu_e)^2. \tag{2.3}$$

Observationally the WD mass distribution is strongly peaked near $\mathcal{M} = 0.6\,\mathcal{M}_\odot$ (Weidemann and Koester 1984) so that a nonrelativistic treatment of the electrons is justified. The low mass of observed WDs is understood by the large rate of mass loss during the red giant and asymptotic giant evolution which can amount to an ejection of the entire envelope and thus to the formation of a planetary nebula (Fig. 2.7). A theoretical evolutionary track for a $3\,\mathcal{M}_\odot$ star from the MS to the WD stage was calculated, e.g. by Mazzitelli and D'Antona (1986). The central stars of planetary nebulae are identified with nascent WDs. The rate of WD formation inferred from the luminosity function discussed below agrees within a factor of about 2 with the observed formation rate of planetary nebulae, which means that both quantities agree to within their statistical and systematic uncertainties.

The hottest and brightest WDs have luminosities of $L \approx 0.5\,L_\odot$ while the faintest ones are observed at $L \approx 0.5\times10^{-4}\,L_\odot$. Thus, because of their small surface area WDs are intrinsically faint (see the Hertzsprung-Russell diagram Fig. 2.9). This implies that they can be observed only in the solar neighborhood, for bright WDs out to about 100 pc (300 lyr). Because their vertical scale height in the galactic disk is about 250 pc (Fleming, Liebert, and Green 1986) the observed WDs homogeneously fill a spherical volume around the Sun. One may then express the observed number of WDs in terms of a volume density; it is of order $10^{-2}\,pc^{-3}$.

The observed luminosity function (the space density of WDs per brightness interval) is shown in Fig. 2.10 and listed in Tab. 2.1 according to Fleming, Liebert, and Green (1986) and Liebert, Dahn, and Monet (1988). The operation of a novel cooling mechanism can be constrained by three important features which characterize the luminosity function: its slope, which signifies the form of the cooling law, its amplitude, which characterizes the cooling time and WD birthrate, and its sudden break at $\log(L/L_\odot) \approx -4.7$, which characterizes the beginning of WD formation. Even the oldest WDs have not had time to cool to lower luminosities. From this break one can infer an age for the galactic disk of 8−10.5 Gyr (Winget et al. 1987; Liebert, Dahn, and Monet 1988; Iben and Laughlin 1989; Wood 1992) while Hernanz et al. (1994) find 9.5−12 Gyr on the basis of their cooling calculations which include crystallization effects.

Because of the WD mass-radius relationship the surface temperature and luminosity are uniquely related for a given WD mass. Therefore, instead of the luminosity function one may consider the temperature distribution which, in principle, is independent of uncertain WD distance determinations. Fleming, Liebert, and Green (1986) gave the distribution of their sample of hot WDs in several temperature bins (Tab. 2.2). Numerical cooling calculations of Blinnikov and Dunina-Barkovskaya (1994) found good agreement with this distribution, as-

Table 2.1. Observed WD luminosity function according to Fleming, Liebert, and Green (1986) and Liebert, Dahn, and Monet (1988). For the hot and bright degenerates (upper part of the table) a large fraction of their spectrum lies in the ultraviolet, causing a large discrepancy between the absolute visual magnitude M_V and the absolute bolometric magnitude M_{bol}. For the hot dwarfs the bins orginally had been chosen on the M_V scale with a width of 0.5 mag, centered on the half-magnitudes; the listed M_{bol} is the mean in these intervals.

M_V	Mean M_{bol}	Mean $\log(L/L_\odot)$	dN/dM_{bol} [pc^{-3} mag^{-1}]	$\log(dN/dM_{bol})$
9.5	5.50	−0.31	1.22×10^{-6}	−5.91 (+0.18,−0.31)
10.0	6.88	−0.86	1.01×10^{-5}	−5.00 (+0.14,−0.21)
10.5	7.84	−1.25	2.16×10^{-5}	−4.67 (+0.13,−0.18)
11.0	8.92	−1.68	9.56×10^{-5}	−4.02 (+0.12,−0.16)
11.5	10.12	−2.16	1.21×10^{-4}	−3.92 (+0.11,−0.15)
12.0	11.24	−2.61	1.51×10^{-4}	−3.82 (+0.11,−0.16)
12.5	11.98	−2.90	2.92×10^{-4}	−3.54 (+0.11,−0.16)
13.0	12.55	−3.13	6.07×10^{-4}	−3.22 (+0.20,−0.39)
	13.50	−3.51	0.89×10^{-3}	−3.05 (+0.14,−0.21)
	14.50	−3.91	1.34×10^{-3}	−2.87 (+0.14,−0.20)
	15.50	−4.31	0.24×10^{-3}	−3.62 (+0.18,−0.31)

Table 2.2. Temperature distribution of hot WDs according to Fleming, Liebert, and Green (1986).

$T_{eff}[10^3\,K]$	Fraction of WDs
40−80	$(7.06 \pm 1.27) \times 10^{-3}$
20−40	0.235 ± 0.026
12−20	0.759 ± 0.083

suming that the WD mass function was peaked around $0.7\,M_\odot$ which is somewhat larger than the canonical value of $0.6\,M_\odot$. For the present purpose, however, the smallness of this difference is taken as a confirmation of the standard WD cooling theory.

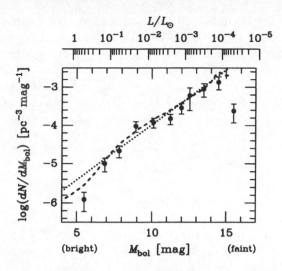

Fig. 2.10. Observed WD luminosity function as in Tab. 2.1. The dotted line represents Mestel's cooling law with a constant WD birthrate of $B = 10^{-3} \, \mathrm{pc^{-3} \, Gyr^{-1}}$. The dashed line is from the numerical cooling curve of a $0.6 \, \mathcal{M}_\odot$ WD (Koester and Schönberner 1986), including neutrino losses and assuming the same constant birthrate.

2.2.2 Cooling Theory

A WD has no nuclear energy sources and so it shines on its residual thermal energy: the evolution of a WD must be viewed as a cooling process (Mestel 1952). Because electron conduction is an efficient mechanism of energy transfer the interior can be viewed, to a first approximation, as an isothermal heat bath with a total amount of thermal energy U. Because the nondegenerate surface layers have a large "thermal resistance," they insulate the hot interior from the cold surrounding space, throttling the energy loss L_γ by photon radiation. Of course, WDs can also lose energy by neutrino volume emission L_ν, and by novel particle emission L_x. Hence, WD cooling is governed by the equation

$$dU/dt = -(L_\gamma + L_\nu + L_x). \tag{2.4}$$

This simple picture ignores the possibility of residual hydrogen burning near the surface, a possibly important luminosity source for young WDs (e.g. Castellani, Degl'Innocenti, and Romaniello 1994; Iben and Tutukov 1984). I will get back to this problem below.

In order to translate this equation into the observable luminosity function I assume a constant WD birthrate B so that the total number density of degenerates is $N = B\,t_{\text{gal}}$ (age of the galactic disk t_{gal}). Taking the above values $N \approx 10^{-2}\,\text{pc}^{-3}$ and $t_{\text{gal}} \approx 9\,\text{Gyr}$ one finds $B \approx 10^{-3}\,\text{pc}^{-3}\,\text{Gyr}^{-1}$. Because the number density of WDs in a given magnitude interval dM_{bol} is proportional to the time interval dt it takes to cool through this magnitude range one readily obtains

$$\frac{dN}{dM_{\text{bol}}} = B\,\frac{dt}{dM_{\text{bol}}} = -B\,\frac{dU/dM_{\text{bol}}}{L_\gamma + L_\nu + L_x}\,. \tag{2.5}$$

The photon luminosity is $L_\gamma = 78.7\,L_\odot\,10^{-2M_{\text{bol}}/5}$ in terms of the bolometric magnitude, equivalent to $\log(L_\gamma/L_\odot) = (4.74 - M_{\text{bol}})/2.5$. L_γ is related to the internal temperature T by the thermal conductance of the surface layers so that one may derive a function $T(L_\gamma)$. The quantities U, L_ν, and L_x are given in terms of T so that they can be expressed in terms of L_γ and hence of M_{bol}.

In hot WDs the thermal energy is largely stored in the nuclei which form a nearly classical Boltzmann gas. At low T the ideal-gas law breaks down and eventually the nuclei arrange themselves in a lattice. The internal energy is then a more complicated function of temperature. The heat capacity per nucleon, which is $\frac{3}{2}$ for an ideal gas, rises to 3 near the Debye temperature Θ_D and then drops approximately as $(16\pi^4/5)\,(T/\Theta_D)^3$ to zero (Shapiro and Teukolsky 1983). However, the observed WDs have a relatively small ρ because of their small mass around $0.6\,\mathcal{M}_\odot$ so that even the oldest WDs have not yet crystallized. Therefore, as a reasonable first approximation the internal energy is $U = C\,T$ with the ideal-gas heat capacity for the entire star of

$$C = \frac{3}{2}\,\frac{\mathcal{M}}{m_u}\,\sum_j \frac{X_j}{A_j} = 3.95\times10^{-2}\,\frac{L_\odot\,\text{Gyr}}{10^7\,\text{K}}\,\frac{\mathcal{M}}{\mathcal{M}_\odot}\,\sum_j \frac{X_j}{A_j}\,, \tag{2.6}$$

where X_j is the mass fraction of the element j, atomic mass A_j, and $m_u = 1.661\times10^{-24}\,\text{g}$ is the atomic mass unit.

The thermal conductance of the surface layers is calculated by solving the stellar structure equations. Using a Kramers opacity law, $\kappa = \kappa_0 \rho\,T^{-7/2}$, one finds (van Horn 1971; Shapiro and Teukolsky 1983)

$$L = 1.7\times10^{-3}\,L_\odot\,\frac{\mathcal{M}}{\mathcal{M}_\odot}\,\left(\frac{T}{10^7\,\text{K}}\right)^{7/2} \equiv K\,T^{7/2}, \tag{2.7}$$

where T is the *internal* temperature. The observed WD luminosities of Tab. 2.1 vary between 0.5×10^{-4} and $0.5\,L_\odot$, corresponding to a range $0.4-6 \times 10^7\,\text{K}$ of internal temperatures.

With these results the luminosity function is found to be

$$\frac{dN}{dM_{\rm bol}} = \frac{4\ln(10)}{35}\, B\, \frac{C\,(L_\gamma/K)^{2/7}}{L_\gamma + L_\nu + L_x}.$$ (2.8)

With $B_3 \equiv B/10^{-3}\,{\rm pc}^{-3}\,{\rm Gyr}^{-1}$ this is numerically

$$\frac{dN}{dM_{\rm bol}} = B_3\, 2.2{\times}10^{-4}\,{\rm pc}^{-3}\,{\rm mag}^{-1}$$

$$\times\; \frac{10^{-4M_{\rm bol}/35}\,L_\odot}{78.7 L_\odot\,10^{-2M_{\rm bol}/5} + L_\nu + L_x}\left(\frac{\mathcal{M}}{\mathcal{M}_\odot}\right)^{5/7}\sum_j \frac{X_j}{A_j}.$$ (2.9)

If one ignores L_ν and L_x this is

$$\frac{dN}{dM_{\rm bol}} = B_3\, 2.9{\times}10^{-6}\,{\rm pc}^{-3}\,{\rm mag}^{-1}\,10^{2M_{\rm bol}/7}\left(\frac{\mathcal{M}}{\mathcal{M}_\odot}\right)^{5/7}\sum_j \frac{X_j}{A_j}.$$

(2.10)

Taking $\mathcal{M} = 0.6\,\mathcal{M}_\odot$ and an equal mixture of ^{12}C and ^{16}O one finds

$$\log(dN/dM_{\rm bol}) = \tfrac{2}{7}\,M_{\rm bol} - 6.84 + \log(B_3),$$ (2.11)

a behavior known as Mestel's cooling law. For $B_3 = 1$ this function is shown as a dotted line in Fig. 2.10. Detailed cooling curves and luminosity functions have been calculated, for example, by Lamb and van Horn (1975), Shaviv and Kovetz (1976), Iben and Tutukov (1984), Koester and Schönberner (1986), Winget et al. (1987), Iben and Laughlin (1989), Segretain et al. (1994), and Hernanz et al. (1994).

From Fig. 2.10 it is evident, however, that Mestel's cooling law provides a surprisingly good representation for intermediate luminosities where it is most appropriate. At the bright end, the luminosity function is slightly depressed, providing evidence for neutrino cooling. It rapidly falls off at the faint end, presumably indicating the beginning of WD formation as discussed above.

2.2.3 Neutrino Cooling

For the hottest WDs volume neutrino emission is more important than surface photon cooling. The photon luminosity of Eq. (2.7) can be expressed as an effective energy-loss rate per unit mass of the star, $\epsilon_\gamma = L_\gamma/\mathcal{M} = 3.3{\times}10^{-3}\,{\rm erg\,g}^{-1}\,{\rm s}^{-1}\,T_7^{3.5}$ with $T_7 = T/10^7\,{\rm K}$. For the upper relevant temperature range, neutrinos are emitted mostly by the plasma process $\gamma \to \nu_e \bar{\nu}_e$ which is studied in detail in Chapter 6; numerical emission rates are discussed in Appendix C. The neutrino energy-loss rate as a function of temperature is shown in Fig. C.6; for a WD the short-dashed curve $(2\rho/\mu_e = 10^6 \, \mathrm{g \, cm}^{-3})$ is most appropriate.

Neutrino cooling causes a depression of the WD luminosity function at the bright end. The dashed line in Fig. 2.10 represents a numerical cooling calculation for a $0.6 \, \mathcal{M}_\odot$ WD which included neutrinos (Koester and Schönberner 1986); the "neutrino dip" is clearly visible. The photon decay $\gamma \to \bar{\nu}\nu$ is a neutrino process made possible by the medium-induced photon dispersion relation and by the medium-induced effective photon-neutrino coupling (Chapter 6). As such this process is not observable in the laboratory, although there is no doubt about its reality. Still, it is encouraging to see it so plainly in the WD luminosity function.

If neutrino emission were much stronger than standard, the neutrino dip would be correspondingly deeper. Stothers (1970) used the observation of several bright WDs in the Hyades cluster to constrain the efficiency of neutrino cooling; at that time the existence and magnitude of a direct neutrino-electron interaction had not yet been experimentally established. Stothers found that an emission rate 300 times larger than standard could be conservatively excluded.

Recently, Blinnikov and Dunina-Barkovskaya (1994) have studied this subject in detail. Their motivation was the possible existence of a neutrino magnetic dipole moment which would enhance the effective neutrino-photon coupling and thus the efficiency of the plasma process. For the present general discussion it is enough to think of their study as an arbitrary variation of the neutrino emissivity, even though the magnetic-moment induced plasma emission rate has a different density dependence—see Eq. (6.94) for details. For the pertinent conditions the energy-loss rate induced by the assumed dipole moment μ_ν is roughly $0.06 \, \mu_{12}^2$ times the standard one where $\mu_{12} \equiv \mu_\nu/10^{-12}\mu_B$ (Bohr magneton μ_B). Blinnikov and Dunina-Barkovskaya (1994) calculated the luminosity function for $0.6 \, \mathcal{M}_\odot$ WDs for the standard case ($\mu_{12} = 0$, dashed line in Fig. 2.11) and for a roughly 25-fold increased rate of neutrino cooling ($\mu_{12} = 20$, dotted line in Fig. 2.11). The birthrate of WDs was assumed constant and adjusted to optimize the agreement with the observations. For $\mu_{12} = 0$ they needed $B = 0.62 \times 10^{-3} \, \mathrm{pc}^{-3} \, \mathrm{Gyr}^{-1}$ while for $\mu_{12} = 20$ the best fit was achieved for 0.67 in these units.

Blinnikov and Dunina-Barkovskaya (1994) pointed out that a particularly sensitive observable for the neutrino dip is the temperature dis-

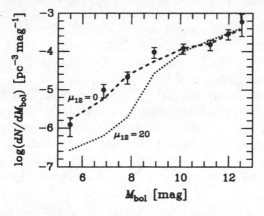

Fig. 2.11. Luminosity function for $0.6\,\mathcal{M}_\odot$ WDs for two values of μ_{12} (Blinnikov and Dunina-Barkovskaya 1994) compared with the observations quoted in the upper part of Tab. 2.1 (Fleming, Liebert, and Green 1986).

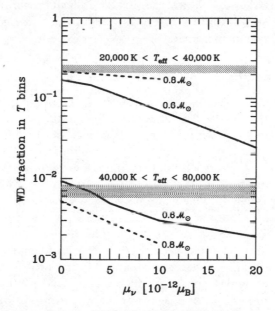

Fig. 2.12. Relative number of 0.6 and $0.8\,\mathcal{M}_\odot$ WDs in the two hot temperature bins of Tab. 2.2 as a function of the anomalous neutrino cooling implied by a magnetic dipole moment μ_ν (Blinnikov and Dunina-Barkovskaya 1994). The number of WDs in the temperature range $12,000-40,000\,\mathrm{K}$ is normalized to unity. The shaded bands correspond to the observations (Tab. 2.2).

tribution of hot WDs derived by Fleming, Liebert, and Green (1986)—
see Tab. 2.2. They calculated cooling sequences for $\mathcal{M} = 0.6\,\mathcal{M}_\odot$ and
$0.8\,\mathcal{M}_\odot$ with varying amounts of nonstandard cooling. The relative
number of WDs in the two hot temperature bins of Tab. 2.2 are shown
as a function of μ_ν in Fig. 2.12.

These results seem to indicate that a significantly enhanced rate of
neutrino emission can be conservatively excluded. However, this view
may be challenged if one includes the possibility of residual hydrogen burning near the WD surface which could mask neutrino cooling
because it would fill in some of the "neutrino dip" in the luminosity
function. Because it is not known how much hydrogen is retained by
a WD after the planetary nebula phase one has an adjustable parameter to provide a desired amount of heating (Castellani, Degl'Innocenti,
and Romaniello 1994). However, preliminary investigations seem to indicate that even when residual hydrogen burning is included the impact
of μ_ν is masked only in one of the temperature bins used by Blinnikov
and Dunina-Barkovskaya (1994) so that a significant deformation of the
luminosity function appears to remain (Blinnikov and Degl'Innocenti
1995, private communication).

2.2.4   Cooling by Boson Emission

Standard or exotic neutrino emission from WDs (or neutron stars) has
the important property that it switches off quickly as the star cools
because of the steep temperature dependence of the emission rates.
Therefore, neutrinos cause a dip at the hot end of the luminosity function while older WDs are left unaffected, even for significantly enhanced
neutrino cooling (Fig. 2.11). One may construct other cases, however,
where this is different. One example is when a putative low-mass boson is emitted in place of a neutrino pair, say, in the bremsstrahlung
process $e + (Z, A) \to (Z, A) + e + \nu\bar{\nu}$. The reduced final-state phase
space then reduces the steepness of the temperature dependence of the
energy-loss rate. The possible existence of such particles is motivated
by theories involving spontaneously broken global symmetries. The
most widely discussed example is the axion which will be studied in
some detail in Chapter 14. For the present discussion all that matters
is the temperature variation of an assumed energy-loss rate.

The bremsstrahlung rate for pseudoscalar bosons will be calculated
in Chapter 3. For the highly degenerate limit the result is given in
Eq. (3.33) where $\alpha' = g^2/4\pi$ is the relevant "fine-structure constant."
Because this rate depends on the density only weakly through a factor

F which includes Coulomb screening by ion correlations and electron relativistic corrections one may easily "integrate" over the entire star. For an assumed equal mixture of carbon and oxygen one finds for the luminosity in pseudoscalar "exotica"

$$L_x = \alpha_{26} \; 2.0 \times 10^{-3} \, L_\odot \, (\mathcal{M}/\mathcal{M}_\odot) \, \langle F \rangle \, T_7^4, \tag{2.12}$$

where $\alpha_{26} = \alpha'/10^{-26}$ and $T_7 = T/10^7 \, \text{K}$ (internal temperature T). Further, $\langle F \rangle \approx 1.0$ within a few 10% (Sect. 3.5.2).

This energy-loss rate varies with internal temperature almost as the surface photon luminosity of Eq. (2.7). If L_x dominates in Mestel's cooling law Eq. (2.11) the slope $\frac{10}{35} M_{\text{bol}}$ is replaced with the almost identical value $\frac{12}{35} M_{\text{bol}}$. Thus, for a given WD birthrate the main impact of pseudoscalars is to reduce the amplitude of the luminosity function. Conversely, the inferred birthrate is larger by about a factor $1 + L_x/L_\gamma \approx 1 + \alpha_{26}$. Because the formation rate of planetary nebulae, the progenitors of WDs, agrees with the standard inferred birthrate to within a factor of about 2, one finds $\alpha_{26} \lesssim 1$.[9]

The observed break of the luminosity function at the faint end (Fig. 2.10) has been interpreted as the beginning of WD formation. If boson cooling were dominant the faintest luminosities would have been reached in a shorter amount of time, reducing the inferred age of the galactic disk; standard cooling implies $t_{\text{gal}} = 8{-}12\,\text{Gyr}$. Even the solar system is 4.5 Gyr old and so a reduction of t_{gal} by more than a factor of 2 is ruled out. This implies $L_x < L_\gamma$ so that $\alpha_{26} < 1.0$ as before.

The galactic age constraint appears to be much more reliable than the one based on the formation rate of planetary nebulae. The former could be avoided if the observations reported by Liebert, Dahn, and Monet (1988) were crudely incomplete at the faint end of the luminosity function, i.e. if many faint WDs had been overlooked so that the break in the luminosity function of Fig. 2.10 were a major observational selection effect. Barring this remote possibility the limit on α_{26} is conservative. Even for a much smaller value of α_{26}, the inferred value for t_{gal} is reduced by an approximate factor $(1 + \alpha_{26})^{-1}$ which may still be significant.

Wang (1992) calculated numerical $1\,\mathcal{M}_\odot$ WD sequences with varying amounts of pseudoscalar cooling while Blinnikov and Dunina-Barkovskaya (1994) performed a more detailed study for $0.6\,\mathcal{M}_\odot$ WDs.

---

[9]In the original derivation (Raffelt 1986b) ion correlations were ignored, leading to $\langle F \rangle \approx 3$ and to the limit $\alpha_{26} \lesssim 0.3$.

They calculated the distribution for the temperature bins of Tab. 2.2 in analogy to the discussion of neutrino dipole moments in the previous section. Their results are shown in Fig. 2.13 as a function of α_{26} where I have corrected from $\langle F \rangle = 3$, which they used in order to compare with Raffelt (1986b), to the more appropriate value $\langle F \rangle = 1$.

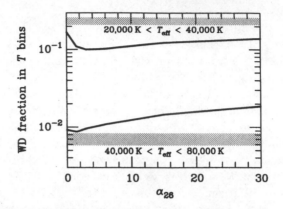

Fig. 2.13. Relative number of $0.6\,\mathcal{M}_\odot$ WDs in the hot and intermediate temperature bins of Tab. 2.2 as a function of the "fine-structure constant" $\alpha_{26} = \alpha'/10^{-26}$ of pseudoscalar bosons (Blinnikov and Dunina-Barkovskaya 1994). I have corrected from $\langle F \rangle = 3$ to the more appropriate value 1. The number of WDs in the temperature range $12{,}000-40{,}000\,\mathrm{K}$ is normalized to unity. The shaded bands correspond to the observations (Tab. 2.2).

This temperature method is now relatively insensitive because axion cooling leaves the shape of the luminosity function nearly unchanged, except that the neutrino dip is washed out for strong cooling. In detail, however, important changes occur. It is noteworthy that boson cooling has a significant impact on the shape of the luminosity function even for $\alpha_{26} < 1$. It appears that the possibility of residual hydrogen burning in young WDs would make the present argument more conservative because it has the same effect as boson emission, namely to reduce or wash out the neutrino dip in the luminosity function.

2.2.5   Period Decrease of Variable White Dwarfs

The luminosity function allows one to determine the WD cooling speed because one is looking at a large WD ensemble of different age. Recently it has become possible for the first time to measure WD cooling directly for a single object by virtue of the period change of a ZZ Ceti star.

White dwarfs have residual atmospheres which may be hydrogen rich (DA white dwarfs) or helium rich (DB); the DA stars are about four times more frequent. The surface layers of DA white dwarfs are not fundamentally different from those of main-sequence or giant stars and so for appropriate conditions one expects Cepheid-type pulsations. Indeed, where the faint continuation of the Cepheid instability strip intersects with the locus of WDs in the Hertzsprung-Russell diagram (Fig. 2.9) one finds the DA variables (DAV), also known as ZZ Ceti stars after their prototype example. They have pulsation periods of a few minutes.

The oscillation period depends on the temperature of the layer which exhibits the κ instability (Sect. 2.1.9) and thus excites the pulsations, and also on the radius of the star. Therefore, the slowing down of the period P is a direct measure of the temperature decrease and thus of the WD cooling speed. Standard pulsation theory yields $\dot{P}/P \approx -a\dot{T}/T + b\dot{R}/R$ where the dimensionless constants a and b are of order unity (Winget, Hansen, and van Horn 1983). Because a WD has an almost fixed radius, the second term may be ignored. The time scale of cooling and of the period change are then related by $T/\dot{T} = -a\,P/\dot{P}$. ZZ Ceti stars have surface temperatures in the neighborhood of 13,000 K where the cooling time scale is of order 1 Gyr. For a period of a few minutes one is talking of a period decrease $\dot{P} = \mathcal{O}(10^{-14}\,\mathrm{s\,s^{-1}})$, not an easy quantity to measure.

After upper limits on $\dot{P}$ had been established over the years for a number of cases, Kepler et al. (1991) succeeded at a measurement for the DAV star G117–B15A (Tab. 2.3) using the Whole Earth Telescope which allows for nearly 24 h a day coverage of a given object.

A variety of model calculations give $\dot{P} = 2-5 \times 10^{-15}\,\mathrm{s\,s^{-1}}$, somewhat smaller than the measured value, i.e. the star appears to cool

Table 2.3. Properties of the DAV star G117–B15A.

Surface temperature	T_{eff}	13,200 K
Luminosity	$\log(L/L_\odot)$	-2.3
Bolometric brightness	M_{bol}	10.49 mag
Mass	$\mathcal{M}$	$(0.49 \pm 0.03)\,\mathcal{M}_\odot$
Pulsation period	P	$(215.197{,}387 \pm 0.000{,}001)\,\mathrm{s}$
Period change	$\dot{P}$	$(12 \pm 4) \times 10^{-15}\,\mathrm{s\,s^{-1}}$
	$P/\dot{P}$	$(0.57 \pm 0.17)\,\mathrm{Gyr}$

faster than predicted by these models (Kepler et al. 1991). It should be noted, however, that part of the observed period change can be attributed to a Doppler shift if G117–B15A is a physical binary with its proper-motion companion G117–B15B.

Isern, Hernanz, and García-Berro (1992) speculated that an additional cooling agent may be operating and notably that this star is cooled by axion emission. For a fiducial WD model ($\mathcal{M} = 0.5\,\mathcal{M}_\odot$, internal temperature $T = 1.8{\times}10^7\,\mathrm{K}$) the cooling time scale is found to be $T/\dot{T} = 1.0\,\mathrm{Gyr}$ while $P/\dot{P} \approx 1.4\,\mathrm{Gyr}$. Therefore, Isern et al. found that $L_x = 0.5\text{–}2.6\,L_\gamma$ was needed to account for the observed $\dot{P}$, although other models needed little or no axion cooling. For their fiducial model the required axion cooling yielded a coupling constant $\alpha_{26} = 0.2 - 0.8$. This interpretation is speculative, of course, but apparently not in conflict with any other constraints on the electron coupling of pseudoscalars (Sect. 3.6.1).

The most conservative interpretation of the $\dot{P}$ measurement of the star G117–B15A is that it agrees with theoretical calculations within the observational and model uncertainties. Therefore, it provides independent evidence that the WD cooling speed is known to within a factor of $\mathcal{O}(1)$ so that any novel cooling agent is constrained to be less efficient than $\mathcal{O}(L_\gamma)$.

2.3 Neutron Stars

2.3.1 Late-Time Cooling

Neutron stars are born when the degenerate iron core of an evolved massive star becomes unstable and collapses to nuclear densities, an implosion which is partly reflected at the core bounce and leads to a type II supernova explosion (Sect. 2.1.8). These events and the first few seconds of neutron star cooling are discussed more fully in Chapter 11. For a few seconds the star emits most of its binding energy in the form of MeV neutrinos which were observed from SN 1987A. Afterward, the temperature at the neutrino sphere (the analogue of the photosphere in ordinary stars) has dropped so much that the detectors are no longer sensitive to the neutrino flux although the star continues to cool by surface neutrino emission.

After $10-100\,\mathrm{yr}$ the internal temperature has dropped to about $10^9\,\mathrm{K} \approx 100\,\mathrm{keV}$ where the neutron star becomes entirely transparent to neutrinos and continues to cool by neutrino volume emission. After about $10^5\,\mathrm{yr}$ it reaches an inner temperature of about $2{\times}10^8\,\mathrm{K}$, a point at which photon emission from the surface becomes the dominant form of cooling. In Fig. 2.14 the central temperature, surface temperature, neutrino luminosity, and photon luminosity are shown as functions of age according to a numerical calculation of Nomoto and Tsuruta (1987).

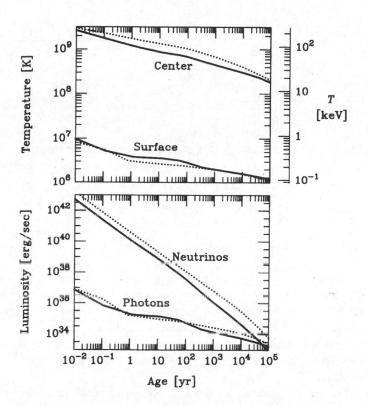

Fig. 2.14. Cooling of a neutron star with baryon mass $1.4\,\mathcal{M}_\odot$ (gravitational mass $1.3\,\mathcal{M}_\odot$) according to Nomoto and Tsuruta (1987). The solid line is for an equation of state of intermediate stiffness (model FP), the dotted line for a stiff model (PS). All "nonstandard" effects were ignored such as nucleon superfluidity, a meson condensate, magnetic fields, and so forth. Temperatures and luminosities are local, ignoring the gravitational redshift.

2.3.2   X-Ray Observations

Because the surface temperature of neutron stars is in the keV regime
they can be observed only by x-ray satellites such as the Einstein Observatory which was launched in 1979, and more recently by EXOSAT and

Table 2.4. Selected x-ray observations of SN remnants and pulsars.[a]

Remnant	Age [yr] (in 1995)	Dist. [kpc]	Pulsar[b]	T_{eff}^{∞} [10^6 K]	[c]	[d]
3C58	(814)	(2.6)	—	2.2 ± 0.2	E	1
Crab	941	1.7–2	0531+21 (r, o, x, γ)	< 1.6	R	2
RCW 103	1500 ± 500	2	—	2.15 ± 0.15	E	1
				< 1.2	R	3
MSH15–52	1850 ± 250	4.2	(r, x)	detected	E	1
Vela X	$\sim 12,000$	0.5	0833–45 (r, o, x, γ)	0.95 ± 0.15 1.6 ± 0.2	E R	1 4
Monogem Ring (?)	$\sim 110,000$	(0.5)	0656+14 (r, x)	0.90 ± 0.04 (0.30 ± 0.05)	R	5
—	$\sim 340,000$	0.15–0.4	Geminga (x, γ)	0.52 ± 0.10	R	6
—	$\sim 540,000$	0.5–1.5	1055–52 (r, x, γ)	0.65 ± 0.15 0.75 ± 0.06	X R	7 8

[a] Adapted from Tsuruta (1986) and updated.
[b] Pulsed radiation: r = radio, o = optical, x = x-rays, γ = γ-rays.
[c] E = Einstein, R = ROSAT, X = EXOSAT observation.
[d] *References:*

1. See the review by Tsuruta (1986) for references to the original literature.
2. Becker and Aschenbach (1995).
3. Becker et al. (1992).
4. Ögelman, Finley, and Zimmermann (1993).
5. Thompson et al. (1991). Finley, Ögelman, and Kiziloğlu (1992).
6. Halpern and Ruderman (1993).
7. Brinkmann and Ögelman (1987).
8. Ögelman and Finley (1993). Anderson et al. (1993).

ROSAT. X-rays have been observed from a number of compact sources in supernova remnants and from several isolated pulsars. However, with the limited spectral resolution of these instruments it is difficult to extract the actual thermal surface emission because there can be significant nonthermal x-ray emission from the magnetosphere.

There are three candidates in the 10^3 yr age category (3C58, the Crab pulsar, RCW 103) from which x-rays have been observed by the Einstein Observatory (Tab. 2.4). However, ROSAT did not detect a compact source in RCW 103; only an upper flux limit has been reported (Tab. 2.4). The Crab pulsar is x-ray bright mostly from non-thermal emission so that ROSAT could only establish an upper limit on its surface temperature. The remaining Einstein source 3C58 probably should also be interpreted as an upper limit on thermal surface emission. These upper limits are in agreement with the standard cooling curves of Nomoto and Tsuruta (1987) shown in Fig. 2.15.

The Einstein data point for the Vela pulsar at an age of about 10^4 yr is somewhat low, a fact which has given rise to the speculation that "exotic" cooling effects may be operating such as neutrino emission by virtue of a meson condensate. However, a blackbody spectral fit to the ROSAT observations yields a much higher temperature (Tab. 2.4,

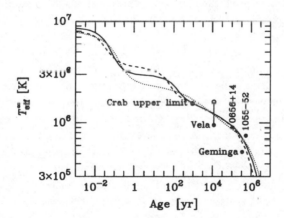

Fig. 2.15. Surface temperature (observed at infinity) of a neutron star with baryon mass $1.4\,\mathcal{M}_\odot$ (gravitational mass $1.3\,\mathcal{M}_\odot$) according to Nomoto and Tsuruta (1987). The solid line is for an equation of state of intermediate stiffness (model FP), the dotted line for a stiff (PS), and the dashed line for a soft model (BPS). All "nonstandard" effects were ignored. The measurements of Tab. 2.4 are also shown.

open circle in Fig. 2.15). Such a high temperature is not compatible with the total x-ray luminosity unless the radius of the neutron star is very small (3–4 km). Either way, it may be premature to reach definite conclusions regarding neutron star cooling on the basis of Vela.

At still larger ages, ROSAT measurements of the surface temperature of the pulsars PSR 0656+14, PSR 1055–52, and Geminga have been reported which lie close to the theoretical standard cooling curves. However, the inferred temperature values depend sensitively on the assumed circumstellar atmospheric models which can modify the spectrum and thus lead to an erroneous temperature assignment. Still, the old isolated pulsars give one a first realistic observational handle at the issue of neutron star cooling.

2.3.3   Nonstandard Cooling and Heating Effects

The so-called "standard" neutron star cooling scenario should be called a "reference" or "minimal" scenario because there are many effects that will alter the cooling history; no doubt at least some of them will be in operation in some or all neutron stars. For reviews see Tsuruta (1986, 1992), for recent numerical cooling curves including various nonstandard effects see Umeda, Tsuruta, and Nomoto (1994).

The occurrence of nucleon superfluidity slows the neutrino emission by the URCA process (Sect. 4.8). However, the cooling curves including superfluidity (Nomoto and Tsuruta 1987) do not seem to differ significantly from the reference curves at ages above a few hundred years unless extreme assumptions are made.

When nucleon superfluidity is important, $\nu\bar{\nu}$ bremsstrahlung emission by electrons in the crust dominates. However, electron band-structure effects may suppress this process, and the crust mass may be smaller than previously thought (Pethick and Thorsson 1994). Therefore, the cooling may be slowed even further. Slowed cooling may also occur by a number of heating effects (accretion, polar cap heating, vortex creep, and others), although such effects become important only for relatively old neutron stars ($t \gtrsim 10^4$ yr). Of course, heating effects related to accretion will not be important in isolated pulsars which thus are preferred laboratories to study neutron star cooling.

The cooling is accelerated if the equation of state provides enough protons to allow for the direct URCA process (Sect. 4.8). In this case the surface temperature drops catastrophically at an age of about 100 yr (Page and Applegate 1992; Lattimer et al. 1994) until superfluidity sets in which essentially stops neutrino cooling from the core. The temperature then stays almost constant for a long time until photon cooling from the surface begins to dominate. In this scenario the cooling curve depends sensitively on the "on switch" set by the occurrence of the direct URCA process anywhere in the star, and by the "off switch" from superfluidity. As the occurrence of these effects depends on fine points of the equation of state as well as on the density and thus the stellar mass there may not be a universal cooling curve for all neutron stars.

Another effect which would accelerate cooling is the occurrence of a meson condensate (Sect. 4.9.1) because of the increased efficiency of neutrino emission. Again, this effect depends sensitively on the equation of state and thus on the density and the stellar mass. The most recent numerical study of neutron-star cooling with a pion condensate was performed by Umeda, Nomoto, and Tsuruta (1994).

2.3.4   Cooling by Particle Emission

The emission of novel particles would also accelerate the cooling of neutron stars. Iwamoto (1984) considered axion emission $e + (Z, A) \rightarrow (Z, A) + e + a$ in the crust and found unacceptably fast cooling unless $\alpha' \lesssim 10^{-25}$ where α' is the axionic fine-structure constant (Chapter 3). However, this result is quite uncertain, notably in view of the above Pethick and Thorsson (1994) band-structure suppression of the bremsstrahlung rate. Moreover, in view of the white-dwarf and globular-cluster bounds of $\alpha' \lesssim 10^{-26}$ it appears that crust cooling by axions is not important in neutron stars.

In the interior of neutron stars, axions can be emitted by the neutron bremsstrahlung process $nn \rightarrow nna$ (Sect. 4.2). With a numerical implementation of Iwamoto's (1984) bremsstrahlung rate Tsuruta and Nomoto (1987) found a limit $g_{an} \lesssim 10^{-10}$ for the axion-neutron Yukawa coupling, based on a comparison with the 10^3 yr old sources. In their calculation the effect of superfluidity apparently was not included which would diminish the bound. Conversely, including protons in a regime where neither protons nor neutrons are superfluid would increase the emission rate (Sect. 4.2.6). Of course, the more recent ROSAT results suggest that thermal surface emission has not been observed for any of the 10^3 yr old sources anyway.

Most recently, Iwamoto et al. (1995) considered the plasma decay process $\gamma \rightarrow \bar{\nu}\nu$ in the crust under the assumption of a large neutrino magnetic dipole moment. They found that for μ_ν of order $10^{-10} \mu_B$ a significant effect would obtain, but that a value as large as $5 \times 10^{-7} \mu_B$ would be consistent with current data. In view of the globular-cluster limit of $\mu_B \lesssim 3\times10^{-12}\,\mu_B$ (Sect. 6.5.6), I interpret these results to mean that a neutrino magnetic dipole moment leaves neutron-star cooling unaffected—one less nonstandard effect to worry about.

The rough agreement between the reference cooling curves and the data points in Fig. 2.15, notably for the old pulsars, suggests that "nonstandard" effects cannot be much more efficient than standard neutrino cooling unless all of the old isolated pulsar surface temperatures have been incorrectly assigned—not a likely scenario. Therefore, it is clear that these and future observations of cooling neutron stars will be pivotal as laboratories to study novel phenomena such as the occurrence of nonstandard phases of nuclear matter.

However, at the present time it is not clear if the emission of weakly interacting particles such as axions could still have an interesting impact on neutron star cooling. At the present time it appears easier to make the reverse statement that such cooling effects likely are not important in view of the restrictive limits on the interaction strength of axions or nonstandard neutrinos set by other astrophysical objects. Therefore, at present it appears that for the more narrowly defined purposes of particle physics the role of old neutron stars as laboratories is less useful than had been thought in the early works discussed above. It also appears that novel weakly interacting particles usually would have a more dramatic impact on the first few seconds of Kelvin-Helmholtz cooling of a protoneutron star (Chapter 11) than they do on the cooling of old pulsars.

2.4   Globular-Cluster Stars

2.4.1   Observables in the Color-Magnitude Diagram

Globular clusters are gravitationally bound associations of typically 10^6 stars (Fig. 2.1); the clusters themselves (about 150 in our galaxy) form an approximately spherical galactic halo. The metallicity is in the range $Z = 10^{-4} - 10^{-2}$; it is usually the same for all stars in one cluster. The low metallicity is one indicator for their great age—like isolated halo stars they belong to the Population II which formed early from a relatively "uncontaminated" hydrogen-helium mixture left over from the big bang of the universe. Stars found in the galactic disk belong to the later Population I which continue to form even today from the interstellar gas. Clusters of disk stars are usually less populous and less tightly bound—the so-called "open clusters."

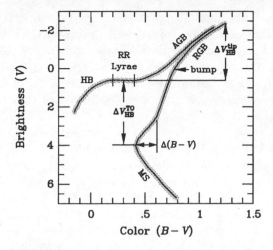

Fig. 2.16. Observables in the color-magnitude diagram of a typical globular cluster (here M15 after Sandage 1986). Depending on the metallicity, the red-giant "bump" can also appear below the HB brightness.

In the color-magnitude diagram of a globular cluster (Fig. 2.3) the stars arrange themselves in a characteristic pattern which is schematically shown in Fig. 2.16. Each branch corresponds to a certain evolutionary phase as discussed in Sect. 2.1. Typically, all stars in a given cluster are coeval and have a fixed chemical composition; they differ only in their mass. After formation, they all began their evolution on the zero-age main-sequence as in Fig. 2.2. The more massive stars evolve faster, become red giants and explode as supernovae ($M \gtrsim 8 \, M_\odot$), or become white dwarfs ($M \lesssim 8 \, M_\odot$). Recently, many pulsars, the remnants of type II supernovae and thus of massive stars, have indeed been found in globular clusters. The upper main sequence (MS) is depleted of stars down to a limiting value below which they have not had time to complete hydrogen burning. Therefore, the stellar mass corresponding to the MS turnoff (TO) is a precise measure of the cluster age. A typical value is $0.7-0.9 \, M_\odot$, depending on the precise age and metallicity.

The color-magnitude diagram of a globular cluster represents an "isochrone" of a stellar population. It shows the locus of coeval stars with different initial masses. It is to be distinguished from the evolutionary track of a single star (Fig. 2.6) which shows a star of a fixed mass at different ages. However, the evolution beyond the MS is very fast. Therefore, the stars along the red-giant branch (RGB), horizontal branch (HB), and asymptotic giant branch (AGB) in a globular cluster have almost identical initial masses whence for these phases a single-star track is practically identical with an isochrone. On the other hand, the TO region requires the construction of detailed theoretical isochrones to compare theory and observations and thus to determine the ages of globular clusters.

In order to associate a certain stellar mass with the TO in a cluster one needs to know the absolute brightness of the stars at the TO, i.e. one needs to know the precise distance. All else being equal, the inferred age varies with the TO luminosity as $\partial \log(\text{age})/\partial \log L_{\text{TO}} = -0.85$ or $\partial \log(\text{age})/\partial V_{\text{TO}} = 0.34$ (Iben and Renzini 1984). Therefore, a 0.1 mag error in V_{TO} leads to an 8% uncertainty in the inferred cluster age. Put another way, because $L \propto (\text{distance})^2$ a 10% uncertainty in cluster distances leads to an 18% age uncertainty. This is the main problem with the age determination of globular clusters.

A particularly useful method to measure the distance is to use RR Lyrae stars as standard candles. As discussed in Sect. 2.1, their luminosity is determined almost entirely by their core mass (apart from a dependence on chemical composition), which in turn is fixed by helium ignition on the RGB which, again, depends only on the chemical composition and not on the red-giant envelope mass. Therefore, the brightness of the HB is nearly independent of stellar mass. Consequently, the brightness difference $\Delta V_{\text{HB}}^{\text{TO}}$ between the HB and the TO is a distance-independent measure of the TO mass and thus of the cluster age. Moreover, because the color of RR Lyrae stars coincides with that of the TO region it is not necessary to convert from the measured brightness with a certain filter (e.g. visual brightness V) to a bolometric brightness, i.e. there is no need for a bolometric correction (BC). Also, RR Lyrae stars are bright and easily identified because of their pulsations. Therefore, $\Delta V_{\text{HB}}^{\text{TO}}$ is one of the most important observables in the color-magnitude diagram of globular clusters (Iben and Renzini 1984; Sandage 1986).

As an example the recent $\Delta V_{\text{HB}}^{\text{TO}}$ determinations of Buonanno, Corsi, and Fusi Pecci (1989) in 19 globular cluster are shown in Fig. 2.17 as a function of metallicity; the logarithmic metallicity measure [Fe/H] is defined in Eq. (2.15). The best linear fit is

$$\Delta V_{\text{HB}}^{\text{TO}} = (3.54 \pm 0.13) - (0.008 \pm 0.078)\,[\text{Fe/H}], \qquad (2.13)$$

so that the HB brightness varies with metallicity almost exactly as the TO brightness. The measured points are in agreement with a Gaussian

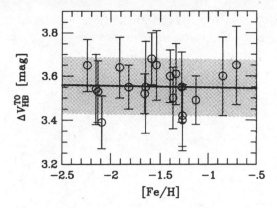

Fig. 2.17. Brightness difference between main-sequence turnoff (TO) and horizontal branch (HB) in 19 galactic globular clusters according to Buonanno, Corsi, and Fusi Pecci (1989). The shaded band indicates the 1σ statistical error of the best fit Eq. (2.13).

distribution about the mean. This result illustrates the level of precision that presently can be achieved at determining $\Delta V_{\mathrm{HB}}^{\mathrm{TO}}$.

In principle, the color of the TO also specifies the location on the MS and thus the TO mass and cluster age. In practice, the color of a star is theoretically less well determined than its luminosity because it depends on the treatment of the photosphere and on the surface area and thus on the radius which is partly fixed by the treatment of convection. Still, the TO color is a useful measure for the relative ages between clusters of identical metallicities. Notably, the distance-independent color difference $\Delta(B - V)$ between the TO and the base of the RGB (Fig. 2.16) has been used to establish an age difference of about (3 ± 1) Gyr between the clusters NGC 288 and 362 (VandenBerg, Bolte, and Stetson 1990; Sarajedini and Demarque 1990).

In order to constrain the operation of a novel energy-loss mechanism the brightness of the RGB tip is particularly useful because particle emission (neutrinos, axions) from a red-giant core delays helium ignition. This delay allows the stars to develop a more massive core and thus to turn brighter before they become HB stars. Again, the inferred luminosity of the brightest star on the RGB depends on the distance whence the most useful observable is the distance-independent brightness difference between the RGB tip and the HB. However, because the color of RR Lyrae stars and red giants is very different one must convert to an absolute bolometric brightness difference $\Delta M_{\mathrm{HB}}^{\mathrm{tip}}$ rather than using the visual brightness difference $\Delta V_{\mathrm{HB}}^{\mathrm{tip}}$.

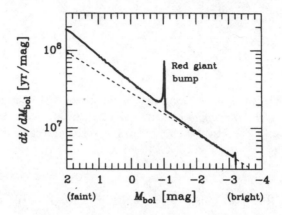

Fig. 2.18. Evolutionary speed on the RGB for a model with $\mathcal{M} = 0.8\,\mathcal{M}_\odot$, metallicity $Z = 10^{-4}$, and initial helium abundance $Y = 0.240$ (Raffelt and Weiss 1992). The dashed line is the tangent near the bright end.

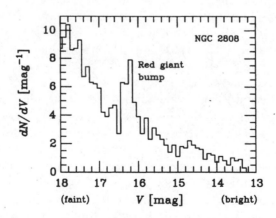

Fig. 2.19. Luminosity function of the RGB of the globular cluster NGC 2808 which has [Fe/H] $= -1.37$ (Fusi Pecci et al. 1990).

A very important measure of the speed of evolution of a single star along the RGB, HB, and AGB is the relative number of stars found on these branches in a given cluster. Because for these advanced evolutionary phases a single-star track is essentially an isochrone, these number ratios R give us directly the relative amounts of time spent on these branches ("R-method"). For example, a novel energy-loss mechanism that operates mostly in the nondegenerate core of an HB star would shorten the helium-burning lifetime. This possibility is constrained by the relative number of stars found on the HB and RGB where the RGB in this context refers to that part which is brighter than the HB.

When a star ascends the RGB, its hydrogen-burning shell encounters at some point a discontinuity in the composition profile left behind by a previous deep penetration of the envelope convection into the region of varying hydrogen content caused by nuclear burning. The ascent is briefly interrupted and the star stays at a fixed luminosity for a brief period of time. Afterward, the hydrogen shell works itself through a constant composition profile which was prepared by the convective envelope. Therefore, at a brightness near the HB one expects to find a "bump" in the number of stars on the RGB, i.e. in the distribution $\partial N/\partial M_{\text{bol}}$ which essentially corresponds to dt/dM_{bol} for a single-star evolutionary track (Fig. 2.18). The bump was recently identified in a number of clusters; for a particularly beautiful example see Fig. 2.19. It has been suggested to use it as a standard candle to calibrate the RR Lyrae brightness-metallicity relation (Fusi Pecci et al. 1990).

An important observable is the absolute brightness of RR Lyrae stars which are not members of globular clusters. No reason is known why these field stars for a given metallicity should be any different from those found in a cluster and so brightness determinations of nearby RR Lyrae stars provide important information about their luminosity calibration.

2.4.2   Theoretical Relations

In order to test the standard stellar-evolution picture against the observables introduced in Sect. 2.4.1 they need to be related to stellar properties such as mass and chemical composition. In the past, extensive grids of stellar evolutionary sequences have been calculated and have been used to derive analytic approximations for the connection between various stellar parameters; as a canonical standard I use the evolutionary HB and RG sequences of Sweigart and Gross (1976, 1978).

a) Composition Parameters

The chemical composition is characterized by the helium content and the metallicity. The convective envelope on the RGB at some point reaches down into the region of a variable composition profile caused by nuclear burning, causing a certain amount of processed material (helium) to be dredged up. Therefore, the helium mass fraction Y_{env} of the envelope is slightly larger than the value Y at formation. The amount of dredge-up is approximately given by (Sweigart, Renzini, and Tornambé 1987)

$$\Delta Y_s \equiv Y_{env} - Y \approx 0.0136 + 0.0055\, Z_{13}, \qquad (2.14)$$

where the metallicity parameter Z_{13} is defined in Eq. (2.16) below. However, because the exact amount of convective dredge-up is somewhat model dependent, and because measurements of $^{12}C/^{13}C$ ratios in metal-poor field red giants seem to indicate that the dredge-up of processed material is more efficient than predicted by standard calculations (Sneden, Pilachewski, and VandenBerg 1986) it is best to use Y_{env} rather than Y as an independent parameter to characterize red giants and HB stars.[10]

Gravitational settling of helium throughout the MS evolution has the opposite effect of reducing the envelope abundance relative to the initial homogeneous value. In recent evolutionary sequences which were calculated to estimate the effect of helium diffusion on the inferred globular-cluster ages a decrease between $\Delta Y_s = -0.009$ and -0.015 was found (Proffitt and Michaud 1991; Chaboyer et al. 1992). Therefore, the effect of MS gravitational settling and that of RG convective dredge-up appear to cancel each other more or less so that Y_{env} appears to be much closer to the initial value Y than had been thought previously.

The metallicity is usually characterized by the mass fraction Z of elements heavier than helium. Because iron is most important for the opacities one often uses the abundance of iron relative to hydrogen as a metallicity measure. It is characterized by the quantity [Fe/H] which is the logarithmic abundance of iron over hydrogen relative to the solar value. If the solar metallicity is taken to be $Z_\odot = 0.02$ so that

---

[10]In their calculation of RG sequences, Sweigart and Gross (1978) used the symbol Y to denote the MS helium abundance; the envelope abundance near the helium flash can be inferred from their tabulation of ΔY_s values for each sequence. In their 1976 study of HB sequences, they used the symbol Y to denote the envelope abundance which is here consistently called Y_{env}.

$\log Z_{\odot} = -1.7$ one finds

$$\log Z = [\text{Fe/H}] + \log Z_{\odot} = [\text{Fe/H}] - 1.7. \qquad (2.15)$$

Because globular-cluster metallicities cover a range of $Z = 10^{-4}$ to 10^{-2} a typical average value is $\log Z = -3$ or $[\text{Fe/H}] = -1.3$. The primordial helium abundance is thought to be about 23%. Therefore, it is useful to employ the "reduced" composition parameters

$$Y_{23} \equiv Y_{\text{env}} - 0.23,$$

$$Z_{13} \equiv \log Z + 3 = [\text{Fe/H}] + 1.3, \qquad (2.16)$$

which are zero for a typical globular cluster.

b) Core Mass at Helium Ignition

One of the most important quantities to be affected by a novel energy-loss mechanism is the core mass at the helium flash because helium ignition is an extremely sensitive function of the temperature. Based on the Sweigart and Gross (1978) models Raffelt (1990b) has derived the analytic approximation

$$\mathcal{M}_c - 0.500 - 0.22\,Y_{23} - 0.011\,Z_{13} - 0.021\,\mathcal{M}_7 + \delta\mathcal{M}_c, \qquad (2.17)$$

where all stellar masses are understood in units of the solar mass $\mathcal{M}_{\odot}$. Here, $\mathcal{M}_c$ is the core mass at helium ignition and $\mathcal{M}_7 \equiv \mathcal{M} - 0.7$ is the "reduced total mass." I have increased $\mathcal{M}_c$ by 0.004 relative to the original calculation to account for the corrected plasma neutrino emission rate (Haft, Raffelt, and Weiss 1994). Within about $0.003\,\mathcal{M}_{\odot}$ Raffelt and Weiss (1992) found the same expression (when corrected for the plasma rates) except for a slightly shallower metallicity dependence.

Recently, Sweigart (1994) has reviewed the core-mass calculations at the helium flash. All workers seem to agree within a few $10^{-3}\,\mathcal{M}_{\odot}$ except for Mazzitelli (1989) who found core masses larger by some $0.020\,\mathcal{M}_{\odot}$. Sweigart (1994) claims that this disagreement cannot be attributed to the algorithm adopted to accomplish the "shell shifting" of the numerical grid which represents the star on a computer.

Because of substantial mass loss on the RGB the meaning of the total mass $\mathcal{M}$ in this equation is not entirely obvious. If there were enough time to relax to equilibrium one would think that it is the instantaneous mass at the helium flash. Indeed, Raffelt and Weiss (1992) found in an evolutionary sequence with mass loss that the end mass determined $\mathcal{M}_c$. However, in this calculation the mass loss was stopped sometime before helium ignition and so this finding is not surprising. Castellani and Castellani (1993) studied RG sequences with mass loss in more detail and found the surprising result that the core mass at helium ignition was determined by the *initial* stellar mass while the envelope structure followed the instantaneous envelope mass. Apparently, in these calculations the core retained memory of a previous configuration. The total amount of mass loss on the RGB may be of order $0.2\,\mathcal{M}_\odot$, causing a maximum discrepancy between the two scenarios of about $0.005\,\mathcal{M}_\odot$ in the expected $\mathcal{M}_c$.

In Eq. (2.17) a deviation $\delta\mathcal{M}_c$ was explicitly included which represents nonstandard changes of $\mathcal{M}_c$. The core-mass increase $\delta\mathcal{M}_c$ is the main quantity to be constrained by observations. It should be thought of as a function of the parameters which govern the physics which causes the delay of helium ignition such as coupling constants of particles which contribute to the energy loss, or a more benign parameter such as the angular frequency of core rotation.

c) Brightness at Helium Ignition

Next, the brightness at helium ignition is needed, identical with the brightness at the tip of the RGB. In the Sweigart and Gross (1978) calculations, both the luminosity and the core mass at the helium flash are functions of Z, $Y_{\rm env}$, and $\mathcal{M}$. For the present purposes, however, the core mass at helium ignition must be viewed as another free parameter which is controlled, for example, by the amount of energy loss by novel particle emission. In order to determine how the luminosity at helium ignition varies with $\mathcal{M}_c$ if all other parameters are held fixed Raffelt (1990b) considered $\partial \log L/\partial\mathcal{M}_c$ for a grid of Sweigart and Gross tracks near the flash. An interpolation yields

$$\log L_{\rm tip} = 3.328 + 0.68\,Y_{23} + 0.129\,Z_{13} + 0.007\,\mathcal{M}_7 + 4.7\,\mathcal{M}_c,$$

$$(2.18)$$

with $L_{\rm tip}$ the luminosity at the RGB tip in units of $L_\odot$. Ignoring the dependence on the total mass here and in Eq. (2.17) one finds for the absolute bolometric brightness of the RGB tip[11]

$$M_{\rm tip} = -3.58 + 0.89\,Y_{23} - 0.19\,Z_{13} - 11.8\,\delta\mathcal{M}_c, \qquad (2.19)$$

slightly different from the results of Raffelt (1990b) who used the mass of RR Lyrae stars for the total mass in Eq. (2.18).

---

[11]Recall that the absolute bolometric brightness is given by $M = 4.74 - 2.5\log L$ for L in units of $L_\odot$, M in magnitudes.

d) Brightness of RR Lyrae Stars

Theoretically, details of the evolution of stars on the HB are difficult to account for. Notably, they may move in and out of the RR Lyrae instability strip so that the stars found there cannot be trivially associated with a specific age after the beginning of helium burning. Therefore, it is easiest to use the absolute bolometric brightness of zero-age HB stars with a mass chosen such that they fall into the RR Lyrae strip (Buzzoni et al. 1983; Raffelt 1990b)

$$M_{RR} = 0.66 - 3.5\,Y_{23} + 0.16\,Z_{13} - \Delta_{RR} - 7.3\,\delta\mathcal{M}_c, \qquad (2.20)$$

where Δ_{RR} is an unknown amount of deviation between real RR Lyrae stars and the zero-age HB models of Sweigart and Gross (1976) that served to derive this relation.

It is expected that Δ_{RR} is a positive number of order 0.1 mag, i.e. on average RR Lyrae stars are thought to be somewhat brighter than zero-age HB star models. This conclusion is supported by Sandage's (1990a) investigation of the vertical height of the HB by means of the pulsational properties of RR Lyrae stars. Sandage found an intrinsic width between 0.2 mag for the most metal-poor and about 0.4 mag for the most metal-rich clusters, i.e. an average deviation between 0.1 and 0.2 mag between zero-age and average HB stars.

Lee, Demarque, and Zinn (1990) have constructed synthetic HBs for a range of metallicities and helium content on the basis of new evolutionary sequences. For a MS helium content of 0.20, which in their calculation amounts to $Y_{env} = 0.22$, they found (see also Lee 1990)

$$M_{RR} = 0.70 + 0.22\,Z_{13}. \qquad (2.21)$$

Comparing this with Eq. (2.20) at $Y_{23} = -0.01$ and $\delta\mathcal{M}_c = 0$ one finds $\Delta_{RR} \approx 0$. This is not in contradiction with the brightening of RR Lyrae stars relative to zero age, it only means that there is a slight offset relative to the analytic representation Eq. (2.20) derived from the Sweigart and Gross (1976) calculations. Δ_{RR} shall always refer to the brightness difference of real RR Lyrae stars relative to Eq. (2.20), it does not refer to an offset relative to real zero-age HB stars.

e) Brightness Difference between HB and RGB Tip

The main observable to constrain a deviation from the standard core mass at the helium flash is the brightness difference between the HB (at the RR Lyrae strip) and the RGB tip for which one finds with Eqs. (2.19) and (2.20)

$$\Delta M_{\mathrm{HB}}^{\mathrm{tip}} \equiv M_{\mathrm{RR}} - M_{\mathrm{tip}}$$

$$= 4.24 - 4.4\,Y_{23} + 0.35\,Z_{13} - \Delta_{\mathrm{RR}} + 4.5\,\delta\mathcal{M}_{\mathrm{c}} \qquad (2.22)$$

which is defined such that it is a positive number.

f) Ratio of HB/RGB Stars

The relative duration of the HB vs. RGB phase is given by the number ratio of the stars on these branches where the RGB is defined as that part which is brighter than the HB. The HB lifetime is found to be (Buzzoni et al. 1983; Raffelt 1990b)

$$\log(t_{\mathrm{HB}}/\mathrm{yr}) = 8.01 + 0.37\,Y_{23} + 0.06\,Z_{13} - 2.9\,\delta\mathcal{M}_{\mathrm{c}}. \qquad (2.23)$$

The RGB lifetime cannot be expressed easily in terms of a simple linear formula because of the RGB bump discussed earlier. A simple approximation to the lifetime ratio $R = t_{\mathrm{HB}}/t_{\mathrm{RGB}}$ is (Buzzoni et al. 1983; Raffelt 1990b)

$$\log R = 0.105 + 2.29\,Y_{23} + 0.029\,Z_{13} + 0.33\,\Delta_{\mathrm{RR}} - 0.70\,\delta\mathcal{M}_{\mathrm{c}}. \qquad (2.24)$$

This quantity is particularly sensitive to the helium content of the stars and almost independent of the core mass at helium ignition.

2.4.3   Observational Results

a) Brightness at the RGB Tip

The brightness at the tip of the RGB can be estimated by the brightest red giant in a given globular cluster. A homogeneous set of observations of the brightest RGs in 33 globular clusters are those of Cohen, Frogel, and Persson (1978), Da Costa, Frogel, and Cohen (1981), Cohen and Frogel (1982) and Frogel, Persson, and Cohen (1981, 1983). According to Frogel, Cohen, and Persson (1983) only in 26 of the 33 clusters the brightest giant was likely observed; for those cases the bolometric brightness difference between the brightest RGs and the HB are shown as a function of metallicity in Fig. 2.20. Also shown is a linear fit $\Delta M_{\mathrm{HB}}^{\mathrm{tip}} = 4.06 + 0.38\,Z_{13}$. The observational errors are thought to be less than about 0.05 mag.

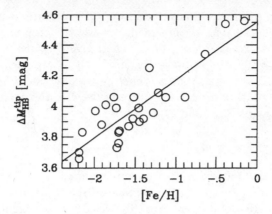

Fig. 2.20. Bolometric brightness difference between the horizontal branch and the brightest red giant in 26 globular clusters from the observations referenced in the text.

However, because there are relatively few stars near the tip on the RGB (on average about $10\,\mathrm{mag}^{-1}$ in the observed clusters), the brightest RG is on average about $0.1\,\mathrm{mag}$ below the actual RGB tip. Raffelt (1990b) estimated the richness of the RGB near the tip for each cluster on the basis of the first few brightest stars provided by the observations and thus estimated the expected brightness difference between the brightest RG and the tip. This yields a linear regression

$$\Delta M_{\mathrm{HB}}^{\mathrm{tip}} = (4.19 \pm 0.03) + (0.41 \pm 0.06)\, Z_{13}, \tag{2.25}$$

about $0.13\,\mathrm{mag}$ brighter than the fit shown in Fig. 2.20. The slope of Eq. (2.25) agrees very well with the theoretical expectation Eq. (2.22), provided that Δ_{RR} does not introduce a large modification.

More recent observations are those of Da Costa and Armandroff (1990) who also found excellent agreement between the theoretical slope of the brightness of the RGB tip luminosity as a function of metallicity.

Because the coefficient of the metallicity dependence agrees well with the predicted value one may restrict a further comparison between theory and observation to $\Delta M_{\mathrm{HB}}^{\mathrm{tip}}$ at a given metallicity for which it is best to use the average value $[\mathrm{Fe/H}] = -1.48$ or $Z_{13} = -0.18$ of the globular clusters used in Fig. 2.20. Inserting these values into Eqs. (2.22) and (2.25) and adding the errors quadratically one finds

$$4.4\, Y_{23} + \Delta_{\mathrm{RR}} - 4.5\, \delta\mathcal{M}_{\mathrm{c}} = 0.06 \pm 0.03. \tag{2.26}$$

If $\Delta_{\mathrm{RR}} = 0.2\,\mathrm{mag}$ this result implies that the envelope helium abundance, which presumably is very close to the primordial value, has to be 0.20 to achieve perfect agreement, or else the core mass at helium ignition has to be $0.030\,\mathcal{M}_\odot$ larger than given by the standard theory. Alternatively, the zero-age HB models may represent the brightness of RR Lyrae stars better than anticipated so that $\Delta_{RR} \approx 0$ in which case there is perfect agreement with the standard values $Y_{env} \approx Y_{primordial} \approx 0.23$ and $\delta\mathcal{M}_c = 0$.

b) Ratio of HB/RGB Stars

Buzzoni et al. (1983) have determined the number of stars on the HB (including RR Lyrae stars),[12] N_{HB}, and on the RGB brighter than the HB, N_{RGB}, in 15 globular clusters. The resulting values for the number ratio $R = N_{HB}/N_{RGB}$ is shown in Fig. 2.21 as a function of metallicity. The individual errors are found by assuming standard deviations of $N^{1/2}$ for the number counts and adding the errors of N_{HB} and N_{RGB} quadratically (Raffelt 1990b). The data are fit by a linear regression

$$\log R = (0.162 \pm 0.016) + (0.065 \pm 0.032)\, Z_{13}, \qquad (2.27)$$

also shown in Fig. 2.21.

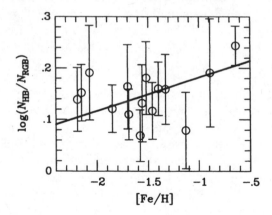

Fig. 2.21. Number ratio of HB/RGB stars for 15 globular clusters according to Buzzoni et al. (1983).

Comparing this result with the theoretical prediction of Eq. (2.24) one finds that the metallicity dependence essentially agrees within the

---

[12]Buzzoni et al. (1983) call this quantity N_{HB+RR}.

stated uncertainty; it is very shallow anyway. Therefore, one compares at the average metallicity of the 15 clusters, $[\text{Fe/H}] = -1.54$ or $Z_{13} = -0.24$, adds the errors quadratically, and finds

$$Y_{23} + 0.14\,\Delta_{\text{RR}} - 0.31\,\delta\mathcal{M}_\text{c} = 0.021 \pm 0.008. \tag{2.28}$$

With $\Delta_{\text{RR}} \approx 0.2\,\text{mag}$ and $\delta\mathcal{M}_\text{c} \approx 0$ this confirms a primordial helium abundance of around 23%.

c) RR Lyrae Absolute Brightness

The RR Lyrae absolute brightness as well as the precise variation of M_{RR} with metallicity is the single most discussed issue about globular-cluster color-magnitude diagrams because the distance and thus the age determination depends critically on this quantity. For example, when using $\Delta V_{\text{HB}}^{\text{TO}}$ measurements like the ones shown in Fig. 2.17, the inferred relative ages of globular clusters depend crucially on the slope a of $M_{\text{RR}} = a\,[\text{Fe/H}] + b$.

One possibility to determine a is the use of the pulsation frequencies of these variable stars. The result is about 0.35, almost twice as large as that obtained by theoretical zero-age HB models or by synthetic HBs, an issue known as the Sandage period shift effect (Sandage 1990 and references therein; Iben and Renzini 1984; Renzini and Fusi Pecci 1988). While this issue is crucial for a relative age determination of globular clusters, it is of relatively minor importance for the present discussion where the zero point b for an intermediate metallicity is the most crucial quantity.

Probably the most direct determination of M_{RR} is to use nearby field RR Lyrae stars for which, in principle, a distance determination by parallax measurements is possible. Barnes and Hawley (1986) applied the method of statistical parallaxes to a sample of 142 stars and found a mean absolute visual brightness of $(0.68 \pm 0.14)\,\text{mag}$. Assuming a bolometric correction for RR Lyrae stars of $-0.06\,\text{mag}$ this leads to

$$\langle M_{\text{RR}} \rangle = (0.62 \pm 0.14)\,\text{mag}. \tag{2.29}$$

The average metallicity of this sample is probably $[\text{Fe/H}] \approx -1.4$.

The Baade-Wesselink method applied to a total of 25 field RR Lyrae stars, and using a bolometric correction of -0.06, leads to

$$M_{\text{RR}} = 0.72 + 0.19\,Z_{13} \tag{2.30}$$

(Sandage and Cacciari 1990, and references therein).

Most recently, the brightness of RR Lyrae stars in the Large Magellanic Cloud was measured; it has a distance which is thought to be well determined by other methods. Walker (1992) found

$$M_{\rm RR} = 0.48 + 0.15\, Z_{13}, \tag{2.31}$$

if the same bolometric correction -0.06 is assumed.

In summary, a reasonably conservative estimate of the absolute RR Lyrae bolometric brightness is

$$M_{\rm RR} = (0.60 \pm 0.15) + 0.17\, Z_{13}. \tag{2.32}$$

A comparison with Eq. (2.20) yields

$$3.5\, Y_{23} + \Delta_{\rm RR} + 7.3\, \delta\mathcal{M}_{\rm c} = 0.06 \pm 0.15. \tag{2.33}$$

This is in good agreement with the standard values $Y_{\rm env} = 0.23$, $\delta\mathcal{M}_{\rm c} = 0$, and $\Delta_{\rm RR} \approx 0.1\,{\rm mag}$.

2.4.4 Interpretation of the Observational Results

In order to interpret the observational results it is first assumed that there is no anomalous core-mass increase. However, considering the uncertainties entering the calculation of $\mathcal{M}_{\rm c}$ such as the precise value of the relevant total stellar mass, uncertainties in the electron conductive opacities, etc., it appears that a plausible range of uncertainty is $\delta\mathcal{M}_{\rm c} = \pm 0.010\,\mathcal{M}_\odot$ even in the absence of any novel phenomena. Adopting this uncertainty and adding it quadratically to the previous uncertainties, the three observables from Eq. (2.26), (2.28), and (2.33) yield

$$Y_{\rm env} = (0.244 \pm 0.012) - 0.23\,\Delta_{\rm RR} \qquad \text{from } \Delta M_{\rm HB}^{\rm tip},$$

$$Y_{\rm env} = (0.251 \pm 0.009) - 0.14\,\Delta_{\rm RR} \qquad \text{from } R,$$

$$Y_{\rm env} = (0.247 \pm 0.048) - 0.29\,\Delta_{\rm RR} \qquad \text{from } M_{\rm RR}. \tag{2.34}$$

The primordial helium abundance likely is in the range $22-24\%$; the envelope abundance in globular clusters is probably slightly larger, depending on details of gravitational settling on the MS and convective dredge-up on the RGB. Therefore, with $\Delta_{\rm RR}$ between 0 and 0.2 mag these results are perfectly consistent.

In order to constrain $Y_{\rm env}$ and $\delta\mathcal{M}_{\rm c}$ simultaneously it is assumed that $\Delta_{\rm RR} = 0.1 \pm 0.1$. Adding this error quadratically in Eq. (2.26),

(2.28), and (2.33) one finds

$$Y_{\text{env}} - 1.0\,\delta\mathcal{M}_{\text{c}} = 0.239 \pm 0.024 \qquad \text{from } \Delta M_{\text{HB}}^{\text{tip}},$$

$$Y_{\text{env}} - 0.3\,\delta\mathcal{M}_{\text{c}} = 0.237 \pm 0.016 \qquad \text{from } R,$$

$$Y_{\text{env}} + 2.1\,\delta\mathcal{M}_{\text{c}} = 0.247 \pm 0.043 \qquad \text{from } M_{\text{RR}}. \tag{2.35}$$

Bands of allowed values for Y_{env} and $\delta\mathcal{M}_{\text{c}}$ are shown in Fig. 2.22.

From Fig. 2.22 one reads off that approximately

$$Y_{\text{env}} = 0.24 \pm 0.02 \qquad \text{and} \qquad |\delta\mathcal{M}_{\text{c}}| < 0.025\,M_{\odot}. \tag{2.36}$$

Therefore, the core mass at the helium flash is tightly constrained to lie within 5% of its standard value.

This result allows one to constrain the operation of a novel energy-loss mechanism in the core of a red giant, i.e. at densities of around $10^6\,\text{g cm}^{-3}$ and at a temperature of about $10^8\,\text{K}$. In some cases, however, an energy-loss mechanism is suppressed by degeneracy effects in a RG core, while it may be fairly efficient in the helium-burning core of an HB star ($\rho \approx 10^4\,\text{g cm}^{-3}$, $T \approx 10^8\,\text{K}$) which is essentially nondegenerate. In this case $\delta\mathcal{M}_{\text{c}} \approx 0$ while the energy loss from the HB core

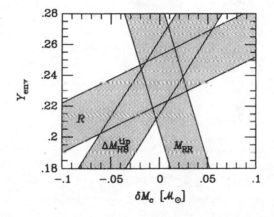

Fig. 2.22. Allowed values for the envelope helium abundance of evolved globular-cluster stars and of an anomalous core-mass excess at helium ignition. The limits were derived from the observed brightness difference $\Delta M_{\text{HB}}^{\text{tip}}$ between the HB and the RGB tip, from the "*R*-method" (number counts on the HB vs. RGB), and from the brightness determination of nearby field RR Lyrae stars (M_{RR}) by statistical parallaxes and the Baade-Wesselink method as well as the brightness of RR Lyrae stars in the LMC.

may be substantial and thus would shorten the helium-burning lifetime as the nuclear fuel would be consumed faster.

The HB/RGB number ratio R indicates that the acceleration of the HB evolution must not be too extreme. From Eq. (2.24) together with the observational result Eq. (2.27) one finds

$$\delta \log R = 0.01 \pm 0.06, \tag{2.37}$$

if one adopts $Y_{\text{env}} = 0.24 \pm 0.02$, $\Delta_{\text{RR}} = 0.1 \pm 0.1$, and $\delta \mathcal{M}_c = \pm 0.010 \, \mathcal{M}_\odot$. Therefore, R cannot be smaller by much more than about 10% of its standard value. Put another way, the helium-burning lifetime of low-mass stars is determined within about 10% from the ratio of HB/RGB stars in globular clusters.

With lesser statistical significance this result is corroborated by number counts of clump giants in open clusters which consist of more recently formed stars in the galactic disk (Population I). Clump giants in open clusters correspond to HB stars in globular clusters; instead of forming a horizontal branch they are concentrated in a clump near the base of the RGB. Cannon (1970) compared the number of clump giants in the open cluster M67 with the number of stars per luminosity interval near the MS turnoff and found $t_{\text{He}} \approx 1.5 \times 10^8 \, \text{yr}$, with a large statistical uncertainty, however, because there were only 5 clump giants. (Open clusters tend to be much less populous than globular ones.) Tinsley and Gunn (1976) derived $t_{\text{He}} = (1.27 \pm 0.29) \times 10^8 \, \text{yr}$ from low-mass giants of the old galactic disk population. These results are in full agreement with Eq. (2.23).

2.4.5 An Alternate Analysis

The above discussion of the globular-cluster limits is a somewhat updated version of my own previous work (Raffelt 1990b). Very recently, Catelan, de Freitas Pacheco, and Horvath (1995) have provided an independent new and extended analysis. While they closely follow the line of reasoning of Raffelt (1990b) they have changed numerous details. Of the 26 globular clusters which enter the $\Delta M_{\text{HB}}^{\text{tip}}$ argument, and of the 15 clusters which enter the R-method, they have discarded several with an extreme HB morphology. For the absolute brightness of RR Lyrae stars they have employed Walker's (1992) values which depend on a precise knowledge of the LMC distance. For the brightness difference between zero-age HB stars and RR Lyrae stars they use $\Delta_{\text{RR}} = 0.31 + 0.10 \, [\text{Fe/H}] = 0.18 + 0.10 \, Z_{13}$.

These authors have introduced a fourth observable, the so-called mass to light ratio A of RR Lyrae stars. This observable has also been used by Castellani and Degl'Innocenti (1993) to constrain a possible core-mass excess. It amounts to a determination of the core mass of an RR Lyrae star on the basis of its luminosity and pulsation period.

Going through similar steps as in the previous sections, Catelan, de Freitas Pacheco, and Horvath (1995) then found the results

$$Y_{\rm env} - 0.9\,\delta\mathcal{M}_{\rm c} = 0.207 \pm 0.039 \qquad \text{from } \Delta M_{\rm HB}^{\rm tip},$$

$$Y_{\rm env} - 0.3\,\delta\mathcal{M}_{\rm c} = 0.218 \pm 0.018 \qquad \text{from } R,$$

$$Y_{\rm env} + 2.1\,\delta\mathcal{M}_{\rm c} = 0.250 \pm 0.049 \qquad \text{from } M_{\rm RR},$$

$$Y_{\rm env} + 1.5\,\delta\mathcal{M}_{\rm c} = 0.238 \pm 0.008 \qquad \text{from } A. \qquad (2.38)$$

Bands of allowed values for $Y_{\rm env}$ and $\delta\mathcal{M}_{\rm c}$ are shown in Fig. 2.23 which is analogous to Fig. 2.22.

From this analysis one infers a best-fit value for the envelope helium abundance which is somewhat low. The primordial abundance probably is not lower than 22%. The only possibility to reduce the envelope abundance from this level is by gravitational settling which is counteracted by convective dredge-up, and perhaps by other effects that might eject helium into the envelope from the core. Therefore,

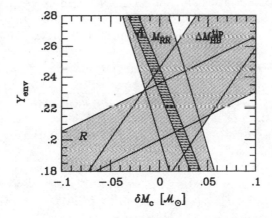

Fig. 2.23. Allowed values for the envelope helium abundance of evolved globular-cluster stars and of an anomalous core-mass excess at helium ignition according to the analysis of Catelan, de Freitas Pacheco, and Horvath (1995). For comparison see Fig. 2.22.

$Y_{env} > 0.22$ is probably a conservative estimate. This takes us back to the conclusion that $\delta \mathcal{M}_c \approx 0$ within at least $0.025\,\mathcal{M}_\odot$. Again, the core mass at helium ignition is found to agree with the standard result to better than 5%.

2.4.6 Systematic Uncertainties

The globular-cluster argument yields some of the most restrictive limits on novel modes of energy loss. Therefore, it is important to understand some of the systematic uncertainties that enter the nominal limit $\delta \mathcal{M}_c \lesssim 0.025\,\mathcal{M}_\odot$.

Core rotation has often been quoted as an effect to change $\mathcal{M}_c$. However, as it would actually delay helium ignition it cannot be invoked to compensate for anomalous cooling effects. In addition, if fast core rotation were an important effect one would expect it to vary from star to star, causing a random broadening of the distribution of ΔM_{HB}^{tip}, an effect not indicated by the observations. The scatter of ΔM_{HB}^{tip} is completely within observational errors and within the scatter caused by the effect that the brightest red giant is not exactly at the tip of the RGB in a given cluster. Further discussions of the rotational impact on $\mathcal{M}_c$ are found in Catelan, de Freitas Pacheco, and Horvath (1995).

The uncertainty of the conductive opacities are relatively large as stressed by Catelan, de Freitas Pacheco, and Horvath (1995). However, conceivable modifications of $\mathcal{M}_c$ do not seem to exceed the $0.010\,\mathcal{M}_\odot$ level even with extreme assumptions.

One of the main theoretical weaknesses of the helium-ignition argument is that the helium flash has never been properly calculated. Because helium ignites off-center one expects that convection plays a major role in the process of heating the entire core and its expansion. One may worry that in the process of the flash, parts of the core are ejected into the stellar envelope, reducing its post-flash size. However, if significant amounts of helium were ejected, the inferred Y_{env} would be changed dramatically. Even the ejection of $0.010\,\mathcal{M}_\odot$ of helium would increase Y_{env} by 0.03 if one assumes an envelope mass of $0.3\,\mathcal{M}_\odot$ and thus would brighten RR Lyrae stars by 0.12 mag. Within the stated limit of $\delta \mathcal{M}_c < 0.025\,\mathcal{M}_\odot$, mass ejection from the core is a dramatic effect that would be hard to hide in the data.

One significant systematic uncertainty arises from the relative abundance of metals among each other which is usually fixed by the solar mixture (Ross and Aller 1976). The assumption of a Ross-Aller mixture for metal-poor systems like globular-cluster stars has been called into question in recent years (e.g. Wheeler, Sneden, and Truran 1989). It is thought that in these systems the "α elements" (mostly oxygen) are enhanced relative to iron. RG sequences calculated by Raffelt and Weiss (1992) with somewhat extreme α enhancements indicated that the red-giant core mass and luminosity at the helium flash are only moderately changed. Most of the change that did occur was due to the reduction of Fe at constant Z because of the α enhancement. Put another way, if $Z_{13} = [\text{Fe/H}] + 1.3$ is used as the defining equation for the above "reduced metallicity" the effect of α enhancements appear to be rather minimal.

Catelan, de Freitas Pacheco, and Horvath (1995) have taken the point of view that for enhanced α elements one should rescale the metallicity according to a recipe given by Chieffi, Straniero, and Salaris (1991). Put another way, the metallicity parameter of Eq. (2.16) that was used in the previous sections should be redefined as

$$Z_{13} \equiv [\text{Fe/H}] + 1.3 + \log(0.579\,f + 0.421). \tag{2.39}$$

Here, f is the enhancement factor of the abundance of α elements relative to the solar value. For $f = 3$ one finds that Z_{13} must be offset by $+0.33$. Catelan, de Freitas Pacheco, and Horvath (1995) went through their analysis with this modification, causing the allowed bands of Fig. 2.23 to be slightly shifted relative to each other. However, the overall change is small, well within the stated upper limit on $\delta \mathcal{M}_c$.

In summary, no effect has been discussed in the literature that would cause the predicted core mass at helium ignition to deviate from its standard value beyond the adopted limit of $+0.025\,\mathcal{M}_\odot$. Therefore, any new energy-loss mechanism that would cause a significantly larger core-mass excess would have to be compensated by a hitherto unidentified other novel effect.

2.5 Particle Bounds from Globular-Cluster Stars

2.5.1 Helium-Burning Lifetime

A particularly simple argument to constrain the properties of novel particles arises from the observed duration of helium burning of low-mass stars, i.e. form the lifetime of stars on the horizontal branch (HB). In Sect. 2.4.4 it was argued that the number ratio R of stars on the HB vs. RGB in globular clusters agreed with standard predictions to within 10%. Therefore, the helium-burning lifetime t_{He} agrees with standard predictions to within this limit. A less significant confirmation arises from the number of clump giants in open clusters and from the old galactic disk population.

In Sect. 1.3.1 it was shown that the main impact of a nonstandard energy-loss rate on a star is an acceleration of the nuclear fuel consumption while the overall stellar structure remains nearly unchanged. The temperature dependence of the helium-burning (triple-α reaction) energy generation rate $\epsilon_{3\alpha} \propto T^{40}$ is much steeper than the case of hydrogen burning discussed in Sect. 1.3.1 and so the adjustment of the stellar structure is even more negligible. With $L_{3\alpha}$ the standard helium-burning luminosity of the core of an HB star and L_x the nonstandard energy-loss rate integrated over the core, t_{He} will be reduced by an approximate factor $L_{3\alpha}/(L_x + L_{3\alpha})$. Demanding a reduction by less than 10% translates into a requirement $L_x \lesssim 0.1\, L_{3\alpha}$. Because this constraint is relatively tight one may compute both L_x and $L_{3\alpha}$ from an unperturbed model. If the same novel cooling mechanism has delayed the helium flash and has thus led to an increased core mass only helps to accelerate the HB evolution. Therefore, it is conservative to ignore a possible core-mass increase.

The standard value for $L_{3\alpha}$ is around $20\, L_\odot$; see Fig. 2.4 for the properties of a typical HB star. Because the core mass is about $0.5\, \mathcal{M}_\odot$ the core-averaged energy generation rate is $\langle \epsilon_{3\alpha} \rangle \approx 80\,\mathrm{erg\,g^{-1}\,s^{-1}}$. Then a nonstandard energy-loss rate is constrained by

$$\langle \epsilon_x \rangle \lesssim 10\,\mathrm{erg\,g^{-1}\,s^{-1}}. \tag{2.40}$$

Previously, this limit had been stated as $100\,\mathrm{erg\,g^{-1}\,s^{-1}}$, overly conservative because it was not based on the observed HB/RGB number ratios in globular clusters. However, in practice Eq. (2.40) does not improve the constraints on a novel energy-loss rate by a factor of 10 because the appropriate average density and temperature are somewhat below the canonical values of $\rho = 10^4\,\mathrm{g\,cm^{-3}}$ and $T = 10^8\,\mathrm{K}$.

For a simple estimate the energy-loss rate may be calculated for average conditions of the core. Typically, ϵ_x will depend on some small power of the density ρ, and a somewhat larger power of the temperature T. For the HB star model of Fig. 2.4 the core-averaged values $\langle \rho^n \rangle$ and $\langle T^n \rangle$ are shown in Fig. 2.24 as a function of n. The dependence on n is relatively mild so that the final result is not sensitive to fine points of the averaging procedure.

In order to test the analytic criterion Eq. (2.40) in a concrete example consider axion losses by the Primakoff process. The energy-loss rate will be derived in Sect. 5.2.1. It is found to be proportional to T^7/ρ and to a coupling constant $g_{10} = g_{a\gamma}/10^{-10}\mathrm{GeV^{-1}}$. For a typical HB

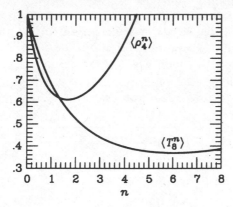

Fig. 2.24. Average values of ρ^n and T^n for the HB star model of Fig. 2.4 where $\rho_4 = \rho/10^4 \, \mathrm{g \, cm^{-3}}$ and $T_8 = T/10^8 \, \mathrm{K}$.

core I find $\langle T_8^7/\rho_4 \rangle \approx 0.3$ which leads to $\epsilon \approx g_{10}^2 \, 30 \, \mathrm{erg \, g^{-1} \, s^{-1}}$. Thus, for $g_{10} = 1$ one concludes that the helium-burning lifetime should be reduced by a factor $80/(80 + 30) = 0.7$.

This result may be compared with numerical evolution sequences for $1.3 \, M_\odot$ stars with an initial helium abundance of 25% and a metallicity of $Z = 0.02$ (Raffelt and Dearborn 1987). The helium-burning lifetime was found to be $1.2 \times 10^8 \, \mathrm{yr}$ which was modified to $0.7 \times 10^8 \, \mathrm{yr}$ for $g_{10} = 1$. This means that the axion losses on the HB led to a t_{He} reduction by a factor 0.6, in good agreement with the analytic estimate. This comparison corroborates that the analytic criterion represents the claimed impact of a novel energy loss on the helium-burning lifetime with a reasonable precision.

Equation (2.40) may now be applied to many different cases. An overview over the most salient results is given in Tab. 2.5 where the original references are given, and the sections of this book are indicated where a more detailed discussion can be found. Apart from the listed examples, the argument has also been applied to low-mass supersymmetric particles (Fukugita and Sakai 1982; Bouquet and Vayonakis 1982; Ellis and Olive 1983). However, it is now thought that supersymmetric particles, if they exist, are probably not light enough to be produced in stars whence the interest in this case has waned.

Table 2.5. Constraints on low-mass particles from the observed duration of helium burning in globular-cluster stars. Details are discussed in the indicated sections of this book.

Particle property	Dominant process	Constraint[a]	[Ref.] Sect.
Yukawa coupling of scalar (vector) boson ϕ to electrons	Bremsstrahlung $e + {}^4\text{He} \to {}^4\text{He} + e + \phi$	$g_S < 1.3 \times 10^{-14}$ $g_V < 0.9 \times 10^{-14}$	[1] 3.5.5
Yukawa coupling to baryons	Compton $\gamma + {}^4\text{He} \to {}^4\text{He} + \phi$	$g_S < 4.3 \times 10^{-11}$ $g_V < 3.0 \times 10^{-11}$	[1] 3.6.2
Photoproduction of X° boson	Photoproduction $\gamma + {}^4\text{He} \to {}^4\text{He} + X^\circ$	$\sigma \lesssim 3 \times 10^{-50}\,\text{cm}^2$	[2]
Yukawa coupling of pseudoscalar boson a to electrons	Compton $\gamma + e \to e + a$	$g < 4.5 \times 10^{-13}$	[3] 3.2.6
Yukawa coupling of "paraphoton" γ'	Compton $\gamma + e \to e + \gamma'$	$g < 3.2 \times 10^{-13}$	[4] 3.2.6
$\gamma\gamma$ coupling of pseudoscalar boson	Primakoff $\gamma + {}^4\text{He} \to {}^4\text{He} + a$	$g_{a\gamma} < 0.6 \times 10^{-10}$ GeV^{-1}	[5] 5.2.5
Neutrino dipole moment	Plasmon decay $\gamma \to \nu\bar{\nu}$	$\mu_{\text{eff}} < 1 \times 10^{-11}\,\mu_\text{B}$	[6] 6.5.6

[a]As derived in this book; may differ from the quoted references.

References:
1. Grifols and Massó (1986); Grifols, Massó, and Peris (1989).
2. van der Velde (1989); Raffelt (1988b).
3. Dicus et al. (1978, 1980); Georgi, Glashow, and Nussinov (1981);
   Barroso and Branco (1982); Fukugita, Watamura, and Yoshimura (1982a,b);
   Pantziris and Kang (1986); Raffelt (1986a).
4. Hoffmann (1987).
5. Raffelt (1986a); Raffelt and Dearborn (1987).
6. Sutherland et al. (1976); Fukugita and Yazaki (1987);
   Raffelt and Dearborn (1988); Raffelt, Dearborn, and Silk (1989).

2.5.2 Helium Ignition

Another powerful constraint arises from the agreement between the predicted and observationally inferred core mass at the helium flash (Sect. 2.4.4). An energy-loss mechanism which is efficient in a degenerate medium ($\rho \approx 10^6 \, \mathrm{g \, cm^{-3}}$, $T \approx 10^8 \, \mathrm{K}$) can delay helium ignition. To establish the core-mass increase as a function of nonstandard particle parameters one needs to evolve red giants numerically to the helium flash.

However, for a simple analytic estimate one observes that the core mass of a red giant grows by hydrogen shell burning. Because it is a degenerate configuration its radius shrinks and so the core releases a large amount of gravitational binding energy which amounts to an average energy source $\langle \epsilon_{\mathrm{grav}} \rangle$. If a novel energy-loss rate $\langle \epsilon_x \rangle$ is of the same order then helium ignition will be delayed.

In order to estimate $\langle \epsilon_{\mathrm{grav}} \rangle$ one may treat the red-giant core as a low-mass white dwarf. Its total energy, i.e. gravitational potential energy plus kinetic energy of the degenerate electrons, is found to be (Chandrasekhar 1939)

$$E = -\frac{3}{7} \frac{G_{\mathrm{N}} \mathcal{M}^2}{R}. \tag{2.41}$$

The radius of a low-mass white dwarf (nonrelativistic electrons!) may be expressed as $R = R_* (\mathcal{M}_\odot / \mathcal{M})^{1/3}$ with $R_* = 8800 \, \mathrm{km}$ so that

$$\langle \epsilon_{\mathrm{grav}} \rangle = -\frac{\dot{E}}{\mathcal{M}} = \frac{G_{\mathrm{N}} \mathcal{M}_\odot}{R_*} \left(\frac{\mathcal{M}}{\mathcal{M}_\odot} \right)^{1/3} \frac{\dot{\mathcal{M}}}{\mathcal{M}_\odot}. \tag{2.42}$$

From the numerical sequences of Sweigart and Gross (1978) one finds that near the helium flash $\mathcal{M} \approx 0.5 \, \mathcal{M}_\odot$ and $\dot{\mathcal{M}} \approx 0.8 \times 10^{-15} \, \mathcal{M}_\odot \, \mathrm{s^{-1}}$ so that $\langle \epsilon_{\mathrm{grav}} \rangle \approx 100 \, \mathrm{erg \, g^{-1} \, s^{-1}}$. Therefore, one must require $\langle \epsilon_x \rangle \ll 100 \, \mathrm{erg \, g^{-1} \, s^{-1}}$ in order to prevent the helium flash from being delayed.

In order to sharpen this criterion one may use results from Sweigart and Gross (1978) and Raffelt and Weiss (1992) who studied numerically the delay of the helium flash by varying the standard neutrino losses with a numerical factor F_ν where $F_\nu = 1$ represents the standard case. The results are shown in Fig. 2.25. Note that for $F_\nu < 1$ the standard neutrino losses are decreased so that helium ignites earlier, causing $\delta \mathcal{M}_c < 0$. It is also interesting that for $F_\nu = 0$ helium naturally ignites at the center of the core while for $F_\nu > 1$ the ignition point moves further and further toward the edge (Fig. 2.26). This behavior is

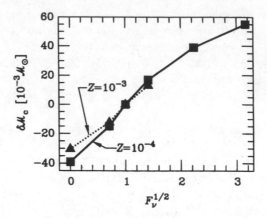

Fig. 2.25. Change of the red-giant core mass at helium ignition, $\mathcal{M}_c$, as a function of a factor F_ν which multiplies the standard neutrino energy-loss rate. *Triangles:* Metallicity $Z = 10^{-3}$ (Sweigart and Gross 1978). *Squares:* $Z = 10^{-4}$ (Raffelt and Weiss 1992).

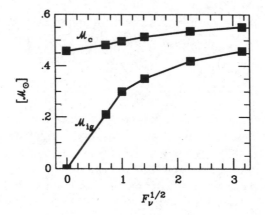

Fig. 2.26. Variation of the red-giant core mass at helium ignition, $\mathcal{M}_c$, as a function of F_ν as in Fig. 2.25 for $Z = 10^{-4}$, $Y = 0.22$, and $\mathcal{M} = 0.8\,\mathcal{M}_\odot$. Also shown is the mass coordinate $\mathcal{M}_{ig}$ where helium ignites (Raffelt and Weiss 1992).

understood from the relatively steep density dependence of the plasma neutrino emission rate which is most efficient at the center.

In order to prevent the core mass from exceeding its standard value by more than 5% ($\delta \mathcal{M}_c < 0.025\,\mathcal{M}_\odot$) one must require $F_\nu \lesssim 3$ or $\langle \epsilon_x \rangle \lesssim 2 \langle \epsilon_\nu \rangle$. According to the Sweigart and Gross (1978) red-giant sequences the neutrino luminosity of the core at helium ignition is approximately $1\,L_\odot$ so that $\langle \epsilon_\nu \rangle \approx 4\,\mathrm{erg\,g^{-1}\,s^{-1}}$. Therefore, an approximate analytic criterion to constrain a nonstandard energy loss is

$$\epsilon_x \lesssim 10\,\mathrm{erg\,g^{-1}\,s^{-1}}, \tag{2.43}$$

where ϵ_x is to be evaluated in a helium plasma at the average density of the core of about $2 \times 10^5\,\mathrm{g\,cm^{-3}}$ (the central density is about $10^6\,\mathrm{g\,cm^{-3}}$) and at the almost constant temperature of $T = 10^8\,\mathrm{K}$. The standard neutrino plasma emission rate evaluated at $2 \times 10^5\,\mathrm{g\,cm^{-3}}$ and $10^8\,\mathrm{K}$ is $3\,\mathrm{erg\,g^{-1}\,s^{-1}}$, in good agreement with the above average neutrino luminosity of the Sweigart and Gross models.

Table 2.6. Increase of the core mass at helium ignition because of the emission of pseudoscalars (Raffelt and Weiss 1995).

$\alpha'\,[10^{-26}]$	$\delta \mathcal{M}_c\,[\mathcal{M}_\odot]$
0.0	0.000
0.5	0.022
1.0	0.036
2.0	0.056

A simple application of Eq. (2.43) is the case of bremsstrahlung emission of pseudoscalars $e + {}^4\mathrm{He} \to {}^4\mathrm{He} + e + a$. For a degenerate medium the emission rate is given in Eq. (3.33); for $T = 10^8\,\mathrm{K}$ it is approximately $\alpha'\,2 \times 10^{27}\,\mathrm{erg\,g^{-1}\,s^{-1}}$ with the "axion fine-structure constant" α'. Then, Eq. (2.43) yields

$$\alpha' \lesssim 0.5 \times 10^{-26}. \tag{2.44}$$

The same case was treated numerically by Raffelt and Weiss (1995) who implemented the energy-loss rate Eq. (3.33) with varying values of α' in several red-giant evolutionary sequences.[13] They found the

---

[13]In a previous numerical treatment by Dearborn, Schramm, and Steigman (1986) the correct emission rate had not yet been available and so they overestimated the energy-loss rate by as much as a factor of 10 at the center of a red-giant core.

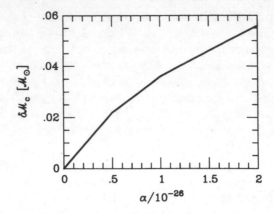

Fig. 2.27. Increase of the core mass of a red giant at helium ignition due to the emission of pseudoscalars according to Tab. 2.6 (Raffelt and Weiss 1995).

core-mass increases given in Tab. 2.6 and shown in Fig. 2.27. The requirement that the core not exceed its standard value by more than 5% reproduces the analytic bound. This example nicely corroborates the surprising precision of the simple criterion Eq. (2.43).

Another important case where a detailed numerical study is available is the emission of neutrinos by the plasma process $\gamma \to \nu\bar{\nu}$ when they have nonstandard magnetic dipole moments μ_ν. The emission rates are derived in Sect. 6.5.5 and the simple criterion Eq. (2.43) is applied in Sect. 6.5.6. It yields a limit $\mu_{12} \lesssim 2$ where $\mu_{12} = \mu_\nu/10^{-12}\,\mu_B$ with the Bohr magneton $\mu_B = e/2m_e$. The numerical variation of the core mass with μ_ν is shown in Fig. 2.28 according to Raffelt and Weiss (1992). An analytic approximation is

$$\delta\mathcal{M}_c = 0.025\,\mathcal{M}_\odot \left[(\mu_{12}^2 + 1)^{1/2} - 1 - 0.17\,\mu_{12}^{3/2}\right]. \qquad (2.45)$$

The requirement $\delta\mathcal{M}_c \lesssim 0.025\,\mathcal{M}_\odot$ then translates into

$$\mu_\nu \lesssim 3\times10^{-12}\,\mu_B, \qquad (2.46)$$

a result which, again, is almost identical with the analytic treatment, supporting the power of the simple criterion stated in Eq. (2.43).

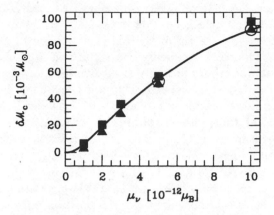

Fig. 2.28. Increase of the core mass of a red giant at helium ignition as a function of an assumed neutrino dipole moment according to Raffelt and Weiss (1992) with a total stellar mass $0.8\,\mathcal{M}_\odot$ and an initial helium abundance of $Y = 0.22$ or 0.24 (the core-mass increase is found to be the same). The triangles refer to the metallicity $Z = 10^{-3}$, the squares to 10^{-4}. The open circles are the corresponding results of Castellani and Degl'Innocenti (1992) with the same stellar mass, $Y = 0.23$, and $Z = 2 \times 10^{-4}$. The solid line is the analytic fit Eq. (2.45).

2.6   Summary

What is the bottom line after studying the impact of an anomalous energy-loss rate on a variety of stellar-evolution phases? While the Sun remains an interesting object from a pedagogical point of view, its main use is that of a distant particle source for terrestrial experimentation, notably to study neutrino properties (Chapters 10 and 12). Supernovae and their collapsed cores (newborn neutron stars) remain very important; they will be discussed in Chapters 11−13. At the present time the cooling of old neutron stars appears to be a useful laboratory to study conventional phenomena (Does the direct URCA process occur? Are there meson condensates or other exotic phases? What is the role of magnetic fields?). The emission of weakly interacting particles other than standard neutrinos appears to play a lesser role in view of other astrophysical limits on their interaction strength.

   Then, apart from the supernova arguments to be discussed later, the most useful and reliable observables to constrain the operation of a nonstandard energy-loss mechanism in stars are the white-dwarf luminosity function, the helium-burning lifetime of horizontal-branch stars, and the nondelay of helium ignition in low-mass red giants as observed by the brightness of the tip of the red-giant branch in globular clusters. It was possible to condense the latter two arguments into two exceedingly simple criteria, namely that an anomalous energy-loss rate in the cores of HB stars as well as in red-giant cores before helium ignition must not exceed about $10 \, \mathrm{erg \, g^{-1} \, s^{-1}}$. The emission rate is to be calculated at the pertinent plasma conditions, i.e. at an approximate temperature of $10^8 \, \mathrm{K} = 8.6 \, \mathrm{keV}$, an electron concentration of $Y_e = 0.5$, and an average density of about $0.6 \times 10^4 \, \mathrm{g \, cm^{-3}}$ (HB stars) or $2 \times 10^5 \, \mathrm{g \, cm^{-3}}$ (red giants). The former case corresponds to roughly nondegenerate conditions, the latter case to degenerate ones so that it depends on the density dependence of the emission rates which of these cases will yield a more restrictive limit.

A red-giant core is essentially a $0.5 \, \mathcal{M}_\odot$ helium white dwarf. The observed "real" white dwarfs have typical masses of about $0.6 \, \mathcal{M}_\odot$; they are thought to consist mostly of carbon and oxygen. Both the helium-ignition argument and the white-dwarf luminosity function allow one to constrain a novel energy-loss mechanism roughly on the level of standard neutrino emission. Therefore, it is no surprise that bounds derived from both arguments tend to be very similar.

There may be other objects or phenomena in the universe that measure novel particle-physics hypotheses even more sensitively than the cases discussed here. They still need to make their way into the particle astrophysics literature.

Chapter 3

Particles Interacting with Electrons and Baryons

The stellar energy-loss argument is applied to weakly interacting particles which couple to electrons and baryons. The emission from a normal stellar plasma can proceed by a variety of reactions, for example the Compton process $\gamma e^- \to e^- \chi$ where χ stands for a single particle (axion, paraphoton, etc.) or a neutrino pair $\nu\bar{\nu}$. Other examples of practical interest are electron bremsstrahlung $e^-(Z, A) \to (Z, A)e^-\chi$ or $e^-e^- \to e^-e^-\chi$, electron free-bound transitions, and pair annihilation $e^-e^+ \to \gamma\chi$ or $e^-e^+ \to \nu\bar{\nu}$. The corresponding energy-loss rates of stellar plasmas are studied for temperatures and densities which are of interest for stellar-evolution calculations. For particles coupled to baryons some of the same processes apply in a normal plasma if one substitutes a proton or a helium nucleus for the electron. Reactions specific to a nuclear medium are deferred to Chapter 4. In addition to draining stars of energy, scalar or vector bosons would mediate long-range forces. Leptonic long-range forces would be screened by the cosmic neutrino background. Thermal graviton emission from stars is mentioned.

3.1 Introduction

If weakly interacting particles couple directly to electrons, they can be produced thermally in stellar plasmas without nuclear processes. For neutrinos, a direct electron coupling was first contemplated after the universal $V-A$ theory for their interactions had been proposed in 1958. It was realized immediately that such a coupling would allow for thermal pair production by bremsstrahlung $e^-(Z, A) \to (Z, A)e^-\nu\bar{\nu}$

(Pontecorvo 1959; Gandel'man and Pinaev 1959) or by photoproduction $\gamma e^- \rightarrow e^- \nu \bar{\nu}$ (Ritus 1961; Chiu and Stabler 1961). The set of processes important for normal stars was completed by Adams, Ruderman, and Woo (1963) who discovered the plasma process $\gamma \rightarrow \nu \bar{\nu}$. Neutrino emission by these reactions is now a standard aspect of stellar-evolution theory.

If other weakly interacting particles were to exist which couple directly to electrons they could essentially play the same role and thus add to the energy loss of stars. The main speculation to be followed up in this chapter is the possible existence of weakly interacting *bosons* that would couple to electrons. Among standard particles the only low-mass bosons are photons and probably gravitons. The former dominate the radiative energy transfer in stars, the latter are so weakly interacting that their thermal emission is negligible (Sect. 3.7). Why worry about others?

Such a motivation arises from several sources. Low-mass bosons could mediate long-range forces between electrically neutral bodies for which gravity is the only standard interaction. It is an interesting end in itself to set the best possible bounds on possible other forces which might arise from the exchange of novel scalar or vector bosons. Their existence seemed indicated for some time in the context of the "fifth-force" episode alluded to in Sect. 3.6.3 below. It is also possible that baryon or lepton number play the role of physical charges similar to the electric charge, and that a new gauge interaction is associated with them. The baryonic or leptonic photons arising from this hypothesis are intriguing candidates for weakly interacting low-mass bosons (Sect. 3.6.4). It will turn out, however, that typically massless bosons which mediate long-range forces are best constrained by experiments which test the equivalence principle of general relativity. Put another way, to a high degree of accuracy gravity is found to be the only long-range interaction between neutral bodies.

Long-range leptonic forces can be screened by the cosmic neutrino background. In this case the stellar energy-loss argument remains of importance to limit their possible strength (Sect. 3.6.4).

The remaining category of interesting new bosons are those which couple to the spin of fermions and thus do not mediate a long-range force between unpolarized bodies. In the simplest case their CP-conserving coupling would be of a pseudoscalar nature. Such particles arise naturally as Nambu-Goldstone bosons in scenarios where a global chiral U(1) symmetry is spontaneously broken at some large energy scale. The most widely discussed example is the Peccei-Quinn symmetry that was proposed as an explanation of CP conservation in strong interactions. It leads to the prediction of axions which will be discussed in some detail in Chapter 14. The most restrictive limits on their coupling strength arise from the stellar energy-loss argument, and it cannot be excluded that in fact they play an important role in the evolution of some stars (Sect. 2.2.5). No wonder that axions have played a primary role in studies concerning the impact of new weakly interacting particles on stellar evolution.

The main focus of this chapter is an application of the stellar energy-loss argument to weakly interacting bosons which couple to electrons by a variety of interaction structures. The relevant processes are entirely analogous to those which emit neutrino pairs except for the plasma process which requires a two-body final state. Therefore, I will presently study photo and bremsstrahlung production of weakly interacting particles, including standard neutrinos.

One may consider the same processes with protons substituted for electrons. For neutrino emission this variation is of no interest because the rate is much smaller. It is significant for low-mass bosons which couple only to baryons.

The energy-loss argument will be systematically applied, yielding restrictive limits on the possible Yukawa and gauge couplings of novel bosons to electrons and baryons. For very low-mass scalar or vector bosons these limits are discussed in the context of those arising from the absence of novel long-range interactions.

3.2 Compton Process

3.2.1 Vector Bosons

The simplest process for the emission of weakly interacting particles from the hot and dense interior of a star is the Compton process where a photon from the heat bath interacts with an electron and is thus converted into a neutrino pair, an axion, or some other boson (Fig. 3.1). These processes are analogous to the usual Compton scattering of photons. Therefore, I begin with this well-known case which is based on the standard electron-photon interaction $\mathcal{L}_{\text{int}} = ie\,\overline{\psi}_e \gamma_\mu \psi_e\, A^\mu$ with the electron charge e, the electron Dirac field ψ_e, and the photon field A. With the fine structure constant $\alpha = e^2/4\pi \approx 1/137$, the electron mass m_e, and $\sigma_0 \equiv \pi\alpha^2/m_e^2$ the total Compton cross section is (e.g. Itzykson and

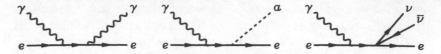

Fig. 3.1. Compton processes for photon scattering as well as for axion and neutrino pair production ("photoneutrino process"). In each case there is another amplitude with the vertices interchanged.

Zuber 1980)

$$\sigma = \sigma_0 \left[\frac{16}{(\hat{s} - 1)^2} + \frac{\hat{s} + 1}{\hat{s}^2} + \frac{2(\hat{s}^2 - 6\hat{s} - 3)}{(\hat{s} - 1)^3} \ln(\hat{s}) \right]. \tag{3.1}$$

Here, $\hat{s} \equiv s/m_e^2$ with $\sqrt{s}$ the CM (center of mass) energy.[14] This cross section is shown in Fig. 3.2 as a function of the CM photon energy ω. For $\omega \ll m_e$ the CM frame is the electron rest frame, $\hat{s} \to 1$, and one recovers the Thomson cross section $\sigma = \frac{8}{3}\sigma_0$. For $\omega \gg m_e$ one has $\hat{s} \gg 1$ and so $\sigma = (\sigma_0/\hat{s})[2\ln(\hat{s}) + 1] = (\pi\alpha^2/\omega^2)[\ln(2\omega) + \frac{1}{4}]$ because in this limit $s = (2\omega)^2$.

The standard Compton cross section can also be used to study the photoproduction of novel low-mass vector particles which couple to electrons in the same way as photons except that the fine-structure constant must be replaced by the new coupling α'. Then

$$\sigma_0 \equiv \pi\alpha\alpha'/m_e^2 \qquad \text{with} \qquad \alpha' \equiv g^2/4\pi, \tag{3.2}$$

a definition that pertains to all bosons which couple to electrons with a dimensionless Yukawa or gauge coupling g.

If novel vector bosons such as "paraphotons" exist (Holdom 1986) they likely couple to electrons by virtue of an induced magnetic moment rather than by a tree-level gauge coupling (Hoffmann 1987),

$$\mathcal{L}_{\text{int}} = (g/4m_e)\,\overline{\psi}_e \sigma_{\mu\nu} \psi_e \, F^{\mu\nu}, \tag{3.3}$$

where g is a dimensionless effective coupling constant and F the para-

---

[14]The square of the CM energy is $s = (P + K)^2$ with $P = (E, \mathbf{p})$ and $K = (\omega, \mathbf{k})$ the four-vectors of the initial-state electron and photon, respectively. The CM frame is defined by $\mathbf{p} = -\mathbf{k}$ so that $s = (E + \omega)^2 = [(m_e^2 + \omega^2)^{1/2} + \omega]^2$ with ω the initial-state photon energy in the CM frame. In the frame where the target electron is at rest ($\mathbf{p} = 0$) one finds $s = 2\omega m_e + m_e^2$ where now ω is the photon energy in the electron frame. Thus, with ω the photon energy in the respective frames,

$$\frac{\omega}{m_e} = \begin{cases} \frac{1}{2}(\hat{s} - 1)/\sqrt{\hat{s}} & \text{in the CM frame,} \\ \frac{1}{2}(\hat{s} - 1) & \text{in the electron rest frame.} \end{cases}$$

photon field tensor. In the nonrelativistic limit this yields the same total cross section as the interaction with pseudoscalars Eq. (3.6) below. This is seen if one compares the matrix elements between two electron states i and f for the two cases. For paraphotons (momentum $\mathbf{k}$, polarization vector $\boldsymbol{\epsilon}$) it is $g\langle f|e^{i\mathbf{k}\cdot\mathbf{r}}(\mathbf{k}\times\boldsymbol{\epsilon})\cdot\boldsymbol{\sigma}|i\rangle$ while for pseudoscalars it is $g\langle f|e^{i\mathbf{k}\cdot\mathbf{r}}\mathbf{k}\cdot\boldsymbol{\sigma}|i\rangle$. After an angular average the two expressions are the same. Of course, one must account for the two paraphoton polarization states by an extra factor of 2.

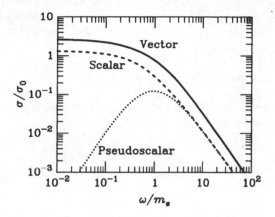

Fig. 3.2. Total cross section for the Compton process with a final-state vector, scalar, or pseudoscalar boson according to Eqs. (3.1), (3.5), and (3.9), respectively, with σ_0 defined in Eq. (3.2) and ω the CM initial photon energy.

3.2.2 Scalars

Grifols and Massó (1986) studied the stellar emission of scalars ψ which couple according to

$$\mathcal{L}_{\text{int}} = g\,\overline{\psi}_e\psi_e\,\phi.\tag{3.4}$$

Integrating their differential cross section I find

$$\sigma = \sigma_0\left[\frac{-16}{(\hat{s}-1)^2} + \frac{1-3\hat{s}}{2\hat{s}^2} + \frac{(\hat{s}+3)^2}{(\hat{s}-1)^3}\ln(\hat{s})\right]\tag{3.5}$$

shown in Fig. 3.2 (dashed line). For small and large photon energies this is half the cross section for massless vector bosons which have two polarization degrees of freedom. For intermediate energies the two results are not related by a simple factor.

3.2.3   Pseudoscalars

Next, turn to the photoproduction of low-mass pseudoscalars ϕ which couple to electrons by the interaction

$$\mathcal{L}_{\text{int}} = (1/2f)\,\overline{\psi}_e \gamma_\mu \gamma_5 \psi_e \,\partial^\mu \phi \quad \text{or} \quad \mathcal{L}_{\text{int}} = -ig\,\overline{\psi}_e \gamma_5 \psi_e \,\phi, \qquad (3.6)$$

where f is an energy scale and g a dimensionless coupling constant. Both interaction laws yield the same Compton cross section with the identification $g = m_e/f$ (see the discussion in Sect. 14.2.3).

If a pseudoscalar (frequency ω, wavevector $\mathbf{k}$) is emitted in a transition between the nonrelativistic electron states $|i\rangle$ and $|f\rangle$, the matrix element is

$$\mathcal{M}_{\text{pseudoscalar}} = \frac{g}{2m_e}\,\frac{1}{\sqrt{2\omega}}\,\big\langle f\big|e^{i\mathbf{k}\cdot\mathbf{r}}\,\boldsymbol{\sigma}\cdot\mathbf{k}\big|i\big\rangle. \qquad (3.7)$$

This is to be compared with the corresponding matrix element for photon transitions,

$$\mathcal{M}_{\text{photon}} = \frac{-e}{2m_e}\,\frac{1}{\sqrt{2\omega}}\,\big\langle f\big|e^{i\mathbf{k}\cdot\mathbf{r}}\,[2\boldsymbol{\epsilon}\cdot\mathbf{p} + \boldsymbol{\sigma}\cdot(\mathbf{k}\times\boldsymbol{\epsilon})]\big|i\big\rangle, \qquad (3.8)$$

with the photon polarization vector $\boldsymbol{\epsilon}$ and the electron momentum operator $\mathbf{p}$. Therefore, transitions involving pseudoscalars closely compare with photonic M1 transitions, a fact that was used to scale nuclear or atomic photon transition rates to those involving axions (Donelly et al. 1978; Dimopoulos, Starkman, and Lynn 1986a,b).

The relativistic Compton cross section for massive pseudoscalars was first worked out by Mikaelian (1978). The most general discussion of the matrix element was provided by Brodsky et al. (1986) and by Chanda, Nieves, and Pal (1988) who also included an effective photon mass relevant for a stellar plasma. For the present purpose it is enough to consider massless photons and pseudoscalars. With σ_0 as defined in Eq. (3.2) one finds

$$\sigma = \sigma_0 \left(\frac{\ln(\hat{s})}{\hat{s}-1} - \frac{3\hat{s}-1}{2\hat{s}^2} \right), \qquad (3.9)$$

shown in Fig. 3.2 (dotted line).

For $\omega \gg m_e$ one finds $\sigma = (\pi\alpha\alpha'/2\omega^2)\,[\ln(2\omega) - \tfrac{3}{4}]$, similar to the scalar case. For $\omega \ll m_e$, however,

$$\sigma = \tfrac{4}{3}\sigma_0\,(\omega/m_e)^2, \qquad (3.10)$$

so that the cross section is suppressed at low energies. This reduction is related to the M1 nature of the transition.

3.2.4 Neutrino Pairs

The photoneutrino process was first studied by Ritus (1961) and by Chiu and Stabler (1961). The effective neutral-current Hamiltonian is

$$\mathcal{H}_{\text{int}} = \frac{G_{\text{F}}}{\sqrt{2}} \, \bar{\psi}_e \gamma_\mu (C_V - C_A \gamma_5) \psi_e \, \bar{\psi}_\nu \gamma^\mu (1 - \gamma_5) \psi_\nu, \tag{3.11}$$

where G_{F} is the Fermi constant, and the dimensionless couplings C_V and C_A are given in Appendix B. Of course, in the early sixties neutral currents were not known—one used Fierz-transformed charged currents which gave $C_V = C_A = 1$. For general C_V's and C_A's the cross section was first calculated by Dicus (1972) who found[15] (Fig. 3.3)

$$\sigma = \sigma_0 \left[(C_V^2 + C_A^2)\,\hat{\sigma}_+ - (C_V^2 - C_A^2)\,\hat{\sigma}_- \right],$$

$$\hat{\sigma}_+ = \frac{49}{12} + \frac{5\,(13\,\hat{s} - 7)}{(\hat{s} - 1)^2} + \frac{15 - 117\,\hat{s} - 55\,\hat{s}^3}{12\,\hat{s}^2}$$

$$+ \frac{25 - 28\,\hat{s} - 27\,\hat{s}^2 - 2\,\hat{s}^3 + 2\,\hat{s}^4}{(\hat{s} - 1)^3} \ln(\hat{s}),$$

$$\hat{\sigma}_- = -39 + \frac{120\,\hat{s}}{(\hat{s} - 1)^2} - \frac{8\,\hat{s} - 1}{\hat{s}^2} + \frac{12\,(2 + 2\,\hat{s} + 5\,\hat{s}^2 + \hat{s}^3)}{(\hat{s} - 1)^3} \ln(\hat{s}),$$

$$\sigma_0 = \frac{\alpha\,G_{\text{F}}^2\,m_e^2}{9\,(4\pi)^2}. \tag{3.12}$$

In the nonrelativistic (NR) limit this is (CM photon energy ω)

$$\sigma_{\text{NR}} = \sigma_0 \tfrac{6}{35} (C_V^2 + 5C_A^2)\,(\hat{s} - 1)^4$$

$$= \sigma_0 \tfrac{96}{35} (C_V^2 + 5C_A^2)\,(\omega/m_e)^4. \tag{3.13}$$

In the extreme relativistic (ER) limit it is

$$\sigma_{\text{ER}} = \sigma_0\,(C_V^2 + C_A^2)\,2\hat{s}\,[\ln(\hat{s}) - \tfrac{55}{24}]$$

$$= \sigma_0\,(C_V^2 + C_A^2)\,16\,(\omega/m_e)^2\,[\ln(2\omega/m_e) - \tfrac{55}{48}], \tag{3.14}$$

where σ_- does not contribute.

---

[15] With $C_V^2 = C_A^2 = 1$ this result agrees with that of Ritus (1961) while Chiu and Stabler (1961) appear to have an extra factor $2\hat{s}/(\hat{s}+1)$. In the nonrelativistic limit with $\hat{s} \to 1$ this deviation makes no difference while in the extreme relativistic limit their result is a factor of 2 larger than that of Ritus (1961) and Dicus (1972).

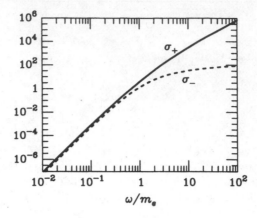

Fig. 3.3. Dimensionless total cross sections $\hat{\sigma}_+$ and $\hat{\sigma}_-$ for the photoneutrino process $\gamma e^- \rightarrow e^- \nu \bar{\nu}$ according to Eq. (3.12) with ω the initial photon energy in the CM frame.

3.2.5 Energy-Loss Rates

The Compton-type processes are typically important when the electrons are nondegenerate (otherwise bremsstrahlung dominates) and nonrelativistic (otherwise e^+e^- annihilation dominates). In these limits one may use the cross sections without Pauli blocking corrections. Because the recoil of the target electron is neglected, the energy ω of a photon impinging on an electron is identical with the energy carried away by the new boson or neutrino pair. Therefore, the energy-loss rate per unit volume is a simple integral over the initial-state photon phase space, weighted with their Bose-Einstein occupation numbers,

$$Q = n_e \int \frac{2\,d^3\mathbf{k}}{(2\pi)^3} \frac{\sigma\,\omega}{e^{\omega/T}-1}, \tag{3.15}$$

where n_e is the number density of electrons, T the temperature, and the factor 2 is for two photon polarization states.

In this expression the photon "plasma mass" $\omega_{\rm P}$ has been neglected. If $\omega_{\rm P} \gtrsim 3T$, corresponding to a typical thermal photon energy, the photon dispersion relation would have to be included properly in both the phase-space integration and in the cross section calculation. However, these are insignificant fine points for the cases to be studied below.

In the nonrelativistic limit it is easy to estimate a suppression factor $F_{\rm deg}$ by electron degeneracy. If recoil effects can be neglected, the initial- and final-state electrons have the same momentum, reducing the calculation to an average of the Pauli blocking factor over all electrons

$$F_{\text{deg}} = \frac{1}{n_e} \int \frac{2\, d^3\mathbf{p}}{(2\pi)^3} \frac{1}{e^{(E-\mu)/T}+1} \left(1 - \frac{1}{e^{(E-\mu)/T}+1}\right), \qquad (3.16)$$

where μ is the electron chemical potential and $E^2 = m_e^2 + \mathbf{p}^2$. Then,

$$F_{\text{deg}} = \frac{1}{n_e\,\pi^2} \int_{m_e}^{\infty} p\, E\, dE\, \frac{e^x}{(e^x+1)^2}, \qquad (3.17)$$

where $x \equiv (E-\mu)/T$. For degenerate conditions the integrand is strongly peaked near $x = 0$ so that one may replace p and E with the values p_{F} and E_{F} at the Fermi surface ($x = 0$), and one may extend the lower limit of integration to $-\infty$. The integral then yields T so that

$$F_{\text{deg}} = 3E_{\text{F}}T/p_{\text{F}}^2, \qquad (3.18)$$

where $n_e = p_{\text{F}}^3/3\pi^2$ was used.

Returning to the nonrelativistic, nondegenerate limit note that the cross sections are of the form $\sigma = \sigma_* \, (\omega/m_e)^p$ so that

$$Q = \frac{\sigma_* n_e\, T^{p+4}}{\pi^2\, m_e^p} \int_0^{\infty} dx\, \frac{x^{p+3}}{e^x-1} = \frac{(p+3)!\,\zeta_{p+4}}{\pi^2} \frac{\sigma_* n_e T^{p+4}}{m_e^p}. \qquad (3.19)$$

Here, $\zeta_n = \zeta(n)$ is the Riemann zeta function which shall be set equal to unity.[16] In a medium of mass density ρ the electron density is $n_e = Y_e\rho/m_u$ where Y_e is the electron number fraction per baryon and m_u the atomic mass unit. Therefore, the energy-loss rate per unit mass is

$$\epsilon = \frac{(p+3)!}{\pi^2} \frac{Y_e\sigma_* T^{p+4}}{m_u m_e^p}. \qquad (3.20)$$

The average energy of the photons which are converted into weakly interacting particles is

$$\langle\omega\rangle = (p+3)\,T. \qquad (3.21)$$

For $p = 0$ one recovers $\langle\omega\rangle = 3T$ for the average energy of blackbody photons.[17]

---

[16] $\zeta_n \to 1$ rather quickly with increasing n; for example $\zeta_4 = \pi^4/90 \approx 1.082$. Therefore, the error is small if one takes $\zeta_n = 1$ which corresponds to using a Maxwell-Boltzmann rather than a Bose-Einstein distribution. They differ at small ω where an effective photon mass in the plasma should be taken into account anyway. Therefore, at the crude level of accuracy where one uses massless photons nothing is gained by using the Bose-Einstein distribution.

[17] With a Bose-Einstein distribution it is $\langle\omega\rangle/T = 3\,\zeta_4/\zeta_3 = \pi^4/30\zeta_3 \approx 2.70$.

Beginning with the case of scalars, the low-energy cross section is constant at $\sigma_* = \frac{4}{3}\pi\alpha\alpha'/m_e^2$, leading to

$$\epsilon_{\text{scalar}} = \frac{8\alpha\alpha'}{\pi}\frac{Y_e T^4}{m_u m_e^2} = \alpha'\, 5.7\times10^{29}\,\text{erg}\,\text{g}^{-1}\,\text{s}^{-1}\,Y_e\,T_8^4, \qquad (3.22)$$

where $T_8 = T/10^8$ K. Vector bosons carry an extra factor of 2 for their polarization states.

Turning to pseudoscalars, the low-energy cross section was found to be $\sigma_* = \frac{4}{3}\left(\pi\alpha\alpha'/m_e^2\right)(\omega/m_e)^2$, i.e. $p=2$, so that

$$\epsilon_{\text{pseudo}} = \frac{160\,\alpha\alpha'}{\pi}\frac{Y_e T^6}{m_u m_e^4} = \alpha'\, 3.3\times10^{27}\,\text{erg}\,\text{g}^{-1}\,\text{s}^{-1}\,Y_e\,T_8^6. \qquad (3.23)$$

The average energy is $\langle\omega\rangle \approx 5T$ or $\langle\omega\rangle/m_e \approx 0.08\,T_8$.

Finally, turn to neutrino pair production for which the NR cross section was given in Eq. (3.13). The energy-loss rate is

$$\epsilon_{\nu\bar{\nu}} = (C_V^2 + 5C_A^2)\,\frac{96\,\alpha}{\pi^4}\frac{G_{\text{F}}^2 m_e^6}{m_u}Y_e\left(\frac{T}{m_e}\right)^8$$

$$= (C_V^2 + 5C_A^2)\,0.166\,\text{erg}\,\text{g}^{-1}\,\text{s}^{-1}\,Y_e\,T_8^8. \qquad (3.24)$$

Because $p=4$ for this process, $\langle\omega\rangle \approx 7T$. Relativistic corrections become important at rather low temperatures. For example, at $T=10^8$ K the true emission rate is about 25% smaller than given by Eq. (3.24).

3.2.6 Applying the Energy-Loss Argument

After the derivation of the energy-loss rates it is now a simple matter to apply the energy-loss argument. In Chapter 2 it was shown that the most restrictive limits obtain from the properties of globular-cluster stars; two simple criteria were derived in Sect. 2.5 which amount to the requirement that a novel energy-loss rate must not exceed $10\,\text{erg}\,\text{g}^{-1}\,\text{s}^{-1}$ for the typical conditions encountered in the core of a horizontal-branch star, and in the core of a red giant just before helium ignition which both have $T \approx 10^8$ K.

The Compton process is suppressed by degeneracy effects in a dense plasma as discussed above—see Eq. (3.18). Therefore, at a fixed temperature the emissivity per unit mass decreases with increasing density. Because a red-giant core is nearly two orders of magnitude denser than the core of an HB star it is enough to apply the argument to the latter case.

The energy-loss rates for scalars and pseudoscalars in Eqs. (3.22) and (3.23) are independent of density, and proportional to T^4 and T^6, respectively. The averages over the core of a typical HB star are $\langle T_8^4 \rangle = 0.40$ and $\langle T_8^6 \rangle = 0.37$ (Fig. 2.24). With $Y_e = 0.5$ appropriate for helium, carbon, and oxygen the $10\,\mathrm{erg\,g^{-1}\,s^{-1}}$ limit yields

$$\alpha' \lesssim \begin{cases} 0.9 \times 10^{-28} & \text{scalar,} \\ 1.6 \times 10^{-26} & \text{pseudoscalar.} \end{cases} \tag{3.25}$$

These bounds apply to bosons with a mass below a few times the temperature, $m \lesssim 20 - 30\,\mathrm{keV}$. For larger masses the limits are significantly degraded because only the high-energy tail of the blackbody photons can produce the particles. For massive pseudoscalars this effect was explicitly studied in Sect. 1.3.5 in the context of solar limits.

For vector bosons which interact by means of a Yukawa coupling, the same limits to α' apply except that they are more restrictive by a factor of 2 because of the two polarization states which increases the emission rate.

For vector bosons which couple by means of a "magnetic moment" as the "paraphotons" in Eq. (3.3), the bound on g is the same as for pseudoscalars apart from an extra factor of 2 in the emission rate from the two polarization states.

3.3 Pair Annihilation

Electron-positron pair annihilation can produce new bosons by the "crossed" version of the Compton amplitude while the conversion into neutrino pairs does not require the participation of a photon (Fig. 3.4). For pseudoscalars the cross section for $e^+ e^- \to \gamma a$ is (Mikaelian 1978)

$$\sigma = \frac{2\pi\alpha\alpha'}{s - 4m_e^2} \ln\left[\frac{s}{4m_e^2}\left(1 + \sqrt{1 - 4m_e^2/s}\right)\right]. \tag{3.26}$$

For $e^+ e^- \to \nu\bar{\nu}$ it is ('t Hooft 1971; Dicus 1972)

$$\sigma = \frac{G_F^2}{12\pi}\frac{(C_V^2 + C_A^2)(s - m_e^2) + 3(C_V^2 - C_A^2)m_e^2}{\sqrt{1 - 4m_e^2/s}}. \tag{3.27}$$

Because pair annihilation requires the presence of positrons it is important only for relativistic plasmas.

In this limit $\sigma \propto s^{-1}\ln(s)$ for the production of bosons (pseudoscalar, scalar, vector) and $\sigma \propto s$ for neutrino pairs. Therefore, the

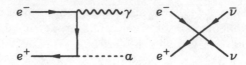

Fig. 3.4. Pair annihilation processes for the production of neutrino pairs or new bosons where a second amplitude with the vertices interchanged is not shown.

importance of stellar energy-loss rates into new scalars relative to neutrino pairs decreases with increasing temperature and density. The impact of new bosons on stellar evolution relative to neutrinos is then expected to be most pronounced for low-mass stars where other processes such as photoproduction dominate. Consequently, the pair process has not played any significant role at constraining the interactions of new bosons.

3.4 Free-Bound and Bound-Free Transitions

Photons, new bosons, or neutrino pairs can be emitted in transitions where a free electron is captured by an ion to form a bound state. For the case of axions this effect was dubbed "axio-recombination" (Dimopoulos et al. 1986). In the Sun, it contributes about 4% of the total axion flux which is mostly from bremsstrahlung. The energy-loss rate scales as $T^{3/2}$, bremsstrahlung as $T^{5/2}$, and Compton emission as T^6. Thus, axio-recombination is of importance in low-mass stars which have low internal temperatures; for main-sequence stars with $M \lesssim 0.2\,M_\odot$ it would be the dominant axion emission process. However, given the limits on the coupling of pseudoscalars to electrons from other arguments, no observable effects can be expected.

Of some practical interest is the inverse process where an axion unbinds an atomic electron, the "axio-electric effect" (Dimopoulos, Starkman, and Lynn 1986a,b). It serves to constrain the solar flux of axions or other pseudoscalars which could produce keV electrons in a Ge spectrometer designed to search for double-β decay (Avignone et al. 1987). Unfortunately, the resulting bound of $\alpha' \lesssim 10^{-21}$ is not very restrictive. Pseudoscalars saturating this limit would be a major energy drain of the Sun and thus not compatible with its observed properties.

For neutrinos, free-bound transitions were first discussed by Pinaev (1963). Recently, Kohyama et al. (1993) studied the corresponding stellar energy-loss rate in detail and found that it dominates the other neutrino emission processes only in such regions of temperature and density where the overall neutrino luminosity is very small. Therefore, free-bound transitions do not seem to be of practical importance as a stellar energy-loss mechanism.

3.5 Bremsstrahlung

3.5.1 Nondegenerate, Nonrelativistic Medium

The last emission process to be discussed is bremsstrahlung (Fig. 3.5) where an electron emits a boson or a neutrino pair when scattering off the Coulomb field of a nucleus. Conceptually, bremsstrahlung is closely related to the Compton process (Fig. 3.1) because in both cases the electron interacts with electromagnetic field fluctuations of the ambient medium which have nonvanishing power for all wavenumbers and frequencies. The Compton process corresponds to wavevectors which satisfy the photon dispersion relation so that a real (on-shell) excitation is absorbed. However, in a typical stellar plasma there is more power in the "off-shell" electromagnetic field fluctuations associated with the charged particles.[18] Moreover, degeneracy effects do not suppress bremsstrahlung at high densities, in contrast with the Compton process.

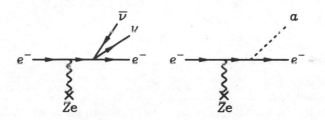

Fig. 3.5. Bremsstrahlung emission of bosons or neutrino pairs by an electron which scatters from the Coulomb field of a nucleus of charge Ze. In each case a second amplitude with the vertices interchanged is not shown.

For neutrino pairs bremsstrahlung dominates over other processes only in the highly degenerate regime (Appendix C). For pseudoscalars,

---

[18]In a typical stellar plasma there are many more charged particles than blackbody photons. In the solar center, for example, the electron density is $6 \times 10^{25}\,\mathrm{cm}^{-3}$ while for photons at a temperature of $1.3\,\mathrm{keV}$ it is $2\zeta(3)\,T^3/\pi^2 = 6 \times 10^{22}\,\mathrm{cm}^{-3}$.

however, it is important even in environments which are approximately nondegenerate and so I begin with this simple case.

The calculation amounts to a straightforward evaluation of the matrix element corresponding to the amplitude of Fig. 3.5 and an integration over the Maxwell-Boltzmann distributions of the electrons. In order to account for screening effects the Coulomb propagator is modified according to Eq. (6.72), $|\mathbf{q}|^{-4} \rightarrow [\mathbf{q}^2(\mathbf{q}^2 + k_S^2)]^{-1}$ where k_S is a screening wave number (Sect. 6.4). For the emission of pseudoscalars one then finds for the energy-loss rate per unit volume to lowest order in k_S^2 (Krauss, Moody, and Wilczek 1984; Raffelt 1986a)

$$Q = \frac{128}{45\sqrt{\pi}} \frac{\alpha^2 \alpha'}{m_e} \left(\frac{T}{m_e}\right)^{5/2} n_e$$

$$\times \sum_j n_j \left[Z_j^2 \sqrt{2} \left(1 - \frac{5}{8} \frac{k_S^2}{m_e T}\right) + Z_j \left(1 - \frac{5}{4} \frac{k_S^2}{m_e T}\right) \right]. \quad (3.28)$$

Here, the sum is extended over all nuclear species with charges $Z_j e$ and number densities n_j; note that $n_e = \sum_j Z_j n_j$. The term quadratic in Z_j corresponds to electron-nucleus collisions while the linear term is from electron-electron scattering which yields a nonnegligible contribution under nondegenerate conditions.[19]

Raffelt (1986a) incorrectly used Eq. (6.61) as a modification of the Coulomb propagator, a procedure which enhances the terms proportional to k_S^2 by a factor of 2. Either way, screening is never an important effect. The screening scale is $k_S^2 = k_D^2 + k_i^2$ because both electrons and nuclei (ions) contribute. Then

$$\frac{k_S^2}{m_e T} = \frac{4\pi\alpha}{m_e T^2} \left(n_e + \sum_j n_j Z_j^2\right), \quad (3.29)$$

which is about 0.12 at the center of the Sun and 0.17 in the cores of horizontal-branch (HB) stars.

Ignoring screening effects, the energy-loss rate per unit mass is

$$\epsilon = \alpha' \, 5.9 \times 10^{22} \, \text{erg g}^{-1} \text{s}^{-1} \, T_8^{2.5} \, Y_e \, \rho \sum_j Y_j \left(Z_j^2 + Z_j/\sqrt{2}\right), \quad (3.30)$$

where $T_8 = T/10^8 \, \text{K}$, ρ is in g cm^{-3}, and $Y_j = X_j/A_j$ is the number fraction of nuclear species j relative to baryons while X_j is the mass fraction, A_j the mass number.

---

[19]Note that $e^- e^- \rightarrow e^- e^- \gamma$ vanishes to lowest order because two particles of equal mass moving under the influence of their Coulomb interaction do not produce a time-varying electric dipole moment because their center of mass and "center of charge" coincide. However, the emission of pseudoscalars corresponds to M1 rather than E1 transitions and so it is not suppressed.

Grifols, Massó, and Peris (1989) have worked out the bremsstrahlung rate for a scalar boson; in this case $e^- e^-$ collisions can be ignored relative to electron-nucleus scattering. They found in the nonrelativistic and nondegenerate limit, ignoring screening effects which are small,

$$\epsilon = \frac{2\alpha^2 \alpha'}{\pi^{3/2}} \frac{n_e T^{1/2}}{m_u m_e^{3/2}} a \sum_j \frac{X_j Z_j^2}{A_j}$$

$$= \alpha' \; 2.8 \times 10^{26} \mathrm{erg\,g^{-1}\,s^{-1}} \, T_8^{0.5} \, Y_e \, \rho \sum_j \frac{X_j Z_j^2}{A_j}, \tag{3.31}$$

where a is an angular integral which numerically is found to be 8.36, m_u is the atomic mass unit, and ρ is in $\mathrm{g\,cm^{-3}}$. For low-mass vector bosons the same result pertains with an extra factor of 2 for the two polarization states.

3.5.2 High Degeneracy: Pseudoscalars

Bremsstrahlung is a particularly important effect under conditions of degeneracy. In this case one neglects $e^- e^-$ collisions entirely which are suppressed by degeneracy relative to the the electron-nucleus process. If the target nuclei are taken to be infinitely heavy, Raffelt (1990) found for the volume emissivity of pseudoscalars ("axions")

$$Q = \frac{4\alpha^2 \alpha'}{\pi^2} \sum_j Z_j^2 n_j \int_{m_e}^{\infty} dE_1 \, f_1 \int_{m_e}^{E_1} dE_2 \, (1 - f_2) \int \frac{d\Omega_2}{4\pi} \int \frac{d\Omega_a}{4\pi}$$

$$\times \, \frac{|\mathbf{p}_1| |\mathbf{p}_2| \omega^2}{\mathbf{q}^2 (\mathbf{q}^2 + \kappa^2)} \left[2\omega^2 \frac{P_1 P_2 - m_e^2 + (P_2 - P_1)K}{(P_1 K)(P_2 K)} + 2 - \frac{P_1 K}{P_2 K} - \frac{P_2 K}{P_1 K} \right], \tag{3.32}$$

where P_1 is the four-vector of the incoming, P_2 of the outgoing electron, $f_{1,2}$ are the electron Fermi-Dirac occupation numbers at energies $E_{1,2}$, temperature T, and chemical potential μ. Further, K is the four-vector of the outgoing pseudoscalar (energy $\omega = E_1 - E_2$, direction Ω_a, a for axion), $\mathbf{q} = \mathbf{p}_1 - \mathbf{p}_2 - \mathbf{k}$ is the momentum transfer to the nucleus, and k_S a screening scale. The Coulomb propagator was modified according to Eq. (6.72). As long as the plasma is not strongly coupled one may use $k_S = k_i$ while the electrons do not contribute because $k_{\mathrm{TF}} \ll k_i$ in a strongly degenerate plasma.

If the electrons are very degenerate, the energy integrals can be done analytically. Moreover, all electron momenta are close to the Fermi surface, $|\mathbf{p}_1| \approx |\mathbf{p}_2| \approx p_F$ with p_F the Fermi momentum defined by $n_e = p_F^3/3\pi^2$. This also implies that $|\mathbf{q}|^2 \approx |\mathbf{p}_1 - \mathbf{p}_2|^2 \approx 2p_F^2(1 - c_{12})$ where c_{12} is the cosine of the angle between $\mathbf{p}_1$ and $\mathbf{p}_2$. With these approximations and the velocity at the Fermi surface $\beta_F \equiv p_F/E_F = p_F/(m_e^2 + p_F^2)^{1/2}$ one finds

$$Q = \frac{\pi^2 \alpha^2 \alpha'}{15} \frac{T^4}{m_e^2} \left(\sum_j n_j Z_j^2 \right) F, \tag{3.33}$$

where

$$F = \int \frac{d\Omega_2}{4\pi} \int \frac{d\Omega_a}{4\pi} \frac{(1 - \beta_F^2)\,[2\,(1 - c_{12}) - (c_{1a} - c_{2a})^2]}{(1 - c_{1a}\beta_F)\,(1 - c_{2a}\beta_F)\,(1 - c_{12})(1 - c_{12} + \kappa^2)} \tag{3.34}$$

with $\kappa^2 \equiv k_S^2/2p_F^2$.

For a single species of nuclei with charge Ze and atomic weight A the energy-loss rate per unit mass is

$$\epsilon = \frac{\pi^2 \alpha^2 \alpha'}{15} \frac{Z^2}{A} \frac{T^4}{m_u\, m_e^2} F$$

$$= \alpha'\, 1.08 \times 10^{27}\, \mathrm{erg\,g^{-1}\,s^{-1}}\, \frac{Z^2}{A}\, T_8^4\, F, \tag{3.35}$$

where again $T_8 = T/10^8\,\mathrm{K}$. Because F is of order unity for all conditions, the bremsstrahlung rate mostly depends on the temperature and chemical composition, and ϵ is not suppressed at high density.

Expanding Eq. (3.34) in powers of β_F for nonrelativistic or partially relativistic electrons one finds

$$F = \frac{2}{3} \ln \left(\frac{2 + \kappa^2}{\kappa^2} \right) + \left[\frac{2 + 5\kappa^2}{15} \ln \left(\frac{2 + \kappa^2}{\kappa^2} \right) - \frac{2}{3} \right] \beta_F^2 + \mathcal{O}(\beta_F^4). \tag{3.36}$$

Therefore, in contrast to the nondegenerate calculation this expression would diverge in the absence of screening.

Another approximation can be made if one observes that Coulomb scattering is mostly forward, i.e. the main contribution to the integral is from $c_{12} \approx 1$ which implies $c_{1a} \approx c_{2a}$. With $c_{2a} = c_{1a}$ only in the denominator one obtains

$$F = \frac{2}{3} \ln \left(\frac{2 + \kappa^2}{\kappa^2} \right) + \left[\frac{2 + 3\kappa^2}{6} \ln \left(\frac{2 + \kappa^2}{\kappa^2} \right) - 1 \right] f(\beta_F) \tag{3.37}$$

with

$$f(\beta_F) = \frac{3 - 2\beta_F^2}{\beta_F^2} - \frac{3\,(1 - \beta_F^2)}{2\beta_F^3} \ln \left(\frac{1 + \beta_F}{1 - \beta_F} \right). \tag{3.38}$$

The function $f(\beta_F)$ is 0 at $\beta_F = 0$ and rises monotonically to 1 for

$\beta_F \to 1$. Hence in the relativistic limit

$$F = \frac{2 + \kappa^2}{2} \ln\left(\frac{2 + \kappa^2}{\kappa^2}\right) - 1, \tag{3.39}$$

somewhat different from what Iwamoto (1984) found who used the Thomas-Fermi wave number as a screening scale.

For a strongly coupled, degenerate plasma typical for white dwarfs the factor F was calculated numerically by Nakagawa, Kohyama, and Itoh (1987) and Nakagawa et al. (1988) who also gave analytic approximation formulae for the axion emission rate, applicable to nonrelativistic and relativistic conditions. For a ^{12}C plasma with densities in the range $10^4 - 10^6 \, \mathrm{g\,cm}^{-3}$ and temperatures of $10^6 - 10^7 \, \mathrm{K}$ it is found that $F = 1.0$ within a few tens of percent. Therefore, for simple estimates this value is a satisfactory approximation.

Altherr, Petitgirard, and del Río Gaztelurrutia (1994) have calculated the bremsstrahlung process with the methods of finite temperature and density (FTD) field theory. The main point is that one computes directly the interaction of the electrons with the electromagnetic field fluctuations which are induced by the ambient charged particles. The result for the emission rate is similar to the one derived above.

3.5.3 High Degeneracy: Neutrino Pairs

Neutrino pair bremsstrahlung (Fig. 3.5) was the first nonnuclear neutrino emission process ever proposed (Pontecorvo 1959; Gandel'man and Pinaev 1959). A detailed calculation of the energy-loss rate in a degenerate medium, relativistic and nonrelativistic, was performed by Festa and Ruderman (1969) while conditions of partial degeneracy were studied by Cazzola, de Zotti, and Saggion (1971). The Festa and Ruderman calculation was extended by Dicus et al. (1976) to include neutral-current interactions. After a calculation very similar to the one presented above for pseudoscalars the emission rate for neutrino pairs is

$$Q = \frac{2\pi\alpha^2}{189} G_F^2 T^6 \left(\sum_j n_j Z_j^2\right) \left[\tfrac{1}{2}(C_V^2 + C_A^2)F_+ + \tfrac{1}{2}(C_V^2 - C_A^2)F_-\right]. \tag{3.40}$$

With a single species of nuclei (charge Z, atomic mass A) the energy-loss rate per unit mass is

$$\epsilon = 0.144 \, \mathrm{erg\,g}^{-1}\,\mathrm{s}^{-1} \, (Z^2/A) \, T_8^6 \, [\ldots], \tag{3.41}$$

where $T_8 = T/10^8 \, \mathrm{K}$ and the square bracket is from Eq. (3.40). The temperature dependence is steeper than for axions by two powers.

The factors F_+ and F_- are of order unity. Dicus et al. (1976) derived analytic expressions in terms of a screening scale and the Fermi velocity of the electrons. In fact, because F_- is always much smaller than F_+ and $C_V^2 - C_A^2$ is much smaller than $C_V^2 + C_A^2$, the "minus" term may be neglected entirely. Therefore, the Dicus et al. (1976) result is identical with that of Festa and Ruderman (1969). Either one is correct only within a factor of order unity because Eq. (6.61) was used as a screening prescription with the Thomas-Fermi wave number as a screening scale. However, in a degenerate medium electrons never dominate screening. The most important effect is from the ion correlations which, in a weakly coupled plasma ($\Gamma \lesssim 1$), can be included by Eq. (6.72) with the Debye scale k_i of the ions as a screening scale. While it is easy to replace k_{TF} with k_i in these results, the modification of the Coulomb propagator according to Eq. (6.72) cannot be implemented without redoing the entire calculation.

A systematic approach to include ion correlations (i.e. screening effects) was pioneered by Flowers (1973, 1974) who showed clearly how to separate the ion correlation effects in the form of a dynamic structure factor from the matrix element of the electrons and neutrinos. This approach also allows one to include lattice vibrations when the ions form a crystal in a strongly coupled plasma. In a series of papers Itoh and Kohyama (1983), Itoh et al. (1984a,b), and Munakata, Kohyama, and Itoh (1987) followed this approach and calculated the emission rate for all conditions and chemical compositions.

As an estimate, good to within a factor of order unity, one may use $F_+ = 1$ and $F_- = 0$. Moreover, inspired by the axion results one can guess a simple expression which can be tested against the numerical rates of Itoh and Kohyama (1983). I find that $F_- = 0$ and

$$F_+ \approx \ln\left(\frac{2 + \kappa^2}{\kappa^2}\right) + \frac{\kappa^2}{2 + \kappa^2} \qquad (3.42)$$

is a reasonable fit even for strongly coupled conditions (Appendix C).

3.5.4   Neutron-Star Crust

The degenerate bremsstrahlung emission of $\nu\bar{\nu}$ pairs is relevant in such diverse environments as the cores of low-mass red giants, white dwarfs, and in neutron-star crusts. Pethick and Thorsson (1994) noted that in the latter case the medium is so dense that band-structure effects of the electrons become important. The band separations can be up to $\mathcal{O}(1\,\mathrm{MeV})$, suppressing electron scattering and bremsstrahlung processes for temperatures below this scale ($T \lesssim 5\times10^9\,\mathrm{K}$). Bremsstrahlung of $\nu\bar{\nu}$ pairs from the crust was thought to dominate neutron-star cooling for some conditions while Pethick and Thorsson (1994) now find that it may never be important. These findings also diminish Iwamoto's (1984) axion bound based on the bremsstrahlung emission from neutron-star crusts.

3.5.5 Applying the Energy-Loss Argument

One may now easily derive astrophysical limits on the Yukawa couplings of scalars and pseudoscalars, in full analogy to Sect. 3.2.6 where the Compton emission rates were used. I begin with the same case that was considered there, namely the restriction $\epsilon \lesssim 10\,\mathrm{erg\,g^{-1}\,s^{-1}}$ in the cores of horizontal-branch stars. For nondegenerate conditions the emission rates are Eq. (3.30) and Eq. (3.31), respectively. They are proportional to $\rho T^{0.5}$ (pseudoscalar) and $\rho T^{2.5}$ (scalar). With $\langle\rho_4\rangle = 0.64$, $\langle T_8^{0.5}\rangle = 0.82$, $\langle T_8^{2.5}\rangle = 0.48$, and a chemical composition of pure helium one finds

$$\alpha' \lesssim \begin{cases} 1.4\times10^{-29} & \text{scalar,} \\ 1.6\times10^{-25} & \text{pseudoscalar.} \end{cases} \tag{3.43}$$

Comparing these limits with those from the Compton process Eq. (3.25) reveals that for scalars the present bremsstrahlung limit is more restrictive, for pseudoscalars the Compton one.

Because bremsstrahlung is not suppressed in a degenerate plasma, one can also apply the helium-ignition argument of Sect. 2.5.2 which again requires $\epsilon \lesssim 10\,\mathrm{erg\,g^{-1}\,s^{-1}}$ at the same $T \approx 10^8\,\mathrm{K}$, however at a density of around $10^6\,\mathrm{g\,cm^{-3}}$. For these conditions the plasma is degenerate but weakly coupled (Appendix D) so that Debye screening should be an appropriate procedure. Therefore, for the emission of pseudoscalars the emission rate Eq. (3.35) with F from Eq. (3.36) should be a reasonable approximation. The screening scale is dominated by the ions, $k_S = k_i = 222\,\mathrm{keV}$, the Fermi momentum is $p_F = 409\,\mathrm{keV}$ so that $\kappa^2 = 0.15$ while $\beta_F = 0.77$, yielding $F \approx 1.8$.

In Fig. 3.6 the energy-loss rate of a helium plasma at $T = 10^8\,\mathrm{K}$ is plotted as a function of density, including the Compton process, and the degenerate (D) and nondegenerate (ND) bremsstrahlung rates. This figure clarifies that bremsstrahlung, of course, is suppressed by degeneracy effects relative to the ND rates, but it is not a significantly decreasing function of density. In this figure a simple interpolation (solid line) between the regimes of high and low degeneracy is shown

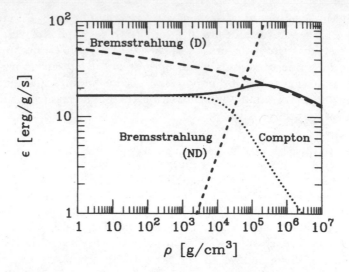

Fig. 3.6. Energy-loss rate of a helium plasma at $T = 10^8$ K as a function of density for pseudoscalars with the coupling $\alpha' = 10^{-26}$. The Compton rate is given in Eq. (3.23), suppressed with the factor F_{deg} of Eq. (3.18) at high density. The nondegenerate (ND) and degenerate (D) bremsstrahlung rates are from Eqs. (3.28) and (3.35), respectively. The solid line is the interpolation formula of Raffelt and Weiss (1995).

that was used in a numerical study of axion emission from red giants by Raffelt and Weiss (1995); details of how it was constructed can be found there.

At $T = 10^8$ K the compound rate (solid line) coincidentally is almost independent of density. The requirement $\epsilon \lesssim 10\,\text{erg}\,\text{g}^{-1}\,\text{s}^{-1}$ then yields a constraint $\alpha' \lesssim 10^{-26}$ for any density whence the helium-burning lifetime of HB stars as well as the core mass at helium ignition yield an almost identical constraint. In detail one needs to apply the helium-ignition argument at an average core density of 2×10^5 g/cm^3 (the central density is about 10^6 g/cm^3) and at the almost constant temperature of $T = 10^8$ K. The degenerate emission rate is then approximately $\alpha'\,2 \times 10^{27}\,\text{erg}\,\text{g}^{-1}\,\text{s}^{-1}$ so that

$$\alpha' \lesssim 0.5 \times 10^{-26}. \tag{3.44}$$

This is the most restrictive available bound on the Yukawa coupling of pseudoscalars to electrons.

The same case was treated numerically by Raffelt and Weiss (1995) who implemented the compound energy-loss rate of Fig. 3.6 with varying values of α' in several red-giant evolutionary sequences. The results of this work have been discussed in Sect. 2.5.2 where it was used as one justification for the simple $10 \, \mathrm{erg} \, \mathrm{g}^{-1} \, \mathrm{s}^{-1}$ energy-loss constraint that was derived there. This detailed numerical study yielded the same limit Eq. (3.44) on a pseudoscalar Yukawa coupling to electrons.

3.6 Astrophysical Bounds on Yukawa and Gauge Couplings

3.6.1 Pseudoscalars (Axions)

The main concern of this chapter was novel bosons which interact with electrons by a dimensionless Yukawa coupling. It will become clear shortly that these results can be easily translated into limits on baryonic couplings as well. It may be useful to pull together the main results found so far, and discuss them in the context of other sources of information on the same quantities.

The best studied case of boson couplings to electrons is that of pseudoscalars because the existence of such particles is motivated by their role as Nambu-Goldstone bosons of a spontaneously broken chiral symmetry of the fundamental interactions. Within this class, axions (Chapter 14) have been most widely discussed; they usually serve as a generic example for low-mass pseudoscalars. For vector bosons which couple by means of a "magnetic moment" (Eq. 3.3) such as "paraphotons" the same limits apply apart from an extra factor of 2 in the emission rate from the two polarization states of these particles.

The simplest constraint on the Yukawa coupling g_a of pseudoscalars (axions) to electrons ($\alpha_a = g_a^2/4\pi$) arises from the argument that the age of the Sun precludes any novel energy-loss mechanism to be more efficient than the surface photon luminosity (Sect. 1.3.2). The relevant emission processes are the Compton reaction with the energy-loss rate given in Eq. (3.23), and bremsstrahlung with electrons scattering on electrons, protons, and helium nuclei; the emission rate was given in Eq. (3.28). An integration over a typical solar model yields an axion luminosity (Raffelt 1986a)

$$L_a = \alpha_a \, 6.0 \times 10^{21} \, L_\odot, \tag{3.45}$$

with about 25% from the Compton process, 25% from ee bremsstrahlung, and 50% from bremsstrahlung by electrons scattering on nuclei.

The requirement $L_a \lesssim L_\odot$ yields a constraint $\alpha_a \lesssim 1.7 \times 10^{-22}$.

A much more restrictive limit arises from the white-dwarf luminosity function as axion emission would accelerate white-dwarf cooling. This argument was studied in detail in Sect. 2.2.4 as a generic case for the use of the white-dwarf luminosity function; the resulting constraint is given in Tab. 3.1. The cooling speed of white dwarfs was also established from a measurement of the period decrease of the DA variable (ZZ Ceti) star G117–B15A which thus yields a similar limit. However, the period decrease of this star may be slightly faster than can be attributed to standard cooling processes; it has been speculated that axions with a coupling strength of about $\alpha_a \approx 0.5 \times 10^{-26}$ could be responsible (Sect. 2.2.5). The limit Eq. (3.44) that was derived in the previous section from the helium-ignition argument in globular clusters is of a similar magnitude, but not restrictive enough to exclude this hypothesis.

All of these constraints apply to low-mass bosons. The most restrictive one is based on the helium ignition argument with $T \approx 10^8\,\mathrm{K} = 8.6\,\mathrm{keV}$. Therefore, these constraints apply if $m \lesssim 10\,\mathrm{keV}$. However, it would be incorrect to think that for larger masses there was no constraint. There is one, but it is degraded because threshold effects limit the particle production to the high-energy tails of the thermal distributions of the plasma constituents. For massive pseudoscalars, this question has been studied in Sect. 1.3.5 in the context of general solar particle constraints. For the more restrictive globular-cluster limits, such a detailed investigation does not exist in the literature.

3.6.2 Energy Loss by Scalar and Vector Bosons

The couplings of low-mass scalar or vector particles ϕ are easy to constrain by the same methods. Because the energy-loss rates have been calculated only for nondegenerate conditions, only the arguments involving the solar age and the helium-burning lifetime can be employed. The latter yields a constraint on the ϕ-e coupling of

$$\alpha_{\phi e} \lesssim 1.4 \times 10^{-29}. \tag{3.46}$$

It is based on the bremsstrahlung process $e + {}^4\mathrm{He} \rightarrow {}^4\mathrm{He} + e + \phi$ as discussed above in Sect. 3.5.5. For vector bosons which couple by a current-current structure analogous to photons the same results apply except for a factor of two in the emission rate which improves the limit by a factor of 2.

Tab. 3.1. Astrophysical bounds on the Yukawa coupling g_a of pseudoscalars (axions) to electrons.

Upper limit $\alpha_a = g_a^2/4\pi$	Astrophysical observable	Dominant emission process	Detailed discussion
1.7×10^{-22}	Solar age	Bremsstrahlung (75%), Compton (25%) $e + X \rightarrow X + e + a$ ($X = e, p, {}^4\mathrm{He}$) $\gamma + e \rightarrow e + a$	here
1.0×10^{-26}	Galactic age inferred from break in white-dwarf luminosity function	Bremsstrahlung (degenerate) $e + {}^{12}\mathrm{C}\,({}^{16}\mathrm{O}) \rightarrow {}^{12}\mathrm{C}\,({}^{16}\mathrm{O}) + e + a$	Sect. 2.2.4
similar	Period decrease in DAV star G117–B15A	same	Sect. 2.2.5
1.6×10^{-26}	Helium-burning lifetime of horizontal-branch stars	Compton $\gamma + e \rightarrow e + a$	Sect. 3.2.6
0.5×10^{-26}	Core mass at helium ignition in low-mass red giants	Bremsstrahlung (degenerate) $e + A \rightarrow A + e + a$ ($A = {}^4\mathrm{He}, {}^{12}\mathrm{C}, {}^{16}\mathrm{O}$)	Sect. 2.5.2 Sect. 3.5.5

One may also consider the Yukawa coupling of scalar (vector) bosons to baryons (Grifols and Massó 1986; Grifols, Massó, and Peris 1989b). In this case one may use the Compton process $\gamma + {}^4\text{He} \to {}^4\text{He} + \phi$ on a helium nucleus for which the emission rate is given *mutatis mutandis* by the same formula as for nonrelativistic electrons. Assuming the same coupling to protons and neutrons the emission rate is coherently enhanced by a factor 4^2. Moreover, in Eq. (3.22) one must replace Y_e with Y_{He} (number of ${}^4\text{He}$ nuclei per baryon) which is $\frac{1}{4}$ for pure helium, m_e is to be replaced by $m_{\text{He}} \approx 4m_u$, and $\alpha \to 4\alpha$ to account for the coherent photon coupling. Pulling these factors together one finds $\epsilon = \alpha_{\phi N}\, 0.7 \times 10^{23}\, \text{erg}\,\text{g}^{-1}\,\text{s}^{-1}$ for pure helium. Because the helium-burning lifetime argument limits this energy-loss rate to $10\,\text{erg}\,\text{g}^{-1}\,\text{s}^{-1}$ one finds

$$\alpha_{\phi N} \lesssim 1.5 \times 10^{-22} \tag{3.47}$$

as a limit on the coupling of a scalar boson to a nucleon N. Again, for vector bosons this limit is a factor of 2 more restrictive.

3.6.3 Long-Range Forces

Scalar or vector particles mediate long-range forces between macroscopic bodies. For pseudoscalars this is not the case because their CP-conserving coupling to fermions has a pseudoscalar structure, i.e. in the nonrelativistic limit they couple to the fermion spin. Therefore, even if the mass of the new particles is very small or exactly zero, they do not mediate a long-range force between unpolarized bodies. The residual force caused by the simultaneous exchange of two pseudoscalars is found to be extremely small (e.g. Grifols and Tortosa 1994).

Consider a scalar of mass $m = \lambda^{-1}$ (Compton wave length λ) which couples to nucleons with a Yukawa strength g. It mediates an attractive force between two nucleons given in terms of the potential $-(g^2/4\pi)\,r^{-1}\,e^{-r/\lambda}$. Two macroscopic test bodies of geometric dimension much below λ are then attracted by virtue of the total potential

$$V(r) = -G_{\text{N}}\,\frac{m_1 m_2}{r}\,(1 + \beta e^{-r/\lambda}), \tag{3.48}$$

where $G_{\text{N}} = m_{\text{Pl}}^{-2}$ is Newton's constant, $m_{\text{Pl}} = 1.22 \times 10^{19}\,\text{GeV}$ is the Planck mass, m_1 and m_2 are the masses of the bodies, and

$$\beta \equiv \frac{g^2}{4\pi}\,\frac{m_{\text{Pl}}^2}{u_1 u_2}. \tag{3.49}$$

Here, $u_{1,2} = m_{1,2}/N_{1,2}$ with $N_{1,2}$ the total number of nucleons in each body. Apart from small binding-energy effects which are different for different materials $u_1 \approx u_2 \approx m_u$ is the atomic mass unit, approximately equal to a nucleon mass m_N. For this force not to compete with gravity one needs $g < \mathcal{O}(10^{-19})$ or $\alpha = g^2/4\pi < \mathcal{O}(10^{-37})$ so that low-mass scalars, if they exist, need to have extremely feeble couplings to matter. Their smallness implies that such particles would not have any impact whatsoever on the energy loss of stars.

Therefore, for boson masses so small that the Compton wave length λ is macroscopic, the most restrictive limits on g obtain from analyzing the forces between macroscopic bodies, not from the energy-loss argument. The existence of a composition-dependent "fifth force" in nature with a strength of about 1% of gravity and a range λ of a few hundred meters seemed indicated by a reanalysis of Eötvös's original data (Fischbach et al. 1986). Subsequently many experiments were carried out to search for this effect, with no believable positive outcome (for a review see Fischbach and Talmadge 1992). However, these investigations produced extremely restrictive limits on β, depending on the assumed range λ of the new force. The limits also depend on the presumed coupling; if the new force couples to baryon number one finds that $\beta \lesssim 10^{-3}$ for λ in the cm range, or $\beta \lesssim 10^{-9}$ for λ of order the Earth-Sun distance and above. Thus, novel long-range forces must be much weaker than gravity. No bounds on β seem to exist for λ below the cm range, i.e. for boson masses of order 10^{-3} eV and above, apart from the stellar energy-loss argument.

The effect of a novel force with intermediate range on the equations of stellar structure was discussed by Glass and Szamosi (1987, 1989). Solar models including such a force and the impact on the solar oscillation frequencies were discussed by Gilliland and Däppen (1987) and Kuhn (1988). For a force so weak or weaker than indicated by the laboratory limits, no observable consequences for stellar structure and evolution seem to obtain. Also, the impact of the new field on the value of fundamental coupling constants even at compact objects such as neutron stars is far below any observable limit (Ellis et al. 1989).

A significant bound obtains from the orbital decay of the Hulse-Taylor binary pulsar. In order for the energy loss in the new scalars to remain below 1% of the gravitational wave emission the Yukawa coupling to baryons must satisfy $g \lesssim 3\times10^{-19}$ (Mohanty and Panda 1994). This translates into $\beta \lesssim 1$ for scalar boson masses below the orbital pulsar frequency of $2\pi/P = 2.251\times10^{-4}\,\mathrm{s}^{-1}$, i.e. for $\lambda^{-1} \lesssim 1.5\times10^{-19}$ eV or $\lambda \gtrsim 1.3\times10^{14}$ cm. This "fifth-force limit," however, is weaker than those derived by terrestrial laboratory methods.

3.6.4   Leptonic and Baryonic Gauge Interactions

The physical motivation for considering long-range interactions mediated by vector bosons, besides the fifth-force episode, is the hypothesis
that baryon number or lepton number could play the role of physical
charges similar to the electric one (Lee and Yang 1955; Okun 1969).
Their association with a gauge symmetry would provide one explanation for the strict conservation of baryon and lepton number which so
far has been observed in nature. In the framework of this hypothesis
one predicts the existence of baryonic or leptonic photons which couple
to baryons or leptons by a charge e_B or e_L, respectively. The novel
gauge bosons would be massless like the ordinary photon.

Therefore, the limits established in the previous sections on the dimensionless couplings of vector bosons can be readily restated as limits
on the values of putative baryonic or leptonic charges. The energy-loss
argument applied to helium-burning stars yields

$$e_L \lesssim 1 \times 10^{-14},$$
$$e_B \lesssim 3 \times 10^{-11}, \qquad\qquad (3.50)$$

according to Eqs. (3.43) and (3.47), respectively. Tests of the equivalence principle (i.e. of a composition-dependent fifth force) on solar-
system scales yield $\beta \lesssim 10^{-9}$ (Sect. 3.6.3) so that

$$e_B \lesssim 1 \times 10^{-23}. \qquad\qquad (3.51)$$

Apparently this is the most restrictive limit on e_B that is currently
available.

One may be tempted to apply the limits from the equivalence principle also to a leptonic charge e_L. However, in this case one has to worry
about the fact that even neutrinos would carry leptonic charges. The
universe is probably filled with a background neutrino sea in the same
way as it is filled with a background of microwave photons. This neutrino medium would constitute a leptonic plasma which screens sources
of the leptonic force just as an electronic plasma screens electric charges
(Zisman 1971; Goldman, Zisman, and Shaulov 1972; Çiftçi, Sultansoi,
and Türköz 1994; Dolgov and Raffelt 1995).

Debye screening will be studied in Sect. 6.4.1. For the screening
wave number one finds the expression

$$k_S^2 = 2\,(e_L/\pi)^2 \int_0^\infty dp\, f_{\mathbf{p}}\, p\,(v + v^{-1}), \qquad\qquad (3.52)$$

where $p = |\mathbf{p}|$ is the momentum (isotropy was assumed), $v = p/E$
the velocity of the charged particles, and $f_{\mathbf{p}}$ the occupation number of

mode **p**. The overall factor of 2 relative to Eq. (6.55) arises because neutrinos and antineutrinos contribute equally.

The smallest k_S (the largest radius over which leptonic fields remain unscreened) arises if the background neutrinos have such small masses that they are still relativistic today ($v = 1$). The neutrino number density is given by $n_{\nu+\bar{\nu}} = 2 \int f_{\mathbf{p}} d^3\mathbf{p}/(2\pi)^3 = \int_0^\infty f_{\mathbf{p}} p^2 dp/\pi^2$. Therefore, in the relativistic limit one finds roughly

$$k_S^{-1} \approx e_L^{-1} n_{\nu+\bar{\nu}}^{-1/3} \approx e_L^{-1} \, 0.2 \, \text{cm}, \tag{3.53}$$

independently of details of the neutrino momentum distribution. In this estimate the predicted number density of background neutrinos in each flavor of $n_{\nu+\bar{\nu}} \approx 100 \, \text{cm}^{-3}$ was used.

The largest conceivable k_S would obtain if neutrinos were nonrelativistic today, and if some of them were bound to the galaxy. The escape velocity from the galaxy is $v_{\text{esc}} \approx 500 \, \text{km s}^{-1}$ so that the maximum momentum of a gravitationally bound neutrino is $p_{\text{max}} = m_\nu v_{\text{esc}}$. Because neutrinos obey Fermi statistics, the largest conceivable galactic neutrino density is $n_{\text{max}} \approx p_{\text{max}}^3 \approx m_\nu^3 v_{\text{esc}}^3$. A typical neutrino velocity is of order the galactic velocity dispersion, i.e. of order v_{esc}. Therefore, from Eq. (3.52) one estimates $k_S^2 \approx e_L^2 n_{\text{max}}^{2/3} v_{\text{esc}}^{-1} \approx e_L^2 m_\nu^2 v_{\text{esc}}$. Because the largest cosmologically allowed neutrino mass is about $30 \, \text{eV}$ one finds that neutrino screening cannot operate on scales below $e_L^{-1} 10^{-5} \, \text{cm}$. Therefore, the screening scale is reduced by no more than six orders of magnitude by the fact that cosmic neutrinos could be nonrelativistic today.

The stellar energy-loss result Eq. (3.50) informs us that for relativistic neutrinos $k_S \gtrsim 10^{13} \, \text{cm} \approx 1 \, \text{AU}$ where $1 \, \text{AU} = 1.5 \times 10^{13} \, \text{cm}$ (astronomical unit) is the distance to the Sun. If neutrinos were nonrelativistic, leptonic forces could be screened over distances six orders of magnitude smaller, i.e. over about $100 \, \text{km}$. However, it is probably safe to assume that leptonic forces with a strength comparable to gravity would have been noticed in terrestrial experiments searching for a composition-dependent fifth force. Therefore, even with neutrino screening it appears inconceivable that e_L could exceed about 10^{-19} In that case the screening scale would always exceed about $0.1 \, \text{AU}$ so that terrestrial limits would easily apply. Thereby one could gain a few orders of magnitude in the limit, and so one could even use solar system constraints, taking one back to a result of order the baryonic one Eq. (3.51). A similar conclusion was reached by Blinnikov et al. (1995).

Finally, the matter of the galaxy would exert a leptonic force on neutrinos propagating, say, from a distant supernova to us. This effect would cause an energy-dependent dispersion of the measurable neutrino burst (Sect. 13.3.3). However, when $e_L \gtrsim 10^{-20}$, which is necessary to cause an interesting effect on supernova neutrinos, then the galactic leptonic charge is completely screened over the relevant length scales, even if the cosmic background neutrinos are relativistic.

3.7 Graviton Emission from Stars

The one nonelectromagnetic long-range force that is actually known to exist is gravity. Whatever the ultimate quantum theory of gravity, there is little doubt that there will be quantized wave excitations, the gravitons. They would be massless spin-2 particles. In fact, gravity is the only possible force that can be mediated by a massless spin-2 boson because as a source it needs a conserved rank-2 tensor. The energy-momentum tensor, which acts as a source for the gravitational field, is the only example.

Classical gravitational waves are an inevitable consequence of Einstein's theory of general relativity. The orbital decay of the binary pulsar PSR 1913+16 (Hulse and Taylor 1975) yields firm evidence for their emission (Taylor and Weisberg 1989; Damour and Taylor 1991).

Gravitons can be produced in hot plasmas in analogy to axions or neutrino pairs; typical processes are bremsstrahlung $e + p \rightarrow e + p + g$ (graviton g) and the Primakoff effect (gravitons have a two-photon coupling). Because of their weak interaction gravitons can freely escape once produced in the interior of stars. Early calculations of the emission rates were summarized by Papini and Valluri (1977). More recent discussions include Schäfer and Dehnen (1983), Gould (1985), and del Campo and Ford (1988). As the graviton coupling involves the inverse of the Planck mass (1.2×10^{19} GeV) the graviton luminosity of stars is inevitably small. For the Sun it is about 10^{15} erg s$^{-1} \approx 10^{-19} L_\odot$, much too small to be of any observational relevance. The same conclusion holds for other stars.

Therefore, gravity itself illustrates that the mediation of long-range forces is a far more important effect of low-mass bosons than their thermal emission from stellar plasmas. Unless, of course, they only couple to the fermion spins rather than to a "charge." Pseudoscalars such as axions are in that category.

Chapter 4

Processes in a Nuclear Medium

The interaction rates of neutrinos and axions with nucleons in a nuclear medium are studied with a focus on neutral-current processes such as bremsstahlung emission of axions $NN \to NNa$ and of neutrino pairs $NN \to NN\nu\bar{\nu}$ as well as neutrino scattering. A severe problem with the perturbative rate calculations at high density is discussed which has a strong impact on axion emissivities and the dominant axial-vector contribution to the neutrino opacities.

4.1 Introduction

New particles which couple to nucleons are emitted from ordinary stars by analogous processes to those discussed in Chapter 3 for electrons. For example, axions can be produced by the Compton process $\gamma p \to pa$; the previous results can be easily adapted to such reactions. Presently I will focus on processes involving neutrinos or axions that are specific to a nuclear medium, i.e. to supernova (SN) cores or neutron stars.

The main focus of the literature which deals with microscopic processes in a nuclear medium was inspired by the problem of late-time neutron-star cooling. The recent progress of x-ray astronomy has led to reasonably safe ROSAT identifications of thermal surface emission from a number of old pulsars (Sect. 2.3). Together with the spin-down age of these objects one can begin to test neutron-star cooling scenarios, notably those that involve novel phases of nuclear matter such as superfluidity, meson condensates, quark matter, etc. (Shapiro and Teukolsky 1983; Tsuruta 1992).

117

From the perspective of particle physics, however, a more interesting nuclear environment is a *young* neutron star for the first few seconds after the progenitor collapsed. The nuclear medium here is so hot that it is essentially nondegenerate, and neutrinos are trapped. Therefore, the production of even more weakly interacting particles such as axions or right-handed neutrinos can compete with neutrino energy transfer which is essentially a diffusion process. For a quantitative understanding of the emissivities of the new particles, but also for the conventional transport of energy and lepton number by neutrinos, a knowledge of the microscopic interaction rates is needed.

The neutrino opacities that went into standard SN collapse and explosion calculations as well as the particle emissivities that went into the derivation of, say, axion bounds from SN 1987A were based on the assumption that the hot nuclear medium can be treated as an ideal Boltzmann gas of free particles, except for degeneracy effects that are easy to include. It turns out, however, that the approximations made are internally inconsistent, notably for the dominant processes which involve couplings to the nucleon spin (axial-vector current interactions). In order for the "naive" neutral-current neutrino opacities to be correct one needs to assume that the nucleon spins do not fluctuate too fast on a time scale set by the temperature. A naive perturbative calculation, however, yields a spin-fluctuation rate which is much larger than this limit. This large rate went into the axion emissivities. Therefore, the existing studies of SN axion bounds are based on microscopic interaction rates which simultaneously make use of the opposite limits of a spin-fluctuation rate very small and very large compared with T.

Beyond the ideal-gas approximation there is virtually no literature on the microscopic interaction rates in a hot nuclear medium, presumably because of the historical focus on old neutron stars, and presumably because degenerate nuclear matter is more reminiscent of actual nuclei. Thus there is a dearth of reliable microscopic input physics for conventional studies of SN evolution, and for variations involving novel particle-physics ideas.

Most of this chapter focusses on the dominant axial-vector current interactions of neutrinos and axions with nucleons in a dense and hot medium. Most of the material is based on a series of papers which I have co-authored (Raffelt and Seckel 1991, 1995; Keil, Janka, and Raffelt 1995; Janka et al. 1995) and as such does not represent a community consensus. On the other hand, I am not aware of a significant controversy. Rather, it appears that very little serious interest has been taken in the difficult question of weakly interacting particles interacting with a hot nuclear medium, even though these issues are of paramount importance for a proper quantitative understanding of SN physics where the interaction of neutrinos with the medium dominates the thermal and dynamical evolution. My discussion can only be a starting point for future work that may actually yield some answers to the questions raised.

4.2 Axionic Bremsstrahlung Process

4.2.1 Matrix Element for $NN \rightarrow NNa$

The simplest neutral-current process of the kind to be discussed in this chapter is bremsstrahlung emission of axions or other pseudoscalars (Fig. 4.1) because the single-particle axion phase space is particularly simple. It will turn out that the result thus derived can be applied to neutrino processes almost without modification. The interaction Hamiltonian with nucleons is of the form

$$\mathcal{H}_{\text{int}} = -\frac{C_N}{2f_a} \overline{\psi}_N \gamma_\mu \gamma_5 \psi_N \, \partial^\mu \phi, \tag{4.1}$$

where f_a is an energy scale (the Peccei-Quinn scale for axions), C_N with $N = n$ or p is a dimensionless, model-dependent coupling constant of order unity, the ψ_N are the proton and nucleon Dirac fields, and ϕ is the axion field or any other pseudoscalar Nambu-Goldstone boson.

Consider a single species of nonrelativistic nucleons interacting by a one-pion exchange (OPE) potential. The spin-summed squared matrix element is (Brinkmann and Turner 1988; Raffelt and Seckel 1995)

$$\sum_{\text{spins}} |\mathcal{M}|^2 = \frac{16\,(4\pi)^3 \alpha_\pi^2 \alpha_a}{3m_N^2} \left[\left(\frac{\mathbf{k}^2}{\mathbf{k}^2 + m_\pi^2} \right)^2 + \left(\frac{\mathbf{l}^2}{\mathbf{l}^2 + m_\pi^2} \right)^2 \right.$$
$$\left. + \frac{\mathbf{k}^2 \, \mathbf{l}^2 - 3\,(\mathbf{k} \cdot \mathbf{l})^2}{(\mathbf{k}^2 + m_\pi^2)(\mathbf{l}^2 + m_\pi^2)} \right]. \tag{4.2}$$

Here, $\alpha_a \equiv (C_N m_N / f_a)^2 / 4\pi$ and $\alpha_\pi \equiv (f\, 2m_N/m_\pi)^2 / 4\pi \approx 15$ with $f \approx 1$ are the axion-nucleon and pion-nucleon "fine-structure constants," respectively. Further, $\mathbf{k} = \mathbf{p}_2 - \mathbf{p}_4$ and $\mathbf{l} = \mathbf{p}_2 - \mathbf{p}_3$ with $\mathbf{p}_i$ the momenta of the nucleons N_i as in Fig. 4.1.

In a thermal medium $\mathbf{k}^2 \approx 3m_N T$ so that $[\mathbf{k}^2/(\mathbf{k}^2 + m_\pi^2)]^2$ varies between 0.86 for $T = 80\,\text{MeV}$ and 0.37 for $T = 10\,\text{MeV}$. Therefore, neglecting the pion mass causes only a moderate error in a SN core (Brinkmann and Turner 1988; Burrows, Ressell, and Turner 1990).

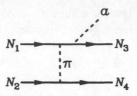

Fig. 4.1. Feynman graph for nucleon-nucleon axion bremsstrahlung. There is a total of eight amplitudes, four with the axion attached to each nucleon line, and an exchange graph each with $N_3 \leftrightarrow N_4$.

Raffelt and Seckel (1995) showed that including m_π causes less than a 30% reduction of typical neutrino or axion rates for $T > 20\,\mathrm{MeV}$. Because this will be a minor error relative to the dominant uncertainties the term in square brackets is approximated as $[3 - (\hat{\mathbf{k}} \cdot \hat{\mathbf{l}})^2]$.

The remaining $(\hat{\mathbf{k}} \cdot \hat{\mathbf{l}})^2$ term is inconvenient without yielding any significant insights. In a degenerate medium it averages to zero in expressions such as the axion emission rate while in a nondegenerate medium it can be as large as about 1.31 (Raffelt and Seckel 1995), leading to an almost 50% reduction of the emissivity. Still, for the present discussion I will neglect this term and use

$$\sum_{\text{spins}} |\mathcal{M}|^2 = 16\,(4\pi)^3 \alpha_\pi^2 \alpha_a m_N^{-2}. \qquad (4.3)$$

While this may seem somewhat arbitrary, it must be stressed that using an OPE potential to model the nucleon interactions in a nuclear medium is in itself an approximation of uncertain precision. For the present discussion a factor of order unity will not change any of the conclusions.

4.2.2 Energy-Loss Rate

The axionic volume energy-loss rate of a medium is the usual phase-space integral,

$$Q_a = \int \frac{d^3 \mathbf{k}_a}{2\omega_a\,(2\pi)^3}\,\omega_a \int \prod_{i=1}^{4} \frac{d^3 \mathbf{p}_i}{2E_i\,(2\pi)^3}\, f_1\, f_2\, (1 - f_3)(1 - f_4)$$

$$\times\,(2\pi)^4\,\delta^4(P_1 + P_2 - P_3 - P_4 - K_a)\,\tfrac{1}{4} \sum_{\text{spins}} |\mathcal{M}|^2, \quad (4.4)$$

where $P_{1,2}$ are the four-momenta of the initial-state nucleons, $P_{3,4}$ are for the final states, and K_a is for the axion. The factor $\tfrac{1}{4}$ is a statistics factor to compensate for double counting of identical fermions in the initial and final state. The occupation numbers $f_{1,2}$ and the Pauli blocking factors $(1 - f_{3,4})$ are for the nucleons while the axions are assumed to escape freely so that a Bose stimulation factor as well as backreactions (axion absorption) can be neglected.

In the nonrelativistic limit the nucleon mass m_N is much larger than all other energy scales such as the temperature or Fermi energies. The nucleon momenta are then much larger than the momentum carried by the radiation. A typical nonrelativistic nucleon kinetic energy is $E_{\text{kin}} = p^2/2m_N$ so that a typical nucleon momentum is $p = (2m_N E_{\text{kin}})^{1/2}$. In a bremsstrahlung process, the radiation typically takes the energy E_{kin} with it, less in a degenerate medium, so that a typical radiation momentum is $k_a = E_{\text{kin}} \ll p$. Therefore, one may ignore the radiation in the law of momentum conservation so that Eq. (4.4) is simplified according to

$$\delta^4(P_1 + P_2 - P_3 - P_4 - K_a) \rightarrow$$

$$\rightarrow \delta(E_1 + E_2 - E_3 - E_4 - \omega_a)\,\delta^3(\mathbf{p}_1 + \mathbf{p}_2 - \mathbf{p}_3 - \mathbf{p}_4). \quad (4.5)$$

In this case, the second integral expression in Eq. (4.4), which "knows" about axions only by virtue of the energy-momentum transfer K_a in the δ function, is only a function of the axion energy ω_a.

Therefore, in terms of a dimensionless function $s(x)$ of the dimensionless axion energy $x = \omega_a/T$ one may write the energy-loss rate in the form (baryon density n_B)

$$Q_a = \left(\frac{C_N}{2f_a}\right)^2 n_B \Gamma_\sigma \int \frac{d^3\mathbf{k}_a}{2\omega_a\,(2\pi)^3}\, \omega_a\, s(\omega_a/T)\, e^{-\omega_a/T}$$

$$= \frac{\alpha_a n_B \Gamma_\sigma T^3}{4\pi\, m_N^2} \int_0^\infty dx\, x^2\, s(x)\, e^{-x}. \quad (4.6)$$

Γ_σ will turn out to represent the approximate rate of change of a nucleon spin under the influence of collisions with other nucleons.

4.2.3 Nondegenerate Limit

To find Γ_σ and $s(x)$ turn first to an evaluation of Eq. (4.4) in the nondegenerate limit. The initial-state nucleon occupation numbers $f_{1,2}$ are given by the nonrelativistic Maxwell-Boltzmann distribution $f_{\mathbf{p}} = (n_B/2)\,(2\pi/m_N T)^{3/2}\, e^{-\mathbf{p}^2/2m_N T}$ so that $\int 2f_{\mathbf{p}}\, d^3\mathbf{p}/(2\pi)^3 = n_B$ gives the nucleon (baryon) density where the factor 2 is for two spin states. Pauli blocking factors are omitted: $(1 - f_{3,4}) \rightarrow 1$.

Because the matrix element has been assumed to be a constant it can be pulled out of the integral which reduces to a phase-space volume. Nonrelativistically, $d^3\mathbf{p}_i/[2E_i(2\pi)^3] = d^3\mathbf{p}_i/[2m_N(2\pi)^3]$ while $E_i = \mathbf{p}_i^2/2m_N$ in the energy δ function. The axion momentum is ignored according to Eq. (4.5). One uses CM momenta $\mathbf{p}_{1,2} = \mathbf{p}_0 \pm \mathbf{p}$ and $\mathbf{p}_{3,4} = \mathbf{p}_0 \pm \mathbf{q}$ where $\mathbf{p}_0 = \frac{1}{2}(\mathbf{p}_1 + \mathbf{p}_2) = \frac{1}{2}(\mathbf{p}_3 + \mathbf{p}_4)$ and defines $u^2 \equiv \mathbf{p}^2/m_N T$ and $y \equiv v^2 \equiv \mathbf{q}^2/m_N T$. Then one finds explicitly

$$\Gamma_\sigma = 4\sqrt{\pi}\, \alpha_\pi^2\, n_B\, T^{1/2} m_N^{-5/2},$$

$$s(x) = 4 \int du\, dv\, u^2\, v^2\, e^{|x|-u^2}\, \delta(u^2 - v^2 - |x|)$$

$$= \int_0^\infty dy\, e^{-y}\, \left(|x|y + y^2\right)^{1/2} \approx \sqrt{1 + |x|\,\pi/4}. \qquad (4.7)$$

The analytic form is accurate to better than 2.2% everywhere; it has the correct asymptotic behavior $s(0) = 1$ and $s(|x|\gg 1) = (|x|\pi/4)^{1/2}$.

We will see in Sect. 4.6.3 that for $|x| \gg 1$ the nondegenerate $s(x)$ must actually *decrease*, in conflict with this explicit calculation. This problem reveals a pathology of the OPE potential which is too singular at short distances. For the present discussion this is of no concern so that I stick to the explicit OPE result in order to facilitate comparison with the existing literature.

To determine the total emission rate one uses the first representation of $s(x)$ and performs $\int dx$ first to remove the δ function; the remaining integrals are easily done. Explicit results for $s_n \equiv \int_0^\infty x^n s(x)\, e^{-x} dx$ are given in Tab. 4.1. The normalized axion energy spectrum $dN_a/dx = x\, s(x) e^{-x}/s_1$ is shown in Fig. 4.2 (solid line). The average axion energy is $\langle \omega_a \rangle/T = s_2/s_1 = 16/7$. The total nondegenerate energy-loss rate is

Table 4.1. $s_n = \int_0^\infty x^n s(x)\, e^{-x} dx$ for nondegenerate (ND) and degenerate (D) conditions.

n	s_n(ND)	s_n(D)
1	8/5	$2\zeta_3 + 6\zeta_5/\pi^2$
2	128/35	$31\pi^4/315$
3	256/21	$24\zeta_5 + (180/\pi^2)\,\zeta_7$
4	4096/77	$82\pi^6/315$

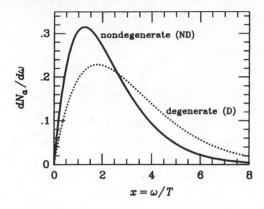

Fig. 4.2. Normalized axion spectrum $x\,s(x)e^{-x}/s_1$ from nucleon-nucleon bremsstrahlung emission. The nondegenerate and degenerate functions $s(x)$ are given in Eqs. (4.7) and (4.9), respectively, while $s_1(\text{ND})$ and $s_1(\text{D})$ are found in Tab. 4.1.

explicitly

$$Q_a^{\text{ND}} = \frac{128\,\alpha_a\alpha_\pi^2}{35\sqrt{\pi}}\,\frac{n_B^2\,T^{7/2}}{m_N^{9/2}},$$

$$\epsilon_a^{\text{ND}} = \alpha_a\,1.69\times10^{35}\,\text{erg g}^{-1}\,\text{s}^{-1}\,\rho_{15}\,T_{\text{MeV}}^{3.5}, \tag{4.8}$$

where $\rho_{15} = \rho/10^{15}\,\text{g cm}^{-3}$, $T_{\text{MeV}} = T/\text{MeV}$, and $\epsilon_a^{\text{ND}} = Q_a^{\text{ND}}/\rho$ is the energy-loss rate per unit mass.

4.2.4 Degenerate Limit

Details of the nucleon phase space in the degenerate limit can be found in Friman and Maxwell's (1979) calculation of $\nu\bar\nu$ emission. The axion energy-loss rate is expressed as in Eq. (4.6). One finds

$$\Gamma_\sigma = \frac{4\alpha_\pi^2}{3\pi}\,\frac{p_F T^3}{n_B} \quad\text{and}\quad s(x) = \frac{(x^2+4\pi^2)\,|x|}{4\pi^2\,(1-e^{-|x|})}, \tag{4.9}$$

where p_F is the nucleon Fermi momentum. As in the nondegenerate case $s(0) = 1$ while $s(|x|\gg1) = |x|^3/4\pi^2$. Values for $s_n = \int_0^\infty x^n s(x)\,e^{-x}dx$ are given in Tab. 4.1. The normalized axion spectrum is shown in Fig. 4.2 (dotted line), the average energy is $\langle\omega_a\rangle/T = s_2/s_1 \approx 3.16$.

The total energy-loss rate is

$$Q_a^D = \alpha_a \alpha_\pi^2 \frac{31\pi^2}{945} \frac{p_F T^6}{m_N^2},$$

$$\epsilon_a^D = \alpha_a \, 1.74 \times 10^{31} \, \text{erg g}^{-1} \text{s}^{-1} \, \rho_{15}^{-2/3} \, T_{\text{MeV}}^6, \tag{4.10}$$

(Iwamoto 1984; Brinkmann and Turner 1988).

The degenerate and nondegenerate rates are best compared in terms of a parameter $\xi \equiv p_F^2/(2\pi \, m_N T)$ which approaches η/π in the degenerate limit (degeneracy parameter η),

$$\frac{Q_a^D}{Q_a^{ND}} = \frac{31\pi^4}{1536\sqrt{2}} \xi^{-5/2} \approx 1.39 \, \xi^{-5/2}. \tag{4.11}$$

Therefore, they are equal for $\xi \approx 1$, or a degeneracy parameter of $\eta \approx 3.5$. This defines the dividing line between the regimes where these approximations can be reasonably used.

In the degenerate limit the nucleon phase-space integrals can be done analytically with the inclusion of a nonzero m_π. The $m_\pi = 0$ rates must be supplemented with a factor[20] (Ishizuka and Yoshimura 1990)

$$G(u) = 1 - \frac{5u}{6} \arctan\left(\frac{2}{u}\right) + \frac{u^2}{3(u^2+4)} +$$

$$+ \frac{u^2}{6\sqrt{2u^2+4}} \arctan\left(\frac{2\sqrt{2u^2+4}}{u^2}\right), \tag{4.12}$$

where $u = m_\pi/p_F$. For only one species of nucleons (as approximately in a neutron star) $p_F = 515 \, \text{MeV} \, \rho_{15}^{1/3}$ so that $u = 0.26 \, \rho_{15}^{-1/3}$ with ρ_{15} the mass density in $10^{15} \, \text{g/cm}^3$. For this case G is shown as a function of ρ in Fig. 4.3 (solid line).

4.2.5 Bremsstrahlung Emission of Scalars

The previous results equally apply to pseudoscalars with a coupling $ig_{aN} \bar\psi_N \gamma_5 \psi_N \phi$ with $\alpha_a \equiv g_{aN}^2/4\pi$ if a derivative pion-nucleon interaction is used (Sect. 14.2.3). However, for scalars which couple according

---

[20]Eq. (4.12) differs from the corresponding result of Friman and Maxwell (1979) which is identical with that of Iwamoto (1984) who apparently did not take the third term of the matrix element Eq. (4.2) properly into account.

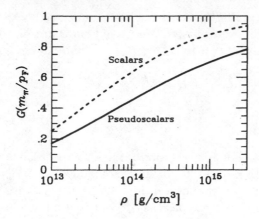

Fig. 4.3. Correction of the degenerate bremsstrahlung rate for a nonzero pion mass. For pseudoscalars, $G(m_\pi/p_F)$ is given explicitly in Eq. (4.12). It is assumed that only one species of nucleons is present.

to $g\overline{\psi}_N\psi_N\phi$ the results are different. The degenerate bremsstrahlung energy-loss rate was worked out by Ishizuka and Yoshimura (1990),

$$Q_{\text{scalar}} = \alpha'\alpha_\pi^2\,\frac{44}{15^3}\left(\frac{T}{m_N}\right)^4 p_F^5\,G_{\text{scalar}}(m_\pi/p_F), \qquad (4.13)$$

with $\alpha' \equiv g^2/4\pi$. The function $G_{\text{scalar}}(u)$ is similar to Eq. (4.12); it is plotted in Fig. 4.3 (dashed line).

4.2.6 Mixture of Protons and Neutrons

For a mixed medium of protons and neutrons one needs to consider the individual Yukawa couplings $g_{an} = C_n m_N/f_a$ and $g_{ap} = C_p m_N/f_a$ as well as the isoscalar and isovector combinations $g_{0,1} = \frac{1}{2}(g_{an} + g_{ap})$ and the "fine-structure constants" $\alpha_j = g_j^2/4\pi$ with $j = n, p, 0, 1$. For equal couplings $\alpha_n = \alpha_p = \alpha_0$ while $\alpha_1 = 0$.

In the nondegenerate limit, the main difference is that np scattering benefits from the exchange of charged pions which couple more strongly by a factor $\sqrt{2}$. Depending on the chemical composition of the medium, the emission rate will be increased by up to a factor of 2. On the other hand, some reduction factors have been ignored such as the $(\hat{\mathbf{k}}\cdot\hat{\mathbf{l}})^2$ term and the pion mass. Therefore, ignoring this enhancement essentially compensates for the previously introduced errors.

In the degenerate limit the changes are more dramatic. The role of the Fermi momentum is played by $p_F \to (3\pi^2 n_B)^{1/3}$. It sets the scale for the proton and neutron Fermi momenta which are $p_F^{n,p} = (3\pi^2 n_B)^{1/3} Y_{n,p}^{1/3}$. According to a result of Brinkmann and Turner (1988) the effective coupling is then

$$\alpha_a \to \alpha_n Y_n^{1/3} + \alpha_p Y_p^{1/3} + \left(\tfrac{28}{3}\alpha_0 + \tfrac{20}{3}\alpha_1 \right) (Y_n^{2/3} + Y_p^{2/3})^{1/2}$$

$$\times \frac{1}{2\sqrt{2}} \left(2 - \frac{|Y_n^{2/3} - Y_p^{2/3}|}{Y_n^{2/3} + Y_p^{2/3}} \right). \quad (4.14)$$

The third term is from np collisions; it has the remarkable feature that it does not vanish as $Y_p \to 0$. For equal couplings ($C_n = C_p$) the variation of the emission rate is shown in Fig. 4.4 as a function of $Y_p = 1 - Y_n$. For all proton concentrations the np contribution dominates.

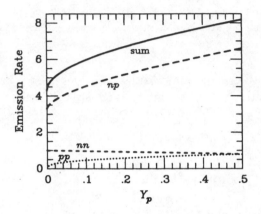

Fig. 4.4. Variation of the degenerate nucleon-nucleon bremsstrahlung rate with Y_p (proton number fraction) according to Eq. (4.14) with equal couplings so that $\alpha_n = \alpha_p = \alpha_0$ and $\alpha_1 = 0$.

4.3   Neutrino Pair Emission

4.3.1   Structure Function

Neutrino pair emission in nucleon-nucleon collisions (Fig. 4.5) is analogous to that of axions. Therefore, instead of embarking on a new calculation it is worth understanding their common features—the neutrino rates can be obtained for free on the basis of the axion ones. This approach amounts to defining the dynamical structure functions

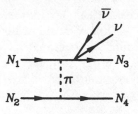

Fig. 4.5. Bremsstrahlung emission of neutrino pairs in nucleon-nucleon collisions. There is a total of eight amplitudes, four with the neutrinos attached to each nucleon line, and an exchange graph each with $N_3 \leftrightarrow N_4$.

of the medium, quantities which are of utmost importance to understand the general properties of the emission, absorption, and scattering rates independently of phase-space details of the neutrinos or axions.

Contrary to the $\nu\bar{\nu}$ bremsstrahlung emission in electron-nucleus collisions (Sect. 3.5.3), nonrelativistically only the nucleon axial-vector coupling contributes in nucleon-nucleon collision (Friman and Maxwell 1979). This difference originates from the interaction potential of the colliding particles which involves a spin-dependent force between nucleons so that the spin fluctuations caused by collisions are more dramatic than those of the velocity—see Sect. 4.6.5 below. Hence, the part of the interaction Hamiltonian relevant for neutrino pair bremsstrahlung is

$$\mathcal{H}_{\text{int}} = \frac{C_A^N G_F}{\sqrt{2}} \, \overline{\psi}_N \gamma_\mu \gamma_5 \psi_N \, \overline{\psi}_\nu \gamma^\mu (1 - \gamma_5) \psi_\nu, \tag{4.15}$$

where $C_A^p \approx -C_A^n \approx \frac{1}{2}$; see Appendix B for a discussion of the appropriate values in a nuclear medium.

The interaction Hamiltonian Eq. (4.15) has the same structure as that for axions Eq. (4.1). The squared matrix elements are then of the general form

$$\sum_{\text{spins}} |\mathcal{M}|^2 = \begin{cases} (C_A^N G_F/\sqrt{2})^2 \, M_{\mu\nu} N^{\mu\nu} & \text{for neutrinos,} \\ (C_N/2f_a)^2 \, M_{\mu\nu} K_a^\mu K_a^\nu & \text{for axions.} \end{cases} \tag{4.16}$$

Here, K_a is the axion four-momentum while

$$N^{\mu\nu} = 8 \left(K_1^\mu K_2^\nu + K_2^\mu K_1^\nu - K_1 \cdot K_2 \, g^{\mu\nu} - i\epsilon^{\alpha\beta\mu\nu} K_{1\alpha} K_{2\beta} \right) \tag{4.17}$$

with the neutrino and antineutrino four momenta K_1 and K_2 (Gaemers, Gandhi, and Lattimer 1989). $N^{\mu\nu}$ and $K_a^\mu K_a^\nu$ are the squared matrix elements of the neutrino and axion current, respectively.

The matrix $M^{\mu\nu}$ is the nuclear part of the squared matrix element. It is exactly the same for axion or neutrino interactions because in Eq. (4.16) the global coupling constants have been explicitly pulled out. Therefore, one may go one step further and perform the entire nucleon phase-space integration for both cases directly on $M^{\mu\nu}$,

$$S^{\mu\nu} \equiv \frac{1}{n_B} \int \prod_{i=1}^{4} \frac{d^3 \mathbf{p}_i}{2E_i (2\pi)^3} \, f_1 f_2 (1 - f_3)(1 - f_4)$$

$$\times (2\pi)^4 \delta^4 (P_1 + P_2 - P_3 - P_4 + K) \, M^{\mu\nu}. \qquad (4.18)$$

Here, the P_i are the nucleon four momenta, $K = -K_a$ for axion emission, and $K = -(K_1 + K_2)$ for neutrino pairs. Thus, K is the energy-momentum transfer from the radiation (axions or neutrino pairs) to the nucleons. Because $S^{\mu\nu}$ knows about the radiation only through the energy-momentum δ function it is only a function of $K = (\omega, \mathbf{k})$, apart from the temperature and chemical potentials of the medium. The slightly awkward definition of the sign of K follows the common definition of the structure function where a positive energy transfer ω refers to energy given to the medium.

The energy-loss rates by $\nu\bar{\nu}$ or axion emission are then the phase-space integrals

$$Q_{\nu\bar{\nu}} = \left(\frac{C_A^N G_F}{\sqrt{2}} \right)^2 n_B \int \frac{d^3 \mathbf{k}_1}{2\omega_1 (2\pi)^3} \frac{d^3 \mathbf{k}_2}{2\omega_2 (2\pi)^3} \, S_{\mu\nu} N^{\mu\nu} \, (\omega_1 + \omega_2),$$

$$Q_a = \left(\frac{C_N}{2f_a} \right)^2 n_B \int \frac{d^3 \mathbf{k}_a}{2\omega_a (2\pi)^3} \, S_{\mu\nu} K_a^\mu K_a^\nu \, \omega_a. \qquad (4.19)$$

Here, it was assumed that both axions and neutrinos can escape freely from the medium so that final-state Pauli blocking or Bose stimulation factors can be ignored.

In the nonrelativistic limit the nucleon current in Eqs. (4.1) and (4.15) reduces to $\chi^\dagger \tau_i \chi$ where χ is a nucleon two-spinor and τ_i ($i = 1, 2, 3$) are Pauli matrices representing the nucleon spin operator. Put another way, in the nonrelativistic limit the axial-vector current represents the nucleon spin density. Therefore, it has only spatial components so that $S^{\mu\nu} \to S^{ij}$ ($i, j = 1, 2, 3$). In order to construct the most general tensorial structure for S^{ij} in an isotropic medium only δ_{ij} is available. Recall that in the nonrelativistic limit $S^{\mu\nu}$ does not know about the momentum transfer $\mathbf{k}$ because of Eq. (4.5). There is then no vector available from which a spatial tensor can be constructed. Thus, the structure function has the most general form

$$S_{ij}(\omega) = S_\sigma(\omega)\, \delta_{ij}. \tag{4.20}$$

For nonrelativistic nucleons all axion or axial-vector neutrino processes involve only one scalar function $S_\sigma(\omega)$ of the energy transfer.

The contraction of δ_{ij} with $K_a^\mu K_a^\nu$ for axion emission yields $\mathbf{k}_a^2 = \omega_a^2$. (The spatial Kronecker δ should be viewed as a Lorentz tensor with a zero in the 00 position.) Therefore, $S_{\mu\nu}K_a^\mu K_a^\nu \to \omega_a^2 S_\sigma(-\omega_a)$ in Eq. (4.19). For neutrino emission, $K_1 \cdot K_2 = \omega_1\omega_2 - \mathbf{k}_1 \cdot \mathbf{k}_2$ and the contraction of δ_{ij} with $g_{\mu\nu}$ is -3. Therefore, the contraction of δ_{ij} with $N^{\mu\nu}$ yields $8\omega_1\omega_2(3 - \cos\theta)$ where θ is the angle between the ν and $\bar{\nu}$ momenta. The neutrino phase-space integration will always average $\cos\theta$ to zero so that it may be dropped. Therefore, $S_{\mu\nu}N^{\mu\nu} \to 24\,\omega_1\omega_2 S_\sigma(-\omega_1-\omega_2)$ in Eq. (4.19). One may then immediately perform the integration over one of the neutrino energies and is left with an integration over the energy transfer.

Thus, in the nonrelativistic limit the energy-loss rates Eq. (4.19) are of the form

$$Q_{\nu\bar{\nu}} = \left(\frac{C_A^N G_F}{\sqrt{2}}\right)^2 \frac{n_B}{20\pi^4} \int_0^\infty d\omega\, \omega^6\, S_\sigma(-\omega),$$

$$Q_a = \left(\frac{C_N}{2f_a}\right)^2 \frac{n_B}{4\pi^2} \int_0^\infty d\omega\, \omega^4\, S_\sigma(-\omega). \tag{4.21}$$

The medium properties are embodied in a common function which may be expressed in the form

$$S_\sigma(\omega) = \frac{\Gamma_\sigma}{\omega^2}\, s(\omega/T) \times \begin{cases} 1 & \text{for } \omega > 0, \\ e^{\omega/T} & \text{for } \omega < 0 \end{cases} \tag{4.22}$$

Because of the "detailed-balance relationship" between positive and negative energy transfers to be discussed more fully below, $s(x)$ must be an even function. In the nondegenerate and degenerate limits Γ_σ and $s(x)$ have been determined above. They allow one to calculate the $\nu\bar{\nu}$ emission rate without any effort.

4.3.2 Bremsstrahlung Emission of Neutrino Pairs

In order to calculate the neutrino pair emission rate explicitly turn first to the nondegenerate limit for which one uses Γ_σ and $s(x)$ given in Eq. (4.7) and the integrals of Tab. 4.1. The average energy of a neutrino pair is $(s_4/s_3)\,T \approx 4.36\,T$ and the total energy-loss rate is

$$Q_{\nu\bar\nu}^{\mathrm{ND}} = \frac{2048}{385\,\pi^{7/2}}\,C_A^2 G_F^2 \alpha_\pi^2\,\frac{n_B^2\,T^{11/2}}{m_N^{5/2}},$$

$$\epsilon_{\nu\bar\nu}^{\mathrm{ND}} = 2.4\times10^{17}\,\mathrm{erg\,g^{-1}\,s^{-1}}\,\rho_{15}\,T_{\mathrm{MeV}}^{5.5}, \tag{4.23}$$

with $\rho_{15} = \rho/10^{15}\,\mathrm{g\,cm^{-3}}$, $T_{\mathrm{MeV}} = T/\mathrm{MeV}$, and $\epsilon_{\nu\bar\nu}^{\mathrm{ND}} = Q_{\nu\bar\nu}^{\mathrm{ND}}/\rho$ is the energy-loss rate per unit mass.

For the degenerate rate one uses Eqs. (4.9). The average energy of a neutrino pair is $(s_4/s_3)\,T \approx 5.78\,T$ and the total energy-loss rate is[21]

$$Q_{\nu\bar\nu}^{\mathrm{D}} = \frac{41\pi}{4725}\,C_A^2\,G_F^2\,\alpha_\pi^2\,p_F\,T^8,$$

$$\epsilon_{\nu\bar\nu}^{\mathrm{D}} = 4.4\times10^{13}\,\mathrm{erg\,g^{-1}\,s^{-1}}\,\rho_{15}^{-2/3}\,T_{\mathrm{MeV}}^8. \tag{4.24}$$

One can correct for a nonvanishing value of the pion mass by virtue of Eq. (4.12).

For a mixture of protons and neutrons the same remarks as in Sect. 4.2.6 apply. Apart from a small correction the neutrino coupling is isovector ($C_A^p \approx -C_A^n$) so that $\alpha_0 \approx 0$ while the other α's are approximately equal (Appendix B). For degenerate conditions, with a small modification the dependence on the proton concentration is the same as that shown in Fig. 4.4. Again, the absolutely dominating contribution is from np collisions unless protons are so rare that they are nondegenerate.[22]

---

[21]Friman and Maxwell's (1979) total energy-loss rate is 2/3 of the one found here. Apparently they did not include the crossterm in the squared matrix element, i.e. the third term in Eq. (4.2).

[22]This conclusion, based on the work of Brinkmann and Turner (1988), is in conflict with the results of Friman and Maxwell (1979). They found that $Q_{\nu\bar\nu}$ was proportional to the proton Fermi momentum which is relatively small in neutron-star matter. On the other hand, for small proton concentrations Brinkmann and Turner's p_F in Eq. (4.24) approaches the *neutron* Fermi momentum. I am in no position to decide between these conflicting results.

4.4 Axion Opacity

When axions or other pseudoscalar bosons interact so "strongly" that they are trapped in a young SN core, they still contribute to the radiative transfer of energy and can thus have an impact on the cooling speed. In order to include axions in a numerical evolution calculation one needs to determine the opacity of the medium to axions (Burrows, Ressell, and Turner 1990).

The "reduced" Rosseland mean opacity relevant for radiative transport by relativistic bosons (Sect. 1.3.4) is defined by

$$\frac{1}{\rho \kappa_a} = \frac{15}{4\pi^2 T^3} \int_0^\infty d\omega\, \ell_\omega \left(1 - e^{-\omega/T}\right)^{-1} \partial_T B_\omega(T), \qquad (4.25)$$

where ρ is the mass density of the medium, ℓ_ω the boson mean free path against absorption, and $B_\omega(T) = (2\pi^2)^{-1}\omega^3(e^{\omega/T} - 1)^{-1}$ is the boson spectral density for one spin degree of freedom. After $\partial_T = \partial/\partial T$ has been taken one finds

$$\frac{1}{\rho \kappa_a} = \frac{15}{8\pi^4} \int_0^\infty dx\, \ell_x \frac{x^4 e^{2x}}{(e^x - 1)^3}, \qquad (4.26)$$

where $x \equiv \omega/T$, and $(1 - e^{-x})^{-1} = e^x(e^x - 1)^{-1}$ was used.

The axion opacity thus defined is to be added to the photon opacity by $\kappa_{\text{tot}}^{-1} = \kappa_\gamma^{-1} + \kappa_a^{-1}$. The energy flux is then given by the usual expression $F = -(3\kappa_{\text{tot}}\rho)^{-1}\nabla a_\gamma T^4$ where $a_\gamma = \pi^2/15$ gives the radiation density of *photons*. Burrows, Ressell, and Turner (1990) have defined an axion opacity which is to be used in conjunction with the axion radiation density which is half as large because there is only one spin degree of freedom, i.e. $a_a = \pi^2/30$. Their κ_a^{-1} is twice that in Eq. (4.26) so that the energy flux is the same. I prefer the present definition because it allows for the usual addition of all opacity contributions.

It is easy to determine the axion absorption rate from the discussion in Sect. 4.3.1. Starting from Q_a in Eq. (4.21) one removes the phase-space integral $\int_0^\infty d\omega\, 4\pi\omega^2/(2\pi)^3$ and a factor ω because Q_a was an energy-loss rate. One includes a factor $e^{\omega/T}$ to account for the detailed-balance relationship (see Eq. 4.43) so that altogether

$$\ell_\omega^{-1} = \left(\frac{C_N}{2f_a}\right)^2 \frac{n_B \Gamma_\sigma}{T} \frac{s(x)}{2x}, \qquad (4.27)$$

where $x = \omega/T$. Therefore, the reduced Rosseland mean opacity is

$$\kappa_a = \left(\frac{C_N}{2f_a}\right)^2 \frac{\Gamma_\sigma}{T m_N} \hat{\kappa}, \qquad (4.28)$$

where $\rho/n_B \approx m_N$ was used. The dimensionless opacity is

$$\hat{\kappa}^{-1} \equiv \frac{15}{8\pi^4} \int_0^\infty dx \, \frac{x^4 e^{2x}}{(e^x - 1)^3} \, \frac{2x}{s(x)}. \tag{4.29}$$

For the nondegenerate case $s(x)$ was given in Eq. (4.7), for the degenerate one in Eq. (4.9). Then one finds numerically $\hat{\kappa}_{\mathrm{ND}} = 0.46$ and $\hat{\kappa}_{\mathrm{D}} = 1.53$. The respective Γ_σ's were given in Eqs. (4.7) and (4.9).

4.5 Neutrino Opacity

4.5.1 Elastic Scattering

The trapping of neutrinos in a hot SN core allows them to escape only by diffusion. This implies that they transport energy in a fashion similar to radiative transfer. Apart from proper spectral weights as in the Rosseland mean opacity, the main figure of merit that determines the transport efficiency is the neutrino mean free path (mfp). While the charged-current absorption of ν_e's is very important for practical SN core-cooling calculations it has no direct bearing on the main issues of interest here.

However, all neutrino flavors interact by neutral-current interactions which allow for the scattering on nucleons. The inverse mean free path is then simply the cross section times the nucleon density. If there is only one species of nondegenerate nucleons one easily finds

$$\lambda^{-1} = \left(C_V^2 + 3C_A^2\right) G_{\mathrm{F}}^2 n_B \, \omega_1^2/\pi, \tag{4.30}$$

where ω_1 is the energy of the incident neutrino, assumed to be much smaller than m_N so that recoil effects can be neglected. For a mixture of protons and neutrons one has to take a proper average with the coupling constants C_V and C_A for protons and neutrons (Appendix B). For ν_e's which have a large chemical potential, a Pauli blocking factor must be included, and the same for nucleons if they are degenerate.

4.5.2 Pair Absorption

Because the neutrinos are assumed to be trapped there is an ambient bath of ν's and $\bar{\nu}$'s which allows for neutrino absorption by the inverse bremsstrahlung process $\nu\bar{\nu} NN \to NN$, i.e. Fig. 4.5 read from right to left. The rate for this process is closely related to that for pair emission:

it is based on the same matrix element with the neutrinos "crossed" into the initial state. Then one finds in analogy to Sect. 4.3.1

$$\lambda^{-1} = \left(\frac{C_A G_F}{\sqrt{2}}\right)^2 \frac{n_B}{2\omega_1} \int \frac{d^3\mathbf{k}_2}{2\omega_2 (2\pi)^3} f_2 \, S_{\mu\nu} N^{\mu\nu}$$

$$= \frac{3C_A^2 G_F^2 n_B}{2\pi^2} \int_0^\infty d\omega_2 \, \omega_2^2 \, f_2 \, S_\sigma(\omega_1 + \omega_2), \qquad (4.31)$$

where 1 refers to the ν for which the mfp is being determined while 2 refers to a $\bar{\nu}$ from the thermal environment (occupation number f_2). With the detailed-balance relationship $S_\sigma(\omega) = S_\sigma(-\omega) \, e^{\omega/T}$ (see Eq. 4.43) and writing $S_\sigma(|\omega|) = (\Gamma_\sigma/\omega^2) \, s(\omega/T)$ as before one finds

$$\lambda^{-1} = 3C_A^2 G_F^2 n_B \Gamma_\sigma T \int_0^\infty dx_2 \, f_2 \, \frac{x_2^2 \, s(x_1 + x_2)}{2\pi^2 \, (x_1 + x_2)^2}, \qquad (4.32)$$

where $x_i = \omega_i/T$. Then one may use the previously determined Γ_σ and $s(x)$ to find the mfp for given conditions (degenerate or nondegenerate).

The ratio between the inverse mfp's from elastic scattering and pair absorption is, ignoring the contribution from C_V^2,

$$\frac{\lambda_{\text{pair}}^{-1}}{\lambda_{\text{scat}}^{-1}} = \frac{\Gamma_\sigma}{T} \int_0^\infty dx_2 \, f_2 \, \frac{x_2^2 \, s(x_1 + x_2)}{2\pi \, x_1^2 (x_1 + x_2)^2}. \qquad (4.33)$$

An average with regard to a thermal x_1 distribution yields about 0.02 for the dimensionless integral where Fermi-Dirac distributions with chemical potentials $\mu_\nu = 0$ were used. The main figure of merit, however, is Γ_σ/T, the ratio of a typical spin-fluctuation rate and the ambient temperature. In the nondegenerate limit one finds with Eq. (4.7)

$$\gamma_\sigma \equiv \frac{\Gamma_\sigma}{T} = 4\pi^{1/2} \alpha_\pi^2 \frac{n_B}{m_N^{5/2} T^{1/2}} \approx 16 \frac{\rho}{\rho_0} \left(\frac{30\,\text{MeV}}{T}\right)^{1/2}, \qquad (4.34)$$

with the nuclear density $\rho_0 = 3 \times 10^{14} \, \text{g/cm}^3$.

4.5.3 Inelastic Scattering

It appears that for typical conditions of a young SN core, pair absorption is almost as important as elastic scattering. However, even though the quantity γ_σ is larger than unity, the pair-absorption rate has an unfavorable phase-space factor from the initial-state $\bar{\nu}$ so that the dimensionless integral in Eq. (4.33) is a small number. This would not be the case for the inelastic scattering process $\nu N N \to N N \nu$ shown in

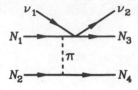

Fig. 4.6. Inelastic neutrino-nucleon scattering. There is a total of eight amplitudes, four with the neutrinos attached to each nucleon line, and an exchange graph each with $N_3 \leftrightarrow N_4$.

Fig. 4.6. In this process, neutrinos can give or take energy even though recoil effects for heavy nucleons are small in the elastic scattering process $\nu N \to N\nu$.

This reaction is identical with pair absorption with the antineutrino line crossed into the final state. The corresponding mfp is given by the same expression as for pair absorption, except that the index 2 now refers to the final-state ν, the initial-state occupation number f_2 is to be replaced with a final-state Pauli blocking factor $(1 - f_2)$, and the energy transfer is $\omega = \omega_2 - \omega_1$,

$$\lambda^{-1} = \frac{3C_A^2 G_F^2 n_B}{2\pi^2} \int_0^\infty d\omega_2\, \omega_2^2\, (1 - f_2)\, S_\sigma(\omega_1 - \omega_2). \tag{4.35}$$

However, because the energy transfer can be zero, and because $S_\sigma(\omega) \propto \omega^{-2}$, this expression diverges. The ω^{-2} behavior seemed harmless before because it was moderated by powers of ω from the phase space of axions or neutrinos.

The occurrence of this divergence could have been predicted without a calculation by inspecting Fig. 4.6. If one cuts the intermediate-state nucleon line, this graph falls into two sub-processes (nucleon-nucleon scattering and nucleon-neutrino scattering) which are each permitted by energy-momentum conservation, allowing the intermediate nucleon in the compound process to go "on-shell." Therefore, the pole of the propagator which corresponds to real particles causes a divergence of the cross section. Physically, the divergence reflects a long-range interaction which occurs because the intermediate nucleon can travel arbitrarily far when it is on its mass shell.

Still, the inelastic scattering process is an inevitable physical possibility. For nonzero energy transfers its differential rate is given by the unintegrated version of Eq. (4.35). For a vanishing energy transfer it is elastic and then its rate should be given by Eq. (4.30).

However, what exactly does one mean by a vanishing energy transfer? The nucleons in the ambient medium constantly scatter with each other so that their individual energies are uncertain to within about $1/\tau_{coll}$ with a typical time between collision τ_{coll}. Therefore, one would expect that the structure of $S_\sigma(\omega)$, which was calculated on the basis of free nucleons which interact only once, is smeared out over scales of order $\Delta\omega \approx \tau_{coll}^{-1}$. In particular, this smearing-out effect naturally regulates the low-ω behavior of the structure function (Raffelt and Seckel 1991).

Because $\Gamma_\sigma \approx \tau_{coll}^{-1}$ is a typical nucleon-nucleon collision rate, a simple ansatz for a modified $S_\sigma(\omega)$ is a Lorentzian (Raffelt and Seckel 1991)

$$S_\sigma(|\omega|) \to \frac{\Gamma_\sigma}{\omega^2 + \Gamma_\sigma^2/4}\, s(\omega/T) = \frac{1}{T}\, \frac{\gamma_\sigma}{x^2 + \gamma_\sigma^2/4}\, s(x), \qquad (4.36)$$

with $x = \omega/T$. For $\gamma_\sigma \ll 1$ one has $\int_{-\infty}^{+\infty} d\omega\, S_\sigma(\omega) = 2\pi + \mathcal{O}(\gamma_\sigma)$ for the modified $S_\sigma(\omega)$. Thus, for $\gamma_\sigma \ll 1$ essentially $S_\sigma(\omega) = 2\pi\delta(\omega)$ so that the total "inelastic" scattering rate Eq. (4.35) reproduces the elastic scattering one. At the same time $S_\sigma(\omega)$ has wings which, for $\omega \gg \Gamma_\sigma$, give the correct inelastic scattering rate which is of order γ_σ.

This simple ansatz can be expected to give a reasonable approximation to the true $S_\sigma(\omega)$ only in the limit $\gamma_\sigma \ll 1$. For practical applications in cooling calculations of young SN cores, however, one has to confront the opposite limit $\gamma_\sigma \gg 1$ causing the smearing-out effect by multiple nucleon collisions to be a dominating feature of $S_\sigma(\omega)$ even for $\omega \gg T$. This implies that for typical thermal energies of neutrinos and nucleons a reasonably clean separation between elastic and inelastic scattering processes is not logically possible—there is only one structure function $S_\sigma(\omega)$, broadly smeared out, which governs all axial-vector scattering, emission, and absorption processes.

This observation has important ramifications not only for neutrino scattering, but also for the bremsstrahlung emission of axions and neutrino pairs from a nuclear medium. Earlier it seemed that one did not have to worry about details of the behavior of $S_\sigma(\omega)$ near $\omega = 0$ because the low-ω part was suppressed by axion or neutrino phase-space factors. However, since the notion of "low energy" presently means $\omega \lesssim \Gamma_\sigma$, and because $T \ll \Gamma_\sigma$, *all* relevant energy transfers are low in this sense, and even the emission processes are dominated by multiple-scattering effects. Consequences of this behavior are explored in more detail below after a formal introduction of the structure functions and their general properties.

4.6   Structure Functions

4.6.1   Formal Definition

To treat axion and neutrino pair emission and absorption on the same footing it became useful in Sect. 4.3.1 to define the quantity $S^{\mu\nu}$ which was the nuclear part of the squared matrix element of nucleon-nucleon collisions, integrated over the nucleon phase space. The structure function thus obtained embodied the medium properties relevant for different processes without involving the radiation phase space. Clearly, this method is not limited to the bremsstrahlung process. For example, one could include the interaction of nucleons with thermal pions or a pion condensate, three-nucleon collisions, and so forth. Whatever the details of the medium physics, in the end one will arrive at some function $S^{\mu\nu}(\omega, \mathbf{k})$ of the energy and momentum transfer which embodies all of its properties. High-density properties of the medium should also appear in the structure function so that multiple-scattering modifications can be consistently applied to all relevant processes such as neutrino scattering, pair emission, axion emission, and others.

A formal definition of the structure function without reference to specific processes begins with a neutral-current interaction Hamiltonian

$$\mathcal{H}_{\text{int}} = (g_V V^\mu + g_A A^\mu) J_\mu, \tag{4.37}$$

where g_V and g_A are (usually dimensionful) coupling constants. The radiation or "probe" is characterized by a current J^μ which for axions is $\partial^\mu \phi$, for neutrino interactions $\overline{\psi}_\nu \gamma^\mu (1 - \gamma_5) \psi_\nu$, and for photons the electromagnetic vector potential. The medium is represented by the vector and axial-vector currents V^μ and A^μ. If the probe couples only to one species N of nucleons, $V^\mu = \overline{\psi}_N \gamma^\mu \psi_N$ and $A^\mu = \overline{\psi}_N \gamma^\mu \gamma_5 \psi_N$.

Next, one imagines that the medium properties are experimentally investigated with a neutrino beam with fixed momentum $\mathbf{k}_1$ which is directed at a bulk sample of the medium, and the distribution of final-state momenta and energies are measured. The transition probability $W(\mathbf{k}_1, \mathbf{k}_2)$ is proportional to $(g_V^2 S_V^{\mu\nu} + g_A^2 S_A^{\mu\nu} + g_V g_A S_{VA}^{\mu\nu}) N_{\mu\nu}$ where $N_{\mu\nu}$ was defined in Eq. (4.17). A standard perturbative expansion (Sect. 9.3) yields for the *dynamical structure functions*

$$S_V^{\mu\nu}(\omega, \mathbf{k}) = \frac{1}{n_B} \int_{-\infty}^{+\infty} dt\, e^{i\omega t} \left\langle V^\mu(t, \mathbf{k}) V^\nu(0, -\mathbf{k}) \right\rangle, \tag{4.38}$$

and analogous expressions for $S_A^{\mu\nu}$ in terms of $\langle A^\mu A^\nu \rangle$ and for $S_{VA}^{\mu\nu}$ involving $\langle V^\mu A^\nu + A^\mu V^\nu \rangle$. The expectation values are to be taken with respect to a thermal ensemble of medium states.

In addition, one frequently uses the *static structure functions* which are functions of the momentum transfer alone. For example,

$$S_V^{\mu\nu}(\mathbf{k}) = \int_{-\infty}^{+\infty} \frac{d\omega}{2\pi}\, S_V^{\mu\nu}(\omega, \mathbf{k}) = \frac{1}{n_B}\left\langle V^\mu(\mathbf{k})V^\nu(-\mathbf{k})\right\rangle. \qquad (4.39)$$

It was used that $\int_{-\infty}^{+\infty} d\omega\, e^{i\omega t} = 2\pi\delta(t)$ so that $\int dt$ in Eq. (4.38) is trivially done and yields the operators at equal times. In Eq. (4.39) $V(\mathbf{k})$ is $V(t, \mathbf{k})$ at an arbitrary time, for example $t = 0$. It only matters that both $V(\mathbf{k})$ and $V(-\mathbf{k})$ are taken at equal times.

In an isotropic medium the tensorial composition of the dynamical structure function can be obtained only from the energy-momentum transfer K and the four-velocity U of the medium; $U = (1, 0, 0, 0)$ in its rest frame. The general form of the vector term is (Kirzhnits, Losyakov, and Chechin 1990)

$$S_V^{\mu\nu} = S_{1,V}\, U^\mu U^\nu + S_{2,V}\left(U^\mu U^\nu - g^{\mu\nu}\right)$$
$$+ S_{3,V}\, K^\mu K^\nu + S_{4,V}\left(K^\mu U^\nu + U^\mu K^\nu\right). \qquad (4.40)$$

An analogous expression pertains to $S_A^{\mu\nu}$ while the mixed term is

$$S_{VA}^{\mu\nu} = i\, S_{VA}\, \epsilon^{\mu\nu\alpha\beta} U_\alpha K_\beta \qquad (4.41)$$

because of its transformation properties under parity. The functions $S_{\ell,V}$ and $S_{\ell,A}$ ($\ell = 1, \ldots, 4$), and S_{VA} depend on medium properties and on the Lorentz scalars K^2 and $U \cdot K$ that can be constructed from U and K; the third possibility $U^2 = 1$ is a constant. Instead of K^2 and $U \cdot K$ one may use the energy and momentum transfer ω and $k = |\mathbf{k}|$ measured in the medium rest frame.

The structure functions are defined for both positive and negative energy transfers because the medium can both give or take energy from a probe. Taking axion emission and absorption as an example, the rate of change of the occupation number of an axion field mode $\mathbf{k}$ is given by

$$\partial_t f_{\mathbf{k}} = \left(\frac{C_A}{2f_a}\right)^2 \frac{n_B}{2\omega}\left[(f_{\mathbf{k}} + 1)\, S_A^{\mu\nu}(-\omega, \mathbf{k}) - f_{\mathbf{k}}\, S_A^{\mu\nu}(\omega, \mathbf{k})\right] K_\mu K_\nu. \qquad (4.42)$$

If axions are trapped and reach thermal equilibrium, $\partial_t f_{\mathbf{k}} = 0$ and $f_{\mathbf{k}} = (e^{\omega/T} - 1)^{-1}$, a Bose-Einstein distribution. This implies the *detailed-balance condition*

$$S_A^{\mu\nu}(\omega, \mathbf{k}) = S_A^{\mu\nu}(-\omega, \mathbf{k})\, e^{\omega/T}. \qquad (4.43)$$

Recall that a positive energy transfer is energy given to the medium.

4.6.2   Nonrelativistic Limit

In a nuclear medium one is interested primarily in the nonrelativistic limit. The vector current is then dominated by $V^0 = \rho = \chi^\dagger \chi$ where χ is a nucleon two-spinor. Here, ρ is the operator for the nucleon number density. The spatial component $\mathbf{V}$ is suppressed by a nonrelativistic velocity factor v. The reverse applies to the axial-vector current where A^0 is suppressed; it is dominated by the spin density $\mathbf{s} = \frac{1}{2}\chi^\dagger \boldsymbol{\tau}\chi$ where $\boldsymbol{\tau}$ is a vector of Pauli matrices.

In Eqs. (4.40) and (4.41) all terms arise from correlators such as $\langle A^0 V^i \rangle$ which are suppressed by v^2, except for the first term of $S_V^{\mu\nu}$ which arises from $\langle V^0 V^0 \rangle$, and the second term of $S_A^{\mu\nu}$ which arises from $\langle A^i A^i \rangle$. Therefore, the only unsuppressed components are

$$S_V^{00}(\omega,\mathbf{k}) = S_\rho(\omega,\mathbf{k}) \quad \text{and} \quad S_A^{ij}(\omega,\mathbf{k}) = S_\sigma(\omega,\mathbf{k})\,\delta^{ij}, \qquad (4.44)$$

where the density and spin-density dynamical structure functions are

$$S_\rho(\omega,\mathbf{k}) = \frac{1}{n_B}\int_{-\infty}^{+\infty} dt\, e^{i\omega t}\,\big\langle \rho(t,\mathbf{k})\rho(0,-\mathbf{k})\big\rangle,$$

$$S_\sigma(\omega,\mathbf{k}) = \frac{4}{3n_B}\int_{-\infty}^{+\infty} dt\, e^{i\omega t}\,\big\langle \mathbf{s}(t,\mathbf{k})\cdot\mathbf{s}(0,-\mathbf{k})\big\rangle \qquad (4.45)$$

(Iwamoto and Pethick 1982).

In order to determine the overall normalization in the nonrelativistic limit consider the static structure function for an ensemble of N_B nucleons enclosed in a large volume V. At a given time the system is characterized by a wavefunction Ψ which depends on the locations $\mathbf{r}_i$ of the nucleons. Then $\rho(\mathbf{r})|\Psi\rangle = \sum_{i=1}^{N_B}\delta(\mathbf{r}-\mathbf{r}_i)|\Psi\rangle$ while the Fourier-transformed operator $V^{-1}\int d^3r\, e^{i\mathbf{k}\cdot\mathbf{r}}\rho(\mathbf{r})$ is

$$\rho(\mathbf{k})|\Psi\rangle = \frac{1}{V}\sum_{i=1}^{N_B} e^{i\mathbf{k}\cdot\mathbf{r}_i}|\Psi\rangle. \qquad (4.46)$$

Therefore, a given configuration of nucleons yields

$$\langle\Psi|\rho(\mathbf{k})\rho(-\mathbf{k})|\Psi\rangle = \frac{1}{V}\sum_{i,j=1}^{N_B} e^{i\mathbf{k}\cdot\mathbf{r}_{ij}} = \frac{N_B}{V} + \frac{1}{V}\sum_{\substack{i,j=1\\i\neq j}}^{N_B} e^{i\mathbf{k}\cdot\mathbf{r}_{ij}}, \qquad (4.47)$$

where $\mathbf{r}_{ij}\equiv\mathbf{r}_i-\mathbf{r}_j$. When averaged over a thermal ensemble the second term will disappear if there are no spatial correlations, and $n_B = N_B/V$ so that $S_\rho(\mathbf{k}) = 1$.

For the static spin-density structure function the same steps can be performed with the inclusion of the spin operators $\frac{1}{2}\boldsymbol{\sigma}_i$ for the individual nucleons. This takes us to the equivalent of Eq. (4.47)

$$\langle\Psi|\mathbf{s}(\mathbf{k})\cdot\mathbf{s}(-\mathbf{k})|\Psi\rangle = \frac{1}{V}\sum_{i,j=1}^{N_B} e^{\mathbf{k}\cdot\mathbf{r}_{ij}}\tfrac{1}{4}\langle\Psi|\boldsymbol{\sigma}_i\cdot\boldsymbol{\sigma}_j|\Psi\rangle =$$

$$= \frac{1}{V}\sum_{i=1}^{N_B}\tfrac{1}{4}\langle\Psi|\boldsymbol{\sigma}_i^2|\Psi\rangle + \frac{1}{V}\sum_{\substack{i,j=1\\i\neq j}}^{N_B} e^{i\mathbf{k}\cdot\mathbf{r}_{ij}}\tfrac{1}{4}\langle\Psi|\boldsymbol{\sigma}_i\cdot\boldsymbol{\sigma}_j|\Psi\rangle. \qquad (4.48)$$

Noting that $\langle(\tfrac{1}{2}\boldsymbol{\sigma}_i)^2\rangle = \tfrac{1}{2}(1+\tfrac{1}{2}) = \tfrac{3}{4}$ the first term is $\tfrac{3}{4}N_B/V = \tfrac{3}{4}n_B$. Therefore, in the absence of correlations one finds $S_\sigma(\mathbf{k}) = 1$.

Even in a noninteracting medium there exist anticorrelations between degenerate nucleons. Standard manipulations yield in this case (e.g. Sawyer 1989)

$$S_{\rho,\sigma}(\mathbf{k}) = \frac{1}{n_B}\int\frac{2d^3\mathbf{p}}{(2\pi)^3}\, f_{\mathbf{p}}\,(1-f_{\mathbf{p}+\mathbf{k}}), \qquad (4.49)$$

with the Fermi-Dirac occupation number $f_{\mathbf{p}}$ for the nucleon mode $\mathbf{p}$. For small temperatures this result can be expanded to yield $S_{\rho,\sigma}(\mathbf{k}) = 3k/2p_F + 3Tm_N/p_F^2$.

4.6.3 The f-Sum Rule

The structure functions have a number of general properties, independently of details of the interactions of the medium constituents. We have already seen that they must obey the normalization condition

$$\int_{-\infty}^{+\infty}\frac{d\omega}{2\pi}\, S_\rho(\omega,\mathbf{k}) = 1 + \frac{1}{n_B}\left\langle\sum_{\substack{i,j=1\\i\neq j}}^{N_B}\cos(\mathbf{k}\cdot\mathbf{r}_{ij})\right\rangle,$$

$$\int_{-\infty}^{+\infty}\frac{d\omega}{2\pi}\, S_\sigma(\omega,\mathbf{k}) = 1 + \frac{4}{3n_B}\left\langle\sum_{\substack{i,j=1\\i\neq j}}^{N_B}\boldsymbol{\sigma}_i\cdot\boldsymbol{\sigma}_j\cos(\mathbf{k}\cdot\mathbf{r}_{ij})\right\rangle. \qquad (4.50)$$

This can be referred to as a "sum rule" because the strength of $S_{\rho,\sigma}$ is "summed" (integrated) over all frequencies ω. Usually we will assume that the correlation expressions on the r.h.s. are negligible.

A more nontrivial sum rule obtains when a factor ω is included under the integral. The definition of the density structure function then yields

$$\int_{-\infty}^{+\infty} \frac{d\omega}{2\pi}\, \omega S_\rho(\omega, \mathbf{k}) = \int_{-\infty}^{+\infty} \frac{d\omega}{2\pi}\, \omega \int_{-\infty}^{+\infty} dt\, e^{i\omega t} \langle \rho(t, \mathbf{k}) \rho(0, -\mathbf{k}) \rangle$$

$$(4.51)$$

and a similar expression for S_σ. Under the integral, a partial integration with suitable boundary conditions allows one to absorb the ω factor, at the expense of $\rho(t, \mathbf{k}) \to \dot{\rho}(t, \mathbf{k})$. Because Heisenberg's equation of motion informs us that $i\dot{\rho} = [\rho, H]$ with H the complete Hamiltonian of the system one finds (Sigl 1995b)

$$\int_{-\infty}^{+\infty} \frac{d\omega}{2\pi}\, \omega\, S_\rho(\omega, \mathbf{k}) = \frac{1}{n_B} \Big\langle \big[\rho(\mathbf{k}), H \big] \rho(-\mathbf{k}) \Big\rangle,$$

$$\int_{-\infty}^{+\infty} \frac{d\omega}{2\pi}\, \omega\, S_\sigma(\omega, \mathbf{k}) = \frac{4}{3 n_B} \Big\langle \big[\boldsymbol{\sigma}(\mathbf{k}), H \big] \cdot \boldsymbol{\sigma}(-\mathbf{k}) \Big\rangle. \qquad (4.52)$$

Here it was used, again, that $\int d\omega\, e^{i\omega t} = 2\pi \delta(t)$ and $\rho(\mathbf{k}) \equiv \rho(0, \mathbf{k})$ and $\boldsymbol{\sigma}(\mathbf{k}) \equiv \boldsymbol{\sigma}(0, \mathbf{k})$.

In order to evaluate this sum rule more explicitly one must assume a specific form for the interaction Hamiltonian. In the simplest case of a medium consisting of only one species of nucleons one may assume that H consists of the kinetic energy for each nucleon, plus a general nonrelativistic interaction potential between all nucleon pairs which depends on the relative distance and the nucleon spins, i.e.

$$H = \sum_{i=1}^{N_B} \frac{\mathbf{p}_i^2}{2m_N} + \frac{1}{2} \sum_{\substack{i,j=1 \\ i \neq j}}^{N_B} V(\mathbf{r}_{ij}, \boldsymbol{\sigma}_i, \boldsymbol{\sigma}_j), \qquad (4.53)$$

where again $\mathbf{r}_{ij} \equiv \mathbf{r}_i - \mathbf{r}_j$. One can then proceed to evaluate the commutators in Eq. (4.52). By virtue of the continuity equation for the particle number one can then show (Pines and Nozières 1966; Sigl 1995b)

$$\int_{-\infty}^{+\infty} \frac{d\omega}{2\pi}\, \omega\, S_\rho(\omega, \mathbf{k}) = \frac{\mathbf{k}^2}{2m_N},$$

$$\int_{-\infty}^{+\infty} \frac{d\omega}{2\pi}\, \omega\, S_\sigma(\omega, \mathbf{k}) = \frac{\mathbf{k}^2}{2m_N}$$

$$+ \frac{4}{3 n_B} \Big\langle \sum_{\substack{i,j=1 \\ i \neq j}}^{N_B} \big[\boldsymbol{\sigma}(\mathbf{k}), V(\mathbf{r}_{ij}, \boldsymbol{\sigma}_i, \boldsymbol{\sigma}_j) \big] \cdot \boldsymbol{\sigma}(-\mathbf{k}) \Big\rangle. \quad (4.54)$$

For the density structure function this exact relationship is known as the f-sum rule.

For the spin-density structure function one can go one step further by expressing the most general nonrelativistic interaction potential as (Sigl 1995b)

$$V(\mathbf{r}_{ij}, \boldsymbol{\sigma}_i, \boldsymbol{\sigma}_j) = U_0(r_{ij}) - U_S(r_{ij})\, \boldsymbol{\sigma}_i \cdot \boldsymbol{\sigma}_j$$

$$- U_T(r_{ij}) \left[3(\boldsymbol{\sigma}_i \cdot \hat{\mathbf{r}}_{ij})(\boldsymbol{\sigma}_j \cdot \hat{\mathbf{r}}_{ij}) - \boldsymbol{\sigma}_i \cdot \boldsymbol{\sigma}_j \right], \qquad (4.55)$$

where U_0 is a spin-independent potential of the interparticle distance $r_{ij} = |\mathbf{r}_{ij}|$, U_S is the scalar, and U_T the tensor part of the spin-dependent potential. Of these terms, only the tensor part does not conserve the total spin in nucleon-nucleon collisions and thus is the only part contributing to spin fluctuations. With $V_{ij}^S \equiv U_S(r_{ij})\, \boldsymbol{\sigma}_i \cdot \boldsymbol{\sigma}_j$ and $V_{ij}^T \equiv U_T(r_{ij}) \left[3(\boldsymbol{\sigma}_i \cdot \hat{\mathbf{r}}_{ij})(\boldsymbol{\sigma}_j \cdot \hat{\mathbf{r}}_{ij}) - \boldsymbol{\sigma}_i \cdot \boldsymbol{\sigma}_j \right]$ one finds

$$\int_{-\infty}^{+\infty} \frac{d\omega}{2\pi}\, \omega\, S_\sigma(\omega, \mathbf{k}) = \frac{\mathbf{k}^2}{2m_N}$$

$$+ \frac{4}{3n_B} \left\langle \sum_{\substack{i,j=1 \\ i \neq j}}^{N_B} \left[1 - \cos(\mathbf{k} \cdot \mathbf{r}_{ij}) \right] V_{ij}^S + \left[1 + \tfrac{1}{2} \cos(\mathbf{k} \cdot \mathbf{r}_{ij}) \right] V_{ij}^T \right\rangle. \quad (4.56)$$

The f-sum of the spin-density structure function is thus closely related to the average spin-spin interaction energy in the medium.

In order for the eigenvalues of the Hamiltonian to be bounded from below one must require that the potentials $U_{0,S,T}(r)$ are not more singular than $1/r^2$. In this case the r.h.s. of the spin-density f-sum rule exists as a nondivergent expression. In our representation $S_\sigma(|\omega|) = (\Gamma_\sigma/\omega^2)\, s(\omega/T)$, the necessary existence of the l.h.s. of Eq. (4.56) implies that $s(x)$ must be a decreasing function of x for large x. This is not the case for the $s(x)$ derived from the OPE potential, indicating that this interaction model is pathological in the sense that it is too singular. Indeed, it corresponds to a dipole potential and thus varies as $1/r^3$. Real nucleon-nucleon interaction potentials have a repulsive core and thus do not exhibit this pathology.

4.6.4 Long-Wavelength Properties

In calculations involving the emission or scattering of neutrinos or axions as in Sect. 4.3.1 one usually neglects the momentum transfer $\mathbf{k}$ because the medium constituents are so heavy that recoil effects are small. In this "long-wavelength limit" one is only interested in the small-$\mathbf{k}$ structure functions

$$S_{\rho,\sigma}(\omega) \equiv \lim_{\mathbf{k}\to 0} S_{\rho,\sigma}(\omega, \mathbf{k}). \tag{4.57}$$

It should be stressed that this quantity is not identical with $S_{\rho,\sigma}(\omega, 0)$. For example, in Eq. (4.47) for $\mathbf{k} = 0$ the interference term does not average to zero. The structure function becomes N_B^2/V and thus coherently enhanced because the momentum transfer is so small that a target consisting of many particles in a volume V cannot be resolved. The limit $\mathbf{k} \to 0$ is understood such that $|\mathbf{k}|^{-1}$ remains much smaller than the geometrical dimension $V^{1/3}$ of the system.

In the long-wavelength limit the normalization Eq. (4.50) and the f-sum rule Eq. (4.56) yield for the spin-density structure function

$$\int_{-\infty}^{+\infty} \frac{d\omega}{2\pi} S_\sigma(\omega) = \int_0^\infty \frac{d\omega}{2\pi} (1 + e^{-\omega/T}) S_\sigma(\omega) = 1 + \frac{4}{3n_B} \left\langle \sum_{\substack{i,j=1 \\ i\neq j}}^{N_B} \boldsymbol{\sigma}_i \cdot \boldsymbol{\sigma}_j \right\rangle,$$

$$\int_{-\infty}^{+\infty} \frac{d\omega}{2\pi} \, \omega \, S_\sigma(\omega) = \int_0^\infty \frac{d\omega}{2\pi} \, \omega \, (1 - e^{-\omega/T}) S_\sigma(\omega) = \frac{4}{3n_B} \left\langle \sum_{\substack{i,j=1 \\ i\neq j}}^{N_B} {\textstyle\frac{3}{2}} V_{ij}^{\mathrm{T}} \right\rangle.$$

$$\tag{4.58}$$

These relations will be of great use to develop a general understanding of the behavior of $S_\sigma(\omega)$ at high densities. The second column of expressions follows from the first by detailed balance. Because $S_\sigma(\omega) \geq 0$ it is evident that all of these expressions are always positive, independently of details of the medium interactions.

In a noninteracting medium the operators $\rho(t, \mathbf{k})$ and $\mathbf{s}(t, \mathbf{k})$ are constant so that $S_{\rho,\sigma}(\omega) = 2\pi\delta(\omega)$, allowing for scattering (zero energy transfer), but not for the emission of radiation. This behavior is familiar from a gas of free particles which can serve as targets for collisons, but which cannot emit radiation because of energy-momentum constraints.

In an interacting medium the density correlator retains this property because in the long-wavelength limit it depends on $\rho(t, \mathbf{k}{\to}0) = V^{-1} \int d^3\mathbf{r}\, \rho(t, \mathbf{r})$ which remains constant. Therefore, even in an interacting medium one expects $S_\rho(\omega) = 2\pi\delta(\omega)$, in agreement with the finding that the neutrino vector current does not contribute to bremsstrahlung in the nonrelativistic limit relative to the axial-vector current (Friman and Maxwell 1979).

The relevant quantity for the latter is $V^{-1} \int d^3\mathbf{r}\, \mathbf{s}(t, \mathbf{r}) = \frac{1}{2} \sum_{i=1}^{N_B} \boldsymbol{\sigma}_i$ with $\boldsymbol{\sigma}_i$ the individual nucleon spins. If the evolution of different spins is uncorrelated one may ignore the cross terms in the correlator. With the single-nucleon spin operator $\boldsymbol{\sigma}$ one finds then

$$S_\sigma(\omega) = \tfrac{1}{3} \int_{-\infty}^{+\infty} dt\, e^{i\omega t} \left\langle \boldsymbol{\sigma}(t) \cdot \boldsymbol{\sigma}(0) \right\rangle. \tag{4.59}$$

With Eq. (4.21) one obtains

$$\frac{d\dot{I}_a}{d\omega} = \left(\frac{C_N}{2f_a}\right)^2 \frac{\omega^4}{12\pi^2} \int_{-\infty}^{+\infty} dt\, e^{-i\omega t} \left\langle \boldsymbol{\sigma}(t) \cdot \boldsymbol{\sigma}(0) \right\rangle \tag{4.60}$$

for the differential axion energy-loss rate (radiation power) per nucleon. Because of collisions with other nucleons, and because of a spin dependent interaction potential caused by pion exchange, the nucleon spins evolve nontrivially so that the correlator has nonvanishing power at $\omega \neq 0$, allowing for axion emission.

4.6.5   Axion Emission in the Classical Limit

The correlator representation Eq. (4.60) of the axion emission rate is extremely useful to develop a general understanding of its main properties without embarking on a quantum-mechanical calculation. To this end the nucleon spin $\boldsymbol{\sigma}$ is approximated by a classical variable, basically a little magnet which jiggles around under the impact of collisions with other nucleons. With an ergodic hypothesis about the spin trajectory on the unit sphere one may replace the ensemble average in Eq. (4.60) by a time average. The radiation intensity (time integrated radiation power) emitted during a long (infinite) time interval is then

$$\frac{dI_a}{d\omega} = \left(\frac{C_N}{2f_a}\right)^2 \frac{\omega^2}{12\pi^2} \left| \int_{-\infty}^{+\infty} dt\, e^{i\omega t}\, \dot{\boldsymbol{\sigma}}(t) \right|^2, \tag{4.61}$$

where two powers of ω were absorbed by a partial integration with suitable boundary conditions at $t = \pm\infty$.

In this form the energy-loss rate is closely related to a well-known expression for the electromagnetic radiation power from a charged particle which moves on a trajectory $\mathbf{r}(t)$. The nonrelativistic limit of a standard result (Jackson 1975) is

$$\frac{dI_\gamma}{d\omega} = \frac{2\alpha}{3\pi} \left| \int_{-\infty}^{+\infty} dt\, e^{i\omega t}\, \mathbf{a}(t) \right|^2, \tag{4.62}$$

where $\mathbf{a}(t) = \ddot{\mathbf{r}}(t)$ is the particle's acceleration on its trajectory.

The same result can be found with the above methods applied to the spatial part of the vector current. Easier still, it can be obtained directly from Eq. (4.61). To this end note that photon emission by an electron involves nonrelativistically $(e/m_e)\,\mathbf{p} = e\mathbf{v}$ so that we must substitute $\dot{\boldsymbol{\sigma}} \to \dot{\mathbf{v}} = \mathbf{a}$. The role of $\mathbf{k}$ (axions) is played by the polarization vector $\boldsymbol{\epsilon}$ (photons) so that $\mathbf{k}^2 = \omega^2$ must be replaced by $\boldsymbol{\epsilon}^2 = 1$. With $(C_A/2f_a) \to e$, using $\alpha = e^2/4\pi$, and inserting a factor of 2 for two photon polarization states completes the translation.

In order to understand the radiation spectrum consider a single "infinitely hard" collision with $\dot{\boldsymbol{\sigma}}(t) = \Delta\boldsymbol{\sigma}\,\delta(t)$. The radiation power is

$$\frac{dI_a}{d\omega} = \left(\frac{C_N}{2f_a}\right)^2 \frac{\omega^2}{12\pi^2}\,|\Delta\boldsymbol{\sigma}|^2. \tag{4.63}$$

For photons one obtains the familiar flat bremsstrahlung spectrum $dI_\gamma/d\omega = (2\alpha/3\pi)|\Delta\mathbf{v}|^2$ which is hardened, for axions, by the additional factor ω^2 from their derivative coupling.

In the form Eq. (4.63) the total amount of energy radiated in a single collision is infinite. In practice, collisions are not arbitrarily hard, and the backreaction of the radiation process on the emitter must be included. This is not rigorously possible in a classical calculation, it requires a quantum-mechanical treatment. Therefore, a classical analysis is useful only for the soft part of the spectrum where backreactions can be ignored, i.e. for radiation frequencies far below the kinetic energy of the emitter. In a thermal environment, a classical treatment then appears reasonable for $\omega \lesssim T$.

Next, consider a large random sequence of n hard collisions with a spin trajectory

$$\dot{\boldsymbol{\sigma}}(t) = \sum_{i=1}^{n} \Delta\boldsymbol{\sigma}_i\,\delta(t - t_i). \tag{4.64}$$

This yields the average radiation intensity per collision of

$$\frac{d\dot{I}_a}{d\omega} = \left(\frac{C_N}{2f_a}\right)^2 \frac{\omega^2}{12\pi^2}\,\Gamma_{\text{coll}}\,\langle(\Delta\boldsymbol{\sigma})^2\rangle\,F(\omega), \tag{4.65}$$

where Γ_{coll} is the average collision rate and $\langle(\Delta\boldsymbol{\sigma})^2\rangle$ is the average squared change of the spin in a collision. Further,

$$F(\omega) = 1 + \lim_{n\to\infty} \frac{1}{n} \sum_{\substack{i,j=1 \\ i\neq j}}^{n} \frac{\Delta\boldsymbol{\sigma}_i \cdot \Delta\boldsymbol{\sigma}_j\,\cos[\omega(t_i - t_j)]}{\langle(\Delta\boldsymbol{\sigma})^2\rangle}, \tag{4.66}$$

where the first term (the "diagonal" part of the double sum) gives the total radiation power as an incoherent sum of individual collisions. The second term takes account of interference effects between the radiation emitted in different collisions.

The interference term yields a suppression of the incoherent summation because the $\Delta\sigma$ in subsequent collisions are anticorrelated. The spin is constrained to move on the unit sphere whence a kick in one direction is more likely than average followed by one in the opposite direction. A similar argument pertains to the velocity; a $\Delta\mathbf{v}$ in one direction is more likely than average to be followed by one in the opposite direction. The radiation spectrum with $\omega \gtrsim \Gamma_{\mathrm{coll}}$ will remain unaffected while for $\omega \lesssim \Gamma_{\mathrm{coll}}$ it is suppressed. The low-ω suppression of bremsstrahlung is known as the Landau-Pomeranchuk-Migdal effect (Landau and Pomeranchuk 1953a,b; Migdal 1956; Knoll and Voskresensky 1995 and references therein).

The summation in Eq. (4.66) can be viewed as an integration over the relative time coordinate $\Delta t = t_i - t_j$ with a certain distribution function $f(\Delta t)$. For a random sequence of "kicks" one expects an exponential distribution of the normalized form $f(\Delta t) = \frac{1}{4}\Gamma_\sigma e^{-\Gamma_\sigma \Delta t/2}$ where Γ_σ is some inverse time-scale. This implies the Lorentzian shape (e.g. Knoll and Voskresensky 1995)

$$F(\omega) = \frac{\omega^2}{\omega^2 + \Gamma_\sigma^2/4}. \tag{4.67}$$

A Lorentzian model is familiar, for example, from the collisional broadening of spectral lines. A comparison with Eq. (4.60) indicates that

$$\frac{\Gamma_{\mathrm{coll}}\langle(\Delta\sigma)^2\rangle}{\omega^2 + \Gamma_\sigma^2/4} = \int_{-\infty}^{+\infty} dt\, e^{i\omega t}\left\langle \sigma(t)\cdot\sigma(0)\right\rangle. \tag{4.68}$$

An integral over $d\omega/2\pi$ reveals a normalization $\langle\sigma^2\rangle$ so that

$$\Gamma_\sigma = \frac{\langle(\Delta\sigma)^2\rangle}{\langle\sigma^2\rangle}\,\Gamma_{\mathrm{coll}}. \tag{4.69}$$

Therefore, Γ_σ is identified with a collisional spin-fluctuation rate.

For nucleons interacting by an OPE potential one may estimate Γ_σ without much effort. The NN cross section is dimensionally α_π^2/m_N^2. A typical thermal nucleon velocity is $v = (3T/m_N)^{1/2}$ yielding for a typical collision rate $\Gamma_{\mathrm{coll}} = \langle v\sigma_{NN}\rangle n_B \approx \alpha_\pi^2 T^{1/2} m_N^{-5/2} n_B$. Because $|\Delta\sigma| \approx 1$ in a collision, $\Gamma_\sigma \approx \Gamma_{\mathrm{coll}}$. This estimate agrees with the detailed result of Eq. (4.7) apart from a numerical factor $4\sqrt{\pi}$.

4.6.6   Classical vs. Quantum Result

In order to make contact with the quantum calculation that led to the axion emission rate in Eq. (4.6) it is useful to juxtapose it with the classical result in terms of the spin-density structure function. The classical calculation led to

$$S_\sigma(\omega) = \frac{\Gamma_\sigma}{\omega^2} \times \frac{\omega^2}{\omega^2 + \Gamma_\sigma^2/4} \qquad (4.70)$$

while the quantum result is

$$S_\sigma(\omega) = \frac{\Gamma_\sigma}{\omega^2} \times s(\omega/T) \times \begin{cases} 1 & \text{for } \omega > 0, \\ e^{\omega/T} & \text{for } \omega < 0. \end{cases} \qquad (4.71)$$

The function $s(x)$ has the property $s(0) = 1$, it is even, and according to the f-sum rule must decrease for large x. In Fig. 4.7 the classical and quantum results are shown as dotted and dashed lines, respectively, where for the purpose of illustration $s(x) = (1 + x^2/4)^{-1/4}$ has been assumed.

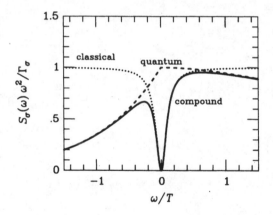

Fig. 4.7.   Classical, quantum, and compound spin-density structure function in the nondegenerate limit according to Eq. (4.70) and Eq. (4.71) with $\Gamma_\sigma/T = 0.2$ and $s(x) = (1 + x^2/4)^{-1/4}$.

This comparison highlights an important weakness of the classical result: it does not obey the detailed-balance requirement $S_\sigma(\omega) = S_\sigma(-\omega)\, e^{\omega/T}$. The classical correlators are invariant under time reflection $t \to -t$ and thus symmetric under $\omega \to -\omega$. Again, the classical result is adequate only for $|\omega| \lesssim T$.

Equation (4.71) also highlights an important weakness of the quantum result: It does not include the interference effect from multiple collisions at $|\omega| \lesssim \Gamma_\sigma$ because the calculation was done assuming individual, isolated collisions. The normalization condition Eq. (4.58) can then be satisfied only by accepting a pathological infrared behavior of $S_\sigma(\omega)$ like the one suggested by Sawyer (1995).

The classical and the quantum results thus both violate fundamental requirements. The quantum calculation applies for $|\omega| \gtrsim \Gamma_\sigma$ while the classical one for $|\omega| \lesssim T$. If $\Gamma_\sigma \ll T$ ("dilute medium") the regimes of validity overlap for $\Gamma_\sigma \lesssim |\omega| \lesssim T$ and the results agree beautifully. In this case the two calculations mutually confirm and complement each other. The compound structure function, where the quantum result is multiplied with $\omega^2/(\omega^2 + \Gamma_\sigma^2/4)$, fulfills the detailed-balance requirement and approximately the normalization condition. Therefore, it probably is a good first guess for the overall shape of $S_\sigma(\omega)$.

4.6.7   High-Density Behavior

In a SN core one is typically in the limit $\Gamma_\sigma \gg T$ so that the regime of overlapping validity $\Gamma_\sigma \lesssim |\omega| \lesssim T$ between a classical and a perturbative quantum calculation no longer exists. Rather, one is in the opposite situation where for $T \lesssim |\omega| \lesssim \Gamma_\sigma$ neither approach appears directly justified. Because the structure function $S_\sigma(\omega)$ determines all axial-vector interaction rates in the long-wavelength limit, it is rather unclear what their high-density behavior might be.

Still, one has important general information about $S_\sigma(\omega)$. The detailed-balance condition $S_\sigma(\omega) = S_\sigma(-\omega)\, e^{\omega/T}$ reveals that it is enough to specify $S_\sigma(\omega)$ for positive energy transfers (energy given to the medium). In the classical limit $S_\sigma(|\omega|) = \Gamma_\sigma/\omega^2$, while the f-sum rule informs us that the quantum version must fall off somewhat faster with large $|\omega|$. Finally, if spin-spin correlations can be neglected, the normalization condition $\int_0^\infty d\omega\, (1 + e^{-\omega/T})\, S_\sigma(\omega) = 2\pi$ obtains.

In order to illustrate the overall impact of the high-density behavior on axion or neutrino pair emission rates and on neutrino scattering rates, it is enough to take the classical limiting case for large $|\omega|$, even though it does not have an integrable f-sum. Thus a simple ansatz is

$$S_\sigma(|\omega|) = \frac{\Gamma_\sigma}{\omega^2 + \Gamma^2/4}, \tag{4.72}$$

where Γ is to be determined by the normalization condition. In the dilute limit ($\Gamma_\sigma \ll T$) this implies $\Gamma \approx \Gamma_\sigma$ while in the dense limit ($\Gamma_\sigma \gg T$) one finds $\Gamma \approx \Gamma_\sigma/2$.

For axion bremsstrahlung, the energy-loss rate of the medium is given by $\epsilon_a = Q_a/\rho \propto \int_0^\infty d\omega\, \omega^4 S_\sigma(\omega)$ so that with Eq. (4.72) one needs to evaluate

$$\epsilon_a \propto \Gamma_\sigma \int_0^\infty d\omega\, \frac{\omega^4 e^{-\omega/T}}{\omega^2 + \Gamma^2/4} \tag{4.73}$$

with Γ determined from the normalization condition. In the dilute limit ($\Gamma \approx \Gamma_\sigma \ll T$) one may ignore Γ in the denominator, so that $\epsilon_a \propto \Gamma_\sigma$. This is indeed what one expects from a bremsstrahlung process for which the volume energy-loss rate is proportional to the density squared, and thus ϵ_a proportional to the density which appears in the spin-fluctuation rate Γ_σ.

In the high-density limit ($\Gamma \approx \Gamma_\sigma/2 \gg T$) the denominator in Eq. (4.73) is dominated by Γ because the exponential factor suppresses the integrand for $\omega \gg T$ so that one expects ϵ_a to be a *decreasing* function of Γ_σ. In Fig. 4.8 the variation of ϵ_a with Γ_σ/T is shown (solid line), taking Eq. (4.72) for the spin-density structure function. The dashed line shows the "naive rate," based on Γ_σ/ω^2 which ignores multiple-scattering effects, and which violates the normalization condition. Fig. 4.8 illustrates that even very basic and global properties of $S_\sigma(\omega)$ reveal an important modification of the axion emission rate at high density: they saturate with an increasing spin-fluctuation rate,

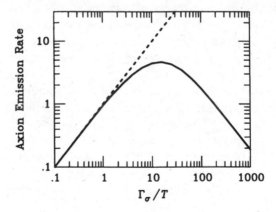

Fig. 4.8. Schematic variation of the axion emission rate per nucleon with Γ_σ/T, taking Eq. (4.72) for the spin-density structure function. The dashed line is the naive rate without the inclusion of multiple-scattering effects, i.e. it is based on $S_\sigma(|\omega|) = \Gamma_\sigma/\omega^2$.

and may even decrease at large densities, although such large values for Γ_σ may never be reached in a nuclear medium as will become clear below.

The high-density downturn of the axion emission rate can be interpreted in terms of the Landau-Pomeranchuk-Migdal effect (Landau and Pomeranchuk 1953a,b; Feinberg and Pomeranchuk 1956; Migdal 1956) as pointed out by Raffelt and Seckel (1991). The main idea is that collisions interrupt the radiation process. The formation of a radiation quantum of frequency ω takes about a time ω^{-1} according to the uncertainty principle and so if collisions are more frequent than this time, the radiation process is suppressed. Classically, this effect was demonstrated in the language of current correlators in Sect. 4.6.5.

The impact of the high-density behavior of $S_\sigma(\omega)$ on the neutral-current neutrino opacity is crudely estimated by the inverse mean free path given in Eq. (4.35), averaged over a thermal energy spectrum of the initial neutrino. Moreover, all expressions become much simpler if one replaces the Fermi-Dirac occupation numbers with the Maxwell-Boltzmann expression $e^{-\omega_i/T}$; the resulting error is small for nondegenerate neutrinos. The relevant quantity is then

$$\left\langle \lambda^{-1} \right\rangle \propto \int_0^\infty d\omega_1 \int_0^\infty d\omega_2\, \omega_1^2\, \omega_2^2\, e^{-\omega_1/T} S_\sigma(\omega_1 - \omega_2). \tag{4.74}$$

One integral can be done explicitly, leaving one with an integral over the energy transfer alone. With Eq. (4.72) for the structure function one finds

$$\left\langle \lambda^{-1} \right\rangle \propto \int_0^\infty d\omega\, \frac{\Gamma_\sigma \left(T^2 + T\omega/2 + \omega^2/12\right) e^{-\omega/T}}{\omega^2 + \Gamma^2/4}. \tag{4.75}$$

This expression is constant for $\Gamma_\sigma \ll T$ where $\Gamma = \Gamma_\sigma$ and thus the structure function is essentially $2\pi\delta(\omega)$. Indeed, the average scattering cross section (or mean free path) is not expected to depend on the density.

For dense media ($\Gamma_\sigma \gg T$), however, the broadening of $S_\sigma(\omega)$ beyond a delta function leads to a decreasing average scattering rate. This means that at a fixed temperature the medium becomes more transparent to neutrinos with increasing density, even without the impact of degeneracy effects. This behavior is shown in Fig. 4.9 in analogy to the axion emission rate Fig. 4.8.

A decreasing cross section is intuitively understood if one recalls that the nucleon spin is typically flipped in a collision with other nucleons because the interaction potential couples to the spin. Neutrino scattering with an energy transfer ω implies that properties of the medium

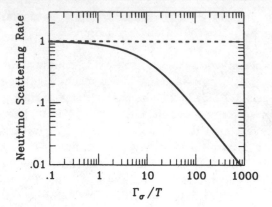

Fig. 4.9. Schematic variation of the neutrino scattering rate per nucleon (axial-vector interaction only) with Γ_σ/T, taking Eq. (4.72) for the spin-density structure function. The dashed line is the naive rate without the inclusion of multiple-scattering effects, i.e. it is based on $S_\sigma(\omega) = 2\pi\delta(\omega)$.

that fluctuate on faster time scales cannot be resolved. Therefore, if a nucleon spin flips many times within the time scale ω^{-1}, the probe "sees" a vanishing average spin and the scattering rate is reduced accordingly.

For the vector current, the relevant zeroth component does not fluctuate so that this contribution to neutrino scattering is not suppressed (Sect. 4.6.4). This observation is intimately tied to the absence of bremsstrahlung emission of neutrino pairs by the vector current. Both of these effects are summarized in the statement that the vector-current structure function $S_\rho(\omega)$ is always given as[23] $2\pi\delta(\omega)$, allowing for scattering (energy transfer $\omega = 0$), but not for bremsstrahlung.

---

[23]This statement is based on the assumption that there are no spatial correlations among the nucleons—possibly a poor approximation in a dense medium. Moreover, collective oscillations may occur so that it can be too simplistic to treat the medium as consisting of essentially free, individual nucleons (Iwamoto and Pethick 1982). Calculations of the long-wavelength structure factor by Sawyer (1988, 1989) in a specific nucleon interaction model showed a substantial suppression of $S_\rho(0)$ and thus, of the neutrino mfp. Other related works are those of Haensel and Jerzak (1987) and of Horowitz and Wehrberger (1991a,b) who calculated the dynamic structure functions within certain interaction models. Unfortunately, these works do not shed much light on the high-density behavior of the quantities which are of prime interest to this book such as the axion emission rate from a hot SN core.

The axion emission and the neutrino scattering rates begin to be suppressed at high density if Γ_σ exceeds a few T. The numerical value Eq. (4.34) that was derived in a simple OPE calculation indicates that such large Γ_σ may be expected in a nuclear medium. Of course, one may well ask if a perturbative calculation of the spin-fluctuation rate is adequate if such a calculation fails for the neutrino scattering rate. Nucleon spins fluctuate because of their spin-dependent interaction with other nucleons and thus are themselves subject to spin averaging effects. The question of the true value of Γ_σ in a nuclear medium (as opposed to the OPE-calculated one) can be addressed empirically by virtue of the SN 1987A neutrino signal and theoretically by the f-sum rule.

Because the axial-vector current contribution to the standard neutral-current scattering rate dominates (the cross section is proportional to $C_V^2 + 3C_A^2$), it would be much easier for neutrinos to diffuse out of the hot SN core if the axial-current scattering rate were significantly suppressed. The observed SN 1987A neutrino signal would have been much shorter than predicted in standard cooling calculations. The impact of reduced neutrino opacities on the SN 1987A signal has been studied by Keil, Janka, and Raffelt (1995); see Sect. 13.6. The main conclusion is that a suppression of the axial-vector current opacity by more than about a factor of 2 is not compatible with the SN 1987A signal duration. This result would indicate that the effective Γ_σ never becomes much larger than a few T.

Theoretically, the f-sum rule Eq. (4.58) allows one to relate Γ_σ to the average spin-spin interaction energy in the medium. According to Sigl's (1995b) estimate one concludes, again, that Γ_σ does not exceed a few T in a nuclear medium.

These results imply that even in a dense medium the axial-vector scattering rate is not suppressed as strongly as one may have expected on the basis of a naive estimate of Γ_σ. It still remains impossible to calculate its exact magnitude from first principles. Therefore, the neutral-current neutrino opacity remains an adjustable function of density for practical SN cooling calculations much as the equation of state.

4.7 Effective Nucleon Mass and Coupling

In a dense medium it is not necessarily possible to use the vacuum masses and coupling constants to determine interaction rates. The axial-vector neutrino couplings are probably suppressed somewhat (Appendix B). For the pion-nucleon coupling, Turner, Kang, and Steigman

(1989) argued on the basis of the nonlinear sigma model that the combination of parameters $\alpha_\pi^2 \alpha_a / m_N^2$ should remain approximately constant. Mayle et al. (1989) similarly found that this parameter should remain somewhere in the range $0.3-1.5$ of its vacuum value.

The nucleon effective mass m_N^* deviates substantially from the vacuum value $m_N = 939\,\text{MeV}$. This shift has an impact on the kinematics of reactions and the nucleon phase-space distribution. A calculation of m_N^* has to rely on an effective theory which describes the interaction of nucleons and mesons. A typical result from a self-consistent relativistic Brueckner calculation including vacuum fluctuations is shown in Fig. 4.10. For conditions relevant for a SN core, an effective value as low as $m_N^*/m_N = 0.5$ is conceivable.

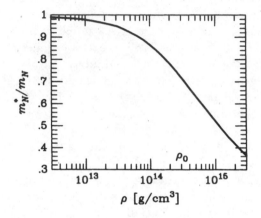

Fig. 4.10. Effective nucleon mass in a nuclear medium according to a self-consistent relativistic Brueckner calculation (Horowitz and Serot 1987). Nuclear density corresponds to about $\rho_0 = 3 \times 10^{14}\,\text{g/cm}^3$.

4.8   The URCA Processes

Neutral-current weak processes are of prime interest for the topics studied in this book because they are closely related to "exotic" reactions involving new particles such as axions. For completeness, however, it must be mentioned that the charged-current reactions depicted in Fig. 4.11 dominate the neutrino energy-loss rate of old neutron stars, and also dominate the ν_e opacity in young SN cores. The processes $n \to p\,e\,\bar{\nu}_e$ (neutron decay) and $e\,p \to n\,\nu_e$ are usually called the URCA

reactions after a casino in Rio de Janeiro. It must have been as easy
to lose money there as it is to lose energy in reactions which produce a
neutrino whether the electron is in the initial or final state.

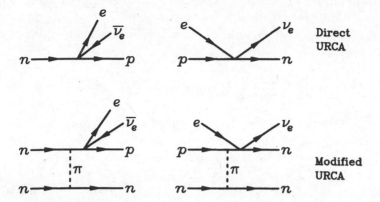

Fig. 4.11. The URCA processes. For the modified versions, there are obvious
other graphs with the leptons attached to other nucleon lines, and exchange
amplitudes.

However, in a neutron star the direct URCA processes can be highly
suppressed by energy-momentum conservation. Because all participat-
ing degenerate fermions are close to their Fermi surface one must require $p_{F,p} + p_{F,e} > p_{F,n}$. Because the proton and electron concentration
is small in a neutron star, it was thought that this "triangle condition"
could not be satisfied. However, the equilibrium proton concentration
depends on details of the equation of state—even in a naive model with
free fermions it is a sensitive function of the effective nucleon mass (Appendix D). Therefore, it is conceivable that the direct URCA processes
actually do take place in neutron stars (Boguta 1981; Lattimer et al.
1991), in which case they provide a cooling mechanism much faster
than all other proposed possibilities.

If the triangle condition is fulfilled, the energy-loss rate is (Lattimer
et al. 1991)

$$Q_{\text{URCA}} = \frac{457\pi}{10080}\, G_F^2 \cos^2\theta_C \left(1 + 3C_A^2\right) m_N^2\, p_{F,e}\, T^6, \qquad (4.76)$$

where $\theta_C \approx 0.24$ is the Cabbibo angle and C_A is the charged-current
axial-vector constant which is -1.26 in vacuum while in nuclear matter
it is suppressed somewhat (Appendix B).

If the triangle condition is not satisfied the URCA processes require bystander particles to absorb momentum, leading to the "modified URCA process" shown in Fig. 4.11 (Chiu and Salpeter 1964). Of course, as below in Sect. 4.9.1, the missing momentum can be provided by pions or a pion condensate (Bahcall and Wolf 1965a,b).

An explicit result for the modified URCA rate in the OPE model for the nucleon interactions is (Friman and Maxwell 1979)

$$Q_{\text{mod. URCA}} = \frac{11513\pi}{120960} \, \alpha_\pi^2 \, G_F^2 \, \cos^2\theta_C \, C_A^2 \, p_{F,e} \, T^8, \qquad (4.77)$$

ignoring factors of order unity to account for a nonzero m_π and nucleon correlations. As in the $\nu\bar{\nu}$ processes, only the axial-vector coupling contributes. Some details of the phase-space integration can be found in Shapiro and Teukolsky (1983).[24]

The modified URCA reactions (Fig. 4.11) are closely related to $\nu\bar{\nu}$ bremsstrahlung and the inelastic scattering process $\nu NN \to NN\nu$ discussed earlier. It is interesting that the rate for the $e\,p\,n \to n\,n\,\nu_e$ process does not diverge, in contrast with $\nu NN \to NN\nu$ where multiple-scattering effects had to be invoked to obtain a sensible result. The divergence was due to the intermediate nucleon going on-shell for a vanishing energy transfer. In $e\,p\,n \to n\,n\,\nu_e$ the electron has the energy $E_{F,e}$, the neutrino T, and because $E_{F,e} \gg T$ (degenerate electrons!), the minimum energy transfer to the leptons is $E_{F,e}$ and thus never zero. Of course, this is the reason why the modified URCA reaction was invoked in the first place: if the intermediate nucleon line is cut, the two sub-processes are suppressed by energy-momentum constraints.

The direct and modified URCA process should be expressed in terms of a common structure function applicable to charged-current processes. One may expect that at high density, spin and isospin fluctuations may suppress these reactions in analogy to the neutral-current processes discussed in the previous section. However, the URCA processes have not been discussed in the literature from this particular perspective.

The equivalent of the modified URCA process for quark matter was calculated by Iwamoto (1980, 1982), more recently by Goyal and Anand (1990), and numerically by Ghosh, Phatak, and Sahu (1994) who claim that Iwamoto's phase-space approximations can lead to substantial errors in the emission rate.

---

[24]Shapiro and Teukolsky's phase-space volume involves a factor $p_{F,e}^3$ because in their treatment of the nuclear matrix element there is no cancellation of $p_{F,e}^2$ with $1/\omega^2$ from the nucleon propagator which occurs in a bremsstrahlung calculation.

4.9   Novel Phases of Nuclear Matter

4.9.1   Pion Condensate

Free particles cannot radiate because of energy-momentum constraints which can be overcome by exchanging momentum with "bystander" particles (bremsstrahlung). No bystander is required if the momentum is taken up by the pion field directly which in Fig. 4.1 only mediated the nucleon interaction (Bahcall and Wolf 1965a,b). This possibility is particularly important if a pion condensate develops so that the medium is characterized by a macroscopic, classical pion field. The pion dispersion relation can be such that the lowest energy state involves a nonvanishing momentum $\mathbf{k}_\pi$ (Baym 1973; Kunihiro et al. 1993; Migdal et al. 1990). Nucleons can exchange pions with the condensate and thereby pick up a momentum $\mathbf{k}_\pi$ which then allows for the radiation of axions or other particles (Fig. 4.12).

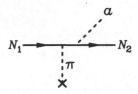

Fig. 4.12.   Axion emission by nucleons scattering off a pion condensate. There is another amplitude with the axion attached to N_1.

The pion condensate causes a periodic potential for the nucleons which thus must be described as quasi-particles or Bloch states with a main momentum component $\mathbf{p}$ and admixtures $\mathbf{p} \pm \mathbf{k}_\pi$. Because an eigenstate of energy is no longer an eigenstate of momentum, these quasi-particles can emit radiation without violating energy-momentum conservation. Therefore, the process of Fig. 4.12 can be equally described as a decay $\widetilde{N}_1 \to \widetilde{N}_2\, a$.

The rate for this reaction was calculated by Muto, Tatsumi, and Iwamoto (1994) who found that the dominant contribution was from a π° condensate. It forms a periodic potential $A \sin(\mathbf{k}_\pi \cdot \mathbf{r})$ where A is a dimensionless amplitude which is small compared to unity for a weakly developed condensate. The Bloch states are to lowest order in A

$$\widetilde{N}_{\mathbf{p}}^{\pm} = N_{\mathbf{p}}^{\pm} \mp \frac{A\kappa_0}{2}\left(\frac{N_{\mathbf{p}+\mathbf{k}_\pi}^{\pm}}{E_{\mathbf{p}+\mathbf{k}_\pi} - E_{\mathbf{p}}} + \frac{N_{\mathbf{p}-\mathbf{k}_\pi}^{\pm}}{E_{\mathbf{p}-\mathbf{k}_\pi} - E_{\mathbf{p}}}\right), \qquad (4.78)$$

where $N_{\mathbf{p}}^{\pm}$ is a plane-wave nucleon state with energy $E_{\mathbf{p}}$ and spin orientation $\pm$ relative to $\mathbf{k}_{\pi}$, and κ_0 is a coupling constant.

In the nonrelativistic limit the axion energy-loss rate corresponding to the decay $\widetilde{N}_1 \to \widetilde{N}_2\, a$ is written as

$$Q_a = \int \frac{d^3\mathbf{k}}{2\omega(2\pi)^3}\, \omega \int \frac{d^3\mathbf{p}_1}{(2\pi)^3} \int \frac{d^3\mathbf{p}_2}{(2\pi)^3}\, f_1(1 - f_2)$$

$$\times\, 2\pi\, \delta(E_1 - E_2 - \omega) \sum_{\text{spins}} |\mathcal{M}|^2, \qquad (4.79)$$

where $\mathbf{k}$ is the axion momentum, $\mathbf{p}_{1,2}$ the nucleon momenta, and $f_{1,2}$ their occupation numbers. The matrix element, averaged over axion emission angles (the condensate is not isotropic!) is found to be

$$\left\langle \sum_{\text{spins}} |\mathcal{M}|^2 \right\rangle = \left(\frac{C_0 - \tilde{g}_A C_1}{2f_a} \right)^2 \tfrac{4}{3} A^2 \kappa_0^2$$

$$\times\, (2\pi)^3 \left[\delta^3(\Delta\mathbf{p} + \mathbf{k}_{\pi}) + \delta^3(\Delta\mathbf{p} - \mathbf{k}_{\pi}) \right], \qquad (4.80)$$

where $\Delta\mathbf{p} = \mathbf{p}_1 - \mathbf{p}_2 - \mathbf{k}$. The isoscalar and isovector axion coupling constants are $C_0 = \frac{1}{2}(C_p + C_n)$ and $C_1 = \frac{1}{2}(C_p - C_n)$ in terms of their couplings to protons and neutrons. The isovector current carries a renormalized axial-vector coupling constant for which Muto, Tatsumi, and Iwamoto (1994) found $\tilde{g}_A \approx 0.43 \times 1.26 = 0.54$.

The phase-space integration was carried out by neglecting the axion momentum in $\delta^3(\Delta\mathbf{p} \pm \mathbf{k}_{\pi})$ and taking the degenerate limit. Then,

$$Q_a = \frac{\pi}{45} \left(\frac{C_0 - \tilde{g}_A C_1}{2f_a} \right)^2 \frac{A^2 \kappa_0^2 m_N^2 T^4}{|\mathbf{k}_{\pi}|}, \qquad (4.81)$$

an emission rate with a relatively soft temperature dependence.

For a numerical estimate Muto, Tatsumi, and Iwamoto (1994) found that in cold neutron-star matter near the critical density (about 2.1 times nuclear) $\mathbf{k}_{\pi} \approx 410\,\text{MeV}$, the neutron and proton Fermi momenta are 410 and 150 MeV, respectively (corresponding to $Y_p = 0.04$), and the coupling strength to the condensate is $\kappa_0 \approx 105\,\text{MeV}$. Therefore,

$$Q_a = \alpha_a A^2 \frac{\pi^2}{45} \frac{\kappa_0^2 T^4}{|\mathbf{k}_{\pi}|} = \alpha_a A^2\, 1.03 \times 10^{44}\, \text{erg cm}^{-3}\, \text{s}^{-1}\, T_9^4, \qquad (4.82)$$

where $\alpha_a \equiv [2m_N(C_0 - \tilde{g}_A C_1)/2f_a]^2/4\pi$ and $T_9 \equiv T/10^9\,\text{K}$.

Instead of an axion one may also emit a neutrino pair in Fig. 4.12. The corresponding energy-loss rate was worked out by Muto and Tatsumi (1988) who found

$$Q_{\nu\bar{\nu}} = \frac{2\pi}{945} \tilde{C}_A^2 \, G_F^2 m_N^2 \, A^2 \, \frac{\kappa_0^2 \, T^6}{|\mathbf{k}_\pi|}. \tag{4.83}$$

Here, $\tilde{C}_A$ is the renormalized neutron neutral-current coupling constant; in vacuum $C_A = -1.26/2$. A somewhat different emission rate was found by Senatorov and Voskresensky (1987).

A process related to that shown in Fig. 4.12 is pionic Compton scattering $\pi N \rightarrow N a$ involving thermal pions. The equivalent of the axion emission rate was calculated by Turner (1992) and by Raffelt and Seckel (1995). This process could be of some importance in the hot nondegenerate medium of a SN core. However, a thermal population of pions yields a rate which is always less important than bremsstrahlung.

For a charged pion or kaon condensate, the equivalent of the modified URCA process can occur by exchanging a meson with the condensate rather than with a bystander nucleon. Recent calculations of such processes include Senatorov and Voskresensky (1987), Muto and Tatsumi (1988), and Thorsson et al. (1995).

4.9.2 Quark Matter

It has been speculated that a "neutron" star may actually undergo a phase transition where the nucleons dissolve in favor of a quark-gluon plasma. Notably, it is possible that the true ground state of nuclear matter consists of "strange quark matter" with about equal numbers of up, down, and strange quarks. Such a system can emit axions by virtue of quark-quark bremsstrahlung (Fig. 4.13).

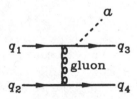

Fig. 4.13. Axion emission by quark-quark bremsstrahlung. There is a total of eight amplitudes, four with the axion attached to each quark line, and an exchange graph each with $q_3 \leftrightarrow q_4$.

The gluon in Fig. 4.13 couples to the quarks with the strong fine-structure constant α_s and the gluon propagator involves an effective mass for which Anand, Goyal, and Iha (1990) used $m_g^2 = (6\alpha_s/\pi)\, p_F^2$ where p_F is the Fermi momentum of the quark sea. The axion coupling to quarks is the usual derivative form $(C_q/2f_a)\, \overline{\psi}_q \gamma_\mu \gamma_5 \psi_q \partial^\mu \phi$ with $q = u, d, s$. After a cumbersome but straightforward calculation Anand et al. found for the energy-loss rate in the degenerate limit

$$Q_a = \frac{62\pi^2 \alpha_s^2}{945} \frac{p_F T^6}{(2f_a)^2} \times \begin{cases} C_q^2\,(I_1 + 2I_2) & \text{for } qq \to qqa, \\ (C_q^2 + C_{q'}^2)\,I_1 & \text{for } qq' \to qq'a, \end{cases} \quad (4.84)$$

where the angular integrals are

$$I_1 \equiv \frac{1}{2\pi^2} \int_0^\pi d\theta \int_0^{2\pi} d\phi\, \frac{s_\theta^5\,(1 + c_\phi^4)}{(s_\theta^2 s_\phi^2 + \kappa^2)^2},$$

$$I_2 \equiv \frac{1}{2\pi^2} \int_0^\pi d\theta \int_0^{2\pi} d\phi\, \frac{s_\theta^5}{(s_\theta^2 s_\phi^2 + \kappa^2)(s_\theta^2 c_\phi^2 + \kappa^2)}, \quad (4.85)$$

where $s_\theta \equiv \sin(\theta/2)$, $s_\phi \equiv \sin(\phi/2)$, $c_\phi \equiv \cos(\phi/2)$, and $\kappa^2 \equiv m_g^2/2p_F^2 = 3\alpha_s/2\pi$. Numerically, $I_1 = 12.3,\ 4.23$, and 2.25 for $\alpha_s = 0.2,\ 0.4$, and 0.6, while $I_2 = 2.34,\ 1.30$, and 0.86, respectively.

The total emission rate involves three processes with equal quarks ($qq = uu,\ dd,\ ss$), and three with different ones ($qq' = ud,\ us,\ ds$) so that Q_a is the prefactor of Eq. (4.84) times $(C_u^2 + C_d^2 + C_s^2)(3I_1 + 2I_2)$, to be compared with Eq. (4.10) for degenerate neutron matter. The ratio between the rates is

$$\frac{Q_a^{qq}}{Q_a^{nn}} = \frac{C_u^2 + C_d^2 + C_s^2}{C_n^2} \frac{p_{F,q}}{p_{F,n}} \frac{\alpha_s^2}{\alpha_\pi^2} \frac{2\pi\,(3I_1 + 2I_2)}{G(m_\pi/p_{F,n})}, \quad (4.86)$$

where the function $G(u)$ was given in Eq. (4.12). The ratio of coupling constants is model dependent (Sect. 14.3.3), the Fermi momenta are approximately equal for equal densities.

For a typical value $\alpha_s = 0.4$ one has $\alpha_s^2/\alpha_\pi^2 = 0.6\times10^{-3}$ while the last factor is about 140 with $G = 0.7$ (Fig. 4.3) so that the last two factors together are about 0.08. Hence, the emission rate from quark matter is much smaller than that from neutron matter because α_π is so large relative to α_s. In the inevitable presence of protons in a neutron star, the np process will be even more important than nn (Fig. 4.4), further enhancing the emission rate of nuclear matter relative to quark matter.

For neutrino pairs the ratio of the emission rates is about the same as for axions and so quark matter is much less effective at $\nu\overline{\nu}$ emission than nuclear matter (Burrows 1979; Anand, Goyal, and Iha 1990). This observation has little effect on the neutrino cooling rate of a quark star because the URCA processes, which are based on charged-current reactions, dominate for both nuclear or quark matter.

For axions, however, it may seem that the emission rate is much suppressed relative to nuclear matter. This is certainly true if "axion" stands for any generic pseudoscalar Nambu-Goldstone boson. The QCD axion, however, which was introduced to solve the CP problem of strong interactions (Chapter 14) necessarily has a two-gluon coupling which allows for the gluonic Primakoff effect (Fig. 4.14) which is analogous to the photon Primakoff effect discussed in Sect. 5.2.

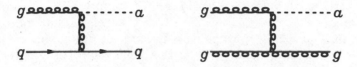

Fig. 4.14. Axion emission by the gluon Primakoff effect.

Altherr (1991) has calculated the emissivity of a hot quark-gluon plasma and found that it was similar to that of a nuclear medium at the same temperature and density. Ellis and Salati (1990) found a much smaller emission rate because they included only the gluonic plasmon decay process $g_T \rightarrow g_L + \gamma$, much in analogy to the corresponding photonic process discussed in Sect. 5.2.2. This decay process, however, is only part of the axion emissivity by the gluon field fluctuations.

For highly degenerate quark matter corresponding to old neutron stars the emission rate has not been calculated as far as I know.

4.9.3 Bubble Phase

In a hot lepton-rich neutron star the nuclear medium in the density regime $\frac{1}{2}\rho_0$ to ρ_0 (nuclear matter density $\rho_0 = 3 \times 10^{14} \, \mathrm{g \, cm^{-3}}$) may form a "bubble phase" with regions of low density embedded in the medium. Leinson (1993) has estimated the bremsstrahlung rate for the neutral-current reactions $N + \mathrm{bubble} \rightarrow \mathrm{bubble} + N + \nu\bar{\nu}$. Naturally, he found that only the axial-vector current contributes. Also, it is easy to translate his results into an axion emission rate. Apparently, the bubble phase can be important for early neutron-star cooling.

4.10 Emission of Right-Handed Dirac Neutrinos

So far in this chapter neutrinos were assumed to be massless particles which interact only by the standard left-handed weak current. It is possible, however, that neutrinos have a Dirac mass in which case any reaction with a final-state (anti)neutrino produces both positive and negative helicity states. Typically, the "wrong-helicity" states will emerge in a fraction $(m_\nu/2E_\nu)^2$ of all cases because of the mismatch between chirality (eigenstates of γ_5) and helicity. For $E_\nu \gg m_\nu$ the wrong-helicity states correspond approximately to right-handed chirality states and so their interaction-rate with the ambient medium is weaker by an approximate factor $(m_\nu/2E_\nu)^2$. This implies that for a sufficiently small mass they would not be trapped in a SN core and thus carry away energy in an "invisible" channel, allowing one to set constraints on a Dirac neutrino mass from the observed neutrino signal of SN 1987A (Sect. 13.8.1). Here, the relationship between the production rate of left- and right-handed neutrinos is explored in some detail because the simple scaling with $(m_\nu/2E_\nu)^2$ is not correct in all cases.

When a neutrino interacts with a medium the transition probability from a state with four-momentum $K_1 = (\omega_1, \mathbf{k}_1)$ to one with $K_2 = (\omega_2, \mathbf{k}_2)$ is written as $W(K_1, K_2)$. The function W is defined for both positive and negative energies. The emission probability for a pair $\overline{\nu}(K_1)\nu(K_2)$ is then $W(-K_1, K_2)$, the absorption probability for a pair is $W(K_1, -K_2)$. The collisional rate of change of the occupation number $f_{\mathbf{k}_1}$ of a neutrino field mode $\mathbf{k}_1$ is then given by

$$\frac{df_{\mathbf{k}_1}}{dt}\bigg|_{\text{coll}} = \int \frac{d^3k_2}{(2\pi)^3} \Big[W_{K_2,K_1} f_{\mathbf{k}_2}(1 - f_{\mathbf{k}_1}) - W_{K_1,K_2} f_{\mathbf{k}_1}(1 - f_{\mathbf{k}_2})$$
$$+ W_{-K_2,K_1}(1 - f_{\mathbf{k}_1})(1 - \overline{f}_{\mathbf{k}_2}) - W_{K_1,-K_2} f_{\mathbf{k}_1}\overline{f}_{\mathbf{k}_2} \Big], \quad (4.87)$$

where the variables of W were written as subscripts. The first term corresponds to neutrino scatterings into the mode $\mathbf{k}_1$ from all other modes, the second term is scattering out of mode $\mathbf{k}_1$ into all other modes, the third term is pair production with a final-state neutrino $\mathbf{k}_1$, and the fourth term is pair absorption of a neutrino of momentum $\mathbf{k}_1$ and an antineutrino of any momentum. $\overline{f}_{\underline{\mathbf{k}}}$ is the occupation number for the antineutrino mode $\mathbf{k}$; $(1-f_{\mathbf{k}})$ or $(1-\overline{f}_{\mathbf{k}})$ represent Pauli blocking factors. This collision integral only includes (effective) neutral-current processes while charged-current reactions were ignored.

According to the discussion of Sect. 4.6 the transition probability is given as

$$W(K_1, K_2) = \frac{G_F^2}{8} \frac{n_B}{2\omega_1 2\omega_2} S^{\mu\nu} N_{\mu\nu}, \tag{4.88}$$

where $S^{\mu\nu}$ is an effective structure function given in terms of the vector, axial-vector and mixed structure function of the medium defined in Eq. (4.38), and n_B is the baryon density. The tensor $N^{\mu\nu}$ is the squared matrix element of the neutrino current; it was given in Eq. (4.17) in terms of K_1 and K_2. In an isotropic medium the tensorial composition of the dynamical structure function can be expressed in terms of the energy-momentum transfer K and the four-velocity U of the medium; $U = (1, 0, 0, 0)$ in its rest frame. Thus, in analogy to Eqs. (4.40) and (4.41) one may write

$$S^{\mu\nu} = S_1 U^\mu U^\nu + S_2 (U^\mu U^\nu - g^{\mu\nu}) + S_3 K^\mu K^\nu$$
$$+ S_4 (K^\mu U^\nu + U^\mu K^\nu) + i S_5 \, \epsilon^{\mu\nu\alpha\beta} U_\alpha K_\beta, \tag{4.89}$$

where the functions S_ℓ ($\ell = 1, \ldots, 5$) depend on medium properties and on the energy-momentum transfer $K = (\omega, \mathbf{k})$. The transition probability is then found to be (Raffelt and Seckel 1995)

$$W(K_1, K_2) = \frac{G_F^2 n_B}{4} \Big[(1 + \cos\theta) S_1 + (3 - \cos\theta) S_2$$
$$- 2(1 - \cos\theta)(\omega_1 + \omega_2) S_5 \Big], \tag{4.90}$$

where θ is the neutrino scattering angle. The terms proportional to S_3 and S_4 vanish identically.

Next, one may consider processes involving massive Dirac neutrinos with specified helicities. Gaemers, Gandhi, and Lattimer (1989) showed that in this case the same expressions apply with $N_{\mu\nu}$ as constructed in Eq. (4.17) if one substitutes $K_i \rightarrow \frac{1}{2}(K_i \pm m_\nu \mathcal{S}_i)$, $i = 1$ or 2, where the plus sign refers to ν, the minus sign to $\bar{\nu}$, and $\mathcal{S}_i$ is the covariant spin vector. For relativistic neutrinos one may consider a noncovariant lowest-order expansion in terms of m_ν. In this limit K_i remains unchanged for left-handed states while

$$K_i = (\omega_i, \mathbf{k}_i) \rightarrow \widetilde{K}_i = (m_\nu / 2\omega_i)^2 \, (\omega_i, -\mathbf{k}_i) \tag{4.91}$$

for right-handed ones. After this substitution has been performed all further effects of m_ν are of higher order so that one may neglect m_ν everywhere except in the global "spin-flip factor."

The dispersion relation of neutrinos in a SN differs markedly from the vacuum form; in the core the "effective m_ν" is several 10 keV. However, m_ν in Eq. (4.91) is the vacuum mass which couples left-handed to right-handed states and thus leads to spin flip while the medium-induced "mass" only affects the dispersion relation of left-handed states. This view is supported by a detailed study of Pantaleone (1991). Of course, for nonrelativistic neutrinos the situation is more complicated because an approximate identification of helicity with chirality is not possible.

Next consider a specific process which involves a left- and a right-handed neutrino such as "spin-flip scattering" $\nu_L(K_L) \rightarrow \nu_R(K_R)$. Then, one needs to construct $N_{\mu\nu}$ from $K_1 = (\omega_L, \mathbf{k}_L)$ and $K_2 = (m_\nu/2\omega_R)^2 (\omega_R, -\mathbf{k}_R)$. The contraction with $S^{\mu\nu}$ leads to (Raffelt and Seckel 1995)

$$\widetilde{W}(K_L, K_R) = \frac{G_F^2 n_B}{4} \left(\frac{m_\nu}{2\omega_R}\right)^2 \Big[(1 - \cos\theta)S_1 + (3 + \cos\theta)S_2$$

$$+ 4\omega_R^2(1 - \cos\theta)S_3 - 4\omega_R(1 - \cos\theta)S_4 + 2(\omega_R - \omega_L)(1 + \cos\theta)S_5\Big].$$

$$(4.92)$$

Following Gaemers, Gandhi, and Lattimer (1989) it must be emphasized that this expression differs in more than the factor $(m_\nu/2\omega_R)^2$ from the nonflip case Eq. (4.90). This difference is due to the changed angular momentum budget of reactions with spin-flipped neutrinos.

This angular-momentum difference between spin-flip and no-flip processes is nicely illustrated by virtue of the pion decay process $\pi^\circ \rightarrow \bar{\nu}\nu$. If both final-state neutrinos are left-handed, i.e. if ν has negative and $\bar{\nu}$ positive helicity, this decay is forbidden by angular momentum conservation. This is seen most easily if one recalls that it is the pion current $\partial_\mu \pi^\circ$ that interacts with the left-handed neutrino current. The squared matrix element of the pion current is thus proportional to $K_{\pi^\circ}^\mu K_{\pi^\circ}^\nu$. For a pion decaying at rest only the 00-component contributes. Contraction with $N_{\mu\nu}$ leads to identically zero if one recalls that for the final-state neutrinos $\omega_1 = \omega_2$ and $\mathbf{k}_1 = -\mathbf{k}_2$. For the spin-flip process one must use the reversed momentum instead so that in $N_{\mu\nu}$ one must use $\omega_1 = \omega_2$ and $\mathbf{k}_1 = \mathbf{k}_2$, leading to a nonvanishing contribution. The decay of thermal pions is an important process for populating the right-handed states of massive Dirac neutrinos in the early universe (Lam and Ng 1991). Contrary to a discussion by Natale (1991), however, pion decays do not seem to provide a particularly strong contribution in SN cores (Raffelt and Seckel 1991).

Right-handed neutrinos can be produced by the interaction with all medium constituents. For a SN core, the production rate from the interaction with charged leptons such as $e^+e^- \to \nu_L\bar{\nu}_R$ or $\nu_L e^- \to e^-\nu_R$ was explicitly studied by Pérez and Gandhi (1990) and by Lam and Ng (1992). Still, in a SN core the dominant contribution to the interaction between neutrinos and the medium is due to the nucleons.

Therefore, one may simplify the expression Eq. (4.92) by applying the same approximations that were used earlier in Sect. 4.6 in the context of purely left-handed neutrinos. Notably, one may use the nonrelativistic and long-wavelength limits which reveal, again, that only the terms proportional to S_1 and S_2 contribute, and that these structure functions can be taken to be functions of the energy transfer ω alone.[25] Moreover, the neutrino phase-space integration will always average the $\cos\theta$ term to zero. Then, the spin-flip reaction rate is indeed simply the nonflip rate times the "spin-flip factor" $(m_\mu/2E_\nu)^2$.

Right-handed neutrinos can be produced by spin-flip scattering of left-handed ones $\nu_L \to \nu_R$, or by the emission of pairs $\nu_L\bar{\nu}_R$ or $\nu_R\bar{\nu}_L$. In a SN core, left-handed neutrinos are trapped while right-handed ones can freely escape whence the quantity of interest is the energy-loss rate of the medium in terms of right-handed states. The total energy-loss rate in ν_R is then $Q_R = Q_{\text{scat}} + Q_{\text{pair}}$ where "scat" refers to spin-flip scattering and "pair" to the pair-emission process. The two contributions are

$$Q_{\text{scat}} = \int \frac{d^3\mathbf{k}_L}{(2\pi)^3} \frac{d^3\mathbf{k}_R}{(2\pi)^3} \widetilde{W}_{K_L,K_R} f_{\mathbf{k}_L} \omega_R,$$

$$Q_{\text{pair}} = \int \frac{d^3\mathbf{k}_L}{(2\pi)^3} \frac{d^3\mathbf{k}_R}{(2\pi)^3} \widetilde{W}_{-K_L,K_R}(1 - \bar{f}_{\mathbf{k}_L})\omega_R. \tag{4.93}$$

An analogous expression pertains to the emission of $\bar{\nu}_R$; if the trapped left-handed neutrinos are nondegenerate this contribution has the same magnitude as that for the emission of ν_R.

Next, consider the limit where the transition probability $\widetilde{W}$ can be represented in terms of a single structure function[26] $S(\omega) = S_1(\omega) + 3S_2(\omega)$. Further, the possibility of neutrino degeneracy is ignored which allows one to approximate the left-handed neutrino Fermi-Dirac distribution by a Maxwell-Boltzmann one, $(e^{\omega/T} + 1)^{-1} \to e^{-\omega/T}$. Then the

---

[25] In the notation of Sect. 4.6 and for a single species of nucleons one has in this limit $S_1 = C_V^2 S_\rho$ and $S_2 = C_A^2 S_\sigma$.

[26] In the notation of Sect. 4.6 this is $S = C_V^2 S_\rho + 3C_A^2 S_\sigma$ in a medium consisting of a single species of nucleons.

energy-loss rates are

$$Q_{\text{scat}} = \frac{G_F^2 m_\nu^2 n_B T^4}{4\pi^4} \int_0^\infty d\omega \left[\left(\frac{\omega}{T}\right)^2 + 6\frac{\omega}{T} + 12 \right] S(\omega),$$

$$Q_{\text{pair}} = \frac{G_F^2 m_\nu^2 n_B T^4}{4\pi^4} \int_0^\infty d\omega \, \frac{1}{12} \left(\frac{\omega}{T}\right)^4 S(\omega), \qquad (4.94)$$

where the detailed-balance condition $S(-\omega) = S(\omega)e^{\omega/T}$ was used. These rates include a factor of 2 for the emission of ν_R and $\bar{\nu}_R$.

As noted by Raffelt and Seckel (1995), unless the functional form of $S(\omega)$ is very bizarre one has $Q_{\text{scat}} \gg Q_{\text{pair}}$. This is easily understood if one observes that in a scattering process the energy available for the final-state ν_R is that of the initial-state ν_L plus energy that is contributed by the medium. For pair emission, the energy for the final-state $\bar{\nu}_L$ and ν_R both have to be provided by the medium. Therefore, the spin-flip scattering process is favored by the neutrino phase space.[27] In a SN core the production of right-handed Dirac neutrinos is given essentially by the standard scattering rate times the factor $(m_\nu/2E_\nu)^2$.

---

[27]This conclusion is in contrast to a suggestion that the pair-emission rates could be relatively important in a SN core (Turner 1992). This statement was not based on a self-consistent treatment of the medium structure functions, but rather on a perturbative treatment of the processes involved. The conflict between Turner's statement and our finding, which essentially was based on phase-space considerations, highlights the inadequacy of naive perturbative results in a nuclear medium.

Chapter 5

Two-Photon Coupling of Low-Mass Bosons

Well-known particles such as neutral pions, or hypothetical ones such as axions or gravitons, each have a two-photon interaction vertex. It allows for radiative decays of the form $\pi^\circ \rightarrow 2\gamma$ as well as for the Primakoff conversion $\pi^\circ \leftrightarrow \gamma$ in the presence of external electric or magnetic fields. The Primakoff conversion of photons into gravitons in putative cosmic magnetic fields could cause temperature fluctuations of the cosmic microwave background radiation. In stars, the Primakoff conversion of photons leads to the production of gravitons and axions, although the graviton luminosity is always negligible. The Primakoff conversion of axions into photons serves as the basis for a detection scheme for galactic axions, and was used in several laboratory axion search experiments. The modified Maxwell equations in the presence of very low mass pseudoscalars would substantially modify pulsar electrodynamics. These physical phenomena are explored and current laboratory and astrophysical limits on the axion-photon coupling are reviewed.

5.1 Electromagnetic Coupling of Pseudoscalars

The idea that symmetries of the fundamental interactions can be broken by the "vacuum" or ground state of the fields plays an important role in modern particle physics. The breakdown of a global symmetry by the vacuum expectation value of a Higgs-like field leads to the existence of massless particles, the "Nambu-Goldstone bosons" of the broken symmetry. One well-known example are the pions which are

165

the Nambu-Goldstone bosons of an approximate $SU(2)$ symmetry of the system of nucleons and pions in the sigma model (e.g. Itzykson and Zuber 1983). Another example is the hypothetical axion which is the Nambu-Goldstone boson of the Peccei-Quinn chiral $U(1)$ symmetry that would solve the CP problem of strong interactions (Chapter 14). Both pions and axions carry a small mass because the underlying symmetry is not exact at low energies; they are sometimes called "pseudo Nambu-Goldstone bosons." An example for a true Nambu-Goldstone boson is the hypothetical majoron which arises from the spontaneous breakdown of a symmetry by a Higgs field which would give the neutrinos Majorana masses (Sect. 15.7).

In these examples the Nambu-Goldstone bosons are pseudoscalars because the underlying symmetry is chiral. Unless the CP symmetry is violated, their possible coupling to photons must be of the form

$$\mathcal{L}_{\text{int}} = -\tfrac{1}{4}\, g_{a\gamma}\, F_{\mu\nu} \tilde{F}^{\mu\nu}\, a = g_{a\gamma}\, \mathbf{E} \cdot \mathbf{B}\, a, \tag{5.1}$$

where a is the pseudoscalar field; the axion will serve as a generic example. Further, $g_{a\gamma}$ is a constant with the dimension $(\text{energy})^{-1}$, F is the electromagnetic field strength tensor, $\tilde{F}$ its dual, and $\mathbf{E}$ and $\mathbf{B}$ are the electric and magnetic fields, respectively. Because $\mathbf{E}$ is a polar and $\mathbf{B}$ an axial vector, $\mathbf{E} \cdot \mathbf{B}$ is a pseudoscalar under a CP transformation and so $\mathcal{L}_{\text{int}}$ remains invariant.

New scalar particles ϕ almost inevitably would also couple to photons with a CP conserving structure $\mathcal{L}_{\text{int}} \propto \tfrac{1}{4}F_{\mu\nu}F^{\mu\nu}\,\phi = \tfrac{1}{2}(E^2 + B^2)\phi$. Therefore, everything that will be said about axions applies *mutatis mutandis* to scalar particles as well.

Gravitons would also have a two-photon vertex. Much of what is said about axions also applies to them, except that their weak couplings render most of the arguments irrelevant—an exception will be mentioned in Sect. 5.5.5.

Returning to the case of pseudoscalar Nambu-Goldstone bosons, by the very construction of the underlying models they interact with certain fermions ψ by a pseudoscalar coupling of the form

$$\mathcal{L}_{\text{int}} = ig\, \overline{\psi}\gamma_5\psi\, a, \tag{5.2}$$

where $g = m/f$ is a Yukawa coupling given in terms of the fermion mass m and an energy scale f related to the vacuum expectation value of the Higgs field which breaks the underlying symmetry—for axions the Peccei-Quinn scale f_a (Chapter 14), for pions the pion decay constant $f_\pi = 93\,\text{MeV}$. In the latter example the relevant fermions are the nucleons, for axions they are the quarks, possibly the charged leptons, and perhaps some exotic heavy quark state not contained in the standard model. For majorons, such couplings exist to neutrinos, and possibly to other fermions.

An interaction of the form Eq. (5.2) with charged fermions automatically leads to an electromagnetic coupling of the form Eq. (5.1) because of the triangle amplitude shown in Fig. 5.1. For one fermion of charge e and mass m an explicit evaluation leads to the relationship (e.g. Itzykson and Zuber 1983)

$$g_{a\gamma} = \frac{\alpha}{\pi} \frac{g}{m} = \frac{\alpha}{\pi f},$$ (5.3)

where m was taken to be much larger than the axion and photon energies. Remarkably, because $g = m/f$ this coupling does not depend on the fermion mass, but only on the scale f of symmetry breaking. In general, one must sum over all possible fermions, taking account of the appropriate charges which are fractional for quarks, and also of the proper pseudoscalar coupling to the individual fermions which may vary from m/f by model-dependent factors of order unity.

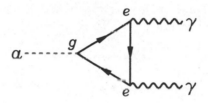

Fig. 5.1. Triangle loop for the coupling of a pseudoscalar a (axion) to two photons.

For massive pseudoscalars the two-photon coupling allows for a decay $a \to 2\gamma$ with a width

$$\Gamma_{a \to 2\gamma} = g_{a\gamma}^2 m_a^3 / 64\pi.$$ (5.4)

For pions in the sigma model, the only charged fermion is the proton. Then $g_{\pi^\circ \gamma} = \alpha/\pi f_\pi$ and $\Gamma_{\pi^\circ \to 2\gamma} = \alpha^2 m_\pi^3 / 64\pi^3 f_\pi^2 = 7.6\,\text{eV}$, in close agreement with the experimental value. (For subtleties of interpretation of this result in the context of current algebra see the standard field theory literature, e.g. Itzykson and Zuber 1983).

The main point for the present discussion is that pseudoscalar massless or low-mass bosons are a natural consequence of certain extensions of the standard model, and that these particles couple to photons according to Eq. (5.1) with a strength

$$g_{a\gamma} = \frac{\alpha}{\pi f_a} C_{a\gamma},$$ (5.5)

where f_a is the energy scale of symmetry breaking and $C_{a\gamma}$ is a model-dependent factor of order unity. (For axions, model-dependent details of the couplings are discussed in Chapter 14.) In the following I will explore a variety of consequences arising from this interaction.

5.2 Primakoff Process in Stars

5.2.1 Screened Cross Section and Emission Rate

The two-photon coupling of pions or other pseudoscalars allows for the conversion $a \leftrightarrow \gamma$ in an external electric or magnetic field by virtue of the amplitude shown in Fig. 5.2. This process was first proposed by Primakoff (1951) to study the π°-γ-coupling which is experimentally difficult to measure in free decays $\pi^\circ \rightarrow 2\gamma$. In stars, this process allows for the production of low-mass pseudoscalars in the electric fields of nuclei and electrons (Dicus et al. 1978).

Fig. 5.2. Primakoff conversion between axions or other pseudoscalars and photons in an external electromagnetic field.

The Primakoff process turns out to be important for nonrelativistic conditions where $T \ll m_e$ so that both electrons and nuclei can be treated as "heavy" relative to typical energies of the ambient photons. Therefore, ignoring recoil effects one finds for the differential cross section in this limit (target charge Ze)

$$\frac{d\sigma_{\gamma \rightarrow a}}{d\Omega} = \frac{g_{a\gamma}^2 Z^2 \alpha}{8\pi} \frac{|\mathbf{k}_\gamma \times \mathbf{k}_a|^2}{\mathbf{q}^4},$$ (5.6)

where $\mathbf{q} = \mathbf{k}_\gamma - \mathbf{k}_a$ is the momentum transfer; the axion and photon energies are the same.

This cross section exhibits the usual forward divergence from the long-range Coulomb interaction. For $m_a \neq 0$ it is cut off in vacuum by the minimum necessary momentum transfer $q_{min} = m_a^2/2\omega$ ($m_a \ll \omega$); the total cross section is then $\sigma_{\gamma \to a} = Z^2 g_{a\gamma}^2 \left[\frac{1}{2} \ln(2\omega/m_a) - \frac{1}{4} \right]$.

In a plasma, the long-range Coulomb potential is cut off by screening effects; according to Sect. 6.4 the differential cross section is modified with a factor $\mathbf{q}^2/(k_S^2 + \mathbf{q}^2)$. In a nondegenerate medium the screening scale is given by the Debye-Hückel formula

$$k_S^2 = \frac{4\pi\alpha}{T} n_B \left(Y_e + \sum_j Z_j^2 Y_j \right), \tag{5.7}$$

where $n_B = \rho/m_u$ (atomic mass unit m_u) is the baryon density while Y_e and Y_j are the number fractions per baryon of the electrons and various nuclear species j. With this modification the total scattering cross section is easily calculated (Raffelt 1986a). Summing over all targets one may derive an expression for the transition rate ("decay rate") of a photon of frequency ω into an axion of the same energy,

$$\Gamma_{\gamma \to a} = \frac{g_{a\gamma}^2 T k_S^2}{32\pi} \left[\left(1 + \frac{k_S^2}{4\omega^2} \right) \ln \left(1 + \frac{4\omega^2}{k_S^2} \right) - 1 \right], \tag{5.8}$$

where the plasma "mass" of the initial-state photon and the axion mass were neglected relative to the energy ω. In the limit $\omega \ll k_S$ this expression expands as $\Gamma_{\gamma \to a} = g_{a\gamma}^2 \omega^2 T/16\pi$ which is entirely independent of the density and chemical composition.

For a stellar plasma, however, this approximation is usually not justified. Ignoring the plasma frequency for the initial-state photons, the energy-loss rate per unit volume is

$$Q = \int \frac{2 \, d^3 \mathbf{k}_\gamma}{(2\pi)^3} \frac{\Gamma_{\gamma \to a} \, \omega}{e^{\omega/T} - 1} = \frac{g_{a\gamma}^2 T^7}{4\pi} F(\kappa^2), \tag{5.9}$$

where $\kappa \equiv k_S/2T$ and

$$F(\kappa^2) = \frac{\kappa^2}{2\pi^2} \int_0^\infty dx \left[(x^2 + \kappa^2) \ln \left(1 + \frac{x^2}{\kappa^2} \right) - x^2 \right] \frac{x}{e^x - 1}, \tag{5.10}$$

with $x = \omega/T$. This function is shown in Fig. 5.3. In a standard solar model $\kappa^2 \approx 12$ throughout the Sun with a variation of less than 15%. In the core of an HB star with $\rho = 10^4 \, \text{g/cm}^3$ and $T = 10^8 \, \text{K}$ it is $\kappa^2 \approx 2.5$. One finds $F = 0.98$ and 1.84 for $\kappa^2 = 2.5$ and 12, respectively.

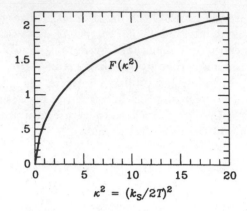

Fig. 5.3. Function $F(\kappa^2)$ according to Eq. (5.10).

5.2.2 Plasmon Decay and Coalescence

Several authors have used the plasma frequency ω_{P} instead of the Debye-Hückel wave number as a screening scale. In a nondegenerate plasma $k_{\mathrm{S}}^2/\omega_{\mathrm{P}}^2 \approx m_e/T$ (Sect. 6.3) and so they overestimated the emission rate. Another source of confusion are statements that the decay $\gamma \rightarrow \gamma a$ was another axion emission process enabled by the 2γ coupling. This issue can be easily clarified, but one needs to draw heavily on the discussion of photon dispersion of Sect. 6.3.

A medium allows for both transverse and longitudinal electromagnetic excitations. In a nondegenerate and nonrelativistic plasma the dispersion relation of the former is $\omega_{\mathrm{T}}^2 - k_{\mathrm{T}}^2 = \omega_{\mathrm{P}}^2$ (plasma frequency ω_{P}) while the latter oscillate essentially with a fixed frequency $\omega_{\mathrm{L}} = \omega_{\mathrm{P}}$, independently of their wave number. Therefore, the plasmon decay process $\gamma_{\mathrm{T}} \rightarrow \gamma_{\mathrm{L}} a$ as well as the plasmon coalescence $\gamma_{\mathrm{T}}\gamma_{\mathrm{L}} \rightarrow a$ are indeed kinematically possible for $k_{\mathrm{L}} > \omega_{\mathrm{P}}$.

It is easy to calculate the inverse lifetime of transverse excitations against these processes

$$\Gamma_{\gamma_{\mathrm{T}} \rightarrow a} = \int \frac{d^3\mathbf{k}_{\mathrm{L}}}{2\omega_{\mathrm{L}}(2\pi)^3} \frac{d^3\mathbf{k}_a}{2\omega_a(2\pi)^3} \frac{1}{2\omega_{\mathrm{T}}} Z_{\mathrm{T}} Z_{\mathrm{L}} |\mathcal{M}|^2 (2\pi)^4$$

$$\times \left[\frac{\delta^4(K_{\mathrm{T}} + K_{\mathrm{L}} - K_a)}{e^{\omega_{\mathrm{L}}/T} - 1} + \frac{\delta^4(K_{\mathrm{T}} - K_{\mathrm{L}} - K_a)}{1 - e^{-\omega_{\mathrm{L}}/T}} \right], \quad (5.11)$$

where $Z_{\mathrm{T,L}}$ are the vertex renormalization factors. Here, the first term in square brackets corresponds to coalescence and so it involves a Bose-Einstein occupation number for the initial-state γ_{L}. The second term is from decay; it involves a Bose stimulation factor $[1 + (e^{\omega_L/T} - 1)^{-1}] = (1 - e^{-\omega_L/T})^{-1}$ for the final-state γ_L. For the axions, a stimulation factor is not included because they are assumed to escape immediately.

Because a longitudinal excitation has no magnetic field, the electric field in the matrix element corresponding to Eq. (5.1) must be associated with a longitudinal, the magnetic field with a transverse excitation. In general $\mathbf{B} = \nabla \times \mathbf{A}$ and $\mathbf{E} = -\nabla A^0 - \partial_t \mathbf{A}$ with the vector potential A. For a propagating mode $A \propto e^{-i(\omega t - \mathbf{k} \cdot \mathbf{x})} \epsilon$ where ϵ is a polarization vector. This implies that $\mathbf{B} \propto \mathbf{k}_T \times \epsilon_T$ where ϵ_T is a polarization vector transverse to $\mathbf{k}_T$ with $\epsilon_T^2 = 1$, and $\mathbf{E} \propto \omega_L \epsilon_L - k_L \epsilon_L^0$ where $\epsilon_L = (k_L^2, \omega_L \mathbf{k}_L) k_L^{-1} (\omega_L^2 - k_L^2)^{-1/2}$ according to Eq. (6.27) so that $\mathbf{E} \propto \hat{\mathbf{k}}_L (\omega_L^2 - k_L^2)^{1/2}$. Then one finds from the usual Feynman rules

$$|\mathcal{M}|^2 = g_{a\gamma}^2 (\omega_L^2 - k_L^2) |\hat{\mathbf{k}}_L \cdot (\mathbf{k}_T \times \epsilon_T)|^2. \tag{5.12}$$

Note that $|\hat{\mathbf{k}}_L \cdot (\mathbf{k}_T \times \epsilon_T)|^2 = |(\hat{\mathbf{k}}_L \times \mathbf{k}_T) \cdot \epsilon_T|^2$. Averaging over the two transverse polarization states yields $\frac{1}{2} |\hat{\mathbf{k}}_L \times \mathbf{k}_T|^2$.

If one writes $Z_L = \tilde{Z}_L \omega_L^2 / (\omega_L^2 - k_L^2)$ as in Sect. 6.3 and performs the $d^3 \mathbf{k}_L$ integration in Eq. (5.11) one finds

$$\Gamma_{\gamma_T \to a} = \frac{g_{a\gamma}^2}{16 (2\pi)^2} \int d\Omega_a d\omega_a \, Z_T \tilde{Z}_L \frac{\omega_a \omega_L}{\omega_T} |\hat{\mathbf{k}}_L \times \mathbf{k}_T|^2$$

$$\times \left[\frac{\delta(\omega_T + \omega_L - \omega_a)}{e^{\omega_L/T} - 1} + \frac{\delta(\omega_T - \omega_L - \omega_a)}{1 - e^{-\omega_L/T}} \right]. \tag{5.13}$$

In a nondegenerate, nonrelativistic plasma typically $\omega_P \ll T$. Because $\omega_L = \omega_P$ one may expand the exponentials so that $(e^{\omega_L/T} - 1)^{-1} \to T/\omega_L$ and also $(1 - e^{-\omega_L/T})^{-1} \to T/\omega_L$. If $\omega_T = \mathcal{O}(T) \gg \omega_P$ one may ignore ω_L in the δ functions which are then trivial to integrate,

$$\Gamma_{\gamma_T \to a} = \frac{g_{a\gamma}^2 T}{8 (2\pi)^2} \int d\Omega_a \, Z_T \tilde{Z}_L |\hat{\mathbf{k}}_L \times \mathbf{k}_T|^2. \tag{5.14}$$

In this limit a has the same energy as γ_T while γ_L has only provided momentum.

To finish up, note that in the nondegenerate, nonrelativistic limit $Z_T = \tilde{Z}_L = 1$. Because $\omega_P \ll T$ and $\omega_T = \mathcal{O}(T)$ we have $\omega_P \ll \omega_T$ so that $k_T \approx \omega_T$. Moreover, $\mathbf{k}_L = \mathbf{k}_T - \mathbf{k}_a$ yielding $|\hat{\mathbf{k}}_L \times \mathbf{k}_T|^2 = |\mathbf{k}_a \times \mathbf{k}_T|^2 / |\mathbf{k}_a - \mathbf{k}_T|^2 = \omega_T^2 (1 - z^2)/2(1 - z) = \frac{1}{2} \omega_T^2 (1 + z)$ where z is the cosine of the angle between $\mathbf{k}_a$ and $\mathbf{k}_T$. The angular integration then averages z to zero and leaves us with

$$\Gamma_{\gamma_T \to a} = \frac{g_{a\gamma}^2 T}{16\pi}. \tag{5.15}$$

Collective longitudinal oscillations only exist for $k_L \lesssim k_D$ (Debye screening scale). Because a momentum transfer $|\mathbf{k}_T - \mathbf{k}_a| = \mathcal{O}(\omega_T)$ is required, this result applies only for $\omega_T \lesssim k_D$. This conversion rate agrees with the low-ω expansion of the Primakoff result Eq. (5.8).

It must be stressed that Eq. (5.15) is not an additional contribution to the conversion rate, it is the same result derived in a different fashion. Here, longitudinal plasmons in the static limit were used as the external electric field in which the Primakoff effect takes place. Before, the static limit was taken from the start; the collective behavior of the electron motion was reflected in the screening of the Coulomb potential. These two paths of performing the calculation in the end extracted the same information from the electromagnetic polarization tensor which defines both screening effects and the dispersion behavior of electromagnetic excitations.

5.2.3 Axion Emission from Electromagnetic Plasma Fluctuations

These simple calculations of the axion emission rate apply only in the classical (nondegenerate, nonrelativistic) limit. Even though this is the most relevant case from a practical perspective it is worth mentioning how one proceeds for a more general evaluation. To this end note that the 2γ interaction Eq. (5.1) corresponds to a source term for the axion wave equation,

$$(\Box + m_a^2)\, a = g_{a\gamma}\, \mathbf{E} \cdot \mathbf{B}, \tag{5.16}$$

where $\Box = \partial^\mu \partial_\mu = \partial_t^2 - \nabla^2$. Axions are then emitted by the $\mathbf{E} \cdot \mathbf{B}$ fluctuations caused by the presence of thermal electromagnetic radiation as well as the collective and random motion of charged particles.

The Primakoff calculation in Sect. 5.2.1 used the (screened) electric field of charged particles and the magnetic field of (transverse) electromagnetic radiation (photons) as a source. Actually, one can include the magnetic field of moving charges for this purpose. Then axions are emitted in the collision of two particles (Fig. 5.4), a process sometimes referred to as the "electro Primakoff effect." Unsurprisingly, the emission rate is much smaller because the magnetic field associated with

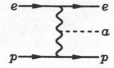

Fig. 5.4. Example for the electro Primakoff effect.

nonrelativistic moving particles is small. (For an explicit calculation see Raffelt 1986a.) Put another way, the dominant contribution to magnetic field fluctuations in a nonrelativistic plasma is from photons, not from moving charges.

In order to illustrate the relationship between field fluctuations and axion emission consider the amplitude for the conversion of a classical transverse electromagnetic wave into a classical axion wave in the presence of an electric field configuration $\mathbf{E}(\mathbf{x})$. One finds from Eq. (5.16)

$$f(\Omega) = \frac{g_{a\gamma}}{4\pi} \left(\boldsymbol{\epsilon} \times \mathbf{k}\right) \cdot \int d^3\mathbf{x}\, e^{-i\mathbf{q}\cdot\mathbf{x}}\, \mathbf{E}(\mathbf{x}) = \frac{g_{a\gamma}}{4\pi} \left(\boldsymbol{\epsilon} \times \mathbf{k}\right) \cdot \mathbf{E}(\mathbf{q}),$$

(5.17)

where $\mathbf{k}$ and $\boldsymbol{\epsilon}$ are the wave and polarization vector of the incident wave, respectively, $\mathbf{q}$ is the "momentum" transfer to the axion, and $\mathbf{E}(\mathbf{q})$ is a Fourier component of $\mathbf{E}(\mathbf{x})$. The differential transition rate is $d\Gamma/d\Omega = |f(\Omega)|^2$ so that

$$\frac{d\Gamma_{\gamma \to a}}{d\Omega} = \frac{g_{a\gamma}^2}{(4\pi)^2} \left(\boldsymbol{\epsilon} \times \mathbf{k}\right)_i \left(\boldsymbol{\epsilon} \times \mathbf{k}\right)_j \mathbf{E}_i(-\mathbf{q})\mathbf{E}_j(\mathbf{q}),$$

(5.18)

where it was used that $\mathbf{E}(\mathbf{x})$ is real so that $\mathbf{E}^*(\mathbf{q}) = \mathbf{E}(-\mathbf{q})$. If $\mathbf{E}(\mathbf{x})$ is a random field configuration one needs to take an ensemble average so that the transition rate is proportional to $\langle \mathbf{E}_i \mathbf{E}_j \rangle_{\mathbf{q}} \equiv \langle \mathbf{E}_i(-\mathbf{q})\mathbf{E}_j(\mathbf{q}) \rangle$.

The electric and magnetic field fluctuations of a plasma are intimately related to the medium response functions to electric and magnetic fields, i.e. to the polarization tensor. For a plasma at temperature T one can show on general grounds (Sitenko 1967)

$$\langle \mathbf{E}_i \mathbf{E}_j \rangle_{\mathbf{q}} = \int_{-\infty}^{+\infty} \frac{d\omega}{2\pi} \frac{2}{e^{\omega/T} - 1}$$

$$\times \left[\frac{q_i q_j}{\mathbf{q}^2} \frac{\mathrm{Im}\,\epsilon_{\mathrm{L}}}{|\epsilon_{\mathrm{L}}|^2} + \left(\delta_{ij} - \frac{q_i q_j}{\mathbf{q}^2} \right) \frac{\mathrm{Im}\,\epsilon_{\mathrm{T}}}{|\epsilon_{\mathrm{T}} - \mathbf{q}^2/\omega^2|^2} \right],$$

(5.19)

where $\epsilon_{\mathrm{L,T}}(\omega, \mathbf{q})$ are the longitudinal and transverse dielectric permittivities of the medium (Sect. 6.3.3). A quantity such as $\mathrm{Im}\,\epsilon_{\mathrm{L}}/|\epsilon_{\mathrm{L}}|^2$ is known as a spectral density—here of the longitudinal fluctuations.

One finds explicitly $\langle \mathbf{E}_i \mathbf{E}_j \rangle_\mathbf{q} = \hat{\mathbf{q}}_i \hat{\mathbf{q}}_j \, T/(1 + \mathbf{q}^2/k_S^2)$ in the classical limit (Sitenko 1967) where k_S^2 is the Debye-Hückel wave number of Eq. (5.7). With this result one easily reproduces the Primakoff transition rate $\Gamma_{\gamma \to a}$ (Raffelt 1988a).

The language of spectral densities for the electromagnetic field fluctuations forms the starting point for a quantum calculation of the axion emission rate in the framework of thermal field theory. This program was carried out in a series of papers by Altherr (1990, 1991), Altherr and Kraemmer (1992), and Altherr, Petitgirard, and del Río Gaztelurrutia (1994). Naturally, in the classical limit they reproduced the Primakoff transition rate $\Gamma_{\gamma_T \to a}$ of Eq. (5.8).

In the degenerate or relativistic limit their results cannot be represented in terms of simple analytic formulae. The most important astrophysical environment to be used for extracting bounds on $g_{a\gamma}$ are low-mass stars before and after helium ignition with a core temperature of about 10^8 K (Sect. 5.2.5). Altherr, Petitgirard, and del Río Gaztelurrutia (1994) gave numerical results for the energy-loss rate for this temperature as a function of density shown in Fig. 5.5 (solid line). The dashed line is the classical limit Eq. (5.9); it agrees well with the general result in the low-density (nondegenerate) limit. In the degen-

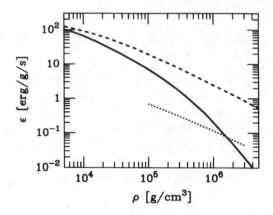

Fig. 5.5. Energy-loss rate of a helium plasma at $T = 10^8$ K by axion emission with $g_{a\gamma} = 10^{-10}$ GeV^{-1}. The solid line is from transverse-longitudinal fluctuations; the dashed line is the corresponding classical limit. The dotted line is from transverse-transverse fluctuations, i.e. in the axion source term $g_{a\gamma} \mathbf{E} \cdot \mathbf{B}$ both fields are from transverse fluctuations. (Adapted from Altherr, Petitgirard, and del Río Gaztelurrutia 1994.)

erate limit, the emission rate drops precipitously, an important feature which will be taken advantage of in Sect. 5.2.5 below.

5.2.4 Solar Axion Spectrum

As a first practical application it is easy to calculate the expected flux of axions at Earth from the Primakoff conversion in the Sun where the classical approximation is well justified. To this end van Bibber et al. (1989) have integrated Eq. (5.9) over a standard solar model which yields an axion luminosity

$$L_a = g_{10}^2 \, 1.7 \times 10^{-3} \, L_\odot, \tag{5.20}$$

with $L_\odot$ the solar luminosity and $g_{10} \equiv g_{a\gamma} \times 10^{10} \, \mathrm{GeV}$. (Recalling that $g_{a\gamma} = (\alpha/\pi f_a) \, C_{a\gamma}$ this corresponds to $f_a/C_{a\gamma} = 2.3 \times 10^7 \, \mathrm{GeV}$.) The differential flux at Earth is well approximated by the formula

$$\frac{dF_a}{d\omega_a} = g_{10}^2 \, 4.02 \times 10^{10} \, \mathrm{cm^{-2} \, s^{-1} \, keV^{-1}} \, \frac{(\omega_a/\mathrm{keV})^3}{e^{\omega_a/1.08\,\mathrm{keV}} - 1} \tag{5.21}$$

which is shown in Fig. 5.6. The average axion energy is $\langle \omega_a \rangle = 4.2 \, \mathrm{keV}$. The total flux at Earth is $F_a = g_{10}^2 \, 3.54 \times 10^{11} \, \mathrm{cm^{-2} \, s^{-1}}$.

The "standard Sun" is about halfway through its main-sequence evolution. Therefore, the solar axion luminosity must not exceed its

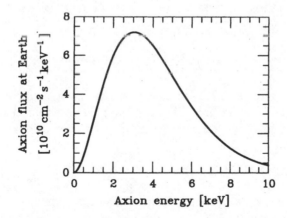

Fig. 5.6. Axion flux at Earth according to Eq. (5.25) from the Primakoff conversion of photons in the Sun.

photon luminosity; otherwise its nuclear fuel would have been spent before reaching an age of 4.5×10^9 yr. This requirement yields a bound

$$g_{a\gamma} \lesssim 2.4 \times 10^{-9} \, \text{GeV}^{-1}. \tag{5.22}$$

Detailed solar evolution calculations of Raffelt and Dearborn (1987) showed that this bound was firm, but also that it could not be improved easily (Sect. 1.3.2). The present-day properties of the Sun could be obtained by a suitable adjustment of the unknown presolar helium abundance.

5.2.5 Globular-Cluster Bound on $g_{a\gamma}$

Armed with the Primakoff emission rate Eq. (5.9) it is an easy task to derive a bound on $g_{a\gamma}$ from the energy-loss argument applied to globular-cluster stars (Sect. 2.5). We need to require that at $T \approx 10^8$ K the axionic energy-loss rate is below $10 \, \text{erg} \, \text{g}^{-1} \, \text{s}^{-1}$ for a density of about $0.6 \times 10^4 \, \text{g} \, \text{cm}^{-3}$, corresponding to a classical plasma, and for about $2 \times 10^5 \, \text{g} \, \text{cm}^{-3}$, corresponding to degeneracy. From Fig. 5.5 it is evident that the emission rate is a steeply falling function of density when degeneracy effects become important. Obviously, the more restrictive limit is found from the low-density case which is based on the helium-burning lifetime of HB stars (Sect. 2.5.1).

In order to calculate the average energy-loss rate of the core of an HB star one needs $\langle T^7/\rho \rangle$ if in Eq. (5.9) one uses a constant $\kappa^2 = 2.5$ or $F = 1.0$. For a typical HB-star model (Fig. 1.4) one finds $\langle T_8^7/\rho_4 \rangle \approx 0.3$ where $T_8 = T/10^8$ K and $\rho_4 = \rho/10^4 \, \text{g} \, \text{cm}^{-3}$. Therefore, $\langle \epsilon_a \rangle \approx g_{10}^2 \, 30 \, \text{erg} \, \text{g}^{-1} \, \text{s}^{-1}$ so that the criterion Eq. (2.40) yields a constraint

$$g_{a\gamma} \lesssim 0.6 \times 10^{-10} \, \text{GeV}^{-1} \quad \text{or} \quad f_a/C_{a\gamma} \gtrsim 4 \times 10^7 \, \text{GeV}. \tag{5.23}$$

The temperature 10^8 K corresponds to 8.6 keV; a typical photon energy is $3T \approx 25$ keV. Therefore, this bound applies to pseudoscalars with a mass $m_a \lesssim 30$ keV while for larger masses it would be degraded.

5.3 Search for Cosmic Axions

One of the most interesting ramifications of the electromagnetic coupling of pseudoscalars is the possibility to search for dark-matter axions. It is briefly explained in Chapter 14 that axions would be produced in the early universe by a nonthermal mechanism which excites classical axion field oscillations, i.e. a highly degenerate axion Bose condensate that would play the role of "cold dark matter," and notably provide the unseen mass necessary to explain the rotation curves of spiral galaxies such as our own (e.g. Kolb and Turner 1990). Because there are uncertainties with regard to details of the primordial axion production mechanism the exact value of the relevant axion mass is not known. However, typically it is of order 10^{-5} eV so that it is a reasonable speculation that the mass of our galaxy is dominated by very low-mass bosons. As these particles are bound to the galaxy they must be nonrelativistic; a typical velocity dispersion corresponding to the galactic gravitational potential is around 10^{-3} in units of the speed of light.

Sikivie (1983) proposed to search for galactic axions by means of a Primakoff-like method. The $a \to \gamma$ conversion of nonrelativistic axions in the μeV mass range produces photons in the microwave (GHz) range. Therefore, the idea is to place a microwave cavity in a strong magnetic field and wait for cavity modes to be excited by the axion field. In this context one may view the electromagnetic modes of the cavity and the free axion field modes as oscillators which are coupled by virtue of the interaction Eq. (5.1) where **B** is the external static field while **E** is from an electromagnetic cavity mode. Then, power is transferred from the axion field to the cavity excitations by virtue of the oscillator beats induced by the coupling; detailed calculations of the conversion rate were performed by Sikivie (1985) and Krauss et al. (1985).

It is worth noting that with a mass of 10^{-5} eV and a velocity of 10^{-3} a typical axion momentum is 10^{-8} eV which corresponds to a wave length of about 20 m. Thus on laboratory scales the axion field is homogeneous. This does not apply to free microwaves—their energy and momentum are the same ($\omega_\gamma = |\mathbf{k}_\gamma|$); for 10^{-5} eV their wave length is 2 cm. The role of the resonant cavity is to overcome this momentum mismatch: on resonance the fundamental cavity frequency is degenerate with nonrelativistic axions of a certain mass for which the energy transfer is maximized. In a search experiment the cavity must be stepped through a range of resonant frequencies which defines the range of axion masses to which a given experimental setup is sensitive.

Two pilot experiments of this sort were completed several years ago (Wuensch et al. 1989; Hagmann et al. 1990). Assuming an axionic darkmatter density at the Earth of 5×10^{-25} g cm^{-3} = 300 MeV cm^{-3} allowed these groups to exclude the range of masses and coupling constants shown in Fig. 5.7. The solid line indicates the relationship between $g_{a\gamma}$ and m_a in axion models where $E/N = 8/3$ or $\xi = 1$ in Eq. (14.24), i.e. where $g_{a\gamma} = (m_a/\mu\text{eV}) (0.69 \times 10^{16} \text{ GeV})^{-1}$.

Most excitingly, the galactic axion search is going to be taken up again with two new experimental setups. The one in Livermore (California) has an increased detection volume and magnetic field ($B^2V = 14\,\text{T}^2\text{m}^3$), and a refined microwave detection method (van Bibber et al. 1992, 1994). Within a running time of two or three years it will be possible to explore the axion mass range $1.3 - 13\,\mu\text{eV}$ down to a coupling strength $g_{a\gamma}$ which is only a factor of about 2.5 shy of the "axion line" in Fig. 5.7. If axions interact with photons somewhat stronger than indicated by this line, or if the local dark-matter density is somewhat larger than assumed in Fig. 5.7, one may already be able to detect axions in this round of measurements.

The second experiment (Kyoto, Japan) will use Rydberg atoms in a novel scheme to detect single microwave quanta (Matsuki et al. 1995). Because of the intrinsic low noise of this detector one can go to lower physical temperatures, thereby reducing thermal noise, and thus allowing one to use smaller cavities. This in turn permits one to search for larger axion masses than is possible with the Livermore-type large cavities. Together, the two experiments can probably cover two decades of axion masses, between about 10^{-6} and $10^{-4}\,\text{eV}$.

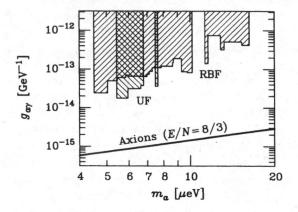

Fig. 5.7. Results of the galactic axion search experiments of the Rochester-Brookhaven-Fermilab (RBF) collaboration (Wuensch et al. 1989) and of the University of Florida (UF) experiment (Hagmann et al. 1990). The hatched areas are excluded, assuming a local dark-matter axion density of $5 \times 10^{-25}\,\text{g cm}^{-3} = 300\,\text{MeV cm}^{-3}$. The "axion line" is the relationship between axion mass and coupling strength for $\xi = 1$ or $E/N = 8/3$ according to Eq. (14.24).

5.4 Axion-Photon Oscillations

5.4.1 Mixing Equations

The "axion haloscope" discussed in the previous section was based on the Primakoff conversion between axions and photons in a macroscopic magnetic or electric field. The same idea can be applied to other axion fluxes such as that expected from the Sun ("axion helioscope," Sikivie 1983). Typical energies of solar axions are in the keV range (Sect. 5.2.4) so that a macroscopic laboratory magnetic field is entirely homogeneous on the scale of the axion wavelength.

In this case the conversion process is best formulated in a way analogous to neutrino oscillations (Anselm 1988; Raffelt and Stodolsky 1988). This approach may seem surprising as the axion has spin zero while the photon is a spin-1 particle. States of different spin-parity can mix, however, if the mixing agent (here the external magnetic field) matches the missing quantum numbers. Therefore, only a transverse magnetic or electric field can mix a photon with an axion; a longitudinal field respects azimuthal symmetry whence it cannot mediate transitions between states of different angular momentum components in the field direction.

The starting point for the magnetically induced mixing between axions and photons is the classical equation of motion for the system of electromagnetic fields and axions in the presence of the interaction Eq. (5.1). In terms of the electromagnetic field-strength tensor F and its dual $\tilde{F}$ one finds, apart from the constraint $\partial_\mu \tilde{F}^{\mu\nu} = 0$,

$$\partial_\mu F^{\mu\nu} = J^\nu + g_{a\gamma} \tilde{F}^{\mu\nu} \partial_\mu a,$$
$$(\Box + m_a^2)\, a = -\tfrac{1}{4}\, g_{a\gamma}\, F_{\mu\nu} \tilde{F}^{\mu\nu}, \qquad\qquad (5.24)$$

where J is the electromagnetic current density.

In a physical situation with a strong external field plus radiation one may approximate $g_{a\gamma} \tilde{F}^{\mu\nu} \partial_\mu a \to g_{a\gamma} \tilde{F}^{\mu\nu}_{\text{ext}} \partial_\mu a$ because a term $g_{a\gamma} \tilde{F}^{\mu\nu}_{\text{rad}} \partial_\mu a$ is of second order in the weak radiation fields. Moreover, if only an external magnetic field is present, the wave equation for the time-varying part of the vector potential $\mathbf{A}$ and for the axion field are

$$\Box \mathbf{A} = g_{a\gamma} \mathbf{B}_{\text{T}} \partial_t a,$$
$$(\Box - m_a^2)\, a = -g_{a\gamma} \mathbf{B}_{\text{T}} \cdot \partial_t \mathbf{A}, \qquad\qquad (5.25)$$

where $\mathbf{B}_{\text{T}}$ is the transverse external magnetic field. If one specializes to a wave of frequency ω propagating in the z-direction, and denoting the components of **A** parallel and perpendicular to $\mathbf{B_T}$ with $A_\parallel$ and $A_\perp$, respectively, one finds (Raffelt and Stodolsky 1988)

$$\left[\omega^2 + \partial_z^2 + 2\omega^2 \begin{pmatrix} n_\perp - 1 & n_R & 0 \\ n_R & n_\parallel - 1 & g_{a\gamma}B_T/2\omega \\ 0 & g_{a\gamma}B_T/2\omega & -m_a^2/2\omega^2 \end{pmatrix}\right] \begin{pmatrix} A_\perp \\ A_\parallel \\ a \end{pmatrix} = 0,$$

(5.26)

where the off-diagonal terms were made real by a suitable global transformation of the fields. Further, a photon index of refraction was included because in practice one never has a perfect vacuum.

The refractive index is generally different for the two linear polarization states parallel and perpendicular to $\mathbf{B_T}$ (Cotton-Mouton effect). Also, there may be mixing between the $A_\perp$ and $A_\parallel$ fields, i.e. the plane of polarization may rotate in optically active media, an effect characterized by n_R. In general, any medium becomes optically active if there is a magnetic field component along the direction of propagation (Faraday effect). Therefore, in general the refractive indices $n_{\perp,\parallel}$ depend on the transverse, the index n_R on the longitudinal magnetic field.

Equation (5.26) is made linear by an approach that will be discussed in more detail for neutrinos in Sect. 8.2. For propagation in the positive z-direction and for very relativistic axions and photons one may expand $(\omega^2 + \partial_z^2) = (\omega + i\partial_z)(\omega - i\partial_z) \rightarrow 2\omega\,(\omega - i\partial_z)$. Then one obtains the usual "Schrödinger equation"

$$\left[\omega + \begin{pmatrix} \Delta_\perp & \Delta_R & 0 \\ \Delta_R & \Delta_\parallel & \Delta_{a\gamma} \\ 0 & \Delta_{a\gamma} & \Delta_a \end{pmatrix} + i\partial_z\right] \begin{pmatrix} A_\perp \\ A_\parallel \\ a \end{pmatrix} = 0,$$

(5.27)

where $\Delta_{\parallel,\perp} = (n_{\parallel,\perp} - 1)\,\omega$, $\Delta_R = n_R\omega$, $\Delta_a = -m_a^2/2\omega$, and $\Delta_{a\gamma} = \frac{1}{2}g_{a\gamma}B_T$.

If one ignores a possible optical activity or Faraday effect ($n_R = 0$), the lower part of this equation represents a 2×2 mixing problem. The matrix is made diagonal by a rotation about an angle

$$\frac{1}{2}\tan 2\theta = \frac{\Delta_{a\gamma}}{\Delta_\parallel - \Delta_a} = \frac{g_{a\gamma}B_T\omega}{(n_\parallel - 1)2\omega^2 + m_a^2}.$$

(5.28)

In analogy to neutrino oscillations (Sect. 8.2.2) the probability for an axion to convert into a photon after travelling a distance ℓ in a transverse magnetic field is

$$\text{prob}(a \rightarrow \gamma) = \sin^2(2\theta)\sin^2(\tfrac{1}{2}\Delta_{\text{osc}}\ell),$$

(5.29)

where $\Delta_{\text{osc}}^2 = (\Delta_\parallel - \Delta_a)^2 + \Delta_{a\gamma}^2$ so that the oscillation length is $\ell_{\text{osc}} = 2\pi/\Delta_{\text{osc}}$.

By adjusting the gas pressure within the magnetic field volume one can make the photon and axion degenerate and thus enhance the transition rate (van Bibber et al. 1989). This applies, in particular, to solar axions which have keV energies so that the corresponding photon dispersion relation in low-Z gases is "particle-like" with the plasma frequency being the effective mass.

If there is a gradient of the gas density, for example near a star, or if the gas density and magnetic field strength change in time as in the expanding universe, suitable conditions allow for resonant axion-photon conversions in the spirit of the neutrino MSW effect (Yoshimura 1988; Yanagida and Yoshimura 1988).

For the magnetic conversion of pseudoscalars in the galactic magnetic field one must worry about density fluctuations of the interstellar medium which can be of order the medium density itself. In this case Eq. (5.29) is no longer valid because it was based on the assumption of spatial homogeneity of all relevant quantities. Carlson and Garretson (1994) have derived an expression for the conversion rate in a medium with large random density variations. They found that it can be significantly suppressed relative to the naive result.

5.4.2 Solar Axions

An axion helioscope experiment was performed by Lazarus et al. (1992) who used a vacuum pipe of 6″ diameter which was placed in the bore of a dipole magnet of 72″ length (1.80 m); the field strength was 2.2 T. The helioscope was oriented so that its long axis pointed along the azimuth of the setting Sun. This provided a time window of approximately 15 min every day during which the line of sight through the vacuum region pointed directly to the Sun. As a detector they used an x-ray proportional chamber at the end of the pipe.

Data were taken on several days with He at different pressures in the pipe. At 1 atm helium provides a plasma mass of about 0.3 keV to x-rays. For vacuum, a 3σ bound of $g_{a\gamma} < 3.6 \times 10^{-9}\,\mathrm{GeV}^{-1}$ for $m_a < 0.050\,\mathrm{eV}$ was found. For a helium pressure of 55 Torr the limit was 3.9 in the same units, applicable to $0.050 < m_a/\mathrm{eV} < 0.086$, and for 100 Torr it was 3.4, applicable to $0.086 < m_a/\mathrm{eV} < 0.110$. These bounds assume an axion flux as given by Eq. (5.21) which in turn assumes an unperturbed Sun. Unfortunately, this assumption is not consistent because the present-day age of the Sun already requires the bound Eq. (5.22).

However, an ongoing experimental project at the Institute for Nuclear Physics in Novosibirsk may be able to improve the helioscope significantly. The conversion magnet has been gimballed so that it can track the Sun, providing much longer exposure times. First results can be expected for late 1995—see Vorobyov and Kolokolov (1995) for a status report.

Another possibility would be to use the straight sections of the beam pipe of the LEP accelerator at CERN as an axion helioscope. Hoogeveen and Stuart (1992) have calculated the times and dates of alignment with the Sun. They proposed an experimental setup that might allow one to reach a sensitivity in $g_{a\gamma}$ down to $4 \times 10^{-10} \, \text{GeV}^{-1}$, which would be very impressive, but still far from the globular-cluster bound Eq. (5.23).

Finally, Paschos and Zioutas (1994) proposed to use a single crystal as a detector where the Primakoff conversion of solar axions is coherently enhanced over the electric fields of many atoms. Put another way, one would expect a strong enhancement via Bragg scattering. Even with this improvement, however, it does not seem possible to beat the bound from globular-cluster stars.

5.4.3   Shining Light through Walls

Instead of using the solar axion flux one can make one's own by shining a laser beam through a long transverse magnetic field region where it develops an axion component. Then the laser beam is blocked while the weakly interacting axions traverse the obstacle. In a second magnet they are back-converted into photons so that one "shines light through walls" (Anselm 1985; Gasperini 1987; van Bibber et al. 1987). Instead of a freely propagating beam one may use resonant cavities on either side of the wall which are coupled by the axion field (Hoogeveen and Ziegenhagen 1991). Another possibility to improve the sensitivity is to use squeezed light (Hoogeveen 1990).

An actual experiment was performed by Ruoso et al. (1992) who used two superconducting magnets of length 440 cm each with a field strength of 3.7 T. The light beam was trapped in a resonant cavity in the first magnet, allowing for about 200 traversals; the incident laser power was 1.5 W. At the end of the second magnet photons were searched for by a photomultiplier. For an axion mass $m_a \lesssim 10^{-3} \, \text{eV}$ an upper bound $g_{a\gamma} < 0.7 \times 10^{-6} \, \text{GeV}^{-1}$ was found.

5.4.4   Vacuum Birefringence

Besides the conversion between photons and axions there is a more subtle effect that can serve to search for the two-photon vertex of low-mass pseudoscalars. In an external transverse **E** or **B** field the mixing between the $A_\parallel$ component with a leads to a backreaction on $A_\parallel$ (Fig. 5.8b) which amounts to a retardation of its phase. Put another way, the $\parallel$ and $\perp$ polarization states have different refractive indices in vacuum with an **E** or **B** field (vacuum Cotton-Mouton effect). Therefore, if one shines a light beam which is linearly polarized at 45° relative to a transverse **B** field, the beam will develop a small degree of elliptic polarization (Maiani, Petronzio, and Zavattini 1986).

A vacuum Cotton-Mouton effect is expected even in the absence of axions from the QED amplitude shown in Fig. 5.8a; in the nonforward direction it describes Delbrück scattering on a charged particle. More generally, an electron loop mediates an effective $\gamma\gamma$ interaction which for low energies can be described by the Euler-Heisenberg Lagrangian,

$$\mathcal{L}_{\gamma\gamma} = \frac{2\alpha^2}{45m_e^4}\Big[(\mathbf{E}^2 - \mathbf{B}^2)^2 + 7(\mathbf{E}\cdot\mathbf{B})^2\Big] \tag{5.30}$$

(Heisenberg and Euler 1936; see also Itzykson and Zuber 1983).

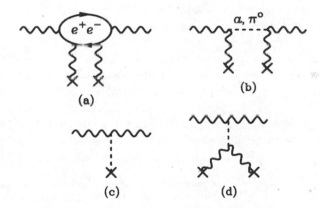

Fig. 5.8. Vacuum birefringence in the presence of external fields. (a) QED contribution according to the Euler-Heisenberg $\gamma\gamma$ interaction. (b) γa or $\gamma\pi^\circ$ oscillations in an external **E** or **B** field. (c) Photon birefringence in an external axion field (axionic domain walls, cosmic axion field). (d) Axion-mediated contribution in a strong **E** · **B** field, e.g. near a pulsar.

The vacuum refractive index in an external magnetic field which results from this interaction was studied by a number of authors—see Tsai and Erber (1975, 1976) for references to the early literature. Adler (1971) provided a comprehensive study and derived the correct expression for the related photon splitting rate $\gamma \to \gamma\gamma$ in an external field.[28] The refractive indices for the $\parallel$ and $\perp$ polarization states are[29]

$$n_\parallel = 1 + 7\,\frac{2\alpha^2}{45}\,\frac{B^2}{m_e^4} \quad\text{and}\quad n_\perp = 1 + 4\,\frac{2\alpha^2}{45}\,\frac{B^2}{m_e^4}, \tag{5.31}$$

where

$$(2\alpha^2/45)\,B^2/m_e^4 = 1.32\times10^{-32}\,(B/\text{Gauss})^2. \tag{5.32}$$

Note that 1 Gauss corresponds to 10^{-4} Tesla, and to 1.95×10^{-2} eV^2 in natural units (Appendix A).

Clearly one needs very strong magnetic fields for vacuum birefringence effects to become important. So far, no positive experimental measurement exists. A proposal to measure the acquired elliptic polarization of a laser beam was put forth by Iacopini and Zavattini (1979). More recently Cantatore et al. (1991) proposed to use polarized light scattered off an electron beam, a method which allows one to obtain polarized GeV photons. As the relative phase shift is $(n_\parallel - n_\perp)\omega\ell$ for a distance of travel ℓ in the magnetic field, high-energy photons show a much stronger effect for otherwise equal conditions.

An experiment to search for the axion contribution of Fig. 5.8b was recently performed (Semertzidis et al. 1990; Cameron et al. 1993). Note that there are *two* axion-induced effects on a laser beam trapped in an optical cavity. One is the birefringence effect analogous to the QED effect which leads to a small amount of elliptical polarization. Another is the loss of $\parallel$ photons into the axion channel which depletes the amplitude of the $\parallel$ mode relative to the $\perp$ one, which in turn leads to a rotation of the plane of polarization. Both effects are of the same order in the coupling constant; experimentally, the rotation effect led to

[28] The photon-splitting box graph with one external field and three real photons attached to an electron loop does not contribute. The lowest-order amplitude is with the external field attached three times, and three real photons (hexagon diagram). For references to the early literature and a discussion of the astrophysical implications of the photon-splitting process see Baring (1991).

[29] $\parallel$, $\perp$ refer to the electric field of the wave relative to the external transverse **B** field while Adler (1971) refers with $\parallel$, $\perp$ to the magnetic field of the wave. Note also that I use rationalized units where $\alpha = e^2/4\pi = 1/137$ while in the literature on photon refraction unrationalized units with $\alpha = e^2 = 1/137$ are often employed.

a more restrictive limit of $g_{a\gamma} < 3.6 \times 10^{-7}\,\mathrm{GeV}^{-1}$ for $m_a \lesssim 7 \times 10^{-4}\,\mathrm{eV}$. For a larger mass an a-γ oscillation pattern develops on the length scale of the optical cavity, leading to an "oscillating limit" as a function of m_a. In this regime the ellipticity measurement was superior.

New experimental efforts in the birefringence category include a proposal by Cooper and Stedman (1995) to use ring lasers. A laser experiment which is actually in the process of being built is PVLAS (Bakalov et al. 1994) which will be able to improve previous laboratory limits on $g_{a\gamma}$ by a factor of 40, i.e. it is expected to be sensitive in the regime $g_{a\gamma} \gtrsim 1 \times 10^{-8}\,\mathrm{GeV}^{-1}$ as long as $m_a \lesssim 10^{-3}\,\mathrm{eV}$. While such strong couplings are astrophysically excluded it is intriguing that this experiment should be able to detect for the first time the standard QED birefringence effect of Fig. 5.8a.

5.5 Astrophysical Magnetic Fields

5.5.1 Transitions in Magnetic Fields of Stars

Certain stars have very strong magnetic fields. For example, neutron stars frequently have fields of 10^{12}–10^{13} G (e.g. Mészáros 1992), and even white dwarfs can have fields of up to 10^9 G. Therefore, one may think that axions produced in the hot interior of neutron stars at a temperature of, say, 50 keV would convert to γ-rays in the magnetosphere (Morris 1986). However, the vacuum refractive term suppresses the conversion rate because the photon momentum for a given frequency is $k_\gamma - n_{\|,\perp}\omega > \omega$ with the refractive indices Eq. (5.31) while for the axions $k_a = \omega - m_a^2/2\omega < \omega$. Therefore, in the presence of a magnetic field axions and photons are less degenerate so that it is more difficult for them to oscillate into each other.

In principle, the refractive index can be cancelled by the presence of a plasma where the photon forward scattering on electrons induces a negative $n-1$, i.e. something like a photon effective mass. In the aligned rotator model for a magnetized neutron star a self-consistent solution of the Maxwell equations with currents requires the presence of an electron density of about $n_e = 7 \times 10^{10}\,\mathrm{cm}^{-3}\,B_{12}P_\mathrm{s}^{-1}$ where B_{12} is the magnetic field along the rotation axis in units of 10^{12} G and P_s is the pulsar period in seconds (Goldreich and Julian 1969). The corresponding plasma frequency is $\omega_\mathrm{P}^2 = 4\pi\alpha n_e/m_e = 0.97 \times 10^{-10}\,\mathrm{eV}^2\,B_{12}P_\mathrm{s}^{-1}$. This implies $k - \omega = -\omega_\mathrm{P}^2/2\omega = -5 \times 10^{-14}\,\mathrm{eV}\,B_{12}P_\mathrm{s}^{-1}\,\omega_\mathrm{keV}^{-1}$ with $\omega_\mathrm{keV} = \omega/\mathrm{keV}$, to be compared with $\Delta_\| = (n_\| - 1)\,\omega = 0.92 \times 10^{-4}\,\mathrm{eV}\,B_{12}^2\omega_\mathrm{keV}$ which is much larger, allowing one to ignore the plasma term.

The off-diagonal term in the mixing matrix Eq. (5.27) is $\Delta_{a\gamma} = \frac{1}{2}g_{a\gamma}B = 0.98\times10^{-9}\,\mathrm{eV}\,g_{10}\,B_{12}$ with $g_{10} = g_{a\gamma}/(10^{-10}\,\mathrm{GeV}^{-1})$. Ignoring m_a the oscillation length is $2\pi\,(\Delta_{\parallel}^2 + \Delta_{a\gamma}^2)^{-1/2} \approx 1.3\,\mathrm{cm}\,B_{12}^{-2}\,\omega_{\mathrm{keV}}^{-1}$ while the geometric dimension of the dipole field is of order the stellar radius, i.e. of order 10 km. Therefore, many oscillations occur within the magnetosphere, and the average transition probability between photons and axions is θ^2 with the mixing angle $\theta \approx \Delta_{a\gamma}/\Delta_{\parallel} = 1.1\times10^{-5}\,g_{10}B_{12}^{-1}\omega_{\mathrm{keV}}^{-1}$. Therefore, the transition rate is very small.

The vacuum refractive index is larger than unity, the plasma contribution less than unity, and so near the stellar surface a crossover must occur where axions and photons are degenerate. However, the length scales do not work out to have a resonant MSW-type transition (Raffelt and Stodolsky 1988; Yoshimura 1988).

At a pulsar, reducing B increases the mixing angle and thus the transition rate while the oscillation length becomes larger. When ℓ_{osc} far exceeds the geometric dimension R of the stellar magnetosphere, and when the mixing angle is small, one can expand the sine functions in Eq. (5.29) so that the transition probability is $(\theta\Delta_{\mathrm{osc}}R)^2 \approx (\Delta_{a\gamma}R)^2$. This transition rate scales with B^2 as expected and becomes smaller for smaller B. Therefore, the optimal situation is when the magnetic field strength and geometric dimensions are matched such that $R \approx \ell_{\mathrm{osc}}$. This condition is approximately met in magnetic white dwarfs with, say, $B = 10^9\,\mathrm{G}$, $\omega = 10\,\mathrm{eV}$, and $R = 10^3\,\mathrm{km}$. In these systems one may even expect a resonant level crossing if they have a dilute atmosphere with an appropriate scale height (Raffelt and Stodolsky 1988; Gnedin and Krasnikov 1992). However, a fortuitous combination of particle and white-dwarf parameters is required, and, even then, *observable* effects apparently have not been proposed.

Most recently, Carlson and Tseng (1995) have performed a study of the conversion of very low-mass pseudoscalars in the magnetic field of sunspots. They find that for certain parameters the x-ray flux from the conversion process could be observable in solar x-ray telescopes such as SXT and Yohkoh.

5.5.2 Birefringence in a Pulsar Magnetosphere

In the previous section it was shown that near a pulsar the QED vacuum Cotton-Mouton effect (Fig. 5.7a) induces a sizeable amount of birefringence between the photon states which are linearly polarized parallel or perpendicular to the transverse component of the magnetic field. Recently, Mohanty and Nayak (1993) showed that in addition there can be a strong and potentially observable *circular* birefringence effect along the polar direction if massless pseudoscalars exist. The masslessness of the pseudoscalars is crucial for this scenario and so axions (Chapter 14) which generically must have a mass do not fulfill this requirement. Therefore, I use "arions" as a generic example which are like axions in all respects except that they are true Nambu-Goldstone bosons of a global chiral $U(1)$ symmetry and thus strictly massless (Anselm and Uraltsev 1982a,b; Anselm 1982).

The main idea of the pulsar birefringence scenario is that in the oblique rotator model a strong $\mathbf{E} \cdot \mathbf{B}$ density exists in the pulsar magnetosphere which serves as a source for the arion field. Therefore, a pulsar would be surrounded by a strong classical arion field density which constitutes an optically active "medium," causing a time delay between the two circular polarization states of the pulsed radio emission from the polar cap region. As a Feynman graph, this situation is represented by Fig. 5.8d.

In detail, Mohanty and Nayak (1993) considered the oblique rotator model where the pulsar magnetic dipole axis is tilted with regard to its rotation axis by an angle α. The instantaneous rotating magnetic dipole field is $\mathbf{B} = (B_0 R^3/r^3)\,[3\hat{\mathbf{r}}\,(\hat{\mathbf{r}} \cdot \hat{\boldsymbol{\mu}}) - \hat{\boldsymbol{\mu}}]$, where B_0 is the magnetic field strength at the poles of the pulsar surface (radius R) and $\hat{\boldsymbol{\mu}}$ is the instantaneous magnetic dipole direction with the angle α relative to the angular velocity vector $\boldsymbol{\Omega}$. The time average of the electric field which is induced by the rotating magnetic dipole, and which matches the boundary condition that the electric field component parallel to the pulsar surface vanishes, is found to be $\langle \mathbf{E} \rangle = B_0 R^5 \Omega \cos\alpha\, r^{-4}\,[3\,(\sin^2\theta - \tfrac{2}{3})\hat{\mathbf{r}} - 2\sin\theta\cos\theta\,\hat{\boldsymbol{\theta}}]$ where θ is the polar angle relative to the rotation axis. The time-averaged value for the pseudoscalar field density is

$$\langle \mathbf{E} \cdot \mathbf{B} \rangle = -B_0^2 \Omega R^8 r^{-7} \cos\alpha\,\cos^3\theta. \tag{5.33}$$

It appears as a source for the arion field on the r.h.s. of Eq. (5.24). Taking account of the relativistic space-time metric outside of the pulsar, Mohanty and Nayak (1993) found for the resulting arion field

$$a = -g_{a\gamma}\frac{2}{575}\,\frac{B_0^2 R^8 \Omega \cos\alpha}{(G_N \mathcal{M})^3}\,\frac{\cos\theta}{r^2} + \mathcal{O}(r^{-3}), \tag{5.34}$$

where G_N is Newton's constant and $\mathcal{M}$ the pulsar mass. The entire magnetosphere contributes coherently to this result. If the pseudoscalars had a mass, only the density $\mathbf{E} \cdot \mathbf{B}$ within a distance of about

m_a^{-1} would effectively act as a source for the local a field and so it would be much smaller.

An inhomogeneous pseudoscalar field configuration represents an optically active medium (Fig. 5.8c) as was noted, for example, in the context of axionic domain wall configurations (Sikivie 1984). To lowest order the dispersion relation for left- and right-handed circularly polarized light is (e.g. Harari and Sikivie 1992)

$$k = \omega \pm \tfrac{1}{2} g_{a\gamma} \, \hat{\mathbf{k}} \cdot \nabla a, \qquad (5.35)$$

so that the momentum is shifted by a frequency-independent amount. The corresponding refractive index is $n = 1 \pm \tfrac{1}{2} g_{a\gamma} \, \omega^{-1} \hat{\mathbf{k}} \cdot \nabla a$.

Taking account of the relativistic metric in the strong gravitational field of a pulsar, Mohanty and Nayak (1993) then found for the time delay between circularly polarized waves which propagate approximately along the polar axis

$$\delta t = \frac{2}{5} \left(\frac{2}{575} \right)^2 g_{a\gamma}^4 \frac{B_0^4 R^{11} \Omega^2 \cos^4 \alpha}{\omega^2 \, (G_N M)^6}. \qquad (5.36)$$

For the pulsar PSR 1937+21 a polarimetric analysis yields a time delay $0.37 \pm 0.67 \, \mu s$ and thus a 1σ upper limit of $\delta t < 1 \mu s$ (Klein and Thorsett 1990). For typical pulsar parameters this allowed Mohanty and Nayak (1993) to place a limit on the arion-photon coupling of $g_{a\gamma} \lesssim 2 \times 10^{-11} \, \text{GeV}^{-1}$.

5.5.3 Conversion of Stellar Arions in the Galactic Field

Stars are powerful sources for pseudoscalars which can be produced in the hot interior by the Primakoff process, i.e. by the conversion $\gamma \to a$ in the electric fields of the charged medium constituents. Outside of the star, the pseudoscalars can be converted back to photons in the galactic magnetic field so that stars would appear to be sources of x- or γ-rays, depending on the characteristic energy of the stellar core (Carlson 1995).

In the galaxy, photons propagate with an effective mass given by the plasma frequency ω_P which for typical electron densities of order $0.1 \, \text{cm}^{-3}$ is of order $10^{-11} \, \text{eV}$. As discussed in Sect. 5.4.1 the pseudoscalar to photon conversion process is an oscillation phenomenon with a mixing angle given by Eq. (5.28); in the present context it is

$$\tfrac{1}{2} \tan 2\theta = \frac{g_{a\gamma} B_T \omega}{m_a^2 - \omega_P^2}. \qquad (5.37)$$

A typical galactic field strength is $1 \, \mu G$, a typical energy at most of order $100 \, \text{MeV}$ for axions from supernovae. With $g_{a\gamma} < 0.6 \times 10^{-10} \, \text{GeV}^{-1}$

one finds $g_{a\gamma} B_T \omega \lesssim 10^{-19} \,\text{eV}^2$. For the allowed range of axion masses this mixing angle is too small to yield a significant conversion effect.

Therefore, this entire line of argument is only relevant for massless (or at least very low-mass) pseudoscalars which again shall be referred to as "arions." Carlson (1995) considered the star α-Ori (Betelgeuse), a red supergiant about 100 pc away from us. He estimated its arion luminosity from the Primakoff process, and compared the expected x-ray flux with data from the HEAO-1 satellite. As a result, a new limit of $g_{a\gamma} \lesssim 2.5 \times 10^{-11} \,\text{GeV}^{-1}$ emerged which is more restrictive than the above bound from globular-cluster stars.

Carlson's argument yields an even more restrictive limit if applied to SN 1987A. One may estimate the arion luminosity of the SN core on the basis of the Primakoff process. If arions couple to quarks or electrons, the luminosity can only be higher because existing axion limits already indicate that arions cannot be trapped by these couplings. In order to evaluate Eq. (5.9) an average temperature of 30 MeV and an average density of $3 \times 10^{14} \,\text{g cm}^{-3}$ with a proton fraction of 0.3 is used (Sect. 13.4.2). The Debye screening scale by the protons is then found to be 36 MeV so that $\kappa^2 = 1.41$ in Eq. (5.10) leading to $F = 0.72$. Therefore, the average energy loss rate is about $g_{10}^2 \, 1.4 \times 10^{16} \,\text{erg g}^{-1}\text{s}^{-1}$. Taking a core mass of $1\mathcal{M}_\odot = 2 \times 10^{33} \,\text{g}$ and a duration of 3 s one expects about $g_{10}^2 \, 10^{50} \,\text{erg}$ to be emitted in arions which is about $10^{-3} g_{10}^2$ of the energy emitted in each neutrino flavor.

Typical arion energies are $3T \approx 100 \,\text{MeV}$ so that Eq. (5.37) together with $\omega_P \approx 10^{-11} \,\text{eV}$ in the interstellar medium reveals that mixing is nearly maximal for the relevant circumstances. Therefore, the oscillation length is given by $\ell_{\text{osc}} = 4\pi/g_{a\gamma} B_T$ which is about 40 kpc for $B_T = 1 \,\mu\text{G}$ and $g_{a\gamma} = 10^{-10} \,\text{GeV}^{-1}$. Therefore, ℓ_{osc} far exceeds the relevant magnetic field region which is of order 1 kpc as discussed in Sect. 13.3.3b. The conversion rate is then

$$\text{prob}(a \to \gamma) = (\tfrac{1}{2} g_{a\gamma} B_T \ell)^2$$

$$= 2.3 \times 10^{-2} \, g_{10}^2 \, (B_T \ell/\mu\text{G kpc})^2, \qquad (5.38)$$

where ℓ is the effective conversion region (distance to source or distance within magnetic field region), and $g_{10} \equiv g_{a\gamma}/10^{-10} \,\text{GeV}^{-1}$.

With Eq. (5.38) and an effective magnetic conversion region of $\ell = 1 \,\text{kpc}$ the expected energy showing up as γ-rays at Earth corresponds to about $10^{-5} g_{10}^4$ relative to the energy in one neutrino species. In Sect. 12.4.3 the radiative decays of low-mass neutrinos from SN 1987A was discussed. On the basis of the SMM data it was found that less than about 10^{-9} of a given neutrino species may show up in the form of decay photons (Fig. 12.9) if the spectral distribution is taken to be characterized by $T \approx 30\,\mathrm{MeV}$. Therefore, one finds a limit of $g_{10}^4 \lesssim 10^{-4}$ or $g_{a\gamma} \lesssim 10^{-11}\,\mathrm{GeV}^{-1}$, applicable if the particle mass is below about $10^{-10}\,\mathrm{eV}$. This is more restrictive than Carlson's original limit, and of the same order as the PSR 1937+21 birefringence limit quoted after Eq. (5.36).

5.5.4   Polarimetry of Distant Radio Sources

Nambu-Goldstone bosons a are by definition the result of a spontaneously broken global symmetry. The cosmic evolution from a very hot initial phase begins with the unbroken symmetry; as the universe expands and cools a phase transition will occur where the field responsible for the spontaneous breakdown must find its new minimum. As this process occurs independently in each causally connected region of the universe at that time, the universe today will be characterized by different orientations of the ground state, i.e. by different values of a classical background a field. If no inflation occurred in the universe after the phase transition, and if the Nambu-Goldstone bosons remain truly massless (in contrast with axions), the background field will not have relaxed to a common ground state everywhere.

In this scenario a radio signal from a distant source travels through regions with different values of the classical a field, and thus through regions of gradients ∇a which act as an optically active medium according to Eq. (5.35). Therefore, linearly polarized light will experience a random rotation of its plane of polarization.

This effect is also expected from the Faraday rotation caused by intervening magnetic fields which induce optical activity in the cosmic background plasma. As this effect is frequency dependent it can be removed by observing a given object at different wavelengths. The effect induced by pseudoscalars, on the other hand, is independent of frequency.

A systematic correlation between the geometric shape of distant radio sources and the linear polarization of the emitted radiation has been observed. This correlation proves that no random rotation of the plane of polarization occurs over cosmic distances, except for the Faraday effect which can be removed from the data. Therefore, the maximum allowed coupling strength of photons to a random cosmic Nambu-Goldstone field can be constrained. Harari and Sikivie (1992) found that $C_{a\gamma} \lesssim 50$ in Eq. (5.5), independently of the symmetry breaking scale f_a. This bound supersedes Sikivie's (1988) previous scenario where he tried to explain the polarization features of certain sources by the conversion of cosmic-string-produced Nambu-Goldstone bosons to photons in cosmic magnetic fields. This scenario would have required $C_{a\gamma} \approx 10^5$.

5.5.5 Temperature Fluctuations in the Cosmic Microwave Background

In the presence of large-scale magnetic fields in the universe, photons of the cosmic microwave background radiation (CMBR) could convert into arions. The angular variations of the CMBR temperature have been measured by the COBE satellite and other instruments to be extremely small; a typical value is $\delta T/T \approx 10^{-5}$. Therefore, the conversion process must not have been very efficient between the surface of last scattering and us.

This argument has been studied in detail by Chen (1995) for photon-graviton conversion which is a very similar effect due to the two-photon coupling vertex which the massless gravitons must have. The coupling constant involves the inverse Planck mass. Therefore, one may also expect interesting effects for hypothetical arions which could couple to photons more strongly than gravitons do.

5.6 Summary of Constraints on $g_{a\gamma}$

The astrophysical and experimental bounds on the photon coupling of arbitrary pseudoscalars are summarized in Fig. 5.9. "Haloscope" refers to the search for galactic axions discussed in Sect. 5.3 and so these constraints (Fig. 5.7) apply only if the pseudoscalars are the dark matter in our galaxy. The dotted line is the search regime for the ongoing experiment mentioned in Sect. 5.3.

"Helioscope" refers to the search for solar axion to x-ray conversion (Sect. 5.4.2). It is shown as a dashed line because it is not self-consistent in that it assumes an unperturbed Sun—the area enclosed by the dashed line is already excluded by the solar age.

"Telescope" refers to the search for decay photons from the cosmic axion background (Sect. 12.7.2, Fig. 12.23). It is assumed that the pseudoscalars were in thermal equilibrium in the early universe.

"Laser" refers to the birefringence and shining-light-through-walls experiments discussed above. The most restrictive such limit is from the rotation of the plane of polarization of a laser beam trapped in an optical cavity in a strong transverse magnetic field (Sect. 5.4.4). The dotted line marks roughly the expected range of sensitivity of the PVLAS experiment (Sect. 5.4.4).

The solar limit (Eq. 5.22) is based on the Primakoff energy loss and the requirement that axions must not exceed the photon luminosity; otherwise the Sun could not have reached its present-day age.

The HB-star limit (Eq. 5.23) comes from the requirement that these objects do not spend their nuclear fuel so fast that their observable number in globular clusters is reduced by more than a factor of ≈ 2.

The "axion line" refers to models where $E/N = 8/3$ or $\xi = 1$ in Eq. (14.24).

For very low-mass bosons ($m_a \lesssim 10^{-10}\,\text{eV}$) the SN 1987A flux of pseudoscalars would be efficiently converted into γ-rays, leading to a limit $g_{a\gamma} \lesssim 10^{-11}\,\text{GeV}^{-1}$ (Sect. 5.5.3).

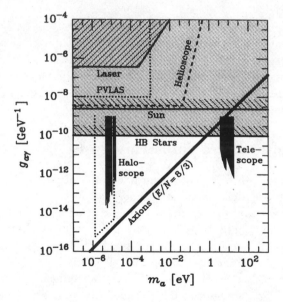

Fig. 5.9. Bounds on the photon coupling $g_{a\gamma}$ as a function of m_a for arbitrary pseudoscalars; see the text for details. (Adapted from Cameron et al. 1993.)

Chapter 6

Particle Dispersion and Decays in Media

Dispersion effects in media have a significant impact on the propagation of some low-mass particles (photons, neutrinos) while others are left unaffected (axions and other Nambu-Goldstone bosons). The relationship between forward scattering and refraction is derived, and the dispersion relations for photons and neutrinos are thoroughly studied. Modified particle dispersion relations allow certain decay processes to occur in media that cannot occur in vacuum, notably the photon decay $\gamma \to \nu\bar{\nu}$ which dominates the neutrino emissivity in a wide range of temperatures and densities ("plasma process"). Other examples are the neutrino and majoron decay $\nu \to \bar{\nu}\chi$ and $\chi \to \nu\bar{\nu}$, respectively. The rates for such processes are derived. The plasma process allows one to derive the most restrictive limits on neutrino magnetic dipole moments. Screening effects in reactions involving Coulomb scattering, and neutrino electromagnetic form factors in media are discussed.

6.1 Introduction

Particles are the quantized excitations of certain fields—photons of the electromagnetic field, electrons of the electron field, and so forth. It is usually convenient to expand these fields in plane waves characterized by frequencies ω and wave vectors $\mathbf{k}$; the excitations of these modes then exhibit a temporal and spatial behavior proportional to $e^{-i(\omega t - \mathbf{k} \cdot \mathbf{x})}$. The frequency for a given wave number is determined by the dispersion relation. Because $(\omega, \mathbf{k})$ is a four-vector, and because of Lorentz invariance, in vacuum the quantity $\omega^2 - \mathbf{k}^2 = m^2$ is the same for all frequencies; m has the usual interpretation of a particle mass. One consequence of this covariant dispersion relation is that decays of the sort $1 \to 2+3$ are only possible if $m_1 > m_2 + m_3$ so that in the rest frame of particle 1 there is enough energy available to produce the final states.

In media the dispersion relations are generally modified by the coherent interactions with the "background." In the simplest case a particle acquires a medium-induced effective mass. For example, photons in a nonrelativistic plasma acquire a dispersion relation $\omega^2 = \omega_P^2 + \mathbf{k}^2$ with the plasma frequency given by $\omega_P^2 = 4\pi\alpha\, n_e/m_e$ (electron density n_e). For $\omega_P > 2m_\nu$ this implies that the decay $\gamma \to \nu\bar{\nu}$ becomes kinematically possible and occurs in stars because the ambient electrons mediate an effective neutrino-photon interaction (Adams, Ruderman, and Woo 1963). In fact, this "plasma process" is the dominant neutrino source in a wide range of temperatures and densities which covers, for example, white dwarfs and red-giant stars (Appendices C and D).

Neutrinos may have nonstandard electromagnetic couplings, notably magnetic dipole moments, which would enhance the plasma process and thus the cooling of stars (Bernstein, Ruderman, and Feinberg 1963). Observational constraints on anomalous cooling rates derived from white dwarfs and globular-cluster stars then provide the most restrictive limits on neutrino electromagnetic couplings (Sect. 6.5.6).

Within the standard model all fermions are fundamentally massless; they acquire an effective mass by their interaction with the vacuum expectation value Φ_0 of a scalar Higgs field (Sect. 8.1.1). Therefore, even vacuum masses can be interpreted as "refractive" phenomena. Because the scalar Φ_0 is Lorentz invariant the dispersion relation thus induced is of the standard form $E^2 = m^2 + \mathbf{p}^2$. "Normal" media, however, single out a preferred Lorentz frame, usually causing $E(\mathbf{p})$ to be a more complicated function than $(m^2 + \mathbf{p}^2)^{1/2}$.

Notably, the dispersion relation can be such that the four-momentum $P = (E, \mathbf{p})$ is "space-like," $P^2 = E^2 - \mathbf{p}^2 < 0$, which amounts to a "negative mass-square" $P^2 = m_{\text{eff}}^2 < 0$. There is nothing wrong with such "tachyons" because the speed of signal propagation safely remains below the speed of light (Sect. 6.2.2). The dispersion relation in isotropic media is often expressed as $k = |\mathbf{k}| = n\omega$ in terms of a refractive index n. Space-like excitations correspond to $n > 1$; examples are photons in water or air. In this case the well-known decay process $e \to e\gamma$ is kinematically allowed for sufficiently fast moving electrons (Cherenkov radiation).

The dispersion relation can also depend on the spin polarization of the radiation. In "optically active" media, the left- and right-handed circular photon polarizations acquire different refractive indices. In this sense all media are optically active for neutrinos where only the left-handed states interact while the right-handed ones are "sterile." For Majorana neutrinos the helicity-plus states are equivalent to $\bar{\nu}$'s which acquire an opposite energy shift from ν so that there is an energy gap between $\nu(\mathbf{p})$ and $\bar{\nu}(\mathbf{p})$. Therefore, in a medium the majoron decays $\nu \to \bar{\nu}\chi$ or $\chi \to \nu\bar{\nu}$ become possible where the majoron χ is a massless particle (Sect. 6.8).

The interaction of ν_μ and ν_τ with a normal medium is different from that of ν_e because of a charged-current $\nu_e\text{-}e^-$ scattering amplitude. Therefore, normal media are "flavor birefringent" in the sense that the medium induces different dispersion relations for neutrinos of different flavors. The importance of this effect for neutrino oscillations, which effectively measure relative phases in the propagation of different-flavored neutrinos, cannot be overstated.

It must be stressed that usually *all* particles acquire nontrivial dispersion relations in media although it depends on the detailed circumstances whether or not the refractive effect is significant. For example, until recently one found statements in the literature that in a sufficiently dense medium where $\omega_{\mathrm{P}} > 2m_e$ photons were damped by electromagnetic pair production $\gamma \to e^+e^-$. However, this is incorrect because the charged leptons also acquire a medium-induced effective mass which is so large that this decay never occurs (Braaten 1991). On the other hand, the above majoron decay is only possible because the majorons χ are Nambu-Goldstone bosons and thus remain massless even in a medium, at least to lowest order (Sect. 6.8).

Besides modifying the dispersion relation of particles it is also possible that the presence of the medium allows for entirely new excitations. The best known example is the longitudinal polarization state of the electromagnetic field which exists in a plasma in addition to the usual states with transverse polarization. These "plasmons" were first discussed by Langmuir (1926). Another example from electromagnetism are the "plasminos," spin-$\frac{1}{2}$ excitations of a plasma that were discussed for the first time only very recently (Klimov 1981; Weldon 1982b, 1989; Pisarski 1989; Braaten 1992). For many purposes such (quantized) collective modes play the same role as the usual particles. For example, in a medium both photons and plasmons can decay into neutrinos and thus contribute to the plasma process of neutrino emission.

In the present chapter I will follow up these questions in detail. While the dispersion relations and couplings of particles in media are formally best dealt with in terms of field theory at finite temperature and density, most of the results relevant for particle physics in stars predate the development of this formalism; they were based on the old-fashioned tools of kinetic theory. Indeed, for simple issues of dispersion or collective effects a kinetic approach seems often physically more transparent while yielding identical results. At any rate, the following discussion is based entirely on kinetic theory.

6.2 Particle Dispersion in Media

6.2.1 Refractive Index and Forward Scattering

How does one go about to calculate the all-important dispersion relation for a given particle in a medium with known properties? Usually it is enough to follow the elementary approach of calculating the forward scattering amplitude of the relevant field excitations with the constituents of the background medium, an approach which has the added advantage of physical transparency over a more formal procedure.[30]

To begin, consider a scalar field Φ which may be viewed as representing one of the photon or electron polarization states. If a plane wave excitation of that field with a frequency ω and a wave vector $\mathbf{k}$ interacts with a scatterer at location $\mathbf{r} = 0$ an additional spherical wave will be created. The asymptotic form of the original plus scattered wave is

$$\Phi(\mathbf{r}, t) \propto e^{-i\omega t} \left(e^{i\mathbf{k}\cdot\mathbf{r}} + f(\omega, \theta) \frac{e^{ikr}}{r} \right) , \qquad (6.1)$$

where $k = |\mathbf{k}|$, $r = |\mathbf{r}|$, and f is the scattering amplitude. It was assumed that it has no azimuthal dependence, something that will always apply on average for a collection of randomly oriented scatterers. The differential scattering cross section is $d\sigma/d\Omega = |f(\omega, \theta)|^2$.

If there is a collection of scattering centers randomly distributed in space, all of the individual scattered waves will interfere. However, because of the random location of the scatterers, constructive and destructive interference terms will average to zero. Thus the total cross section of the ensemble is the (incoherent) sum of the individual ones.

In the forward direction, however, the scattered waves add up coherently with each other and with the parent wave, leading to a phase shift and thus to refraction. This is seen if one considers a plane wave in the z-direction incident on an infinitesimally thin slab (thickness δa) at

---

[30]The derivation below follows closely the exposition of Sakurai (1967).

$z = 0$ which contains n scattering centers per unit volume, and which is infinite in the x- and y-directions. At a distance z from the slab, large compared with k^{-1}, the asymptotic form of the parent plus scattered wave is, ignoring the temporal variation $e^{-i\omega t}$,

$$\Phi(z) \propto e^{i\omega z} + n\delta a \int_0^\infty \frac{e^{ik(\rho^2+z^2)^{1/2}}}{(\rho^2 + z^2)^{1/2}} f(\omega,\theta)\, 2\pi\rho\, d\rho, \tag{6.2}$$

where $\rho \equiv (x^2 + y^2)^{1/2}$ and $\theta = \arctan(\rho/z)$. Moreover, it was assumed that in vacuum the wave propagates relativistically so that $k = \omega$.

The integral in Eq. (6.2) is ill defined because the integrand oscillates with a finite amplitude even for large values of ρ. It is made convergent by substituting $k \to k + i\kappa$ with $\kappa > 0$ an infinitely small real parameter. Integration by parts then yields

$$\Phi(z) \propto e^{i\omega z} \left[1 + i\, \frac{2\pi n \delta a}{\omega}\, f_0(\omega) \right], \tag{6.3}$$

where a term of order $(\omega z)^{-1}$ was neglected which becomes small for large z. Here, $f_0(\omega) \equiv f(\omega, 0)$ is the *forward scattering amplitude*.

Turn next to a slab of finite thickness a. The phase change of the transmitted wave is obtained by compounding infinitesimal ones with $\delta a = a/j$ and taking the limit $j \to \infty$,

$$\lim_{j \to \infty} \left[1 + i\, \frac{2\pi n a}{j\omega}\, f_0(\omega) \right]^j = e^{i(2\pi n/\omega)f_0 a}. \tag{6.4}$$

Inserting this result in Eq. (6.3) reveals that over a distance a in the medium the wave accumulates a phase $e^{i n_{\mathrm{refr}}\omega a}$ where

$$n_{\mathrm{refr}} = 1 + \frac{2\pi}{\omega^2}\, n\, f_0(\omega) \tag{6.5}$$

is recognized as the index of refraction.

If the relativistic approximation $|n_{\mathrm{refr}} - 1| \ll 1$ is not valid one must treat the wave self-consistently in the medium and distinguish carefully between frequency and wavenumber. In this case one finds (Foldy 1945)

$$n_{\mathrm{refr}}^2 = 1 + \frac{4\pi}{k^2}\, n\, f_0(k), \tag{6.6}$$

where the forward scattering amplitude must be calculated taking the modified dispersion relation into account.

For the propagation of a field with several spin or flavor components the same result applies if one remembers that "forward scattering" not

only refers to scattering in the forward direction, but that all properties
of the wave and the scatterer are left unchanged. If the medium particles have a distribution of momenta, spins, etc. the forward scattering
amplitude must be averaged over those quantities, and different species
of medium particles must be summed over.

For a practical calculation it helps to recall that $d\sigma/d\Omega = |f(\theta)|^2$
so that $|f_0|$ is the square root of the forward differential cross section.
For example, the Thomson cross section for photons interacting with
nonrelativistic electrons is $d\sigma/d\Omega = (\alpha/m_e)^2\,|\epsilon \cdot \epsilon'|^2$ with the polariza-
tion vectors ϵ and ϵ' of the initial- and final-state photon. Forward
scattering implies $|\epsilon \cdot \epsilon'|^2 = 1$ so that $|f_0| = \alpha/m_e$. The dispersion rela-
tion is then $\omega^2 = k^2 + \omega_P^2$ with the plasma frequency $\omega_P^2 = 4\pi\alpha\,n_e/m_e$.
Of course, the absolute sign of f_0 has to be derived from some other
information—for photon dispersion see Sect. 6.3.

The forward scattering amplitude and the refractive index are generally complex numbers. Physically it is evident that in a medium the
intensity of a beam is depleted as $e^{-z/\ell}$. The mean free path is given by
$\ell^{-1} = \sigma n v$ where σ is the total scattering cross section, n is the number
density of scatterers, and v is the velocity of propagation. Thus the
amplitude of a plane wave varies as $e^{ikz-z/2\ell}$. Moreover, the derivation
of the refractive index indicates that the amplitude varies according to
$e^{in_{\mathrm{refr}}\omega z}$, yielding $k = \mathrm{Re}\,n_{\mathrm{refr}}\omega$ and $(2\ell)^{-1} = \mathrm{Im}\,n_{\mathrm{refr}}\omega$. For relativistic
propagation $(v = 1)$ the last equation implies $\sigma(\omega) = (4\pi/\omega)\,\mathrm{Im}\,f_0(\omega)$,
a relationship known as the optical theorem.

For the applications discussed in this book specific interaction models between the propagating particles and the medium will be assumed
so that it is usually straightforward to calculate the dispersion relation
according to Eq. (6.5). One should keep in mind, however, that n_{refr} as
a function of ω has a number of general properties, independently of the
interaction model. For example, its real and imaginary part are connected by the Kramers-Kronig relations (Sakurai 1967; Jackson 1975).

6.2.2   Particle Momentum and Velocity

The four-vector $(\omega, \mathbf{k})$ which governs the spatial and temporal behavior
of a plane wave can be time-like $(\omega^2 - \mathbf{k}^2 > 0)$ as for massive particles
in vacuum, it can be light-like $(\omega^2 - \mathbf{k}^2 = 0)$ as for photons in vacuum,
or it can be space-like $(\omega^2 - \mathbf{k}^2 < 0)$ as for visible light in water or air.
Because the quantized excitations of such field modes are interpreted
as particles, $E = \hbar\omega$ is the particle's energy. (I have temporarily restored $\hbar$ even though it is 1 in natural units.) Similarly one may be

tempted to interpret $\mathbf{p} = \hbar\mathbf{k}$ as the particle's momentum. In vacuum a particle's velocity is $\mathbf{p}/E$, a quantity which exceeds the speed of light for space-like excitations. Occasionally one reads in the literature that for this reason only those branches of a particle dispersion relation were physical where $|\mathbf{p}| < E$. Such statements are incorrect, however, and the underlying concern about tachyonic propagation is unfounded.

The quantity $\mathbf{p}/E$ has no general physical relevance. Two significant velocity definitions are the *phase velocity* and the *group velocity* of a wave (Jackson 1975). The former is the speed with which the crest of a plane wave propagates, i.e. it is given by the condition $\omega t - kz = 0$ or $v_{\text{phase}} = \omega/k = n_{\text{refr}}^{-1}$. For a massive particle in vacuum $v_{\text{phase}} > 1$. However, the phase velocity can drop below the speed of light in a medium. When this occurs for electromagnetic excitations in a plasma, electrons can "surf" in the wave which thus transfers energy at a rate proportional to the fine structure constant α (Landau 1946), an effect known as Landau damping. As long as $v_{\text{phase}} > 1$ the photon propagation is damped only by Thomson scattering which is an effect of order α^2.

The group velocity $v_{\text{group}} = d\omega/dk$ is the speed with which a wave packet or pulse propagates. In terms of the refractive index it is

$$v_{\text{group}}^{-1} = n_{\text{refr}}(\omega) + \omega\, dn_{\text{refr}}/d\omega \qquad (6.7)$$

(Jackson 1975). For a massive particle in vacuum with $\omega^2 = (k^2+m^2)^{1/2}$ it is $v_{\text{group}} - k/\omega < 1$, and also in a medium normally $v_{\text{group}} < 1$. Near a resonance it can happen that $v_{\text{group}} > 1$, but there is still no reason for alarm. The fast variation of $n_{\text{refr}}(\omega)$ as well as the presence of a large imaginary part near a resonance imply that the issue of signal propagation is much more complicated than indicated by the simple approximations which enter the definition of the group velocity. For a detailed discussion of electromagnetic signal propagation in dispersive media see Jackson (1975).

Evidently a naive interpretation of $\hbar\mathbf{k}$ as a particle momentum can be quite misleading. Another example relates to the difficulty of separating the momentum flow of a (light) beam in a medium into one part carried by the wave and one carried by the medium. There was a long-standing dispute in the literature with famous researchers on different sides of an argument that was eventually resolved by Peierls (1976); see also Gordon (1973). Experimentally, it was addressed by shining a laser beam vertically through a water-air interface and measuring the deformation of the surface due to the force which must occur because of a photon's change of momentum between the two media (Ashkin and Dziedzic 1973).

The problem of the physical momentum flow associated with a wave will be of no concern to the issues addressed in this book. In microscopic reactions the quantity which appears in the law of "energy-momentum conservation" is the wave vector. For example, in the plasma process $\gamma \to \nu\bar{\nu}$ the momenta of the outgoing neutrinos must balance against the wave vector of the decaying electromagnetic excitation. In this book dispersion effects will be important only for pulse propagation from distant sources, for particle oscillation effects, and for energy-momentum conservation in microscopic reactions. In these cases the naive interpretation of $\hbar\mathbf{k}$ as a particle's momentum is safe. For the remainder of this book the wave number (or pseudomomentum) and the momentum of a field excitation will not be distinguished.

6.2.3   Wave-Function Renormalization

In particle reactions the main impact of medium-induced modifications of the dispersion relations is on the kinematics, notably if a threshold condition is involved. One is thus tempted to proceed with the usual Feynman rules and take account of the dispersion relations only in the phase-space integration, notably in the law of energy-momentum conservation. In most practical cases this approach causes no problems, although an exception are interactions involving longitudinal plasmons (Sect. 6.3). Therefore, one should be aware that the matrix element also must be modified because of the subtle issue of what one means with a "particle" in a medium.

After a spatial Fourier transform the equations of motion for the Fourier components $\phi_{\mathbf{k}}$ of a free field are those of a harmonic oscillator. Interpreting the amplitude $\phi_{\mathbf{k}}$ and its velocity $\dot{\phi}_{\mathbf{k}}$ as conjugate variables, the canonical quantization procedure leads to quantized energy levels $\hbar\omega_{\mathbf{k}}$, where $\omega_{\mathbf{k}}$ is the classical frequency of the mode $\mathbf{k}$ according to its dispersion relation. Conversely, a quantized excitation with energy $\hbar\omega_{\mathbf{k}}$ has a certain field strength which determines its coupling strength to a source, e.g. the coupling strength of a photon to an electron.

In a medium, the energy associated with a frequency $\omega_{\mathbf{k}}$ is still $\hbar\omega_{\mathbf{k}}$. However, because of the presence of interaction energy between ϕ and the medium, the field strength associated with a quantized excitation is modified. For example, photons with a given frequency couple with a different strength to electrons in a medium than they do in vacuum. This modification can be lumped into a "renormalization factor" $\sqrt{Z}$ of the coupling strength of external photon lines in a Feynman graph.

In order to determine this factor from the dispersion relation consider a scalar field ϕ in the presence of a medium which induces a refractive index. This means that the Klein-Gordon equation in Fourier space, including a source term ρ, is of the form

$$[-K^2 + \Pi(K)]\phi(K) = g\rho(K), \qquad (6.8)$$

where $K = (\omega, \mathbf{k})$ is a four-vector in Fourier space and g is a coupling constant. $\Pi(K)$ is the "self-energy" which includes a possible vacuum mass m^2 and medium-induced contributions which are calculated from the forward scattering amplitude. The homogeneous equation with $\rho = 0$ has nonvanishing solutions only for $K^2 = \Pi(K)$ which defines the dispersion relation

$$\omega_{\mathbf{k}}^2 - \mathbf{k}^2 = \Pi(\omega_{\mathbf{k}}, \mathbf{k}). \qquad (6.9)$$

This equation determines implicitly the frequency $\omega_{\mathbf{k}}$ related to a wave number $\mathbf{k}$ for a freely propagating mode.

A problem with Eq. (6.8) is the general dependence of Π on ω and $\mathbf{k}$ which implies "dispersion," i.e. in coordinate space it is not a simple second-order differential equation. Otherwise the equation of motion for a single field mode $\phi_{\mathbf{k}}$ would be $(\partial_t^2 + \mathbf{k}^2 + \Pi_{\mathbf{k}})\phi_{\mathbf{k}} = g\rho_{\mathbf{k}}$. Apart from the source term this is a simple harmonic oscillator with frequency $\omega_{\mathbf{k}}^2 = \mathbf{k}^2 + \Pi_{\mathbf{k}}$. The canonical quantization procedure then leads to quantized excitations with energy $\hbar\omega_{\mathbf{k}}$—the usual "particles."

In a medium one follows this procedure in an approximate sense by expanding $\Pi_{\mathbf{k}}(\omega) = \Pi(\omega, \mathbf{k})$ to lowest order around $\omega_{\mathbf{k}}$,

$$\Pi_{\mathbf{k}}(\omega) = \Pi_{\mathbf{k}}(\omega_{\mathbf{k}}) + \Pi_{\mathbf{k}}'(\omega_{\mathbf{k}})(\omega - \omega_{\mathbf{k}}), \qquad (6.10)$$

where $\Pi_{\mathbf{k}}'(\omega) \equiv \partial_\omega \Pi_{\mathbf{k}}(\omega)$. To this order the Klein-Gordon equation is

$$\left[-\omega^2 + \omega_{\mathbf{k}}^2 + \Pi_{\mathbf{k}}'(\omega_{\mathbf{k}})(\omega - \omega_{\mathbf{k}}) \right] \phi_{\mathbf{k}}(\omega) = g\rho_{\mathbf{k}}(\omega), \qquad (6.11)$$

where the dispersion relation Eq. (6.9) was used. To first order in $\omega - \omega_{\mathbf{k}}$ one may use $2\omega = 2\omega_{\mathbf{k}} = \omega + \omega_{\mathbf{k}}$ which allows one to write

$$Z^{-1} \left(\omega^2 - \omega_{\mathbf{k}}^2 \right) \phi_{\mathbf{k}}(\omega) = g\rho_{\mathbf{k}}(\omega), \qquad (6.12)$$

where

$$Z^{-1} \equiv \frac{2\omega_{\mathbf{k}} - \Pi_{\mathbf{k}}'(\omega_{\mathbf{k}})}{2\omega_{\mathbf{k}}} = 1 - \left. \frac{\partial \Pi(\omega, \mathbf{k})}{\partial \omega^2} \right|_{\omega^2 - \mathbf{k}^2 = \Pi(\omega, \mathbf{k})} \qquad (6.13)$$

Because Z is a constant for a fixed $\mathbf{k}$ the approximate equation of motion Eq. (6.12) corresponds to a Hamiltonian

$$H = H_0 + H_{\text{int}} = \tfrac{1}{2}Z^{-1}(\dot{\phi}_{\mathbf{k}}^2 + \omega_{\mathbf{k}}^2\phi_{\mathbf{k}}^2) + g\phi_{\mathbf{k}}\rho_{\mathbf{k}}. \tag{6.14}$$

The free-field term is of the standard harmonic-oscillator form if one substitutes $\phi_{\mathbf{k}} = \sqrt{Z}\,\tilde{\phi}_{\mathbf{k}}$, i.e. free particles are excitations of the field $\tilde{\phi}_{\mathbf{k}}$ which has a renormalized amplitude relative to $\phi_{\mathbf{k}}$.

In terms of the renormalized field the interaction Hamiltonian is now of the form $H_{\text{int}} = \sqrt{Z}g\tilde{\phi}_{\mathbf{k}}\rho_{\mathbf{k}}$ which means that particles in the medium interact with an external source with a modified strength $\sqrt{Z}g$. Therefore, in Feynman graphs one must include one factor of $\sqrt{Z}$ for each external line of the ϕ field. Equivalently, the squared matrix element involves a factor Z for each external ϕ particle.

For relativistic modes where $|\omega_{\mathbf{k}}^2 - \mathbf{k}^2| \ll \omega_{\mathbf{k}}^2$ the modification is inevitably small, $|Z-1| \ll 1$. For (longitudinal) plasmons, however, the dispersion relation in a nonrelativistic plasma is approximately $\omega = \omega_{\mathrm{P}}$ with the plasma frequency ω_{P}, i.e. they are far away from the light cone, and then Z is a nonnegligible correction (Sect. 6.3). In the original calculation of the plasma decay process $\gamma \to \nu\bar{\nu}$ an incorrect Z was used for the longitudinal excitations (Adams, Ruderman, and Woo 1963). The correct factor was derived by Zaidi (1965).

6.3 Photon Dispersion

6.3.1 Maxwell's Equations

For the astrophysical applications relevant to this book the photon refractive index in a fully ionized plasma consisting of nuclei and electrons will be needed. On the quantum level, this system is entirely described by quantum electrodynamics (QED). It is sometimes referred to as a QED plasma—in contrast with a quark-gluon plasma which is described by quantum chromodynamics (QCD). The calculation of the refractive index amounts to an evaluation of the forward scattering amplitude of photons on electrons, a simple task except for the complications from the statistical averaging over the electrons which are partially or fully relativistic and exhibit any degree of degeneracy. Recently Braaten and Segel (1993) have found an astonishing simplification of this daunting problem (Sect. 6.3.4).

A more conceptual complication is the occurrence of a third photon degree of freedom in a medium (Langmuir 1926), sometimes referred to as Langmuir waves or plasmons.[31] Still, one should not think of photons as becoming literally massive like a massive vector boson which also carries three polarization states. A photon mass is prohibited by gauge invariance which remains intact. However, the medium singles out an inertial frame and thus breaks Lorentz invariance, an effect which is ultimately responsible for the possibility of a third polarization state.

It is useful, then, to begin with some general aspects of photon propagation in a medium which are unrelated to specific assumptions about the medium constituents. Notably, begin with the classical Maxwell equations for the electric and magnetic fields

$$\nabla \cdot \mathbf{E} = \rho, \qquad \nabla \times \mathbf{B} - \dot{\mathbf{E}} = \mathbf{J},$$
$$\nabla \cdot \mathbf{B} = 0, \qquad \nabla \times \mathbf{E} + \dot{\mathbf{B}} = 0. \tag{6.15}$$

An additional condition is that the electric charge density ρ and current density $\mathbf{J}$ obey the continuity equation

$$\partial \cdot J = \dot{\rho} - \nabla \cdot \mathbf{J} = 0, \tag{6.16}$$

where $J = (\rho, \mathbf{J})$ and $\partial = (\partial_t, \nabla)$. Covariantly, Maxwell's equations are

$$\partial_\mu F^{\mu\nu} = J^\nu, \qquad \epsilon^{\mu\nu\rho\sigma} \partial_\mu F_{\rho\sigma} = 0, \tag{6.17}$$

where $F^{\mu\nu}$ is the antisymmetric field-strength tensor with the nonvanishing components $F^{0i} = -F^{i0} = -\mathbf{E}^i$, and $F^{ij} = -F^{ji} = -\epsilon^{ijk}\mathbf{B}_k$. Applying ∂_μ to the inhomogeneous equation and observing that $F^{\mu\nu}$ is antisymmetric and $\partial_\mu\partial_\nu$ symmetric under $\mu \leftrightarrow \nu$ reveals that for consistency J must obey the continuity equation.

An equivalent formulation arises from expressing the field strengths in terms of a four-potential $A = (\Phi, \mathbf{A})$ by virtue of

$$F^{\mu\nu} = \partial^\mu A^\nu - \partial^\nu A^\mu, \tag{6.18}$$

which amounts to $\mathbf{E} = -\nabla\Phi - \dot{\mathbf{A}}$ and $\mathbf{B} = \nabla \times \mathbf{A}$. This representation is enabled by the homogeneous set of Maxwell equations which are then automatically satisfied. The inhomogeneous set now takes the form

$$\Box A - \partial (\partial \cdot A) = J, \tag{6.19}$$

where $\Box = \partial \cdot \partial = \partial_\mu \partial^\mu = \partial_t^2 - \nabla^2$.

---

[31] They are sometimes called "longitudinal plasmons" in contrast to "transverse plasmons." In this nomenclature the term "plasmon" refers to any excitation of the electromagnetic field in a medium while "photon" refers to an excitation in vacuum.

The field-strength tensor contains six independent degrees of freedom, the **E** and **B** fields. The redundancy imposed by the constraint of the homogeneous equations was removed by introducing the vector potential. There still remains one redundant degree of freedom related to a constraint imposed by current conservation. The Maxwell equations remain invariant under a "gauge transformation" $A \to A - \partial\alpha$ where α is an arbitrary scalar function. The modified A yields the same fields **E** and **B** which are the physically measurable quantities.

The relationship to current conservation is easiest recognized if one recalls that Maxwell's equations can be derived from a Lagrangian $-\frac{1}{4}F^2 - J \cdot A$ where $F^2 = F_{\mu\nu}F^{\mu\nu}$. A gauge transformation introduces an additional term $J \cdot \partial\alpha$ which is identical to a total divergence $\partial \cdot (\alpha J)$ if $\partial \cdot J = 0$ and thus leaves the Euler-Langrange equations unchanged. Indeed, current conservation is a necessary and sufficient condition for the gauge invariance of the theory (Itzykson and Zuber 1983).

A judicious choice of gauge can simplify the equations enormously. Two important possibilities are the Lorentz gauge and the Coulomb, transverse, or radiation gauge, based on the conditions

$$\partial \cdot A = 0, \qquad \text{Lorentz gauge,}$$

$$\nabla \cdot \mathbf{A} = 0, \qquad \text{Coulomb gauge.} \tag{6.20}$$

Maxwell's equations are then found to be (Jackson 1975)

$$\Box\,\Phi = \rho, \qquad \Box\,\mathbf{A} = \mathbf{J}, \qquad \text{Lorentz gauge,}$$

$$-\nabla^2\Phi = \rho, \qquad \Box\,\mathbf{A} = \mathbf{J}_\mathrm{T}, \qquad \text{Coulomb gauge,} \tag{6.21}$$

where $\mathbf{J}_\mathrm{T}$ is the transverse part of **J** characterized by $\nabla \cdot \mathbf{J}_\mathrm{T} = 0$.

In the absence of sources ($\rho = 0$ and $\mathbf{J} = 0$) the potential Φ vanishes in the Coulomb gauge while **A** obeys a wave equation. A Fourier transformation leads to $(-\mathbf{k}^2 + \omega^2)\mathbf{A} = 0$ whence the propagating modes have the dispersion relation $\mathbf{k}^2 = \omega^2$ corresponding to massless particles. Because of the transversality condition $\mathbf{k} \cdot \mathbf{A} = 0$ there are only two polarizations, the usual transverse electromagnetic waves. They are characterized by an electric field **E** transverse to **k** and a magnetic field of the same magnitude transverse to both.

6.3.2   Linear Response of the Medium

Maxwell's equations allow one to calculate the electromagnetic fields in the presence of prescribed external currents. However, the charged particles which constitute the currents move themselves under the influence of electromagnetic fields. Therefore, the interaction between fields and currents must be calculated self-consistently. If the fields are sufficiently weak one may assume that the reaction of the currents to the fields can be described as a linear response. (For a general review of linear-response theory in electromagnetism see Kirzhnits 1987.)

In general this statement cannot be made locally in the sense that the currents at space-time point $(t, \mathbf{x})$ were only linear functions of $A(t, \mathbf{x})$. Within the restrictions imposed by causality the relationship between fields and currents is nonlocal; for example, a solution of Maxwell's equations with prescribed currents requires integrations over the sources in space and time. After a Fourier transformation, however, the assumption of a linear response can be stated as

$$J^\mu_{\text{ind}} = -\Pi^{\mu\nu} A_\nu. \tag{6.22}$$

The polarization tensor $\Pi(K)$ with $K = (\omega, \mathbf{k})$ is a function of the medium properties.

Besides the induced current there may be an externally prescribed one J_{ext} which is unrelated to the response of the microscopic medium constituents to the fields; the total current is $J = J_{\text{ind}} + J_{\text{ext}}$. Maxwell's equations (6.19) are then in Fourier space

$$(-K^2 g^{\mu\nu} + K^\mu K^\nu + \Pi^{\mu\nu}) A_\nu = J^\mu_{\text{ext}}. \tag{6.23}$$

Invariance under a gauge transformation $A_\nu \to A_\nu + K_\nu \alpha$ requires that $\Pi^{\mu\nu} K_\nu = 0$. Because the external and total currents are conserved the induced current is conserved as well, leading to $K \cdot J_{\text{ind}} = 0$ or $K_\mu \Pi^{\mu\nu} = 0$. Altogether

$$K_\mu \Pi^{\mu\nu} = \Pi^{\mu\nu} K_\nu = 0 \tag{6.24}$$

which is an important general property of the polarization tensor.

Considering the Maxwell equations in Coulomb gauge in the absence of external currents, the transversality of $\mathbf{A}$ still implies that it provides only two wave polarization states, albeit with modified dispersion relations due to the presence of Π. With regard to the Φ equation note that in an isotropic medium the induced charge density ρ_{ind} must be a spatial scalar and so can depend only on Φ and the combination $\mathbf{k} \cdot \mathbf{A} = 0$ which is the only available scalar linear in $\mathbf{A}$. Therefore, the homogeneous equation for Φ is

$$(\mathbf{k}^2 + \Pi^{00}) \Phi = 0. \tag{6.25}$$

Because Π^{00} is a function of ω and $\mathbf{k}$ this is a wave equation with the dispersion relation $\mathbf{k}^2 + \Pi^{00}(\omega, \mathbf{k}) = 0$. The electric field associated with this third polarization degree of freedom is proportional to $\mathbf{k}\Phi$, along the direction of propagation—hence the term longitudinal excitation. There is no magnetic field associated with it. Physically, it corresponds to a density wave of the electrons much like a sound wave. Obviously this mode requires being carried by a medium, as opposed to the transverse waves which propagate in vacuum as well.

The wave equation Eq. (6.23) corresponds to a Langrangian density in Fourier space which involves a new term $-V$ with $V = \frac{1}{2} A_\mu \Pi^{\mu\nu} A_\nu$ which plays the role of a medium-induced potential energy for the field A. In vacuum $\Pi^{\mu\nu}$ can be constructed only from $g^{\mu\nu}$ and $K^\mu K^\nu$ which both violate the gauge condition $\Pi K = 0$. Notably, this forbids a photon mass term which would have to be of the form $\Pi^{\mu\nu}_{\text{mass}} = m^2 g^{\mu\nu}$. In a medium an inertial frame is singled out, allowing one to construct Π from the medium four-velocity U and to find a structure which obeys the gauge constraint. Strictly speaking, however, the medium does not induce an effective-mass term which would be of the form $m^2_{\text{eff}} g^{\mu\nu}$ and which remains forbidden by gauge invariance.

6.3.3   Isotropic Polarization Tensor in the Lorentz Gauge

The Coulomb gauge is well suited to treat radiation in vacuum because the propagating modes are neatly separated from the scalar potential, and the gauge component is easily identified with the longitudinal part of $\mathbf{A}$. In a medium, however, the different appearance of Φ and $\mathbf{A}$ in their respective wave equations is cumbersome. The Maxwell equations in Lorentz gauge are symmetric between Φ and $\mathbf{A}$ which allows one to treat all polarization states on the same footing.

In order to construct the most general $\Pi(K)$ for an isotropic medium it is useful to define four basis vectors for Minkowski space which are adapted to the symmetry of the medium as well as to the Lorentz condition (Weldon 1982a; Haft 1993). For that purpose one may use the preferred directions in Minkowski space, namely K and the four-velocity of the medium U which is $(1, \mathbf{0})$ in its inertial frame. Moreover, the notation ω and $\mathbf{k}$ is used for the frequency and wave vector of K in the medium frame; they are covariantly given by $\omega = U \cdot K$ and $k^2 = \mathbf{k}^2 = (U \cdot K)^2 - K^2$.

In Lorentz gauge the physical A fields obey $K \cdot A = 0$. Therefore, one defines a basis vector for the gauge degree of freedom by

$$e_{\text{g}} \equiv K/\sqrt{K^2}. \qquad (6.26)$$

Next, one chooses a vector which is longitudinal relative to the spatial part of K and which obeys $K \cdot e_L = 0$,

$$e_L \equiv \frac{\omega K - K^2 U}{k\sqrt{K^2}} = \frac{(k^2, \omega\mathbf{k})}{k\sqrt{K^2}}, \tag{6.27}$$

where the second expression refers to the medium rest frame. There remain two directions orthogonal to e_g and e_L, or equivalently, to K and U. If $\mathbf{k}$ is taken to point in the z-direction two possible choices are the unit vectors $\mathbf{e}_x$ and $\mathbf{e}_y$, respectively. However, in order to retain the azimuthal symmetry around the $\mathbf{k}$ direction the "circular polarization vectors" $\mathbf{e}_\pm = (\mathbf{e}_x \pm i\mathbf{e}_y)/\sqrt{2}$ are needed. Then

$$e_\pm \equiv (0, \mathbf{e}_\pm) \tag{6.28}$$

in the rest frame of the medium; a suitable covariant formulation is also possible. The basis vectors obey $e_\pm^* = e_\mp$ while $e_{g,L}$ are real for $K^2 > 0$ (time-like) and $e_{g,L}^* = -e_{g,L}$ for $K^2 < 0$ (space-like). They are normalized according to $e_\pm^* \cdot e_\pm = e_\mp \cdot e_\pm = -1$ and $e_L^* \cdot e_L = \mp 1$ and $e_g^* \cdot e_g = \pm 1$, depending on K^2 being time- or space-like. Evidently, e_g and e_L switch properties between a time- and space-like K.

This choice of basis vectors is only possible if $K^2 \neq 0$. If K is light-like one must make some other choice, for example $(1, \mathbf{0})$ and $(0, \hat{\mathbf{k}})$ in the rest frame of the medium. Because the goal is to describe electromagnetic excitations in a medium, and because usually $K^2 \neq 0$ for such waves, this is no serious limitation. It will turn out that the dispersion relation of (longitudinal) plasmons crosses the light-cone, i.e. there is a wave number for which $\omega^2 - k^2 = 0$. The degeneracy of e_g with e_L at this single point will cause no trouble.

The most general polarization tensor compatible with the gauge condition Eq. (6.24) must be constructed from $e_\pm$ and e_L alone. Azimuthal symmetry about the $\mathbf{k}$ direction requires that they occur only in the scalar combinations $e_a^\mu e_a^{*\nu}$. Therefore, one defines the projection operators on the basis vectors

$$P_a^{\mu\nu} \equiv -e_a^\mu e_a^{*\nu}, \qquad (a = \pm, L). \tag{6.29}$$

The most general polarization tensor is then given as

$$\Pi^{\mu\nu} = \sum_{a=\pm,L} \pi_a P_a^{\mu\nu}, \tag{6.30}$$

where the π_a are functions of the Lorentz scalars K^2 and $U \cdot K$ or equivalently of ω and k in the medium frame. They represent the medium response to circularly and longitudinally polarized A's.

The homogeneous Maxwell equations in Lorentz gauge in an isotropic medium then have the most general form

$$\left(-K^2 g^{\mu\nu} + \sum_{a=\pm,L} \pi_a P_a^{\mu\nu}\right) A_\nu = 0. \tag{6.31}$$

The metric tensor is $g = P_{\rm g} + P_{\rm L} + P_+ + P_-$ where $P_{\rm g}^{\mu\nu} = e_{\rm g}^\mu e_{\rm g}^{*\nu}$. Hence one obtains decoupled wave equations $[-\omega^2 + \mathbf{k}^2 + \pi_a(\omega, \mathbf{k})] \tilde{A}_a = 0$ for the physical degrees of freedom $A_a = P_a A$ with $a = \pm, {\rm L}$. The corresponding dispersion relation is

$$-\omega^2 + k^2 + \pi_a(\omega, k) = 0. \tag{6.32}$$

It yields the frequency ω_k for modes with a given polarization and wave number. The so-called effective mass is then $m_{\rm eff}^2 = \pi_a(\omega_k, k)$. This expression is different for different polarizations and wave numbers, and may even be negative.

Generally, an isotropic medium is characterized by *three* different response functions because the left- and right-handed circular polarization states may experience different indices of refraction (Nieves and Pal 1989a,b). Such optically active media are not symmetric under a parity transformation. For example, a sugar solution changes under a spatial reflection because the sugar molecules have a definite handedness.

If the medium and all relevant interactions are even under parity the circular polarization states have the same refractive index. Then one needs to distinguish only between transverse and longitudinal modes; one defines $\pi_{\rm T} \equiv \pi_+ = \pi_-$ and $P_{\rm T} = P_+ + P_-$ which projects on the plane transverse to K and U in Minkowski space.

In macroscopic electrodynamics the medium effects are frequently stated in the form of response functions to applied electric and magnetic fields instead of a response to A. The displacement induced by an applied electric field is $\mathbf{D} = \epsilon \mathbf{E}$ with ϵ the dielectric permittivity. Similarly, the magnetic field is $\mathbf{H} = \mu^{-1} \mathbf{B}$ for an applied magnetic induction where μ is the magnetic permeability. For time-varying and/or inhomogeneous fields these relationships are understood in Fourier space where the response functions depend on ω and k.

The magnetic field $\mathbf{H}$ and the transverse part of $\mathbf{D}$, characterized by $\mathbf{k} \cdot \mathbf{D}_{\rm T} = 0$, do not have independent meaning (Kirzhnits 1987). Therefore, among other possibilities one may choose $\mathbf{H} = \mathbf{B}$, $\mathbf{D}_{\rm T} = \epsilon_{\rm T} \mathbf{E}_{\rm T}$, and $\mathbf{D}_{\rm L} = \epsilon_{\rm L} \mathbf{E}_{\rm L}$. In this case $\epsilon_{\rm L} \equiv \epsilon$ is the longitudinal and $\epsilon_{\rm T} \equiv \epsilon_{\rm L} + (1 - \mu^{-1}) k^2/\omega^2$ the transverse dielectric permittivity.

The relationship to the transverse and longitudinal components of the polarization tensor is (Weldon 1982a)

$$\epsilon_{\rm L} = 1 - \pi_{\rm L}/(\omega^2 - k^2) \quad \text{and} \quad \epsilon_{\rm T} = 1 - \pi_{\rm T}/\omega^2. \tag{6.33}$$

This yields the well-known dispersion relations (Sitenko 1967)

$$\epsilon_{\rm L}(\omega, k) = 0 \quad \text{and} \quad \omega^2 \epsilon_{\rm T}(\omega, k) = k^2 \tag{6.34}$$

for the longitudinal and transverse modes.

Calculations of the polarization tensor from the forward scattering amplitudes on microscopic medium constituents are usually performed in a cartesian basis and thus yield an expression for $\Pi^{\mu\nu}$. The longitudinal and transverse components are projected out by virtue of $\pi_{\rm L} = e_{\rm L}^{*\mu} \Pi_{\mu\nu} e_{\rm L}^\nu$ and $\pi_{\rm T} = e_{\pm}^{*\mu} \Pi_{\mu\nu} e_{\pm}^\nu$, or explicitly in the medium frame (Weldon 1982a)

$$\pi_{\rm L} = (1 - \omega^2/k^2)\, \Pi^{00} \quad \text{and} \quad \pi_{\rm T} = \tfrac{1}{2}(\mathrm{Tr}\,\Pi - \pi_{\rm L}), \tag{6.35}$$

with $\mathrm{Tr}\,\Pi = g_{\mu\nu}\Pi^{\mu\nu}$.

Recall that the dispersion relations are given by $\omega^2 - k^2 = \pi_{\rm T,L}(\omega, k)$. Thus the frequency $\omega(k)$ and the "effective mass" of a given mode are generally complicated functions of k, notably in a medium involving bound electrons where various resonances occur. It can be shown on general grounds (Jackson 1975), however, that for frequencies far above all resonances the transverse mode has a particle-like dispersion relation $\omega^2 - k^2 = m_{\rm T}^2$ where $m_{\rm T}$ is the "transverse photon mass" which is a constant independent of the wave number or frequency.

6.3.4 Lowest-Order QED Calculation of Π

On the level of quantum electrodynamics (QED) the potential $V = \tfrac{1}{2} A_\mu \Pi^{\mu\nu} A_\nu$ which modifies the free Lagrangian is interpreted as the self-energy of the photons in the medium. As a Feynman graph, it corresponds to an insertion of $\Pi^{\mu\nu}$ into a photon line of four-momentum K and thus corresponds to forward scattering on the medium constituents (Fig. 6.1), entirely analogous to the interpretation of the refractive index in terms of a forward scattering amplitude in Sect. 6.2.1. One then concludes that $\Pi^{\mu\nu}(K)$ is the truncated matrix element for the forward scattering of a photon with momentum K, i.e. it is the matrix element of the medium constituents alone, uncontracted with the photon polarization vectors ϵ^μ and ϵ^ν.

In general, the calculation of Π requires the methods of field theory at finite temperature and density. To lowest order, however, this

Fig. 6.1. Polarization tensor as photon self-energy insertion.

formalism is not required because the only contribution is from lowest-order forward scattering on charged particles. Moreover, because the scattering amplitude involves nonrelativistically the inverse mass of the targets one may limit one's attention to the electrons.

Then one takes the standard (truncated) Compton-scattering matrix element (e.g. Bjorken and Drell 1964; Itzykson and Zuber 1983) and takes an average over the Fermi-Dirac distributions of the electrons. To lowest order in $\alpha = e^2/4\pi$ this yields (Altherr and Kraemmer 1992; Braaten and Segel 1993)

$$\Pi^{\mu\nu}(K) = 16\pi\alpha \int \frac{d^3\mathbf{p}}{2E(2\pi)^3} \left(\frac{1}{e^{(E-\mu)/T} + 1} + \frac{1}{e^{(E+\mu)/T} + 1} \right)$$

$$\times \frac{(P \cdot K)^2 g^{\mu\nu} + K^2 P^\mu P^\nu - P \cdot K \left(K^\mu P^\nu + K^\nu P^\mu \right)}{(P \cdot K)^2 - \frac{1}{4}(K^2)^2},$$

(6.36)

where $P = (E, \mathbf{p})$ and $E = (\mathbf{p}^2 + m_e^2)^{1/2}$, apart from refractive effects for the electrons and positrons. The phase-space distributions represent electrons and positrons at temperature T and chemical potential μ.

Over the years, the phase-space integration has been performed in various limits (Silin 1960; Tsytovich 1961; Jancovici 1962; Klimov 1982; Weldon 1982a; Altherr, Petitgirard, and del Río Gaztelurrutia 1993). The most comprehensive analytic result is that of Braaten and Segel (1993) which contains all previous cases in the appropriate limits.

The main simplification occurs from neglecting the $(K^2)^2$ term in the denominator of Eq. (6.36). For light-like K's this is exactly correct, and in the nonrelativistic limit where m_e is much larger than all other energy scales the approximation is also trivially justified. In the relativistic limit it is only justified if one is interested in $\Pi(K)$ near the light cone ($\omega = k$) in Fourier space. In the relativistic limit both transverse and longitudinal excitations have dispersion relations which are approximately $(\omega^2 - k^2)^{1/2} \approx eT$ or eE_{F} in the nondegenerate and degenerate limits, respectively. As detailed by Braaten and Segel (1993), this deviation from masslessness is small enough to justify the approximation if one aims at the dispersion relations. Including $\frac{1}{4}(K^2)^2$ yields an $\mathcal{O}(\alpha^2)$ correction—it can be ignored in an $\mathcal{O}(\alpha)$ result.

In a higher-order calculation one has to include the proper $e^\pm$ dispersion relations which imply that electromagnetic excitations are never damped by $\gamma \to e^+ e^-$ decay (Braaten 1991), in contrast with statements found in the previous literature. Dropping the $(K^2)^2$ term in the denominator of Eq. (6.36) prevents Π from developing an imaginary part from this decay, even with the vacuum $e^\pm$ dispersion relations. Therefore, the "approximate" integral actually provides a better representation of the $\mathcal{O}(\alpha)$ dispersion relations than the "exact" one.

With the approximation $K^2 = 0$ in the denominator of Eq. (6.36) the angular integral is trivial.[32] With Eq. (6.35) one finds

$$\pi_{\mathrm{L}} = \frac{4\alpha}{\pi} \frac{\omega^2 - k^2}{k^2} \int_0^\infty dp\, f_p \frac{p^2}{E} \left[\frac{\omega}{kv} \log\left(\frac{\omega + kv}{\omega - kv} \right) - \frac{\omega^2 - k^2}{\omega^2 - k^2 v^2} - 1 \right],$$

$$\pi_{\mathrm{T}} = \frac{4\alpha}{\pi} \frac{\omega^2 - k^2}{k^2} \int_0^\infty dp\, f_p \frac{p^2}{E} \left[\frac{\omega^2}{\omega^2 - k^2} - \frac{\omega}{2kv} \log\left(\frac{\omega + kv}{\omega - kv} \right) \right], \quad (6.37)$$

where $v = p/E$ is the $e^\pm$ velocity and f_p represents the sum of their phase-space distributions.

The remaining integration can be done analytically in the classical, degenerate, and relativistic limits where one finds

$$\pi_{\mathrm{L}} = \omega_{\mathrm{P}}^2 \left[1 - G(v_*^2 k^2/\omega^2) \right] + v_*^2 k^2 - k^2,$$

$$\pi_{\mathrm{T}} = \omega_{\mathrm{P}}^2 \left[1 + \tfrac{1}{2} G(v_*^2 k^2/\omega^2) \right]. \quad (6.38)$$

Here, v_* is a "typical" electron velocity defined by

$$v_* \equiv \omega_1/\omega_{\mathrm{P}}. \quad (6.39)$$

The plasma frequency ω_{P} and the frequency ω_1 are

$$\omega_{\mathrm{P}}^2 \equiv \frac{4\alpha}{\pi} \int_0^\infty dp\, f_p\, p \left(v - \tfrac{1}{3} v^3 \right),$$

$$\omega_1^2 \equiv \frac{4\alpha}{\pi} \int_0^\infty dp\, f_p\, p \left(\tfrac{5}{3} v^3 - v^5 \right). \quad (6.40)$$

The function G (Fig. 6.2) is defined by

$$G(x) = \frac{3}{x} \left[1 - \frac{2x}{3} - \frac{1-x}{2\sqrt{x}} \log\left(\frac{1 + \sqrt{x}}{1 - \sqrt{x}} \right) \right]$$

$$= 6 \sum_{n=1}^\infty \frac{x^n}{(2n+1)(2n+3)}. \quad (6.41)$$

Note that $G(0) = 0$, $G(1) = 1$, and $G'(1) = \infty$.

---

[32]In the degenerate limit, Jancovici (1962) has calculated analytically the full integral without the $K^2 = 0$ approximation.

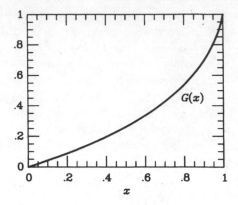

Fig. 6.2. Function $G(x)$ according to Eq. (6.41).

The most astonishing observation of Braaten and Segel (1993) is that Eq. (6.38) is a good approximation for *all* conditions, not only for the limiting cases for which it was derived. As the approximation is much better than 1%, which is the approximate accuracy of an $\mathcal{O}(\alpha)$ result, these representations can be taken to be exact to this order.

6.3.5 Dispersion Relations

In order to determine the photon dispersion relation for specific conditions one must determine $\omega_{\rm P}$ and v_* corresponding to the temperature T and chemical potential μ of the electrons. In Fig. 6.3 contours for v_* and $\gamma \equiv \omega_{\rm P}/T$ are shown in the T-ρ-plane of a plasma. Analytic limiting cases are (Braaten and Segel 1993)

$$v_* = \begin{cases} (5T/m_e)^{1/2} & \text{Classical,} \\ v_{\rm F} & \text{Degenerate,} \\ 1 & \text{Relativistic,} \end{cases} \qquad (6.42)$$

$$\omega_{\rm P}^2 = \begin{cases} \dfrac{4\pi\alpha\, n_e}{m_e}\left(1 - \dfrac{5}{2}\dfrac{T}{m_e}\right) & \text{Classical,} \\[2mm] \dfrac{4\pi\alpha\, n_e}{E_{\rm F}} = \dfrac{4\alpha}{3\pi}\, p_{\rm F}^2 v_{\rm F} & \text{Degenerate,} \\[2mm] \dfrac{4\alpha}{3\pi}\left(\mu^2 + \tfrac{1}{3}\pi^2 T^2\right) & \text{Relativistic,} \end{cases} \qquad (6.43)$$

where $v_{\rm F} = p_{\rm F}/E_{\rm F}$ is the velocity at the Fermi surface, "classical" refers to the nondegenerate and nonrelativistic limit, and "relativistic" is for any degree of degeneracy.

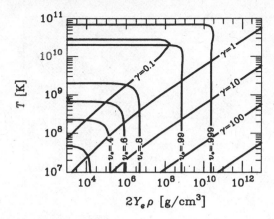

Fig. 6.3. Contours for v_* and $\gamma = \omega_P/T$ as defined in Eqs. (6.39) and (6.40) where Y_e is the number of electrons per baryon.

Next, with Eq. (6.38) one must solve the transcendental equations $\pi_{T,L}(\omega, k) = \omega^2 - k^2$ which are explicitly

$$\omega^2 - k^2 \quad = \omega_P^2 \left[1 + \tfrac{1}{2}G(v_*^2 k^2/\omega^2)\right] \quad \text{Transverse,}$$

$$\omega^2 - v_*^2 k^2 = \omega_P^2 \left[1 - G(v_*^2 k^2/\omega^2)\right] \quad \text{Longitudinal.} \tag{6.44}$$

In the classical limit this is to lowest order in T/m_e

$$\omega^2 = k^2 + \omega_P^2 \left(1 + \frac{k^2}{\omega^2}\frac{T}{m_e}\right) \quad \text{Transverse,}$$

$$\omega^2 = \omega_P^2 \left(1 + 3\frac{k^2}{\omega^2}\frac{T}{m_e}\right) \quad \text{Longitudinal.} \tag{6.45}$$

For small temperatures the longitudinal modes oscillate with an almost fixed frequency, independently of momentum, while the transverse modes behave almost like massive particles (Fig. 6.4).

The general result Eq. (6.38) and the behavior of the function $G(x)$ reveal that for transverse excitations $\omega^2 - k^2$ can vary only between ω_P^2 and $\tfrac{3}{2}\omega_P^2$. Also, $\omega^2 - k^2 > 0$ so that K^2 is always time-like. For $k \gg \omega_P$ the transverse dispersion relation approaches that of a massive particle with a fixed mass m_T, the "transverse photon mass". With Eq. (6.38) and because $k/\omega \to 1$ for $k \gg \omega_P$ one finds $m_T^2 = \omega_P^2[1 + \tfrac{1}{2}G(v_*^2)]$, or

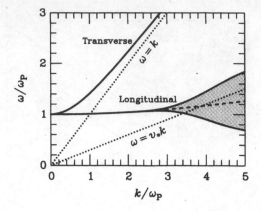

Fig. 6.4. Electromagnetic dispersion relations in the classical limit according to Eq. (6.45) for $v_* = (5T/m_e)^{1/2} = 0.2$. The shaded area indicates the "width" of $\omega(k)$ in the longitudinal case due to Landau damping.

with $\omega = k$ directly from Eq. (6.37)

$$m_{\rm T}^2 = \frac{4\alpha}{\pi} \int_0^\infty dp \, f_p \, \frac{p^2}{E}. \tag{6.46}$$

Limiting cases are

$$\frac{m_{\rm T}^2}{\omega_{\rm P}^2} = \begin{cases} 1 & \text{Classical,} \\[2mm] \dfrac{3}{2v_{\rm F}^2} \left[1 - \dfrac{1 - v_{\rm F}^2}{2v_{\rm F}} \log\left(\dfrac{1 + v_{\rm F}}{1 - v_{\rm F}} \right) \right] & \text{Degenerate,} \\[2mm] \dfrac{3}{2} & \text{Relativistic.} \end{cases} \tag{6.47}$$

In Fig. 6.5 $(\omega^2 - k^2)$ is shown for several values of v_* as a function of k. It is quite apparent how the transverse mass is asymptotically approached.

The dispersion relation for longitudinal modes is more interesting in several regards. First, according to Eq. (6.44) the oscillation frequency is only a function of $v_* k$ and so the natural scale for k is $\omega_{\rm P}/v_*$. In Fig. 6.6 I show $\omega^2 - v_*^2 k^2$ as a function of $v_* k$.

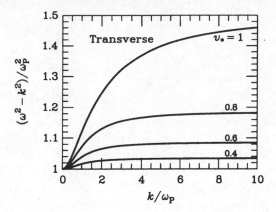

Fig. 6.5. Dispersion relation for transverse modes according to Eq. (6.44).

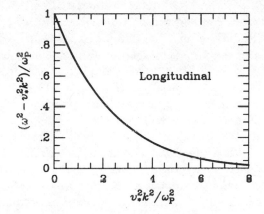

Fig. 6.6. Dispersion relation for longitudinal modes according to Eq. (6.44).

Second, for $v_* < 1$ there is always a wave number k_1 where $\omega(k)$ "crosses the light cone" ($\omega/k = 1$),

$$
\begin{aligned}
k_1^2 &= \frac{4\alpha}{\pi} \int_0^\infty dp\, f_p \frac{p^2}{E} \left[\frac{1}{v} \log\left(\frac{1+v}{1-v} \right) - 1 \right] \\
&= \omega_{\mathrm{P}}^2 \frac{3}{v_*^2} \left[\frac{1}{2v_*} \log\left(\frac{1+v_*}{1-v_*} \right) - 1 \right].
\end{aligned} \tag{6.48}
$$

The second identity (Braaten and Segel 1993) applies at the same level of approximation as $\pi_{\mathrm{T,L}}$ in Eq. (6.38). Some analytic limiting cases are

$$\frac{k_1^2}{\omega_{\rm P}^2} = \begin{cases} 1 + 3T/m_e & \text{Classical,} \\ \dfrac{3}{v_{\rm F}^2}\left[\dfrac{1}{2v_{\rm F}}\log\left(\dfrac{1+v_{\rm F}}{1-v_{\rm F}}\right)-1\right] & \text{Degenerate,} \\ \infty & \text{Relativistic.} \end{cases} \qquad (6.49)$$

Then, for $k > k_1$ the four-momentum is space-like, $\omega^2 - k^2 < 0$.

As discussed in Sect. 6.2.2 there is nothing wrong with a space-like four-momentum of an excitation. In media with electron resonances such as water or air even (transverse) photons exhibit this behavior which allows kinematically for their Cherenkov emission $e \to e\gamma$ or absorption $\gamma e \to e$. In a plasma, transverse excitations are always time-like and thus cannot be Cherenkov absorbed. Their lowest-order damping mechanism is Thomson scattering $\gamma e \to e\gamma$ which is not included because it is an $\mathcal{O}(\alpha^2)$ effect. Longitudinal excitations with $k > k_1$, in contrast, can and will be Cherenkov absorbed by the ambient electrons, leading to an $\mathcal{O}(\alpha)$ damping rate. It corresponds to an imaginary part of the dispersion relation (an imaginary part of $\pi_{\rm L}$).

In the expression Eq. (6.36) this damping effect corresponds to a vanishing denominator, essentially to $P\cdot K = 0$, which occurs when the intermediate electron in Compton scattering "goes on-shell." Evidently, $P\cdot K = E\omega - \mathbf{p}\cdot\mathbf{k}$ can never vanish for $k < \omega$ while for $k > \omega$ there are always *some* electrons, even in a nonrelativistic plasma, which satisfy this condition. When the phase velocity ω/k becomes of the order of a typical thermal velocity the number of electrons which match the Cherenkov condition becomes large, and then the damping of plasmons becomes strong. Because v_* measures a typical electron velocity this occurs for $k \gtrsim \omega/v_*$ (Fig. 6.4). Therefore, while nothing dramatic happens where the dispersion relation crosses the light cone, it fizzles out near the "electron cone." For $k \gtrsim \omega/v_*$ there are no organized oscillations of the electrons—longitudinal modes no longer exist.

This damping mechanism of plasma waves was first discussed by Landau (1946) and is named after him. A calculation in terms of Cherenkov absorption was performed by Tsytovich (1961). In the classical limit the Landau damping rate (the imaginary part of the frequency) is

$$\frac{\Gamma_{\rm L}}{\omega_{\rm P}} = \sqrt{\frac{\pi}{8}}\left(\frac{k_{\rm D}}{k}\right)^3 e^{-\frac{k_{\rm D}^2}{2k^2}} = \sqrt{\pi}\left(\frac{5}{2}\right)^{3/2}\left(\frac{\omega_{\rm P}}{v_*k}\right)^3 e^{-\frac{5}{2}\frac{\omega_{\rm P}^2}{v_*^2k^2}}, \quad (6.50)$$

where $k_{\rm D} = 4\pi\alpha\, n_e/T$ is the Debye screening scale. (Note that $\omega_{\rm P}^2/k_{\rm D}^2 = T/m_e = v_*^2/5$.) For a given wave number a plasmon must be viewed as a resonance with a finite width Γ_L. The approximate uncertainty $\pm\Gamma_L(k)$ of the energy $\omega(k)$ is shown in Fig. 6.4 as a shaded area.

The damping rate Eq. (6.50) does not show a threshold effect at $k = \omega$ because it was calculated nonrelativistically so that the high-energy tail of the electron distribution contains particles with velocities exceeding the speed of light. Tsytovich (1961) calculated a relativistic result with the correct threshold behavior. However, because the damping rate is exceedingly small for $k \ll k_D$, the correction is very small if $\omega_P \ll k_D$, equivalent to $T \ll m_e$. For the degenerate case, explicit expressions for the imaginary parts of $\pi_{T,L}(\omega, k)$ were derived by Altherr, Petitgirard, and del Río Gaztelurrutia (1993).

The general expressions Eq. (6.38) for the real parts of π_L and π_T were derived without the need to assume that K was time-like. Therefore, they should also apply "below the light cone," even though Braaten and Segel (1993) confined their discussion to the region $k < \omega$. As expected, these expressions break down for $k > \omega/v_*$ where Landau damping becomes strong.

6.3.6 Renormalization Constants $Z_{T,L}$

Armed with the dispersion relation one may determine the vertex renormalization constants $Z_{T,L}$ relevant for the coupling of external photons or plasmons to an electron in the medium (Sect. 6.2.3),

$$Z_{T,L}^{-1} = 1 - \left.\frac{\partial \pi_{T,L}(\omega, \mathbf{k})}{\partial \omega^2}\right|_{\omega^2 - \mathbf{k}^2 = \pi_{T,L}(\omega, \mathbf{k})}. \tag{6.51}$$

With the same approximations as before, Braaten and Segel (1993) found an analytic representation accurate to $\mathcal{O}(\alpha)$,

$$Z_T = \frac{2\omega^2(\omega^2 - v_*^2 k^2)}{\omega^2[3\omega_P^2 - 2(\omega^2 - k^2)] + (\omega^2 + k^2)(\omega^2 - v_*^2 k^2)},$$

$$Z_L = \frac{2(\omega^2 - v_*^2 k^2)}{3\omega_P^2 - (\omega^2 - v_*^2 k^2)}\frac{\omega^2}{\omega^2 - k^2}. \tag{6.52}$$

In each case ω and k are "on shell," i.e. they are related by the dispersion relation relevant for the T and L case, respectively.

Inspection of Eq. (6.52) reveals that Z_T is always very close to unity, as expected for excitations with only a small deviation from a massive-particle dispersion relation. The contours in Fig. 6.7 confirm that Z_T never deviates from unity by more than a few percent.

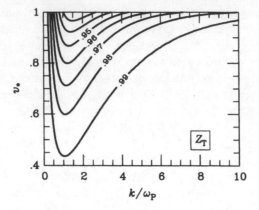

Fig. 6.7. Contours for the vertex renormalization factor Z_T for transverse electromagnetic excitations in a medium according to Eq. (6.52).

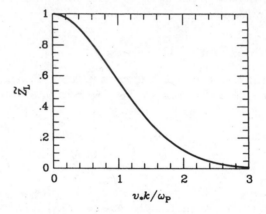

Fig. 6.8. Modified vertex renormalization factor $\tilde{Z}_L$ for longitudinal electromagnetic excitations in a medium according to Eq. (6.53).

The longitudinal case is more complicated. Z_L is a product of two factors $\tilde{Z}_L$ and $\omega^2/(\omega^2 - k^2)$ where the former,

$$\tilde{Z}_L \equiv \frac{2\left(\omega^2 - v_*^2 k^2\right)}{3\omega_P^2 - \left(\omega^2 - v_*^2 k^2\right)}, \tag{6.53}$$

is a function of the variable $v_* k$ alone because for plasmons ω is a function of $v_* k$ alone. The function $\tilde{Z}_L(v_* k)$ is shown in Fig. 6.8; for $v_* k \gg \omega_P$ it quickly drops to zero. However, in a relativistic plasma with $v_* = 1$ the complete factor is $Z_L = 2\omega^2/[3\omega_P^2 - (\omega^2 - k^2)]$ and thus rises quickly with k because $\omega \approx k$.

As discussed in the previous section, the dispersion relation crosses the light cone at $k = k_1$, a point at which Z_L diverges and changes sign. The sign change is compensated by the change of the polarization vector e_L (Eq. 6.27) at the light cone where it becomes imaginary. Because in the squared matrix element a factor $e_L^{*\mu} e_L^\nu$ appears, and because this expression changes sign at the light cone, the expression $Z_L e_L^{*\mu} e_L^\nu$ remains positive.

As for the divergence, it is harmless in reactions of the sort $\gamma_L \to \nu\bar\nu$ (plasmon decay), $e \to e\gamma_L$ (Cherenkov emission), $\gamma_L e \to e$ (Cherenkov absorption), and $\gamma_T \gamma_L \to a$ (plasmon coalescence into axions) which are of interest in this book. These reactions involving three particles are constrained by their phase space to either time-like excitations (plasmon decay), or to space-like ones (Cherenkov and coalescence process). Therefore, the threshold behavior of the phase space moderates the divergence in these cases.

6.4 Screening Effects

6.4.1 Debye Screening

Scattering processes in the Coulomb field of charged particles such as Rutherford scattering, bremsstrahlung, or the Primakoff effect typically lead to cross sections which diverge in the forward direction because of the long-range nature of the electrostatic interaction. In a plasma this divergence is moderated by screening effects which thus are crucial for a calculation of the cross sections or energy-loss rates.

Screening effects are revealed by turning to the static limit of Maxwell's equations in a medium,

$$\left[k^2 + \pi_L(0,k)\right]\Phi(k) = \rho(k),$$

$$\left[k^2 + \pi_T(0,k)\right]\mathbf{A}(k) = \mathbf{J}(k), \tag{6.54}$$

where the current must be transverse in both Coulomb and Lorentz gauge as $\partial_t \rho = 0$ in the static limit. Notably, the equation for Φ in vacuum is the Fourier transform of Poisson's equation and thus gives rise to a $1/r$ Coulomb potential if the source is point-like, $\rho(r) = e\delta(r)$.

In a QED plasma $\pi_{L,T}(\omega,k)$ are given by the integrals Eq. (6.37). In the static limit ($\omega = 0$) one finds $\pi_T(0,k) = 0$ because all terms in the integrand involve factors of ω. Therefore, stationary currents

($\partial_t \mathbf{J} = 0$) are not screened. The magnetic field associated with a stationary current is the same at a distance whether or not the plasma is present.

Not so for the electric field associated with a charge. In the static limit one finds

$$\pi_{\rm L}(0, k) = \frac{4\alpha}{\pi} \int_0^\infty dp\, f_p\, p\, (v + v^{-1}). \tag{6.55}$$

Because this expression does not depend on k it can be identified with the square of a fixed wave number $k_{\rm S}$, leading to Poisson's equation in the form

$$\left(k^2 + k_{\rm S}^2\right)\Phi(k) = \rho(k). \tag{6.56}$$

Because for a point-like source this gives a Yukawa potential

$$\Phi(r) \propto r^{-1} \to r^{-1}\, e^{-k_{\rm S}r} \tag{6.57}$$

electric charges are screened for distances exceeding about $k_{\rm S}^{-1}$.

Evaluating Eq. (6.55) explicitly in the classical limit reproduces the well-known Debye screening scale (Debye and Hückel 1923; for a textbook discussion see Landau and Lifshitz 1958)

$$k_{\rm S}^2 = k_{\rm D}^2 = \frac{4\pi\alpha\, n_e}{T} = \frac{m_e}{T}\, \omega_{\rm P}^2. \tag{6.58}$$

At this point one recognizes that $k_{\rm D}^2$ is independent of the electron mass, in contrast with the plasma frequency $\omega_{\rm P}^2$. Therefore, it is no longer justified to ignore the ions or nuclei; they contribute little to dispersion because of their reduced Thomson scattering amplitude, but they contribute equally to screening. Therefore, one finds $k_{\rm S}^2 = k_{\rm D}^2 + k_{\rm i}^2$ with

$$k_{\rm i}^2 = \frac{4\pi\alpha}{T} \sum_j n_j Z_j^2, \tag{6.59}$$

where the sum is over all species j with charge $Z_j e$.

Comparing Eq. (6.55) with the corresponding expression for the plasma frequency Eq. (6.40) reveals that in the relativistic limit ($v = 1$) $k_{\rm D}^2 = 3\omega_{\rm P}^2$. It is clear that apart from a numerical factor they must be the same because a relativistic plasma has only one natural scale, namely a typical electron energy.

In the limit of degenerate electrons the integral is also easily solved and leads to the familiar Thomas-Fermi wave number[33] (Jancovici 1962)

$$k_S^2 = k_{TF}^2 = \frac{4\alpha}{\pi} E_F p_F = \frac{3\omega_P^2}{v_F^2}. \tag{6.60}$$

However, k_D^2 always exceeds k_{TF}^2 so that in a medium of degenerate electrons and nondegenerate ions the main screening effect is from the latter. Recall that the Fermi momentum is related to the electron density by $n_e = p_F^3/3\pi^2$ and the Fermi energy is $E_F = (p_F^2 + m_e^2)^{1/2}$.

To compare the Thomas-Fermi with the Debye scale take the non-relativistic limit $(k_{TF}/k_D)^2 = \frac{3}{2} T/(E_F - m_e)$. This is much less than 1 or the medium would not be degenerate whence $k_{TF} \ll k_D$. Therefore, if the electrons are degenerate and the ions nondegenerate, a test charge is mostly screened by the polarization of the ion "fluid" because the electrons form a "stiff" background. Unfortunately, one often finds calculations in the literature which include screening by the electrons (screening scale k_{TF}) but ignore the ions. The resulting error need not be large because the screening scale typically appears logarithmically in the final answer (see below).

Screening effects in Coulomb processes are often found to be implemented by a modified Coulomb propagator

$$\frac{1}{|\mathbf{q}|^4} \rightarrow \frac{1}{(\mathbf{q}^2 + k_S^2)^2}, \tag{6.61}$$

where $\mathbf{q}$ is the momentum transfer carried by the intermediate photon. This substitution arises if one considers Coulomb scattering from a Yukawa-like charge distribution. It corresponds to the substitution Eq. (6.57), i.e. to a single charge with an exponential screening cloud. This picture is appropriate if the Coulomb scattering process itself is so slow that the charged particles move around and rearrange themselves so much that the probe, indeed, sees an average screening cloud.

In the opposite limit, a given probe sees a certain configuration, a different probe a different one, etc., and one has to average over all of these possibilities. In this case one needs to square the matrix element first, and then take an average over different medium configurations. For Eq. (6.61) one averages first, obtains an average scattering amplitude or matrix element, and squares afterward.

---

[33]For a textbook derivation from a Thomas-Fermi model see Shapiro and Teukolsky (1983). Note that they work in the nonrelativistic limit: their $E_F = p_F^2/2m_e$.

It depends on the physical circumstances which procedure is a better approximation. If one considers bremsstrahlung processes with degenerate electrons scattering off nondegenerate nuclei, the crossing time of an electron of a region the size k_S^{-1} is short compared to the crossing time of nuclei. Hence, the latter can be viewed as static, the probe sees one configuration at a time, and one certainly should use the "square first" procedure instead of Eq. (6.61) to account for screening. This is achieved by the following consideration of correlation effects.

6.4.2   Correlations and Static Structure Factor

The screening of electric fields in a plasma is closely related to correlations of the positions and motions of the charged particles. If a negative test charge is known to be in a certain position, the probability of finding an electron in the immediate neighborhood is less than average, while the probability of finding a nucleus is larger than average. It is this polarization of the surrounding plasma which screens a charge.

Take one particle of a given species to be the origin of a coordinate system, and take their average number density to be n. The electrostatic repulsion of the test charge causes a deviation of the surrounding charges from the average density by an amount

$$S(\mathbf{r}) = \delta^3(\mathbf{r}) + n\, h(\mathbf{r}), \tag{6.62}$$

where $h(\mathbf{r})$ measures the particle correlations. They vanish in an ideal Boltzmann gas: $h(\mathbf{r}) = 0$. The Fourier transform

$$S(\mathbf{q}) = \int d^3\mathbf{r}\, S(\mathbf{r})\, e^{-i\mathbf{q}\cdot\mathbf{r}} \tag{6.63}$$

is the *static structure factor* of the electron distribution. In the absence of correlations ($h = 0$) one has trivially $S(\mathbf{q}) = 1$.

In order to make contact with Debye screening consider the Yukawa potential of Eq. (6.57) which represents a charge density

$$\rho(\mathbf{r}) = \delta^3(\mathbf{r}) - \frac{k_S^2}{4\pi}\frac{e^{-k_S r}}{r}. \tag{6.64}$$

The volume integral of $\rho(\mathbf{r})$ vanishes, giving zero total charge, i.e. complete screening at infinity. If one imagines that only one species of charged particles is mobile on a uniform background of the opposite charge, then Eq. (6.64) implies correlations between the mobile species of $n\, h(\mathbf{r}) = -(k_S^2/4\pi r)\, e^{-k_S r}$. As expected, Debye screening corresponds to spatial anticorrelations of like-charged particles. Fourier transforming Eq. (6.64) yields the important result

$$S(\mathbf{q}) = \frac{\mathbf{q}^2}{\mathbf{q}^2 + k_S^2} \tag{6.65}$$

for the structure factor.

The assumption of one mobile species of particles on a uniform background corresponds to the model of a "one-component plasma." It is approximately realized in the interior of hot white dwarfs or the cores of red-giant stars where the degenerate electrons form a "stiff" background of negative charge in which the nondegenerate ions move.

In a nondegenerate situation, however, there are at least two mobile species, ions of charge Ze and electrons. For this two-component plasma Salpeter (1960) derived the structure functions

$$S_{ee}(\mathbf{q}) = \frac{\mathbf{q}^2 + Zk_D^2}{\mathbf{q}^2 + (1 + Z)k_D^2},$$

$$S_{ii}(\mathbf{q}) = \frac{\mathbf{q}^2 + k_D^2}{\mathbf{q}^2 + (1 + Z)k_D^2},$$

$$S_{ei}(\mathbf{q}) = \frac{k_D^2}{\mathbf{q}^2 + (1 + Z)k_D^2}. \tag{6.66}$$

The Fourier transform of the screening cloud around an electron is

$$S(\mathbf{q}) - S_{ee}(\mathbf{q}) - ZS_{ei}(\mathbf{q}) - \frac{\mathbf{q}^2}{\mathbf{q}^2 + k_S^2}, \tag{6.67}$$

with $k_S^2 = k_D^2 + k_i^2 = (1 + Z)\, k_D^2$. (Note that for only one species of ions $k_i^2 = Zk_D^2$.) Hence one reproduces a screened charge distribution which causes a Yukawa potential. However, the small-q behavior of S_{ee} or S_{ii} is very different: $S_{ee}(0) = Z/(1 + Z)$ in a two-component plasma while $S_{ee}(0) = 0$ for only one component.

6.4.3 Strongly Coupled Plasma

For low temperatures, the screening will not be of Yukawa type and the structure factor will deviate from the simple Debye formula. A plasma can be considered cold if the average Coulomb interaction energy between ions is much larger than typical thermal energies. To quantify this measure, one introduces the "ion-sphere radius" a_i by virtue of $n_i^{-1} = 4\pi a_i^3/3$ where n_i is the number density of the mobile particle species. Hence, a measure for the Coulomb interaction energy is $Z^2\alpha/a_i$, assuming the ions have charge Ze. One usually introduces the parameter

$$\Gamma \equiv \frac{Z^2\alpha}{a_i T} = \frac{(k_i a_i)^2}{3} \tag{6.68}$$

as a measure for how strongly the plasma is coupled, where $k_i^2 = 4\pi Z^2\alpha/T$. For $\Gamma \ll 1$ it is weakly coupled and approaches an ideal Boltzmann gas.

The Debye structure factor of a one-component plasma can be written as

$$S_D(\mathbf{q}) = \frac{|a_i \mathbf{q}|^2}{|a_i \mathbf{q}|^2 + 3\Gamma}. \tag{6.69}$$

This result applies even for large Γ if $|a_i \mathbf{q}| \ll 1$. For $\Gamma \gg 1$, the plasma is strongly coupled, and for $\Gamma \gtrsim 178$ the ions will arrange themselves in a body centered cubic lattice (Slattery, Doolen, and DeWitt 1980, 1982).

In Fig. 6.9 I show S and S_D as functions of $a_i q = |a_i \mathbf{q}|$ for $\Gamma = 2$, 10 and 100 where S was numerically determined (Hansen 1973; Galam and Hansen 1976). The emerging periodicity for a strongly coupled plasma is quite apparent. It is also clear that for $\Gamma \lesssim 1$ the Debye formula gives a fair representation of the structure factor while for a strongly coupled plasma it is completely misleading. The interior of white dwarfs is in the regime of large Γ, and old white dwarfs are believed to crystallize. (See Appendix D for an overview over the conditions relevant for stellar plasmas.)

6.4.4 Screened Coulomb Scattering

Armed with these insights one may turn to the issue of Coulomb scattering processes in a plasma. In the limit of nonrelativistic and essentially static sources for the electric fields the relevant quantity entering the matrix element is the Fourier component $\rho(\mathbf{q})$ of the charge distribution $\rho(\mathbf{r})$ where $\mathbf{q}$ is the momentum transferred by the Coulomb field to the sources. The squared matrix element thus involves the quantity $\rho(\mathbf{q})\rho^*(\mathbf{q})$ which is $\rho(\mathbf{q})\rho(-\mathbf{q})$ because $\rho(\mathbf{r})$ is real. Taking a statistical average over all possible configurations of the charge distribution leads to a rate proportional to

$$S(\mathbf{q}) = \langle \rho(\mathbf{q})\rho(-\mathbf{q}) \rangle. \tag{6.70}$$

This is the static structure factor introduced earlier as a measure of the correlation between the charged particles of the medium.

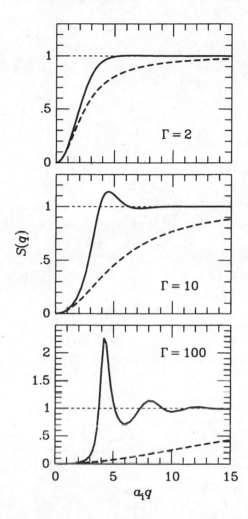

Fig. 6.9. Static structure factor for a one-component plasma according to the numerical calculations of Hansen (1973) and Galam and Hansen (1976). The dashed lines correspond to the Debye structure factor Eq. (6.69).

Without correlations, the squared matrix element involves $|\mathbf{q}|^{-4}$ from the Coulomb propagator. In order to account for screening effects one should substitute

$$|\mathbf{q}|^{-4} \rightarrow |\mathbf{q}|^{-4} S(\mathbf{q}) \tag{6.71}$$

which implies

$$\frac{1}{|\mathbf{q}|^4} \rightarrow \frac{1}{\mathbf{q}^2 (\mathbf{q}^2 + k_S^2)} \tag{6.72}$$

in the weak-screening limit (Debye screening).

The difference in a scattering cross section implied by Eq. (6.72) relative to (6.61) is easily illustrated. Observe that a cross section involving a Coulomb divergence is typically of the form

$$\int_{-1}^{+1} dx \, \frac{(1 - x) \, f(x)}{(1 - x)^2}, \tag{6.73}$$

where x is the cosine of the scattering angle of the probe. Here, $f(x)$ is a slowly varying function which embodies the details of the scattering or bremsstrahlung process. If this function is taken to be a constant, the two screening prescriptions amount to the two integrals

$$\int_{-1}^{+1} dx \, \frac{1}{(1 - x + \kappa^2)} = \log \left(\frac{2 + \kappa^2}{\kappa^2} \right),$$

$$\int_{-1}^{+1} dx \, \frac{(1 - x)}{(1 - x + \kappa^2)^2} = \log \left(\frac{2 + \kappa^2}{\kappa^2} \right) - \frac{2}{2 + \kappa^2}, \tag{6.74}$$

where $\kappa^2 \equiv k_S^2/2\mathbf{p}^2$ is the screening scale expressed in units of the initial-state momentum of the probe. Usually, it far exceeds the screening scale whence $\kappa^2 \ll 1$. Then Eq. (6.72) yields a cross section proportional to $\log(4\mathbf{p}^2/k_S^2)$ while Eq. (6.61) gives $[\log(4\mathbf{p}^2/k_S^2) - 1]$.

Thus, if one is only interested in a rough estimate, either screening prescription and any reasonable screening scale yield about the same result. For an accurate calculation, however, one needs to identify the dominant source of screening (for example, the nondegenerate ions in a degenerate plasma and not the electrons), and the appropriate moderation of the Coulomb propagator, usually Eq. (6.72).

6.5 Plasmon Decay in Neutrinos[34]

6.5.1 Millicharged Neutrinos

Transverse and longitudinal electromagnetic excitations in a plasma are both kinematically able to decay into neutrino pairs (Fig. 6.10) of sufficiently small mass, namely $2m_\nu < K^2$ where K is the plasmon[35] four-momentum. In the following, the neutrinos are always taken to be massless relative to the plasma frequency and so $K^2 > 0$ is required which restricts longitudinal excitations to $k < k_1$, the wave number where their dispersion relation crosses the light cone.

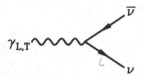

Fig. 6.10. Plasmon decay in neutrinos.

In addition, a ν-γ-interaction is required which does not exist in the standard model. Still, plasmon decays occur because the medium itself mediates an effective coupling as will become clear below. As an easy start, however, consider the hypothesis that neutrinos carry small electric charges ("millicharges"). Interestingly, this possibility is not excluded by the structure of the standard model and has received some recent attention in the literature (Sect. 7.3.2).

With a neutrino millicharge e_ν the interaction with the electromagnetic vector potential A is the standard expression

$$\mathcal{L}_{\text{int}} = -ie_\nu \, \overline{\psi}_\nu \gamma^\alpha \psi_\nu \, A_\alpha. \tag{6.75}$$

The spin-summed squared matrix element is of the form

$$\sum_{\text{spins}} |\mathcal{M}|^2 = M_{\alpha\beta} P^\alpha \overline{P}^\beta \tag{6.76}$$

where explicitly

$$M_{\alpha\beta} = 4e_\nu^2 \, Z \, (g_{\alpha\beta} + 2\epsilon_\alpha^* \epsilon_\beta). \tag{6.77}$$

Here, Z is the renormalization constant (Sect. 6.2.3 and 6.3.6), P and $\overline{P}$ are the ν and $\overline{\nu}$ four-momenta, and ϵ is the plasmon polarization vector for which one uses the basis vectors of Eqs. (6.27) and (6.28).

---

[34]I closely follow Haft (1993).

[35]In this section the term "plasmon" refers to both transverse and longitudinal electromagnetic excitations in a medium.

The decay width of a plasmon with four-momentum $K = (\omega, \mathbf{k})$ in the medium frame, and with a definite polarization is

$$\Gamma = \int \frac{d^3\mathbf{p}}{2E_{\mathbf{p}}(2\pi)^3} \frac{d^3\overline{\mathbf{p}}}{2E_{\overline{\mathbf{p}}}(2\pi)^3} (2\pi)^4 \delta^4(K - P - \overline{P}) \frac{1}{2\omega} \sum_{\text{spins}} |\mathcal{M}|^2.$$

(6.78)

Because of Eq. (6.76) one may use Lenard's (1953) formula

$$\int \frac{d^3\mathbf{p}}{2E_{\mathbf{p}}} \frac{d^3\overline{\mathbf{p}}}{2E_{\overline{\mathbf{p}}}} P^\alpha \overline{P}^\beta \, \delta^4(K - P - \overline{P}) = \frac{\pi}{24} \left(K^2 g^{\alpha\beta} + 2K^\alpha K^\beta \right).$$

(6.79)

With $\alpha_\nu \equiv e_\nu^2/4\pi$ this leads to

$$\Gamma = \alpha_\nu \, Z \, (\omega^2 - k^2)/3\omega,$$

(6.80)

where the normalization $\epsilon_\alpha^* \epsilon^\alpha = -1$ for transverse and time-like longitudinal plasmons was used as well as $\epsilon \cdot K = 0$. Γ applies to both transverse and longitudinal plasmons with the appropriate $Z_{\text{T,L}}$. For a chosen three-momentum $k = |\mathbf{k}|$ the quantities Z, ω, and $K^2 = \omega^2 - k^2$ are all functions of k by virtue of the dispersion relation $K^2 = \pi_{\text{T,L}}(K)$.

In the classical limit transverse plasmons propagate like massive particles with $K^2 = \omega_{\text{P}}^2$ and $Z_{\text{T}} = 1$. Then $\Gamma_{\text{T}} = \frac{1}{3}\alpha_\nu \, \omega_{\text{P}}^3 \, (\omega_{\text{P}}/\omega)$ where the last factor is recognized as a Lorentz time-dilation factor. For a general dispersion relation which is not Lorentz covariant it makes little sense, of course, to express the decay rate in the plasmon frame. An example is the classical limit for the longitudinal mode for which to zeroth order in T/m_e the frequency is $\omega = \omega_{\text{P}}$, $Z_{\text{L}} = \omega^2/K^2$, and then $\Gamma_{\text{L}} = \frac{1}{3}\alpha_\nu\omega_{\text{P}}$ with the restriction $k < \omega_{\text{P}}$.

6.5.2 Neutrino Dipole Moments

Another direct coupling between neutrinos and photons arises if the former have electric or magnetic dipole or transition moments (Sects. 7.2.2 and 7.3.2)

$$\mathcal{L}_{\text{int}} = \frac{1}{2} \sum_{a,b} \left(\mu_{ab} \, \overline{\psi}_a \sigma_{\mu\nu} \psi_b + \epsilon_{ab} \, \overline{\psi}_a \sigma_{\mu\nu} \gamma_5 \psi_b \right) F^{\mu\nu}.$$

(6.81)

Here, F is the electromagnetic field tensor, $\sigma_{\mu\nu} = \gamma_\mu \gamma_\nu - \gamma_\nu \gamma_\mu$, and ψ_a with $a = \nu_{1,2,3}$ or $\nu_{e,\mu,\tau}$ are the neutrino fields.

The squared matrix element is of the form Eq. (6.76); for a magnetic dipole coupling μ of a single flavor one finds

$$M_{\alpha\beta} = 4\mu^2 Z \left(2K_\alpha K_\beta - 2K^2 \epsilon_\alpha^* \epsilon_\beta - K^2 g_{\alpha\beta}\right). \tag{6.82}$$

This leads to a decay rate

$$\Gamma = \left(\mu^2/24\pi\right) Z \left(\omega^2 - k^2\right)^2/\omega, \tag{6.83}$$

applicable to either plasmon polarization with the appropriate Z and dispersion relation.

In the presence of electric and magnetic transition moments one obtains the same result with

$$\mu^2 \to \sum_{a,b} \left(|\mu_{ab}|^2 + |\epsilon_{ab}|^2 \right). \tag{6.84}$$

There is no interference term between the electric and magnetic couplings. The presence of transition moments allows for plasmon decays of the sort $\gamma \to \nu_a \bar{\nu}_b$ with different flavors $a \neq b$, doubling the final states which may be $\nu_a \bar{\nu}_b$ or $\nu_b \bar{\nu}_a$.

6.5.3   Standard-Model Couplings

In a medium there is an effective coupling between electromagnetic fields and neutrinos mediated by the ambient electrons. For the purpose of a plasmon-decay calculation it can be visualized with the Feynman graph Fig. 6.11 which corresponds to the "photoneutrino process" or Compton production of neutrino pairs (Sect. 3.2.4). However, the final-state electron can have the same four-momentum as the initial state which amounts to forward scattering of the electrons. In this case the energy-momentum transfer to the electron vanishes, allowing for a coherent superposition of these amplitudes from all electrons.

Fig. 6.11. Compton production of neutrino pairs (nonzero momentum transfer to electrons) or plasmon decay (electron forward scattering).

With the dimensionless coupling constants C_V and C_A given in Appendix B the neutral-current ν-e-interaction is

$$\mathcal{L}_{\text{int}} = -\frac{G_{\text{F}}}{\sqrt{2}}\, \overline{\psi}_e \gamma_\alpha (C_V - C_A \gamma_5)\psi_e\, \overline{\psi}_\nu \gamma^\alpha (1 - \gamma_5)\psi_\nu. \qquad (6.85)$$

The vector-current has the same structure that pertains to the electron interaction with photons, $\mathcal{L}_{\text{int}} = -ie\,\overline{\psi}_e \gamma_\alpha \psi_e\, A^\alpha$. Therefore, after performing a thermal average over the electron forward scattering amplitudes the plasmon decay is represented by the Feynman graph Fig. 6.12 which is identical with Fig. 6.1 with one photon line replaced by a neutrino pair. As far as the electrons are concerned, photon forward scattering $\gamma \to \gamma$ is the same as the conversion $\gamma \to \nu\overline{\nu}$.

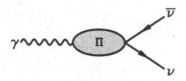

Fig. 6.12. Photon-neutrino coupling by the photon polarization tensor; the second photon line in Fig. 6.1 was replaced by a neutrino pair.

Put another way, in a plasma the propagation of an electromagnetic excitation is accompanied by an organized oscillation of the electrons. This is particularly obvious for longitudinal modes which *are* the collective oscillation of the electron gas, but it also applies to transverse modes. The collective motion is the medium response $J_{\text{ind}} = \Pi A$ to an electromagnetic excitation, a response which is characterized by the polarization tensor. The coherent electron oscillations serve as sources for the neutrino current whence they emit neutrino pairs.

The electron collective motion is an oscillation of their location or density while their spins remain unaffected apart from relativistic corrections. Because the axial-vector current represents the electron spin density, and because to lowest order no collective spin motion is expected as a response to an electromagnetic excitation, the axial-vector neutrino coupling to electrons will contribute very little to plasmon decay. This remains true in a relativistic plasma. For example, Koyama, Itoh, and Munakata (1986) found numerically that the axial vector contributes less than 0.01% to neutrino emission by the plasma process for all conditions of astrophysical interest. A detailed study of the axial response function can be found in Braaten and Segel (1993). For the present discussion I will not worry any further about the axial-vector contribution.

The matrix element for the interaction between neutrinos and photons can then be read from the effective vertex

$$i \frac{C_V G_F}{e\sqrt{2}} A^\alpha \Pi_{\alpha\beta} \overline{\psi}_\nu \gamma^\alpha (1 - \gamma_5) \psi_\nu. \tag{6.86}$$

The electric charge in the denominator removes one such factor contained in Π which was calculated for photon forward scattering. In the matrix element, Π is to be taken at the four-momentum K of the photon. Besides plasmon decay $\gamma \to \nu\bar{\nu}$, this interaction also allows for processes such as Cherenkov absorption $\gamma\nu \to \nu$ or emission $\nu \to \gamma\nu$.

For the decay of a plasmon with a polarization vector ϵ and four-momentum K the squared matrix element has the form Eq. (6.76) with

$$M_{\alpha\beta} = 8 \frac{G_F^2}{2e^2} \pi_{\mathrm{T,L}}^2 \left(g_{\alpha\beta} + 2\epsilon_\alpha^* \epsilon_\beta \right). \tag{6.87}$$

Because the plasmon is a propagating mode it obeys its dispersion relation, i.e. $\pi_{\mathrm{T,L}} = \omega^2 - k^2$. The decay rate is then

$$\Gamma = \frac{C_V^2 G_F^2}{48\pi^2 \alpha} Z_{\mathrm{T,L}} \frac{(\omega^2 - k^2)^3}{\omega}. \tag{6.88}$$

This result was first derived by Adams, Ruderman, and Woo (1963), the correct Z for the longitudinal case was first derived by Zaidi (1965). This equation is understood "on shell" where ω depends on k through the dispersion relation $\omega^2 - k^2 = \pi_{\mathrm{T,L}}(\omega, k)$.

6.5.4 Summary of Decay Rates

In order to express the decay rates in a compact form, recall that on shell the scale for K^2 is set by the plasma frequency ω_{P}. Therefore, it is useful to define

$$\hat{\pi}_{\mathrm{T,L}}(k) \equiv \frac{\pi_{\mathrm{T,L}}(\omega_k, k)}{\omega_{\mathrm{P}}^2}, \tag{6.89}$$

where ω_k is the frequency related to k as a solution of the dispersion equation $\omega^2 - k^2 = \pi(\omega, k)$. Recall that $1 \leq \hat{\pi}_{\mathrm{T}} < \frac{3}{2}$ while $0 \leq \hat{\pi}_{\mathrm{L}} \leq 1$ for a time-like K^2 ($k < k_1$). At $k = k_1$ the L dispersion relation crosses the light cone.

The decay rates of Eqs. (6.80), (6.83) and (6.88) of a plasmon with three-momentum k are then expressed as

$$
\Gamma_k = \frac{4\pi}{3} \frac{Z_k}{\omega_k} \times
\begin{cases}
\alpha_\nu & \dfrac{\omega_{\rm P}^2 \,\hat\pi_k}{4\pi} & \text{Millicharge,} \\[3ex]
\dfrac{\mu^2}{2} & \left(\dfrac{\omega_{\rm P}^2 \,\hat\pi_k}{4\pi}\right)^2 & \text{Dipole Moment,} \\[3ex]
\dfrac{C_V^2 G_{\rm F}^2}{\alpha} & \left(\dfrac{\omega_{\rm P}^2 \,\hat\pi_k}{4\pi}\right)^3 & \text{Standard Model,}
\end{cases}
\qquad (6.90)
$$

where for Z_k and $\hat\pi_k$ the T or L value appropriate for the chosen polarization must be used.

6.5.5   Energy-Loss Rates

It is now an easy task to calculate stellar energy-loss rates for the plasma process. An integration over the Bose-Einstein distributions of the transverse and longitudinal plasmons yields for the energy-loss rate per unit volume

$$
Q_{\rm T} = \frac{2}{2\pi^2} \int_0^\infty dk\, k^2 \, \frac{\Gamma_{\rm T}\, \omega}{e^{\omega/T} - 1},
$$

$$
Q_{\rm L} = \frac{1}{2\pi^2} \int_0^{k_1} dk\, k^2 \, \frac{\Gamma_{\rm L}\, \omega}{e^{\omega/T} - 1}.
\qquad (6.91)
$$

In $Q_{\rm T}$ the factor 2 counts two polarization states. In $Q_{\rm L}$ the integration can be extended only to the wave number k_1 where the L dispersion relation crosses the light cone—for $k > k_1$ decays are kinematically forbidden. In either case $\Gamma_{\rm T,L}$ and ω are functions of k, the latter given by the dispersion relation.

For the specific neutrino interaction models discussed in the previous section one obtains with the decay rates of Eq. (6.90)

$$
Q = \frac{8\zeta_3}{3\pi} T^3 \times
\begin{cases}
\alpha_\nu & \dfrac{\omega_{\rm P}^2}{4\pi} \, Q_1 & \text{Millicharge,} \\[3ex]
\dfrac{\mu^2}{2} & \left(\dfrac{\omega_{\rm P}^2}{4\pi}\right)^2 Q_2 & \text{Dipole Moment,} \\[3ex]
\dfrac{C_V^2 G_{\rm F}^2}{\alpha} & \left(\dfrac{\omega_{\rm P}^2}{4\pi}\right)^3 Q_3 & \text{Standard Model,}
\end{cases}
\qquad (6.92)
$$

where $\zeta_3 \approx 1.202$ refers to the Riemann Zeta function. The dimensionless emission rates Q_n for the three cases are each a sum of a transverse and longitudinal term, $Q_n = Q_{L,n} + Q_{T,n}$ where

$$Q_{L,n} = \frac{1}{4\zeta_3 T^3} \int_0^{k_1} dk\, k^2 \frac{Z_L\, \hat{\pi}_L^n}{e^{\omega/T} - 1} = \frac{1}{4\zeta_3 T^3} \int_0^{k_1} dk\, k^2 \frac{\omega^2}{\omega_P^2} \frac{\tilde{Z}_L\, \hat{\pi}_L^{n-1}}{e^{\omega/T} - 1},$$

$$Q_{T,n} = \frac{1}{2\zeta_3 T^3} \int_0^\infty dk\, k^2 \frac{Z_T\, \hat{\pi}_T^n}{e^{\omega/T} - 1}. \tag{6.93}$$

The second equation for $Q_{L,n}$ relies on the definition Eq. (6.53), i.e. $Z_L = \tilde{Z}_L \omega^2/(\omega^2 - k^2)$, and $\omega^2 - k^2 = \pi_L$ was used.

The normalization factors were chosen such that $Q_{T,n} = 1$ if the plasmons are treated as effectively massless particles for the phase-space integration. Then $Z_T = \hat{\pi}_T = 1$ which is a reasonable approximation in a nondegenerate, nonrelativistic plasma. In that limit to lowest order $k_1 = \omega_P$, $\tilde{Z}_L = 1$, and $\pi_L = \omega_P^2 - k^2$. Therefore, in this limit $Q_{L,n} \ll Q_{T,n}$. In fact, the longitudinal emission rate is of comparable importance to the transverse one only in a narrow range of parameters of astrophysical interest (Haft, Raffelt, and Weiss 1994).

These simple approximations, however, are not adequate for most of the conditions where the plasma process is important. In Appendix C the numerical neutrino emission rates are discussed; a comparison between Fig. C.1 and Fig. 6.3 reveals that the plasma process is important for $0.3 \lesssim \omega_P/T \lesssim 30$, i.e. transverse plasmons can be anything from relativistic to entirely nonrelativistic. For a practical stellar evolution calculation one may use the analytic approximation formula for the plasma process discussed in Appendix C, based on the representation of the dispersion relations of Sect. 6.3.

The main issue at stake in this book, however, is nonstandard neutrino emission from the direct electromagnetic couplings discussed in Sect. 6.5. Instead of constructing new numerical emission rate formulae one uses the existing ones for the standard-model (SM) couplings and scales them to the novel cases. Numerically, one finds

$$\frac{Q_{\text{charge}}}{Q_{\text{SM}}} = \frac{\alpha_\nu \alpha\, (4\pi)^2}{C_V^2 G_F^2 \omega_P^4} \frac{Q_1}{Q_3} = 0.664\, e_{14}^2 \left(\frac{10\,\text{keV}}{\omega_P}\right)^4 \frac{Q_1}{Q_3},$$

$$\frac{Q_{\text{dipole}}}{Q_{\text{SM}}} = \frac{\mu^2 \alpha\, 2\pi}{C_V^2 G_F^2 \omega_P^2} \frac{Q_2}{Q_3} = 0.318\, \mu_{12}^2 \left(\frac{10\,\text{keV}}{\omega_P}\right)^2 \frac{Q_2}{Q_3}, \tag{6.94}$$

where $e_{14} = e_\nu/10^{-14} e$ and $\mu_{12} = \mu/10^{-12}\mu_B$ with $\mu_B = e/2m_e$.

Contours for Q_1/Q_3 and Q_2/Q_3 are shown in Fig. 6.13 according to Haft, Raffelt, and Weiss (1994). Replacing these ratios by unity in a practical stellar evolution calculation introduces only a small error.

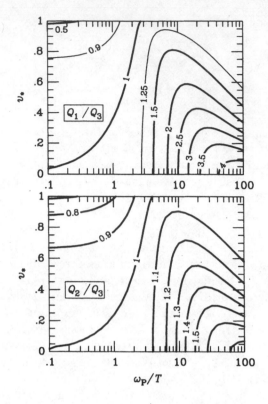

Fig. 6.13. Contours of Q_1/Q_3 and Q_2/Q_3 defined by Eq. (6.93) in the plane defined by the plasma frequency ω_P and a "typical" electron velocity v_* discussed in Sect. 6.3. See Fig. 6.3 for contours of v_* and ω_P in the T-ρ-plane. (Adapted from Haft, Raffelt, and Weiss 1994.)

6.5.6 Astrophysical Bounds on Neutrino Electromagnetic Properties

The plasma decay process is the most important neutrino emission process for a large range of temperatures and densities. Moreover, it has an observable impact on the cooling of hot white dwarfs, and on the core mass at helium ignition in low-mass red giants. For a neutrino millicharge $e_\nu \gtrsim 10^{-14}e$ and a dipole moment $\mu \gtrsim 10^{-12}\mu_B$ the nonstandard plasmon decay rates in Eq. (6.94) begin to compete with the standard one if the plasma frequency is around 10 keV. Therefore, neutrino millicharges or dipole moments of this magnitude will have observable effects on these stars and thus can be excluded.

In Sect. 2.2.3 it has been discussed that neutrino emission cools hot white dwarfs so fast that there is a clear depression of the white-dwarf luminosity function at the hot and bright end relative to the simple Mestel law which takes only surface photon emission into account. This depression would be enhanced by additional cooling caused by dipole moments or millicharges, allowing one to derive a limit of about $\mu_{12} < 10$, although even for $\mu_{12} \approx 3$ a nonnegligible effect is apparent.

A more restrictive and probably more reliable limit can be derived from the properties of globular-cluster stars (Raffelt 1990b). To this end one may use the simple criteria derived in Sect. 2.5 which state that a novel energy-loss rate is constrained by $\langle \epsilon_x \rangle \lesssim 10 \, \text{erg} \, \text{g}^{-1} \, \text{s}^{-1}$ for the average core-conditions of a horizontal branch star, and for those of a low-mass red giant before the helium flash. In both cases $T \approx 10^8 \, \text{K}$. For this temperature, the plasma loss rates are shown in Fig. 6.14 as a function of density. The anomalous rates were obtained from the standard one according to Eq. (6.94), taking $Q_1/Q_3 = Q_2/Q_3 = 1$ and using the zero-temperature plasma frequency given in Eq. (D.12) as a function of density.

The first criterion of Sect. 2.5, based on the helium-burning lifetime of HB stars, requires calculating the energy loss rate at an average density which is below $10^4 \, \text{g} \, \text{cm}^{-3}$. Therefore, the medium is so dilute

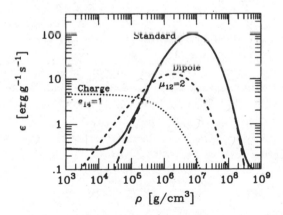

Fig. 6.14. Neutrino energy-loss rate in helium at $T = 10^8 \, \text{K}$. *Solid line:* Total standard rate. *Long dashes:* Standard plasma rate. *Short dashes:* Plasma rate induced by a dipole moment $\mu_\nu = 2 \times 10^{-12} \, \mu_{\text{B}}$. *Dots:* Plasma rate induced by a neutrino "millicharge" $e_\nu = 10^{-14} \, e$.

that one may employ the simple analytic form Eq. (6.92) of the emission rate with $Q_n = 1$. The core of HB stars consists at first of helium, later also of carbon and oxygen, for all of which $Y_e = 0.5$. Then,

$$\epsilon_x = 1\,\mathrm{erg\,g^{-1}\,s^{-1}} \times T_8^3 \times \begin{cases} 5.0\,e_{14}^2 & \text{Millicharge,} \\ 0.098\,\mu_{12}^2\,\rho_4 & \text{Dipole Moment,} \\ 0.0127\,\rho_4^2 & \text{Standard Model,} \end{cases} \quad (6.95)$$

where $T_8 = T/10^8\,\mathrm{K}$ and $\rho_4 = \rho/10^4\,\mathrm{g\,cm^{-3}}$. The core averages for a typical HB star are $\langle T_8^3 \rangle = 0.44$, $\langle T_8^3\rho_4 \rangle = 0.47$ and $\langle T_8^3\rho_4^2 \rangle = 0.57$. The requirement $\langle \epsilon_x \rangle < 10\,\mathrm{erg\,g^{-1}\,s^{-1}}$ then gives the limits

$$e_\nu \lesssim 2\times10^{-14}\,e \quad \text{and} \quad \mu_\nu \lesssim 14\times10^{-12}\mu_B. \quad (6.96)$$

Of course, for such large dipole moments the core would grow far beyond its standard value before helium ignites, causing an additional acceleration of the HB lifetime. In fact, this indirect impact on the HB lifetime would be the dominant effect as shown, for example, by the numerical calculations of Raffelt, Dearborn, and Silk (1989).

From Fig. 6.14 it is clear that the dipole-induced emission rate is larger for the conditions of the second criterion, based on the helium-ignition argument where $\langle \rho \rangle \approx 2\times10^5\,\mathrm{g\,cm^{-3}}$. According to Eq. (D.12) the relevant plasma frequency is $\omega_P = 8.6\,\mathrm{keV}$ so that $Q_{\mathrm{charge}}/Q_{\mathrm{SM}} \approx 1.2\,e_{14}^2$ and $Q_{\mathrm{dipole}}/Q_{\mathrm{SM}} \approx 0.4\,\mu_{12}^2$ in Eq. (6.94). The average total emission rate is then given by the standard rate times $F_\nu = 1 + Q_j/Q_{\mathrm{SM}}$ where j stands for "charge" or "dipole." In order to prevent the core mass at helium ignition from exceeding its standard value by more than 5% one must require $F_\nu < 3$. Then one finds $e_\nu \lesssim 1.3\times10^{-14}\,e$ and $\mu_\nu \lesssim 2\times10^{-12}\mu_B$. For the dipole case, a detailed numerical implementation yielded $\mu_\nu \lesssim 3\times10^{-12}\mu_B$ (Sect. 2.5.2), nearly identical with this simple analytic estimate. The limit on the charge could also be slightly degraded and so I adopt

$$e_\nu \lesssim 2\times10^{-14}\,e \quad \text{and} \quad \mu_\nu \lesssim 3\times10^{-12}\mu_B \quad (6.97)$$

as the final globular-cluster limits.

6.6 Neutrino Form Factors in Media

In a medium, neutrinos can interact with photons using electrons or
other charged particles as go-betweens. The basic idea is to consider the
Compton process of Fig. 6.11 with the initial- and final-state electrons
in the same state, i.e. forward scattering for the electrons. Then one
may sum over all electrons of the medium. This coherent superposition
of the amplitudes from all electrons was used in the previous section to
calculate the standard-model plasmon decay rate. There, only on-shell
(propagating) photons were considered. In general one may consider
other cases, for example electromagnetic scattering by the exchange of
a space-like photon, or the behavior of neutrinos in an external electric
or magnetic field.

The neutrino electromagnetic form factors in vacuum will be studied
in Sect. 7.3.2. They can be classified as a charge radius, an anapole
moment, and an electric and a magnetic dipole moment. They are
induced by intermediate (virtual) charged particles such as charged
leptons or W bosons. In the present case the form factors are induced
by the real particles of the ambient heat bath. The effective Lagrangian
Eq. (7.19) is fundamentally Lorentz covariant, a fact which reduces
the number of possible form factors to four. While in a medium the
couplings may also be written in a nominally Lorentz covariant form,
the medium singles out an inertial frame, leading to more complicated
structures. This is analogous to dispersion which is simple in vacuum
(a mass term is the only possibility) while in a medium the dispersion
relations can be excruciatingly complicated.

Limiting the couplings to the ones mediated by electrons and protons, the induced photon coupling to the neutrino is proportional to A^μ
because these fermions couple by the usual $e\overline{\psi}\gamma_\mu\psi A^\mu$ interaction. The
neutrinos couple by their standard effective neutral-current interaction
$(G_F/\sqrt{2})\,\overline{\psi}_\nu\gamma_\alpha(1-\gamma_5)\psi_\nu\,\overline{\psi}_e\gamma^\alpha(C_V-C_A\gamma_5)\psi_e$ with the weak coupling
constants C_V and C_A given in Appendix B. Therefore, after summing
over all intermediate electron states the effective neutrino-photon interaction may be written in the form

$$\mathcal{L}_{\text{eff}} = -\sqrt{2}\,G_F\,\overline{\psi}_\nu\gamma_\alpha\tfrac{1}{2}(1-\gamma_5)\psi_\nu\,\Lambda^{\alpha\beta}A_\beta, \tag{6.98}$$

where $\Lambda^{\alpha\beta}$ is a matrix which depends on the medium properties and
on the energy-momentum transfer, i.e. the energy momentum K of the
photon line. It consists of a symmetric piece $\Lambda_V^{\alpha\beta}$ which is proportional
to C_V, and an antisymmetric piece $\Lambda_A^{\alpha\beta}$ which is proportional to C_A.

Explicit expressions in terms of the electron phase-space integrals were derived, e.g. by D'Olivo, Nieves, and Pal (1989) and by Altherr and Salati (1994),

$$\Lambda_V^{\alpha\beta} = 4e\, C_V \int \frac{d^3\mathbf{p}}{2E(2\pi)^3} \Big[f_{e^-}(\mathbf{p}) + f_{e^+}(\mathbf{p}) \Big]$$

$$\times \frac{(P\cdot K)^2 g^{\alpha\beta} + K^2 P^\alpha P^\beta - P\cdot K\,(K^\alpha P^\beta + K^\alpha P^\beta)}{(P\cdot K)^2 - \frac{1}{4}(K^2)^2}\,,$$

$$\Lambda_A^{\alpha\beta} = 2ie\, C_A \epsilon^{\alpha\beta\mu\nu} \int \frac{d^3\mathbf{p}}{2E(2\pi)^3} \Big[f_{e^-}(\mathbf{p}) - f_{e^+}(\mathbf{p}) \Big]$$

$$\times \frac{K^2\, P_\mu K_\nu}{(P\cdot K)^2 - \frac{1}{4}(K^2)^2}\,, \tag{6.99}$$

where $f_{e^\pm}(\mathbf{p})$ is the electron and positron phase-space distribution with $P = (E,\mathbf{p})$ the electron or positron four-momentum.

Instead of a neutrino pair, another photon can be thought of as being coupled to the electron line in Fig. 6.11 or 6.12, a process which represents photon forward scattering. Therefore, apart from overall coupling constants $\Lambda_V^{\alpha\beta}$ is identical with the electronic contribution to the photon polarization tensor $\Pi^{\alpha\beta}$ studied earlier in this chapter

$$\Lambda_V^{\alpha\beta} = (C_V/e)\, \Pi^{\alpha\beta}. \tag{6.100}$$

In an isotropic plasma, $\Pi^{\alpha\beta}$ is characterized by the two medium characteristics $\pi_T(\omega, k)$ and $\pi_L(\omega, k)$ which are functions of the photon four-momentum $K = (\omega, \mathbf{k})$ with $k = |\mathbf{k}|$. For the antisymmetric piece, in an isotropic medium the phase-space integration averages the spatial part of P^α to zero so that

$$\Lambda_A^{\alpha\beta} = 2ieC_A \epsilon^{\alpha\beta\mu 0} K_\mu\, a(\omega, k) \tag{6.101}$$

in terms of a single medium characteristic a.

The single most important application of the effective neutrino electromagnetic coupling is the photon decay process $\gamma \to \nu\bar{\nu}$ that was studied in the previous section. It turns out that $\Lambda_A^{\alpha\beta}$ contributes very little to the decay process so that $\Lambda_V^{\alpha\beta}$, or rather the polarization tensor $\Pi^{\alpha\beta}$ determines all aspects of the plasma process.

Another possible process is the Cherenkov emission of photons by neutrinos. Of course, because in a plasma transverse photons acquire an "effective mass," i.e. their dispersion relation is time-like ($\omega^2 - \mathbf{k}^2 > 0$), this process is kinematically forbidden. However, longitudinal electromagnetic excitations (plasmons), which exist only in the medium, propagate such that for some momenta $\omega^2 - \mathbf{k}^2 < 0$ (space-like four-momentum), allowing for Cherenkov emission. One finds statements in the literature that the neutrino energy transfer to the medium by this process exceeded the transfer by (incoherent) ν-e scattering (e.g. Oraevskiĭ and Semikoz 1984; Oraevskiĭ, Semikoz, and Smorodinskiĭ 1986; Semikoz 1987a). This is in conflict with the discussion of Kirzhnits, Losyakov, and Chechin (1990) who found on general grounds that the energy loss of a neutrino propagating in a stable medium was always bounded from above by the collisional energy loss, apart from a factor of order unity. Granting this, the Cherenkov process does not seem to be of great practical importance.

If neutrinos have masses and mix, decays of the form $\nu_2 \to \nu_1 \gamma$ are possible in vacuum, and can be kinematically possible in a medium if the photon "effective mass" does not exceed the neutrino mass difference $m_2 - m_1$. If kinematically allowed, this decay receives a contribution from the medium-induced coupling which may far exceed the vacuum decay rate. Explicit calculations were performed by a number of authors[36] who unfortunately ignored the kinematic constraint imposed by the photon dispersion relation. This is not a reasonable approximation in view of the relatively small neutrino masses that remain of practical interest. Further, in order to judge the importance of the medium-induced decay it is not relevant to compare with the vacuum decay rate, but rather one should compare with the collisional transition rate $\nu_2 e \to e \nu_1$ (mediated by photon exchange) which is the process with which the coherent reaction directly competes.

The photon decay as well as the Cherenkov process and the medium induced neutrino decay all have in common that the neutrino couples to an electromagnetic field which is a freely propagating wave, obeying the dispersion relation in the medium which is $\omega^2 - k^2 = \pi_{T,L}(\omega, k)$ for transverse and longitudinal excitations, respectively. However, one may also consider the effect of a static external electric or magnetic field.[37] To this end, one must take the static limit $\omega \to 0$ of the vertex functions $\Lambda_{V,A}^{\alpha\beta}(\omega, k)$. For an external static electric field the only nonvanishing component of the vector potential A^μ is A^0. Then

---

[36]D'Olivo, Nieves, and Pal (1990); Kuo and Pantaleone (1990); Giunti, Kim, and Lam (1991).

[37]This issue has been investigated in many works, e.g. Oraevskiĭ and Semikoz (1985, 1987), Semikoz (1987a,b), Nieves and Pal (1989c, 1994), Semikoz and Smorodinskiĭ (1988, 1989), and D'Olivo, Nieves, and Pal (1989).

$\Lambda_A^{\alpha\beta}$ does not contribute so that only the Λ_V^{00} component remains of interest. In the static limit Π^{00} is simply given by $\pi_L(0,k)$ which in turn can be identified with the square of the screening scale k_S^2 in a medium (Sect. 6.4.1). This implies that the neutrino interacts with the external electric field as if it had a charge

$$e_\nu = -(C_V/e)\sqrt{2}\,G_F k_S^2. \tag{6.102}$$

In a classical (nondegenerate, nonrelativistic) hydrogen plasma the screening scale is given by the Debye scale through $k_S^2 = 2k_D^2 = 2e^2 n_e/T$ with the electron density n_e so that the induced neutrino charge is $e_\nu = C_V e\, 2\sqrt{2}\,G_F n_e/T$. This induced charge is explained by the medium polarization caused by the weak force exerted by the presence of the neutrino.

While this induced charge is conceptually very interesting it does not seem to have any immediate practical consequences. Notably, it is not the relevant quantity for the interaction with a static magnetic field, i.e. one may not infer that neutrinos move on curved paths in magnetic fields. The presence of a "neutrino charge" was derived for the interaction with a static *electric* field! The relevant form factor for the interaction with a static magnetic field is identified by noting that now only the spatial components of A^μ are nonzero. In the static limit only the component Π^{00} of the polarization tensor survives when contracted with A^μ. Then there is no contribution from $\Lambda_S^{\alpha\beta}$ for the neutrino interaction with a magnetic field. The contribution from $\Lambda_A^{\alpha\beta}$ can be interpreted as a "normal" or "Dirac magnetic moment" induced by the medium (Semikoz 1987a; D'Olivo, Nieves, and Pal 1989)

$$\mu_\nu = -eC_A\sqrt{2}G_F\,4\pi\int_0^\infty dp\left[f_{e^-}(p)-f_{e^+}(p)\right]. \tag{6.103}$$

In the limit of a classical plasma this is $\mu_\nu = (e_\nu/2m_e)(2C_A/C_V)$ where e_ν is the induced electric charge of Eq. (6.102).

This induced Dirac magnetic moment is to be compared with the electron's Dirac moment $e/2m_e$, not with an anomalous moment. The former arises from the $e\bar\psi_e\gamma_\mu\psi_e A^\mu$ coupling, the latter is described by $\frac{1}{2}\mu_e\bar\psi_e\sigma_{\mu\nu}\psi_e F^{\mu\nu}$. This means that the induced dipole moment does not lead to neutrino spin precession—it only couples to left-handed states. It entails an energy difference between neutrinos moving in opposite directions along a magnetic field. The transverse part of the field has no impact on the neutrino—there is no spin precession, and no curvature of the trajectory. (These conclusions pertain to the limit of weak magnetic fields. For strong fields the modification of the electron wavefunctions, i.e. Landau levels rather than plane waves, would have to be used for a self-consistent treatment of the photon polarization tensor and thus, for the neutrino coupling to a magnetic field. For a first discussion see Oraevskiĭ and Semikoz 1991.)

In summary, on the basis of the existing literature it appears that the medium-induced electromagnetic form factors of neutrinos are of practical importance only for the photon decay process that was discussed in the previous section.

6.7 Neutrino Refraction

6.7.1 Neutrino Refractive Index

When neutrinos propagate in a medium they will experience a shift of their energy, similar to photon refraction, due to their coherent interaction with the medium constituents (Wolfenstein 1978). The neutrino refractive index can be calculated in the same way as that for any other particle which propagates in a medium, namely on the basis of the forward scattering amplitudes as discussed in Sect. 6.2.1. As one needs only forward scattering, and as the relevant medium constituents are protons, neutrons, electrons, and possibly other neutrinos, only the Feynman graphs of Fig. 6.15 need to be considered.[38]

In most situations of practical interest the energies of the neutrinos and of the medium particles are much smaller than the W and Z mass (80.2 and 91.2 GeV) so that the energy and momentum transferred by the gauge bosons is always much less than their mass.[39] This justifies to expand their propagators (energy-momentum transfer Q) as

$$D_{\mu\nu}(Q) = \frac{g_{\mu\nu}}{m_{Z,W}^2} + \frac{Q^2 g_{\mu\nu} - Q_\mu Q_\nu}{m_{Z,W}^4} + \dots \qquad (6.104)$$

and keep only the first term. (The second term is needed if the contribution of the first one cancels as in a CP symmetric medium—see

---

[38]In the formalism of finite temperature and density (FTD) field theory the amplitudes may be written in a more compact form so that the relevant Feynman graphs reduce to a tadpole and a bubble graph (Nötzold and Raffelt 1988; Nieves 1989; Pal and Pham 1989). Apart from a more compact notation, however, the FTD formalism leads to the same expressions as the "pedestrian" approach chosen here.

[39]See however Learned and Pakvasa (1995) as well as Domokos and Kovesi-Domokos (1995) for a discussion of the oscillations of very high-energy cosmic neutrinos for which this approximation is not adequate.

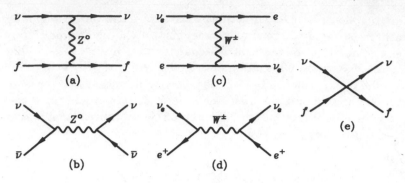

Fig. 6.15. Amplitudes contributing to forward scattering: (a) Neutral-current scattering for any ν or $\bar{\nu}$ on any f or $\bar{f}$ as target. (b) ν_e-$\bar{\nu}_e$ scattering. (c) ν_e-e charged-current scattering. (d) ν_e-e^+ charged-current scattering. (e) Effective four-fermion vertex in the low-energy limit.

Sect. 6.7.2.) This amounts to reducing the weak interaction to the low-energy Fermi effective Hamiltonian represented by graph (e) in Fig. 6.15. For the neutral-current processes (a) and (b) it is explicitly

$$\mathcal{H}_{\text{int}} = \frac{G_{\text{F}}}{\sqrt{2}} \, \bar{\psi}_f \gamma_\mu (C_V - C_A \gamma_5) \psi_f \, \bar{\psi}_{\nu_\ell} \gamma^\mu (1 - \gamma_5) \psi_{\nu_\ell}, \qquad (6.105)$$

where ψ_{ν_ℓ} is a neutrino field ($\ell = e, \mu, \tau$) while ψ_f represents fermions of the medium ($f = e, p, n$, or even neutrinos $\nu_{\ell'}$). Here, $G_{\text{F}} = 1.166 \times 10^{-5} \, \text{GeV}^{-2}$ is the Fermi constant. The relevant values of the vector and axial-vector weak charges C_V and C_A are given in Appendix B. In the low-energy limit the charged-current reactions (c) and (d) can also be represented as an effective neutral-current interaction of the same form with $C_V = C_A = 1$.

It is now straightforward to work out the forward scattering amplitudes. The axial-vector piece represents the spin of f and so it averages to zero if the medium is unpolarized. Then one finds for the refractive index of a neutrino (upper sign) or antineutrino (lower sign) with energy ω

$$n_{\text{refr}} - 1 = \mp C_V' \, G_{\text{F}} \, \frac{n_f - n_{\bar{f}}}{\omega \sqrt{2}}, \qquad (6.106)$$

where n_f and $n_{\bar{f}}$ are the number densities of fermions f and antifermions $\bar{f}$, respectively. The effective weak coupling constants C_V' are identical with the C_V given in Appendix B except for neutrinos as medium particles[40] which are left-handed and thus polarized. Therefore, $(1 - \gamma_5)\psi_\nu = 2\psi_\nu$ and $C_V' = 2C_V$.

Because we are dealing with forward scattering where recoil effects do not occur, the contributions from free or bound nucleons are the same. Therefore, Eq. (6.106) allows one to determine the refractive index of any normal medium. Because electric neutrality implies an excess density of electrons over positrons which balances against the protons, their neutral-current contributions cancel. (A possible exception is a π^- condensate that may exist in neutron stars.) An excess of ν_e over $\bar{\nu}_e$ appears to occur only in a young supernova core where neutrinos have a large chemical potential for the first few seconds after collapse.

All told, the dispersion relation for unmixed neutrinos, valid even in the nonrelativistic limit (Chang and Zia 1988), can be written in terms of a potential energy as

$$(\omega - V)^2 = k^2 + m^2, \tag{6.107}$$

where $V = -(n_{\text{refr}} - 1)\,\omega$. For all practical cases

$$V = \pm\sqrt{2}G_{\text{F}}n_B \times \begin{cases} (-\tfrac{1}{2}Y_n + Y_e + 2Y_{\nu_e}) & \text{for } \nu_e, \\ (-\tfrac{1}{2}Y_n + Y_{\nu_e}) & \text{for } \nu_{\mu,\tau}, \end{cases} \tag{6.108}$$

(upper sign ν, lower sign $\bar{\nu}$). Here, n_B is the baryon density and

$$Y_f \equiv \frac{n_f - n_{\bar{f}}}{n_B} \tag{6.109}$$

are the particle number fractions commonly used in astrophysics. Numerically,

$$\sqrt{2}\,G_{\text{F}}n_B = 0.762{\times}10^{-13}\,\text{eV}\,\frac{\rho}{\text{g cm}^{-3}} \tag{6.110}$$

with the mass density ρ.

A remark concerning the absolute sign of V is in order. The relative signs between the different C_V's can be worked out easily from the weak interaction structure of the standard model. Also, the relative sign of the effective neutral-current amplitudes which follow from Z° and W exchange follows directly, for example, from the FTD approach (Nötzold and Raffelt 1988). Thus to fix the overall sign it is enough

---

[40]In a supernova core or in the early universe it is not possible to distinguish between a "test neutrino" and a "medium neutrino." There, one has to study the nonlinear evolution of the entire ensemble self-consistently (Sect. 9.3.2).

to understand the absolute sign of neutrino-neutrino scattering. It is a case where identically "charged" fermions scatter by the exchange of a vector boson. The structure of this process is analogous to Coulomb scattering of like-charged particles which experience a repulsive force. (The exchange of a spin-0 or spin-2 boson leads to an attractive force.) Thus a neutrino of given momentum in a region of space filled with other neutrinos will have a positive potential energy V in addition to its kinetic energy. The correct absolute sign was first pointed out by Langacker, Leveille, and Sheiman (1983).

The deviation from relativistic propagation described by Eq. (6.107) can be expressed as an effective refractive index which includes the vacuum neutrino mass,

$$n_{\text{refr}} = \left[\left(1 - \frac{V}{\omega} \right)^2 - \frac{m^2}{\omega^2} \right]^{1/2} \quad \to \quad 1 - \frac{V}{\omega} - \frac{m^2}{2\omega^2} \qquad (6.111)$$

(relativistic limit). Therefore, the effect of a medium can be expressed as an effective mass

$$m_{\text{eff}}^2 = m^2 + 2\omega V. \qquad (6.112)$$

A numerical comparison with the vacuum mass is achieved by

$$\left(2\omega \sqrt{2} G_{\text{F}} n_B \right)^{1/2} = 3.91 \times 10^{-4} \, \text{eV} \left(\frac{\rho}{\text{g cm}^{-3}} \right)^{1/2} \left(\frac{\omega}{\text{MeV}} \right)^{1/2}. $$

$$(6.113)$$

Of course, the term "effective mass" is a misnomer because m_{eff} depends on the energy ω, and m_{eff}^2 can be negative depending on the vacuum mass, the medium composition, the flavor of the neutrino, and whether it is ν or $\bar{\nu}$.

The phase velocity $v_{\text{phase}} = \omega/k = n_{\text{refr}}^{-1}$ can be larger or less than the speed of light, depending on those parameters. However, the group velocity $v_{\text{group}} = d\omega/dk$ remains at its vacuum value $(1 + m^2/k^2)^{-1/2}$ for a given momentum k because ω is shifted by a constant amount V, independently of k.

Because normal media contain about equal numbers of protons and neutrons $Y_n \approx Y_e \approx \frac{1}{2}$ and so $V_{\nu_e} \approx \frac{1}{4}\sqrt{2} G_{\text{F}} n_B \approx -V_{\nu_{\mu,\tau}}$. Therefore, ν_e and $\nu_{\mu,\tau}$ are shifted by almost exactly opposite amounts. An exception is the proton-rich material of normal stars which initially contain about 75% hydrogen. Another exception is the neutron-rich matter of neutron stars where even V_{ν_e} is negative.

The absolute shift of the neutrino "masses" is rather negligible because we are dealing with highly relativistic particles. Even in this limit, however, the *difference* between the dispersion relation of different flavors is important for oscillation effects. Hence the most noteworthy medium effect is its *flavor birefringence*: ν_e and $\nu_{\mu,\tau}$ acquire different effective masses because of the charged-current contribution from ν_e-e scattering. The difference of their potentials is $V_{\nu_e} - V_{\nu_{\mu,\tau}} = \sqrt{2} G_F n_L$ with $n_L = Y_L n_B$ the lepton-number density where the number fraction of leptons is $Y_L = Y_e + Y_{\nu_e}$. Of course, neutrinos as a background medium contribute only in a young supernova core.

6.7.2 Higher-Order Effects

In the early universe one has nearly equal densities of particles and antiparticles with an asymmetry of about 10^{-9}, leading to a near cancellation of the refractive terms Eq. (6.106). One may think that the next most important contribution is from ν-γ scattering, a process closely related to the $\nu \to \nu' \gamma \gamma$ decay briefly discussed in Sect. 7.2.2. If one approximates the weak interactions by an effective four-fermion coupling the relevant amplitude is given by the graph Fig. 6.16 which on dimensional grounds should be of order αG_F. However, electromagnetic gauge invariance together with the left-handedness of the weak interaction implies that it vanishes identically (Gell-Mann 1961). For massive neutrinos the amplitude is proportional to $\alpha G_F m_\nu$, but even in this case it vanishes in the forward direction (Langacker and Liu 1992).

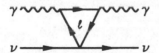

Fig. 6.16. Neutrino-photon scattering with an effective four-fermion weak interaction and a charged lepton ℓ in the loop. This amplitude vanishes entirely for massless neutrinos, and for massive ones it still vanishes in the forward direction.

For the lowest-order ν-γ contribution to the refractive index one must then use the full gauge-boson propagator and include all one-loop amplitudes required by the standard model. Such calculations were performed by Levine (1966) and by Cung and Yoshimura (1975) who found that the scattering amplitude was proportional to $\alpha G_F s / m_W^2$ (center of mass energy $\sqrt{s}$). Recently this problem was revisited by

Dicus and Repko (1993) who worked out explicitly the matrix elements and cross sections. From their results one can extract the forward scattering amplitude which leads to a refractive index for ν_ℓ ($\ell = e, \mu, \tau$) in a photon bath,

$$n_{\text{refr}} - 1 = \frac{\alpha}{4\pi} \frac{G_{\text{F}}}{m_W^2} \left[1 + \frac{4}{3} \ln \left(\frac{m_W^2}{m_\ell^2} \right) \right] \langle E_\gamma \rangle n_\gamma, \tag{6.114}$$

where $\langle \ldots \rangle$ means an average. Numerically, the term in square brackets is 32.9 for $\ell = e$ and thus not small, but with 4π in the denominator the whole expression is still of order $\alpha G_{\text{F}}/m_W^2$, i.e. of order[41] G_{F}^2. (See also Nieves, Pal, and Unger 1983; Nieves 1987; Langacker and Liu 1992.)

Because the ν-γ term is so small a larger refractive index arises if one includes the second term in the expansion Eq. (6.104) of the gauge-boson propagators. Of course, in graphs (a) and (c) of Fig. 6.15 forward scattering implies $Q = 0$ so that only the first term contributes, except when $f = \nu$ where the exchange graph has $Q \neq 0$. This case and graphs (b) and (d) yield a second-order contribution (Nötzold and Raffelt 1988). For a neutrino of flavor ℓ which is either e, μ, or τ it is

$$n_{\text{refr}} - 1 = \frac{8\sqrt{2}\,G_{\text{F}}}{3m_Z^2} \left(\langle E_{\nu_\ell} \rangle n_{\nu_\ell} + \langle E_{\bar{\nu}_\ell} \rangle n_{\bar{\nu}_\ell} \right)$$
$$+ \frac{8\sqrt{2}\,G_{\text{F}}}{3m_W^2} \left(\langle E_{\ell^-} \rangle n_{\ell^-} + \langle E_{\ell^+} \rangle n_{\ell^+} \right). \tag{6.115}$$

In this case the contributions from background fermions and antifermions add with the same sign, and the global sign remains the same for ν and $\bar{\nu}$ as test particles. In practice, only an electron-positron background is of relevance in the early universe so that ν_τ, for example, only feels a second-order contribution from other ν_τ's and $\bar{\nu}_\tau$'s. These results are of order G_{F}^2/α and thus they are the dominant contribution in a CP-symmetric plasma.

One-loop corrections to the amplitudes of Fig. 6.15 yield other higher-order terms which are of order G_{F}^2 like Eq. (6.114). They are still interesting because the loops involve charged leptons with a mass depending on their flavor. Therefore, the universality of the effective neutral-current interaction is broken on this level, leading to different refractive indices for different ν_ℓ. Between ν_e and ν_μ or ν_τ the medium is already birefringent to lowest order from ν_e-e charged-current interactions. Between ν_μ and ν_τ the one-loop correction dominates. Assum-

---

[41]Note that $m_Z^{-2} = \cos^2 \Theta_W \sin^2 \Theta_W \sqrt{2}\,G_{\text{F}}/\pi\alpha$ and $m_W^{-2} = \sin^2 \Theta_W \sqrt{2}\,G_{\text{F}}/\pi\alpha$.

ing electric neutrality it is (Botella, Lim, and Marciano 1987; see also Semikoz 1992 and Horvat 1993)

$$V_{\nu_\tau} - V_{\nu_\mu} = \frac{3G_F^2 m_\tau^2}{2\pi^2}\, n_B \left[\ln\left(\frac{m_W^2}{m_\tau^2}\right) - 1 + \frac{Y_n}{3}\right], \qquad (6.116)$$

with a sign change for $V_{\bar{\nu}_\tau} - V_{\bar{\nu}_\mu}$. The shift of the "effective mass" m_{eff}^2 is numerically

$$2\omega(V_{\nu_\tau} - V_{\nu_\mu}) = \left(2.06{\times}10^{-6}\,\text{eV}\right)^2 \frac{\rho}{\text{g cm}^{-3}} \frac{\omega}{\text{MeV}} \frac{6.61 + Y_n/3}{7},$$
$$(6.117)$$

much smaller than the corresponding difference between ν_e and $\nu_{\mu,\tau}$.

6.7.3 The Sun a Neutrino Lens?

The most important consequence of neutrino refraction in media is its impact on neutrino oscillations because different flavors experience a different index of refraction. In optics, the most notable consequence of refraction is the possibility to deflect light and thus to use lenses and other optical instruments. In principle, the same is also possible for neutrinos. The Sun, for example, could act as a gigantic neutrino lens.

One may easily calculate the deflection caused by a given body. If s is a unit vector along the direction of a propagating wave, and if s is a coordinate along the beam, the deflection is given by (Sommerfeld 1958)

$$|d\mathbf{s}/ds| = n_{\text{refr}}^{-1}|\mathbf{s} \times \nabla n_{\text{refr}}|, \qquad (6.118)$$

where n_{refr} is the refractive index. One may equally write

$$|d\alpha/ds| = n_{\text{refr}}^{-1}\,|\nabla_\perp n_{\text{refr}}|, \qquad (6.119)$$

where α is the angle relative to the local tangential vector, i.e. $d\alpha/ds$ is the local curvature of the beam, and $\nabla_\perp$ is the transverse gradient. The total angle of deflection is

$$|\Delta\alpha| = \int_{-\infty}^{+\infty} ds\, |\nabla_\perp n_{\text{refr}}| \qquad (6.120)$$

if the curvature is small which is the case for $|n_{\text{refr}} - 1| \ll 1$.

If the beam hits a spherically symmetric body (radius R) at an impact parameter $b < R$ its angle against the radial direction at a radius r from the center is $\sin\beta = b/r$ so that $\nabla_\perp n_{\text{refr}} = (b/r)\,\partial_r n_{\text{refr}}$.

Moreover, if s is measured from the point of closest approach one has $s = (r^2 - b^2)^{1/2}$ and so $ds = (r^2 - b^2)^{-1/2}\, r\, dr$. Altogether one finds

$$\Delta\alpha = -2\int_b^R dr\, \frac{b\,\partial_r n_{\mathrm{refr}}}{\sqrt{r^2 - b^2}}. \qquad (6.121)$$

The refraction by the surface of the body was ignored: it is assumed that it is "soft" with $\partial_r n_{\mathrm{refr}} = 0$ at $r = R$. The absolute sign was chosen such that $\Delta\alpha$ is positive if the spherical body acts as a "focussing lens." This is the case, for example, for light passing through the atmosphere of the Earth where $\partial_r n_{\mathrm{refr}} < 0$ and so the Sun near the horizon appears "lifted" (Sommerfeld 1958).

With the results for the neutrino refractive index of Sect. 6.7.1 one has for a beam of energy ω

$$\Delta\alpha = 2\,\frac{\sqrt{2}\,G_{\mathrm{F}} n_{B,c}}{\omega}\int_b^R dr\, \frac{b\,\partial_r[n_B(-\tfrac{1}{2}Y_n + Y_e)]}{n_{B,c}\sqrt{r^2 - b^2}}, \qquad (6.122)$$

where $n_{B,c}$ is the baryon density at the center of the lens. This expression applies to ν_e while for ν_μ or ν_τ the term Y_e is absent, and for antineutrinos the overall sign changes.

For the Sun with a central density of about $150\,\mathrm{g\,cm^{-3}}$ the overall coefficient is $2.3\times10^{-18}\,(10\,\mathrm{MeV}/\omega)$. The integral expression is dimensionless and thus of order unity. Therefore, the focal length of the Sun as a neutrino lens is of order $10^{18}\,R_\odot$ (solar radius) for $10\,\mathrm{MeV}$ neutrinos, or about the radius of the visible universe!

6.8 Majoron Decay

As a first application for the neutrino dispersion relation I consider the interaction of neutrinos with the hypothetical majorons (Sect. 15.7). These particles are Nambu-Goldstone bosons of a symmetry which is spontaneously broken by a Higgs fields which gives the neutrinos Majorana masses. For the present purposes it is enough to specify a pseudoscalar interaction

$$\mathcal{L}_{\mathrm{int}} = ih\,\overline{\psi}_\nu\gamma_5\psi_\nu\,\chi \qquad (6.123)$$

between the neutrinos of a given family and the massless majorons. Here, ψ_ν is a Majorana neutrino field while h is a dimensionless Yukawa coupling constant.

Majorana neutrinos are fermions with only two degrees of freedom. They correspond to a helicity-minus ν and a helicity-plus $\overline{\nu}$. In fact, if they are massless there is no operational distinction between a Majorana neutrino and the two active degrees of freedom of a Dirac neutrino. Therefore, according to Eq. (6.107) the dispersion relation for the helicity $\pm$ states of a Majorana neutrino is

$$E_\pm = (m^2 + \mathbf{p}^2)^{1/2} \mp V, \tag{6.124}$$

where the medium-induced potential V was given in Eq. (6.108). For ν_e it is $V = \sqrt{2} G_F(n_e - \frac{1}{2}n_n)$ with the electron and neutron densities n_e and n_n, respectively.

This dispersion relation implies that the medium is "optically active" with regard to the neutrino helicities, just as some media are birefringent with regard to the photon circular polarization. In the optical case the left-right symmetry (parity) is broken by the medium constituents which must have a definite handedness; sugar molecules are a well-known example. In the neutrino case parity is broken by the structure of the interaction; the medium itself is unpolarized.

Because there is an energy difference between the Majorana helicity states $\nu_\pm$ for a given momentum, decays $\nu_- \to \nu_+ \chi$ are kinematically allowed. For relativistic neutrinos the squared matrix element is found to be $|\mathcal{M}|^2 = 4h^2 P_1 \cdot P_2$ with the four-momenta $P_{1,2}$ of the initial and final neutrino state. The differential decay rate is then

$$d\Gamma = \frac{4h^2}{2E_1} \frac{d^3 \mathbf{p}_2}{2E_2 (2\pi)^3} \frac{d^3 \mathbf{k}}{2\omega (2\pi)^3} P_1 \cdot P_2 \, (2\pi)^4 \, \delta^4(P_1 - P_2 - K) \tag{6.125}$$

with the majoron four-momentum K. Integrating out the $d^3 \mathbf{k}$ variable removes the momentum δ function. The remaining differential decay is

$$\frac{d\Gamma}{dE_2} = \alpha_\chi \int_{-1}^{+1} dx \, \frac{P_1 \cdot P_2 \, p_2^2}{E_1 E_2 \omega} \, \delta(E_1 - E_2 - \omega), \tag{6.126}$$

where $x = \cos\theta$ for the angle between $\mathbf{p}_1$ and $\mathbf{p}_2$ and $\alpha_\chi \equiv h^2/4\pi$ is the majoron "fine-structure constant." In the δ function one must use $\omega = k = |\mathbf{p}_1 - \mathbf{p}_2| = (p_1^2 + p_2^2 - 2p_1 p_2 x)^{1/2}$. With $\int dx \, \delta[f(x)] = |df/dx|^{-1} = \omega/p_1 p_2$ and with energy-momentum conservation which yields $P_1 - P_2 = K$ and thus $P_1 \cdot P_2 = \frac{1}{2}(P_1^2 + P_2^2)$ one finds

$$\frac{d\Gamma}{dE_2} = \alpha_\chi \frac{(E_1^2 + E_2^2 - p_1^2 - p_2^2) \, p_2}{2 E_1 E_2 \, p_1}. \tag{6.127}$$

If one ignores the vacuum mass relative to V one has $p_{1,2} = E_{1,2} \mp V$ so that to lowest order in V

$$\frac{d\Gamma}{dE_2} = \alpha_\chi \frac{V(E_1 - E_2)}{E_1^2}. \tag{6.128}$$

Therefore, the final-state neutrino spectrum has a triangular shape where E_2 varies between 0 and E_1. The integrated decay rate is

$$\Gamma = \tfrac{1}{2}\alpha_\chi V \tag{6.129}$$

as first shown by Berezhiani and Vysotsky (1987). Various subleties were covered in the detailed discussion by Giunti et al. (1992).

Interestingly, under the same circumstances the decay $\chi \to \nu_+ \nu_-$ is equally possible (Rothstein, Babu, and Seckel 1993) and proceeds with the same rate Eq. (6.129)—see Berezhiani and Rossi (1994).

In these calculations it was assumed that the majoron is massless as it behooves a Nambu-Goldstone boson. Does this remain true in a medium? The majorons could have a Yukawa coupling g to electrons in which case one would expect on dimensional grounds that they develop a medium-induced "mass" of order $g(n_e/m_e)^{1/2}$ in analogy to the photon plasma mass. Indeed, if one works with a pseudoscalar coupling analogous to Eq. (6.123) one finds such a result. Even if they did not couple to electrons, in a supernova core there is a background of neutrinos to which majorons couple by assumption.

However, a pseudoscalar coupling is not appropriate for a Nambu-Goldstone boson as it is not invariant under a shift $\chi \to \chi + \chi_0$. The pseudoscalar expression is only the lowest-order expansion of an exponential coupling which respects the symmetry. Equivalently, a derivative coupling of the sort $(1/2f)\,\overline{\psi}\gamma_\mu\gamma_5\psi\,\partial^\mu\chi$ can be used which satisfies the symmetry explicitly (Sect. 14.2.3). Either way one finds that the forward scattering amplitude between Nambu-Goldstone bosons and fermions vanishes—there is no refractive index. The same conclusion was reached by Flynn and Randall (1988) on more general grounds.

For a suitable choice of parameters the medium-induced decay of electron neutrinos can deplete the solar neutrino flux before it leaves the Sun. However, it is doubtful if this effect could explain all current solar neutrino measurements (Sect. 10.8). For $h \gtrsim 10^{-6}$ a radical modification of the neutrino signal from a supernova collapse is expected (Sect. 15.7.2).

Chapter 7

Nonstandard Neutrinos

The phenomenological consequences of nonvanishing neutrino masses and mixings and of electromagnetic couplings are explored. The decay channels and electromagnetic properties of mixed neutrinos are discussed. Experimental, astrophysical, and cosmological limits on neutrino masses, decays, and electromagnetic properties are summarized.

7.1 Neutrino Masses

7.1.1 The Fermion Mass Problem

In the physics of elementary particles one currently knows of two categories of apparently fundamental fields: the spin-$\frac{1}{2}$ quarks and leptons on the one-hand side, and the spin-1 gauge bosons on the other. The former constitute "matter" while the latter mediate the electromagnetic, weak, and strong forces. The gauge-theory description of the interactions among these particles is renowned for its elegance and stunning in its success at accounting for all relevant measurements. At the same time it has many entirely loose ends. Perhaps the most puzzling problem is that of fermion masses and the related issue of the threefold replication of families: The electron and neutrino as well as the up and down quarks (which make up protons and neutrons) each come in two additional "flavors" or families which seem to differ from the first one only in their masses (Fig. 7.1).

The standard model of particle physics holds that all fermions and gauge bosons are fundamentally massless. The gauge symmetry forbids a fundamental mass for the latter while the masslessness of the former is indicated by the handedness of the weak interaction: only left-handed (l.h.) fermions feel this force while the right-handed (r.h.)

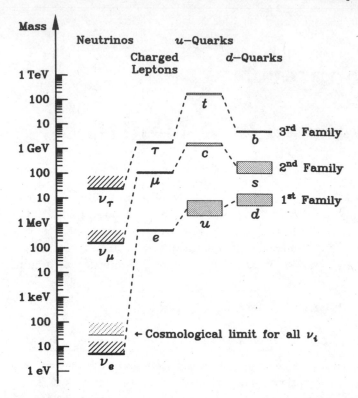

Fig. 7.1. Mass spectrum of elementary fermions according to the Review of Particle Properties (Particle Data Group 1994). For the top quark see CDF Collaboration (1995) and D0 Collaboration (1995). For neutrinos the experimental upper limits of Tab. 7.2 are shown.

ones are "sterile." Handedness, however, is not a Lorentz-invariant concept because a particle with a spin opposite to its momentum (l.h. or negative helicity) is r.h. when viewed by a sufficiently fast-moving observer. Only massless particles cannot be "overtaken" because they move with the speed of light and so they can be classified into l.h. and r.h. states without reference to a specific Lorentz frame.

Any deviation from relativistic propagation is thought to be a "refractive" effect, much as a photon acquires a nontrivial dispersion in a medium. Here, the "medium" is the spin-0 Higgs field Φ, a hypothetical third category of fundamental objects, which is believed to take on the nonzero classical value $\Phi_0 \approx 246\,\mathrm{GeV}$ in vacuum due to self-interactions ("spontaneous symmetry breaking"). Fermion fields ψ interact with the Higgs field by virtue of a Lagrangian $g\Phi\bar{\psi}\psi$ where g is a dimensionless (Yukawa) coupling constant. In vacuum, this coupling leads to an interaction term $g\Phi_0\bar{\psi}\psi$ which has the form of a standard Dirac mass term $m\bar{\psi}\psi$. Different fermion masses thus arise from different Yukawa couplings. Φ_0 does not depend on the Lorentz frame because of the scalar nature of the Higgs field and so $g\Phi_0\bar{\psi}\psi$ is the same in all frames, unlike a refractive photon "mass" in a medium.

It is not known whether the Higgs mechanism is the true source for the masses of the fundamental fermions. Experimentally, the Higgs particle (excitations of the Higgs field) has not yet been discovered, while theoretically the fermion masses are the least appealing aspect of the standard model because they require a host of ad hoc coupling constants which must be experimentally determined.

7.1.2 Dirac and Majorana Masses

Neutrinos break the pattern of Fig. 7.1 in that they are much lighter than the other members of a given family, a discrepancy which is most severe for the third family where cosmologically $m_{\nu_\tau} \lesssim 30\,\text{eV}$, eight and ten orders of magnitude less than m_τ and m_t, respectively! Moreover, neutrinos are different in that their r.h. chirality states are sterile because of the handedness of the weak interaction. The r.h. states interact with the rest of the world only by gravity and by a possible Yukawa coupling to the Higgs field.

It is frequently assumed that neutrinos do not couple to the Higgs field, and that the r.h. components do not even exist, assumptions which are part of the particle-physics standard model. In this case there are only two neutrino states for a given family as opposed to four states for the charged leptons. Actually, one may interpret the two components of such a neutrino as the spin states of a *Majorana fermion* which is defined to be its own antiparticle. Fermions with four distinct states are known as *Dirac fermions*. Naturally, a Majorana fermion cannot carry a charge as that would allow one to distinguish it from its antiparticle. A magnetic or electric dipole moment is equally forbidden: its orientation relative to the spin is reversed for antiparticles. For example, the neutron cannot be a Majorana fermion among other reasons because it carries a magnetic moment.

Because the r.h. neutrino components of a given family, if they exist, are sterile anyway, there is no practical distinction between massless Dirac and Majorana neutrinos except in a situation where gravitational interactions dominate. For example, even otherwise sterile neutrinos should be thermally emitted from black holes which are thought to emit blackbody radiation of all physical fields. Equally, they would have been produced in the very early universe when quantum gravitational effects dominate. However, their present-day cosmic density, like that of primordial gravitons, would be very dilute relative to microwave background photons.

If neutrinos were Majorana particles they could still have a mass even though it could not arise from the usual Higgs field which induces Dirac masses. However, a Majorana mass could arise from the interaction with a second Higgs field which also develops a vacuum expectation value. Then the smallness of the neutrino masses could be due to a small vacuum value of the new Higgs field while the Yukawa couplings would not need to be anomalously small.

It is also possible that the r.h. components of the neutrinos do exist, but are themselves (sterile) Majorana fermions with large masses. It should be noted that any Dirac fermion (four components) can be viewed as a combination of two Majorana fermions (two components each) with degenerate masses. Certain variations of such models ("see-saw mass models") predict for the light, interacting neutrinos

$$m_1 : m_2 : m_3 = m_e^2 : m_\mu^2 : m_\tau^2 \quad \text{or} \quad m_u^2 : m_c^2 : m_t^2. \tag{7.1}$$

The smallness of the neutrino masses is then a suppression effect by the large mass scale of the heavy sterile state whose mass would arise, for example, at the grand unification scale of $10^{15}-10^{16}$ GeV.

For an elementary introduction to the most common models for neutrino masses see, for example, Mohapatra and Pal (1991). The dizzying variety of such models alone attests to the fact that even very basic questions about the nature of neutrinos remain unanswered. While some mass schemes like the see-saw relationship Eq. (7.1) are intriguing, they have no predictive power because there are many other possibilities. Therefore, it is best to remain open to all possibilities which are not excluded by experimental or astrophysical arguments.

7.1.3   Kinematical Mass Bounds

Unsurprisingly, much experimental effort goes into attempts to measure or narrow down the range of possible neutrino masses, an area where astrophysics and cosmology have made their most renowned contributions to particle physics. Direct laboratory experiments rely on the kinematical impact of a mass on certain reactions such as nuclear decays of the form $(A, Z) \rightarrow (A, Z+1)\, e^- \bar{\nu}_e$ where the continuous energy spectrum of the electrons originally revealed the emission of another particle that carried away the remainder of the available energy. The minimum amount of energy taken by the neutrino is the equivalent of its mass so that the upper endpoint of the electron spectrum is a sensitive measure for m_{ν_e}. Actually, the most sensitive probe is the *shape* of the electron spectrum just below its endpoint, not the value of the endpoint itself. The best constraints are based on the tritium decay $^3\mathrm{H} \rightarrow {}^3\mathrm{He}^+\, e^-\, \bar{\nu}_e$ with a maximum amount of kinetic energy for the electron of $Q = 18.6\,\mathrm{keV}$. This unusually small Q-value ensures that a large fraction of the electron counts appear near the endpoint (Boehm and Vogel 1987; Winter 1991).

In Tab. 7.1 the results from several recent experiments are summarized which had been motivated by the Moscow claim of $17\,\mathrm{eV} < m_{\nu_e} < 40\,\mathrm{eV}$ (Boris et al. 1987). This range is clearly incompatible with the more recent data which, however, find *negative* mass-squares. This means that the endpoint spectra tend to be slightly deformed in the opposite direction from what a neutrino mass would do. This effect is particularly striking and significant for the Livermore experiment where it is nearly impossible to blame it on a statistical fluctuation. Therefore, at the present time one cannot escape the conclusion that this experimental technique suffers from some unrecognized systematic effect. This problem must be resolved before it will become possible to extract a reliable bound on m_{ν_e} although it appears unlikely that an m_{ν_e} in excess of 10 eV could be hidden by what-

Table 7.1. Summary of tritium β decay experiments.

Experiment	$m_{\nu_e}^2 \pm \sigma_{\mathrm{stat}} \pm \sigma_{\mathrm{syst}}$ [eV^2]	Reference
Los Alamos	$-147 \pm 68 \pm 41$	Robertson et al. (1991)
Tokyo	$-65 \pm 85 \pm 65$	Kawakami et al. (1991)
Zürich	$-24 \pm 48 \pm 61$	Holzschuh et al. (1992)
Mainz	$-39 \pm 34 \pm 15$	Weinheimer et al. (1993)
Livermore	$-130 \pm 20 \pm 15$	Stoeffl and Decman (1994)
Troitsk	-18 ± 6	Belesev et al. (1994)[a]

[a]See Otten (1995) for a published description.

ever effect causes the "wrong" deformation of the end-point spectrum. In spite of apparent systematic problems, the Troitsk group (Belesev et al. 1994) claims a limit of $m_{\nu_e} < 4.5\,\mathrm{eV}$ at 95% CL; see also Otten (1995).

One has attempted to determine m_{ν_μ} by measuring the muon momentum from the decay of stopped pions, $\pi^+ \to \mu^+\nu_\mu$, leading to $m_{\nu_\mu}^2 = m_{\pi^+}^2 + m_\mu^2 - 2m_{\pi^+}(m_\mu^2 + p_\mu^2)^{1/2}$. A recent p_μ measurement (Daum et al. 1991) implied a negative squared mass of $m_{\nu_\mu}^2 = -(0.154 \pm 0.045)\,\mathrm{MeV}^2$, probably due to large systematic uncertainties in the determination of m_{π^+}. Hence, it seemed that the often-quoted bound of $m_{\nu_\mu} < 0.27\,\mathrm{MeV}$ did not apply. An older experiment studied the in-flight decay of pions with a result $m_{\nu_\mu}^2 = -(0.14 \pm 0.20)\,\mathrm{MeV}^2$, largely independent of the pion mass (Anderhub et al. 1982). This implies a 90% CL upper limit of $m_{\nu_\mu} < 0.50\,\mathrm{MeV}$. Most recently, the mass of the negative pion was reconsidered by Jeckelmann, Goudsmit, and Leisi (1994). Their previous experiment allows for two mass assignments, $m_\pi = 139.56782 \pm 0.00037\,\mathrm{MeV}$ or $139.56995 \pm 0.00035\,\mathrm{MeV}$. The larger value had previously been rejected on the basis of evidence which now appears questionable. Together with a new p_μ measurement (Assamagan et al. 1994) one finds $m_{\nu_\mu}^2 = -0.148 \pm 0.024\,\mathrm{MeV}^2$ or $-0.022 \pm 0.023\,\mathrm{MeV}^2$. The first value is negative by 6.2 standard deviations and thus may be rejected as unphysical. The second solution is compatible with zero and gives a 90% CL upper limit of $m_{\nu_\mu} < 0.16\,\mathrm{MeV}$.

For ν_τ the best bounds also come from limits on missing energy in certain reactions, the only form in which ν_τ has ever been "observed." The ARGUS Collaboration (1988, 1992) studied the decay $\tau^- \to 3\pi^- 2\pi^+ \nu_\tau$ with a total of 20 events with good energy determinations for all five pions, leading to $m_{\nu_\tau} < 31\,\mathrm{MeV}$ at 95% CL. A similar experiment by the CLEO Collaboration (1993) based on a much larger data sample gave $m_{\nu_\tau} < 32.6\,\mathrm{MeV}$ at 95% CL. Most recently, the ALEPH Collaboration (1995) at CERN has reported a new

Table 7.2. Experimental neutrino mass limits.

Flavor	Limit	CL	Reference
ν_e	(5 eV)	—	See Tab. 7.1
ν_μ	0.16 MeV	90%	Assamagan et al. (1994)
ν_τ	23.8 MeV	95%	ALEPH Collaboration (1995)

95% CL mass limit of 23.8 MeV on the basis of 25 events of the form $\tau \to 5\pi\nu_\tau$ and $5\pi\pi^\circ\nu_\tau$ where π stands for a charged pion.

Another kinematical method to be discussed in Sect. 11.3.4 uses the neutrino pulse dispersion from a distant supernova (SN). For ν_e the observed neutrinos from SN 1987A gave $m_{\nu_e} \lesssim 20\,\mathrm{eV}$, less restrictive than the tritium experiments. However, if the neutrino pulse from a future galactic SN will be detected one may be able to probe even a ν_τ mass down to the cosmologically interesting 30 eV range (Sect. 11.6)!

For Dirac neutrinos there is another essentially kinematical constraint from the SN 1987A neutrino observations. The sterile ν_{Dirac} components can be produced in scattering processes by helicity flips. In a supernova core this effect leads to an anomalous energy drain, limiting a Dirac mass to be less than a few 10 keV (Sect. 13.8.1).

7.1.4 Neutrinoless Double-Beta Decay

If neutrinos have Majorana masses, lepton number is not conserved as one cannot associate a conserved "charge" with a Majorana particle. One observable consequence would be the occurrence of neutrinoless nuclear decay modes of the form $(A, Z) \to (A, Z+2)\, 2e^-$ which would violate lepton number by two units. There are several isotopes which can decay only by the simultaneous conversion of two neutrons. Recently it has become possible to observe the electron spectra from the standard two-neutrino mode $(A, Z) \to (A, Z+2)\, 2e^-\, 2\bar{\nu}_e$; for a recent review see Moe (1995). The decay $^{76}\mathrm{Ge} \to {}^{76}\mathrm{Se}\, 2e^-2\bar{\nu}_e$, for example, is found to have a half life of $(1.43\pm0.04_{\mathrm{stat}}\pm0.13_{\mathrm{syst}})\times10^{21}$ yr (Beck 1993). The age of the universe, by comparison, is about 10^{10} yr.

In the 0ν decay mode, loosely speaking, one of the emitted Majorana neutrinos would be reabsorbed as an antineutrino with an amplitude proportional to $m_{\nu_e,\mathrm{Majorana}}$ and thus a rate proportional to $m_{\nu_e,\mathrm{Majorana}}^2$. In a measurement of the combined energy spectrum of both electrons the 0ν mode would show up as a peak at the endpoint. The best current upper bound is from the Heidelberg-Moscow $^{76}\mathrm{Ge}$ experiment which yields $m_{\nu_e,\mathrm{Majorana}} < 0.65\,\mathrm{eV}$ (Balysh et al. 1995), a number which will likely improve to 0.2 eV over the next few years. This nominal limit must be relaxed by as much as a factor of $2-3$ for the uncertainty in the nuclear matrix elements which are needed to translate an experimental limit on the neutrinoless decay rate into a mass limit.

With neutrino mixing (Sect. 7.2) the other flavors also contribute so that the bound is really on the quantity $\langle m_\nu \rangle \equiv \sum_j \lambda_j |U_{ej}|^2 m_j$ where λ_j is a CP phase equal to ± 1, and the sum is to be extended over all two-component Majorana neutrinos. In this language a four-component Dirac neutrino consists of two degenerate two-component Majorana ones with $\lambda = +1$ and -1 so that their contributions cancel exactly, reproducing the absence of lepton number violation for Dirac neutrinos. If one takes the largest cosmologically allowed value $m_{\nu_\tau} = 30\,\text{eV}$ and the largest experimentally allowed mixing amplitude (Sect. 8.2.4) of $|U_{e,3}| \approx 0.16$ one may have a contribution as large as $0.8\,\text{eV}$ from ν_τ.

7.1.5 Cosmological Mass Bounds

Cosmology arguably yields the most important neutrino mass bounds (Kolb and Turner 1990; Börner 1992). In the framework of the big-bang scenario of the early universe one expects about as many "blackbody neutrinos" in the universe as there are cosmic microwave photons. In detail, the cosmic energy density in massive neutrinos is found to be

$$\rho_\nu = \frac{3}{11}\, n_\gamma \sum_{i=1}^{3} m_i, \tag{7.2}$$

with n_γ the present-day density of microwave background photons and m_i the neutrino masses. In units of the cosmic critical density this is

$$\Omega_\nu h^2 = \sum_{i=1}^{3} \frac{m_i}{93\,\text{eV}}, \tag{7.3}$$

where h is the Hubble constant in units of $100\,\text{km}\,\text{s}^{-1}\,\text{Mpc}^{-1}$. The observed age of the universe together with the measured expansion rate yields $\Omega h^2 \lesssim 0.4$ so that for any of the known families

$$m_\nu \lesssim 30\,\text{eV}\,. \tag{7.4}$$

If one of the neutrinos had a mass near this bound it would be the main component of the long-sought dark matter of the universe.

Certain scenarios of structure formation currently favor "hot plus cold dark matter" where neutrinos with $m_{\nu_e} + m_{\nu_\mu} + m_{\nu_\tau} \approx 5\,\text{eV}$ play a sub-dominant dynamical role but help to shape the required spectrum of primordial density perturbations (Pogosyan and Starobinsky 1995 and references therein). Preferably, the three neutrino masses should be degenerate rather than one dominating flavor.

If neutrinos were unstable and if they decayed so early that their decay products were sufficiently redshifted by the expansion of the universe, the cosmological mass bound can be violated without running into direct conflict with observations. The excluded range of masses

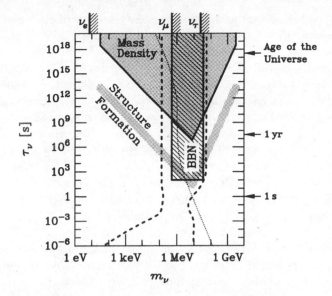

Fig. 7.2. Cosmological bounds on neutrino masses and lifetimes as described in the text. The experimental limits are shown above the main panel. If the dominant decay channel is the majoron mode $\nu \to \nu'\chi$ the BBN-excluded range extends between the dashed lines. The dotted line is $\tau_\nu |U_{e3}|^2$ for standard-model decays $\nu_3 \to \nu_1$ according to Eq. (7.17).

and lifetimes according to Dicus, Kolb, and Teplitz (1977)[42] is shown in Fig. 7.2 as a shaded area marked "Mass Density."

Decaying neutrinos would cause a second cosmic epoch of radiation domination, suppressing the growth of density fluctuations and thus the formation of structure (Steigman and Turner 1985; Krauss 1991; Bond and Efstathiou 1991). Somewhat schematically, the area above the shaded band in Fig. 7.2 marked "Structure Formation" is excluded by this more model-dependent argument. For masses and lifetimes on this band, neutrinos would actually have the beneficial effect of modifying the primordial spectrum of density fluctuations such as to avoid the problem of too much small-scale power in cold dark matter universes (Bardeen, Bond, and Efstathiou 1987; Bond and Efstathiou 1991; Dodelson, Gyuk, and Turner 1994; White, Gelmini, and Silk 1995).

---

[42]Note that the corresponding limits discussed in the book by Kolb and Turner (1990) are somewhat less restrictive because their treatment does not seem to be entirely self-consistent (G. Gelmini, private communication).

The expansion rate and thus the energy density of the universe are well "measured" at the epoch of nucleosynthesis ($T \approx 0.3\,\text{MeV}$) by the primordial light-element abundances (Yang et al. 1984). This big-bang nucleosynthesis (BBN) argument has been used to constrain the number of light neutrino families to $N_\nu \lesssim 3.4$ (Yang et al. 1984; Olive et al. 1990). Even though the measured $Z°$ decay width has established $N_\nu = 3$ (Particle Data Group 1994) the BBN bound remains of interest as a *mass* limit because massive neutrinos contribute more than a massless one to the expansion rate at BBN. For a lifetime exceeding about $100\,\text{s}$ this argument excludes $500\,\text{keV} \lesssim m_\nu \lesssim 35\,\text{MeV}$ (Kolb et al. 1991; Dolgov and Rothstein 1993; Kawasaki et al. 1994), with even more restrictive limits for Dirac neutrinos (Fuller and Malaney 1991; Enqvist and Uibo 1993; Dolgov, Kainulainen, and Rothstein 1995). In Fig. 7.2 the region thus excluded is hatched and marked "BBN."

Kawasaki et al. (1994) have considered the majoron mode $\nu \to \nu'\chi$ (Sect. 15.7) as a specific model for the neutrino decay. Including the energy density of the scalar χ they find even more restrictive limits which exclude the region between the dashed lines in Fig. 7.2.

7.2   Neutrino Mixing and Decay

7.2.1   Flavor Mixing

One of the most mysterious features of the particle zoo is the threefold repetition of families (or "flavors") shown in Fig. 7.1. The fermions in each column have been arranged in a sequence of increasing mass which appears to be the only significant difference between them. There is no indication for higher sequential families; the masses of their neutrinos would have to exceed $\frac{1}{2}m_Z = 46\,\text{GeV}$ according to the CERN and SLAC measurements of the $Z°$ decay width (Particle Data Group 1994). If the origin of masses is indeed the interaction with the vacuum Higgs field, the only difference between the fermions of a given column in Fig. 7.1 is their Yukawa coupling to Φ.

If the only difference between, say, an electron and a muon is the vacuum refraction, any superposition between them is an equally legitimate charged lepton except for the practical difficulty of preparing it experimentally. When such a mixed state propagates, the two components acquire different phases along the beam exactly like two photon helicities in an optically active medium, leading to a rotation of the plane of polarization. Of course, now this "polarization" is understood in the abstract flavor space rather than in coordinate space.

In three-dimensional flavor space one is free to choose any superposition of states as a basis. It is convenient and common practice to use the mass eigenstates (vacuum propagation eigenstates) for each column of Fig. 7.1. Thus by definition the electron is the charged lepton with the smallest mass eigenvalue, the muon the second, and the tau the heaviest, and similarly for the quarks.

All fermions interact by virtue of the weak force and thus couple to the $W^{\pm}$ and Z° gauge bosons, the quarks and charged leptons in addition couple to photons, while only the quarks interact by the strong force and thus couple to gluons. The $W^{\pm}$ (charged current) interaction has the important property of changing, for example, a charged lepton into a neutrino as in the reaction $p + e^{-} \rightarrow n + \nu_{e}$ (Fig. 7.3). If the initial charged lepton was an electron (the lightest charged lepton mass eigenstate), the outgoing neutrino state is defined to be an "electron neutrino" or ν_{e} which in general will be a certain superposition of neutrino mass eigenstates.

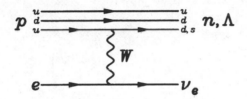

Fig. 7.3. Typical charged-current reaction.

This phenomenon of *Cabbibo mixing* is well established among the quarks. For example, in the process of Fig. 7.3 the transition among the quarks is between u and $\cos\theta_{C}\, d + \sin\theta_{C}\, s$ where $\cos\theta_{C} = 0.975$ refers to the Cabbibo angle. Kinematics permitting, the final-state hadron will sometimes be uds which constitutes the Λ particle with a mass of 1.116 GeV compared with 0.934 GeV for the neutron (udd).

The superposition of quark states into which u transforms by a charged-current interaction is commonly denoted by d', charm couples to s', and top to b' while the unprimed states refer to the first, second, and third mass eigenstates in the d-column of Fig. 7.1. Ignoring the third family one has

$$\begin{pmatrix} d' \\ s' \end{pmatrix} = \begin{pmatrix} \cos\theta_{C} & \sin\theta_{C} \\ -\sin\theta_{C} & \cos\theta_{C} \end{pmatrix} \begin{pmatrix} d \\ s \end{pmatrix}. \tag{7.5}$$

It is only by convention that the mixing is applied to the d-column of the quarks rather than the u-column or both.

Including the third generation, the mixing is induced by the three-dimensional Cabbibo-Kobayashi-Maskawa (CKM) matrix V. After removing all unphysical phases by an appropriate redefinition of the quark fields this unitary matrix is given in terms of four significant parameters. The Particle Data Group (1994) recommends a standard parametrization in terms of three two-family mixing angles $\theta_{ij} < \pi/2$ and one phase $0 \leq \delta < 2\pi$. With $C_{ij} \equiv \cos\theta_{ij}$ and $S_{ij} \equiv \sin\theta_{ij}$ this standard form is (Fritzsch and Plankl 1987)

$$
V = \begin{pmatrix} 1 & 0 & 0 \\ 0 & C_{23} & S_{23} \\ 0 & -S_{23} & C_{23} \end{pmatrix} \begin{pmatrix} C_{13} & 0 & S_{13}e^{-i\delta} \\ 0 & 1 & 0 \\ -S_{13}e^{i\delta} & 0 & C_{13} \end{pmatrix} \begin{pmatrix} C_{12} & S_{12} & 0 \\ -S_{12} & C_{12} & 0 \\ 0 & 0 & 1 \end{pmatrix}
$$

$$
= \begin{pmatrix} C_{12}C_{13} & S_{12}C_{13} & S_{13}e^{-i\delta} \\ -C_{23}S_{12} - C_{12}S_{23}S_{13}e^{i\delta} & C_{12}C_{23} - S_{12}S_{23}S_{13}e^{i\delta} & C_{13}S_{23} \\ S_{12}S_{23} - C_{12}C_{23}S_{13}e^{i\delta} & -C_{12}S_{23} - C_{23}S_{12}S_{13}e^{i\delta} & C_{13}C_{23} \end{pmatrix}
$$

$$
\approx \begin{pmatrix} 1 & S_{12} & S_{13}e^{-i\delta} \\ -S_{12} & 1 & S_{23} \\ S_{12}S_{23} - S_{13}e^{i\delta} & -S_{23} & 1 \end{pmatrix}. \tag{7.6}
$$

Experimentally one has the 90% CL ranges (Particle Data Group 1994)

$$
0.218 < S_{12} < 0.224,
$$
$$
0.032 < S_{23} < 0.048,
$$
$$
0.002 < S_{13} < 0.005. \tag{7.7}
$$

The approximation in Eq. (7.6) is justified by the small mixing angles. They and the CP-violating phase $\delta = 3.3 \times 10^{-3}$ (Wolfenstein 1986) are measured parameters of the standard model which, like the fermion masses, are not theoretically accounted for at the present time.

If neutrinos have masses one naturally expects that they follow a similar scheme and so the "weak interaction eigenstates" ν_ℓ ($\ell = e, \mu, \tau$) are thought to be given as linear superpositions of the mass eigenstates ν_i by virtue of

$$
\nu_\ell = \sum_{i=1}^{3} U_{\ell i} \nu_i, \tag{7.8}
$$

where the unitary matrix U plays the role of the CKM matrix. Unless otherwise stated ν_1 will always refer to the dominant mass admixture of ν_e and so forth. It seems plausible that $m_1 < m_2 < m_3$, a hierarchy that is often assumed.

Flavor mixing is the only possibility for members of one family (one row in Fig. 7.1) to transform into those of a different family. This phenomenon is known as *the absence of flavor-changing neutral currents*; it means that the Z° coupling to quarks and leptons, like the photon coupling, leaves a given superposition of fermions unaltered. For example, muons decay only by the flavor-conserving mode $\mu^- \to \nu_\mu e^- \overline{\nu}_e$ (Fig. 7.4); the experimental upper limits on the branching ratios for $\mu^- \to e^- \gamma$ and $\mu^- \to e^- e^+ e^-$ are 5×10^{-11} and 1.0×10^{-12}, respectively. In the absence of neutrino masses and mixing the individual lepton flavor numbers are conserved: a lepton can be transformed only into its partner of the same family, or it can be created or annihilated together with an antilepton of the same family.

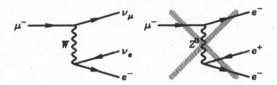

Fig. 7.4. Allowed and forbidden μ decays.

When neutrinos have masses and mixings, flavor-violating lepton decays become possible, but their rate would be so small that their experimentally observed absence does not yield interesting constraints on neutrino parameters. Because neutrino masses must be very small if they exist at all, the most significant observable effect is that of neutrino oscillations.

7.2.2 Standard-Model Decays of Mixed Neutrinos

For massive neutrinos it is kinematically possible to decay according to $\nu \to \nu' \gamma$, $\nu \to \nu' \gamma \gamma$, or $\nu \to \nu' \overline{\nu}'' \nu''$. In the absence of mixing, of course, all of these modes are forbidden. Even in the presence of mixing,

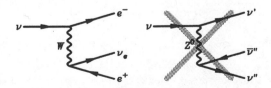

Fig. 7.5. Allowed and forbidden tree-level decays of mixed neutrinos.

however, the three-neutrino channel remains forbidden by the absence of flavor-changing neutral currents because the flavor content of ν and ν' in Fig. 7.5 must remain unaltered by the $Z°$ vertex. Therefore, in the standard model low-mass neutrinos can decay only by higher-order (radiative) amplitudes.

"Heavy" neutrinos ν_h with $m_h > 2m_e \approx 1\,\mathrm{MeV}$ may decay at tree-level through the channel $\nu_h \to \nu_e e^+ e^-$ at a rate

$$\frac{1}{\tau_{e^+e^-}} = |U_{eh}|^2 \frac{G_\mathrm{F}^2}{3\,(4\pi)^3}\, m_h^5\, \Phi(m_h)$$

$$= |U_{eh}|^2\, 3.5\times10^{-5}\,\mathrm{s}^{-1}\, m_\mathrm{MeV}^5\, \Phi(m_h), \qquad (7.9)$$

where U_{eh} is the mixing amplitude between ν_h and ν_e, G_F is the Fermi constant, and $m_\mathrm{MeV} \equiv m_h/\mathrm{MeV}$. The phase-space factor is (Shrock 1981)

$$\Phi(m_h) = (1 - 4a)^{1/2}\,(1 - 14a - 2a^2 - 12a^3)$$

$$+ 24a^2(1 - a^2)\ln\frac{1 + (1 - 4a)^{1/2}}{1 - (1 - 4a)^{1/2}} \qquad (7.10)$$

with $a \equiv m_e^2/m_h^2$; it is shown in Fig. 7.6.

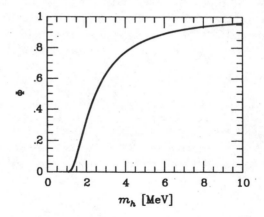

Fig. 7.6. Phase space factor for $\nu_h \to \nu_e e^+ e^-$ according to Eq. (7.10).

The one- and two-photon decay modes arise in the standard model with mixed neutrinos from the amplitudes shown in Fig. 7.7. Turn first to the one-photon decay $\nu_i \to \nu_j\gamma$ with the neutrino masses $m_i > m_j$.

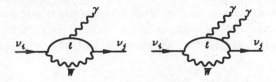

Fig. 7.7. Feynman graphs for neutrino radiative decays. There are other similar graphs with the photon lines attached to the intermediate W boson.

In general the matrix element can be thought of as arising from an effective interaction Lagrangian of the form

$$\mathcal{L}_{\text{int}} = \tfrac{1}{2}\overline{\psi}_i \sigma_{\alpha\beta}(\mu_{ij} + \epsilon_{ij}\gamma_5)\psi_j \, F^{\alpha\beta} + \text{h.c.}, \tag{7.11}$$

where $F^{\alpha\beta}$ is the electromagnetic field tensor, ψ_i and ψ_j are the neutrino fields, and μ_{ij} and ϵ_{ij} are magnetic and electric transition moments which are usually expressed in units of Bohr magnetons $\mu_{\text{B}} = e/2m_e$. The decay rate is

$$\frac{1}{\tau_\gamma} = \frac{|\mu_{ij}|^2 + |\epsilon_{ij}|^2}{8\pi}\left(\frac{m_i^2 - m_j^2}{m_i}\right)^3 = 5.308\,\text{s}^{-1}\left(\frac{\mu_{\text{eff}}}{\mu_{\text{B}}}\right)^2 \delta_m^3\, m_{\text{eV}}^3, \tag{7.12}$$

where $\mu_{\text{eff}}^2 \equiv |\mu_{ij}|^2 + |\epsilon_{ij}|^2$, $m_{\text{eV}} \equiv m_i/\text{eV}$, and $\delta_m \equiv (m_i^2 - m_j^2)/m_i^2$.

An explicit evaluation of the one-photon amplitude of Fig. 7.7 yields for Dirac neutrinos (Pal and Wolfenstein 1982)

$$\left.\begin{array}{c}\mu_{ij}^{\text{D}}\\ \epsilon_{ij}^{\text{D}}\end{array}\right\} = \frac{e\sqrt{2}\,G_{\text{F}}}{(4\pi)^2}\,(m_i \pm m_j)\sum_{\ell=e,\mu,\tau} U_{\ell j}U_{\ell i}^* \, f(r_\ell). \tag{7.13}$$

For Majorana neutrinos one has instead $\mu_{ij}^{\text{M}} = 2\mu_{ij}^{\text{D}}$ and $\epsilon_{ij}^{\text{M}} = 0$ or $\mu_{ij}^{\text{M}} = 0$ and $\epsilon_{ij}^{\text{M}} = 2\epsilon_{ij}^{\text{D}}$, depending on the relative CP phase of ν_i and ν_j.

In Eq. (7.13) $r_\ell \equiv (m_\ell/m_W)^2$ where the charged-lepton masses are $m_e = 0.511\,\text{MeV}$, $m_\mu = 105.7\,\text{MeV}$, and $m_\tau = 1.784\,\text{GeV}$ while the $W^\pm$ gauge boson mass is $m_W = 80.2\,\text{GeV}$. Thus for all charged leptons $r_\ell \ll 1$; in this limit

$$f(r_\ell) \to -\tfrac{3}{2} + \tfrac{3}{4}\,r_\ell. \tag{7.14}$$

If one inserts the leading term $-\tfrac{3}{2}$ into the sum in Eq. (7.13) one finds that its contribution vanishes because the unitarity of U implies that its rows or columns represent orthogonal vectors. Because the first nonzero contribution is from $\frac{3}{4}r_\ell$, the transition moments are suppressed by $(m_\ell/m_W)^2$, an effect known as *GIM cancellation* after Glashow, Iliopoulos, and Maiani (1970). Explicitly, the transition moments are

$$\left.\begin{array}{r}\mu_{ij}^{\rm D}/\mu_{\rm B}\\ \epsilon_{ij}^{\rm D}/\mu_{\rm B}\end{array}\right\} = \frac{3G_{\rm F}m_e}{\sqrt{2}\,(4\pi)^2}\,(m_i\pm m_j)\left(\frac{m_\tau}{m_W}\right)^2\sum_{\ell=e,\mu,\tau}U_{\ell j}U_{\ell i}^*\left(\frac{m_\ell}{m_\tau}\right)^2$$

$$= 3.96\times10^{-23}\,\frac{m_i\pm m_j}{1\,{\rm eV}}\sum_{\ell=e,\mu,\tau}U_{\ell j}U_{\ell i}^*\left(\frac{m_\ell}{m_\tau}\right)^2. \qquad (7.15)$$

These small numbers imply that neutrino radiative decays are exceedingly slow in the standard model.

Dirac neutrinos would have static or diagonal $(i = j)$ magnetic dipole moments while the electric dipole moments vanish according to Eq. (7.13). Their presence would require CP-violating interactions. Majorana neutrinos, of course, cannot have any diagonal electromagnetic moments. For $\mu_{ii}^{\rm D}$ the leading term of Eq. (7.14) in Eq. (7.13) does not vanish because the unitarity of U implies that the sum equals unity for $i = j$. Therefore,

$$\frac{\mu_{ii}^{\rm D}}{\mu_{\rm B}} = \frac{6\sqrt{2}\,G_{\rm F}m_e}{(4\pi)^2}\,m_i = 3.20\times10^{-19}\,m_{\rm eV}, \qquad (7.16)$$

much larger than the transition moments because it is not GIM suppressed.

The two-photon decay rate $\nu_i \to \nu_j\gamma\gamma$ is of higher order and thus may be expected to be smaller by a factor of $\alpha/4\pi$. However, it is not GIM suppressed so that it is of interest for a certain range of neutrino masses (Nieves 1983; Ghosh 1984). Essentially, the result involves another factor $\alpha/4\pi$ relative to the one-photon rate, and $f(r_\ell)$ in Eq. (7.13) is replaced by $(m_i/m_\ell)^2$.

As an example consider the different decay modes for $\nu_3 \to \nu_1$, assuming that $m_3 \gg m_1$ and that the mixing angles are small so that $\nu_3 \approx \nu_\tau$ and $\nu_1 \approx \nu_e$. Then one has explicitly

$$\frac{1}{\tau} \approx |U_{e3}|^2\,\frac{G_{\rm F}^2m_3^5}{3\,(4\pi)^3}\times\begin{cases}\Phi(m_3), & \nu_3\to\nu_1e^+e^-,\\[2mm]\dfrac{27}{8}\dfrac{\alpha}{4\pi}\left(\dfrac{m_\tau}{m_W}\right)^4, & \nu_3\to\nu_1\gamma,\\[2mm]\dfrac{1}{180}\left(\dfrac{\alpha}{4\pi}\right)^2\left(\dfrac{m_3}{m_e}\right)^4, & \nu_3\to\nu_1\gamma\gamma,\end{cases} \qquad (7.17)$$

where $\Phi(m_h)$ was given in Eq. (7.10) and shown in Fig. 7.6. The $\gamma\gamma$ decay dominates in a small range of m_3 just below $2m_e$.

The quantity $\tau \, |U_{e3}|^2$ is shown as a dotted line in Fig. 7.2. Even without nonstandard physics the decay rate is fast on cosmological scales if $m_3 \gtrsim 2m_e$. Because experimentally m_3 may be as large as 24 MeV the cosmological mass bound of 30 eV does not automatically apply. Experimental limits on U_{e3} together with the BBN mass bound, however, exclude a heavy standard ν_3. Even without reference to BBN it can be excluded on the basis of the SN 1987A neutrino radiative lifetime limits (Sect. 12.5.2).

7.3 Neutrino Electromagnetic Form Factors

7.3.1 Overview

When Wolfgang Pauli in 1930 first postulated the existence of neutrinos he speculated that they might interact like a magnetic dipole of a certain moment μ_ν. If that were the case they could be measured by their ionizing power when they move through a medium; this ionizing power was first calculated by Bethe (1935). Nahmias (1935) measured the event rates in a Geiger-Müller counter in the presence and absence of a radioactive source and interpreted his null result as a limit $\mu_\nu \lesssim 2\times10^{-4}\,\mu_{\rm B}$ (Bohr magneton $\mu_{\rm B} = e/2m_e$) on the neutrino dipole moment. He concluded that "since this limit is already smaller than a nuclear magneton, it seems probable that the neutrino has no moment at all." Subsequent attempts to measure ever smaller neutrino dipole moments have consistently failed.

The main difference between then and now is the advanced theoretical understanding of neutrino interactions in the context of the standard model of electroweak gauge interactions. A magnetic dipole interaction couples l.h. with r.h. states so that the latter would not be strictly sterile. This would be in conflict with the standard model where neutrinos interact only by their l.h. coupling to W and Z gauge bosons. Thus neutrino dipole moments must vanish identically because weak interactions violate parity maximally.

This picture changes when neutrinos have masses because even the r.h. components of a Dirac neutrino are then not strictly sterile as they couple to the Higgs field—or else they would not have a mass. Indeed, an explicit calculation in the standard model with neutrino masses gave a magnetic dipole moment $\mu_\nu = 3.20\times10^{-19}\mu_{\rm B}\,(m_\nu/{\rm eV})$ (Eq. 7.16). If neutrinos mix, they also obtain transition magnetic and electric moments. However, they are even smaller because of the GIM suppression effect—see Eq. (7.15).

Much larger values would obtain with direct r.h. neutrino interactions. For example, in left-right symmetric models there exist heavier gauge bosons which mediate r.h. interactions; parity violation would occur because of the mass difference between the l.h. and r.h. gauge bosons. For a neutrino ν_ℓ (flavor $\ell = e$, μ or τ) the dipole moment in such models is (Kim 1976; Marciano and Sanda 1977; Bég, Marciano, and Ruderman 1978)

$$\mu_\nu = \frac{eG_{\rm F}}{2\sqrt{2}\,\pi^2} \left[m_\ell \left(1 - \frac{m_{W_1}^2}{m_{W_2}^2} \right) \sin 2\zeta + \tfrac{3}{4}\, m_{\nu_\ell} \left(1 + \frac{m_{W_1}^2}{m_{W_2}^2} \right) \right],$$

(7.18)

where ζ is the left-right mixing angle between the gauge bosons W_L and W_R; $W_{1,2}$ are their mass eigenstates.

Because of the smallness of the mass-induced standard dipole moments any evidence for neutrino electromagnetic interactions would represent evidence for interactions beyond the standard model. Therefore, the quest for neutrino electromagnetic interactions is more radical than that for masses and mixings.

In a dense medium even standard massless neutrinos interact with photons by an effective coupling which is mediated by the ambient electrons. This coupling can be expressed in terms of an effective neutrino charge radius. Presently I focus on neutrino interactions in vacuum, leaving a discussion of their properties in media to Sect. 6.6.

7.3.2   Single-Photon Coupling

There are many possible extensions of the standard model which would give sizeable neutrino dipole and transition moments by some novel r.h. interaction. For the purposes of this book the underlying new physics is of no concern; all we need is a generic representation of its observable effects in terms of neutrino electromagnetic form factors.

The most general interaction structure of a fermion field ψ with the electromagnetic field can be expressed as an effective Lagrangian

$$\mathcal{L}_{\rm int} = -F_1\, \overline{\psi}\gamma_\mu\psi A^\mu - G_1\, \overline{\psi}\gamma_\mu\gamma_5\psi\, \partial_\mu F^{\mu\nu}$$

$$- \tfrac{1}{2}\overline{\psi}\sigma_{\mu\nu}(F_2 + G_2\gamma_5)\psi F^{\mu\nu},$$

(7.19)

where A^μ is the electromagnetic vector potential and $F^{\mu\nu}$ the field strength tensor. The interpretation of the coupling constants is that of an electric charge for F_1, an anapole moment for G_1, a magnetic dipole moment for F_2, and an electric dipole moment for G_2. In the matrix element derived from this Lagrangian these couplings should be viewed as form factors which are functions of Q^2 where Q is the energy-momentum transfer to the fermion, i.e. the energy momentum of the photon line attached to the fermion current. The interpretation of a charge etc. then pertains to the $Q^2 \to 0$ limit.

It is usually assumed that neutrinos are electrically neutral, i.e. that $F_1(0) = 0$ because electric charge quantization implies that elementary particles carry only charges in multiples of $\frac{1}{3}e$ where e is the electron charge. In recent discussions of electric charge quantization[43] it was stressed, however, that the standard model of electroweak interactions without grand unification requirements does allow neutrinos to carry small electric charges. Their possible magnitude is thus an experimental issue; existing limits are reviewed in Sect. 15.8. Because these limits are very restrictive, i.e. because neutrino electric charges must be very small, it appears likely that electric charge is quantized after all so that neutrino electric charges vanish identically.

Even if neutrinos are electrically neutral, as shall be assumed henceforth, they can virtually dissociate into charged particles and so they will have a form factor $F_1(Q^2)$ which does not vanish for $Q^2 \neq 0$. One may visualize the neutral object as a superposition of two charge distributions of opposite sign with different spatial extensions. In terms of a power series expansion of $F_1(Q^2)$ one usually defines the charge radius by virtue of

$$\langle r^2 \rangle = 6 \left. \frac{\partial F_1(Q^2)}{e \, \partial Q^2} \right|_{Q^2=0} \tag{7.20}$$

where $\langle r^2 \rangle$ may be both positive or negative.

For neutrinos, the interpretation of the charge radius as an observable quantity is a rather subtle issue as it is probed by "off-shell" photons ($Q^2 \neq 0$), i.e. by intermediate photons in processes such as scattering by photon exchange. Because the form factor is proportional to Q^2 such scattering processes do not exhibit a Coulomb divergence. The charge radius induces a short-range or contact interaction similar to processes involving Z° exchange. Therefore, the charge radius represents a correction to the standard tree-level electroweak scattering amplitude between neutrinos and charged particles. This tree-level

---

[43]Babu and Mohapatra 1990; Babu and Volkas 1992; Takasugi and Tanaka 1992; Foot, Lew, and Volkas 1993; Foot 1994. References to earlier works are given in these papers.

amplitude will receive radiative corrections from a variety of diagrams, including photon exchange, which must be considered simultaneously so that it is not at all obvious that one can extract a finite, gauge-invariant, observable quantity that can be physically interpreted as a charge radius.[44]

The charge radius, even if properly defined, represents only a correction to the tree-level weak interaction and as such it is best studied in precision accelerator experiments. In the astrophysical context, weak interaction rates involving standard l.h. neutrinos cannot be measured with the level of precision required to test for small deviations from the standard model. Indeed, a recent compilation (Salati 1994) reveals that experimental bounds on $\langle r^2 \rangle$ are more sensitive than astrophysical limits, except perhaps for ν_τ for which experimental data are scarce and so its standard-model neutral-current interactions are not well tested.

The matrix element for the anapole interaction in the Lorentz gauge is proportional to Q^2. Therefore, it vanishes in the limit $Q^2 \to 0$, i.e. for real photons coupled to the neutrino current. The role of the anapole form factor G_1 is thus very similar to a charge radius: it represents a correction to the standard tree-level weak interaction and as such does not seem to be of astrophysical interest.

The form factors F_2 and G_2 are of much greater importance because they may obtain nonvanishing values even in the $Q^2 \to 0$ limit. Henceforth I shall refer to $\mu \equiv F_2(0)$ as a magnetic dipole moment, to $\epsilon \equiv iG_2(0)$ as an electric dipole moment, respectively. This identification is understood if one derives the Dirac equation of motion $i\partial_t \psi = H\psi$ for a neutrino field ψ (mass m) in the presence of an external, weak, slowly varying electromagnetic field $F^{\mu\nu}$. From Eq. (7.19) one finds for the Hamiltonian

$$H = -i\boldsymbol{\alpha} \cdot \boldsymbol{\nabla} + \beta\Big[m - (\mu + i\epsilon\gamma_5)(i\boldsymbol{\alpha} \cdot \mathbf{E} + \boldsymbol{\Sigma} \cdot \mathbf{B})\Big], \qquad (7.21)$$

where $\frac{1}{2}\sigma_{\mu\nu}F^{\mu\nu} = i\boldsymbol{\alpha} \cdot \mathbf{E} + \boldsymbol{\Sigma} \cdot \mathbf{B}$ was used. In the Dirac representation one has

$$\boldsymbol{\alpha} = \begin{pmatrix} 0 & \boldsymbol{\sigma} \\ \boldsymbol{\sigma} & 0 \end{pmatrix}, \qquad \beta = \begin{pmatrix} I & 0 \\ 0 & -I \end{pmatrix}, \qquad \boldsymbol{\Sigma} = \begin{pmatrix} \boldsymbol{\sigma} & 0 \\ 0 & \boldsymbol{\sigma} \end{pmatrix}, \quad (7.22)$$

where $\boldsymbol{\sigma}$ is a vector of Pauli matrices while I is the 2×2 unit matrix. For a neutrino at rest the Dirac spinor is characterized by its large

[44]For recent discussions of these matters see Lucio, Rosado, and Zepeda (1985), Auriemma, Srivastava, and Widom (1987), Degrassi, Sirlin, and Marciano (1989), Musolf and Holstein (1991), and Góngora-T. and Stuart (1992).

component, a Pauli two-spinor ϕ, which is then found to evolve as

$$i\partial_t\phi = (\mu\mathbf{B}_0 + \epsilon\mathbf{E}_0) \cdot \boldsymbol{\sigma}\,\phi, \qquad (7.23)$$

where the index 0 refers to quantities in the neutrino rest frame. A neutrino polarized in or opposite to the field direction has the energy $\pm\mu B_0$ or $\pm\epsilon E_0$, respectively, so that μ and ϵ are indeed magnetic and electric dipole moments, respectively.

The electromagnetic form factors obey certain constraints for Dirac and Majorana neutrinos; detailed discussions were provided by a number of authors.[45] For Dirac neutrinos, all form factors must be real relative to each other (no relative phases) if CP invariance holds. For the diagonal case (coupling to one neutrino species) all form factors must be real, and CP invariance implies that the electric dipole moment must vanish. For Majorana neutrinos, a magnetic transition moment (F_2) must be imaginary, an electric transition moment (G_2) real. If CP invariance holds, in addition one of them must vanish, i.e. there is either a transition electric, or a transition magnetic moment, but not both. Majorana neutrinos cannot have diagonal electric nor magnetic moments, nor can they have a charge or charge radius; they may have an anapole form factor.

7.3.3   Two-Photon Coupling

Discussions of neutrino electromagnetic form factors are usually restricted to the effective neutrino coupling to an electromagnetic wave or static field. However, a two-photon coupling of neutrinos is also possible and of some interest. Historically, it was thought for some time that the process $\gamma\gamma \to \nu\bar{\nu}$ could be of great importance for the emission of neutrinos from stars until it was shown by Gell-Mann (1961) that the amplitude for this process vanishes identically if neutrinos have only l.h. local interactions with electrons. Several authors discussed the $\gamma\gamma \to \nu\bar{\nu}$ process when neutrinos are massive, or when they have more general interaction structures (Halprin 1975; Fischbach et al. 1976, 1977; Natale, Pleitez, and Tacla 1987; Gregores et al. 1995). However, there does not seem to be a plausible scenario where this process would be of serious astrophysical interest.

In the standard model, there is an effective two-photon coupling to neutrinos because the interaction is not local; rather, it is mediated by finite-mass gauge bosons. Early calculations of the effective

---

[45]For example Nieves (1982), Kayser (1982), Shrock (1982), and Li and Wilczek (1982).

coupling in gauge theories were performed by Levine (1966) and Cung
and Yoshimura (1975); for more recent discussions in the framework of
the standard model see Nieves, Pal, and Unger (1983), Nieves (1987),
Dodelson and Feinberg (1991), Liu (1991), Langacker and Liu (1992),
Kuznetsov and Mikheev (1993), and Dicus and Repko (1993). This
two-photon coupling leads to a higher-order contribution to the neutrino refractive index in a bath of photons (Sect. 6.7.2). However, this
and other consequences do not seem to be important in any astrophysical or laboratory setting of practical interest.

For neutrinos with masses and mixings there is a decay amplitude
$\nu \to \nu'\gamma\gamma$ which can dominate for a small range of neutrino masses
below $m_\nu = 2m_e$ as discussed by Nieves (1983) and Ghosh (1984).
Explicit results were given in Sect. 7.2.2 above.

One of the photons may represent an external electric or magnetic
field, i.e. one may consider the neutrino decay $\nu \to \nu'\gamma$ in the presence of a strong external field which would modify the propagators
of the intermediate charged leptons. For very strong fields it appears
that a substantial decay rate can obtain (Gvozdev, Mikheev, and Vassilevskaya 1992a,b, 1993, 1994a,b). From this literature it does not
seem to become entirely clear under which if any circumstances these
results might be of practical interest for, say, the decay of supernova or
neutron-star neutrinos.

7.4   Electromagnetic Processes

The presence of electromagnetic dipole and transition moments implies
that neutrinos couple directly to the electromagnetic field, allowing for
a variety of nonstandard processes (Fig. 7.8). Most obviously, neutrinos
can scatter on electrons by photon exchange. The ν-e scattering cross
section (electrons at rest in the laboratory frame) was given, e.g. by
Vogel and Engel (1989)

$$\frac{d\sigma}{dT} = \frac{G_F^2 m_e}{2\pi} \left[(C_V + C_A)^2 + (C_V - C_A)^2 \left(1 - \frac{T}{E_\nu}\right)^2 \right.$$
$$\left. + (C_A^2 - C_V^2) \frac{m_e T}{E_\nu^2} \right] + \alpha\mu_\nu^2 \left[\frac{1}{T} - \frac{1}{E_\nu} \right], \qquad (7.24)$$

where C_V and C_A are the weak coupling constants given in Appendix B,
T is the electron recoil energy with the limits $0 \le T \le 2E_\nu^2/(2E_\nu + m_e)$,
and μ_ν is the neutrino dipole moment.

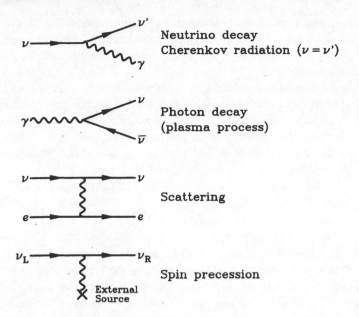

Fig. 7.8. Important processes involving direct neutrino electromagnetic couplings, notably magnetic dipole moments and magnetic or electric transition moments.

Because an anomalous neutrino charge radius causes a contact interaction it would manifest itself by the modification $C_V \rightarrow C_V + \frac{1}{3}\sqrt{2}\,\pi\alpha\langle r^2\rangle/G_F$, i.e. there would be a small correction to the overall cross section. A dipole moment, on the other side, modifies the energy spectrum of the recoil electrons, notably at low energies, because of its forward-peaked nature from the Coulomb divergence.

If neutrinos had electric dipole moments, or electric or magnetic transition moments, these quantities would also contribute to the scattering cross section. Therefore, the quantity measured in electron recoil experiments involving relativistic neutrinos is (e.g. Raffelt 1989)

$$\mu_\nu^2 = \sum_{j=\nu_e,\nu_\mu,\nu_\tau} |\mu_{ij} - \epsilon_{ij}|^2, \tag{7.25}$$

where i refers to the initial-state neutrino flavor. Therefore, in principle there is a possibility of destructive interference between the magnetic and electric transition moments of Dirac neutrinos. (Majorana neutrinos have only magnetic or electric transition moments, but not both if CP is conserved.)

A related process to scattering by photon exchange is the spin-precession in a macroscopic magnetic or electric field into r.h. states of the same or another flavor, i.e. electromagnetic oscillation into "wrong-helicity" states. Such effects will be studied in Sect. 8.4 in the general context of neutrino oscillations. In principle, this process can be important at modifying the measurable l.h. solar neutrino flux as discussed in Sect. 10.7 in the context of the solar neutrino problem. Spin oscillations can also be important in supernovae where strong magnetic fields exist, although a detailed understanding remains elusive at the present time (Sect. 11.4).

The most interesting process caused by dipole moments is the photon decay into neutrino pairs, $\gamma \to \bar\nu\nu$, which is enabled in media where the photon dispersion relation is such that $\omega^2 - \mathbf{k}^2 > 0$. It is the most interesting process because it occurs even in the absence of dipole moments due to a medium-induced neutrino-photon coupling (Chapter 6). This is the dominant standard neutrino emission process from stars for a wide range of temperatures and densities. Details of both the standard and the dipole-induced "plasma process" depend on complicated fine points of the photon dispersion relation in media that were taken up in Chapter 6.

The salient features, however, can be understood in an approximation where photons in a medium ("transverse plasmons") are treated as particles with an effective mass equal to the plasma frequency ω_P which in a nonrelativistic medium is $\omega_P^2 = 4\pi\alpha n_e/m_e$ with the electron density n_e and electron mass m_e. The decay rate of these electromagnetic excitation in their own rest frame is then

$$\Gamma_\gamma = \mu_\nu^2 \, \frac{\omega_P^3}{24\pi} \quad \text{with} \quad \mu_\nu^2 = \sum_{i,j}\Big(|\mu_{ij}|^2 + |\epsilon_{ij}|^2\Big), \tag{7.26}$$

while in the frame of the medium where the photon has the energy ω a Lorentz factor ω_P/ω must be included. The sum includes all final-state neutrino flavors with $m_i \ll \omega_P$; otherwise phase-space modifications occur, and even a complete suppression of the decay by a neutrino mass threshold. In contrast with Eq. (7.25) relevant for the scattering rate, no destructive interference effects between magnetic and electric dipole amplitudes occur.

Transition moments would allow for the radiative decay $\nu_i \to \nu_j\gamma$. Again, because a neutrino charge radius or anapole moment vanish in the $Q^2 \to 0$ limit relevant for free photons, radiative neutrino decays are most generally characterized by their magnetic and electric transition moments. The decay rate is found to be

$$\Gamma_{\nu_i} = \mu_\nu^2 \frac{m_i^3}{8\pi} \qquad \text{with} \qquad \mu_\nu^2 = \sum_j \left(|\mu_{ij}|^2 + |\epsilon_{ij}|^2 \right). \qquad (7.27)$$

The sum is extended over all ν_j with $m_j \ll m_i$, otherwise phase-space corrections as given in Eq. (7.12) must be included. There is no destructive interference between electric and magnetic amplitudes. Because of the long decay path available in the astronomical environment, the search for decay photons from known astrophysical neutrino sources is by far the most efficient method to set limits on radiative decays. This method and its results are studied in detail in Chapter 12.

Neutrino radiative decays, the plasma process, and scatterings by photon exchange all depend on the same electromagnetic form factors. Therefore, a limit on the radiative decay time can be expressed as a limit on neutrino transition moments, and a limit on a transition moment from, say, a scattering experiment can be translated into a limit on a radiative decay time. In much of the literature this simple connection has not been made; the search for neutrino decays and that for dipole moments were strangely dealt with as separate and unrelated efforts.

There are a number of interesting neutrino electromagnetic processes which I will not discuss in any detail because they either yield very small rates or have not so far led to any new results. Among them are the bremsstrahlung emission of photons in neutrino electron collisions $\nu e \to \nu e \gamma$ (Mourão, Bento, and Kerimov 1990; Bernabéu et al. 1994). Another possibility is nuclear excitation due to neutrino dipole moments (Dodd, Papageorgiu, and Ranfone 1991; Sehgal and Weber 1992) or the disintegration of deuterons (Akhmedov and Berezin 1992). Also, Cherenkov radiation (Grimus and Neufeld 1993) or transition radiation (Sakuda 1994) of neutrinos with magnetic moments have been studied in the literature.

7.5 Limits on Neutrino Dipole Moments

7.5.1 Scattering Experiments

In principle, there are a number of possibilities to search for direct neutrino electromagnetic couplings in the laboratory. In practice, the only method that has so far yielded significant limits is a study of the neutrino-electron scattering cross section that would receive a forward-peaked contribution from a dipole moment.

The best limits on μ_{ν_e} are based on the use of reactor neutrinos as a source. The measurement of the $\bar{\nu}_e$-e scattering cross section by Reines, Gurr, and Sobel (1976) was interpreted by Kyuldjiev (1984) to yield a bound of $\mu_{\nu_e} < 1.5 \times 10^{-10} \mu_B$. Since then, the reactor $\bar{\nu}_e$ spectrum has been much better understood. Vogel and Engel (1989) stressed that a literal interpretation of the old results by Reines, Gurr, and Sobel (1976) would actually yield evidence for a dipole moment of about $2-4 \times 10^{-10} \mu_B$. There is, however, a more recent limit of

$$\mu_{\nu_e} < 2.4 \times 10^{-10} \mu_B \tag{7.28}$$

from the Kurchatov Institute (Vidyakin et al. 1992).

When interpreting this bound as a limit on the ν_e-ν_τ transition moment, recall that for an experimentally allowed ν_τ mass in the 10 MeV regime this limit would not apply as reactor neutrinos have relatively low energies (below about 10 MeV). Therefore, it remains useful to consider the bounds from the neutrino beam at LAMPF with a ν_e endpoint energy of 52.8 MeV

$$\mu_{\nu_e} < 10.8 \times 10^{-10} \mu_B,$$
$$\mu_{\nu_\mu} < 7.4 \times 10^{-10} \mu_B \tag{7.29}$$

(Krakauer et al. 1990). A similar limit of $\mu_{\nu_\mu} < 8.5 \times 10^{-10} \mu_B$ was obtained by Ahrens et al. (1990).

An improvement of the bound on μ_{ν_e} by an order of magnitude or more can be expected from a new reactor experiment currently in preparation ("MUNU experiment," Broggini et al. 1990). However, results will not become available before a few years from now.

The transition moments between ν_τ and other sequential neutrinos are bounded by the above experiments involving initial-state ν_e's or ν_μ's or their antiparticles. The diagonal ν_τ magnetic dipole moment, however, is much less constrained because no strong ν_τ sources are available in the laboratory; the ν_τ-e cross section has never been measured. However, the calculated ν_τ flux produced in a proton beam dump from the decay of D_s mesons can be used to derive an upper limit on the ν_τ-e cross section which can be translated into a bound

$$\mu_{\nu_\tau} < 5.4 \times 10^{-7} \mu_B, \tag{7.30}$$

(Cooper-Sarkar et al. 1992). Of course, in this range the ν_τ electromagnetic cross section would far exceed a typical weak interaction one.

7.5.2 Spin-Flip Scattering in Supernovae

Neutrino scattering by photon exchange can also be important in astrophysical settings, notably if neutrinos are Dirac particles. The magnetic or electric dipole coupling is such that it flips the helicity of relativistic neutrinos, i.e. the final state is r.h. for an initial l.h. neutrino. This spin flip is of no importance in experiments where the electron recoil is measured, but it can have dramatic consequences in supernovae where l.h. neutrinos are trapped by the standard weak interactions. The spin-flip scattering by a electromagnetic dipole interaction would produce "wrong-helicity" states that could freely escape unless they scattered again electromagnetically. The SN 1987A neutrino signal indicates that this anomalous cooling channel cannot have been overly effective, yielding a constraint of around $\mu_\nu \lesssim 3 \times 10^{-12} \mu_B$ on all Dirac diagonal or transition moments in the sense of Eq. (7.25); see Sect. 13.8.3 for a more detailed discussion.

One should keep in mind, however, that for dipole moments in this range the spin precession in the strong macroscopic magnetic fields that are believed to exist in and near SN cores could also cause significant left-right transitions. Notably, the back conversion of r.h. neutrinos could cause a transfer of energy between widely separated regions of the SN core, and might even help at the explosion (Sect. 13.8.3). The role of relatively large neutrino dipole moments in SN physics has not been elaborated in enough depth to arrive at reliable regions of parameters that are ruled out or ruled in by SN physics and the SN 1987A neutrino signal.

7.5.3 Spin-Flip Scattering in the Early Universe

Neutrino spin-flip scattering has important consequences in the early universe as it can bring some or all of the "wrong-helicity" Dirac neutrino degrees of freedom into thermal equilibrium. The usual big bang nucleosynthesis (BBN) argument previously mentioned in Sect. 7.1.5 allows one to exclude this possibility because even one additional thermally excited neutrino degree of freedom appears to be forbidden by the spectacular agreement between the predicted and observed primordial light-element abundances.

This argument was first advanced by Morgan (1981a,b). Unfortunately, he used an unrealistically small cutoff for the Coulomb divergence of the spin-flip scattering cross section, leading to an overestimate of the efficiency by which r.h. Dirac neutrinos can be brought into thermal equilibrium. A reasonable cutoff by the Debye screening scale was used by Fukugita and Yazaki (1987) who found

$$\mu_\nu \lesssim 0.5 \times 10^{-10} \mu_{\rm B}, \tag{7.31}$$

about a factor of 3.5 looser than Morgan's original constraint.

This limit can be avoided for the diagonal ν_τ magnetic moment which would contribute to the $\nu_\tau \bar{\nu}_\tau \to e^- e^+$ annihilation process. If the ν_τ also had a mass in the 10 MeV regime the spin-flip excitation of the r.h. degrees of freedom would be compensated by the annihilation depletion of the ν_τ and $\bar{\nu}_\tau$ population before nucleosynthesis (Giudice 1990). A detailed analysis (Kawano et al. 1992) reveals that a μ_{ν_τ} larger than about $0.7 \times 10^{-8} \mu_{\rm B}$ is allowed in the mass range between a few and about 30 MeV.

Like in a SN core, there could exist large magnetic fields in the early universe that would allow for left-right transitions by the magnetic dipole induced spin precession (Sect. 8.4); for early discussions of this possibility see Lynn (1981) and Shapiro and Wasserman (1981). Arguments of this sort naturally depend on assumptions concerning the primordial magnetic field distribution; such fields may be required as seeds for the dynamo mechanism to create present-day galactic magnetic fields. A quantitative kinetic understanding of the process of populating the r.h. neutrinos requires a simultaneous treatment of the neutrino spin-precession and scattering much along the lines of Chapter 9 where flavor oscillations are studied in an environment where neutrinos scatter frequently. The most recent investigation of primordial neutrino magnetic oscillations is Enqvist, Rez, and Semikoz (1995); see their work for references to the previous literature. In certain plausible scenarios of primordial magnetic field distributions neutrino Dirac dipole moments as small as $10^{-20} \mu_{\rm B}$ seem to be in conflict with BBN.

7.5.4   Search for Radiative Neutrino Decays

The search for radiative decays of reactor, beam, solar, supernova, and cosmic neutrinos will be discussed at length in Chapter 12. For electron neutrinos, the effective transition moment in the sense of Eq. (7.27) will be found to be limited by

$$\mu_\nu \lesssim \begin{cases} 0.9 \times 10^{-1} \mu_{\rm B} \, ({\rm eV}/m_\nu)^2 & \text{Reactors,} \\ 0.5 \times 10^{-5} \mu_{\rm B} \, ({\rm eV}/m_\nu)^2 & \text{Sun,} \\ 1.5 \times 10^{-8} \mu_{\rm B} \, ({\rm eV}/m_\nu)^2 & \text{SN 1987A,} \\ 3 \times 10^{-10} \, \mu_{\rm B} \, ({\rm eV}/m_\nu)^{2.3} & \text{Cosmic background,} \end{cases} \tag{7.32}$$

where m_ν is the mass of the decaying parent neutrino. Of these limits, all except the cosmic one are based on measured neutrino fluxes. For nonelectron neutrinos, the cosmic and SN 1987A limits apply equally because these sources emit neutrinos of all flavors.

The SN 1987A constraints apply in this form only for $m_\nu \lesssim 40\,\text{eV}$, and for total lifetimes which in the laboratory exceed the transit time between the SN and Earth. The cosmological limit assumes that the radiative channel dominates and as such it requires $m_\nu \lesssim 30\,\text{eV}$; otherwise the universe would be "overclosed" by neutrinos. Because $m_{\nu_e} \lesssim 5\,\text{eV}$ these conditions are plausibly satisfied for ν_e. For ν_μ and ν_τ one may contemplate masses in excess of $30\,\text{eV}$ if one simultaneously contemplates novel interactions which allow for invisible fast decays. Therefore, radiative decay limits for ν_μ and ν_τ must be derived as a function of the assumed mass and of the assumed total decay time. For SN 1987A this exercise will be performed in Sect. 12.4.5, the results are displayed in Fig. 12.17. For cosmic neutrinos, I do not know of a published comparable contour plot.

7.5.5 Plasmon Decay in Stars

The last and most interesting constraint arises from the energy-loss argument applied to globular cluster stars. The neutrino emissivity by the plasma process $\gamma \to \nu\bar{\nu}$ would be too large unless

$$\mu_\nu \lesssim 3\times 10^{-12}\,\mu_B. \tag{7.33}$$

This bound, which applies for $m_\nu \lesssim 5\,\text{keV}$, has been derived in detail in Sect. 6.5.6.

Chapter 8

Neutrino Oscillations

The phenomenon of neutrino oscillations in vacuum and in media as well as in magnetic fields is studied. Experimental constraints on neutrino mixing parameters are reviewed.

8.1 Introduction

If neutrinos do not have novel interactions that allow them to decay fast then they must obey the cosmological mass limit of $m_\nu \lesssim 30\,\text{eV}$. This is even true for ν_τ although it could decay sufficiently fast into the $e^+e^-\nu_e$ channel if it had a mass in the 10 MeV range. However, the absence of γ rays from SN 1987A in conjunction with the neutrino signal (Sect. 12.5.2) and independently arguments of big-bang nucleosynthesis (Fig. 7.2) exclude this option. If neutrino masses are indeed so small then there is no hope for a direct experimental measurement at the present time, with the possible exception of m_{ν_e} which could still show up in tritium β decay or neutrinoless $\beta\beta$ decay experiments as discussed in Chapter 7.

Pontecorvo (1967) was the first to realize that the existence of several neutrino flavors (two were known at the time) allows even very small masses to become visible.[46] A "weak-interaction eigenstate" which is produced, say, in the neutron decay $n \to pe^-\bar{\nu}_e$ is in general expected to be a mixture of neutrino mass eigenstates. The phenomenon of particle mixing (Sect. 7.2.1) is familiar from the quarks

---

[46]Pontecorvo's (1957, 1958) original discussion referred to $\nu \leftrightarrow \bar{\nu}$ oscillations in analogy to the experimentally observed case of $K^\circ \leftrightarrow \overline{K}^\circ$. For a historical overview see Pontecorvo (1983). In this book I will not discuss $\nu \leftrightarrow \bar{\nu}$ oscillations any further—see Akhmedov, Petcov, and Smirnov (1993) for a recent reexamination of "Pontecorvo's original oscillations."

and hence does not appear to be an exotic assumption. Whatever the physical cause of particle masses, it seems unrelated to their gauge interactions! Expanding the neutrino state in plane waves, each mass eigenstate propagates as $e^{-i(\omega t - \mathbf{k}_i \cdot \mathbf{x})}$ where $\mathbf{k}_i^2 = \omega^2 - m_i^2$. Therefore, the different mass components develop phase differences, causing the original superposition which formed a $\bar{\nu}_e$ to turn partially into other flavors. Therefore, one can search for the disappearance of neutrinos of a given flavor from a beam, or one can search for the appearance of "wrong-flavored" states in a beam. The measured deficit of solar ν_e's (Chapter 10) has long been attributed to the oscillation phenomenon even though a definitive proof is still missing.

Neutrino oscillations effectively measure a phase difference between two components of a beam, much as the rotation of the plane of polarization of linearly polarized light represents a phase difference between the circularly polarized components of a beam in an optically active medium. This method is sensitive to small differences in the refractive index of the two components. For example, the Faraday rotation effect can be used to measure very weak interstellar magnetic fields even though the interstellar medium is quite dilute. Both for neutrinos and photons, fine points of the dispersion relation have a significant impact on the oscillation effect.

Wolfenstein (1978) was the first to recognize that the medium-induced modification of the neutrino dispersion relations (Sect. 6.7.1) is not an academic affair, but rather of immediate relevance for some neutrino oscillation experiments. In Mikheyev and Smirnov's (1985) seminal paper it was shown that oscillations can be "resonant" when a beam passes through such a density gradient that the flavor branches of the dispersion relation cross. This Mikheyev-Smirnov-Wolfenstein (MSW) effect is very important in astrophysics because neutrinos are naturally produced in the interior of stars and stream through a density gradient into empty space.

An adiabatic crossing of the dispersion relations has the effect of interchanging the flavor content of the neutrino flux even if the mixing angle is very small. This effect is one version of the oscillation solution of the solar neutrino problem. For suitable parameters it is also significant in supernovae where different-flavored neutrinos are thought to be produced with different energy spectra. The MSW effect could swap the spectral characteristics of the neutrinos emerging from a newborn neutron star, allowing for a number of fascinating novel effects.

If neutrinos had large magnetic dipole moments they could spin-precess in magnetic fields. This effect is completely analogous to flavor oscillations, except that here the two helicity components rather than the flavor components get transformed into each other. Again, this is a standard effect familiar from the behavior of electrons in magnetic fields. In astrophysical bodies large magnetic fields exist, especially in supernovae, so that magnetic helicity oscillations are potentially interesting. However, much larger magnetic dipole moments are required than are predicted for standard massive neutrinos. Thus, flavor oscillations have rightly received far more attention.

Presently I will develop the theoretical tools for neutrino oscillations, and summarize the current experimental situation. In Chapter 9 I will discuss the more complicated phenomena that obtain when oscillating neutrinos are trapped in a supernova core. The story of solar neutrinos (Chapter 10) is inextricably intertwined with that of neutrino oscillations, especially of the MSW variety. Finally, oscillations may also play a prominent role for supernova neutrinos and the interpretation of the SN 1987A signal (Chapter 11).

8.2 Vacuum Oscillations

8.2.1 Equation of Motion for Mixed Neutrinos

In order to derive a formal equation for the oscillation of mixed neutrinos I begin with the equation of motion of a Dirac spinor ν_i which describes the neutrino mass eigenstate i. It obeys the Dirac and thus the Klein-Gordon equation $(\partial_t^2 - \nabla^2 + m_i^2)\,\nu_i = 0$. One may readily combine all mass eigenstates in a single equation

$$(\partial_t^2 - \nabla^2 + M^2)\,\Psi = 0, \tag{8.1}$$

where

$$M^2 \equiv \begin{pmatrix} m_1^2 & 0 & 0 \\ 0 & m_2^2 & 0 \\ 0 & 0 & m_3^2 \end{pmatrix} \quad \text{and} \quad \Psi \equiv \begin{pmatrix} \nu_1 \\ \nu_2 \\ \nu_3 \end{pmatrix}. \tag{8.2}$$

Eq. (8.1) may be written in any desired flavor basis, notably in the basis of weak-interaction eigenstates to which one may transform by virtue of Eq. (7.8).

$$\begin{pmatrix} \nu_e \\ \nu_\mu \\ \nu_\tau \end{pmatrix} = U \begin{pmatrix} \nu_1 \\ \nu_2 \\ \nu_3 \end{pmatrix} \tag{8.3}$$

The mass matrix transforms according to $M^2 \to U M^2 U^\dagger$ and is no longer diagonal. (Note that $U^{-1} = U^\dagger$ because it is a unitary matrix.)

As usual one expands the neutrino fields in plane waves of the form $\Psi(t, \mathbf{x}) = \Psi_{\mathbf{k}}(t)\, e^{i\mathbf{k}\cdot\mathbf{x}}$ for which Eq. (8.1) is

$$(\partial_t^2 + \mathbf{k}^2 + M^2)\, \Psi_{\mathbf{k}}(t) = 0. \tag{8.4}$$

In general one cannot assume a temporal variation $e^{-i\omega t}$ because there are three different branches of the dispersion relation with $\omega_i^2 = \mathbf{k}^2 + m_i^2$. A mixed neutrino cannot simultaneously have a fixed momentum *and* a fixed energy!

In practice one has always to do with very relativistic neutrinos for which $k = |\mathbf{k}| \gg m_i$. In this limit one may linearize Eq. (8.4) by virtue of $\partial_t^2 + \mathbf{k}^2 = (i\partial_t + k)(-i\partial_t + k)$. For each mass eigenstate $i\partial_t \rightarrow \omega_i \approx k$ and one needs to keep the exact expression only in the second factor where the difference between energy and momentum appears. Thus $\partial_t^2 + \mathbf{k}^2 \approx 2k(-i\partial_t + k)$, leading to the Schrödinger-type equation

$$i\partial_t \Psi_{\mathbf{k}} = \Omega_{\mathbf{k}} \Psi_{\mathbf{k}} \quad \text{where} \quad \Omega_{\mathbf{k}} \equiv \left(k + \frac{M^2}{2k} \right). \tag{8.5}$$

The vector Ψ originally consisted of neutrino Dirac spinors but it was reinterpreted as a vector of (positive-energy) probability amplitudes. For negative-energy states (antineutrinos) a global minus sign appears in Eq. (8.5).

The Schrödinger equation (8.5) describes a spatially homogeneous system with a nonstationary temporal evolution. In practice one usually deals with the opposite situation, namely a stationary neutrino flux such as that from a reactor or the Sun with a nontrivial spatial variation. Then it is useful to expand $\Psi(t, \mathbf{x})$ in components of fixed frequency $\Psi_\omega(\mathbf{x})e^{-i\omega t}$, yielding

$$(-\omega^2 - \nabla^2 + M^2)\, \Psi_\omega(\mathbf{x}) = 0. \tag{8.6}$$

In the relativistic limit and restricting the spatial variation to the z-direction one obtains in full analogy to the previous case

$$i\partial_z \Psi_\omega = -K_\omega \Psi_\omega \quad \text{where} \quad K_\omega \equiv \left(\omega - \frac{M^2}{2\omega} \right). \tag{8.7}$$

This equation describes the spatial variation of a neutrino beam propagating in the positive z-direction with a fixed energy ω.

Ultimately one is not interested in amplitudes but in the observable probabilities $|\Psi_\ell|^2 = \Psi_\ell^* \Psi_\ell$ with $\ell = e$, μ, or τ. One may derive an equation of motion for these quantities which is most compact in terms of a density matrix

$$\rho_{ab} = \Psi_b^* \Psi_a \,. \tag{8.8}$$

Then $i\partial_t \rho_{\mathbf{k}} = [\Omega_{\mathbf{k}}, \rho_{\mathbf{k}}]$ or $i\partial_z \rho_\omega = -[K_\omega, \rho_\omega]$ where $[A, B] = AB - BA$ is a commutator of matrices. Therefore,

$$i\partial_t \rho = (2k)^{-1}[M^2, \rho] \quad \text{or} \quad i\partial_z \rho = (2\omega)^{-1}[M^2, \rho]\,, \tag{8.9}$$

where the indices ω or k have been dropped.

A beam evolves from the $z = 0$ state or density matrix as

$$\Psi_\omega(z) = e^{iKz}\Psi_\omega(0) \quad \text{or} \quad \rho_\omega(z) = e^{-iKz}\rho_\omega(0)\,e^{iKz}\,. \tag{8.10}$$

An analogous result applies to the case of temporal rather than spatial oscillations. In the weak-interaction basis e^{iKz} will be denoted by W,

$$W(z) \equiv (e^{iKz})_{\text{weak}} = U(e^{iKz})_{\text{mass}} U^\dagger \,. \tag{8.11}$$

If the neutrino is known to be a ν_e at the source $(z = 0)$ its probability for being measured as a ν_e at a distance z ("survival probability") is $|W_{ee}(z)|^2$.

One may be worried that the simple-minded derivation and interpretation of these results is problematic because the neutrino wave function is never directly observed. What is observed are the charged leptons absorbed or emitted in conjunction with the neutrino production and detection. However, if one performs a fully quantum-mechanical calculation of the probability (or cross section) for the compound process of neutrino production, propagation, and absorption, the naive oscillation probability described by the elements of the W matrix factors out for all situations of practical interest, and notably in the relativistic limit (Giunti et al. 1993; Rich 1993).

8.2.2 Two-Flavor Oscillations

The neutrino mixing matrix U can be parametrized exactly as the Cabbibo-Kobayashi-Maskawa (CKM) matrix in the quark sector in Eq. (7.6). If one of the three two-family mixing angles is much larger than the others (as for the quarks) one may study oscillations between the dominantly coupled families as a two-flavor mixing problem. Moreover, because so far all experiments—with the possible exception of solar and certain atmospheric neutrino observations—yield only upper limits on oscillation parameters one usually restricts the analysis to a two-flavor scenario. Of course, if one were to observe the appearance of a certain flavor—rather than the disappearance of ν_e as for solar neutrinos or of ν_μ in the atmospheric case—one would have to consider the possibility that they arise from sequential transitions of the sort $\nu_e \to \nu_\mu \to \nu_\tau$. Note that in the bottom-left entry of the approximate expression for the CKM matrix Eq. (7.6) the term $S_{12}S_{23}$ had to be kept because it is larger than the direct term S_{13}. The neutrino mixing angles could show a similar hierarchy.

With these caveats in mind we turn to the two-flavor mixing case where U has the 2×2 Cabbibo form Eq. (7.5),

$$U = \cos\theta\, I + i\sin\theta\, \sigma_2, \tag{8.12}$$

with the mixing angle θ, the 2×2 unit matrix I, and the Pauli matrix[47] σ_2. The mass matrix may be written in the form

$$M^2/2\omega = b_0 - \tfrac{1}{2}\mathbf{B}\cdot\boldsymbol{\sigma}, \tag{8.13}$$

where $b_0 = (m_1^2 + m_2^2)/4\omega$. In the weak-interaction basis

$$\mathbf{B} = \frac{2\pi}{\ell_{\mathrm{osc}}} \begin{pmatrix} \sin 2\theta \\ 0 \\ \cos 2\theta \end{pmatrix}, \tag{8.14}$$

a vector which is tilted with regard to the 3-axis by twice the mixing angle (Fig. 8.2). Further,

$$\ell_{\mathrm{osc}} \equiv \frac{4\pi\,\omega}{m_2^2 - m_1^2} \tag{8.15}$$

is the *oscillation length*. Its meaning will presently become clear.

In this representation it is straightforward to work out the spatial behavior of a stationary neutrino beam. From $K = \omega - b_0 + \tfrac{1}{2}\mathbf{B}\cdot\boldsymbol{\sigma}$ one finds

$$W = e^{i(\omega - b_0)z} \left[\cos\left(\frac{\pi z}{\ell_{\mathrm{osc}}}\right) - i\sin\left(\frac{\pi z}{\ell_{\mathrm{osc}}}\right) \begin{pmatrix} -\cos 2\theta & \sin 2\theta \\ \sin 2\theta & \cos 2\theta \end{pmatrix} \right]. \tag{8.16}$$

Assuming that the oscillations are among the first two families, the appearance probability for a ν_μ and the ν_e survival probability are for an initial ν_e

$$\mathrm{prob}\,(\nu_e \to \nu_\mu) = |W_{e\mu}|^2 = \sin^2(2\theta)\,\sin^2(\pi z/\ell_{\mathrm{osc}}),$$
$$\mathrm{prob}\,(\nu_e \to \nu_e) = |W_{ee}|^2 = 1 - \mathrm{prob}\,(\nu_e \to \nu_\mu). \tag{8.17}$$

The oscillation behavior is shown in Fig. 8.1.

---

[47]The Pauli matrices are $\sigma_1 = \begin{pmatrix} 0 & 1 \\ 1 & 0 \end{pmatrix}$, $\sigma_2 = \begin{pmatrix} 0 & -i \\ i & 0 \end{pmatrix}$, and $\sigma_3 = \begin{pmatrix} 1 & 0 \\ 0 & -1 \end{pmatrix}$.

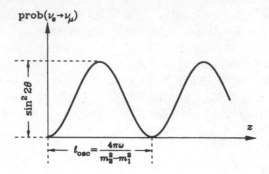

Fig. 8.1. Oscillation pattern for two-flavor oscillations (neutrino energy ω).

The flavor oscillations described by Eq. (8.16) are fully analogous to the rotation of the plane of polarization in an optically active medium or to the spin precession in a magnetic field. This analogy is brought out more directly if one starts with the equation of motion for the density matrix Eq. (8.9). Suppressing the index ω the matrices can be expressed as

$$\rho = \tfrac{1}{2}\left(1 + \mathbf{P} \cdot \boldsymbol{\sigma}\right) \quad \text{and} \quad K = \omega - b_0 + \tfrac{1}{2}\mathbf{B} \cdot \boldsymbol{\sigma}, \tag{8.18}$$

and a similar representation for Ω where $\mathbf{B}$ is expressed as a function of k by virtue of $\omega \to k$ to lowest order for relativistic neutrinos.

The vector $\mathbf{P}$ is a *flavor polarization vector*. In the weak-interaction basis $|\Psi_e|^2 = \tfrac{1}{2}\left(1 + P_3\right)$ and $|\Psi_\mu|^2 = \tfrac{1}{2}\left(1 - P_3\right)$ give the probability for the neutrino to be measured as ν_e or ν_μ, respectively. P_1 and P_2 contain phase information and thus reveal the degree of coherence between the flavor states. For a pure state $|\mathbf{P}| = 1$ while in general $|\mathbf{P}| < 1$. For $\mathbf{P} = 0$ one has a completely incoherent equal mixture of both flavors. In optics, $\mathbf{P}$ describes the degree of polarization of a light beam in the Poincaré sphere representation of the Stokes parameters (Poincaré 1892; Born and Wolf 1959).

The equation of motion for the polarization vector in any flavor basis is found to be (Stodolsky 1987; Kim, Kim, and Sze 1988)

$$\partial_z \mathbf{P} = \mathbf{B} \times \mathbf{P} \quad \text{or} \quad \partial_t \mathbf{P} = \mathbf{B} \times \mathbf{P}. \tag{8.19}$$

Here, $\mathbf{B}$ plays the role of a "magnetic field" and $\mathbf{P}$ that of a "spin vector." The precession of $\mathbf{P}$ for an initial ν_e where $\mathbf{P}(0) = (0, 0, 1)$ is shown in Fig. 8.2.

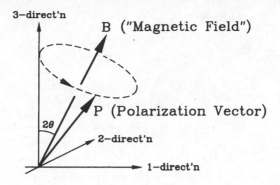

Fig. 8.2. Flavor oscillation as a "spin precession." (After Stodolsky 1987.)

8.2.3 Distribution of Sources and Energies

If the neutrino source region is not point-like relative to the oscillation length, one has to average the appearance or survival probabilities accordingly. If the source locations z_0 are distributed according to a normalized function $f(z_0)$ the ν_μ appearance probability is

$$\text{prob}\,(\nu_e \to \nu_\mu) = \sin^2 2\theta \int dz_0 \, f(z_0) \, \sin^2 \frac{\pi\,(z - z_0)}{\ell_{\text{osc}}}. \tag{8.20}$$

For example, consider a Gaussian distribution $f(z_0) = e^{-z_0^2/2s^2}/s\sqrt{2\pi}$ of size s for which

$$\text{prob}\,(\nu_e \to \nu_\mu) = \tfrac{1}{2} \sin^2 2\theta \left[1 - e^{-2\pi^2(s/\ell_{\text{osc}})^2} \cos(2\pi z/\ell_{\text{osc}}) \right]. \tag{8.21}$$

For $s = \tfrac{1}{5}\ell_{\text{osc}}$ this result is shown in Fig. 8.3. For $s = 0$ Eq. (8.21) is identical with Eq. (8.17) while for $s \gg \ell_{\text{osc}}$ it is $\tfrac{1}{2}\sin^2 2\theta$ which reflects that the beam is an incoherent mixture: the relative phases between different flavor components have been averaged to zero.

No source is exactly monochromatic; usually the neutrino energies are broadly distributed. With a point source and a normalized distribution $g(\omega)$ one finds

$$\text{prob}\,(\nu_e \to \nu_\mu) = \sin^2 2\theta \int d\omega \, g(\omega) \, \sin^2 \frac{(m_2^2 - m_1^2)\,z}{4\omega}. \tag{8.22}$$

As an example let $g(\omega)$ such that $\Delta = 2\pi/\ell_{\text{osc}}$ follows a Gaussian distribution $e^{-(\Delta-\Delta_0)^2/2\delta^2}/\delta\sqrt{2\pi}$ of width δ and with $\Delta_0 = 2\pi/\ell_0$. Then

$$\text{prob}\,(\nu_e \to \nu_\mu) = \tfrac{1}{2} \sin^2 2\theta \left[1 - e^{-\delta^2 z^2/2} \cos(2\pi z/\ell_0) \right] \tag{8.23}$$

which is shown in Fig. 8.4 for $\delta = \tfrac{1}{10}\Delta_0$. For $\delta = 0$ Eq. (8.23) reproduces Eq. (8.17) while for $z \gg \delta^{-1}$ it approaches $\tfrac{1}{2}\sin^2 2\theta$.

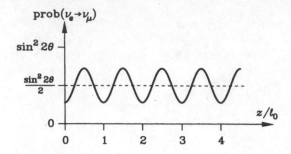

Fig. 8.3. Oscillation pattern for a Gaussian source distribution with $s = \frac{1}{5}\ell_{\text{osc}}$ according to Eq. (8.21).

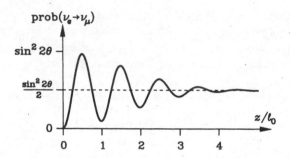

Fig. 8.4. Oscillation pattern for a mixture of neutrino energies with $\delta = \frac{1}{10}\pi/\ell_0$ according to Eq. (8.23).

These phenomena are well described in the picture of a precessing polarization vector which represents a density matrix and thus is designed to deal with incoherent or partially coherent beams. Notably, for a distribution of energies the polarization vector is $\mathbf{P} = \int d\omega\, g(\omega)\, \mathbf{P}_\omega$. Because the components $\mathbf{P}_\omega$ precess with different frequencies about a common "magnetic field" direction the component of $\mathbf{P}$ transverse to $\mathbf{B}$ disappears as the $\mathbf{P}_\omega$ approach a uniform distribution on the precession cone in Fig. 8.2. The projection of $\mathbf{P}$ on $\mathbf{B}$, however, is conserved so that $\mathbf{P} \rightarrow (\mathbf{P} \cdot \hat{\mathbf{B}})\, \hat{\mathbf{B}}$ for $t \rightarrow \infty$. If originally $\mathbf{P} = (0,0,1)$ for initial ν_e's the geometry of Fig. 8.2 indicates that $\mathbf{P}_3 \rightarrow \cos^2 2\theta$ for $t \rightarrow \infty$ and so $\text{prob}(\nu_e \rightarrow \nu_\mu) = \frac{1}{2}\sin^2 2\theta$.

8.2.4 Experimental Oscillation Searches

Because neutrino masses must be very small the oscillation length involves macroscopic scales. Numerically, it is

$$\ell_{\mathrm{osc}} = 2.48\,\mathrm{m}\,\frac{E_\nu}{1\,\mathrm{MeV}}\,\frac{1\,\mathrm{eV}^2}{\Delta m^2}\,. \tag{8.24}$$

Therefore, the modulation of the flavor content of a neutrino beam can occur on large, even astronomical length scales.

There exists a large number of oscillation searches using terrestrial neutrino sources (reactors, accelerators); for detailed references see Particle Data Group (1994). In Fig. 8.5 (curves a–f) I show the most restrictive limits on oscillations between the known neutrinos where

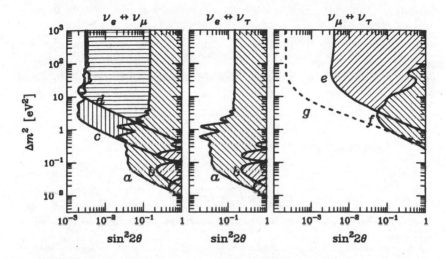

Fig. 8.5. Experimental limits on neutrino masses and mixing angles. *Reactors, ν_e disappearance:* (a) Bugey 4 (Achkar et al. 1995), superseding the Gösgen limits (Zacek et al. 1986); (b) Kurchatov Institute (Vidyakin et al. 1987, 1990, 1991). *Accelerator experiments:* (c) BNL Experiment 776, wideband beam, ν_e and $\bar{\nu}_e$ appearance (Borodovsky et al. 1992). (d) BNL Experiment 734, measurement of ν_e/ν_μ ratio (Ahrens et al. 1985). (e) Fermilab Experiment 531, ν_τ appearance (Ushida et al. 1986); similar constraints were reported by the CHARM II Collaboration (1993). (f) CDHS Experiment, ν_μ disappearance (Dydak et al. 1984). (g) Anticipated range of sensitivity for the CHORUS and NOMAD experiments which are currently taking data at CERN (DiLella 1993; Winter 1995).

the analysis was always based on the assumption that two-flavor oscillations dominate. The disappearance experiments, of course, also constrain oscillations into hypothetical sterile neutrinos.

Even though the experimental results look very impressive, a glance on the CKM matrix Eq. (7.6) reveals that one could not yet have expected to see oscillations in the $\nu_e \leftrightarrow \nu_\tau$ or $\nu_\mu \leftrightarrow \nu_\tau$ channel if the neutrino mixing angles are comparably small. It is very encouraging that the NOMAD and CHORUS experiments which are currently taking data at CERN (DiLella 1993; Winter 1995) anticipate a range of sensitivity (curve g in Fig. 8.5) which is promising both in view of the possible cosmological role of a m_ν in the 10 eV range and the small mixing angles probed. Other future but less advanced projects for terrestrial oscillation searches were reviewed by Schneps (1993, 1995).

At the time of this writing the LSND Collaboration has reported a signature that is consistent with the occurrence of $\bar{\nu}_\mu \to \bar{\nu}_e$ oscillations (Athanassopoulos et al. 1995). If this interpretation is correct, the corresponding Δm^2 would exceed about $1\,\mathrm{eV}^2$, while $\sin^2 2\theta$ would be a few 10^{-3}. The status of this claim is controversial at the present time—see, e.g. Hill (1995). No doubt more data need to be taken before one can seriously begin to believe that neutrino oscillations have indeed been observed.

8.2.5 Atmospheric Neutrinos

Besides reactors and accelerators, one may also use atmospheric neutrinos as a source to search for oscillations. Primary cosmic ray protons produce hadronic showers when interacting with atmospheric nuclei (A). Neutrinos are subsequently produced according to the simple scheme

$$
\begin{aligned}
p + A &\to n + \pi/K + \ldots \\
\pi/K &\to \mu^+(\mu^-) + \nu_\mu(\bar{\nu}_\mu) \\
\mu^+(\mu^-) &\to e^+(e^-) + \nu_e(\bar{\nu}_e) + \bar{\nu}_\mu(\nu_\mu).
\end{aligned}
$$

$$(8.25)$$

Therefore, one expects twice as many ν_μ's as ν_e's, and equally many neutrinos as antineutrinos of both flavors. At a detector, the neutrino flux is approximately isotropic except at energies below about 1 GeV where geomagnetic effects become important. Because the neutrinos come from anywhere in the atmosphere, from directly overhead or from as far as the antipodes, oscillation lengths between about 10

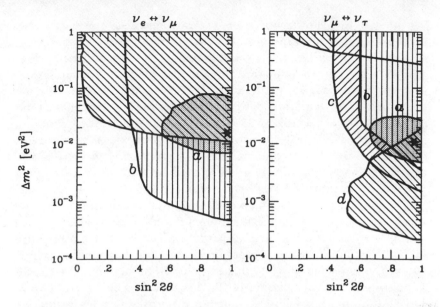

Fig. 8.6. Limits on neutrino masses and mixing angles from atmospheric neutrinos. (a) The shaded area is the range of masses and mixing angles required to explain the ν_e/ν_μ anomaly at Kamiokande (Fukuda et al. 1994); the star marks the best-fit value for the mixing parameters. The hatched areas are excluded by: (b) ν_e/ν_μ ratio at Fréjus (Fréjus Collaboration 1990, 1995; Daum 1994). (c) Absolute rate and (d) stopping fraction of upward going muons at IMB (Becker-Szendy et al. 1992). Also shown are the excluded areas from the experimental limits of Fig. 8.5.

and 13000 km are available.[48] The energy spectrum and absolute normalization of the flux must be determined by calculations and thus is probably uncertain to within about $\pm 30\%$ while the ν_e/ν_μ flavor ratio is likely known to within, say, $\pm 5\%$.

Several underground proton decay experiments have reported measurements of atmospheric neutrinos. The Fréjus detector (an iron calorimeter) saw the expected ν_e/ν_μ flavor ratio and thereby excluded the range of masses and mixing angles marked b in Fig. 8.6 for ν_e-ν_μ and ν_μ-ν_τ oscillations (Fréjus Collaboration 1990, 1995; Daum 1994).

Instead of measuring the neutrinos directly one may also study the flux of secondary muons produced by interactions in the rock surround-

---

[48]The effect of matter must be included for ν_e-ν_μ atmospheric neutrino oscillations. For a recent detailed analysis see Akhmedov, Lipari, and Lusignoli (1993).

ing the detector. (Electrons from ν_e interactions range out much faster
in the rock and so one expects mostly muons from ν_μ's.) This method is
sensitive to the high-energy spectral regime of the atmospheric ν_μ flux.
Moreover, one may select upward going muons which are produced from
ν_μ's which traversed the entire Earth and thus have a large oscillation
length available. The IMB detector excludes range c by this method
(Becker-Szendy et al. 1992). Also, one may determine the fraction of
muons stopped within the detector to those which exit, allowing one to
constrain a spectral deformation caused by the energy dependence of
the oscillation length. Range d is excluded by this method according
to the IMB detector (Becker-Szendy et al. 1992).

However, several detectors see a substantial deficit of atmospheric
ν_μ's relative to ν_e's, a finding usually expressed in terms of a "ratio of
ratios," i.e. the measured over the expected ratio of e-like over μ-like
events (Fig. 8.7). While this procedure is justified because it is largely
free of the uncertain absolute flux normalization, one must be careful
at interpreting the significance of the flux deficit. The error of a measured ratio does not follow a Gaussian distribution; a representation
like Fig. 8.7 tends to overemphasize the significance of the discrepancy
(Fogli and Lisi 1995).

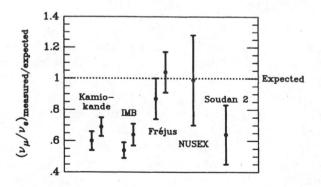

Fig. 8.7. Measured ratio of the atmospheric ν_μ/ν_e fluxes relative to the
expected value ("ratio of ratios") in five detectors. Where two results are
shown they refer to different signatures or data samples. (See Goodman
1995 for references.)

Apparently, the anomaly observed at the Kamiokande water Cherenkov detector (Hirata et al. 1992; Fukuda et al. 1994) can be explained in terms of oscillations for neutrino parameters in the shaded area in Fig. 8.6; the best-fit value is indicated by a star. These results are a combined fit for the sub-GeV and multi-GeV data as published by the Kamiokande collaboration (Fukuda et al. 1994). The oscillation hypothesis appears to be buttressed by a zenith-angle variation of the effect observed in Kamiokande's multi-GeV data sample although the claimed significance of this effect has been critiqued, e.g. by Fogli and Lisi (1995) and by Saltzberg (1995).

The required large mixing angle as well as the exclusion regions of the other experiments make it appear dubious that the anomaly is caused by oscillations. Still, it is a serious effect that cannot be blamed easily on problems with the Kamiokande detector. Also, the reliability of some of the exclusion areas in Fig. 8.6 may be called into question, notably because of their dependence on absolute flux normalizations. The intuition against a large ν_μ-ν_τ mixing angle may be misguided. In the future, it will be possible to test the relevant regime of mixing parameters in long-baseline laboratory experiments (e.g. Schneps 1995). At the time of this writing, the possibility that the atmospheric neutrino anomaly may be revealing neutrino oscillations remains a lively-debated possibility.

8.3   Oscillations in Media

8.3.1   Dispersion Relation for Mixed Neutrinos

The neutrino refractive index of a normal medium is extremely small and so its only potentially observable effect occurs in neutrino oscillations. The refractive index is different for different flavors—the medium is "flavor birefringent"—and so neutrinos from different families which propagate with the same energy through the same medium acquire different phases. If in addition these flavors mix, the medium-induced phase shift between them shows up in the interference between the mixed states. Without a medium the phase difference between mixed states arises from their mass difference. Hence, medium effects will be noticable only if the induced "effective mass" is of the same order as the vacuum masses. Therefore, medium refraction is important for neutrino oscillations in certain situations because the vacuum masses are very small.

The dispersion relation for a single flavor in a medium was given by Eq. (6.107). For three flavors which mix according to Eq. (8.3) the Klein-Gordon equation in Fourier space is

$$\left\{ \left[\omega - \frac{G_{\mathrm{F}} n_B}{\sqrt{2}} \begin{pmatrix} 3Y_e - 1 & 0 & 0 \\ 0 & Y_e - 1 & 0 \\ 0 & 0 & Y_e - 1 \end{pmatrix} \right]^2 \right.$$

$$\left. - k^2 - U \begin{pmatrix} m_1^2 & 0 & 0 \\ 0 & m_2^2 & 0 \\ 0 & 0 & m_3^2 \end{pmatrix} U^\dagger \right\} \begin{pmatrix} \nu_e \\ \nu_\mu \\ \nu_\tau \end{pmatrix} = 0, \quad (8.26)$$

where a possible neutrino background was ignored and $Y_n = 1 - Y_p = 1 - Y_e$ was used. Also, the higher-order difference between ν_μ and ν_τ was ignored. This equation has nonzero solutions only if $\det\{\ldots\}$ vanishes, a condition that gives us the dispersion relation for the three normal modes. For unmixed neutrinos where U is the unit matrix one recovers Eq. (6.107) for each flavor.

For all practical cases the neutrinos are highly relativistic so that one may linearize this equation,

$$\left(\omega - k - M_{\mathrm{eff}}^2 / 2k \right) \Psi = 0 \tag{8.27}$$

where it is easy to read the matrix M_{eff}^2 from Eq. (8.26). Because to lowest order $\omega = k$ one may equally use $M_{\mathrm{eff}}^2 / 2\omega$ in order to derive the dispersion relation, depending on whether one wishes to write ω as a function of k or vice versa.

For two-flavor mixing between ν_e and ν_μ or ν_τ (mixing angle θ_0) the effective mass matrix may be written in the same form as in vacuum

$$M_{\mathrm{eff}}^2 / 2\omega = b_0 - \tfrac{1}{2} \mathbf{B} \cdot \boldsymbol{\sigma}, \tag{8.28}$$

where $b_0 = (m_1^2 + m_2^2)/4k + \sqrt{2}\, G_{\mathrm{F}} n_B \left(Y_e - \tfrac{1}{2} \right)$ and

$$\mathbf{B} = \frac{2\pi}{\ell_{\mathrm{osc}}} \begin{pmatrix} \sin 2\theta \\ 0 \\ \cos 2\theta \end{pmatrix} = \frac{m_2^2 - m_1^2}{2\omega} \begin{pmatrix} \sin 2\theta_0 \\ 0 \\ \cos 2\theta_0 \end{pmatrix} - \sqrt{2}\, G_{\mathrm{F}} n_e \begin{pmatrix} 0 \\ 0 \\ 1 \end{pmatrix} \tag{8.29}$$

This equation defines implicitly the mixing angle θ as well as the oscillation length ℓ_{osc} in the medium in terms of the masses, the vacuum mixing angle θ_0, and the electron density n_e.

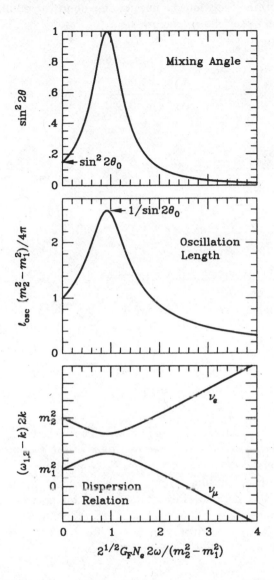

Fig. 8.8. Mixing angle, oscillation length, and neutrino dispersion relation as a function of the electron density. The medium was taken to have equal numbers of protons and neutrons ($Y_e = \frac{1}{2}$), the ratio of neutrino masses was taken to be $m_1 : m_2 = 1 : 2$, and $\sin^2 2\theta_0 = 0.15$.

Explicitly one finds for the mixing angle and the oscillation length in the medium the following expressions,

$$\tan 2\theta = \frac{\sin 2\theta_0}{\cos 2\theta_0 - \xi},$$

$$\sin 2\theta = \frac{\sin 2\theta_0}{[\sin^2 2\theta_0 + (\cos 2\theta_0 - \xi)^2]^{1/2}}, \qquad (8.30)$$

where

$$\xi \equiv \frac{\sqrt{2}\, G_F n_e\, 2\omega}{m_2^2 - m_1^2} = 1.53 \times 10^{-7}\, \frac{Y_e\,\rho}{\mathrm{g\,cm^{-3}}}\, \frac{\omega}{\mathrm{MeV}}\, \frac{\mathrm{eV}^2}{m_2^2 - m_1^2}, \qquad (8.31)$$

and

$$\ell_{\mathrm{osc}} = \frac{4\pi\,\omega}{m_2^2 - m_1^2}\, \frac{\sin 2\theta}{\sin 2\theta_0}. \qquad (8.32)$$

For $m_2 > m_1$ these functions are shown in Fig. 8.8; they exhibit a "resonance" for $\cos 2\theta_0 = \xi$.

The dispersion relation has two branches which in vacuum correspond to $\omega_{1,2} = (k^2 - m_{1,2}^2)^{1/2}$. In the relativistic limit they are

$$\omega_{1,2} - k = \frac{m_1^2 + m_2^2}{4k} + \sqrt{2}\, G_F n_B \left(Y_e - \tfrac{1}{2}\right)$$

$$\pm \frac{m_2^2 - m_1^2}{4k} \left[\sin^2 2\theta_0 + (\cos 2\theta_0 - \xi)^2\right]^{1/2}, \qquad (8.33)$$

a result schematically shown in Fig. 8.8. The "resonance" of the mixing angle corresponds to the crossing point of the two branches of the dispersion relation. Of course, the levels do not truly cross, but rather show the usual "repulsion."

Because the medium effect changes sign for antineutrinos, a resonance occurs between ν_e and ν_μ if $m_2 > m_1$, while none occurs between $\bar\nu_e$ and $\bar\nu_\mu$. If the mass hierarchy is the other way round, a resonance occurs for $\bar\nu_e$ and $\bar\nu_\mu$, but not for ν_e and ν_μ.

8.3.2   Oscillations in Homogeneous Media

In a homogeneous medium the treatment of neutrino oscillations is exactly as in vacuum except that one must use the effective medium mixing angle and oscillation length given above. If the medium is sufficiently dilute it will not affect the oscillations at all. This is the case when the quantity ξ of Eq. (8.31) is much smaller than unity. In the opposite limit one finds for the mixing angle in the medium

$$\sin 2\theta = \frac{m_2^2 - m_1^2}{\sqrt{2}\, G_F n_e\, 2\omega}\, \sin 2\theta_0. \tag{8.34}$$

The oscillation length becomes

$$\ell_{\text{osc}} = \frac{2\pi}{\sqrt{2}\, G_F n_e} = 1.63 \times 10^4\, \text{km}\ \frac{\text{g cm}^{-3}}{Y_e \rho}, \tag{8.35}$$

independent of the neutrino masses or energy.

Typically the effect of the medium is, therefore, to suppress the mixing angle and thus the possibility to observe oscillations. Of course, for normal materials with a density of a few g cm^{-3} and neutrino masses in the eV range one needs TeV neutrino energies for the medium to be relevant at all. For a review of the impact of oscillations in a medium on neutrino experiments or the observation of atmospheric, solar, or supernova neutrinos see, for example, Kuo and Pantaleone (1989).

8.3.3 Inhomogeneous Medium: Adiabatic Limit

In an inhomogeneous medium the oscillation problem is much more complicated. Recall that the spatial variation of a stationary neutrino beam in the z-direction is given by $i\partial_z \Psi = -K\Psi$ according to Eq. (8.7) if one drops the index ω. The matrix $K = \omega - M_{\text{eff}}^2/2\omega$ is now a function of z. Formally, the solution is $\Psi(z) = W\Psi(0)$ with

$$W = \mathcal{S}\, \exp\left(i \int_0^z K(z')\, dz' \right), \tag{8.36}$$

where $\mathcal{S}$ is the space-ordering operator. An explicit solution is not available because the matrices $K(z)$ generally do not commute for different z. Of course, for a constant K one recovers the previous result $W = e^{iKz}$.

In certain limits one may still find simple solutions. The most interesting case of neutrinos moving through an inhomogeneous medium is the emission from stars, notably the Sun, where they are produced in a relatively high-density region and then escape into vacuum. The density at the center of the Sun is about 150 g cm^{-3}, the solar radius 6.96×10^{10} cm, yielding an extremely shallow density variation by terrestrial standards! Therefore, consider the adiabatic limit where the density of the medium varies slowly over a distance ℓ_{osc} which is the characteristic length scale for the oscillation problem.

This case is best understood for two-flavor mixing if one studies the (temporal) evolution of the neutrino flavor polarization vector $\mathbf{P}$. Recall that it evolves according to the spin-precession formula $\dot{\mathbf{P}} = \mathbf{B} \times \mathbf{P}$ where the "magnetic field" is now a function of time. If $\mathbf{B}$ varies slowly relative to the precession frequency the "spin" follows the magnetic field in the sense that it moves on a precession cone which is "attached" to $\mathbf{B}$. If the spin is oriented essentially along the magnetic field direction it stays pinned to that direction. Therefore, it can be entirely reoriented by slowly turning the external magnet.

For the case of neutrino oscillations this means that in the adiabatic limit a state can be entirely reoriented in flavor space, i.e. an initial ν_e can be turned almost completely into a ν_μ even though the vacuum mixing angle may be small. Consider $\theta_0 \ll 1$, an initial density so large that the medium effects dominate, begin with a ν_e, and let $m_1 < m_2$. This means that in Fig. 8.8 (lowest panel) begin on the upper branch of the dispersion relation far to the right of the crossover. Then let the neutrino propagate toward vacuum through an adiabatic density gradient. This implies that it stays on the upper branch and ends up at $n_e = 0$ (vacuum) as the mass eigenstate of the *upper* eigenvalue m_2 which corresponds approximately to a ν_μ. This behavior is known as "resonant neutrino oscillations" or MSW effect after Mikheyev, Smirnov, and Wolfenstein. Mikheyev and Smirnov (1985) first discovered this effect when they studied the oscillation of solar neutrinos while Wolfenstein (1978) first emphasized the importance of refraction for neutrino oscillations. A simple interpretation of resonant oscillations in terms of an adiabatic "level crossing" was first given by Bethe (1986).

In order to quantify the adiabatic condition return to the picture of a spin precessing around a magnetic field. For flavor oscillations the precession frequency is $2\pi/\ell_{\rm osc}$. The "magnetic field" is tilted with an angle 2θ (medium mixing angle) against the 3-direction (Fig. 8.2) and so its speed of angular motion is $2\,d\theta/dt$. For spatial rather than temporal oscillations the adiabatic condition is

$$|\nabla \theta| \ll \pi/\ell_{\rm osc}. \tag{8.37}$$

This translates into $\nabla\theta = \frac{1}{2}\,\xi\,(\sin^2 2\theta/\sin 2\theta_0)\,\nabla \ln n_e$ while $\ell_{\rm osc}$ is given by Eq. (8.32) and ξ by Eq. (8.31). Thus, the adiabatic condition is

$$\xi \sin^3 2\theta\,|\nabla \ln n_e| \ll \sin^2 2\theta_0\,|m_2^2 - m_1^2|/2\omega. \tag{8.38}$$

This condition must be satisfied along the entire trajectory.

If the neutrino crosses a density region such that a resonance occurs, this part of the trajectory yields the most restrictive adiabaticity requirement. On resonance $\xi = \cos 2\theta_0$ and $\sin 2\theta = 1$. In this case one defines an adiabaticity parameter

$$\gamma \equiv \frac{m_2^2 - m_1^2}{2\omega} \frac{\sin 2\theta_0 \tan 2\theta_0}{|\nabla \ln n_e|_{\text{res}}}, \tag{8.39}$$

where the denominator is to be evaluated at the resonance point. The adiabatic condition is $\gamma \gg 1$. It establishes a relationship between vacuum mixing angles and neutrino masses for which resonant oscillations occur.

8.3.4 Inhomogeneous Medium: Analytic Results

For practical problems, notably the oscillation of solar neutrinos on their way out of the Sun, one may easily solve the equation $i\partial_z \Psi = -K\Psi$ numerically for prescribed profiles of the electron density and neutrino production rates. The main features of such calculations, however, can be understood analytically because for certain simple density profiles one can find analytic representations of Eq. (8.36) independently of the adiabatic approximation.

Consider the situation where a ν_e is produced in a medium and subsequently escapes into vacuum through a monotonically decreasing density profile. Because in the adiabatic limit the production and detection points are separated by many oscillation lengths, the oscillation pattern will be entirely washed out and one may use average probabilities for the flavor content. Initially, the projection of the polarization vector on the "magnetic field" direction is $\cos 2\theta$ where θ is the medium mixing angle at the production point. Because the component of **P** in the **B** direction is conserved when **B** changes adiabatically, the final average projection of **P** on the 3-axis is $\cos 2\theta_0 \cos 2\theta$. Then the survival probability is $\text{prob}(\nu_e \to \nu_e) = \frac{1}{2}(1 + \cos 2\theta_0 \cos 2\theta)$. This result applies in the adiabatic limit whether or not a resonance occurs.

Next, drop the adiabatic condition but assume that the production and detection points are many oscillation lengths away on opposite sides of a resonance so that the oscillation pattern remains washed out; then it remains sufficient to consider average probabilities. In this case it is useful to write

$$\text{prob}(\nu_e \to \nu_e) = \tfrac{1}{2} + (\tfrac{1}{2} - p) \cos 2\theta_0 \cos 2\theta, \tag{8.40}$$

where the correction p to the adiabatic approximation is the probability that the neutrino jumps from one branch of the dispersion relation to the other (Fig. 8.8, lowest panel) when it moves across the resonant density region.

A linear density profile near the resonance region, which is always a first approximation, yields the "Landau-Zener probability"

$$p = e^{-\pi\gamma/2} \tag{8.41}$$

which was first derived in 1932 for atomic level crossings. The adiabaticity parameter γ was defined in Eq. (8.39); in the adiabatic limit $\gamma \gg 1$ one recovers $p = 0$.

For a variety of other density profiles and without the assumption of a small mixing angle one finds a result of the form

$$p = \frac{e^{-(\pi\gamma/2)F} - e^{-(\pi\gamma/2)F'}}{1 - e^{-(\pi\gamma/2)F'}}, \tag{8.42}$$

where $F' = F/\sin^2\theta_0$ and F is an expression characteristic for a given density profile. For a linear profile $n_e \propto r$ one has $F = 1$ so that for a small mixing angle one recovers the Landau-Zener probability. The profile $n_e \propto r^{-1}$ leads to $F = \cos^2 2\theta_0/\cos^2\theta_0$. Of particular interest is the exponential $n_e \propto e^{-r/R_0}$ which yields

$$F = 1 - \tan^2\theta_0. \tag{8.43}$$

All of these results are quoted after the review by Kuo and Pantaleone (1989) where other special cases and references to the original literature can be found.

Going beyond the Landau-Zener approximation requires assuming one of the above specific forms for $n_e(r)$ for which analytic results exist. Recently, Guzzo, Bellandi, and Aquino (1994) used a somewhat different approach which is free of this limitation. They derived an approximate solution to the equivalent of Eq. (8.36) by the method of stationary phases for the space-ordered exponential. They found

$$p = \left(\frac{1-\gamma'}{1+\gamma'}\right)^2 \sin^2(\theta_0 - \theta) + \frac{2\gamma'}{(1+\gamma')^2}\left[\cos^2(\theta_0 - \theta) + \cos^2(\theta_0 + \theta)\right], \tag{8.44}$$

where $\gamma' \equiv \pi\gamma/16$. It was assumed that a resonance occurs between the production point (mixing angle θ) and the detection point which is taken to lie in vacuum (mixing angle θ_0). Eq. (8.44) can be used only in the nonadiabatic regime as it works only for $\gamma < 16/\pi$.

8.3.5   The Triangle and the Bathtub

The most important application of resonant neutrino oscillations is the possible reduction of the solar ν_e flux, an issue to be discussed more fully in Chapter 10. Here, I will use the above simple analytic results to discuss schematically the survival probability $\mathrm{prob}(\nu_e \to \nu_e)$ of neutrinos produced in the Sun.

To this end I use an exponential profile for the electron density, $n_e = n_c\, e^{-r/R_0}$, which is a reasonable first approximation with $R_0 = R_\odot/10.54$ (Bahcall 1989). Then $|\nabla \ln n_e| = R_0^{-1}$ is a quantity independent of location so that there is no need to evaluate it specifically on resonance. Thus $|\nabla \ln n_e|_{\mathrm{res}} = 3 \times 10^{-15}\,\mathrm{eV}$ is independent of neutrino parameters. Therefore, the adiabaticity parameter is

$$\gamma = \tfrac{1}{6} \times 10^3 \sin 2\theta_0 \tan 2\theta_0 \, \frac{\Delta m^2}{\mathrm{meV}^2} \frac{\mathrm{MeV}}{E_\nu}. \tag{8.45}$$

An electron density at the center of $n_c = 1.6 \times 10^{26}\,\mathrm{cm}^{-3}$ yields

$$\xi_c = 40 \, \frac{\mathrm{meV}^2}{\Delta m^2} \frac{E_\nu}{\mathrm{MeV}}. \tag{8.46}$$

The quantity ξ was defined in Eq. (8.31).

It is further assumed that all neutrinos are produced with a fixed energy by a point-like source at the solar center. They will encounter a resonance on their way out if $\xi_c > \cos 2\theta_0$. In this case one may calculate the "jump probability" p according to Eq. (8.42) with F from Eq. (8.43) for the exponential profile. The survival probability is then given by Eq. (8.40) with the mixing angle at the solar center

$$\cos 2\theta = \frac{\cos 2\theta_0 - \xi_c}{[(\cos 2\theta_0 - \xi_c)^2 + \sin^2 2\theta_0]^{1/2}}. \tag{8.47}$$

If $\xi_c < \cos 2\theta_0$ no resonance is encountered; the vacuum parameters dominate throughout the Sun. In this case one may use $p = 0$.

In the framework of these approximations it is straightforward to evaluate $\mathrm{prob}(\nu_e \to \nu_e)$ as a function of Δm^2 and $\sin^2 2\theta_0$. In Fig. 8.9 contours for the survival probability are shown for $E_\nu = 1\,\mathrm{MeV}$. Because the energy always appears in the combination $\Delta m^2/2E_\nu$ one may obtain an analogous plot for other energies by an appropriate vertical shift. This kind of plot represents the well-known MSW triangle. The dashed line marks the condition $\gamma = 1$ and thus divides the parameter plane into a region where the oscillations are adiabatic and one where they are not. The dotted line marks $\xi_c = \cos 2\theta_0$ and thus indicates for which neutrino parameters a resonance occurs on the way out of the Sun.

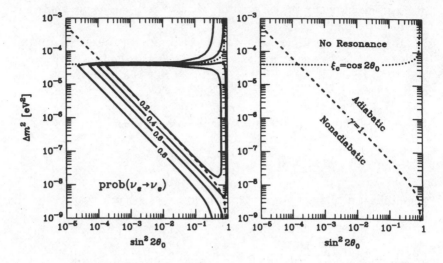

Fig. 8.9. The MSW triangle for the simplified solar model discussed in the text with $E_\nu = 1\,\mathrm{MeV}$.

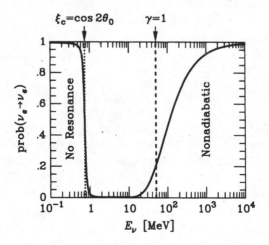

Fig. 8.10. The MSW bathtub for the simplified solar model discussed in the text with $\Delta m^2 = 3\times10^{-5}\,\mathrm{eV}^2$ and $\sin^2 2\theta_0 = 0.01$.

It is also instructive to look at a vertical cut through this contour plot, except that it is more useful to represent it as a function of E_ν for a fixed Δm^2 rather than the reverse. Both possibilities are equivalent because what appears is the combination $\Delta m^2/2E_\nu$. The result is the MSW bathtub shown in Fig. 8.10 for $\Delta m^2 = 3 \times 10^{-5}\,\text{eV}^2$ and $\sin^2 2\theta_0 = 0.01$ which is one set of parameters that might solve the solar neutrino problem (Chapter 10).

The main point of Fig. 8.10 is that resonant neutrino oscillations modify the spectrum of observable ν_e's differently for different energies. Low-energy neutrinos remain entirely unaffected, intermediate-energy ones are strongly suppressed, and the effect on high-energy states varies over a broad interval. There, the shape of the observable spectrum is modified relative to the source spectrum.

8.3.6 Neutrino Oscillations without Vacuum Mixing

Neutrino oscillations in vacuum or in a medium arise because in the weak-interaction basis the matrix of effective neutrino masses, M_{eff}^2, has off-diagonal elements which induce transitions between, say, an initial ν_e and a ν_μ. The effect of the medium is to modify the diagonal elements of this matrix, possibly such that different flavor states become degenerate, causing the effect of resonant oscillations. Conceivably, oscillations could occur even if $M_{\text{eff}}^2 = 0$ in vacuum so that in the standard model there are no off-diagonal terms.

The absence of off-diagonal medium contributions to M_{eff}^2 within the standard model reflects the absence of flavor-changing neutral currents (Sect. 7.2). It cannot be excluded experimentally, however, that for neutrinos such currents exist on some level, implying that a ν_e, for example, sometimes emerges as a ν_μ from a collision with another particle. In the forward direction this would cause an "off-diagonal refractive index" so that mixing and thus oscillations would be induced by the medium (Valle 1987; Fukugita and Yanagida 1988). Certain supersymmetric extensions of the standard model with R-parity breaking predict such effects (Guzzo, Masiero, and Petcov 1991; Kapetanakis, Mayr, and Nilles 1992). Therefore, if neutrino oscillations indeed explain the solar neutrino problem (Chapter 10) this does not inevitably imply neutrino vacuum masses and mixings—it could also point to the existence of flavor-changing neutral currents. An analytic description of two-flavor oscillations for this case was given by Guzzo and Petcov (1991) while a detailed parameter study for the solution of the solar neutrino problem was provided by Barger, Phillips, and Whisnant (1991).

8.4   Spin and Spin-Flavor Oscillations

8.4.1   Vacuum Spin Precession

According to Sect. 7.2 neutrinos may have magnetic dipole moments. In particular, if neutrinos have masses they inevitably have small but nontrivial electromagnetic form factors. Novel interactions can induce large dipole moments even for massless neutrinos. In the presence of magnetic fields such a dipole moment leads to the familiar spin precession which causes neutrinos to oscillate into opposite helicity states.

If a state started out as a helicity-minus neutrino, the outgoing particle is a certain superposition of both helicities. The "wrong-helicity" (right-handed) component does not interact by the standard weak interactions, diminishing the *detectable* neutrino flux. As the Sun has relatively strong magnetic fields it is conceivable that magnetic spin oscillations could be responsible for the measured solar neutrino deficit (Werntz 1970; Cisneros 1971)—see Sect. 10.7. Magnetic spin precessions could also affect supernova neutrinos, and in the early universe it could help to bring the "wrong-helicity" Dirac neutrino degrees of freedom into thermal equilibrium. Presently I will focus on some general aspects of neutrino magnetic spin oscillations rather than discussing specific astrophysical scenarios.

In the rest frame of a neutrino the evolution of its spin operator $\mathbf{S} = \frac{1}{2}\boldsymbol{\sigma}$ (Pauli matrices $\boldsymbol{\sigma}$) is governed by the Hamiltonian $H_0 = 2\mu\mathbf{B}_0 \cdot \mathbf{S}$; this follows directly from Eq. (7.23). Here, $\mathbf{B}_0$ refers to the magnetic field in the neutrino's rest frame as opposed to $\mathbf{B}$ in the laboratory frame. The equation of motion of the spin operator in the Heisenberg picture is then given by $i\partial_t\mathbf{S} = [\mathbf{S}, H_0]$ or

$$\dot{\mathbf{S}} = 2\mu\mathbf{B}_0 \times \mathbf{S}, \tag{8.48}$$

which represents the usual precession with frequency $2\mu\mathbf{B}_0$ around the magnetic field direction. Equivalently, one may consider Eq. (7.23) for the neutrino spinor which has the form of a Schrödinger equation.

Usually one will be concerned with very relativistic neutrinos so that the magnetic field $\mathbf{B}_0$ in the rest frame must be obtained from the electric and magnetic fields in the laboratory frame $\mathbf{E}$ and $\mathbf{B}$ by virtue of the usual Lorentz transformation

$$\mathbf{B}_0 = (\mathbf{B} \cdot \hat{\mathbf{v}})\hat{\mathbf{v}} + \gamma[\hat{\mathbf{v}} \times (\mathbf{B} \times \hat{\mathbf{v}}) + \mathbf{E} \times \mathbf{v}], \tag{8.49}$$

where $\mathbf{v}$ is the velocity of the neutrino, $\hat{\mathbf{v}}$ the velocity unit vector, and $\gamma = (1 - \mathbf{v}^2)^{-1/2} = E_\nu/m$ the Lorentz factor with E_ν the neutrino energy and m its mass. The spin precession as viewed from the laboratory system, however, involves a time-dilation factor γ^{-1} so that the Hamiltonian for the spin evolution in the laboratory system is $H = \mu\gamma^{-1}\mathbf{B}\cdot\boldsymbol{\sigma}$ or explicitly

$$H = \mu\Big[(m/\omega)(\mathbf{B}\cdot\hat{\mathbf{v}})\hat{\mathbf{v}} + \hat{\mathbf{v}}\times(\mathbf{B}\times\hat{\mathbf{v}}) + \mathbf{E}\times\mathbf{v}\Big]\cdot\boldsymbol{\sigma}. \qquad (8.50)$$

If the spin is quantized along the direction of motion (helicity states) and if there is only a magnetic field in the laboratory ($\mathbf{E} = 0$), the Hamiltonian is

$$H = \mu B \begin{pmatrix} \gamma^{-1}\cos\theta & e^{-i\varphi}\sin\theta \\ e^{+i\varphi}\sin\theta & -\gamma^{-1}\cos\theta \end{pmatrix}, \qquad (8.51)$$

with θ the magnetic field direction relative to the neutrino direction of motion and φ its azimuthal direction which may be chosen as $\varphi = 0$.

The bottom line is that very relativistic neutrinos ($\gamma \to \infty$) spin-precess like nonrelativistic ones, except that the relevant magnetic field is the transverse component B_T; the effective laboratory Hamiltonian is

$$H = \mu B_T \begin{pmatrix} 0 & 1 \\ 1 & 0 \end{pmatrix}. \qquad (8.52)$$

Therefore, if one begins with left-handed neutrinos a complete precession into right-handed ones will always occur (assuming a sufficient path length), independently of the direction between the laboratory magnetic field and the neutrino direction of motion (Fujikawa and Shrock 1980).

If there is only an electric field in the laboratory frame, magnetic oscillations will nevertheless occur because Eq. (8.50) implies that the neutrino sees an effective magnetic field $\mathbf{B}_T = \mathbf{E}\times\mathbf{v}$ (Okun 1986). The precession thus takes place in the plane of $\mathbf{E}$ and $\mathbf{v}$.

Contrary to flavor oscillations, the oscillation length does not depend on the energy. Therefore, a broad energy spectrum would not cause a depolarization of the neutrino flux from an astrophysical source (Sun, supernovae). Rather, neutrinos of all energies would spin-precess in step with each other.

Because the precession frequency is $2\mu B_T$ the neutrinos will have reversed their spin after a distance

$$\tfrac{1}{2}\ell_{\mathrm{osc}} = \frac{\pi}{2\mu B_T} = 5.36\times10^{13}\,\mathrm{cm}\,\frac{10^{-10}\,\mu_B}{\mu}\,\frac{1\,\mathrm{G}}{B_T}, \qquad (8.53)$$

with $\mu_B = e/2m_e$ the Bohr magneton. With a solar B_T in the kG regime the oscillation length is of order a solar radius $R_\odot = 6.96\times10^{10}\,\mathrm{cm}$. A substantial reduction of the left-handed solar neutrino flux then requires magnetic dipole moments of order $10^{-10}\,\mu_{\rm B}$. Typical galactic magnetic fields are of order $10^{-6}\,{\rm G}$ with coherence scales of order kpc. Therefore, neutrinos from a galactic source such as a supernova could be "flipped" before reaching Earth if their dipole moment is of order $10^{-12}\,\mu_{\rm B}$. These values are in the neighborhood of what is experimentally and observationally allowed so that there remains interest in the possibility that magnetic spin oscillations could have a significant impact on the neutrino fluxes from the Sun or from supernovae.

8.4.2 Spin Precession in a Medium

The picture of magnetic spin oscillations thus developed is incomplete. If neutrinos propagate in a medium as in the Sun they are subject to medium-induced refractive effects which shift the energy of left-handed states relative to right-handed ones which propagate as in vacuum because they are sterile. Therefore, including medium effects the effective Hamiltonian for the evolution of the two helicity states of ν_e is

$$H = \begin{pmatrix} 0 & \mu B_{\rm T} \\ \mu B_{\rm T} & \sqrt{2}G_{\rm F}(n_e - \tfrac{1}{2}n_n) \end{pmatrix}, \qquad (8.54)$$

where n_e and n_n are the electron and neutron densities, respectively. Here, "spin up" refers to the direction of motion, i.e. to right-handed states, "spin down" to left-handed ones. The two helicity states are no longer degenerate, and the spin precession will no longer lead to a complete reversal of the spin. One may write $H = b + \mu\mathbf{B}\cdot\boldsymbol{\sigma}$ so that there is effectively a longitudinal magnetic field, i.e. parallel to the direction of motion or z-direction

$$B_{\rm L} = \frac{G_{\rm F}(n_e - \tfrac{1}{2}n_n)}{\sqrt{2}\,\mu} = 66.6\,{\rm kG}\ \frac{10^{-10}\mu_{\rm B}}{\mu}\ \frac{\rho\,(Y_e - \tfrac{1}{2}Y_n)}{{\rm g\,cm^{-3}}}. \qquad (8.55)$$

Here, ρ is the mass density and $Y_{e,n}$ the electron and neutron number per baryon. If $B_{\rm L} \gg B_{\rm T}$ the precession is around a direction close to the direction of motion, and so the spin reversal will always be far from complete. Hence, the presence of the medium suppresses magnetic spin oscillations (Voloshin, Vysotskiĭ, and Okun 1986).

8.4.3 Spin-Flavor Precession

If neutrinos were Majorana particles they could not have magnetic dipole moments, but they could still possess transition moments between different flavors. In this case the magnetic oscillation would be, say, between a helicity-minus ν_e to a helicity-plus ν_μ which, because of its assumed Majorana nature, is identical to a $\bar\nu_\mu$. Because typically the ν_e and ν_μ masses will be different they must be included in the phase evolution of the neutrinos. Thus, for relativistic states (momentum p) one arrives at a two-level equation of motion of the form

$$i\partial_t \begin{pmatrix} \nu_e \\ \bar\nu_\mu \end{pmatrix} = \begin{pmatrix} m_{\nu_e}^2/2p & \mu B_T \\ \mu B_T & m_{\nu_\mu}^2/2p \end{pmatrix} \begin{pmatrix} \nu_e \\ \bar\nu_\mu \end{pmatrix}. \tag{8.56}$$

The transition magnetic moment μ thus leads to simultaneous spin and flavor oscillations (Schechter and Valle 1981). Moreover, neutrinos will be converted to antineutrinos (and the reverse). If such spin-flavor oscillations occurred, say, in the Sun one might actually be able to measure a flux of *antineutrinos* from that source.

A transition magnetic moment implies that the individual flavor lepton numbers are not conserved. Thus most likely there is also standard flavor mixing which allows for, say, $\nu_e \leftrightarrow \nu_\mu$ oscillations. If the mass eigenstates are m_1 and m_2, respectively, and if the vacuum mixing angle is θ, the four-level equation of motion in vacuum is

$$i\partial_t \begin{pmatrix} \nu_e \\ \nu_\mu \\ \bar\nu_e \\ \bar\nu_\mu \end{pmatrix} = \begin{pmatrix} \Delta_m c_{2\theta} & \Delta_m s_{2\theta} & 0 & \mu B_T \\ \Delta_m s_{2\theta} & -\Delta_m c_{2\theta} & \mu B_T & 0 \\ 0 & \mu B_T & \Delta_m c_{2\theta} & \Delta_m s_{2\theta} \\ \mu B_T & 0 & \Delta_m s_{2\theta} & -\Delta_m c_{2\theta} \end{pmatrix} \begin{pmatrix} \nu_e \\ \nu_\mu \\ \bar\nu_e \\ \bar\nu_\mu \end{pmatrix}, \tag{8.57}$$

where $\Delta_m \equiv (m_2^2 - m_1^2)/4p$, $c_{2\theta} \equiv \cos 2\theta$, and $s_{2\theta} \equiv \sin 2\theta$. Any of ν_e, ν_μ, $\bar\nu_e$, and $\bar\nu_\mu$ can oscillate into any of the others.

Including medium effects further complicates this matrix because the energies of ν_e and ν_μ are shifted by different amounts, and each is shifted in the opposite directions from its antiparticle. With $V_{\nu_e} = \sqrt{2}G_F(n_e - \frac{1}{2}n_n)$ and $V_{\nu_\mu} = \sqrt{2}G_F(-\frac{1}{2}n_n)$ the matrix becomes

$$\begin{pmatrix} \Delta_m c_{2\theta} + V_{\nu_e} & \Delta_m s_{2\theta} & 0 & \mu B_T \\ \Delta_m s_{2\theta} & -\Delta_m c_{2\theta} + V_{\nu_\mu} & \mu B_T & 0 \\ 0 & \mu B_T & \Delta_m c_{2\theta} - V_{\nu_e} & \Delta_m s_{2\theta} \\ \mu B_T & 0 & \Delta_m s_{2\theta} & -\Delta_m c_{2\theta} - V_{\nu_\mu} \end{pmatrix}. \tag{8.58}$$

If there are density gradients (solar neutrinos!) one may have resonant magnetic conversions between, say, ν_e and $\bar\nu_\mu$ where the barrier to spin-precessions caused by the mass difference is compensated by matter effects, i.e. one may have resonant spin-flavor oscillations (Akhmedov 1988a,b; Barbieri and Fiorentini 1988; Lim and Marciano 1988).

For Dirac neutrinos, transitions to antineutrinos are not possible and so one needs to consider oscillations separately in the four-level system $(\nu_e^L, \nu_\mu^L, \nu_e^R, \nu_\mu^R)$ and $(\bar{\nu}_e^L, \bar{\nu}_\mu^L, \bar{\nu}_e^R, \bar{\nu}_\mu^R)$. Moreover, one may have both diagonal and transition magnetic moments so that for neutrinos the r.h.s. of the equation of motion is

$$
\begin{pmatrix}
\Delta_m c_{2\theta} + V_{\nu_e} & \Delta_m s_{2\theta} & \mu_{ee} B_{\mathrm{T}} & \mu_{\mu e} B_{\mathrm{T}} \\
\Delta_m s_{2\theta} & -\Delta_m c_{2\theta} + V_{\nu_\mu} & \mu_{e\mu} B_{\mathrm{T}} & \mu_{\mu\mu} B_{\mathrm{T}} \\
\mu_{ee} B_{\mathrm{T}} & \mu_{\mu e} B_{\mathrm{T}} & \Delta_m c_{2\theta} & \Delta_m s_{2\theta} \\
\mu_{e\mu} B_{\mathrm{T}} & \mu_{\mu\mu} B_{\mathrm{T}} & \Delta_m s_{2\theta} & -\Delta_m c_{2\theta}
\end{pmatrix}
\begin{pmatrix}
\nu_e^L \\ \nu_\mu^L \\ \nu_e^R \\ \nu_\mu^R
\end{pmatrix},
\tag{8.59}
$$

where the right-handed neutrinos do not experience a medium-induced energy shift.

For certain values of the parameters the oscillations between two states may dominate which are then described by a certain 2×2 submatrix. Then the general treatment is much like the flavor mixing problems studied earlier. In general, a much richer collection of solutions obtains—for a review see Pulido (1992).

Electric transition moments can also play a role, and precessions in electric fields may be important. Majorana neutrinos have either electric or magnetic transition moments, but not both as long as CP remains conserved. Still, even electric transition moments would lead to spin-flavor oscillations in macroscopic magnetic fields.

8.4.4   Twisting Magnetic Fields

The problem of spin oscillations is further complicated by the possibility that the magnetic field changes its direction along the neutrino trajectory, i.e. that it may be "twisting" (Aneziris and Schechter 1991). In this case the equation of motion of a two-level spin-precession problem is based on the Hamiltonian

$$
H = \mu \begin{pmatrix} B_{\mathrm{L}} & B_{\mathrm{T}} e^{-i\varphi} \\ B_{\mathrm{T}} e^{i\varphi} & -B_{\mathrm{L}} \end{pmatrix},
\tag{8.60}
$$

where the three parameters B_{L}, B_{T}, and φ vary along the neutrino trajectory. The transverse field strength B_{T} is a physical magnetic field while the longitudinal one B_{L} is an effective magnetic field which represents the neutrino mass difference (spin-flavor oscillations!) and refractive medium effects. As long as $\mathbf{B}_{\mathrm{T}}$ maintains its direction along the trajectory, the phase φ can be globally chosen to be zero.

If φ varies along the trajectory so that the neutrino experiences it to be a function of time, it is useful to transform the equation of motion to a coordinate system which corotates with $\mathbf{B}_{\mathrm{T}}$ so that in the new frame one is back to the old situation of a fixed direction for $\mathbf{B}_{\mathrm{T}}$ (Smirnov

1991). This transformation is achieved by $\nu' = U\nu$ where ν represents a two-level wave function and

$$U = \begin{pmatrix} e^{-i\varphi/2} & 0 \\ 0 & e^{i\varphi/2} \end{pmatrix}, \tag{8.61}$$

so that the spin-up and down states acquire opposite, time-dependent phases. However, because the transformation does not mix spin up with spin down, the transition rate between the two levels is the same in both coordinate systems. The equation of motion $i\partial_t \nu' = H'\nu'$ involves the Hamiltonian

$$H' = \mu \begin{pmatrix} B_\mathrm{L} & B_\mathrm{T} \\ B_\mathrm{T} & -B_\mathrm{L} \end{pmatrix} + \tfrac{1}{2}\dot{\varphi} \begin{pmatrix} -1 & 0 \\ 0 & 1 \end{pmatrix}. \tag{8.62}$$

Put another way, the effective longitudinal magnetic field in the rotating frame is $B_\mathrm{L}' = B_\mathrm{L} - \tfrac{1}{2}\dot{\varphi}/\mu$ while B_T remains unchanged. $|B_\mathrm{L}'|$ can be less or larger than $|B_\mathrm{L}|$ so that the transition rate between the two levels can be increased or decreased by a twist. B_L may even be cancelled entirely, enabling resonant oscillations.

For a systematic discussion of the full four-level spin-flavor problem see, for example, Akhmedov, Petcov, and Smirnov (1993).

Chapter 9

Oscillations
of Trapped Neutrinos

In a supernova (SN) core or in the early universe neutrinos scatter on the background medium and on each other. Therefore, oscillations of mixed neutrinos are frequently interrupted, leading to flavor equilibrium if there is enough time. A Boltzmann-type kinetic equation is derived that accounts simultaneously for neutrino oscillations and collisions for arbitrary neutrino degeneracy. It is applied to a SN core, yielding an estimate of the time scale to reach flavor equilibrium. On the basis of the SN 1987A neutrino observations a limit on the mixing of sequential neutrinos with a hypothetical sterile flavor is derived.

9.1 Introduction

So far oscillations were discussed in the context of "beam experiments" where neutrinos of a known flavor are produced at a certain location, for example in a power reactor or in the Sun, then propagate over a distance, and are then detected with a device that allows one to distinguish between different flavors. It was assumed that there were no interactions between the production and detection point, with the possible exception of decays, allowing one to treat neutrino oscillations as a simple propagation phenomenon fully analogous to light propagation in an optically active medium.

This approach is not adequate for the early universe and a young SN core. In both cases there are frequent neutrino collisions which affect the free evolution of the phases. The impact of these collisions can be understood if one assumes for the purpose of illustration that one

310

neutrino flavor (say ν_e) scatters with a rate Γ while the other (say ν_μ) does not. An initial ν_e will begin to oscillate into ν_μ. The probability for finding it in one of the two flavors evolves as previously discussed and as shown in Fig. 9.1 (dotted line). However, in each collision the momentum of the ν_e component of the superposition is changed, while the ν_μ component remains unaffected. Thus, after the collision the two flavors are no longer in the same momentum state and so they can no longer interfere: each of them begins to evolve separately. This allows the remaining ν_e to develop a new coherent ν_μ component which is made incoherent in the next collision, and so forth. This process will come into equilibrium only when there are equal numbers of ν_e's and ν_μ's.

This decoherence effect is even more obvious when one includes the possibility of ν_e absorption and production by charged-current reactions $\nu_e n \leftrightarrow pe$. Because of oscillations an initial ν_e is subsequently found to be a ν_μ with an average probability of $\frac{1}{2}\sin^2 2\theta$ (mixing angle θ) and as such cannot be absorbed, or only by the reaction $\nu_\mu n \leftrightarrow p\mu$ if it has enough energy. The continuous emission and absorption of ν_e's spins off a ν_μ with an average probability of $\frac{1}{2}\sin^2 2\theta$ in each collision! Chemical relaxation of the neutrino flavors will occur with an approximate rate $\frac{1}{2}\sin^2 2\theta\,\Gamma$ where Γ is a typical weak interaction rate for the ambient physical conditions. An initial ν_e population turns into an equal mixture of ν_e's and ν_μ's as shown schematically in Fig. 9.1 (solid line).

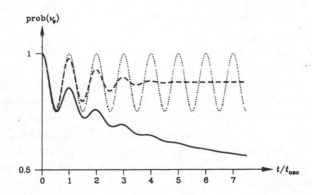

Fig. 9.1. Neutrino oscillations with collisions (solid line). In the absence of collisions and for a single momentum one obtains periodic oscillations (dotted line), while for a mixture of energies the oscillations are washed out by "dephasing" (dashed line).

A certain "damping" of flavor oscillations occurs even without collisions when the neutrinos are not monochromatic because then different modes oscillate with different frequencies. This "dephasing" effect was shown in Fig. 8.4 and is repeated in Fig. 9.1 (dashed line). While the dephasing washes out the oscillation pattern, it does not lead to flavor equilibrium: the probability for ν_e ends at a constant of $1 - \frac{1}{2}\sin^2 2\theta$. For the rest of this chapter "damping of oscillations" never refers to this relatively trivial dephasing effect.

The interplay of collisions and oscillations leads to flavor equilibrium between mixed neutrinos. In a SN core the concentration of electron lepton number is initially large so that the ν_e form a degenerate Fermi sea. The other flavors ν_μ and ν_τ are characterized by a thermal distribution at zero chemical potential. However, if they mix with ν_e they will achieve the same large chemical potential. In a SN core heat and lepton number are transported mostly by neutrinos; the efficiency of these processes depends crucially on the degree of neutrino degeneracy for each flavor. Therefore, it is of great interest to determine the time it takes a non-ν_e flavor to equilibrate with ν_e under the assumption of mixing (Maalampi and Peltoniemi 1991; Turner 1992; Pantaleone 1992a; Mukhopadhyaya and Gandhi 1992; Raffelt and Sigl 1993).

Moreover, if ν_e mixes with a sterile neutrino species, conversion into this inert state leads to the loss of energy and lepton number from the inner core of a SN. The observed SN 1987A neutrino signal may thus be used to constrain the allowed range of masses and mixing angles (Kainulainen, Maalampi, and Peltoniemi 1991; Raffelt and Sigl 1993; see also Shi and Sigl 1994).

These applications are discussed in Sects. 9.5 and 9.6 below. A simple estimate of the rate of flavor conversion and the emission rate of sterile neutrinos from a SN core requires not much beyond the approximate rate $\frac{1}{2}\sin^2 2\theta\,\Gamma$. However, a proper kinetic treatment of the evolution of a neutrino ensemble under the simultaneous action of oscillations and collisions is an interesting theoretical problem in its own right. Notably, it is far from obvious how to treat degenerate neutrinos in a SN core because the different flavors will suffer different Pauli blocking factors. Does this effect break the coherence between mixed flavors in neutral-current collisions?

The bulk of this chapter is devoted to the derivation and discussion of a general kinetic equation for mixed neutrinos (Dolgov 1981; Rudzsky 1990; Raffelt, Sigl, and Stodolsky 1993; Sigl and Raffelt 1993). This equation provides a sound conceptual and quantitative framework for dealing with various aspects of coherent and incoherent neutrino interactions with a medium, whether the neutrinos are degenerate or not. As a free spin-off it will provide a proper formalism to deal with the refractive effects of neutrinos propagating in a bath of neutrinos, a problem that is of interest in the early universe where interactions among neutrinos produce the dominant medium effect, and in the neutrino flow from a SN core where the density of neutrinos is larger than that of nonneutrino background particles (Sect. 11.4).

Besides neutrino flavor oscillations, magnetically induced spin or spin-flavor oscillations are also of potential interest because in supernovae and the early universe strong magnetic fields are believed to exist. Spin relaxation (the process of populating the r.h. degrees of freedom by the simultaneous action of spin oscillations and collisions) is a very similar problem to that of achieving chemical equilibrium between different flavors which is discussed here. Therefore, similar kinetic methods can be applied (Enqvist, Rez, and Semikoz 1995 and references therein).

9.2 Kinetic Equation for Oscillations and Collisions

9.2.1 Stodolsky's Formula

The loss of coherence between mixed neutrinos in collisions cannot be properly understood on the amplitude level because it is not the amplitudes, but only their relative coherence that is damped—the flavor states "decohere," they do not disappear. This is different from the decay of mixed particles where one of the amplitudes can be viewed as decreasing exponentially so that the total number of particles is not conserved. A natural description of decoherence is achieved by a density matrix as in Eq. (8.8); for a two-flavor mixing problem it was expressed in terms of a polarization vector $\mathbf{P}$ according to Eq. (8.18), $\rho = \frac{1}{2}(1 + \mathbf{P} \cdot \boldsymbol{\sigma})$. In the weak interaction basis its diagonal elements are the probabilities for measuring ν in, say, the ν_e or ν_μ state, respectively, while the off-diagonal elements contain relative phase information.

In a two-level system, the length of the polarization vector measures the degree of coherence: length 1 corresponds to a pure state, shorter P's to some degree of incoherence, and length zero is the completely mixed or incoherent state (Stodolsky 1987). In this latter case ρ is proportional to the unit matrix which is invariant under a transformation of basis. This state of "chemical" or "flavor equilibrium" has no off-diagonal elements in any basis.

Even coherent or partially coherent density matrices can be diago-
nalized in *some* basis. The interactions with the background medium
will also be diagonal in some basis; for neutrinos, this is the weak interaction basis. If the density matrix and the interactions are diagonal in
the same basis, there is no decoherence effect. For example, a density
matrix diagonal in the weak interaction basis implies that there is no
relative phase information between, say, a ν_e and a ν_μ and so collisions
which affect ν_e's and ν_μ's separately have no impact on the density
matrix. However, a density matrix which is diagonal in the mass basis,
assumed to be different from the weak interaction basis, will suffer a
loss of coherence in the same medium.

All told, the loss of coherence is given by a shrinking of the length
of $\mathbf{P}$. More precisely, only the component $\mathbf{P_T}$ is damped which represents the part "transverse" to the interaction basis, i.e. which in the
interaction basis represents the off-diagonal elements of ρ. Thus, in the
presence of collisions the evolution of $\mathbf{P}$ is given by (Stodolsky 1987)

$$\dot{\mathbf{P}} = \mathbf{V} \times \mathbf{P} - D\mathbf{P_T}. \tag{9.1}$$

The first part is the previous precession formula, except that here a
temporal evolution is appropriate since one has in mind the evolution
of a spatially homogeneous ensemble rather than the spatial pattern of
a stationary beam. The "magnetic field" $\mathbf{V}$ is

$$\mathbf{V} = \frac{2\pi}{t_{\text{osc}}} \begin{pmatrix} \sin 2\theta \\ 0 \\ \cos 2\theta \end{pmatrix}, \tag{9.2}$$

where θ is the mixing angle in the medium and t_{osc} the oscillation
period. They are given in terms of the neutrino masses and momen-
tum, the vacuum mixing angle, and the medium density by Eqs. (8.30)
and (8.32) where strictly speaking the neutrino energy is to be replaced
by its momentum as we have turned to temporal rather than spatial
oscillations. The damping parameter D is determined by the scattering
amplitudes on the background.

The evolution described by Eq. (9.1) is a precession around the
"magnetic field" $\mathbf{V}$, combined with a shrinking of the length of $\mathbf{P}$ to
zero. This final state corresponds to $\rho = \frac{1}{2}$ where both flavors are
equally populated, and with vanishing coherence between them. For
$D = t_{\text{osc}}^{-1}$ and $\sin 2\theta = \frac{1}{2}$ the evolution of the flavors is shown in Fig. 9.1
(solid line). Ignoring the wiggles in this curve it is an exponential as
can be seen by multiplying both sides of Eq. (9.1) with $\mathbf{P}/P^2$ which

leads to $\dot{P}/P = -D\,(P_{\mathrm{T}}/P)^2$ with $P = |\mathbf{P}|$ and $P_{\mathrm{T}} = |\mathbf{P}_{\mathrm{T}}|$. For a small collision rate relative to the oscillation frequency one may use the precession-averaged P_{T} which is found by taking the transverse part of the projection of $\mathbf{P}$ on $\mathbf{V}$. Elementary geometry in Fig. 8.2 yields $\langle P_{\mathrm{T}}\rangle/P = \cos 2\theta \sin 2\theta$ so that $\langle \dot{P}/P\rangle = -D\cos^2 2\theta \sin^2 2\theta$. If one neutrino interacts while the other is sterile (for example ν_μ with regard to charged-current absorption) D is half the collision rate Γ of the active flavor (Stodolsky 1987). For small mixing angles one recovers the previous intuitive relaxation rate $\frac{1}{2}\sin^2 2\theta\,\Gamma$.

Equation (9.1) is based on a single-particle wave function picture of neutrino oscillations and thus it is applicable if effects nonlinear in the neutrino density matrices can be ignored. It does not allow one to include the effect of Pauli blocking of neutrino phase space, which is undoubtedly important in a SN core where the ν_e Fermi sea is highly degenerate. In this case it is rather unclear what one is supposed to use for the damping parameter D. If ν_e and ν_μ scatter with equal amplitudes, there is no damping at all because the collisions do not distinguish between flavors, preserving the coherence between them. Does this remain true if ν_e collisions are Pauli blocked by their high Fermi sea while those of ν_μ are not? Such and other related questions can be answered if one abandons a single-particle approach to neutrino oscillations, i.e. if one moves to a field-theoretic framework which includes many-body effects from the start.

9.2.2 Matrix of Densities

Which quantity is supposed to replace the previous single-particle density matrix as a means to describe a possibly degenerate neutrino ensemble? For unmixed neutrinos the relevant observables are time-dependent occupation numbers $f_{\mathbf{p}}$ for a given mode $\mathbf{p}$ of the neutrino field. They are given as expectation values of number operators $n_{\mathbf{p}} = a_{\mathbf{p}}^\dagger a_{\mathbf{p}}$ where $a_{\mathbf{p}}$ is a destruction operator for a neutrino in mode $\mathbf{p}$ and $a_{\mathbf{p}}^\dagger$ the corresponding creation operator. The expectation value is with regard to the state $|\,\rangle$ of the entire ensemble. For several flavors it is natural to generalize the $f_{\mathbf{p}}$'s to matrices $\rho_{\mathbf{p}} = \rho(\mathbf{p})$ of the form[49]

---

[49]Strictly speaking $\rho(\mathbf{p})$ is defined by $\langle a_j^\dagger(\mathbf{p})a_i(\mathbf{p}')\rangle = (2\pi)^3\delta^{(3)}(\mathbf{p} - \mathbf{p}')\rho_{ij}(\mathbf{p})$ and similar for $\bar{\rho}(\mathbf{p})$. Therefore, the expectation values in Eq. (9.3) diverge because they involve an infinite factor $(2\pi)^3\delta^{(3)}(0)$ which is related to the infinite quantization volume necessary for continuous momentum variables. In practice, this factor always drops out of final results so that one may effectively set $(2\pi)^3\delta^{(3)}(0)$ equal to unity.

(Dolgov 1981)

$$\rho_{ij}(\mathbf{p}) = \left\langle a_j^\dagger(\mathbf{p})a_i(\mathbf{p})\right\rangle \quad \text{and} \quad \overline{\rho}_{ij}(\mathbf{p}) = \left\langle b_i^\dagger(\mathbf{p})b_j(\mathbf{p})\right\rangle, \qquad (9.3)$$

where $a_i(\mathbf{p})$ and $a_i^\dagger(\mathbf{p})$ are the destruction and creation operators for neutrinos of flavor i in mode $\mathbf{p}$ while b is for antineutrinos which otherwise are referred to by overbarred quantities. The reversed order of the flavor indices in the definition of $\overline{\rho}(\mathbf{p})$ guarantees that both matrices transform in the same way under a unitary transformation in flavor space. Also, for brevity $\langle\,|\ldots|\,\rangle$ is always written as $\langle\ldots\rangle$.

The diagonal elements of $\rho_\mathbf{p}$ and $\overline{\rho}_\mathbf{p}$ are the usual occupation numbers while the off-diagonal ones represent relative phase information. In the nondegenerate limit, up to a normalization $\rho_\mathbf{p}$ plays the role of the previously defined single-particle density matrix. Therefore, the $\rho_\mathbf{p}$'s and $\overline{\rho}_\mathbf{p}$'s are well suited to account simultaneously for oscillations and collisions. In fact, one can argue that a homogeneous neutrino ensemble is completely characterized by these "matrices of densities" (Sigl and Raffelt 1993). It remains to derive an equation of motion which in the appropriate limits should reduce to the previous precession equation, to a Boltzmann collision equation, and to Stodolsky's damping equation (9.1), respectively.

9.2.3 Free Evolution: Flavor Oscillations

The creation and annihilation operators which appear in the definition of $\rho_\mathbf{p}$ are the time-dependent coefficients of a spatial Fourier expansion of the neutrino field [notation $d\mathbf{p} \equiv d^3\mathbf{p}/(2\pi)^3$]

$$\Psi(t,\mathbf{x}) = \int d\mathbf{p}\left[a_\mathbf{p}(t)u_\mathbf{p} + b_{-\mathbf{p}}^\dagger(t)v_{-\mathbf{p}}\right]e^{i\mathbf{p}\cdot\mathbf{x}}. \qquad (9.4)$$

More precisely, $a_\mathbf{p}$ is an annihilation operator for negative-helicity neutrinos of momentum $\mathbf{p}$ while $b_\mathbf{p}^\dagger$ is a creation operator for positive-helicity antineutrinos. The Dirac spinors $u_\mathbf{p}$ and $v_\mathbf{p}$ refer to massless negative-helicity particles and positive-helicity antiparticles, respectively; the spinor normalization is taken to be unity. For n flavors, $a_\mathbf{p}$ and $b_\mathbf{p}^\dagger$ are column vectors of components $a_i(\mathbf{p})$ and $b_i^\dagger(\mathbf{p})$, respectively. They satisfy the anticommutation relations $\{a_i(\mathbf{p}),a_j^\dagger(\mathbf{p}')\} = \{b_i(\mathbf{p}),b_j^\dagger(\mathbf{p}')\} = \delta_{ij}(2\pi)^3\delta^{(3)}(\mathbf{p}-\mathbf{p}')$.

In the massless limit and when only left-handed (l.h.) interactions are present one may ignore the right-handed (r.h.) field entirely. However, in order to include flavor mixing one needs to introduce a $n\times n$

mass matrix M which is nondiagonal in the interaction basis. Even in this case the "wrong" helicity states will be ignored because in the ultrarelativistic regime spin-flip reactions are suppressed by an approximate factor $(m_\nu/2E_\nu)^2 \ll 1$. In this limit lepton number violating effects from possible Majorana masses are also ignored.

In the absence of interactions Ψ satisfies the free Dirac equation, implying $a_{\mathbf{p}}(t) = a_{\mathbf{p}}(0)\exp(-i\Omega^0_{\mathbf{p}}t)$ and $b_{\mathbf{p}}(t) = b_{\mathbf{p}}(0)\exp(-i\Omega^0_{\mathbf{p}}t)$, where

$$\Omega^0_{\mathbf{p}} \equiv \left(\mathbf{p}^2 + M^2\right)^{1/2} \tag{9.5}$$

is a $n \times n$ matrix of "vacuum oscillation frequencies." In the mass basis it has only diagonal elements which are the energies $E_i = (\mathbf{p}^2 + m_i^2)^{1/2}$. Therefore, one may use

$$H_0 = \int d\mathbf{p} \sum_{i,j=1}^{n} \left[a_i^\dagger(\mathbf{p})\Omega^0_{ij}(\mathbf{p})a_j(\mathbf{p}) + b_j^\dagger(\mathbf{p})\Omega^0_{ij}(\mathbf{p})b_i(\mathbf{p})\right] \tag{9.6}$$

as a free neutrino Hamiltonian.

In order to find the evolution of $\rho_{\mathbf{p}}$ and $\bar{\rho}_{\mathbf{p}}$ one needs to study the equations of motion of the $n \times n$ operator matrices

$$\hat{\rho}_{ij}(\mathbf{p}, t) = a_j^\dagger(\mathbf{p}, t)a_i(\mathbf{p}, t) \quad \text{and} \quad \hat{\bar{\rho}}_{ij}(\mathbf{p}, t) \equiv b_i^\dagger(\mathbf{p}, t)b_j(\mathbf{p}, t). \tag{9.7}$$

With a Hamiltonian H their evolution is given by Heisenberg's equation,

$$i\partial_t\,\hat{\rho} = [\hat{\rho}, H], \tag{9.8}$$

and similar for $\hat{\bar{\rho}}_{\mathbf{p}}$. With $H = H_0$ from Eq. (9.6) one finds

$$i\partial_t\,\hat{\rho}_{\mathbf{p}} = [\Omega^0_{\mathbf{p}}, \hat{\rho}_{\mathbf{p}}] \quad \text{and} \quad i\partial_t\,\hat{\bar{\rho}}_{\mathbf{p}} = -[\Omega^0_{\mathbf{p}}, \hat{\bar{\rho}}_{\mathbf{p}}]. \tag{9.9}$$

Taking expectation values on both sides yields equations of motion for $\rho_{\mathbf{p}}$ and $\bar{\rho}_{\mathbf{p}}$. For two flavors one may use the representation $\Omega^0_{\mathbf{p}} = \omega^0_{\mathbf{p}} + \frac{1}{2}\mathbf{V}_{\mathbf{p}} \cdot \boldsymbol{\sigma}$ and $\rho_{\mathbf{p}} = \frac{1}{2}f_{\mathbf{p}}(1 + \mathbf{P}_{\mathbf{p}} \cdot \boldsymbol{\sigma})$. Then Eq. (9.9) leads to the precession formulas $\dot{\mathbf{P}}_{\mathbf{p}} = \mathbf{V}_{\mathbf{p}} \times \mathbf{P}_{\mathbf{p}}$ and $\dot{\bar{\mathbf{P}}}_{\mathbf{p}} = -\mathbf{V}_{\mathbf{p}} \times \bar{\mathbf{P}}_{\mathbf{p}}$.

9.2.4 Interaction with a Background Medium

Interactions with a medium are introduced by virtue of a general interaction Hamiltonian $H_{\text{int}}(B, \Psi)$ which is a functional of the neutrino field Ψ and a set B of background fields; specific cases will be discussed in Sects. 9.3 and 9.4 below. The equation of motion for $\hat{\rho}_{\mathbf{p}}$ is found from

Heisenberg's equation with $H = H_0 + H_{\text{int}}$. Taking an expectation value with regard to the initial state yields

$$\dot\rho_{\mathbf{p}}(t) = -i\left[\Omega^0_{\mathbf{p}}, \rho_{\mathbf{p}}(t)\right] + i\left\langle\left[H_{\text{int}}(B(t), \Psi(t)), \hat\rho_{\mathbf{p}}(t)\right]\right\rangle, \qquad (9.10)$$

and an analogous equation for $\bar\rho_{\mathbf{p}}(t)$. These equations are exact, but they are not a closed set of differential equations for the $\rho_{\mathbf{p}}$ and $\bar\rho_{\mathbf{p}}$. To this end one needs to perform a perturbative expansion.

To first order one may set the interacting fields $B(t)$ and $\Psi(t)$ on the r.h.s. of Eq. (9.10) equal to the free fields[50] $B_0(t)$ and $\Psi_0(t)$. Under the assumption that the original state contained no correlations between the neutrinos and the background the expectation value factorizes into a medium part and a neutrino part. With Wick's theorem and ignoring fast-varying terms such as $b^\dagger b^\dagger$ it can be reduced to an expression which contains only $\rho_{\mathbf{p}}$'s and $\bar\rho_{\mathbf{p}}$'s. The result gives the forward-scattering or refractive effect of the interaction.

To include nonforward collisions one needs to go to second order in the perturbation expansion. At a given time t a general operator $\xi(t) = \xi(B(t), \Psi(t))$ which is a functional of B and Ψ is to first order

$$\xi(t) = \xi_0(t) + i\int_0^t dt'\left[H^0_{\text{int}}(t - t'), \xi_0(t)\right], \qquad (9.11)$$

where ξ_0 and H^0_{int} are functionals of the freely evolving fields $B_0(t)$ and $\Psi_0(t)$. Applying this general iteration formula to the operator $\xi = [H_{\text{int}}(B, \Psi), \hat\rho_{\mathbf{p}}]$ which appears on the r.h.s. of Eq. (9.10) one arrives at

$$\dot\rho_{\mathbf{p}}(t) = -i\left[\Omega^0_{\mathbf{p}}, \rho_{\mathbf{p}}(t)\right] + i\left\langle\left[H^0_{\text{int}}(t), \hat\rho^0_{\mathbf{p}}\right]\right\rangle$$
$$- \int_0^t dt'\left\langle\left[H^0_{\text{int}}(t - t'), \left[H^0_{\text{int}}(t), \hat\rho^0_{\mathbf{p}}\right]\right]\right\rangle, \qquad (9.12)$$

and similar for $\bar\rho_{\mathbf{p}}(t)$. The second term on the r.h.s. is the first-order refractive part associated with forward scattering. The second-order term contains both forward- as well as nonforward-scattering effects.

---

[50]These free operators are the solutions of the equations of motion in the absence of H_{int}. However, internal interactions of the medium such as nucleon-nucleon scattering are not excluded. Moreover, $\Psi(0) = \Psi_0(0)$ etc. are taken as initial conditions for the interacting fields. Also, the mass term is ignored in the definition of Ψ_0; its effect is included only in the first term on the r.h.s. of Eq. (9.10), the "vacuum oscillation term." Therefore, the free creation and annihilation operators vary as $a^0_j(\mathbf{p}, t) = a_j(\mathbf{p}, 0)e^{-ipt}$ etc. for all flavors with $p = |\mathbf{p}|$. This implies that the operators $\hat\rho^0(\mathbf{p})$ and $\hat{\bar\rho}^0(\mathbf{p})$, which are constructed from the free a's and b's, are time independent.

Because all operators on the r.h.s. of Eq. (9.12) are free the expectation values in the first- and second-order term factorize between the neutrinos and the medium. This leads one to equations for $\dot{\rho}_{\mathbf{p}}(t)$ and $\dot{\bar{\rho}}_{\mathbf{p}}(t)$ which on the r.h.s. involve only $\rho_{\mathbf{p}}(t)$ and $\bar{\rho}_{\mathbf{p}}(t)$ as well as $\langle \hat{\rho}_{\mathbf{p}}^0 \rangle$ and $\langle \hat{\bar{\rho}}_{\mathbf{p}}^0 \rangle$ besides expectation values of B operators.

The interactions described by H_{int} are taken as individual, isolated collisions where the neutrinos go from free states to free states as in ordinary scattering theory. The duration of one collision (the inverse of a typical energy transfer) is assumed to be small relative to the time scale over which the density matrices vary substantially, i.e. small relative to the oscillation time and the inverse collision frequency. Physically this amounts to the restriction that the neutrino collision rate is small enough that multiple-scattering effects can be ignored. Further, it is assumed that the medium is not changed much by the interactions with the neutrino ensemble, allowing one to neglect evolution equations for the medium variables which can thus be taken to be externally prescribed, usually by conditions of thermal equilibrium. If the medium is not stationary it is assumed that the time scale of variation is large compared to the duration of typical neutrino-medium collisions.

One may then choose the time step of iteration t in Eq. (9.12) both small relative to the evolution time scale and large relative to the duration of one collision. Under these circumstances the time integral can be extended to infinity while setting $\rho_{\mathbf{p}}(t)$ equal to $\rho_{\mathbf{p}}(0) = \langle \hat{\rho}_{\mathbf{p}}^0 \rangle$. This leads to

$$\dot{\rho}_{\mathbf{p}}(0) = -i\left[\Omega_{\mathbf{p}}^0, \rho_{\mathbf{p}}(0)\right] + i\left\langle\left[H_{\text{int}}^0(0), \hat{\rho}_{\mathbf{p}}^0\right]\right\rangle$$
$$-\frac{1}{2}\int_{-\infty}^{+\infty} dt\,\left\langle\left[H_{\text{int}}^0(t), \left[H_{\text{int}}^0(0), \hat{\rho}_{\mathbf{p}}^0\right]\right]\right\rangle, \quad (9.13)$$

and similar for $\bar{\rho}_{\mathbf{p}}$. Here, $\int_0^\infty dt\,\langle\ldots\rangle$ was replaced by $\frac{1}{2}\int_{-\infty}^{+\infty} dt\,\langle\ldots\rangle$. The difference between these expressions corresponds to a principle-part integral which leads to a second-order correction to the refractive term which is ignored. In the form of Eq. (9.13) the time integral leads to energy conservation in individual collisions.

An explicit evaluation of the r.h.s. of Eq. (9.13) for a given interaction model yields the desired set of differential equations for the $\rho_{\mathbf{p}}$'s and $\bar{\rho}_{\mathbf{p}}$'s at time $t = 0$. It will be valid at all times if the correlations built up by neutrino collisions are "forgotten" before the next collision occurs. This assumption corresponds to "molecular chaos" in the derivation of the usual Boltzmann equation.

9.3 Neutral-Current Interactions

9.3.1 Hamiltonian

In order to make Eq. (9.13) explicit one must use a specific model for
the interactions between neutrinos and the medium. To this end I begin
with fermions which interact by virtue of an effective neutral-current
(NC) Hamiltonian,

$$H_{\rm NC} = \frac{G_{\rm F}}{\sqrt{2}} \sum_a \int d^3{\bf x}\, B_a^\mu({\bf x})\, \overline{\Psi}({\bf x}) G_a \gamma_\mu (1 - \gamma_5)\Psi({\bf x})\,. \qquad (9.14)$$

Here, B_a^μ typically is also a bilinear of the form $\overline{\psi}_a\gamma^\mu\psi_a$ or $\overline{\psi}_a\gamma_5\gamma^\mu\psi_a$
with a Dirac field ψ_a which describes fermions of the medium. It is
assumed that all neutrino flavors scatter on a given species a in the
same way apart from overall factors which are given as a hermitian
$n \times n$ matrix G_a of dimensionless coupling constants. In the absence of
flavor-changing neutral currents it is diagonal in the weak interaction
basis.

As a concrete example consider the ν_e and ν_μ flavors in a medium
of ultrarelativistic electrons which may be classified into a l.h. and a
r.h. "species," $a = L$ or R, so that $B_{L,R}^\mu = \frac{1}{2}\overline{\psi}_e\gamma^\mu(1 \mp \gamma_5)\psi_e$. With the
standard-model couplings given in Sect. 6.7.1 one finds in the weak
interaction basis $G_L = 2\sin^2\theta_W + \sigma_3$ and $G_R = 2\sin^2\theta_W$.

For the calculations it is convenient to write Eq. (9.14) in momentum space,

$$H_{\rm NC} = \frac{G_{\rm F}}{\sqrt{2}} \sum_a \int d{\bf p}\, d{\bf p}'\, B_a^\mu({\bf p} - {\bf p}')\, \overline{\Psi}_{\bf p} G_a \gamma_\mu (1 - \gamma_5)\Psi_{\bf p'}\,, \qquad (9.15)$$

where $B_a^\mu({\bf \Delta}) = \int d^3{\bf x}\, B_a^\mu({\bf x})e^{-i{\bf \Delta}\cdot{\bf x}}$ is the Fourier transform of $B_a^\mu({\bf x})$
and $\Psi_{\bf p} = a_{\bf p}u_{\bf p} + b_{-\bf p}^\dagger v_{-\bf p}$ in terms of the annihilation and creation
operators of Eq. (9.4).

A special case of NC interactions are those among the neutrinos
themselves with a Hamiltonian that is quartic in Ψ. In momentum
space these "self-interactions" are given by

$$H_{\rm S} = \frac{G_{\rm F}}{\sqrt{2}} \int d{\bf p}\, d{\bf p}'\, d{\bf q}\, d{\bf q}'\, (2\pi)^3\delta^{(3)}({\bf p} + {\bf q} - {\bf p}' - {\bf q}')$$
$$\times\, \overline{\Psi}_{\bf q} G_{\rm S}\gamma^\mu(1 - \gamma_5)\Psi_{\bf q'}\, \overline{\Psi}_{\bf p} G_{\rm S}\gamma_\mu(1 - \gamma_5)\Psi_{\bf p'}\,. \qquad (9.16)$$

In the standard model with three sequential neutrino families $G_{\rm S}$ is the
3×3 unit matrix. For the evolution of a normal and a hypothetical
sterile flavor one would have $G_{\rm S} = {\rm diag}(1, 0)$.

9.3.2 Neutrino Refraction

As a first simple application one may recover neutrino refraction by a medium that was previously discussed in Sect. 6.7.1. An explicit evaluation of the second term of Eq. (9.13) with $H_{int} = H_{NC}$ yields

$$i\dot{\rho} = [\Omega_{\mathbf{p}}^0, \rho_{\mathbf{p}}] + \sum_a n_a [G_a, \rho_{\mathbf{p}}], \qquad (9.17)$$

where $n_a \equiv \langle B_a^\mu \rangle P_\mu / P_0$ where P is the neutrino four momentum and thus P/P_0 their four velocity. In an isotropic medium the spatial parts of $\langle B_a^\mu \rangle$ vanish so that n_a is the number density of fermions a. If the medium is unpolarized, axial currents do not contribute.

In the standard model with the coupling constants of Appendix B one finds for an isotropic, unpolarized medium of protons, neutrons, and electrons,

$$i\dot{\rho}_{\mathbf{p}} = \left[(\Omega_{\mathbf{p}}^0 + \sqrt{2}\, G_F N_\ell), \rho_{\mathbf{p}}\right],$$
$$-i\dot{\bar{\rho}}_{\mathbf{p}} = \left[(\Omega_{\mathbf{p}}^0 - \sqrt{2}\, G_F N_\ell), \bar{\rho}_{\mathbf{p}}\right], \qquad (9.18)$$

where $N_\ell = \text{diag}(n_e, 0, 0)$ in the flavor basis (electron density n_e). For two flavors this is equivalent to the previous precession formula. Notably, the ν and $\bar{\nu}$ oscillation frequencies are shifted in opposite directions relative to the vacuum energies.

Neutrino-neutrino interactions make an additional contribution to the refractive energy shifts, i.e. to the first-order term in Eq. (9.13). After the relevant contractions one finds[51] (Sigl and Raffelt 1993)

$$\Omega_{\mathbf{p}}^S = \sqrt{2}\, G_F \int d\mathbf{q} \left\{ G_S(\rho_{\mathbf{q}} - \bar{\rho}_{\mathbf{q}}) G_S + G_S \text{Tr}\left[(\rho_{\mathbf{q}} - \bar{\rho}_{\mathbf{q}}) G_S\right] \right\}.$$
$$(9.19)$$

$\bar{\Omega}_{\mathbf{p}}^S$ is given by the same formula with $\rho_{\mathbf{q}}$ and $\bar{\rho}_{\mathbf{q}}$ interchanged. The trace expression implies the well-known result that neutrinos in a bath of their own flavor experience twice the energy shift relative to a bath of another flavor.

The early universe is essentially matter-antimatter symmetric so that higher-order terms to the refractive index must be included as discussed in Sect. 6.7.2; see also Sigl and Raffelt (1993). In stars,

---

[51]If the neutrino ensemble is not isotropic one has to include a factor $(1 - \cos\theta_{\mathbf{pq}})$ under the integral, where $\theta_{\mathbf{pq}}$ is the angle between $\mathbf{p}$ and $\mathbf{q}$.

neutrinos are important only in young SN cores where one may ignore antineutrinos. With the total neutrino matrix of densities

$$\rho \equiv \int d\mathbf{p} \, \rho_{\mathbf{p}}, \qquad (9.20)$$

and with $G_S = 1$ in the standard model, the neutrino contribution to the refractive energy shift is $\Omega_{\mathbf{p}}^S = \sqrt{2} \, G_F \, \rho$. The trace term was dropped because it does not contribute to the commutator in the equation of motion. The diagonal entries of ρ are the neutrino densities. However, in the presence of mixing and oscillations ρ also has off-diagonal elements, i.e. there are "off-diagonal refractive indices" as first realized by Pantaleone (1992b).

In a SN core the complete first-order equation of motion for $\rho_{\mathbf{p}}$ is then

$$i\dot{\rho}_{\mathbf{p}} = \left[\Omega_{\mathbf{p}}^0, \rho_{\mathbf{p}} \right] + \sqrt{2} \, G_F \left[(N_\ell + \rho), \rho_{\mathbf{p}} \right], \qquad (9.21)$$

which is intrinsically nonlinear. Interestingly, if one integrates both sides over $d\mathbf{p}$ one obtains an equation for ρ which is linear as the neutrino term drops out from the commutator. Therefore, even though individual modes of the neutrino field oscillate differently in the presence of other neutrinos, the instantaneous rate of change of the overall flavor polarization is as if they were absent.

In a SN core the refractive effects are dominated by nonneutrino particles, notably by electrons. However, above the neutrino sphere the flow of neutrinos itself represents a particle density exceeding that of the background medium (Sect. 11.4). Also, the medium of the early universe is dominated by neutrinos so that self-interactions and the corresponding nonlinearities of the neutrino flavor oscillations must be carefully included. For recent studies of primordial neutrino oscillations see Samuel (1993), Kostolecký, Pantaleone, and Samuel (1993), and references to the earlier literature given there.

9.3.3 Kinetic Terms

The refractive term (first order) of Eq. (9.13) is just a sum over different medium components, whereas the collision term (second order) in general contains interference terms between different target species. However, if they are uncorrelated, corresponding to $\langle B_a^\mu B_b^\nu \rangle = \langle B_a^\mu \rangle \langle B_b^\nu \rangle$ for $a \neq b$, these interference terms only contribute to second-order forward-scattering effects which are neglected. The collision term is then an incoherent sum over all target species so that in the following one may suppress the subscript a for simplicity.

After a lengthy but straightforward calculation one arrives at the NC collision term (Sigl and Raffelt 1993)

$$\dot{\rho}_{\mathbf{p},\text{coll}} = \tfrac{1}{2} \int d\mathbf{p}' \Big[W_{P',P} G \rho_{\mathbf{p}'} G (1 - \rho_{\mathbf{p}}) - W_{P,P'} \rho_{\mathbf{p}} G (1 - \rho_{\mathbf{p}'}) G$$
$$+ W_{-P',P} (1 - \rho_{\mathbf{p}}) G (1 - \bar{\rho}_{\mathbf{p}'}) G - W_{P,-P'} \rho_{\mathbf{p}} G \bar{\rho}_{\mathbf{p}'} G + \text{h.c.} \Big],$$

$$(9.22)$$

where P and P' are neutrino four momenta with physical (positive) energies $P_0 = |\mathbf{p}|$ and $P_0' = |\mathbf{p}'|$. The nonnegative transition probabilities $W_{K',K} = W(K', K)$ are Wick contractions of medium operators of the form

$$W(K', K) = \tfrac{1}{8} G_{\text{F}}^2 \, S^{\mu\nu}(K' - K) \, N_{\mu\nu}(K', K),$$

$$(9.23)$$

where K and K' correspond to neutrino four-momenta with K_0 and K_0' positive or negative. The "medium structure function" is

$$S^{\mu\nu}(\Delta) \equiv \int_{-\infty}^{+\infty} dt \, e^{i\Delta_0 t} \, \langle B^{\mu}(t, \Delta) B^{\nu}(0, -\Delta) \rangle,$$

$$(9.24)$$

where the energy transfer Δ_0 can be both positive and negative. In the ultrarelativistic limit the neutrino tensor can be written as

$$N^{\mu\nu} = \tfrac{1}{2} (U^{\mu} U'^{\nu} + U'^{\mu} U^{\nu} - U \cdot U' g^{\mu\nu} - i\epsilon^{\mu\nu\alpha\beta} U_{\alpha} U'_{\beta}),$$

$$(9.25)$$

where $U \equiv K/K^0$ and $U' \equiv K'/K^0$ are the neutrino four velocities. Therefore, $N^{\mu\nu}$ is an even function of K and K'. Note that the definition Eq. (9.25) differs slightly from the corresponding Eq. (4.17).

The first two terms of the collision integral Eq. (9.22) are due to neutrino scattering off the medium. The positive term represents gains from scatterings $\nu_{\mathbf{p}'} \to \nu_{\mathbf{p}}$ while the negative one is from losses by the inverse reaction. The third and fourth expressions account for pair processes, i.e. the creation or absorption of $\nu_{\mathbf{p}} \bar{\nu}_{\mathbf{p}'}$ by the medium. The pair terms are found by direct calculation or from the scattering ones by "crossing,"

$$P \to -P \quad \text{and} \quad \rho_{\mathbf{p}} \to (1 - \bar{\rho}_{\mathbf{p}}).$$

$$(9.26)$$

For example, the reaction $\nu_{\mathbf{p}} X \to X' \nu_{\mathbf{p}'}$ transforms to $X \to X' \bar{\nu}_{\mathbf{p}} \nu_{\mathbf{p}'}$ under this operation where X and X' represent medium configurations.

The collision integral for $\bar{\rho}_{\mathbf{p}}$ is found by direct calculation or by applying the crossing operation Eq. (9.26) to all neutrinos and antineutrinos appearing in Eq. (9.22). The neutrino gain terms then transform to the antineutrino loss terms and vice versa.

Equation (9.22) and the corresponding result for $\bar{\rho}_{\mathbf{p}}$ were derived by Sigl and Raffelt (1993) whose exposition I have closely followed. In the nondegenerate limit where $(1 - \rho_{\mathbf{p}}) \to 1$ it agrees with a kinetic equation of Dolgov (1981) and Barbieri and Dolgov (1991). Moreover, a similar equation was derived by Rudzsky (1990) which can be shown to be equivalent to Eq. (9.22) in the appropriate limits.

The relatively complicated collision term that follows from the neutrino-neutrino Hamiltonian Eq. (9.16) has been worked out by Sigl and Raffelt (1993). However, in a SN core the collisions of neutrinos with each other are negligible relative to interactions with nucleons and electrons.

In the limit of a single neutrino flavor, or several unmixed flavors, the role of $\rho_{\mathbf{p}}$ is played by the usual occupation numbers $f_{\mathbf{p}}$ while the matrix G is unity, or the unit matrix. Then Eq. (9.22) is

$$\dot{f}_{\mathbf{p},\text{coll}} = \int d\mathbf{p}' \left[W_{P',P} \, f_{\mathbf{p}'}(1 - f_{\mathbf{p}}) - W_{P,P'} \, f_{\mathbf{p}}(1 - f_{\mathbf{p}'}) \right.$$
$$\left. + W_{-P',P}(1 - f_{\mathbf{p}})(1 - \bar{f}_{\mathbf{p}'}) - W_{P,-P'} \, f_{\mathbf{p}}\bar{f}_{\mathbf{p}'} \right] \quad (9.27)$$

which is the usual Boltzmann collision integral. The main difference to Eq. (9.22) is the appearance there of "nonabelian Pauli blocking factors" which involve noncommuting matrices of neutrino occupation numbers and coupling constants.

9.3.4 Recovering Stodolsky's Formula

The damping of neutrino oscillations becomes particularly obvious in the limit where a typical energy transfer Δ_0 in a neutrino-medium interaction is small relative to the neutrino energies themselves. This would be the case for "heavy" and thus nonrelativistic background fermions. Then pair processes may be ignored and neutrinos change their direction of motion in a collision, but not the magnitude of their momentum, which also implies $W(P, P') \approx W(P', P)$. If the neutrino ensemble is isotropic one then has $\rho_{\mathbf{p}} = \rho_{\mathbf{p}'}$ under the integral in Eq. (9.22). For the matrix structure of the collision term this leaves $2G\rho_{\mathbf{p}}G - GG\rho_{\mathbf{p}} - \rho_{\mathbf{p}}GG = -[G, [G, \rho_{\mathbf{p}}]]$ which puts the nature of the collision term as a double commutator in evidence—see also Eq. (9.13). One may define a total scattering rate for nondegenerate neutrinos of momentum $\mathbf{p}$ by virtue of

$$\Gamma_{\mathbf{p}} = \int d\mathbf{p}' \, W(P', P) \,. \quad (9.28)$$

In the present limit this yields a collision "integral"

$$\dot{\rho}_{\mathbf{p},\text{coll}} = -\tfrac{1}{2}\Gamma_{\mathbf{p}}\left[G,[G,\rho_{\mathbf{p}}]\right],\tag{9.29}$$

where terms nonlinear in $\rho_{\mathbf{p}}$ have disappeared even though the neutrinos may still be degenerate.

Eq. (9.29) is more transparent in the case of two-flavor mixing where one may write $\rho_{\mathbf{p}} = \tfrac{1}{2}f_{\mathbf{p}}(1 + \mathbf{P}_{\mathbf{p}} \cdot \boldsymbol{\sigma})$ and $G = \tfrac{1}{2}(g_0 + \mathbf{G} \cdot \boldsymbol{\sigma})$. The total occupation number $f_{\mathbf{p}}$ is conserved while the polarization vector is damped according to

$$\dot{\mathbf{P}}_{\mathbf{p},\text{coll}} = -\tfrac{1}{2}\Gamma_{\mathbf{p}}\,\mathbf{G} \times (\mathbf{G} \times \mathbf{P}_{\mathbf{p}}).\tag{9.30}$$

The r.h.s. is a vector transverse to $\mathbf{G}$, allowing one to write

$$\dot{\mathbf{P}}_{\mathbf{p},\text{coll}} = -\tfrac{1}{2}\Gamma_{\mathbf{p}}\,|\mathbf{G}|^2\,\mathbf{P}_{\mathbf{p},\text{T}}.\tag{9.31}$$

Thus one naturally recovers Stodolsky's damping term Eq. (9.1) with $D = \tfrac{1}{2}\Gamma_{\mathbf{p}}|\mathbf{G}|^2$. For ν_e and ν_μ and if one writes $G = \text{diag}(g_{\nu_e}, g_{\nu_\mu})$ in the weak interaction basis $D = \tfrac{1}{2}\Gamma_{\mathbf{p}}(g_{\nu_e} - g_{\nu_\mu})^2$. This representation reflects that the damping of neutrino oscillations depends on the difference of the scattering *amplitudes:* D is the square of the amplitude difference, not the difference of the squares. If one flavor does not scatter at all, D is half the scattering rate of the active flavor.

Collisions thus lead to chemical equilibrium as discussed in the introduction to this chapter and as shown in Fig. 9.1. However, the simple exponential damping represented by Stodolsky's formula can be reproduced only in the limit of vanishing energy transfers in collisions, an assumption which amounts to separating the neutrino momentum degrees of freedom from the flavor ones. In a more general case the evolution is more complicated. In particular, collisions usually lead to a transient flavor polarization in an originally unpolarized ensemble if the momentum degrees of freedom were out of equilibrium. Still, the neutrinos always move toward kinetic and chemical equilibrium under the action of the collision integral Eq. (9.22) in the sense that the properly defined free energy never increases (Sigl and Raffelt 1993).

9.3.5 Weak-Damping Limit

Even for two-flavor mixing the general form of the collision integral Eq. (9.22) remains rather complicated. However, for the conditions of a SN core one may apply two approximations which significantly simplify the problem. First, one is mostly concerned with the evolution of ν_e's because initially they have a large chemical potential and thus are far away from chemical equilibrium with the other flavors. Their high degree of degeneracy implies that $\bar\nu_e$'s may be ignored and with them all pair processes. Therefore, the evolution of ν_e's mixed with one other flavor (standard or sterile) is given by

$$\dot\rho_{\mathbf p} = i[\rho_{\mathbf p},\Omega_{\mathbf p}] + \tfrac12 \int d\mathbf p' \left[W_{P'P}\, G\rho_{\mathbf p'}G(1-\rho_{\mathbf p}) \right.$$
$$\left. -W_{PP'}\,\rho_{\mathbf p}G(1-\rho_{\mathbf p'})G + \text{h.c.}\right]. \qquad (9.32)$$

The matrix of oscillation frequencies includes vacuum and first-order medium contributions. With the momentum-dependent oscillation period $t_{\rm osc}$ it is

$$\Omega_{\mathbf p} = (2\pi/t_{\rm osc})\,\tfrac12\,\mathbf v_{\mathbf p}\cdot\boldsymbol\sigma \quad \text{with} \quad \mathbf v_{\mathbf p} \equiv (s_{\mathbf p},0,c_{\mathbf p}). \qquad (9.33)$$

Here,

$$s_{\mathbf p} \equiv \sin 2\theta_{\mathbf p} \quad \text{and} \quad c_{\mathbf p} \equiv \cos 2\theta_{\mathbf p} \qquad (9.34)$$

with the momentum-dependent mixing angle in the medium $\theta_{\mathbf p}$.

The second approximation for the conditions of a SN core is the weak-damping limit or limit of fast oscillations. It is easy to show that for the relevant physical conditions $2\pi/t_{\rm osc}$ is typically much faster than the scattering rate. Therefore, it is justified to consider density matrices $\tilde\rho_{\mathbf p}$ averaged over a period of oscillation. While the $\rho_{\mathbf p}$'s are given by four real parameters which are functions of time, the $\tilde\rho_{\mathbf p}$'s require only two, for example the occupation numbers of the two mixed flavors. It is straightforward to show that in the weak interaction basis

$$\tilde\rho_{\mathbf p} = \begin{pmatrix} f_{\mathbf p}^e & 0 \\ 0 & f_{\mathbf p}^x \end{pmatrix} + \tfrac12\, t_{\mathbf p}\,(f_{\mathbf p}^e - f_{\mathbf p}^x)\begin{pmatrix} 0 & 1 \\ 1 & 0 \end{pmatrix}, \qquad (9.35)$$

where $t_{\mathbf p} \equiv \tan 2\theta_{\mathbf p} = s_{\mathbf p}/c_{\mathbf p}$, $f_{\mathbf p}^e$ is the occupation number of ν_e, not of electrons, and $f_{\mathbf p}^x$ refers to a standard or sterile flavor ν_x.

One way of looking at the weak-damping limit is that between collisions neutrinos are best described by "propagation eigenstates," i.e. in a basis where the $\tilde\rho_{\mathbf p}$ are diagonal. Then the matrix of coupling constants G is no longer diagonal and so flavor conversion is understood as the result of "flavor-changing neutral currents" where "flavor" refers to the propagation eigenstates. However, because in general the effective mixing angle is a function of the neutrino momentum one would have to use a different basis for each momentum, an approach that complicates rather than simplifies the equations. Therefore, it is easiest and physically most transparent to work always in the weak interaction basis.

In order to derive an equation of motion for $\tilde{\rho}_{\mathbf{p}}$ one evaluates the collision term in Eq. (9.32) by inserting $\tilde{\rho}_{\mathbf{p}}$'s under the integral. Expanding the result in Pauli matrices leads to an expression of the form $\frac{1}{2}(a_{\mathbf{p}} + \mathbf{A}_{\mathbf{p}} \cdot \boldsymbol{\sigma})$. In general, the polarization vector $\mathbf{A}_{\mathbf{p}}$ produced by the collision term is not parallel to $\mathbf{v}_{\mathbf{p}} = (s_{\mathbf{p}}, 0, c_{\mathbf{p}})$ because the collision term couples modes with different mixing angles. However, the assumed fast oscillations average to zero the $\mathbf{A}_{\mathbf{p}}$ component perpendicular to $\mathbf{v}_{\mathbf{p}}$. Therefore, the r.h.s. of Eq. (9.32) is of the form $\frac{1}{2}[a_{\mathbf{p}} + (\mathbf{v}_{\mathbf{p}} \cdot \mathbf{A}_{\mathbf{p}})(\mathbf{v}_{\mathbf{p}} \cdot \boldsymbol{\sigma})]$. With a matrix of coupling constants in the weak basis of

$$G = \begin{pmatrix} g_e & 0 \\ 0 & g_x \end{pmatrix} \tag{9.36}$$

the collision integral Eq. (9.32) becomes explicitly

$$\dot{f}_{\mathbf{p}}^x = \frac{1}{4} \int d\mathbf{p}' \left\{ w(\nu_{\mathbf{p}}^x \to \nu_{\mathbf{p}'}^x) \left[(4 - s_{\mathbf{p}}^2) g_x^2 + (2 - s_{\mathbf{p}}^2) t_{\mathbf{p}} t_{\mathbf{p}'} g_e g_x \right] \right.$$
$$+ w(\nu_{\mathbf{p}}^e \to \nu_{\mathbf{p}'}^x) \left[-s_{\mathbf{p}}^2 g_x^2 - s_{\mathbf{p}}^2 t_{\mathbf{p}} t_{\mathbf{p}'} g_e g_x \right]$$
$$+ w(\nu_{\mathbf{p}}^x \to \nu_{\mathbf{p}'}^e) \left[s_{\mathbf{p}}^2 g_e^2 - (2 - s_{\mathbf{p}}^2) t_{\mathbf{p}} t_{\mathbf{p}'} g_e g_x \right]$$
$$\left. + w(\nu_{\mathbf{p}}^e \to \nu_{\mathbf{p}'}^e) \left[s_{\mathbf{p}}^2 g_e^2 + s_{\mathbf{p}}^2 t_{\mathbf{p}} t_{\mathbf{p}'} g_e g_x \right] \right\}, \tag{9.37}$$

where

$$w(\nu_{\mathbf{p}}^a \to \nu_{\mathbf{p}'}^b) \equiv W_{P'P} f_{\mathbf{p}'}^b (1 - f_{\mathbf{p}}^a) - W_{PP'} f_{\mathbf{p}}^a (1 - f_{\mathbf{p}'}^b) \tag{9.38}$$

(Raffelt and Sigl 1993). The equation for $f_{\mathbf{p}}^e$ is the same if one exchanges $e \leftrightarrow x$ everywhere. In the absence of mixing $s_{\mathbf{p}} = t_{\mathbf{p}} = 0$, leading to the usual collision integral for each species separately.

If ν_x is neither ν_μ nor ν_τ but rather some hypothetical sterile species its coupling constant is $g_x = 0$ by definition. In this case the collision integral simplifies to

$$\dot{f}_{\mathbf{p}}^x = \frac{1}{4} s_{\mathbf{p}}^2 g_e^2 \int d\mathbf{p}' \left[W_{P'P} f_{\mathbf{p}'}^e (2 - f_{\mathbf{p}}^e - f_{\mathbf{p}}^x) \right.$$
$$\left. - W_{PP'} (f_{\mathbf{p}}^e + f_{\mathbf{p}}^x)(1 - f_{\mathbf{p}'}^e) \right]. \tag{9.39}$$

If the ν_e stay approximately in thermal equilibrium, detailed balance yields

$$\dot{f}_{\mathbf{p}}^x = \frac{1}{4} s_{\mathbf{p}}^2 g_e^2 \int d\mathbf{p}' \left[W_{P'P} f_{\mathbf{p}'}^e (1 - f_{\mathbf{p}}^x) - W_{PP'} f_{\mathbf{p}}^x (1 - f_{\mathbf{p}'}^e) \right]. \tag{9.40}$$

If in addition the mixing angle is so small that the ν_x freely escape one may set $f_{\mathbf{p}}^x = 0$ on the r.h.s. so that the integral expression becomes $\int d\mathbf{p}' \, W_{P'P} f_{\mathbf{p}'}^e$.

9.3.6   Small Mixing Angle

In practice the mixing angle is usually small, allowing for substantial further simplifications. In this limit the approach to flavor equilibrium is much slower than that to kinetic equilibrium for each flavor separately (ν_x is taken to be one of the active flavors ν_μ or ν_τ). Therefore, each flavor is characterized by a Fermi-Dirac distribution so that it is enough to specify the total number density n_{ν_x} rather than the occupation numbers of individual modes. Integrating Eq. (9.37) over all modes, using detailed balance to lowest order in $s_\mathbf{p}^2$, and with $t_\mathbf{p} = s_\mathbf{p}$ one finds for the evolution of the ν_x number density

$$\dot{n}_{\nu_x} = \tfrac{1}{4} \int d\mathbf{p}\, d\mathbf{p}'\, W_{PP'} \Big[(g_x s_\mathbf{p} - g_e s_{\mathbf{p}'})^2 f_\mathbf{p}^e (1 - f_{\mathbf{p}'}^x)$$
$$- (g_e s_\mathbf{p} - g_x s_{\mathbf{p}'})^2 f_\mathbf{p}^x (1 - f_{\mathbf{p}'}^e) \Big], \quad (9.41)$$

and a similar equation for $\dot{n}_{\nu_e}$. Together with the condition of β equilibrium, $\mu_n - \mu_p = \mu_e - \mu_{\nu_e}$, that of charge neutrality, $n_p = n_e$, and the conservation of the trapped lepton number, $d(n_e + n_{\nu_e} + n_{\nu_x})/dt = 0$, these equations represent differential equations for the chemical potentials $\mu_{\nu_x}(t)$ and $\mu_{\nu_e}(t)$ if the temperature is fixed.

9.3.7   Flavor Conversion by Neutral Currents?

Next I turn to the conceptually interesting question whether flavor conversion (or the damping of neutrino oscillations) is possible by NC collisions alone. Considering only standard flavors the matrix of coupling constants G is then proportional to the unit matrix. In this case Stodolsky's damping formula in the form Eq. (9.29) gives $\dot{\rho}_{\mathbf{p},\text{coll}} = 0$.

This formula applies in the limit when the neutrino energies do not change in collisions (a medium of "heavy" fermions). If one lifts this restriction the situation is more complicated, but it simplifies again for weak damping and a small mixing angle. Then one may apply Eq. (9.41) with $g_e = g_x = 1$,

$$\dot{n}_{\nu_x} = \tfrac{1}{4} \int d\mathbf{p}\, d\mathbf{p}'\, W_{PP'} (s_\mathbf{p} - s_{\mathbf{p}'})^2 \Big[f_\mathbf{p}^e (1 - f_{\mathbf{p}'}^x) - f_\mathbf{p}^x (1 - f_{\mathbf{p}'}^e) \Big]. (9.42)$$

If in a collision $|\mathbf{p}| = |\mathbf{p}'|$ and thus $s_\mathbf{p} = s_{\mathbf{p}'}$ one recovers the previous result $\dot{n}_{\nu_x} = 0$. However, if the mixing angle is a function of the neutrino momentum, NC collisions do lead to flavor conversion and thus to the damping of oscillations.

Of course, if only true NC interactions existed, the mixing angle in the medium would be fixed at its vacuum value and so no flavor conversion could occur. The deviation of θ from its vacuum value in a medium is entirely from *charged-current* interactions with electrons, even though they may be written in an effective NC form. The coherent neutrino energy shifts by an electron background are enough to allow true NC collisions with, say, neutrons to achieve flavor equilibrium!

9.4 Charged-Current Interactions

9.4.1 Hamiltonian

Besides neutrino scattering or pair processes one must also include charged-current (CC) reactions where neutrinos are absorbed or produced by the medium (converted into or from charged leptons) such that the total lepton number of the neutrino ensemble changes by one unit. The corresponding interaction Hamiltonian can be written in the form

$$H_{CC} = \frac{G_F}{\sqrt{2}} \int d^3\mathbf{x}\, \overline{\Upsilon}(\mathbf{x})\Psi(\mathbf{x}) + \text{h.c.}, \qquad (9.43)$$

where the neutrino field Ψ is, again, a column vector in flavor space with the entries Ψ_ℓ, $\ell = e, \mu, \tau$ in the standard model. Further, $\overline{\Upsilon}$ is a row of Dirac operators representing the medium. In the interaction basis Υ_ℓ carries the lepton number corresponding to the flavor ℓ. For example, in a medium of nucleons and electrons the field Υ_e corresponding to the electron lepton number can be written for standard-model couplings as

$$\Upsilon_e = \gamma^\mu(1 - \gamma_5)\psi_e \overline{\psi}_n \gamma_\mu (C_V - C_A\gamma_5)\psi_p, \qquad (9.44)$$

where ψ_p, ψ_n, and ψ_e are the proton, neutron, and electron Dirac fields, respectively, while $C_V = 1$ and $C_A = 1.26$ are the dimensionless CC vector and axial-vector nucleon coupling constants.

9.4.2 Kinetic Terms

One may now insert H_{CC} into Eq. (9.13) in order to derive the explicit CC collision integral for the evolution of $\rho_\mathbf{p}$ and $\overline{\rho}_\mathbf{p}$. The operators Υ_ℓ violate the lepton number L_ℓ corresponding to flavor ℓ. Therefore, $\langle \Upsilon_\ell \rangle = 0$ at all times if the medium is in an eigenstate of L_ℓ ($\ell = e, \mu, \tau$, or additional exotic flavors). This assumption implies that the CC interaction Eq. (9.43) does not contribute to refractive effects given by the first-order term in Eq. (9.13).

In the second-order term H_{int} appears quadratic so that one obtains expressions like $\langle \overline{\Upsilon}_\ell \Upsilon_k \rangle$. However, because the medium is assumed to be in an eigenstate of L_ℓ they do not contribute for $\ell \neq k$. Thus, in the final result the contributions of different flavors can be added incoherently.

The rates of production $\mathcal{P}^\ell_\Delta$ and absorption $\mathcal{A}^\ell_\Delta$ of a ν_ℓ are functions of the energy-momentum transfer Δ to the medium,

$$\mathcal{P}^\ell_\Delta = \tfrac{1}{2} G_{\text{F}}^2 \int_{-\infty}^{+\infty} dt\, e^{-i\Delta_0 t} \langle \overline{\Upsilon}_\ell(\boldsymbol{\Delta}, t)\gamma_\mu \Delta^\mu \Upsilon_\ell(\boldsymbol{\Delta}, 0)\rangle,$$

$$\mathcal{A}^\ell_\Delta = \tfrac{1}{2} G_{\text{F}}^2 \int_{-\infty}^{+\infty} dt\, e^{-i\Delta_0 t} \langle \text{Tr}\big[\gamma_\mu \Delta^\mu \Upsilon_\ell(\boldsymbol{\Delta}, 0)\overline{\Upsilon}_\ell(\boldsymbol{\Delta}, t)\big]\rangle. \qquad (9.45)$$

These expressions are defined for both positive and negative energy transfer Δ_0 because $\mathcal{P}^\ell_{-P}$ plays the role of an absorption rate for antineutrinos with physical ($P_0 > 0$) four momentum while $\mathcal{A}^\ell_{-P}$ plays that of a production rate. Put another way, $\mathcal{A}^\ell_\Delta$ and $\mathcal{P}^\ell_\Delta$ represent the rate of absorption or production of lepton number of type ℓ, independently of the sign of Δ_0.

It is useful to define a flavor matrix of production rates which in the weak basis has the form

$$\mathcal{P}_\Delta \equiv \frac{1}{2} \begin{pmatrix} \mathcal{P}^e_\Delta & 0 & 0 \\ 0 & \mathcal{P}^\mu_\Delta & 0 \\ 0 & 0 & \mathcal{P}^\tau_\Delta \end{pmatrix}. \qquad (9.46)$$

An analogous definition pertains to $\mathcal{A}_\Delta$. Then one finds for the CC collision integrals (Sigl and Raffelt 1993)

$$\dot{\rho}_{\mathbf{p},\text{CC}} = \{\mathcal{P}_P, (1 - \rho_{\mathbf{p}})\} - \{\mathcal{A}_P, \rho_{\mathbf{p}}\},$$

$$\dot{\overline{\rho}}_{\mathbf{p},\text{CC}} = \{\mathcal{A}_{-P}, (1 - \overline{\rho}_{\mathbf{p}})\} - \{\mathcal{P}_{-P}, \overline{\rho}_{\mathbf{p}}\}, \qquad (9.47)$$

where $\{\cdot, \cdot\}$ is an anticommutator. The kinetic term for $\overline{\rho}_{\mathbf{p}}$ is related to that for $\rho_{\mathbf{p}}$ by the crossing relation Eq. (9.26).

The r.h.s. of Eq. (9.47) for $\rho_{\mathbf{p}}$ is the difference between a gain and a loss term corresponding to the production or absorption of a $\nu_{\mathbf{p}}$. For a single flavor they take on the familiar form $\mathcal{P}_P(1 - f_{\mathbf{p}})$ and $\mathcal{A}_P f_{\mathbf{p}}$ where $(1 - f_{\mathbf{p}})$ is the usual Pauli blocking factor.

9.4.3   Weak-Damping Limit

The meaning of Eq. (9.47) becomes more transparent if one makes various approximations which are justified for the conditions of a SN core. As discussed in Sect. 9.3.5 one may ignore the antineutrino degrees of freedom, and one may use the weak-damping limit where neutrino oscillations are much faster than their rates of collision or absorption. Then one finds for two flavors (Raffelt and Sigl 1993)

$$
\begin{aligned}
\dot{f}_{\mathbf{p}}^{x} &= (1 - f_{\mathbf{p}}^{x})\mathcal{P}_{P}^{x} - f_{\mathbf{p}}^{x}\mathcal{A}_{P}^{x} \\
&+ \tfrac{1}{4}s_{\mathbf{p}}^{2}\big[(2 - f_{\mathbf{p}}^{x} - f_{\mathbf{p}}^{e})(\mathcal{P}_{P}^{e} - \mathcal{P}_{P}^{x}) - (f_{\mathbf{p}}^{x} + f_{\mathbf{p}}^{e})(\mathcal{A}_{P}^{e} - \mathcal{A}_{P}^{x})\big],
\end{aligned}
\tag{9.48}
$$

where $f_{\mathbf{p}}^{e}$ and $f_{\mathbf{p}}^{x}$ are the occupation numbers for ν_{e} and ν_{x} as in Sect. 9.3.5. The corresponding equation for ν_{e} is found by exchanging $e \leftrightarrow x$ everywhere.

For $\nu_{x} = \nu_{\mu}$ flavor conversion can build up a nonvanishing muon density in a SN core because they are light enough to be produced initially when the electron chemical potential is on the order of $200-300$ MeV. For $\nu_{x} = \nu_{\tau}$ or some sterile flavor, the direct production or absorption is not possible, $\mathcal{A}_{P}^{x} = \mathcal{P}_{P}^{x} = 0$. This simplifies Eq. (9.48) considerably,

$$
\dot{f}_{\mathbf{p}}^{x} = \tfrac{1}{4}s_{\mathbf{p}}^{2}\big[(2 - f_{\mathbf{p}}^{x} - f_{\mathbf{p}}^{e})\mathcal{P}_{P}^{e} - (f_{\mathbf{p}}^{x} + f_{\mathbf{p}}^{e})\mathcal{A}_{P}^{e}\big].
\tag{9.49}
$$

If neither ν_{e} nor ν_{τ} are occupied because, for example, the medium is transparent to neutrinos so that they escape after production, one has $\dot{f}_{\mathbf{p}}^{x} = \tfrac{1}{2}s_{\mathbf{p}}^{2}\mathcal{P}_{P}^{e}$ so that the production rate of ν_{x} is that of ν_{e} times $\tfrac{1}{2}\sin^{2}2\theta_{\mathbf{p}}$ as one would have expected.

In a SN core where normal neutrinos are trapped Eq. (9.49) is more complicated as backreaction and Pauli blocking effects must be included. It becomes simple again if $s_{\mu}^{2} \ll 1$, because then the production of ν_{x} causes only a small perturbation of β equilibrium. Therefore, one may use the detailed-balance condition $(1 - f_{\mathbf{p}}^{e})\mathcal{P}_{P}^{e} - f_{\mathbf{p}}^{e}\mathcal{A}_{P}^{e} = 0$. Inserting this into Eq. (9.49) leads to

$$
\dot{f}_{\mathbf{p}}^{x} = \tfrac{1}{4}s_{\mathbf{p}}^{2}\big[(1 - f_{\mathbf{p}}^{x})\mathcal{P}_{P}^{e} - f_{\mathbf{p}}^{x}\mathcal{A}_{P}^{e}\big],
\tag{9.50}
$$

so that now the ν_{x} follow a Boltzmann collision equation with rates of gain and loss given by those of ν_{e} times $\tfrac{1}{4}\sin^{2}2\theta_{\mathbf{p}}$. If the ν_{x} are sterile they escape without building up so that their production rate is $\tfrac{1}{4}\sin^{2}2\theta_{\mathbf{p}}$ that of ν_{e}.

9.5   Flavor Conversion in a SN Core

9.5.1   Rate Equation

As a first application of the formalism developed in the previous sections consider two-flavor mixing between ν_e and another active neutrino species $\nu_x = \nu_\mu$ or ν_τ (vacuum mixing angle θ_0). In a SN core immediately after collapse electron lepton number is trapped and so the ν_e's have a large chemical potential on the order of 200 MeV. The trapped energy and lepton number diffuses out of the SN core and is radiated away within a few seconds. Will ν_x achieve equilibrium with ν_e on this time scale and thus share the large chemical potential? For the μ flavor this would also imply the production of muons by the subsequent charged-current absorption of ν_μ so that the lepton number would be shared between e, μ, ν_e, and ν_μ.

If the mixing angle in the medium were not small, flavor conversion would occur about as fast as it takes to establish β equilibrium. In this case a detailed calculation is not necessary so that one may focus on the limit of small mixing angles. In addition the oscillations are fast which allows one to use Eqs. (9.41) and (9.50). Moreover, the medium properties are assumed to be isotropic so that the production and absorption rates $\mathcal{P}^e$ and $\mathcal{A}^e$ of ν_e's depend only on their energy E. Also, the transition rate $W_{PP'}$ for the scattering of a neutrino with four momentum P to one with P' may be replaced by an angular average which depends only on the energies E and E'. Altogether one finds a rate of change for the ν_x density of

$$\dot{n}_{\nu_x} = \tfrac{1}{4}\int d\mathbf{p}\, s_\mathbf{p}^2\big[(1-f_\mathbf{p}^x)\,\mathcal{P}_E^e - f_\mathbf{p}^x\,\mathcal{A}_E^e\big]$$
$$+ \tfrac{1}{4}\sum_a \int d\mathbf{p}\,d\mathbf{p}'\,W_{EE'}^a\big[(g_x^a s_\mathbf{p} - g_e^a s_{\mathbf{p}'})^2 f_\mathbf{p}^e (1-f_{\mathbf{p}'}^x)$$
$$- (g_e^a s_\mathbf{p} - g_x^a s_{\mathbf{p}'})^2 f_\mathbf{p}^x (1-f_{\mathbf{p}'}^e)\big], \quad (9.51)$$

where $f_\mathbf{p}^e$ and $f_\mathbf{p}^x$ are the occupation numbers of ν_e and ν_x which are given by Fermi-Dirac distributions because *kinetic* equilibrium was assumed for both flavors. Also, $\tfrac{1}{4}s_\mathbf{p}^2 = \tfrac{1}{4}\sin^2 2\theta_E = \theta_E^2$ for small mixing angles.

A summation over different species a of medium fermions was restored; g_e^a and g_x^a are dimensionless effective NC coupling constants of ν_e and ν_x to fermion species a. For $a = n$ or p these constants are the same for all active neutrino species. Electrons as scattering targets are very relativistic so that they may be classified into a l.h. and a r.h. "species." The scattering with ν_e and ν_x is then described by the effective NC Hamiltonians

$$H_{L,R} = \frac{G_F}{2\sqrt{2}} \bar{\psi}_e \gamma_\mu (1 \mp \gamma_5) \psi_e \, \bar{\Psi} G_{L,R} \gamma^\mu (1 - \gamma_5) \Psi, \qquad (9.52)$$

where Ψ is again a neutrino column vector in flavor space. Further,

$$G_L = \begin{pmatrix} 2\sin^2\theta_W + 1 & 0 \\ 0 & 2\sin^2\theta_W - 1 \end{pmatrix},$$

$$G_R = \begin{pmatrix} 2\sin^2\theta_W & 0 \\ 0 & 2\sin^2\theta_W \end{pmatrix}, \qquad (9.53)$$

where $\sin^2\theta_W \approx \frac{1}{4}$ will be used. Hence the effective NC coupling constants are different for ν_e and ν_x interacting with l.h. electrons while they are the same for r.h. ones.

The ν_e Fermi sea is very degenerate. With regard to the neutrino distributions one may thus use the approximation $T = 0$ so that neutrino occupation numbers are 1 below their Fermi surface, and 0 above. Because the chemical potential μ_{ν_e} of the ν_e population exceeds μ_{ν_x}, and because neutrinos can only down-scatter in the $T = 0$ limit, the term proportional to $f_{\mathbf{p}}^x(1 - f_{\mathbf{p}'}^e)$ vanishes. Moreover, the detailed-balance requirement $(1 - f_{\mathbf{p}}^e)\mathcal{P}_E^e = f_{\mathbf{p}}^e \mathcal{A}_E^e$ implies $\mathcal{P}_E^e = 0$ for $E > \mu_{\nu_e}$ where $f_{\mathbf{p}}^e = 0$. Then altogether

$$\dot{n}_{\nu_x} = \int_{\mu_{\nu_x}}^{\mu_{\nu_e}} dE \left(\frac{\theta_E^2 \mathcal{P}_E^e E^2}{2\pi^2} + \int_{\mu_{\nu_x}}^{E} dE' \sum_a W_{EE'}^a \frac{(g_x^a \theta_E - g_e^a \theta_{E'})^2 E^2 E'^2}{4\pi^4} \right). \qquad (9.54)$$

Because for degenerate neutrinos $n_\nu = \mu_\nu^3/6\pi^2$ Eq. (9.54) can be written as a differential equation for μ_{ν_x}.

According to Eq. (8.30) the mixing angle in a medium which is dominated by protons, neutrons, and electrons is given by

$$\tan 2\theta_E = \frac{\sin 2\theta_0}{\cos 2\theta_0 - E/E_\rho}, \qquad (9.55)$$

where the density-dependent "resonance energy" is

$$E_\rho \equiv \frac{\Delta m^2}{2\sqrt{2}\,G_F n_e}, \qquad (9.56)$$

with the electron density n_e. For $E = \mu_{\nu_e}$ one finds

$$\frac{E}{E_\rho} = \frac{(68\,\text{keV})^2}{\Delta m^2} Y_e Y_{\nu_e}^{1/3} \rho_{14}^{4/3}, \qquad (9.57)$$

where ρ_{14} is the density in units of $10^{14}\,\text{g cm}^{-3}$ and as usual Y_j gives the abundance of species j relative to baryons. The approximate parameter range where resonance is important is shown in Fig. 9.3 as diagonal shaded band.

For conditions near resonance there is no need for a detailed calculation because then the mixing angle is large. Therefore, one may focus on the limiting cases where Δm^2 is either so small or so large that $\theta_E^2 \ll 1$,

$$|\theta_E| = \theta_0 \times \begin{cases} 1 & \text{if } E_\rho/E \gg 1 \text{ (large } \Delta m^2), \\ E_\rho/E & \text{if } E_\rho/E \ll 1 \text{ (small } \Delta m^2). \end{cases} \qquad (9.58)$$

Thus, for large Δm^2 the ν_x production rate is

$$\dot{n}_{\nu_x} = \theta_0^2 \int_{\mu_{\nu_x}}^{\mu_{\nu_e}} dE \left(\frac{\mathcal{P}_E^e E^2}{2\pi^2} + \int_{\mu_{\nu_x}}^E dE' \sum_a W_{EE'}^a \frac{(g_x^a - g_e^a)^2 E^2 E'^2}{4\pi^4} \right), \quad (9.59)$$

while for small Δm^2 it is

$$\dot{n}_{\nu_x} = \left(\frac{\theta_0 \Delta m^2}{2\sqrt{2} G_{\mathrm{F}} n_e} \right)^2 \int_{\mu_{\nu_x}}^{\mu_{\nu_e}} dE \left(\frac{\mathcal{P}_E^e}{2\pi^2} + \int_{\mu_{\nu_x}}^E dE' \sum_a W_{EE'}^a \frac{(g_x^a E' - g_e^a E)^2}{4\pi^4} \right).$$
$$(9.60)$$

9.5.2 Neutrino Interaction Rates

In order to evaluate these integrals one first needs the production rate $\mathcal{P}_E^e$ of ν_e's with energy E due to the CC reaction $p + e \to n + \nu_e$. The leptons are taken to be completely degenerate, the nucleons to be completely nondegenerate. Because they are also nonrelativistic the absorption of an e^- produces a ν_e of the same energy. Therefore, $\mathcal{P}_E^e = \sigma_E n_p$ where n_p is the proton density and $\sigma_E = (C_V^2 + 3C_A^2) G_{\mathrm{F}}^2 E^2/\pi$ is the CC scattering cross section for electrons of energy E. Here, $C_V = 1$ and $C_A = 1.26$ are the usual vector and axial-vector weak couplings. Altogether one finds

$$\mathcal{P}_E^e = \frac{C_V^2 + 3C_A^2}{\pi} G_{\mathrm{F}}^2 E^2 n_p. \qquad (9.61)$$

In practice this rate is reduced by various factors. First, Pauli blocking of nucleons cannot be neglected entirely. Second, the degeneracy of electrons is not complete. Third, the axial-vector scattering rate may be suppressed in a medium at nuclear densities (Sect. 4.6.7).

For NC scattering, nucleons may be neglected entirely. In Eq. (9.59) their contribution vanishes identically because $g_e = g_x$. In Eq. (9.60) it is suppressed because they are relatively heavy so that $E' \approx E$.

For l.h. electrons the coupling strengths are different, and they are relativistic so that recoil effects are not small. However, they are degenerate so that their contribution is expected to be smaller than the ep process. It is not entirely negligible, however, especially if the nucleon contribution is partly suppressed by many-body effects. The transition rates were worked out in detail by Raffelt and Sigl (1993) for entirely degenerate leptons. They found

$$
\begin{aligned}
W_{EE'}^R &\approx \frac{G_F^2 \mu_e^2}{3\pi} \frac{(E - E')E'}{E^2}, \\
W_{EE'}^L &\approx \frac{G_F^2 \mu_e^2}{3\pi} \frac{(E - E')E}{E'^2},
\end{aligned}
\tag{9.62}
$$

a result which is the lowest-order term of an expansion in powers of E/μ_e (electron chemical potential μ_e).

9.5.3 Time Scale for Flavor Conversion

Given enough time the ν_τ's will reach the same number density as the ν_e's. Therefore, it is most practical to discuss the approach to chemical equilibrium in terms of a time scale

$$
\tau^{-1} \equiv -\frac{d}{dt} \ln \left(\frac{n_{\nu_e} - n_{\nu_x}}{n_{\nu_e}} \right) .
\tag{9.63}
$$

If τ were a constant independent of n_{ν_x}/n_{ν_e} the difference between the number densities would be damped exponentially.

Collecting the results of the previous section one then finds easily for the case of a "large" Δm^2 of Eq. (9.59)

$$
\tau^{-1} = \theta_0^2 \frac{3 G_F^2}{5\pi} n_p \mu_{\nu_e}^2 \left[(C_V^2 + 3C_A^2) F_p + \frac{\mu_{\nu_e}}{\mu_e} F_e \right],
\tag{9.64}
$$

where F_p and F_e give the contributions of protons (CC process) and electrons (effective NC process). They are functions of μ_{ν_x} which is parametrized by

$$
\eta \equiv \mu_{\nu_x}/\mu_{\nu_e}.
\tag{9.65}
$$

One finds (Fig. 9.2, left panel)

$$
\begin{aligned}
F_p &= (1 - \eta^5)/(1 - \eta^3), \\
F_e &= (\tfrac{5}{6} + \tfrac{1}{2}\eta + \tfrac{1}{4}\eta^2 + \tfrac{1}{12}\eta^3)(1 - \eta)^3/(1 - \eta^3),
\end{aligned}
\tag{9.66}
$$

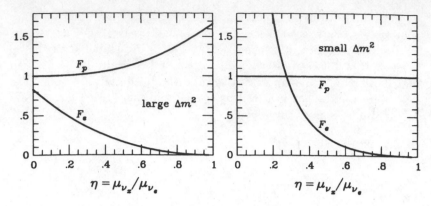

Fig. 9.2. Dependence of the flavor relaxation rate on $\eta = \mu_{\nu_x}/\mu_{\nu_e}$ for the case of a "large" Δm^2 Eq. (9.66) and a "small" Δm^2 Eq. (9.69). F_e is the contribution of electron targets ($\nu e \rightarrow e\nu$) while F_p is from protons ($ep \leftrightarrow n\nu_e$).

and numerically

$$\frac{3G_F^2}{5\pi}\,n_p\mu_{\nu_e}^2 = 1.0\times10^9\,\mathrm{s}^{-1}\left(\frac{Y_p\rho}{10^{14}\,\mathrm{g\,cm^{-3}}}\right)^{5/3}\left(\frac{\mu_{\nu_e}}{\mu_e}\right)^2, \qquad (9.67)$$

which sets the time scale for flavor conversion.

For the case of a "small" Δm^2 one finds from Eq. (9.60) and from the production and scattering rates of the previous section

$$\tau^{-1} = \theta_0^2\,\frac{(\Delta m^2)^2}{8\pi n_p}\left[(C_V^2 + 3C_A^2)\,F_p + \frac{\mu_{\nu_e}}{\mu_e}\,F_e\right], \qquad (9.68)$$

where $F_p = 1$ and

$$F_e = \frac{\frac{9}{5}\eta^{-1} - \frac{137}{40} + \frac{3}{4}(1+\eta^4)\log\eta + \frac{5}{3}\eta + \eta^3 - \frac{101}{120}\eta^4 - \frac{1}{5}\eta^5}{\frac{8}{3}\left(1-\eta^3\right)} \qquad (9.69)$$

(Fig. 9.2, right panel). Numerically,

$$\frac{(\Delta m^2)^2}{8\pi n_p} = 1.4\times10^2\,\mathrm{s}^{-1}\left(\frac{10^{14}\,\mathrm{g\,cm^{-3}}}{Y_p\rho}\right)\left(\frac{(\Delta m^2)^{1/2}}{1\,\mathrm{keV}}\right)^4. \qquad (9.70)$$

This time scale is independent of Fermi's constant because a factor G_F^2 from the scattering rate cancels against G_F^{-2} from θ_E^2.

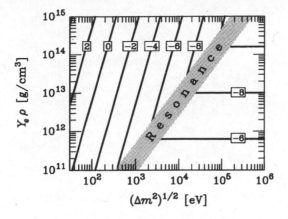

Fig. 9.3. Contour plot for $\log(\tau\theta_0^2)$ with τ in seconds according to Eqs. (9.64) and (9.68), taking $F_e = 0$, $F_p = 1$, $C_V^2 + 3C_A^2 = 4$, and $\mu_{\nu_e}/\mu_e = 1$. (Adapted from Raffelt and Sigl 1993.)

From Fig. 9.2 it is clear that effective NC scattering on electrons slightly accelerates the initial rate of flavor conversion, but it does not dramatically affect the overall time scale for achieving equilibrium. This time scale is crudely estimated by ignoring F_e entirely in Eqs. (9.64) and (9.68) and by setting $F_p = 1$ and $C_V^2 + 3C_A^2 = 4$. With $\mu_{\nu_e} \approx \mu_e$ one then finds results for $\tau\theta_0^2$ shown as contours in Fig. 9.3. The diagonal band refers to the resonance condition of Eq. (9.57); there flavor equilibrium would be established on a time scale nearly independent of the vacuum mixing angle. However, the detailed behavior in this range of parameters has not been determined.

Armed with these results it is straightforward to determine the range of masses and mixing angles where ν_x would achieve flavor equilibrium and thus would effectively participate in β equilibrium $ep \leftrightarrow n\nu$. The initially trapped lepton number escapes within a few seconds. Therefore, $\tau \lesssim 1\,\mathrm{s}$ is adopted as a criterion for ν_x to have any novel impact on SN cooling or deleptonization. The relevant density is about three times nuclear while $Y_p \approx 0.35$ so that $Y_p\rho = 3\times10^{14}\,\mathrm{g\,cm^{-3}}$ is adopted. Otherwise the same parameters are used as in Fig. 9.3. Then one finds

$$\sin^2 2\theta_0 \gtrsim \begin{cases} 0.02\,(\mathrm{keV^2}/\Delta m^2)^2 & \text{for } \Delta m^2 \lesssim (100\,\mathrm{keV})^2 \\ 2\times10^{-10} & \text{for } \Delta m^2 \gtrsim (100\,\mathrm{keV})^2 \end{cases} \qquad (9.71)$$

as a requirement for ν_x to reach chemical equilibrium. This range of parameters is shown as a hatched region in Fig. 9.4.

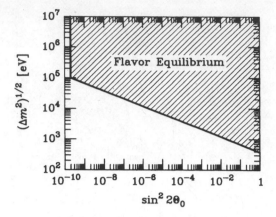

Fig. 9.4. In the hatched parameter range ν_μ or ν_τ would achieve chemical equilibrium with ν_e in a SN core within about one second after collapse.

The cosmological mass limit of about 30 eV prevents ν_μ or ν_τ from playing any novel role in the cooling or deleptonization of a SN core. However, only a very small mixing angle is required to achieve equilibrium if one of the neutrinos defied either standard particle physics or standard cosmology and had a mass in the keV range or above. Of course, if the mass were of Dirac type the cooling effect from the production of spin-flipped neutrinos would be too large to be compatible with the SN 1987A neutrino signal, yielding a bound on Dirac neutrino masses in the 10 keV range (Sect. 13.8.1). It is in this context that flavor conversion was first discussed by Maalampi and Peltoniemi (1991), Turner (1992), and Pantaleone (1992a).

9.6   Sterile Neutrinos and SN 1987A

If a hypothetical sterile neutrino ν_x existed, it would be produced in the inner core of a SN by virtue of its assumed mixing with ν_e. The ν_x would escape directly from the inner SN core, carrying away energy and lepton number. If this process occurred too fast the observed neutrino signal of SN 1987A would have been unduly shortened, allowing one to exclude a certain range of ν_x masses and mixing angles with ν_e (Kainulainen, Maalampi, and Peltoniemi 1991).

For small mixing angles one may use Eqs. (9.40) and (9.50) as a starting point for the rate of change of the ν_x occupation numbers $f_\mathbf{p}^x$.

Because the ν_x are assumed to escape freely one may set $f_{\mathbf{p}}^x = 0$ on the r.h.s. of these equations. Moreover, for the ν_x coupling constants one may use $g_x = 0$ because it is sterile. For an isotropic medium and taking an angular average of the scattering rate as in the previous section one finds

$$\dot{f}_{\mathbf{p}}^x = \tfrac{1}{4}s_{\mathbf{p}}^2 \left(\mathcal{P}_E^e + \sum_a (g_e^a)^2 \int d\mathbf{p}' W_{E'E}^a f_{\mathbf{p}'}^e \right). \tag{9.72}$$

A summation over different target species for effective NC scattering was restored. Taking the ν_e's to be completely degenerate one has $f_{\mathbf{p}}^e = 1$ for $E = |\mathbf{p}| < \mu_{\nu_e}$ and $f_{\mathbf{p}}^e = 0$ otherwise so that

$$\dot{f}_{\mathbf{p}}^x = \tfrac{1}{4}s_{\mathbf{p}}^2 \left(\mathcal{P}_E^e + \sum_a (g_e^a)^2 \int dE' \, \frac{E'^2 \, W_{E'E}^a}{2\pi^2} \right). \tag{9.73}$$

With this result and

$$\dot{n}_L = -\int_0^{\mu_{\nu_e}} dE \, \dot{f}_{\mathbf{p}}^x \qquad \text{and} \qquad \dot{Q} = -\int_0^{\mu_{\nu_e}} dE \, \dot{f}_{\mathbf{p}}^x \, E \tag{9.74}$$

the volume loss rates for lepton number and energy can be easily determined.

Turn first to the case $\Delta m^2 \gtrsim (100\,\text{keV})^2$ so that one may use the vacuum mixing angle in a SN core. One must now include both the CC process $ep \to n\nu_x$ as well as the NC process $\nu_e N \to N\nu_x$ which is not suppressed because for a sterile ν_x there is no destructive interference effect. For nondegenerate nucleons and using $C_V^2 + 3C_A^2 \approx 4$ for both CC and NC processes one easily finds from the results of Sect. 9.5.2

$$\dot{n}_L = -\tfrac{1}{4} \sin^2 2\theta_0 \, \frac{2G_F^2}{5\pi^3} \, n_B \left(Y_p + \tfrac{1}{4} \right) \mu_{\nu_e}^5, \tag{9.75}$$

and the same for $\dot{Q}$ with $\tfrac{1}{5}\mu_{\nu_e}^5 \to \tfrac{1}{6}\mu_{\nu_e}^6$. Electrons as NC scattering targets may be neglected because of their degeneracy. Lepton number is thus lost at a rate

$$\dot{Y}_L = -\sin^2 2\theta_0 \, \tau^{-1} \left(Y_e + \tfrac{1}{4} \right) Y_{\nu_e}^{5/3}, \tag{9.76}$$

where

$$\tau^{-1} = \tfrac{3}{5} (36\pi)^{1/3} \, G_F^2 n_B^{5/3} = 7.7 \times 10^{10} \, \text{s}^{-1} \, \rho_{15}^{5/3}. \tag{9.77}$$

Here, $Y_p = Y_e$, $Y_L \equiv n_L/n_B$, and $Y_\nu n_B = n_\nu = \mu_\nu^3/6\pi^2$ was used, and ρ_{15} is ρ in units of $10^{15}\,\text{g cm}^{-3}$.

Because Y_{ν_e} is about $Y_L/4$, lepton number and energy are lost at about a rate of $\sin^2 2\theta_0 \, 10^{10}\,\mathrm{s}^{-1}$. Because the SN 1987A signal lasted for several seconds, a conflict with these observations is avoided if

$$\sin^2 2\theta_0 \lesssim 10^{-10} \tag{9.78}$$

(Kainulainen, Maalampi, and Peltoniemi 1991; Raffelt and Sigl 1993). If $m_{\nu_x} \gtrsim 1\,\mathrm{MeV}$ the assumed mixing with ν_e allows for decays $\nu_x \to \nu_e e^- e^+$ and $\nu_x \to \nu_e e^- e^+ \gamma$. The resulting γ signal from the SN 1987A ν_x flux (Sect. 12.4.7) does not allow for a dramatic improvement of the bound Eq. (9.78). However, the decay argument does exclude the possibility of a mixing angle so large that even the "sterile" ν_x would be trapped by virtue of its mixing with ν_e.

The case of $\Delta m^2 \lesssim (100\,\mathrm{keV})^2$ is complicated because neutrinos in a certain energy range below their Fermi surface encounter a ν_e-ν_x mixing resonance. This implies that during the SN infall phase a large amount of lepton number can be lost. The impact on the equation of state can be strong enough to prevent a subsequent explosion. Shi and Sigl (1994) found that this argument requires $\sin^2 2\theta_0 \lesssim 10^{-8}\,\mathrm{keV}^2/\Delta m^2$ for $\Delta m^2 \gtrsim (1\,\mathrm{keV})^2$. They found additional constraints from the anomalous contribution to the cooling by ν_x emission.

These are all limits on the mixing of a sterile neutrino with ν_e, the only case that has been studied in the literature. Historically, this is related to the now forgotten episode of the 17 keV neutrino which for some time seemed to exist and which could have been a sterile neutrino mixed with ν_e. However, similar limits can be derived for ν_μ-ν_x or ν_τ-ν_x mixing. The main difference is that the nonelectron neutrinos would not normally obtain a chemical potential so that only a thermal ν_μ and ν_τ population can be converted. A typical temperature is 30 MeV or more, the average energy of a thermal population of relativistic fermions is about $3T \gtrsim 100\,\mathrm{MeV}$, so there will be a significant thermal muon population ($m_\mu = 106\,\mathrm{MeV}$). Thus sterile states can be produced in charged-current muon scatterings in analogy to the above discussion of ν_x production involving electron scattering. The resulting limit on the mixing angle may be slightly weaker than in the ν_e-ν_x case, but it will be of the same general order of magnitude.[52] For ν_τ-ν_x mixing the situation is different in that there are no thermally excited τ leptons. Still, any reaction that produces $\nu_\tau \bar\nu_\tau$ pairs can also produce ν_x and $\bar\nu_x$ particles.

---

[52] This remark is relevant in the context of recent speculations about the existence of a 34 MeV sterile neutrino (Barger, Phillips, and Sarkar 1995) as an explanation of an anomaly observed in the KARMEN experiment (KARMEN Collaboration 1995).

Chapter 10

Solar Neutrinos

The current theoretical and experimental status of the Sun as a neutrino source is reviewed. Particle-physics interpretations of the apparent deficit of measured solar neutrinos are discussed, with an emphasis on an explanation in terms of neutrino oscillations.

10.1 Introduction

The Sun, like other hydrogen-burning stars, liberates nuclear binding energy by the fusion reaction

$$4p + 2e^- \rightarrow {}^4\text{He} + 2\nu_e + 26.73\,\text{MeV} \tag{10.1}$$

which proceeds through a number of different reaction chains and cycles (Fig. 10.2). With a total luminosity of $L_\odot = 3.85 \times 10^{33}\,\text{erg s}^{-1} = 2.4 \times 10^{39}\,\text{MeV s}^{-1}$, the Sun produces about $1.8 \times 10^{38}\,\text{s}^{-1}$ neutrinos, or at Earth (distance $1.50 \times 10^{13}\,\text{cm}$) a flux of $6.6 \times 10^{10}\,\text{cm}^{-2}\,\text{s}^{-1}$. While this is about a hundred times less than the $\bar{\nu}_e$ flux near a large nuclear power reactor it is still a measurable flux which can be used for experimentation just like the flux from any man-made source.

The most straightforward application of the solar neutrino flux is a search for radiative decays by measurements of x- and γ-rays from the quiet Sun. Because of the long decay path relative to laboratory experiments one obtains a limit which is about 9 orders of magnitude more restrictive (Sect. 12.3.1).

A more exciting application is a search for neutrino oscillations. In fact, the current measurements of the solar neutrino flux are neither compatible with theoretical predictions nor with each other ("solar neutrino problem"); all discrepancies disappear with the assumption of neutrino oscillations. At the present time, however, this interpretation is not established "beyond reasonable doubt"—a final verdict can be expected from the new experiments currently in preparation. They may be able to discover a characteristic distortion of the neutrino spectrum, or they may actually measure the "wrong-flavored" neutrinos that were produced by oscillations from the ν_e originating in the Sun.

Contrary to the $\bar{\nu}_e$ flux from a power reactor, the solar neutrino spectrum arises from a small number of specific reactions (Fig. 10.1). The main contribution (91%) is from the reaction $p + p \rightarrow d + e^+ + \nu_e$ with a maximum neutrino energy 0.420 MeV ("pp neutrinos"). Second at about 7% of the flux are the "beryllium neutrinos" from the electron-capture reaction $e^- + {}^7\text{Be} \rightarrow {}^7\text{Li} + \nu_e$ with a fixed energy 0.862 MeV. Finally, very small fraction ($\approx 10^{-4}$) are the "boron neutrinos" from ${}^8\text{B} \rightarrow {}^8\text{Be} + e^+ + \nu_e$. Still, they are of major importance because their large energies of up to 15 MeV allow for a less difficult detection procedure than is required for the soft part of the spectrum.

The first solar neutrino experiment is based on the nuclear reaction $\nu_e + {}^{37}\text{Cl} \rightarrow {}^{37}\text{Ar} + e^-$. With a threshold of 0.814 MeV it picks up both beryllium and boron neutrinos, although the argon production rate is dominated by the latter. The target consists of about 4×10^5 liters (615 tons) of perchloroethylene (C_2Cl_4) in a huge tank which is located in the Homestake Mine in South Dakota (U.S.A.), about 1.5 km underground for protection against the cosmic-ray background. After an exposure of a few months to the solar neutrino flux a few argon atoms have been produced (about 0.4 atoms per day). They are chemically extracted and counted by their subsequent decays (half-life 35.0 days).

When this pioneering experiment first produced data (Davis, Harmer, and Hoffman 1968) there appeared a deficit relative to the theoretically expected flux, a discrepancy which has persisted ever since—the experiment is still taking data today! On the theoretical side, the ever refined predictions of Bahcall and his collaborators (e.g. Bahcall 1989) were instrumental at establishing the notion that this discrepancy—a factor of around 3—was to be taken seriously. However, in spite of the acknowledged experimental care of Davis and his collaborators, doubt has always lingered about the reliability of the data because this detector has never been subject to an on-off test as the Sun is the only available neutrino source powerful enough to cause a detectable signal. On the solar side, the boron neutrino flux depends crucially on the reaction rate $p + {}^7\text{Be} \rightarrow {}^8\text{B} + \gamma$ with a cross section that is relatively poorly known.

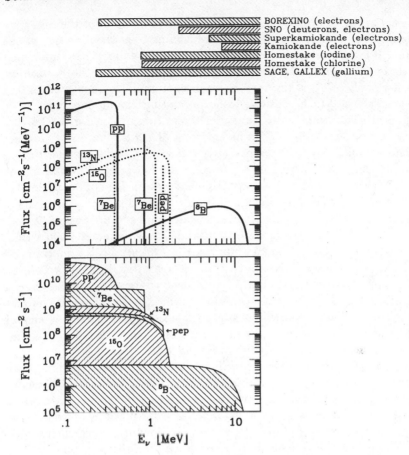

Fig. 10.1. Solar neutrino flux at Earth according to the Bahcall and Pinsonneault (1995) solar model. *Upper panel:* Continuum spectra in $cm^{-2}s^{-1}MeV^{-1}$, line spectra in $cm^{-2}s^{-1}$. Solid lines are the sources of dominating experimental significance. *Above:* Range of sensitivity of current and near-future solar neutrino experiments. *Lower panel:* Cumulative spectrum integrated from a given energy to infinity.

The situation changed radically when the Kamiokande detector, originally built to search for proton decay, began in 1987 to measure the solar neutrino flux by virtue of the elastic scattering reaction $\nu_e + e^- \rightarrow e^- + \nu_e$ which is detected by the Cherenkov light emitted by the kicked electron. With a threshold of about 9 MeV (later 7 MeV) it is exclusively sensitive to the boron neutrino flux. Because of its directional sensitivity it is a true "neutrino telescope" and for the first time established that indeed neutrinos are coming from the direction of the Sun. It is perplexing, however, that the measured flux is less suppressed relative to solar-model predictions than that found in the Homestake experiment. This is the reverse from what would be expected on the grounds that ^{37}Cl is sensitive to boron *and* beryllium neutrinos.

The situation changed yet again when the experiments SAGE (Soviet-American Gallium Experiment) and GALLEX began to produce data in 1990 and 1991, respectively. They are radiochemical experiments using the reaction $\nu_e + {}^{71}\text{Ga} \rightarrow {}^{71}\text{Ge} + e^-$ which has a threshold of 233 keV. Therefore, these experiments pick up the dominant *pp* neutrino flux which can be calculated from solar models with a precision of a few percent unless something is radically wrong with our understanding of the Sun. Therefore, a substantial deficit of measured *pp* neutrinos would have been a "smoking gun" for the occurrence of neutrino oscillations. While the first few exposures of SAGE seemed to indicate a low flux, the good statistical significance of the data that have since been accumulated by both experiments indicate a flux which is high enough so that no *pp* neutrinos are reported missing, but low enough to confirm the existence of a significant problem with the high-energy part of the spectrum (beryllium and boron neutrinos).

With four experiments reporting data, which represent three different spectral responses to the solar neutrino flux, the current attention has largely shifted from a comparison between experiments and theoretical flux predictions to a "model-independent analysis" which is based on the small number of possible source reactions each of which produces neutrinos of a well-defined spectral shape. This sort of analysis currently indicates a lack of consistency among the experiments which can be brought to perfect agreement if neutrinos are assumed to oscillate.

Even though the attention has currently shifted away from theoretical solar neutrino flux predictions it should be noted that in recent years there has been much progress in a quantitative theoretical treatment of the Sun. Independently of the interest in the Sun as a neutrino source it serves as a laboratory to test the theory of stellar structure and evolution. Particularly striking advances have been made in the field of helioseismology. There are two basic vibration patterns for the Sun, one where gravity represents the restoring force (g-modes), and normal "sound" or pressure (p) modes. The former are evanescent in the solar convection zone (depth about $0.3\,R_\odot$ from the surface) and have never been unambiguously observed. The oscillation period of the highest-frequency g-modes would be about 1 h.

There exist vast amounts of data concerning p-modes (periods between 2 min and 1 h) which can be measured from the Doppler shifts of spectral lines on the solar surface. The main point is that one can establish a relationship between the multipole order of the oscillation pattern and the frequency. Because different vibration modes probe the sound speed at different depths one can invert the results to derive an empirical profile for the sound speed in the solar interior. The agreement with theoretical expectations for the square of the sound speed is better than about 0.3% except in the inner $0.2\,R_\odot$ which are not probed well by p-modes (Christensen-Dalsgaard, Proffitt, and Thompson 1993; Dziembowski et al. 1994). Such results make it very difficult to contemplate the possibility that the Sun is radically different from a standard structure. On the other hand, these results are not precise enough to reduce the uncertainty of solar neutrino predictions which arise from the uncertainty of the opacity coefficients.

The differences between the solar neutrino predictions of different authors are minimal when identical input parameters are used. Therefore, the expected error resulting from solar modelling is very small, except that some key input parameters remain uncertain. The dominant source of uncertainty for the boron neutrino flux is the cross section for the reaction $p + {}^{7}\text{Be} \rightarrow {}^{8}\text{B} + \gamma$ which appears as a multiplicative factor for the solar flux prediction and thus is unrelated to solar modelling.

A more astrophysical uncertainty are the opacity coefficients. Although they are thought to be well known for the conditions in the deep solar interior, even a relatively small error translates into a non negligible uncertainty of the boron flux prediction because of its steep temperature dependence. Crudely, a 10% error of the opacity coefficients translates into a 1% error of the central solar temperature and then into a 20% uncertainty of the boron flux. Indeed, a measurement of the solar neutrino flux was originally envisaged as a method to measure precisely the inner temperature of the Sun.

The Sun as a neutrino source naturally has received much attention in the literature because of its outstanding potential to finally confirm the existence of neutrino oscillations in nature. Because this book is not primarily on solar neutrinos I will limit my discussion to what I consider the most important features of this unique neutrino source, and the role it plays for particle physics. A lot of key material can be found in Bahcall's (1989) book on solar neutrinos, in the more recent Physics Report on the solar interior by Turck-Chièze et al. (1993), and on the Sun in general in the book by Stix (1989).

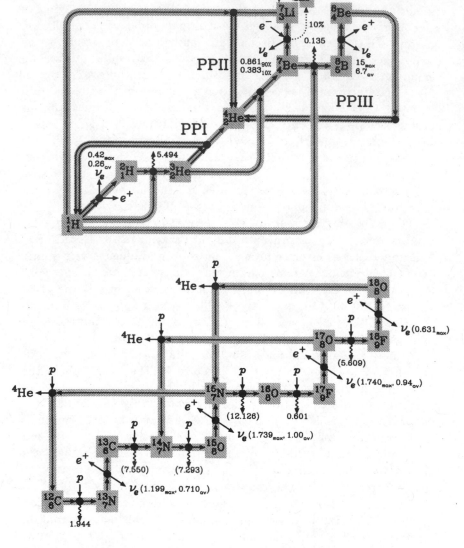

Fig. 10.2. Reaction chains PPI–PPIII and CNO tri-cycle. Nuclear reactions, including decays, are marked with a bullet (•). Average (av) and maximum (max) energies in MeV are given for the neutrinos. For photons (wavy arrows) numbers in brackets refer to the total energy of a cascade; otherwise it is the energy of a monochromatic γ line.

10.2 Calculated Neutrino Spectrum

10.2.1 Individual Sources

A hydrogen-burning star like the Sun liberates nuclear binding energy by helium fusion from hydrogen which proceeds by virtue of the *pp* chains and the CNO cycle (Fig. 10.2). The energy-generation rate for the CNO cycle is a much steeper function of temperature whence it dominates in hot stars. For a 2% mass fraction of CN elements, typical for population I stars like the Sun, the crossover temperature is at about 1.8×10^7 K (Clayton 1968). The central temperature of the Sun is about 1.56×10^7 K and so the CNO reactions contribute only about 1% to the total energy budget. In fact, of the CNO tri-cycle shown in Fig. 10.2, in practice only the first loop (the CN cycle) is of importance in the Sun as the branching rates into the second or even third loop are extremely low.

In the Sun, the *pp* chains terminate in about 85% of all cases via PPI, i.e. by the fusion of two ^{3}He nuclei. In this case the only neutrino-producing reactions are *pp* ($pp \rightarrow de^+\nu_e$) and pep ($pe^-p \rightarrow d\nu_e$), the latter occurring very rarely. In about 15% of all cases the termination is via PPII where ^{7}Be is formed from ^{4}He $+ ^3$He. Because of the small energy difference between the ground states of the ^{7}Be and ^{7}Li

Table 10.1. Source reactions for solar neutrinos.

Reaction	Q (a) [MeV]		Flux at Earth[b] $[\text{cm}^{-2}\,\text{s}^{-1}]$	Uncertainty[b] [%]	
$pp \rightarrow {}^2\text{H}\,e^+\nu_e$	0.420	c	5.9×10^{10}	$+1$	-1
$pe^-p \rightarrow {}^2\text{H}\,\nu_e$	1.442		1.4×10^8	$+1$	-2
${}^3\text{He}\,p \rightarrow {}^4\text{He}\,e^+\nu_e$	18.77	c	1.2×10^3	Factor 6	
${}^7\text{Be}\,e^- \rightarrow {}^7\text{Li}\,\nu_e$	0.862		4.6×10^9	$+6$	-7
${}^7\text{Be}\,e^- \rightarrow {}^7\text{Li}^*\,\nu_e$	0.384		5.2×10^8	$+6$	-7
${}^8\text{B} \rightarrow {}^8\text{Be}^*\,e^+\nu_e$	≈ 15	c	6.6×10^6	$+14$	-17
${}^{13}\text{N} \rightarrow {}^{13}\text{C}\,e^+\nu_e$	1.199	c	6.2×10^8	$+17$	-20
${}^{15}\text{O} \rightarrow {}^{15}\text{N}\,e^+\nu_e$	1.732	c	5.5×10^8	$+19$	-22
${}^{17}\text{F} \rightarrow {}^{17}\text{O}\,e^+\nu_e$	1.740	c	6.5×10^6	$+15$	-19
Total			6.5×10^{10}	$+1$	-1

[a]Maximum ν_e energy for continuum (c) sources.
[b]Bahcall and Pinsonneault (1995).

nuclei, the usual weak decay by $e^+ \nu_e$ emission is not possible and so the conversion proceeds by electron capture, leading to the emission of an almost monochromatic neutrino. In about 10% of all cases the capture reaction goes to the first excited state (478 keV) of ^{7}Li so that there are two neutrino lines.

Instead of an electron, ^{7}Li very rarely captures a proton and forms ^{8}B which subsequently decays into ^{8}Be, a nucleus unstable against spontaneous fission into ^{4}He + ^{4}He. This PPIII termination occurs in about 0.02% of all cases, too rare to be of importance for nuclear energy generation. Its importance arises entirely from the high energy of the ^{8}B neutrinos which are the ones least difficult to measure.

The neutrino-producing reactions are summarized in Tab. 10.1 with their maximum energies, and with the resulting neutrino flux at Earth found in the solar model of Bahcall and Pinsonneault (1995).

Apart from small screening and thermal broadening effects, the spectral shape for each individual source is independent of details of the solar model. The pp, ^{13}N, ^{15}O, and ^{17}F reactions are allowed or superallowed weak transitions so that their spectra are

$$dN/dE_\nu = A\left(Q + m_e - E_\nu\right)\left[(Q + m_e - E_\nu)^2 - m_e^2\right]^{1/2} E_\nu^2 \, F \,,$$

$$(10.2)$$

where Q is the maximum e^+ kinetic energy and also the maximum ν_e energy, A is a normalization constant, and F is a function of E_{e^+} which takes the e^+ final-state interactions into account. For the low-Z nuclei under consideration this correction is small for most of the neutrino spectrum. With $F = 1$ the normalization constants are given in Tab. 10.2. In Fig. 10.3 the normalized spectra from the pp and the ^{15}O processes are shown where the pp spectrum was taken from the tabulation of Bahcall and Ulrich (1988). Using Eq. (10.2) instead would cause a change so small that it would be nearly hidden by the line width of the curve in Fig. 10.3.

Table 10.2. Normalization of the spectrum Eq. (10.2) with $F = 1$.

Source	Q [MeV]	A [MeV^{-5}]
pp	0.420	193.9
^{13}N	1.199	3.144
^{15}O	1.732	0.668

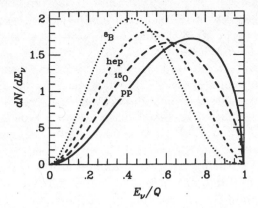

Fig. 10.3. Normalized spectra of neutrino source reactions in the Sun.

It is much more difficult to obtain the spectrum from ^{8}B decay because the final-state ^{8}Be* nucleus is unstable against spontaneous fission into two α particles. Even though several states of ^{8}Be contribute to the transition, it is dominated by the 2.9 MeV excitation and so the neutrino spectrum can be determined with relatively little ambiguity by folding Eq. (10.2) with the experimental α spectrum (Kopysov and Kuzmin 1968). A more recent and more detailed analysis was performed by Bahcall and Holstein (1986). An analytic approximation to their tabulated spectrum is

$$dN/dE_\nu = 8.52 \times 10^{-6} (15.1 - E_\nu)^{2.75} E_\nu^2 , \qquad (10.3)$$

where the neutrino energies are in MeV. This normalized spectrum is also shown in Fig. 10.3 where, again, the difference between the tabulated values and the analytic approximation would be hidden by the line width (maximum deviation less than 0.02 in units of the vertical axis in Fig. 10.3).

Neutrinos from the hep reaction extend to the highest energies of all solar sources, but their overall flux is very small and very uncertain because of large uncertainties in the low-energy ^{3}Hep cross section— see Bahcall and Pinsonneault (1992) for a detailed discussion. The tabulated spectrum (Bahcall and Ulrich 1988) can be represented by

$$dN/dE_\nu = 2.33 \times 10^{-5} (18.8 - E_\nu)^{1.80} E_\nu^{1.92} , \qquad (10.4)$$

where the quality of the fit is equally good as that for the ^{8}B neutrinos.

In principle, any of the neutrino-producing reactions in the Sun with a final-state positron can also occur with an initial-state electron such as $pe^-p \to d\nu_e$ instead of $pp \to de^+\nu_e$. In most cases, however, the electron-capture process is strongly suppressed relative to positron emission because of an unfavorable phase-space factor, or equivalently because of the relatively small electron density. (For a general comparison between these two reaction channels see Bahcall 1990.) The only exception is pep which yields a nonnegligible contribution because the Q value of the pp reaction and thus the available positron phase space are rather small. Another exception are the beryllium neutrinos because positron emission is inhibited entirely by the small energy difference between the ground states of ^{7}Be and ^{7}Li.

The shapes of the neutrino spectra from the individual source reactions are determined entirely by the matrix element and phase space of the microscopic reactions apart from small broadening effects by the thermal motion of the reaction participants. As typical thermal energies in the Sun are 1 keV the spectral modification by such effects is entirely negligible. (For a detailed discussion of this point see Bahcall 1991.) The high-energy part of the solar neutrino spectrum between about $2-15$ MeV is dominated by the boron neutrinos (Fig. 10.1), i.e. by a single source reaction. Therefore, this energy range is a clean example to test the spectral shape. The confirmation of the expected spectral shape above 7 MeV by the Kamiokande II detector excludes

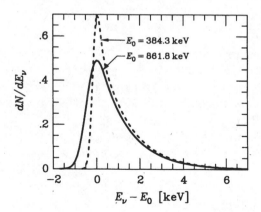

Fig. 10.4. Normalized spectra of the thermally and Doppler broadened beryllium neutrino lines where E_0 is the corresponding laboratory energy (Bahcall 1994). The line shapes involve an integral over a solar model.

a certain range of neutrino masses and mixing angles (Sect. 10.3.4). Of course, the Superkamiokande detector with its much improved sensitivity and lower threshold could still detect a deviation; this would be a clear indication for neutrino oscillations or other neutrino-related novel phenomena.

For the neutrino lines from $e^- + {}^7\mathrm{Be} \rightarrow {}^7\mathrm{Li} + \nu_e (861.8\,\mathrm{keV})$ and $e^- + {}^7\mathrm{Be} \rightarrow {}^7\mathrm{Li}^* + \nu_e (384.3\,\mathrm{keV})$ thermal broadening effects are of some interest because they dominate the line shape. A detailed discussion is found in Bahcall (1994) who calculated the observable spectra at Earth shown in Fig. 10.4 where E_0 is the energy of the transition in the laboratory. The peak of the lines is shifted to higher energies by 0.43 and 0.19 keV, respectively, while the average energy is shifted by an even larger amount because of the asymmetric form. Perhaps in some next-century detector this line shape could be used to measure the central temperature of the Sun.

10.2.2 Standard Solar Models

In order to calculate the expected solar neutrino flux at Earth one needs to construct a model of the Sun. A "standard solar model" is obtained by solving the stellar structure equations discussed in Sect. 1.2 in several time steps to evolve it to the solar age of 4.5×10^9 yr. At this point it must produce the observed present-day luminosity of the Sun of $1\,L_\odot = 3.85 \times 10^{33}\,\mathrm{erg\,s^{-1}}$ (it is about 30% brighter than a zero-age model). This agreement is enforced by tuning the unknown presolar helium abundance Y_initial to the required value which is usually found to be about 27%.

Another present-day boundary condition is the measured solar radius of $1\,R_\odot = 6.96 \times 10^{10}\,\mathrm{cm}$ which is adjusted by the mixing-length parameter which enters the standard treatment of convection. In the Sun, the outer layers (depth about 0.3 $R_\odot$) are found to be convective; in lower-mass main-sequence stars convection reaches deeper, in higher-mass ones it disappears entirely near the surface while the central region becomes convective. The Sun calibrates the mixing-length parameter which is then used in evolutionary calculations of other stars.

In order to calculate a standard solar model one needs a variety of calculated or measured input information, notably the photon opacities, the equation of state, nuclear cross sections with an appropriate screening prescription, diffusion coefficients, a prescription to treat convection, the abundances of metals (elements heavier than helium), and the solar age, luminosity, and radius. Because of the large number of details that have to be minded it is sometimes difficult to compare exactly the solar models of different workers. However, wherever a detailed comparison has been performed, different results for such output quantities as predicted neutrino fluxes usually can be understood in terms of different choices regarding some of the input physics (e.g. Bahcall and Pinsonneault 1992; Turck-Chièze and Lopes 1993). Thus, there is very little ambiguity in solar modelling for a common choice of input physics.

Nonstandard solar models are ones where standard input parameters have been chosen outside the range of recognized uncertainties (for example the opacities), or which include entirely new physical effects such as strong magnetic fields, fast rotation in the deep interior, nonstandard nuclear reaction rates (involving free quarks, for example), energy transfer by a nonstandard mechanism (e.g. by trapped massive weakly interacting particles), and others. Some such possibilities can be excluded by the helioseismologically determined sound-speed profile of the Sun. I will not discuss nonstandard solar models any further—for an overview see Bahcall (1989). Naturally, it is always possible that a new "nonstandard" effect is recognized which could be important for the solar structure and neutrino fluxes.

One standard physical effect that has made its way into standard solar models only recently is the gravitational settling of helium and metals. This effect leads to a stronger concentration of helium in the central region than is caused by nuclear burning alone so that the present-day Sun is "more evolved." Consequently, the central temperature is slightly higher, leading to an increased flux of boron neutrinos. The settling of metals leads to an increased opacity, causing a further increase of the temperature and of the neutrino fluxes by a similar amount. Solar models with gravitational settling of helium significantly improve the already good agreement with the helioseismologically inferred sound-speed profile (Christensen-Dalsgaard, Proffitt, and Thompson 1993) while the settling of metals does not seem to have a strong additional impact on the p-mode frequencies (Proffitt 1994).

Bahcall and Pinsonneault (1992) were the first to include helium settling in a standard solar model; in their 1995 paper they included metal settling as well. They find that the compound effect is to change the ^{8}B flux by +36%, the ^{7}Be flux by +14%, and the *pp* flux by −1.7%. Similar changes (+31%, +13%, and −1.7%) were found by Proffitt (1994) while Kovetz and Shaviv (1994), who included only helium settling, obtained smaller effects relative to the corresponding case of Bahcall and Pinsonneault (1992).

While helium and metal diffusion increases the neutrino fluxes, the changes are roughly within the claimed errors of previous standard solar model predictions. Many improvements of input physics over the years have left the neutrino flux predictions of Bahcall and his collaborators surprisingly stable over 25 years (Bahcall 1989, 1995). Bahcall (1994) has compiled the central temperature predictions from a heterogeneous set of 12 standard solar models without diffusion calculated by different authors since 1988. The temperature predictions are almost uniformly distributed on the interval $15.40 - 15.72 \times 10^6$ K, i.e. these authors agree with each other on the value 15.56×10^6 K within $\pm 1\%$.

Thus, in spite of differences in detail there exists a broad consensus on what one means with a standard solar model. Therefore, the neutrino fluxes of the Bahcall and Pinsonneault (1995) model with element diffusion (Tab. 10.1 and Fig. 10.1) can be taken to be representative. The general agreement on a standard solar model does not guarantee, of course, that there might not exist problems related to incorrect standard assumptions or incorrect input parameters common to all such models.

10.2.3 Uncertainties of Standard Neutrino Predictions

a) Opacities

In order to compare the neutrino flux predictions with the experimental measurements one needs to develop a sense for the reliability of the calculations. Naturally, it is impossible to quantify the probability for the operation of some hitherto unknown physical effect that might spoil the predictions; an error analysis can only rely on the recognized uncertainties of standard input physics. Very detailed error analyses can be found for the standard solar models of Bahcall and Pinsonneault (1992) and of Turck-Chièze and Lopes (1993).

As for solar modelling, the dominating uncertainty arises from the radiative opacities which largely determine the temperature profile of the Sun. A reduction of the Rosseland mean opacity in the central region by 10% reduces the temperature by about 1%.

There is not a one-to-one correspondence between the central solar temperature T_c and the neutrino fluxes because it is not possible to adjust T_c and leave all else equal. Conversely, one must modify some input parameters and evolve a self-consistent solar model. Still, if one allows all input parameters to vary according to a distribution determined by their measured or assumed uncertainties one finds a strong correlation between T_c and the neutrino fluxes which allows one to understand the impact of certain modifications of a solar model on the neutrino fluxes via their impact on T_c. Bahcall (1989) found

$$\text{Neutrino Flux} \propto \begin{cases} T_c^{-1.2} & \text{for } pp, \\ T_c^{8} & \text{for } {}^{7}\text{Be}, \\ T_c^{18} & \text{for } {}^{8}\text{B}. \end{cases} \tag{10.5}$$

It is noteworthy that the pp flux decreases with increasing T_c because of the constraint imposed by the solar luminosity. One concludes that a 1% uncertainty in T_c translates roughly into a 20% uncertainty of the boron flux.

There are two sources of uncertainty for the opacity. First, for an assumed chemical composition, the actual opacity calculation which involves complicated details of atomic and plasma physics. Second, there is the uncertain metal content in the central region of the Sun (Z/X). Because iron retains several bound electrons even for the conditions at the solar center it contributes substantially to the Rosseland mean opacity; removing iron entirely would reduce it by 25–30%.

For a recent calculation and detailed discussion of solar opacities see Iglesias and Rogers (1991a). In view of the relatively small deviations between different opacity calculations for the relevant conditions Turck-Chièze and Lopes (1993) as well as Bahcall and Pinsonneault (1992) agree that the radiative opacities are likely calculated with a precision of better than a few percent.[53]

The actual amount of heavy elements in the Sun, notably iron, is determined by spectroscopic measurements in the photosphere, and by the abundance in meteorites which are assumed to represent the presolar material. The results from both methods seem to agree essentially on a common value—see Bahcall and Pinsonneault (1995) for a summary of the recent status. They believe that the uncertainty in the ${}^{8}\text{B}$ flux caused by the uncertainty of Z/X is about 8%. Of course, in the central regions of the Sun the metal abundance is also determined by gravitational settling as discussed above. The uncertainty of the ${}^{8}\text{B}$ flux inherent in the treatment of gravitational settling is thought to be 8% as well.

---

[53]However, most recently Tsytovich et al. (1995) have claimed that relativistic corrections to the electron free-free opacity as well as a number of other hitherto ignored effects reduce the standard total opacity by as much as 5%.

Bahcall and Pinsonneault (1995) then find that the opacity-related uncertainty of the ^{8}B neutrino flux is about 12%. For the ^{7}Be flux the errors are thought to be roughly half as large, in agreement with Eq. (10.5). However, as stressed by Bahcall and Pinsonneault (1995), the interpretation as an effective 1σ error is misleading as the main source of uncertainty is of a systematic and theoretical nature. Previously, these authors had stated "theoretical 3σ errors;" in that sense an opacity-related uncertainty of the ^{8}B flux of about 30% was found in Bahcall and Pinsonneault (1992). Turck-Chièze and Lopes (1993) adopted 15% without a commitment to a specific number of sigmas. None of these errors can be interpreted in a strict statistical sense. Rather, they give one an idea of what the workers in that field consider a plausible range of possibilities.

b) Beryllium-Proton Reaction

A dominating uncertainty for the important flux of boron neutrinos arises from the cross section ^{7}Be $+ p \rightarrow {}^8$B $+ \gamma$ which plays no role whatsoever for the energy generation in the Sun because the PPIII termination of the pp chain is extremely rare. Therefore, a modification of this cross section has no impact on the structure of the Sun and thus no other observable consequence but to modify the high-energy neutrino flux.

The cross section for this reaction is parametrized for low energies in the usual form with an astrophysical S-factor

$$\sigma(E) = S(E) E^{-1} e^{-2\pi\eta(E)}, \tag{10.6}$$

with the Sommerfeld parameter

$$\eta = Z_1 Z_2 e^2 v^{-1}. \tag{10.7}$$

Here, $Z_{1,2}e$ are the charges of the reaction partners, v their relative velocity, and E their CM kinetic energy. The S-factor is expected to be essentially constant at low energies unless there is a resonance near threshold.

The six "classical" measurements of the ^{7}Be $+ p \rightarrow {}^8$B $+ \gamma$ reaction are referenced in Tab. 10.3. In the Sun, the most effective energy range is around $E = 20\,\mathrm{keV}$, far in the tail of the thermal distributions of the reaction partners, but still far below the lowest laboratory energies of around $120\,\mathrm{keV}$ in the experiments of Kavanagh et al. (1969) and Filippone et al. (1983). Therefore, one must extrapolate the factor

Table 10.3. Extrapolation of S_{17} to $E = 0$ for measurements of $^7\text{Be}\,p \to {}^8\text{B}\,\gamma$ according to Johnson et al. (1992).

Experiment	$S_{17}(0)$ [eV b]
Kavanagh et al. (1960)	15 ± 6
Parker (1966, 1968)	27 ± 4
Kavanagh et al. (1969)	25.2 ± 2.4
Vaughn et al. (1970)	19.4 ± 2.8
Wiezorek et al. (1977)	41.5 ± 9.3
Filippone et al. (1983)	20.2 ± 2.3

S_{17} down to the astrophysically interesting regime. The most recent comprehensive reanalysis was performed by Johnson et al. (1992) who found the values listed in Tab. 10.3. They tend to be smaller by around 10% relative to previous extrapolations which did not take into account that for laboratory energies there is a contribution from d-waves in the entrance channel. Johnson et al. used two different interaction models for the extrapolation which yielded identical results within 2%. This does not necessarily imply that the theoretical extrapolation is known with this precision; for example, Riisager and Jensen (1993) suggest much smaller values for $S_{17}(0)$.

As stressed by Johnson et al. (1992) it is problematic to combine the results of Tab. 10.3 to a "world average" because they show systematic discrepancies. In fact, the data of the two "low-energy" experiments (Kavanagh et al. 1969; Filippone et al. 1983) agree very well with each other over a wide range of energies with a systematic offset by a factor 1.34. A similar offset exists between the "high-energy" data of Parker (1966, 1968) and Vaughn et al. (1970) with a factor 1.42 (Gai 1995).

Because it is unknown which of the experiments is right (if any) Johnson et al. (1992) combined the data according to a prescription adopted by the Particle Data Group (1994) for such cases. It amounts to the usual weighted average, but increasing the error by a certain factor derived from the statistical significance of the discrepancies. Johnson et al. then arrive at a world average

$$S_{17}(0) = (22.4 \pm 2.1)\,\text{eV b}. \tag{10.8}$$

Bahcall and Pinsonneault (1992, 1995) used this value, i.e. they used a 1σ uncertainty of 9%. Turck-Chièze and Lopes (1993) used $(22.4 \pm 1.3_{\text{stat}} \pm 3.0_{\text{syst}})\,\text{eV b}$, i.e. they adopted an uncertainty of $\pm 15\%$.

Xu et al. (1994) have discussed a unique relationship between $S_{17}(0)$ and the nuclear vertex constant of the overlap wave function for the virtual decay $^8B \rightarrow {}^7Be+p$; the nuclear vertex constant can be predicted from other nuclear data. These authors' calculation of $S_{17}(E)$ agrees remarkably well with the data points of Filippone et al. (1983) at low energies, and with Vaughn et al. (1970) at higher energies; these are the experiments which gave a low $S_{17}(0)$. Xu et al. (1994) find an even lower value of $S_{17}(0) \approx 17.6 \, \text{eV b}$.

On the other hand, Brown, Csótó, and Sherr (1995) studied the relationship between the Coulomb displacement energy for the $A = 8$, $J = 2^+$, $T = 1$ state and S_{17}. They found a high value $S_{17}(20 \, \text{keV}) = (26.5 \pm 2.0) \, \text{eV b}$.

The $S_{17}(0)$ factor was recently determined from the Coulomb dissociation of ^8B, i.e. by the Primakoff-type process $^8B \rightarrow {}^7Be + p$ in the electric field of a ^{208}Pb nucleus (Motobayashi et al. 1994). These authors find a very low preliminary value of $S_{17}(0) = (16.7 \pm 3.2) \, \text{eV b}$. Langanke and Shoppa (1994) think that the true value may be significantly lower still if allowance is made for a possible E2 amplitude in the Primakoff reaction. However, this possibility is heavily disputed by Gai and Bertulani (1995); see also the reply by Langanke and Shoppa (1995).

All of these new results are somewhat preliminary at the present time. Independently of their ultimate status it is evident that the S_{17} factor must be considered the weakest link in the prediction of the boron neutrino flux.

10.3 Observations

10.3.1 Absorption Reactions for Radiochemical Experiments

The longest-running solar neutrino experiment is the Homestake chlorine detector which is based on a reaction proposed by Pontecorvo (1948) and Alvarez (1949),

$$\nu_e + {}^{37}Cl \rightarrow {}^{37}Ar + e \qquad \text{(threshold } 0.814 \, \text{MeV)}. \qquad (10.9)$$

The two other data-producing radiochemical experiments use gallium as a target according to the reaction

$$\nu_e + {}^{71}Ga \rightarrow {}^{71}Ge + e \qquad \text{(threshold } 0.233 \, \text{MeV)}. \qquad (10.10)$$

Because of its low threshold, the gallium experiments can pick up the solar pp neutrino flux. The absorption cross sections as a function of

neutrino energy are shown in Fig. 10.5 according to the tabulation of
Bahcall and Ulrich (1988).

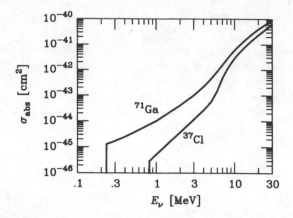

Fig. 10.5. Neutrino absorption cross section on ^{37}Cl and ^{71}Ga according to
Bahcall and Ulrich (1988).

Because the absorption cross sections are steeply increasing functions of energy while the predicted solar neutrino spectrum (Fig. 10.1)
steeply decreases, the low flux of boron neutrinos yields the dominant
contribution to the expected counting rate for chlorine, and a sizeable
contribution to gallium. This is illustrated in Fig. 10.6 where the predicted counting rates from solar neutrinos, integrated between energy
E_ν and infinity, are shown as a function of E_ν. These plots correspond
to the lower panel of Fig. 10.1 if the differential flux is weighted with
the relevant absorption cross section. It is customary to express the
absorption rate per nucleus in "solar neutrino units"

$$1\,\mathrm{SNU} = 10^{-36}\,\mathrm{s}^{-1}, \tag{10.11}$$

not to be confused with 1 SNu, the supernova unit, which quantifies
the rate of supernova occurrences in a galaxy. Because one measures
a fully integrated flux one only needs the absorption cross sections
folded with the spectra of the individual source reactions. For ^{37}Cl and
^{71}Ga they are given in Tab. 10.4. Multiplying the predicted fluxes with
these cross sections and applying a factor 10^{36} gives the absorption rate
in SNUs.

Table 10.4. Spectrally averaged absorption cross sections on ^{37}Cl and ^{71}Ga for different solar source reactions according to Bahcall (1989), except for ^{8}B neutrinos on ^{37}Cl which is according to Garcia et al. (1991). The unit is 10^{-46} cm^2.

	pp	pep	^{7}Be	^{8}B	^{13}N	^{15}O
^{37}Cl	0	16	2.4	10,900	1.7	6.8
^{71}Ga	11.8	215	73.2	24,300	61.8	116

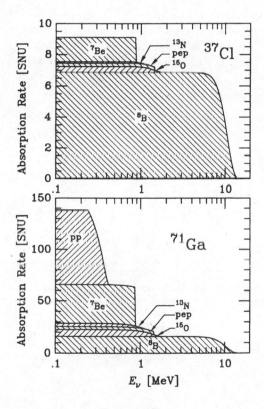

Fig. 10.6. Absorption rate by ^{37}Cl and ^{71}Ga of the solar neutrino flux at Earth, integrated from a given neutrino energy to infinity. The predicted flux is according to Bahcall and Pinsonneault (1995), the absorption cross sections according to Bahcall and Ulrich (1988)—see Fig. 10.5.

10.3.2   Chlorine Detector (Homestake)

Solar neutrino observations were pioneered by the chlorine detector of Davis (1964) which is located in the Homestake Gold Mine at Lead, South Dakota (U.S.A.). The target consists of about 615 tons of perchloroethylene (C_2Cl_4), a cleaning fluid, from which argon is extracted every few months by an intricate chemical procedure. Because ^{37}Ar has a half-life of 35.0 days the amount of neutrino-produced ^{37}Ar begins to saturate after a couple of months whence a longer exposure time is not warranted. The extracted argon—only a few atoms in a small amount of carrier gas—is then viewed by a proportional counter which registers the Auger electrons which are ejected when ^{37}Ar decays by electron capture. The introduction in 1970 of an electronic system which analyzes the pulse rise time greatly enhanced the sensitivity. Therefore, usually only the results after 1970 are quoted, beginning with run 18. Except for a period from May 1985 to October 1986 where the experiment was down due to successive electrical failures of the circulation pumps, data have been taken continuously since 1967.

The known backgrounds for the experiment are cosmic-ray produced ^{37}Ar atoms which correspond to (0.29 ± 0.08) SNU, and an average neutron background corresponding to (0.13 ± 0.13) SNU. Because these average backgrounds as well as the backgrounds of the proportional

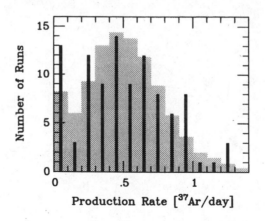

Fig. 10.7. Distribution of the counting rates of 99 runs (18−117). The black bars represent the data shown in Fig. 10.7, the shaded histogram the expected distribution (adapted from Lande 1995). The absorption rate in SNU is obtained by multiplying the argon production rate by 5.31.

Table 10.5. Predicted absorption rate (in SNU) by ^{37}Cl for different solar source reactions. BP92 give "theoretical 3σ" uncertainties.

	Diffusion	pp	pep	^{7}Be	^{8}B	^{13}N	^{15}O	Total
BP95	He, metals	0.0	0.2	1.2	7.2	0.1	0.4	$9.3^{+1.2}_{-1.4}$
BP92	He	0.0	0.2	1.2	6.2	0.1	0.3	8.0 ± 3.0
BP92	—	0.0	0.2	1.2	5.5	0.1	0.2	7.2 ± 2.7
TL93	—	0.0	0.2	1.1	4.6	0.1	0.2	6.4 ± 1.4

BP92 = Bahcall and Pinsonneault (1992).
BP95 = Bahcall and Pinsonneault (1995).
TL93 = Turck-Chièze and Lopes (1993).

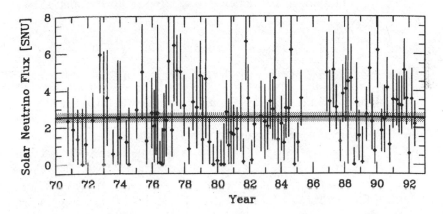

Fig. 10.8. Solar neutrino flux at the Homestake experiment ($\nu_e\ ^{37}$Cl $\rightarrow$ ^{37}Ar e) for the past quarter century. The shaded band indicates the 1σ uncertainty of the global best-fit average neutrino flux (Lande 1995).

counters must be subtracted, the counting rate attributed to solar neutrinos in a given run is sometimes found to be formally negative. In those cases, a zero counting rate is adopted. The distribution of counting rates for 99 runs (18−117) is shown in Fig. 10.7. The global average given in Eq. (10.12) is based on a maximum-likelihood analysis which includes the background. It would not be correct to average the data points with individual background subtractions to obtain a global average signal.

The recognized systematic errors are 1.5% for the extraction efficiency, 3% for the proportional counter efficiency, 3% for the cosmic-ray background, 5% for the neutron background, and 2% for the proportional counter background, which amounts to a total of 7% or 0.18 SNU.

The measured counting rate for the individual runs at Homestake is shown in Fig. 10.8. The current global best-fit average for the solar neutrino flux measurement, derived from runs 18−124 is (Lande 1995)

$$(2.55 \pm 0.17_{\text{stat}} \pm 0.18_{\text{syst}}) \, \text{SNU} = (2.55 \pm 0.25) \, \text{SNU} \qquad (10.12)$$

where the errors were combined in quadrature. The average argon production rate per day in the detector is obtained by dividing the SNUs by 5.31 so that it is found to be 0.48 ^{37}Ar/day. Eq. (10.12) is to be compared with the predictions from different solar models shown in Tab. 10.5.

10.3.3   Gallium Detectors (SAGE and GALLEX)

The gallium experiments involve far more complicated chemical extraction procedures for the neutrino-produced ^{71}Ge which is not a noble gas. The Soviet-American (now Russian-American) Gallium Experiment (SAGE) used at first 27 tons, later 55 tons of metallic gallium while the European GALLEX collaboration uses 100 tons of an aqueous gallium chloride solution, corresponding to 30.3 tons of gallium. The SAGE experiment is located in the Baksan Neutrino Observatory in Mount Andyrchi, Caucasus Mountains (Russia) while GALLEX is located in the Gran Sasso tunnel near Rome (Italy). SAGE has been taking data since January 1990, GALLEX since May 1991. The current results of SAGE were published by Abdurashitov et al. (1994) and Gavrin (1995), those of GALLEX by the GALLEX Collaboration (1994, 1995b) and Kirsten (1995).

The half-life of ^{71}Ge is 11.43 days so that one needs only about a three-week exposure, allowing for frequent extractions. SAGE has already accumulated a total of 21 analyzed runs, GALLEX a total of 39. In Fig. 10.9 the individual counting rates are shown as well as the global best-fit averages and the distribution of counting rates in 25 SNU bins. For the SAGE data, formally negative rates after background subtraction are forced to zero. In both cases, recognized background signals of 7−8 SNU were subtracted. The average counting rates attributed to solar neutrinos are found to be (in SNU)

SAGE: $74 + 13/-12_{stat} + 5/-7_{syst} = 74 \pm 14,$

GALLEX: $77 \pm 9_{stat} + 4/-5_{syst} = 77 \pm 10,$ (10.13)

where systematic and statistical errors were added in quadrature. These results are to be compared with the predictions from standard solar models shown in Tab. 10.6.

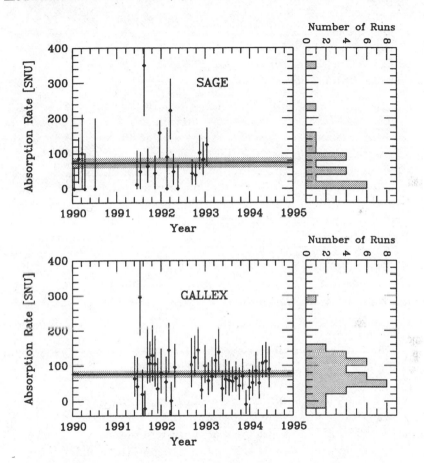

Fig. 10.9. Solar neutrino flux at the gallium detectors ($\nu_e\,^{71}\mathrm{Ga} \rightarrow {}^{71}\mathrm{Ge}\,e$). The shaded bands indicate a 1σ uncertainty of the global best-fit average neutrino fluxes (statistical and systematic errors added in quadrature). In the SAGE data, formally negative fluxes are forced to zero. In the right panels, the bin size is 20 SNU.

Table 10.6. Predicted absorption rate (in SNU) by ^{71}Ga for different solar source reactions. BP92 give "theoretical 3σ" uncertainties.

	Diffusion	pp	pep	^{7}Be	^{8}B	^{13}N	^{15}O	Total
BP95	He, metals	69.7	3.0	37.7	16.1	3.8	6.3	137^{+8}_{-7}
BP92	He	70.8	3.1	35.8	13.8	3.0	4.9	131.5^{+21}_{-17}
BP92	—	71.3	3.1	32.9	12.3	2.7	4.3	127^{+19}_{-16}
TL93	—	71.1	3.0	30.9	10.8	2.4	3.7	122.5 ± 7

BP92 = Bahcall and Pinsonneault (1992).
BP95 = Bahcall and Pinsonneault (1995).
TL93 = Turck-Chièze and Lopes (1993).

The GALLEX experiment was subjected to an "on-off test" by virtue of a laboratory ν_e source strong enough to "outshine" the Sun in its local neutrino flux (GALLEX collaboration 1995a). To this end a container with activated chromium was inserted in the center of the tank containing the target fluid. The relevant isotope is ^{51}Cr which decays to ^{51}V with a half-life of 27.71 d by electron capture. The ν_e spectrum consists of four monoenergetic lines of energies 426 keV (9%), 431 keV (1%), 746 keV (81%), and 751 keV (9%). The dominating line is very close to the solar beryllium line. An analysis of the first seven exposures reveals a ratio between the measured and expected counting rate of 1.04 ± 0.12. At the present time, four further extractions are still being analyzed. Meanwhile, the source is being reactivated at the Siloé nuclear reactor in Grenoble (France) in order to perform further exposures.

The GALLEX collaboration has interpreted the source experiment as a global test of their detector efficiency. Most recently, Hata and Haxton (1995) have advocated a somewhat different view. They argue that the gallium absorption cross section for the 746 keV line is poorly known because of excited-state contributions which have not been directly measured, and which are more uncertain than had been acknowledged in the previous literature. According to Hata and Haxton the source experiment should not be taken as measuring, say, the GALLEX extraction efficiency but rather as measuring the excited-state contributions to the absorption cross section for beryllium neutrinos.

10.3.4 Water Cherenkov Detector (Kamiokande)

A water Cherenkov detector like the one located in the Kamioka metal
mine (Gifu prefecture, Japan) is a large body of water, surrounded
by photomultipliers which register the Cherenkov light emitted by relativistic charged particles. Solar neutrinos are detected by virtue of
their elastic scattering on the electrons bound in the water molecules,

$$\nu + e \rightarrow e + \nu. \tag{10.14}$$

This process has no significant threshold, although in practice the detection of the electrons is background-limited to relatively large energies.
The Kamiokande detector began its measurement of solar neutrinos
in January 1987 with an effective analysis threshold of about 9 MeV
which was reduced to about 7 MeV in mid 1988. The much larger
Superkamiokande detector, which is scheduled to begin data-taking in
April 1996, will have a threshold of about 5 MeV.

In contrast with the radiochemical detectors which are effectively
based on the charged-current reaction $\nu_e + n \rightarrow p + e$, the elastic
scattering on electrons is sensitive to all (left-handed) neutrinos and
antineutrinos. The cross section is given by the well-known formula
(e.g. Commins and Bucksbaum 1983)

$$\frac{d\sigma_{\nu e}}{dy} = \frac{G_F^2 m_e E_\nu}{2\pi} \left[A + B(1-y)^2 - C y \frac{m_e}{E_\nu} \right], \tag{10.15}$$

where G_F is the Fermi constant and the coefficients A, B, and C are
tabulated in Tab. 10.7. Further,

$$y \equiv \frac{E_\nu - E_\nu'}{E_\nu} - \frac{T_e}{E_\nu} \quad \text{and} \quad 0 < y < \frac{2E_\nu}{2E_\nu + m_e}, \tag{10.16}$$

where E_ν and E_ν' are the initial- and final-state neutrino energies while
T_e is the kinetic energy of the final-state electron.

Table 10.7. Coefficients in Eq. (10.15) for elastic neutrino electron scattering.

Flavor	A	B	C
ν_e	$(C_V + C_A + 2)^2$	$(C_V - C_A)^2$	$(C_V + 1)^2 - (C_A + 1)^2$
$\bar{\nu}_e$	$(C_V - C_A)^2$	$(C_V + C_A + 2)^2$	$(C_V + 1)^2 - (C_A + 1)^2$
$\nu_{\mu,\tau}$	$(C_V + C_A)^2$	$(C_V - C_A)^2$	$C_V^2 - C_A^2$
$\bar{\nu}_{\mu,\tau}$	$(C_V - C_A)^2$	$(C_V + C_A)^2$	$C_V^2 - C_A^2$

$C_V = -\frac{1}{2} + \sin^2 \Theta_W \approx -0.04, \quad C_A = -\frac{1}{2}.$

For $E_\nu > 5\,\text{MeV}$, i.e. for energies above the detection threshold of Kamiokande and Superkamiokande, the total cross section is well approximated by

$$\sigma_{\nu e} \approx \frac{G_F^2 m_e E_\nu}{2\pi}\left(A + \tfrac{1}{3}B\right) = 9.5 \times 10^{-44}\,\text{cm}^2\,\frac{E_\nu}{10\,\text{MeV}} \times \begin{cases} 1 & \text{for } \nu_e, \\ \frac{1}{2.4} & \text{for } \bar{\nu}_e, \\ \frac{1}{6.2} & \text{for } \nu_{\mu,\tau}, \\ \frac{1}{7.1} & \text{for } \bar{\nu}_{\mu,\tau}. \end{cases}$$

$$(10.17)$$

Of course, for a flux of $\bar{\nu}_e$'s such as that from a supernova collapse, the dominant signal in a water Cherenkov detector is from the charged-current process $\bar{\nu}_e + p \to n + e^+$ (Chapter 11).

The νe elastic scattering cross section is strongly forward peaked for $E_\nu \gg m_e$. One easily finds

$$\frac{d\sigma}{d\cos\theta} = 4\,\frac{m_e}{E_\nu}\,\frac{(1 + m_e/E_\nu)^2 \cos\theta}{[(1 + m_e/E_\nu)^2 - \cos^2\theta]^2}\,\frac{d\sigma}{dy}, \qquad (10.18)$$

where $d\sigma/dy$ was given in Eq. (10.15) with

$$y = \frac{2\,(m_e/E_\nu)\cos^2\theta}{(1 + m_e/E_\nu)^2 - \cos^2\theta}. \qquad (10.19)$$

Here, θ is the angle between the direction of motion of the final-state electron relative to the incident neutrino; one finds $0 \leq \cos\theta \leq 1$. For $E_\nu = 5\,\text{MeV}$ and $10\,\text{MeV}$ this cross section is shown in Fig. 10.10, normalized to unity for $\theta = 0$. The electron keeps the direction of the incident neutrino within $\theta \lesssim (2m_e/E_\nu)^{1/2}$.

In a water Cherenkov detector, the direction of the charged particles can be reconstructed from the ring of Cherenkov light hitting the photomultipliers. Therefore, the direction of the incident neutrino is known within an uncertainty which is determined by the angular distribution of Fig. 10.10, and by the random motion of the electron due to multiple scattering in the Coulomb fields of the medium constituents. At an electron energy of $10\,\text{MeV}$, the Kamiokande detector has an angular resolution of about $28°$ for the direction of the electron. For this energy, the electron keeps the neutrino direction within about $18°$ so that the uncertainty is dominated by the electron Coulomb scattering.

Apart from these effects which blur the reconstruction of the incident neutrino direction, a water Cherenkov detector is an imaging

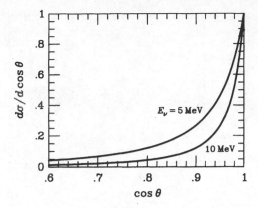

Fig. 10.10. Differential scattering cross section for $\nu_e e \to e\nu_e$ according to Eq. (10.18), normalized to unity for $\theta = 0$, where θ is the angle between the incident neutrino and final-state electron.

device and thus a true "neutrino telescope." Thus, for the first time the Kamiokande detector actually proved that neutrinos are coming from the direction of the Sun. The first main publication of these results was by Hirata et al. (1991) from a data sample of 1040 live detector days, taken between 1987 and 1990; an update including data until July 1993 (a total of 1670 live detector days) was given by Suzuki (1995). The angular distribution of the registered electrons relative to the direction of the Sun is shown in Fig. 10.11. Even though there remains an isotropic background, probably from radioactive impurities in the water, the solar neutrino signal beautifully shows up in these measurements.

In elastic neutrino-electron scattering, the energy distribution of the kicked electrons is nearly flat so that the spectral shape of the incident neutrino flux is only indirectly represented by the measured electron spectrum. In Fig. 10.12 the electron recoil spectrum is shown for an incident spectrum of ^{8}B neutrinos. A water Cherenkov detector can resolve the energy of the charged particles from the intensity of the measured light which for low-energy electrons is roughly proportional to the energy. Therefore, the electron recoil spectrum from the interaction with solar neutrinos can be resolved at Kamiokande. The measured shape relative to the theoretically expected one is shown in Fig. 10.13, arbitrarily normalized at $E_\nu = 9.5\,\text{MeV}$ (Suzuki 1995). Within statistical fluctuations the agreement is perfect.

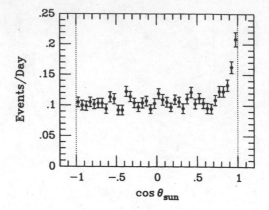

Fig. 10.11. Angular distribution of the solar neutrino events at Kamiokande for 1670 live detector days (January 1987–July 1993), according to Suzuki (1995). There remains an isotropic background, probably from radioactive impurities in the water.

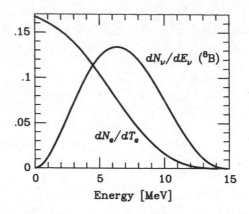

Fig. 10.12. Normalized spectra of ^{8}B neutrinos and of recoil electrons (kinetic energy T_e) from the elastic scattering process $\nu_e e \to e \nu_e$.

Even though the Sun is thought to produce only ν_e's one may speculate that some of them are converted into $\bar{\nu}_e$'s on their way to Earth by spin-flavor oscillations in magnetic fields (Sect. 8.4) or by matter-induced majoron decays (Sect. 6.8). The $\bar{\nu}_e$'s would cause a signal by the reaction $\bar{\nu}_e p \to n e^+$ with an isotropic angular distribution. The remaining measured isotropic background shown in Fig. 10.11 can thus

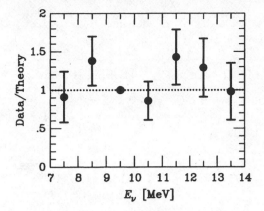

Fig. 10.13. Measured over expected spectrum of recoil electrons from solar neutrinos at Kamiokande, arbitrarily normalized to unity at $E_\nu = 9.5\,\text{MeV}$ (Suzuki 1995).

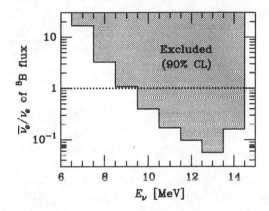

Fig. 10.14. Limit on a solar $\bar{\nu}_e$ flux relative to ^{8}B solar ν_e's from the isotropic background shown in Fig. 10.11. The data used for this result extend until August 1992 (Suzuki 1993).

be used to constrain the $\bar{\nu}_e$ flux. At different energies a 90% CL upper limit on the solar $\bar{\nu}_e$ flux relative to ^{8}B solar neutrinos is shown in Fig. 10.14 according to Suzuki (1993). The best relative limit is at $E_\nu = 13\,\text{MeV}$ where the $\bar{\nu}_e$ flux is less than 5.8% of ^{8}B solar ν_e's at 90% CL. If the cause of the isotropic background could be reliably identified these upper limits could be improved accordingly.

Table 10.8. Predicted flux of ^{8}B neutrinos at Earth. For BP92, the uncertainty is a "theoretical 3σ error."

Model	Diffusion	^{8}B Flux $[10^6\,\mathrm{cm}^{-2}\,\mathrm{s}^{-1}]$
BP95	He, metals	$6.6^{+0.9}_{-1.1}$
BP92	He	5.7 ± 2.4
BP92	—	5.1 ± 2.2
TL93	—	4.4 ± 1.1

BP95 = Bahcall and Pinsonneault (1995).
BP92 = Bahcall and Pinsonneault (1992).
TL93 = Turck-Chièze and Lopes (1993).

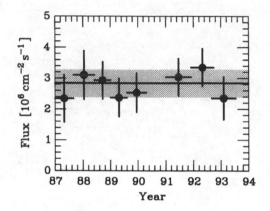

Fig. 10.15. Measured ^{8}B solar neutrino flux at Kamiokande (Suzuki 1993). The shaded band gives the average with 1σ statistical and systematic errors added in quadrature.

Next, one may determine the absolute flux of solar ^{8}B neutrinos. As a function of time it is shown in Fig. 10.15. Assuming that the spectrum is indeed that of ^{8}B neutrinos, the measured flux from the data of January 1987 until July 1993 is (units $10^6\,\mathrm{cm}^{-2}\,\mathrm{s}^{-1}$)

$$2.89 + 0.22/-0.21_{\mathrm{stat}} \pm 0.35_{\mathrm{syst}} = 2.89 \pm 0.41 \qquad (10.20)$$

where the errors were added in quadrature. This measurement is to be compared with the predictions shown in Tab. 10.8

Table 10.9. Current and near-future solar neutrino experiments.

Experiment	Operation[a] from	until	Detection	Thresh. [MeV]	Solar flux	Measurement[c]	Prediction BP95	TL93
Homestake	1967	(?)	Radiochemical $\nu_e \,^{37}Cl \to \,^{37}Ar\, e$	0.814	not pp	2.55 ± 0.25 SNU	$9.3^{+1.2}_{-1.4}$	6.4 ± 1.4
SAGE	1990	(?)	Radiochemical $\nu_e \,^{71}Ga \to \,^{71}Ge\, e$	0.233	all	74 ± 14 SNU	137^{+8}_{-7}	123 ± 7
GALLEX	1991	(?)	...	...	...	77 ± 10 SNU	...	...
Kamiokande II	1987	(1996)	H_2O Cherenkov $\nu_x e \to e \nu_x$ (b)	7	^8B	2.89 ± 0.41 $\times 10^6$ cm^{-2} s^{-1}	$6.6^{+0.9}_{-1.1}$	4.4 ± 1.1
Superkamiok.	(1996)	(?)	...	5	...	—	...	...
SNO	(1996)	(?)	D_2O Cherenkov $\nu_e d \to p p e$ $\nu_x e \to e \nu_x$ (b) $\nu_x d \to p n \nu_x$ (b)	(1.44) 5 2.225	^8B	—	...	...
BOREXINO	(1997)	(?)	Scintillator $\nu_x e \to e \nu_x$ (b)	0.25	^7Be	— $\times 10^9$ cm^{-2} s^{-1}	$5.2^{+0.3}_{-0.4}$	—

[a] In brackets anticipated.   [b] $\nu_x = \nu_e, \nu_\mu,$ or ν_τ.   [c] Statistical and systematic 1σ errors added in quadrature.
BP95 = Bahcall and Pinsonneault (1995), with helium and metal diffusion.
TL93 = Turck-Chièze and Lopes (1993), no diffusion.

10.3.5   Summary

The main features and results of the solar neutrino experiments discussed in this section, and of near-future experiments to be discussed in Sect. 10.9, are summarized in Tab. 10.9. More technical aspects can be found in the original papers quoted in this section and in previous papers by the referenced authors. Overviews of many experimental aspects can be found in Bahcall (1989), Davis, Mann, and Wolfenstein (1989), and Koshiba (1992). The two theoretical predictions shown are somewhat extreme in that TL93 yields lowish neutrino fluxes when compared with BP92 (no diffusion), while BP95 with the inclusion of helium and metal diffusion is presently at the upper end of what is being predicted on the basis of standard solar models. It is clear, of course, that including diffusion would also increase the TL93 fluxes.

10.4   Time Variations

10.4.1   Day-Night Effect

It is commonly assumed that the Sun is in a stationary state so that the solar neutrino flux should be constant in time. One may still expect certain temporal variations of the measured flux. Between day and night the line of sight between the Kamiokande detector and the Sun intersects with different parts of the Earth. If neutrinos oscillate, for certain masses and mixing angles the Earth's matter would alter the oscillation pattern such that the counting rate at Kamiokande would be expected to vary between day and night. The day rate is found to be $0.90 \pm 0.10_{stat} \pm 0.12_{syst}$ times the average, the night rate $1.04 \pm 0.10_{stat} \pm 0.12_{syst}$, i.e. there is no significant difference (Suzuki 1995). The radiochemical detectors with their long exposure times cannot resolve a possible day-night difference.

10.4.2   Seasonal Variation

Because of the ellipticity of the Earth's orbit the distance to the Sun varies during the year from a minimum of 1.471×10^{13} cm in January to a maximum of 1.521×10^{13} cm in July, i.e. it varies by $\pm 1.67\%$ from its average during the year. Therefore, the solar neutrino flux varies by $\pm 3.3\%$ from average between January and July. This effect is too small to be observed by any of the present-day experiments.

This variation can be amplified if neutrinos oscillate. Notably, if the vacuum oscillation length of the monochromatic ^{7}Be neutrinos is of order the annual distance variation, the ^{7}Be flux measured in the chlorine and gallium detectors could vary between zero and its predicted full rate. Moreover, at different times of the year the neutrinos have to traverse on average a different amount of terrestrial matter which could affect neutrino oscillations and thus lead to an annual variation.

There is no evidence for such an effect in any of the detectors. GALLEX reports a counting rate of $(82 \pm 15_{stat})$ SNU for October–March and $(78 \pm 14_{stat})$ SNU for April–September, i.e. there is no significant difference (GALLEX collaboration 1994).

A semiannual variation could be caused by nonstandard neutrino interactions with the solar magnetic field (Sect. 10.7). The solar equatorial plane is at an angle of $7°15'$ relative to the Earth's orbital plane (the ecliptic). Around 7 June and 8 December, the solar core is viewed from Earth through the solar equator where the magnetic field is thought to be weaker than at higher latitudes. The Kamiokande II data (1040 live detector days in 1987–1990) were subdivided into three-months periods which include (rate Γ_I) or exclude (Γ_{II}) the intersection points, respectively. The relative difference in counting rate was found to be $(\Gamma_I - \Gamma_{II})/(\Gamma_I + \Gamma_{II}) = -0.06 \pm 0.11_{stat} \pm 0.02_{syst}$, i.e. there was no indication for a time variation (Hirata et al. 1991). The Homestake data also do not show any evidence for a semiannual variation.

10.4.3 Correlation with Solar Cycle at Homestake

The Sun shows a prominent magnetic activity cycle which is thought to be due to dynamo action within the convective surface layers, driven by the nonuniform rotation of the Sun (Stix 1989). One of the best-known manifestations of this activity is the cycle of sunspots, measured by their total number appearing on the solar disk.[54] A given solar cycle lasts for about 11 years from one minimum of sunspot number to the following; the first recorded cycle begins with the minimum around A.D. 1755. Sunspots are caused by magnetic flux tubes which break through the surface; they are thought to be manifestations of a subsurface toroidal magnetic field with opposite directions between the southern and northern hemisphere. In addition, the Sun has a poloidal (dipole) field. The fields reverse polarity after 11 years so that the full magnetic cycle lasts 22 years.

---

[54]In the following, the "number of sunspots" refers to the international sunspot index according to the Zürich system where both spots and spot groups are counted and averaged from the reports of many solar observatories. The sunspot index is regularly published in *Solar Geophysical Data*.

The solar neutrino measurements of the chlorine experiment seem to anticorrelate with solar activity as first suggested by Subramanian (1979) and Bazilevskaya, Stozhkov, and Charakhch'yan (1982). These latter authors also noted a correlation with the primary cosmic proton flux which, however, is known to correlate with the solar magnetic cycle because the solar magnetic field affects the flux of charged cosmic rays hitting the Earth. The cosmic-ray flux is not expected to affect the chlorine experiment directly; the average cosmic-ray induced ^{37}Ar production rate is about $0.055\,\mathrm{day}^{-1}$, compared with an average solar neutrino signal of $0.48\,\mathrm{day}^{-1}$. Figure 10.16 shows the temporal variation from 1970–1989 of the Homestake argon production rate, the integral number of sunspots, and the neutron counts at the McMurdo station in Antarctica. The latter are indicative of the primary cosmic-ray proton flux with energies above $0.4\,\mathrm{GeV}$.

While a certain degree of anticorrelation between the number of sunspots and the argon production rate is plainly visible in Fig. 10.16, there are two important questions: How significant is the effect? If it is significant, what are its physical origins?

As for the latter question, the only plausible explanation[55] that has been put forth over the years is the hypothesis that neutrinos possess a small magnetic moment which allows them to spin-precess into sterile right-handed states in the presence of magnetic fields (Voloshin and Vysotskiĭ 1986; Voloshin, Vysotskiĭ, and Okun 1986a,b), and variations of this scheme which allow for simultaneous spin and flavor oscillations (Sect. 10.7). In this case one would expect some degree of a semiannual variation as explained above, but none is found in the Homestake data.

The issue of statistical significance was addressed on the basis of the 1970–1989 data by Filippone and Vogel (1990), Bieber et al. (1990), and Bahcall and Press (1991). These latter authors argued for rank-ordering as a statistical method because of the likely nonlinear relationship (if any) with indicators of solar activity, and because of the possibly non-Gaussian nature of the experimental errors. They produced pairs of an observed argon production rate and the corresponding sunspot index. Next, the Homestake data and sunspot indices were replaced by the ordinal rank in their respective data sets, i.e. the smallest value has rank 1, the second-largest rank 2, and so forth. In order to discover correlations they applied two standard rank statistical tests, the Spear-

---

[55] The refractive term in the MSW conversion probability from convective currents could play a role and could be coupled to solar activity (Haxton and Zhang 1991). However, extreme conditions are required to explain the observed effect.

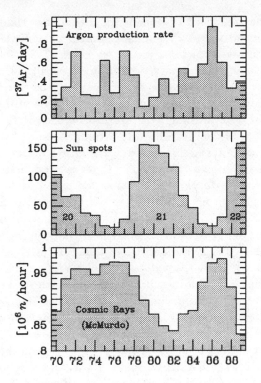

Fig. 10.16. ^{37}Ar production rate at Homestake, integral number of sunspots, and neutron counts at McMurdo (Antarctica) which are indicative of the primary cosmic-ray proton flux above 0.4 GeV; annual bins each. In the center panel, the number of the solar cycle is also indicated. (Adapted from Bieber et al. 1990.)

man rank-order correlation coefficient and Kendall's tau (e.g. Press et al. 1986). They found a moderate statistical significance level of 1.3% (Spearman) and 0.9% (Kendall) for the argon-production rate to be correlated with the sunspot index.[56]

Most recently, a correlation between the 1970–1991 argon production rate and the solar activity cycle was investigated by Oakley et al. (1994). They used the Mt. Wilson 150-ft. tower magnetograms as an indicator for the solar magnetic surface flux, both for the full solar disk

---

[56]The significance level gives approximately the probability that a random shuffling of the data gives an equally good or better (anti)correlation. A significance level smaller than 1% is considered "highly significant."

and for the central $14° \times 14°$ which is the region most significant for the neutrino flight path to us. The disk-centered magnetic-field cycle lags full-disk indicators by about 1 y. Oakley et al. (1994) found a significance level of 0.001% for an anticorrelation between the Homestake rate and the disk-centered magnetic flux. Including the flux from increasingly higher latitudes reduced the significance level of the correlation. With only high-latitude magnetic information the correlation was lost.

Another test of time variation is a comparison as in Fig. 10.7 between the expected and measured distribution of argon production rates. A time variation would broaden the measured distribution (black histogram) relative to the expectation for a constant neutrino flux (shaded histogram). A certain degree of broadening is certainly compatible with Fig. 10.7, but a precise statistical analysis does not seem to be available at the present time. At any rate, the rank-ordering result does not specify the amplitude of a time varying signal relative to a constant base rate while the width of the distribution in Fig. 10.7 would be sensitive mostly to this amplitude. Therefore, the two methods yield rather different information concerning a possible time variation.

The gallium experiments have not been running long enough to say much about a time variation on the time scale of several years. Also, there does not seem to be a significant time variation in the Kamiokande data between 1987 and 1992 which covers a large fraction of the current solar cycle No. 22. Notably, there is no apparent (anti)correlation with sunspot number (Suzuki 1993). However, a correlation with disk-centered magnetic indicators may yield a different result—according to the analysis of Oakley et al. (1994) the use of a disk-centered indicator with its inherent time-lag relative to full-disk spot counts is crucial for the strong anticorrelation at Homestake.

The constancy of the Kamiokande rate, if confirmed over a more extended period, severely limits a conjecture that the nuclear energy generating region in the Sun was not constant (e.g. Raychaudhuri 1971, 1986). Because the Kamiokande detector is mostly sensitive to the boron neutrinos with their extreme sensitivity to temperature, they would be expected to show the most extreme time variations if the solar cycle was caused by processes in the deep interior rather than by dynamo action in the convective surface layers.

10.4.4 Summary

There is no indication for a day-night variation of the solar neutrino flux at Kamiokande and none for a seasonal variation at Kamiokande,

GALLEX, or Homestake. The Homestake argon production rate appears to be strongly anticorrelated with solar disk-centered indicators of magnetic activity, with no evidence for a long-term variation at Kamiokande. However, the period covered there (1987–1992) may be too short, and the Kamiokande rate has not been correlated with disk-centered indicators. No homogeneous analysis of both Kamiokande and Homestake data relative to solar activity exists at the present time.

It is very difficult to judge what the apparent Homestake anticorrelation with solar activity means. Is there an unrecognized background which correlates with solar activity such as cosmic rays? Is there a long-term drift in some aspect of the experiment which then naturally correlates with other causally unrelated long-term varying phenomena such as solar activity? Is it all some statistical fluke? Or is it an indication of nonstandard neutrino properties, notably magnetic dipole moments which spin precess in the solar magnetic field into sterile states? This latter possibility will be elaborated further in Sect. 10.7.

10.5 Neutrino Flux Deficits

10.5.1 Boron Flux

Even if one ignores for the moment a possible time variation of the Homestake neutrino measurements there remain several "solar neutrino problems." All of the solar neutrino flux measurements discussed in Sect. 10.3 exhibit a deficit relative to theoretical predictions. The simplest case to interpret is that of Kamiokande because it is sensitive only to the boron flux. In this case the most uncertain input parameter is the cross section for the reaction $p^7\mathrm{Be} \to {}^8\mathrm{B}\gamma$ which enters the flux prediction as a multiplicative factor, independently of other details of solar modelling. As discussed in Sect. 10.2.3, the astrophysical S-factor for this reaction depends on theoretical extrapolations to low energies of experimental data which themselves seem to exhibit relatively large systematic uncertainties. Therefore, one may turn the argument around and consider the solar neutrino flux measurement at Kamiokande as another determination of $S_{17}(0)$.

According to Eq. (10.20) the measured flux is $2.89 \times (1 \pm 0.14)$ in units of $10^6\,\mathrm{cm}^{-2}\,\mathrm{s}^{-1}$ while the prediction is $4.4 \times (1 \pm 0.21) \times S_{17}(0)/22.4$ for TL93 (Turck-Chièze and Lopes 1993) where S_{17} is understood in units of eV b. For the BP95 (Bahcall and Pinsonneault 1995) it is $6.6 \times (1 \pm 0.15) \times S_{17}(0)/22.4$ where I have used an average symmetric error for simplicity. This yields

$$S_{17}(0)/\text{eV b} = \begin{cases} 14.7 \pm 3.7 & \text{TL93,} \\ 9.8 \pm 2.0 & \text{BP95,} \end{cases} \qquad (10.21)$$

to be compared with the world average 22.4 ± 2.1 (Sect. 10.2.3). While the discrepancy is severe, one would still be hard-pressed to conclude with a reasonable degree of certainty that the solar neutrino problem is not just a nuclear physics problem.

10.5.2   Beryllium Flux

The Kamiokande solar neutrino problem may be explained by a low $S_{17}(0)$ factor, perhaps in conspiration with lower-than-standard opacities that cause the central solar temperature to be lower than expected. In this case the spectral shape would remain unchanged, leading to an unambiguous prediction for the neutrino signal that must be caused by ^8B neutrinos in the chlorine and gallium detectors. This contribution may be subtracted from the Homestake and SAGE/GALLEX data, respectively, to obtain a measurement of the remaining neutrino sources (Tab. 10.10)—see Kwong and Rosen (1994). Moreover, the predicted pp and pep signals in these detectors do not seem to involve any significant uncertainties so that they may be subtracted as well, effectively leading to a measurement of the flux of ^7Be and CNO-neutrinos (Tab. 10.10). These fluxes are then found to be formally negative; they are consistent with zero within the experimental measurement errors.

These remaining fluxes inferred from the chlorine and gallium experiments are to be compared with the predictions for the ^7Be and CNO neutrinos. In Tab. 10.10 the predictions of BP95 are shown; those of TL93 and other authors are similar. Bahcall (1994b) has compiled the predictions for the ^7Be flux from a heterogeneous set of 10 solar models by different authors whose predictions agree to within $\pm 10\%$—there is a broad consensus on the standard value of this flux. Therefore, the errors of the predictions are much smaller than the experimental ones, leaving one with a very significant discrepancy between the predicted and observed flux both at Homestake and at SAGE/GALLEX, even if one were to ignore the CNO contributions entirely.

In the analysis of Hata and Haxton (1995) which uses the GALLEX source experiment as a constraint on the gallium absorption cross section, it is found that at the 99% CL the measured ^7Be flux is less than 0.3 of the BP95 prediction. This high significance of the beryllium problem relies on the assumption that all solar neutrino experi-

Table 10.10. Measured vs. predicted ^{7}Be and CNO neutrino fluxes (in SNU).

	Homestake (^{37}Cl)	GALLEX/ SAGE (^{71}Ga)
Measurements:		
Total	2.55 ± 0.25	77 ± 10
^{8}B (inferred from Kamiokande)	3.15 ± 0.44	7 ± 1
pp + pep (calculated)	0.20 ± 0.01	74 ± 1
Remainder	-0.8 ± 0.5	-4 ± 10
BP95 prediction:[a]		
^{7}Be	1.24	37.7
CNO	0.48	10.1
Total	1.72	47.8

[a]Bahcall and Pinsonneault (1995).

ments are correct within the acknowledged uncertainties. If one ignored Homestake the 99% CL limit would be at 0.5 of the BP95 prediction.

Therefore, the "new solar neutrino problem" consists in the measured absence of the beryllium neutrino flux. This is a far more serious deficit than that of the boron flux which has been measured to be a significant fraction of its expected value, and for which the prediction is very uncertain anyway.

10.5.3   An Astrophysical Solution?

The beryllium problem is very significant because its predicted flux is closely related to that of the boron flux. A glance at the solar reaction chains (Fig. 10.2) reveals that both fluxes start with ^{7}Be,

$$^7\text{Be} + e \rightarrow {}^7\text{Li} + \nu_e,$$

$$^7\text{Be} + p \rightarrow {}^8\text{B} \rightarrow {}^8\text{Be} + e^+ + \nu_e. \tag{10.22}$$

Therefore, the flux ratio of ^{7}Be/^{8}B neutrinos essentially measures the branching ratio between the electron and proton capture on ^{7}Be, admittedly averaged over slightly different regions of the solar core. The boron and beryllium fluxes respond very differently to a modification of the solar central temperature (Eq. 10.5). Because the boron flux has been measured, even at a reduced strength, it is not possible to explain the missing beryllium neutrinos by a low solar temperature unless one is willing to contemplate simultaneously an extremely reduced temperature and an extremely enhanced S_{17} factor.

Many authors[57] have recently investigated the question of the significance of the solar neutrino problem, and if it can be solved by a plausible, or even implausible, combination of erroneous nuclear cross sections, solar opacities, and so forth. The consensus is that an astrophysical solution is not possible, even if one ignores part of the experimental data. There is no obvious uncertain input quantity into solar modelling or the flux predictions that could be tuned to obtain the measured fluxes. Therefore, something mysterious is wrong with the calculation of the solar neutrino fluxes and/or the solar neutrino experiments.

However, if our understanding of the solar neutrino source and of the detection experiments is not totally wrong, an attractive alternative is to contemplate the option that something happens to the neutrinos as they propagate from the solar core to us.

10.6   Neutrino Oscillations

10.6.1   Which Data to Use?

The idea to test the hypothesis of neutrino oscillations by means of the solar neutrino flux goes back to Pontecorvo (1957, 1958, 1967) and Maki, Nakagawa, and Sakata (1962). After the first measurements in 1968 it was revived by Gribov and Pontecorvo (1969) as well as Bahcall and Frautschi (1969). If some of the ν_e's produced in the Sun transformed into $\bar{\nu}_e$'s, into other sequential neutrinos (ν_μ, ν_τ), or into hypothetical sterile states ν_s, the *measurable* flux at Earth would be depleted. Even the Kamiokande detector which responds to ν_μ and ν_τ because of the possibility of neutral-current ν-e scattering would show a reduced flux because of the smaller cross section. The measured flux

---

[57]These include Bludman, Kennedy, and Langacker (1992a,b); Bludman et al. (1993); Bahcall and Bethe (1993); Castellani, Degl'Innocenti, and Fiorentini (1993); Castellani et al. (1994b); Hata, Bludman, and Langacker (1994); Shi and Schramm (1994); Shi, Schramm, and Dearborn (1994); Kwong and Rosen (1994); Bahcall (1994b); Berezinski (1994); Degl'Innocenti, Fiorentini, and Lissia (1995); Parke (1995); Hata and Haxton (1995); Haxton (1995).

deficits in all detectors may give us the first indication for neutrino oscillations and thus for nonvanishing neutrino masses and mixings.

Because of the statistically high significance of the anticorrelations with solar activity of the Homestake results it is not entirely obvious how to proceed with a quantitative test of the oscillation hypothesis which can cause only a day-night or semiannual time variation. In the present section a possible long-term variability of the solar neutrino flux or its detection methods is ignored, i.e. the apparent anticorrelation with solar activity at Homestake is considered to be a statistical fluctuation. In Sect. 10.7 a possible explanation in terms of neutrino interactions with the solar magnetic field is considered.

One may take the opposite point of view that the variability at Homestake is a real effect, and that it is not related to neutrino magnetic moments because that hypothesis requires fairly extreme values for the dipole moments and solar magnetic fields (Sect. 10.7). An interpretation of the data in terms of neutrino oscillations then becomes difficult because one admits from the start that unknown physical effects are either operating in the Sun, in the intervening space, or in the detectors. One could argue perhaps that the apparent variability of the Homestake data in itself was evidence that something was wrong with this experiment. In this case one could still test the hypothesis of neutrino oscillations under the assumption that the signal recorded at Homestake was spurious in which case one must discard the entire data set, not only parts of it as has sometimes been done. Ignoring the Homestake data does not solve the solar neutrino problem, but its significance is reduced.

Still, as no one has put forth a plausible hypothesis for a specific problem with the Homestake experiment it is arbitrary to discard the data. Admittedly, it is also arbitrary to consider the time variation spurious even though standard statistical methods seem to reveal a highly significant anticorrelation with solar activity.

10.6.2 Vacuum Oscillations

A formal treatment of vacuum neutrino oscillations was presented in Sect. 8.3. If the observer is many oscillation lengths away from the source, and if the neutrinos are produced with a broad spectrum of energies from an extended source, one will observe an incoherent mixture of the mass eigenstates. If one considers two-flavor oscillations $\nu_e \leftrightarrow \nu_a$ with $\nu_a = \nu_\mu$, ν_τ, or some hypothetical sterile flavor, the ν_e flux is reduced by a factor $1 - \frac{1}{2}\sin^2 2\theta$ (vacuum mixing angle θ). Therefore, one needs a large mixing angle to achieve a substantial suppression of the measurable flux. If one includes the possibility of three-flavor oscillations one can achieve, in principle, a maximum ν_e reduction by a factor $\frac{1}{3}$.

In this picture the ν_e flux is reduced by the same factor for all energies, contrary to what is indicated by the observations. Therefore, one needs an oscillation length of order the Earth-Sun distance (astronomical unit $1\,\text{AU} = 1.496 \times 10^{-13}\,\text{cm}$) because in this case the oscillation pattern is not completely smeared out. Then one may measure a different flux suppression for different energies; the wiggles in the reduction factor as a function of neutrino energy can be resolved (Fig. 10.17).

For two-flavor oscillations the relationship Eq. (8.15) between neutrino energy E_ν, the mass-square difference Δm_ν^2, and the oscillation length ℓ_{osc} can be written as

$$\Delta m_\nu^2 = 1.66 \times 10^{-10}\,\text{eV}^2\,\frac{1\,\text{AU}}{\ell_{\text{osc}}}\,\frac{E_\nu}{10\,\text{MeV}}. \tag{10.23}$$

Therefore, the neutrino masses are very small if these "long-wavelength oscillations" are the explanation for the observed flux deficits.

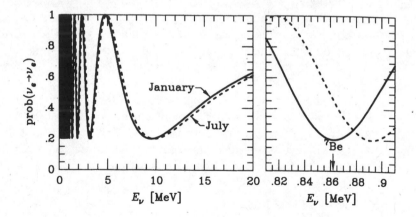

Fig. 10.17. Survival probability of solar ν_e's for $\Delta m_\nu^2 = 0.8 \times 10^{-10}\,\text{eV}^2$ and $\sin^2 2\theta = 0.8$, parameters which reconcile all measurements with the standard solar flux predictions. The solid line is for the minimum annual distance to the Sun of $1.471 \times 10^{13}\,\text{cm}$ (January), the dashed line for the maximum distance $1.521 \times 10^{13}\,\text{cm}$ (July). The right panel is an enlarged section around the line of beryllium neutrinos. Between January and July, their flux is suppressed by very different amounts.

The ν_e survival probability as a function of energy is shown in Fig. 10.17 for $\Delta m_\nu^2 = 0.8 \times 10^{-10} \, \mathrm{eV}^2$ and $\sin^2 2\theta = 0.8$. The Sun is treated as a point-like source because the extension of its neutrino-producing core is minute relative to the oscillation length of order the Earth-Sun distance. The solid line is for the minimum, the dashed line for the maximum annual distance to the Sun. The nearly monochromatic beryllium neutrinos ($E_\nu = 0.862 \, \mathrm{MeV}$) easily retain their phase relationship because they are distributed in energy only over a width of about $2 \, \mathrm{keV}$ (Fig. 10.4). Therefore, one would expect a pronounced annual flux variation which is not observed, thereby reducing the allowed range of masses and mixing angles where long-wavelength oscillations could solve the solar neutrino problem. The analysis of Barger, Phillips, and Whisnant (1992) and more recently of Krastev and Petcov (1994) reveals that there remain a few small parameter pockets with $\sin^2 2\theta$ around $0.8-1$ and Δm_ν^2 around $0.6-0.9 \times 10^{-10} \, \mathrm{eV}^2$. Krastev and Petcov (1994) also conclude that oscillations into sterile neutrinos are slightly disfavored.

While the spectral deformation of the ^{8}B neutrino spectrum is quite dramatic, the spectrum of recoil electrons smears out most of this effect (Fig. 10.18). Notably, above the Kamiokande detection threshold of about $7 \, \mathrm{MeV}$ the expected distortion is too small to be detected and

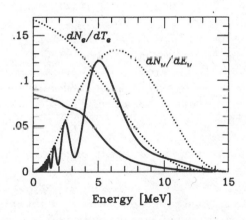

Fig. 10.18. Spectrum of ^{8}B neutrinos and of recoil electrons. *Dotted lines:* Standard spectrum from Fig. 10.12. *Solid lines:* Surviving ν_e spectrum after oscillations (suppression factor from Fig. 10.17) and modified electron recoil spectrum. For oscillations into $\nu_{\mu,\tau}$ one would have to include their neutral-current scattering on electrons with a cross section about $\frac{1}{6}$ that of ν_e.

so the spectral shape measured at Kamiokande (Fig. 10.13) does not further constrain the vacuum oscillation hypothesis.

Barger, Phillips, and Whisnant (1992) as well as Krastev and Petcov (1994) gave precise confidence contours in the $\sin^2 2\theta$-Δm_ν^2-plane where the solar neutrino problems are solved. However, some of the input information involves large systematic uncertainties so that it is well possible that, for example, a revised S_{17} factor would shift the allowed regions beyond their stated confidence contours—see, for example, Berezhiani and Rossi (1995). The main message is that there remain parameters in the quoted range where vacuum oscillations ("just so oscillations") could reconcile all solar flux measurements with the standard flux predictions, except for the apparent anticorrelation of the Homestake data with solar activity.

10.6.3 Resonant Oscillations (MSW Effect)

The solar neutrino problem has become tightly intertwined with the issue of neutrino oscillations thanks to the work of Mikheyev and Smirnov (1985) who showed that even for small mixing angles one can achieve a large rate of flavor conversion because of medium-induced "resonant oscillations." This "MSW effect" was conceptually and quantitatively discussed in Sect. 8.3.

For a practical application to the solar neutrino problem two main features of the MSW effect are of great relevance. One is the bathtub-shaped suppression function of Fig. 8.10 which replaces a constant reduction factor (short-wavelength vacuum oscillations) or the wiggly shape of Fig. 10.17 (long-wavelength vacuum oscillations). It implies that ν_e's of intermediate energy can be reduced while low- and high-energy ones are left relatively unscathed. This is what appears to be indicated by the observations with the beryllium neutrinos more strongly suppressed than the pp or boron ones.

Another important feature are the triangle-shaped suppression contours in the $\sin^2 2\theta$-Δm_ν^2-plane for a fixed neutrino energy (Fig. 8.9). Each experiment produces its own triangular band of neutrino parameters where its measured rate is reconciled with the standard flux prediction. As the signal in different detectors is dominated by neutrinos of different energies, these triangles are vertically offset relative to one another and so there remain only a few intersection points where all experimental results are accounted for. Therefore, as experiments with three different spectral responses are now reporting data the MSW solution to the solar neutrino problem is very well constrained.

The allowed ranges from the different experiments as well as the combined allowed range are shown in Fig. 10.19 according to the analysis of Hata and Haxton (1995). They used the Bahcall and Pinsonneault (1995) solar model with helium and metal diffusion and its uncertainties. The black areas represent the 95% CL allowed range if one includes all the experimental information quoted in Sect. 10.3, including the GALLEX source experiment. The center of the allowed area for the "Large-Angle Solution" is roughly

$$\sin^2 2\theta = 0.6,$$
$$\Delta m_\nu^2 = 2 \times 10^{-5} \, \text{eV}^2, \tag{10.24}$$

while for the "Nonadiabatic Solution" or "Small-Angle Solution" it is

$$\sin^2 2\theta = 0.6 \times 10^{-2},$$
$$\Delta m_\nu^2 = 0.6 \times 10^{-5} \, \text{eV}^2. \tag{10.25}$$

Both cases amount essentially to the same neutrino mass-square difference. Other recent and very detailed analyses are by Fiorentini et al. (1994), Hata and Langacker (1994), and Gates, Krauss, and White (1995) who find similar results. In detail, their confidence contours differ because they used other solar models or a different statistical analysis.

A certain region of parameters (dotted area in Fig. 10.19) can be excluded by the absence of a day/night difference in the Kamiokande counting rates which are expected for the matter-induced back oscillations into ν_e when the neutrino path intersects with part of the Earth at night. (Even though the detector is deep underground in a mine the overburden of material during daytime is negligible.) The excluded range of masses and mixing angles is independent of solar modelling as it is based only on the relative day/night counting rates.

Except for the Kamiokande day/night exclusion region one should be careful at taking the confidence contours in a plot like Fig. 10.19 too seriously. The solar model uncertainties are dominated by systematic effects which cannot be quantified in a statistically objective sense. One of the main uncertain input parameters is the astrophysical S-factor for the reaction $p\,^7\text{Be} \rightarrow {}^8\text{B}\,\gamma$. Speculating that all or most of the missing boron neutrino flux is accounted for by a low S_{17} has the effect of reducing the best-fit $\sin^2 2\theta$ of the small-angle solution by about a factor of 3 while the large-angle solution disappears, at least in the analysis of Krastev and Smirnov (1994). The best-fit Δm^2, however,

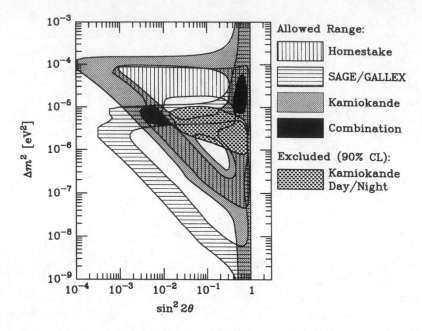

Fig. 10.19. Allowed range of neutrino masses and mixing angles in the neutrino experiments if the flux deficit relative to the Bahcall and Pinsonneault (1995) solar model is interpreted in terms of neutrino oscillations. The experimental data include all summarized in Sect. 10.3. (Plot adapted from Hata and Haxton 1995.)

is rather stable against variations of S_{17} (Krastev and Smirnov 1994; Berezinsky, Fiorentini, and Lissia 1994).

Another approach is to allow S_{17} to float freely when performing a maximum-likelihood analysis, i.e. to fit it simultaneously with the neutrino parameters from all solar neutrino experiments (Hata and Langacker 1994). The best-fit value is found to be $1.43^{+0.65}_{-0.42}$ times the standard 22.4 eV b. Of course, the 95% CL range for the best-fit neutrino parameters is now much larger, allowing any $\sin^2 2\theta$ between about 10^{-3} and 0.8. Notably the range of allowed large-angle solutions is vastly increased.

In summary, the hypothesis of neutrino oscillations can beautifully explain almost all experimental results to date. Only the curious anticorrelation of the Homestake data with solar activity remains unaccounted for. It is interpreted as a statistical fluctuation.

10.7 Spin and Spin-Flavor Oscillations

If the anticorrelation between disk-centered magnetic indicators of solar activity and the event rate at Homestake is taken seriously, the only plausible explanation put forth to date is that of magnetically induced neutrino spin or spin-flavor transitions that were discussed in Sect. 8.4. It was first pointed out by Voloshin and Vysotskiĭ (1986) that the varying magnetic-field strength in the solar convective surface layers could cause a time-varying depletion of the left-handed solar neutrino flux. A refined discussion was provided by Voloshin, Vysotskiĭ, and Okun (1986a,b) after whom this mechanism is called the VVO solution to the solar neutrino problem.

Strong magnetic fields may also exist in the nonconvective interior of the Sun. In principle, they could also cause magnetic oscillations and thus reduce the solar neutrino flux (Werntz 1970; Cisneros 1971). However, the amount of reduction would not be related to the magnetic activity cycle which is confined to the convective surface layers with an approximate depth of $0.3R_\odot \approx 2 \times 10^{10}$ cm $= 200,000$ km. The oscillation length is given by Eq. (8.53) so that over this distance a complete spin reversal is achieved for

$$\mu_\nu B_{\rm T} \approx 3 \times 10^{-10} \, \mu_{\rm B} \, {\rm kG}, \qquad (10.26)$$

where $\mu_{\rm B} = e/2m_e$ is the Bohr magneton and $B_{\rm T}$ is the magnetic-field strength perpendicular to the neutrino trajectory. Because the magnetic field is mainly toroidal, the condition of transversality is automatically satisfied.

Magnetic spin oscillations in vacuum do not have any energy dependence and so the solar neutrino flux would be reduced by a common factor for the entire spectrum. However, a large conversion rate is only achieved if the spin states are nearly degenerate which is not the case in media where refractive effects change the dispersion relation of left-handed neutrinos. In the context of spin-flavor oscillations, degeneracy can be achieved by a proper combination of medium refraction and mass differences (resonant spin-flavor oscillations). In this case the diagonal elements of the neutrino oscillation Hamiltonian involve energy-dependent terms of the form $(m_2^2 - m_1^2)/2E_\nu$, causing a strong energy dependence of the conversion probability. Therefore, the measured signals in the detectors as well as the time variation at Homestake and the absence of such a variation at Kamiokande can all be explained by a suitable choice of neutrino parameters and magnetic-field profiles of the Sun.[58] The simple estimate Eq. (10.26) remains approximately valid. A neutrino mass-square difference $m_2^2 - m_1^2$ below about 10^{-5} eV^2 is required because of the small matter densities encountered in the convective layers.

The bounds on neutrino dipole and transition moments discussed in Sect. 6.5.6 yield $\mu_\nu \lesssim 3\times10^{-12}\mu_B$ so that a magnetic-field strength $B \gtrsim 100$ kG is required in the solar convection zone. Typical field strengths measured in sunspots where the flux breaks through the surface are of a few kG. While this may not be representative of the large-scale toroidal field, several general arguments suggest that 10 kG is a generous upper limit (Shi et al. 1993). If there were much stronger fields they would have to be confined to flux ropes which would not be effective at inducing neutrino spin oscillations. Therefore, the spin oscillation scenario would require anomalously large magnetic fields, in conflict with the arguments presented by Shi et al., or dipole moments in excess of what is allowed by the bounds of Sect. 6.5.6. Even smaller dipole moments than $3\times10^{-12}\mu_B$ require novel neutrino interactions.

The solar magnetic field is believed to consist of two opposite flux tori, separated by the solar equatorial plane. In the course of a year the line of sight from Earth to the solar core varies between $\pm7°15'$ solar latitude and so the neutrinos measured here traverse a field configuration which varies in the course of a year. The predicted semiannual flux variation (Voloshin, Vysotskiĭ, and Okun 1986a), however, has not been confirmed by any of the experiments.

In the spin-flavor oscillation scenario involving Majorana neutrinos the Sun would be a source for antineutrinos, some of which could be $\bar{\nu}_e$'s by a combination of spin-flavor ($\nu_e \to \bar{\nu}_\mu$) and flavor ($\bar{\nu}_\mu \to \bar{\nu}_e$) oscillations. However, the Kamiokande data already yield restrictive limits on a solar $\bar{\nu}_e$ flux (Barbieri et al. 1991; see also Fig. 10.14).

In summary, the magnetic spin oscillation scenario requires new neutrino interactions to generate large magnetic dipole moments, and new astrophysics to allow for sufficiently strong magnetic fields in the solar convection zone. Neither a semiannual variation of the flux nor $\bar{\nu}_e$'s have been observed; each would have been a smoking gun for the occurrence of this effect. Therefore, one is led to disfavor the magnetic spin oscillation scenario. Then, of course, one is back to a statistical fluctuation as an explanation for the time structure of the Homestake data.

---

[58]Recent detailed investigations were performed by Akhmedov, Lanza, and Petcov (1993, 1995), Krastev (1993), Nunokawa and Minakata (1993), Guzzo and Pulido (1993), and Pulido (1993, 1994). For a review of earlier works see Pulido 1992.

10.8 Neutrino Decay

A deficit of solar neutrinos measured at Earth can be related to neutrino decay. However, the in-flight decay of neutrinos does not provide the required deformation of the spectrum because the decay rate in the laboratory system involves a Lorentz factor m_ν/E_ν so that low-energy neutrinos decay faster. If ν_e is a mixture of mass eigenstates only the heavier one decays. It is possible that the heavy admixture decays fast so that the spectrum is reduced by a constant factor. Even this extreme case does not provide a good fit to the data. A detailed analysis of different cases was performed by Acker and Pakvasa (1994) who found that the in-flight decay solution was ruled out at the 98% CL, even when allowing for the solar model uncertainties.[59]

Neutrino decays can yield a solar $\bar{\nu}_e$ flux which is, in principle, measurable. Such a flux can be produced if neutrinos are Majorana particles, and if they couple to majorons χ (Sect. 15.7). Some fraction of $\nu \to \nu' + \chi$ decays flip the helicity of the neutrino so that the ν' is effectively a $\bar{\nu}'$. Thus after an MSW conversion $\nu_e \to \nu_{\mu,\tau}$ one could have decays $\nu_{\mu,\tau} \to \bar{\nu}_e + \chi$ (Raghavan, He, and Pakvasa 1988). Even without oscillations one can have matter-induced decays of the form $\nu_e \to \bar{\nu}_e + \chi$ as discussed in Sect. 6.8. Detailed predictions for the $\bar{\nu}_e$ flux for this type of scenario were worked out by Berezhiani et al. (1992) and by Berezhiani, Moretti, and Rossi (1993). The Kamiokande detector has already produced limits on solar $\bar{\nu}_e$'s (Fig. 10.14), with much better limits to be expected from Superkamiokande. However, in view of other limits on the neutrino majoron coupling Berezhiani, Moretti, and Rossi (1993) found that it seemed unrealistic to hope for a detectable solar $\bar{\nu}_e$ signal.

Malaney, Starkman, and Butler (1994) showed that in decays of the form $\nu \to \nu' + $ boson, final-state stimulation effects ("neutrino lasing") could enhance the decay rate. However, the best-motivated case is that of majoron decays which involve a γ_5 coupling. The shape of the resulting majoron spectrum is such that the crucial emission of low-momentum bosons and thus the lasing effect is suppressed whence Acker and Pakvasa's conclusions remain valid. For models involving scalar or vector bosons a detailed new analysis is required.

At the present time it looks rather unconvincing that the solar neutrino problem is related to some form of neutrino decays.

---

[59] See Acker and Pakvasa (1994) for references to earlier discussions of neutrino decay as a potential solution to the solar neutrino problem.

10.9   Future Experiments

10.9.1   Superkamiokande

If some form of neutrino oscillations are the explanation for the measured solar flux deficits relative to standard predictions, how are we ever going to know for sure? One needs to measure a signature which is characteristic only for neutrino oscillations. The most convincing case would be a measurement of the "wrong-flavored" neutrinos, i.e. the ν_μ or ν_τ appearance rather than the ν_e disappearance. Other clear signatures would be a deformation of the ^{8}B spectrum or a diurnal or seasonal flux variation.

The latter cases can be very well investigated with the Superkamiokande detector which is scheduled to begin its operation in April of 1996. It is a water Cherenkov detector like Kamiokande, with about 20 times the fiducial volume. With about twice the relative coverage of the surface area with photocathodes and a detection threshold as low as 5 MeV it will count about 30 events/day from the solar boron neutrino flux, as opposed to about 0.3 events/day at Kamiokande.

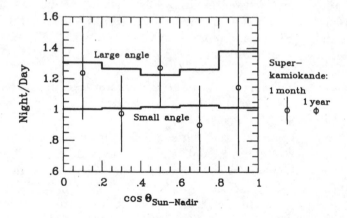

Fig. 10.20. Expected signal at night relative to the average daytime signal in a water Cherenkov detector as a function of the angle between the Sun and the detector nadir. The large-angle example is for $\sin^2 2\theta = 0.7$ and $\Delta m_\nu^2 = 1 \times 10^{-5}\,\text{eV}^2$, the small-angle case for $\sin^2 2\theta = 0.01$ and $\Delta m_\nu^2 = 0.3 \times 10^{-5}\,\text{eV}^2$. Also shown are the existing Kamiokande measurements and the expected Superkamiokande error bars after 1 month and 1 year of running, respectively. (Adapted from Suzuki 1995.)

The small- and large-angle MSW solutions suggested by Fig. 10.19 would cause a day/night variation of the solar neutrino signal as indicated in Fig. 10.20. The expected signal is shown in bins for the angle between the Sun and the nadir of the detector, i.e. in bins of the intersection length of the neutrino flight path with the Earth. Also shown are the current Kamiokande measurements, and the expected error bars after 1 month and 1 year of Superkamiokande running time, respectively. Shortly after Superkamiokande starts taking data one should be able to decide whether the large-angle MSW solution applies!

The MSW solutions would also cause a spectral distortion of the recoil electron spectrum from the primary boron neutrinos. The expected spectral shape relative to the standard one, arbitrarily normalized at an electron kinetic energy of $T_e = 10\,\mathrm{MeV}$, is shown in Fig. 10.21 for several values of the assumed mixing angle. After several years of running, Superkamiokande should be able to identify clearly the small-angle solution if it applies.

If Superkamiokande measures neither a spectral distortion nor a day/night effect, the deficiency of the boron flux probably would have

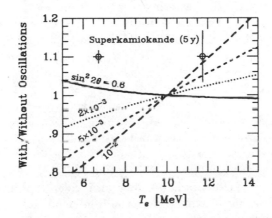

Fig. 10.21. Spectral distortion of the recoil electrons from the primary boron neutrinos in a water Cherenkov detector. The ratio relative to the standard spectrum is arbitrarily normalized at an electron kinetic energy of $T_e = 10\,\mathrm{MeV}$. For the large-angle example (solid line) the assumed mass-square difference is $\Delta m_\nu^2 = 2\times10^{-5}\,\mathrm{eV}^2$, for the small-angle examples (broken lines) it is $\Delta m_\nu^2 = 0.6\times10^{-5}\,\mathrm{eV}^2$. The anticipated error bars after 5 years of running Superkamiokande are also indicated for two energies. (Adapted from Krastev and Smirnov 1994.)

to be attributed to a small astrophysical S_{17} factor. In this case, the explanation for the deficiency of beryllium neutrinos could not be resolved by this detector.

10.9.2 Sudbury Neutrino Observatory (SNO)

The Sudbury Neutrino Observatory (SNO), also scheduled to take up operation in 1996, is a heavy-water Cherenkov detector which is expected to be able to measure the appearance of "wrong-flavored" neutrinos if the MSW effect solves the solar neutrino problems (Sudbury Neutrino Observatory Collaboration 1987; Lesko et al. 1993). The SNO detector consists of 1000 tons of heavy water (D_2O) in a spherical acrylic vessel of 12 m diameter, immersed in an outer vessel of ultrapure light water (H_2O), surrounded by about 9600 photomultiplier tubes of 20 cm diameter each. The detector is located 2000 m underground in the Creighton mine, an operating Nickel mine, near Sudbury in Ontario (Canada). Neutrinos can be detected by three different reactions in this detector: by electron elastic scattering $\nu + e \to e + \nu$ and by the deuterium dissociation reactions $\nu_e + d \to p + p + e$ and $\nu + d \to p + n + \nu$.

The electron elastic scattering reaction is analogous to the Kamiokande and Superkamiokande detectors: the recoiling electron is measured by the detection of its Cherenkov light. The effective detection threshold is expected to be at 5 MeV as in Superkamiokande. This reaction is sensitive to both ν_e and $\nu_{\mu,\tau}$, albeit with a reduced cross section for the latter (Eq. 10.17).

The charged-current deuterium dissociation $\nu_e d \to ppe$ has a threshold of 1.44 MeV, i.e. the final-state electron kinetic energy is essentially $T_e = E_\nu - 1.44$ MeV. The electron is detected by its Cherenkov light; the effective energy resolution is about 20%. The angular distribution relative to the incident neutrino is given by $1 - \frac{1}{3}\cos\Theta$. The cross section for this reaction is large. For an incident spectrum of boron neutrinos one expects 9 times more electron counts above 5 MeV than from electron elastic scattering; above 9 MeV even 13 times as many.

One can search for neutrino oscillations by a spectral distortion of the electron spectrum, similar to Fig. 10.21 for Superkamiokande. In fact, the spectral distortion is more pronounced as it is not washed out by a broad final-state distribution of electron energies. Also, one can search for a day/night effect.

The most important detection reaction, however, is the neutral-current deuteron disintegration $\nu d \to pn\nu$ which has the same cross section for all flavors and so it measures the total (left-handed) neutrino flux above its threshold of 2.2 MeV, independently of the occurrence of oscillations. The main problem here is the measurement of the final-state neutron by the detection of γ rays from the subsequent neutron capture, or by a neutron detector array in the heavy water. Ultrapure water, acrylic, and other materials are needed to prevent an excessive radioactive background that would spoil this measurement. One anticipates to obtain the full unsuppressed ^{8}B flux with a precision of about 1% after 5 years of operation. The measured ratio between the charged-current and neutral-current deuterium disintegration will give an immediate measure of the electron survival probability and thus of the occurrence of neutrino oscillations.

If spin or spin-flavor oscillations occur such that a sizeable $\bar{\nu}_e$ flux is produced, it can be detected by the reaction $\bar{\nu}_e + d \rightarrow n + n + e^+$ which produces three detectable particles (Balantekin and Loreti 1992). Of course, a possible $\bar{\nu}_e$ flux is already constrained by Kamiokande (Fig. 10.14), and can be detected or constrained by Superkamiokande.

Suggestions for solar model independent methods of analyzing future solar neutrino data were made, for example, by Spiro and Vignaud (1990), Bilenky and Giunti (1993, 1994), and Castellani et al. (1994).

10.9.3 BOREXINO

Superkamiokande and SNO are both limited to a measurement of the boron neutrino flux because of their relatively high detection thresholds. If the boron flux is partly or mostly suppressed by a low S_{17} factor or a low central solar temperature instead of neutrino oscillations these experiments may have difficulties at identifying oscillations which would still be indicated by the missing beryllium neutrino flux. Therefore, it is interesting that another experiment (BOREXINO[60]) is being prepared which would be sensitive dominantly to the beryllium neutrinos.

The main detection reaction is elastic ν-e scattering as in the light-water Cherenkov detectors. However, the kicked electron is detected by virtue of scintillation light rather than Cherenkov radiation, the former

---

[60]The name of this experiment is derived from BOREX (boron solar neutrino experiment), a proposed detector that was to use ^{11}B as a target (Raghavan, Pakvasa, and Brown 1986; see also Bahcall 1989). For a practical implementation it was envisaged to use a boron loaded liquid scintillator (e.g. Raghavan 1990); because of the relatively small size of this detector the Italian diminutive BOREXINO (baby BOREX) emerged. Ultimately, the idea of using a borated scintillator was dropped entirely, leaving boron only in the name of the experiment. Confusingly, then, BOREXINO is unrelated to a boron target, and also unrelated to the solar boron neutrinos because the experiment is designed to hunt the beryllium ones.

yielding about 50 times more light at the relevant energies below about 1 MeV. Naturally, as a target one needs to use an appropriate scintillator rather than water. The advantage of a lowered threshold is bought at the price of losing all directional information. However, because of the good energy resolution the monochromatic beryllium neutrinos at 862 keV should be clearly detectable as a distinct shoulder in the energy spectrum of the recoil electrons. Optimistically, a scintillation detector could have a threshold as low as $E_\nu = 250$ keV.

The main challenge at implementing this method is to lower the radioactive contamination of the scintillator, its vessel, and the surrounding water bath to an unprecedented degree of purity. For example, the allowed mass fraction of ^{238}U of the scintillator is less than about 10^{-16}g/g. The feasibility of this method is currently being studied at the CTF (Counting Test Facility) experiment, located in the Gran Sasso underground laboratory. Assuming a positive outcome, BOREXINO would be built, consisting of 300 tons of scintillator, surrounded by 3000 tons of water. Optimistically, data taking with this facility could commence in 1997.

10.9.4   Homestake Iodine Detector

Currently, a modular 100-ton iodine detector is under construction in the Homestake mine. It is similar to the Homestake chlorine detector, except that it uses the ^{127}I $\rightarrow$ ^{127}Xe transition to measure the ν_e flux. It has an effective threshold of 0.789 MeV, similar to the chlorine detector. However, for a standard solar neutrino flux the detection rate should be about four times higher if the cross section calculations are correct. The currently built detector should take up operation in mid-1995. It may be expanded at a later time after running experience has been obtained, and after the neutrino cross sections have been measured (Bahcall et al. 1995; Engel, Krastev, and Lande 1995).

10.9.5   Summary

The hypothesis of neutrino flavor oscillations is strongly supported by the results of all existing solar neutrino experiments. With the new generation of detectors which will begin to take up operation in 1996 it looks plausible that nonstandard neutrino properties can be firmly established on the basis of the solar neutrino flux before the millenium ends.

Chapter 11

Supernova Neutrinos

The general physical picture of stellar collapse and supernova (SN) explosions is described with an emphasis on the properties of the observable neutrino burst from such events. The measurements of the neutrino burst from SN 1987A are reviewed. Its lessons for particle physics are deferred to Chapter 13 except for the issue of neutrino masses and mixings. Future possibilities to observe SN neutrinos are discussed.

11.1 Stellar Collapse and Supernova Explosions

11.1.1 Stellar Collapse

A massive star ($\mathcal{M} \gtrsim 8\,\mathcal{M}_\odot$) inevitably becomes unstable at the end of its life. It collapses and ejects its outer mantle in a SN explosion as briefly described in Sect. 2.1.8. Within fractions of a second the collapsing core forms a compact object at supranuclear density which radiates its gravitational binding energy $E_b \simeq 3 \times 10^{53}$ erg within a few seconds in the form of neutrinos. Gamow and Schoenberg (1940, 1941) were the first to speculate that neutrino emission would be a major effect in the collapse of a star. The only direct observation of neutrinos from such an event occurred on 23 February 1987 when the blue supergiant Sanduleak −69 202 in the Large Magellanic Cloud exploded in what became known as SN 1987A. The neutrinos, and possibly other low-mass particles, emitted from a collapsing star are the main topic of this chapter, with SN 1987A playing a primary role.

Before an evolved massive star collapses, its core is a degenerate configuration made up of iron-group elements. They cannot release nuclear energy by fusion as they are already the most tightly bound nuclei so that no further nuclear burning stage can be ignited. The precollapse inner "iron white dwarf" has a mass of about $1.5\,\mathcal{M}_\odot$, a central density of about $3.7\times10^9\,\mathrm{g\,cm^{-3}}$, a central temperature of about $0.69\,\mathrm{MeV}$, and a number fraction of electrons per baryon of $Y_e \approx 0.42$ (Brown, Bethe, and Baym 1982). As the mass of this object grows and its radius shrinks it reaches its Chandrasekhar limit: relativistic electrons cannot support a self-gravitating body.

In practice, a thermal pressure contribution cannot be neglected. The collapse is triggered when the temperature has become so high that the photodissociation of iron commences, $\gamma + {}^{56}\mathrm{Fe} \to 13\alpha + 4n$, a reaction which consumes $124.4\,\mathrm{MeV}$ of energy. This energy loss reduces the thermal contribution of the electron pressure. Therefore, compression yields a lesser pressure increase than would occur in the absence of photodissociation. A star near its Chandrasekhar mass is close to a point where the increased gravitational pull caused by a small contraction is no longer overcompensated by a large enough push from the corresponding pressure increase. Therefore, a small reduction of the adiabatic index $\Gamma \equiv (\partial\ln p/\partial\ln\rho)_s$ is enough to cause an instability. Once the collapse has begun, pressure support is also lost by the capture of electrons on heavy nuclei which amounts to the reaction $e^- + p \to n + \nu_e$. It converts electrons to neutrinos which escape freely.

The inner part of the core ($\mathcal{M} \approx 0.6\,\mathcal{M}_\odot$) has $\Gamma \approx \frac{4}{3}$. Its collapse is homologous, i.e. it maintains its relative density profile (Goldreich and Weber 1980). The collapse velocity is proportional to the radius with $v/r = 400 - 700\,\mathrm{s^{-1}}$, yet it remains subsonic and so this part of the core is in good communication with itself. The nearly free fall of the outer part is supersonic.

At a certain density the neutrinos will no longer be able to stream freely from the core. When their diffusion time exceeds a dynamical collapse time scale they will be trapped (Mazurek 1974, 1975, 1976; Sato 1975). Neutral-current scatterings on large nuclei are particularly effective at trapping neutrinos because the cross section is coherently enhanced (Freedman 1974).[61] One finds a trapping density of around $10^{12}\,\mathrm{g\,cm^{-3}}$ for $10\,\mathrm{MeV}$ neutrinos (Brown, Bethe, and Baym 1982). The neutrino trapping radius as a function of time is shown as a dotted line in Fig. 11.1.

---

[61] Neutrinos with $10\,\mathrm{MeV}$ energies cannot "resolve" the nucleus, causing it to act as a single scattering center. Because the neutral-current interaction with protons is reduced by a factor $1 - 4\sin^2\Theta_W$ (weak mixing angle Θ_W) with $\sin^2\Theta_W \approx 0.23$ (Appendix B), the elastic scattering cross section of a nucleus scales with the square of the neutron number.

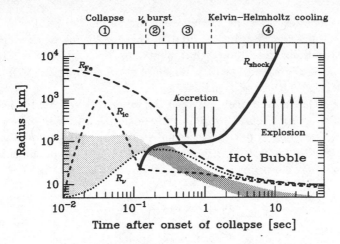

Fig. 11.1. Schematic picture of the core collapse of a massive star ($\mathcal{M} \gtrsim$ $8\,\mathcal{M}_\odot$), of the formation of a neutron-star remnant, and the beginning of a SN explosion. There are four main phases numbered 1–4 above the plot: 1. Collapse. 2. Prompt-shock propagation and break-out, release of prompt ν_e burst. 3. Matter accretion and mantle cooling. 4. Kelvin-Helmholtz cooling of "protoneutron star." The curves mark the time evolution of several characteristic radii: The stellar iron core (R_{Fe}). The "neutrino sphere" (R_ν) with diffusive transport inside, free streaming outside. The "inner core" (R_{ic}) which for $t \lesssim 0.1\,$s is the region of subsonic collapse, later it is the settled, compact inner region of the nascent neutron star. The SN shock wave (R_{shock}) is formed at core bounce, stagnates for several 100 ms, and is revived by neutrino heating—it then propagates outward and ejects the stellar mantle. The shaded area is where most of the neutrino emission comes from; between this area and R_ν neutrinos still diffuse, but are no longer efficiently produced. (Adapted from Janka 1993.)

Neutrino trapping has the effect that the lepton number fraction Y_L is nearly conserved at the value Y_e which obtains at the time of trapping. However, electrons and electron neutrinos still interconvert (β equilibrium), causing a degenerate ν_e sea to build up. The core of a collapsing star is the only known astrophysical site apart from the early universe where neutrinos are in thermal equilibrium. It is the only site where neutrinos occur in a degenerate Fermi sea as the early universe is thought to be essentially CP symmetric with equal numbers of neutrinos and antineutrinos to within one part in 10^9. When neutrino trapping becomes effective, the lepton fraction per baryon is $Y_L \approx 0.35$,

not much lower than the initial iron-core value, i.e. not much of the
lepton number is lost during infall. The conditions of chemical and
thermal equilibrium dictate that neutrinos take up only a relatively
small part (about $\frac{1}{4}$) of the lepton number (Appendix D.2). A typical
Y_L profile after collapse is shown in Fig. 11.2

The collapse is intercepted when the inner core reaches nuclear den-
sity ($\rho_0 \approx 3 \times 10^{14}\,\mathrm{g\,cm^{-3}}$), a point where the equation of state stiffens.
Because the inner core collapse is subsonic, the information about the
central condition spreads throughout, i.e. the collapse of the entire homologous core slows down. However, this information cannot propagate

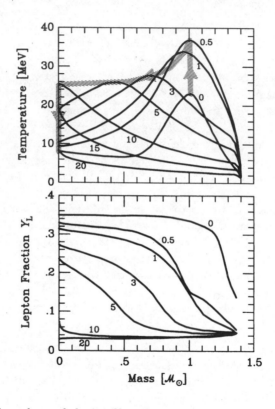

Fig. 11.2. Snapshots of the profiles of temperature T and lepton number
fraction Y_L in the collapsed core of a massive star. The indicated times are
in seconds after collapse. The shaded arrows in the upper panel indicate
the motion of the temperature maximum. For a soft equation of state the
maximum temperature can be up to about 70 MeV. (Adapted from Burrows
and Lattimer 1986.)

beyond the sonic point at the edge of the inner core which now encompasses about $0.8\,\mathcal{M}_\odot$. As material continues to fall onto the inner core at supersonic velocities a shock wave builds up at the sonic point which is at the edge of the inner core, not at its center. As more material moves in, more and more energy is stored in this shock wave which almost immediately begins to propagate outward into the collapsing outer part of the iron core—see the thick solid line in Fig. 11.1. Assuming that enough energy is stored in the shock wave it will eventually eject the stellar mantle outside of what was the iron core. The rebound or "bounce" of the collapse turns the implosion of the core into an explosion of the outer star—a SN occurs.

This "bounce and shock" scenario of SN explosions was first proposed by Colgate and Johnson (1960) and then elaborated by a number of authors (see Brown, Bethe, and Baym 1982 and references therein). In practice, however, the story of SN explosions appears to be more complicated than this "prompt explosion scenario." Neutrino losses and the dissociation of the iron material through which the shock wave propagates dissipate much of the shock's energy so that in typical calculations it stalls and eventually recollapses. It is currently believed that the energy deposition by neutrinos revives the shock wave, leading to the "delayed explosion scenario" detailed in Sect. 11.1.3 below.

11.1.2 Deleptonization and Cooling

After core bounce and the formation of a shock wave the next dramatic step in the evolution of the core is when the outward propagating shock breaks through the "neutrino sphere," i.e. the shell within which neutrinos are trapped, most effectively by the coherent scattering on heavy nuclei. As the passage of the shock wave dissociates these nuclei, it is easier for neutrinos to escape. Moreover, the protons newly liberated from the iron nuclei allow for quick neutronization by virtue of $e^- + p \to n + \nu_e$, causing a short ν_e burst which is often called the "prompt ν_e burst" or "deleptonization burst" (phase No. 2 in Figs. 11.1 and 11.3). However, the material which is quickly deleptonized encompasses only a few tenths of a solar mass so that most of the leptons remain trapped in the inner core (Fig. 11.2).

At this stage, the object below the shock has become a "protoneutron star." It has a settled inner core within the radius where the shock wave first formed and which consists of neutrons, protons, electrons, and neutrinos (lepton fraction $Y_L \approx 0.35$). The protoneutron star also has a bloated outer part which has lost a large fraction of

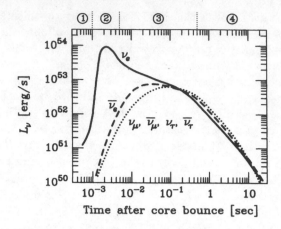

Fig. 11.3. Schematic neutrino "lightcurves" during the phases of (1) core collapse, (2) shock propagation and shock breakout, (3) mantle cooling and accretion, and (4) Kelvin-Helmholtz cooling. (Adapted from Janka 1993.)

its lepton number during the ν_e burst at shock break-out. This outer part settles within the first $0.5-1\,\mathrm{s}$ after core bounce, emitting most of its energy in the form of neutrinos. Also, more material is accreted while the shock wave stalls. As much as a quarter of the expected total amount of energy in neutrinos is liberated during this phase (No. 3 in Figs. 11.1 and 11.3).

Meanwhile, the stalled shock wave has managed to resume its outward motion and has begun to eject the overburden of matter. Therefore, the protoneutron star after about $0.5-1\,\mathrm{s}$ can be viewed as a star unto itself with a radius of around $30\,\mathrm{km}$ which slowly contracts and cools by the emission of (anti)neutrinos of all flavors, and at the same time deleptonizes by the loss of ν_e's. After $5-10\,\mathrm{s}$ it has lost most of its lepton number, and slightly later most of its energy. This is the "Kelvin-Helmholtz cooling phase," marked as No. 4 in Figs. 11.1 and 11.3. Afterward, the star has become a proper neutron star whose small lepton fraction is determined by the condition of a vanishing neutrino chemical potential (Appendix D.2), and whose further cooling history has been discussed in Sect. 2.3.

Immediately after collapse the protoneutron star is relatively cold (see the $t = 0$ curve in Fig. 11.2). Half or more of the energy to be radiated later is actually stored in the degenerate electron Fermi sea with typical Fermi momenta of order $300\,\mathrm{MeV}$. The corresponding degenerate neutrino sea has its Fermi surface at around 200 MeV. The lepton number profile (lower panel of Fig. 11.2) has a step-like form, with the step moving inward very quickly when the shock breaks through the neutrino sphere. The quick recession of the lepton profile during the first 0.5 s represents deleptonization during the mantle cooling phase.

After this initial phase, however, the bloated outer part of the star has settled; it is more difficult for neutrinos to escape from this compact object. Still, the steep gradient of lepton number drives an outward diffusion of neutrinos which move toward regions of lower Fermi momentum and thus, of lower degeneracy energy. Therefore, they "downscatter," releasing most of the previous electron and neutrino degeneracy energy as heat. Hence near the edge of the lepton number step the medium is heated efficiently. In Fig. 11.2 it is plainly visible that the temperature maximum of the medium is always in the region of the steepest lepton number gradient.[62] Therefore, the medium first heats near the core surface, and then the temperature maximum moves inward until it has reached the center. At this time the core is entirely deleptonized; it continues to cool, the temperature maximum at the center drops to obscurity.

The neutrino radiation leaving the star has typical energies in the 10 MeV range, compared with a 200 MeV neutrino Fermi energy in the interior. Therefore, the loss of lepton number by itself is associated with relatively little energy. Put another way, only a small excess of ν_e over $\bar{\nu}_e$ is needed to carry away the lepton number. The total energy is carried away in almost equal parts by each (anti)neutrino flavor.

11.1.3 Supernova Explosions

How do supernovae explode? For some time it was thought that the outer layers of the star were ejected by the momentum transferred from the outward neutrino flow which is released after the core collapse (Colgate and White 1966). This scenario had to be abandoned after the discovery of neutral-current neutrino interactions which trap these particles so that they are released only relatively slowly. With the demise of the neutrino explosion scenario the earlier suggestion of Colgate and Johnson (1960) of a hydrodynamic shock wave driving the explosion became the standard, the so-called "prompt explosion scenario" or "direct mechanism." It continued to malfunction, however, because in numerical calculations the shock tended to stall because of energy dissipation

---

[62] I thank David Seckel for explaining this point to me which is not usually stressed in the pertinent literature.

associated with the dissociation of the remaining iron shell it had to work through before it could fly. Recall that the shock wave forms relatively deep inside of the collapsing iron core of the progenitor star.

A possible solution is the "delayed mechanism" (Wilson 1983; Bethe and Wilson 1985) where the shock wave lingers at a constant radius for a few 100 ms and then takes off again (Fig. 11.1), powered both by the accretion of material and by the energy deposition of the neutrino flow. In this regard the dilute hot region (ρ of order $10^6-10^8\,\mathrm{g\,cm^{-3}}$, T of order 1 MeV) below the stalled shock plays a major role at absorbing neutrino energy.[63] Unfortunately, it is not certain that the neutrino flux can deposit enough energy to revitalize the shock as the energy transfer is relatively inefficient. However, if one adjusts the amount of neutrino energy transfer to a value above a rather well defined threshold one obtains beautiful explosions (Fig. 11.4). It should be noted that the spectacular explosion of a SN—which at the peak of its lightcurve outshines an entire galaxy—is only a "dirt effect" relative to the release of neutrino energy which equals the gravitational binding energy of the newborn neutron star of about $G_N \mathcal{M}/R \approx 3\times10^{53}$ erg with $\mathcal{M} \approx 1.4\,\mathcal{M}_\odot$ and $R \approx 10$ km. The total energy released in the kinetic energy of the ejecta and in electromagnetic radiation is a few[64] 10^{51} erg, on the order of 1% of the total neutron-star binding energy. Therefore, on energetic grounds alone there is no problem at tapping the neutrino flux for explosion energy.

A variety of schemes are currently being discussed to achieve a successful shock revival. It is thought that convection may play a major role at transporting energy to the surface of the bloated protoneutron star which has formed after the break-out of the shock (e.g. Burrows and Lattimer 1988; Wilson and Mayle 1988; Mayle and Wilson 1993; Burrows and Fryxell 1992, 1993; Janka and Müller 1993b). It may be that this mechanism can boost the effective neutrino luminosity for a few hundred milliseconds, enough in some calculations to trigger an explosion (see however Bruenn and Mezzacappa 1994). At any rate, in the absence of a fundamental treatment of convection this method is essentially one way of parametrizing the initial amount of neutrino heating below the shock.

---

[63]Goodman, Dar, and Nussinov (1987) proposed that the pair-annihilation process $\nu\bar\nu \to e^+ e^-$ might be the dominant mode of energy transfer. However, Cooperstein, van den Horn, and Baron (1987) critizised the neutrino emission parameters of that study while Janka (1991) found that a proper treatment of the phase space renders this process less efficient than had been originally thought.

[64]This unit is sometimes referred to as 1 foe, for "ten to the fifty one ergs."

Convection is also important at transporting energy within the dilute region between the protoneutron star and the stalling shock wave (Herant, Benz, and Colgate 1992; Herant et al. 1994; Janka and Müller 1993a, 1994, 1995a,b; Sato, Shimizu, and Yamada 1993; Burrows,

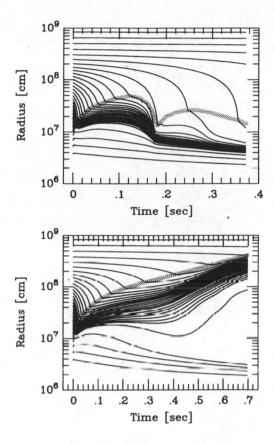

Fig. 11.4. Unsuccessful (upper panel) and successful (lower panel) SN explosion. In each case, the location of several mass shells is shown as a function of time. The thick shaded line indicates the location of the shock. The only difference between the two cases is the adjusted neutrino luminosity from a central source; it was chosen as $L_\nu = 2.10 \times 10^{52} \, \mathrm{erg \, s^{-1}}$ (upper panel) and $L_\nu = 2.20 \times 10^{52} \, \mathrm{erg \, s^{-1}}$ (lower panel), respectively. In the upper case, which is just below threshold for a successful explosion, the shock displays an interesting oscillatory behavior. (Curves courtesy of H.-T. Janka, taken from Janka and Müller 1993a.)

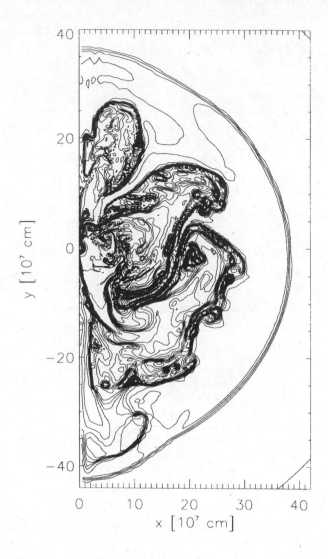

Fig. 11.5. Entropy contours between neutron star and shock wave in a 2-dimensional calculation of a SN explosion. The entropy per nucleon is shown in contours at equal steps of 0.5 k_B between 5 and 16 k_B, and in steps of 1 k_B between 16 and 23 k_B. This snapshot represents model T2c of Janka and Müller (1995b) at $t = 377$ ms after bounce. (Original of the figure courtesy of H.-T. Janka.)

Hayes, and Fryxell 1995). For the first time 2- and 3-dimensional calculations have become possible. They reveal a large-scale convective overturn (Fig. 11.5) which helps at revitalizing the shock because it brings hot material from depths near the neutrino sphere quickly up to the region immediately behind the shock, and cooler material down to the neutrino sphere where it absorbs energy from the neutrino flow. Successful explosions can be obtained for amounts of neutrino heating where 1-dimensional calculations did not succeed. The sharp transition between failed and successful explosions as a function of neutrino heating that was found in 1-dimensional calculations (Fig. 11.4) is smoothed out, but neutrino heating still plays a pivotal role at obtaining a successful and sufficiently energetic explosion.

In summary, then, the current standard picture of SN explosions is a modification and synthesis of the Colgate and Johnson (1960) shock-driven and the Colgate and White (1966) neutrino-driven explosions. At the present time there may still remain a quantitative problem at obtaining enough neutrino energy deposition behind the shock wave to guarantee a successful and sufficiently energetic explosion. It remains to be seen if this scenario withstands the test of time, or if a novel ingredient will have to be invoked in the future.

11.1.4 Nucleosynthesis

The universe began in a hot "big bang" which allowed for the formation of nuclei from the protons and neutrons originally present in thermal equilibrium with the ambient heat bath. The primordial abundances "froze out" at about 22−24% helium, the rest hydrogen, and a small trace of other light elements such as lithium. The present-day distribution of elements was bred from this primeval mix mostly by nuclear processes in stars; they eject some of their mass at the end of their lives (Chapter 2), returning processed material to the interstellar medium from which new stars and planets are born.

However, the normal stellar burning processes can produce elements only up to the iron group which have the largest binding energy per nucleon. Thus, the heavy elements must have been produced by different processes at different sites. It has long been thought that nuclei with $A \gtrsim 70$ were predominantly made by neutron capture, notably the s- and r- (slow and rapid) processes (Burbidge et al. 1957; Cameron 1957; Clayton 1968; Meyer 1994). The site for the occurrence of the r-process has remained elusive for the past three decades, although many different suggestions have been made. The crux is that one needs to produce the heavy elements in the observed proportions, and with a total amount compatible with a plausible galactic history. It has long been held that the r-process elements were made in SN explosions; in this case one needs a yield of about $10^{-4}\,\mathcal{M}_\odot$ of heavy elements per SN.

Perhaps the first realistic scenario that appears to meet these requirements is r-process nucleosynthesis in the hot bubble between a protoneutron star and the escaping shock wave in a core-collapse SN explosion at a time of a few seconds after core bounce (Woosley and Hoffmann 1992; Meyer et al. 1992; Woosley et al. 1994; Witti, Janka, and Takahashi 1994; Takahashi, Witti, and Janka 1994; Meyer 1995). The material in this region is very dilute because of the successful explosion, yet very hot—around 10^9 K or $100\,\mathrm{keV}$ in the region where the r-process is thought to occur. This hot bubble is not entirely empty because of a neutrino-driven stellar wind. Therefore, one is talking about a high-entropy environment (a few hundred k_B per baryon), i.e. a large number of photons per baryon (a few ten). For such conditions the required neutron/proton ratio is achieved even for electron fractions of $Y_e \approx 0.40$ typical for the material outside of a collapsed SN core.

This scenario appears to be qualitatively and quantitatively almost perfect except that the necessary combination of entropy, electron fraction Y_e, and expansion time scale do not seem to be quite born out by current numerical calculations. Whatever the explanation of this problem, it is fascinating that both the occurrence of a successful and sufficiently energetic SN explosion as well as the occurrence of the r-process in the high-entropy environment of the "hot bubble" seem to depend crucially on the neutrino energy transfer which thus plays a dominant role in this scenario. One may expect that r-process nucleosynthesis will turn into a tool to calibrate the neutrino flux from a nascent neutron star, and perhaps into a tool to study nonstandard neutrino properties (for a first example see Sect. 11.4.5). In effect, the distribution and quantity of r-process elements gives us a measure of SN neutrino fluxes, independent of direct observations! This is not unlike big-bang nucleosynthesis where the primeval light-element abundances have been an extremely useful tool to study the properties of the primordial neutrino heat bath (Kolb and Turner 1990). At the present time, of course, a quantitative understanding of SN nucleosynthesis in conjunction with a quantitative understanding of SN explosions is a field in its infancy—it remains to be seen if it grows up to be as beautiful as big-bang nucleosynthesis.

11.2 Predicted Neutrino Signal

11.2.1 Overall Features

One of the most important aspects of SN physics relevant to particle astrophysics is the immense flux of neutrinos liberated after the core collapse. This flux has been measured from SN 1987A, and with luck will be measured again from a galactic SN in the future. Therefore, it is important to understand the neutrino signal to be expected from this sort of event.

On a crude level of approximation one can understand the main features of the overall neutrino signal on the basis of very simple physical principles. The overall amount of energy to be expected is given by the binding energy of the compact star that formed after collapse

$$E_b \approx \frac{3}{5}\frac{G_N\mathcal{M}^2}{R} = 1.60\times 10^{53}\,\text{erg}\ \left(\frac{\mathcal{M}}{\mathcal{M}_\odot}\right)^2\left(\frac{10\,\text{km}}{R}\right). \qquad (11.1)$$

It is reasonable to expect the energy to be equipartitioned among the different neutrino flavors and so to expect about $\frac{1}{6}E_b$ in each of the six standard (anti)neutrino degrees of freedom. (Here and in the following Newtonian physics is used; general relativistic corrections to energies, temperatures, etc. as viewed from a distant observer can be as large as several 10% due to gravitational redshifts.)

Neutrinos are trapped in the interior of the high-density neutron star. Therefore, they are emitted from the relatively well defined surface at a radius of 10–20 km, depending on the mass and the nuclear equation of state. As long as the material near the surface is nondegenerate it must support itself against the local gravitational field by normal thermal pressure. One may apply the virial theorem (Chapter 1) which informs us that the average kinetic energy of a typical nucleon near the neutron-star surface must be half of its gravitational potential, i.e. $2\langle E_{\text{kin}}\rangle \approx G_N\mathcal{M}\,m_N/R$ (nucleon mass m_N). With a neutron-star mass of $\mathcal{M} = 1.4\,\mathcal{M}_\odot$ and a radius $R = 15\,\text{km}$ one finds $\langle E_{\text{kin}}\rangle \approx 25\,\text{MeV}$ or $T = \frac{2}{3}\langle E_{\text{kin}}\rangle \approx 17\,\text{MeV}$. Therefore, thermal neutrinos emitted from the neutron-star surface are characterized by a temperature of order 10 MeV.

The duration of neutrino emission is a multiple of the neutrino diffusion time scale over the dimension of the neutron-star radius,

$$t_{\text{diff}} \approx R^2/\lambda, \qquad (11.2)$$

where λ is a typical mean free path. A typical neutral-current weak scattering cross section on nonrelativistic nucleons is given by

$$\sigma \approx G_F^2 E_\nu^2 / \pi = 1.7 \times 10^{-42} \, \text{cm}^2 \, (E_\nu / 10 \, \text{MeV})^2. \qquad (11.3)$$

Nuclear density ($\rho_0 \approx 3 \times 10^{14} \, \text{g cm}^{-3}$) corresponds to a nucleon density of about $1.8 \times 10^{38} \, \text{cm}^{-3}$ so that $\lambda \approx 300 \, \text{cm}$ for 30 MeV neutrinos. This yields a diffusion time scale $t_{\text{diff}} = \mathcal{O}(1 \, \text{s})$.

In summary, one expects an energy of about $0.5 \times 10^{53} \, \text{erg}$ to be emitted in each (anti)neutrino degree of freedom over a time scale of order 1 sec with typical energies of order several 10 MeV.

11.2.2 Energies and Spectra

These global properties of the expected neutrino signal are broadly confirmed by detailed numerical calculations of neutrino transport.[65] However, there are a number of important "fine points" to keep in mind. First, the nonelectron neutrino degrees of freedom $\nu_{\mu,\tau}$ and $\bar{\nu}_{\mu,\tau}$ have smaller opacities; their energies are too low for charged-current reactions of the sort $\bar{\nu}_\mu + p \rightarrow n + \mu^+$ because of the large masses of the μ and τ leptons. These flavors decouple at higher densities and temperatures than ν_e and $\bar{\nu}_e$ and so they are emitted with higher average energies. Equally important, ν_e's have lower energies than $\bar{\nu}_e$'s because the opacities are dominated by $\nu_e + n \rightarrow p + e^-$ and $\bar{\nu}_e + p \rightarrow n + e^+$, respectively, and because there are fewer protons than neutrons. Typically one finds (Janka 1993)

$$\langle E_\nu \rangle = \begin{cases} 10-12 \, \text{MeV} & \text{for } \nu_e, \\ 14-17 \, \text{MeV} & \text{for } \bar{\nu}_e, \\ 24-27 \, \text{MeV} & \text{for } \nu_{\mu,\tau} \text{ and } \bar{\nu}_{\mu,\tau}, \end{cases} \qquad (11.4)$$

i.e. typically $\langle E_{\nu_e} \rangle \approx \frac{2}{3} \langle E_{\bar{\nu}_e} \rangle$ and $\langle E_\nu \rangle \approx \frac{5}{3} \langle E_{\bar{\nu}_e} \rangle$ for the other flavors.

The number fluxes of the nonelectron flavors are smaller than those of $\bar{\nu}_e$ because the energy is found to be approximately equipartitioned between the flavors: the total $E_{\nu_e + \bar{\nu}_e}$ lies between $\frac{1}{3}$ and $\frac{1}{2}$ of E_b. Similarly, the number flux of ν_e is larger than that of $\bar{\nu}_e$ (the lepton number is carried away in ν_e's!) so that, again, the energy is approximately equipartitioned between ν_e and $\bar{\nu}_e$. The total $E_{\bar{\nu}_e}$ is found to lie between $\frac{1}{6}$ and $\frac{1}{4}$ of E_b. The SN 1987A observations were almost exclusively sensitive to the $\bar{\nu}_e$ flux. The total $E_{\bar{\nu}_e}$ inferred from these measurements

---

[65]See, e.g., Burrows and Lattimer (1986); Bruenn (1987); Mayle, Wilson, and Schramm (1987); Burrows (1988); Janka and Hillebrandt (1989a,b); Myra and Bludman (1989); Myra and Burrows (1990). For reviews see Cooperstein (1988) and Burrows (1990a,b).

must then be multiplied with a factor between 4 and 6 to obtain an estimate of E_b.

Even though the emission of neutrinos is a quasithermal process their energies are not set at the "neutrino sphere" which is defined to be the approximate shell from where they can escape without substantial further diffusion. Of course, even the notion of a neutrino sphere is a crude concept because of the E_ν^2 dependence of the scattering cross section on nonrelativistic nucleons which implies that there is a separate neutrino sphere for each energy group. The scattering with nucleons does not allow for much energy transfer apart from recoil effects. What is relevant for determining the neutrino energies is their "energy sphere" where they last exchanged energy by the scattering on electrons, by pair processes, and by charged-current absorption. Naturally, this region lies interior to the neutrino sphere—see the shaded areas in Fig. 11.1 as opposed to the dotted line which represents the neutrino sphere.

The concept of an "energy sphere" (where neutrinos last exchanged energy with the medium) and of a "transport sphere" (beyond which they can stream off without further scattering) helps to explain the apparent paradox that the $\bar{\nu}_\mu$ spectrum is, say, twice as hard as that of $\bar{\nu}_e$, yet the same amount of energy is radiated. Both fluxes originate from the same radius of about 15 km so that the Stefan-Boltzmann law ($L \propto R^2 T^4$) would seem to indicate that the $\bar{\nu}_\mu$ flux should carry 16 times as much energy. However, the place to which the Stefan-Boltzmann law should be applied is the energy sphere, yet the neutrinos cannot escape from there because the flow is impeded by neutral-current scattering on an overburden of nucleons. One may crudely think of the energy sphere being covered with a skin that does not allow the radiation to stream off except through some holes. Thus the effectively radiating surface is smaller than $4\pi R^2$. (For a more technical elaboration of this argument see Janka 1995a.)

Evidently, neutrino transport is a rather complicated problem, especially in the transition region between diffusion and free escape. The most accurate numerical way to implement it would be a Monte Carlo integration of the Boltzmann collision equation (Janka and Hillebrandt 1989a,b). In practice, this is not possible because of the constraints imposed by the limited speed of present-day computers so that a variety of approximation methods are used to solve this problem. While there is broad agreement on the general features of the expected neutrino signal, there remain differences between the predicted spectra and lightcurves of different authors.

Apart from the highly nontrivial problem of neutrino transport, the expected signal depends on a variety of physical assumptions such as the nuclear equation of state which determines the stellar equilibrium configuration and the amount of energy that is liberated, on the properties of the progenitor star (notably the iron core mass), on the duration of the accretion phase while the shock stalls, on the treatment of convection during the first few 100 ms, and others.

In the Monte Carlo integrations of Janka and Hillebrandt (1989a,b) the neutrino spectra at a given time are found to be reasonably well described by the Fermi-Dirac shape

$$\frac{dL_\nu}{dE_\nu} \propto \frac{E_\nu^3}{1 + e^{E_\nu/T_\nu - \eta_\nu}}, \tag{11.5}$$

where T_ν is an effective neutrino temperature and η_ν an effective degeneracy parameter. This ansatz allows one to fit the overall luminosity by a global normalization factor as well as the energy moments $\langle E_\nu \rangle$ and $\langle E_\nu^2 \rangle$; finer details of the spectrum are probably not warranted anyway. Throughout the emission process, η_{ν_e} decreases from about 5 to 3, $\eta_{\bar{\nu}_e}$ from about 2.5 to 2, and $\eta_{\nu_{\mu,\tau},\bar{\nu}_{\mu,\tau}}$ from 2 to 0. This effective degeneracy parameter is the same for $\nu_{\mu,\tau}$ and $\bar{\nu}_{\mu,\tau}$, in contrast with a real chemical potential which changes sign between particles and antiparticles.

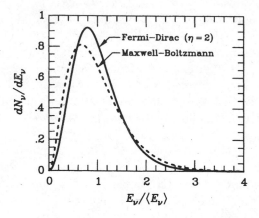

Fig. 11.6. Normalized neutrino spectral distribution according to a Maxwell-Boltzmann distribution and a Fermi-Dirac distribution with an effective degeneracy parameter $\eta = 2$, typical for the Monte Carlo transport calculations of Janka and Hillebrandt (1989a,b). The temperatures are $T_\nu = \frac{1}{3}\langle E_\nu \rangle$ (Maxwell-Boltzmann) and $T_\nu = 0.832\,\frac{1}{3}\langle E_\nu \rangle$ (Fermi-Dirac with $\eta = 2$).

Figure 11.6 shows a normalized Maxwell-Boltzmann spectrum and a Fermi-Dirac spectrum with $\eta = 2$, both for the same average energy $\langle E_\nu \rangle$. This choice implies $T_\nu = \frac{1}{3} \langle E_\nu \rangle$ (Maxwell-Boltzmann) and $T_\nu = 0.832 \frac{1}{3} \langle E_\nu \rangle$ (Fermi-Dirac with $\eta = 2$). What is shown in each case is the normalized number spectrum, i.e. $dN_\nu/dE_\nu \propto E_\nu^2 e^{-E_\nu/T_\nu}$ (Maxwell-Boltzmann) and $E_\nu^2/(1 + e^{E_\nu/T_\nu - \eta_\nu})$ (Fermi-Dirac). It is apparent that for a fixed average energy the Fermi-Dirac spectrum is "pinched," i.e. it is suppressed at low and high energies relative to the Maxwell-Boltzmann case.

The most important difference between the two cases is the suppressed high-energy tail of the pinched spectra which causes a significant reduction of neutrino absorption rate. This may be important for neutrino detection in terrestrial detectors as well as for neutrino-induced nuclear reactions in the SN mantle and envelope. However, for the sparse SN 1987A signal the differences would not have been overly dramatic. Therefore, in view of the many other uncertainties at predicting the spectrum most practical studies of neutrino emission and possible detector signals used simple Maxwell-Boltzmann spectra, or Fermi-Dirac spectra with a vanishing chemical potential which differ from the former only in minor detail.

11.2.3 Time Evolution of the Neutrino Signal

The schematic time evolution of the (anti)neutrino luminosities of the different flavors was shown in Fig. 11.3. The prompt ν_e burst has relatively high energies ($\langle E_{\nu_e} \rangle \approx 15\,\text{MeV}$), but the total energy content of a few 10^{51} erg renders it negligible relative to the integrated luminosity of the subsequent emission phases.

The average energies and luminosities of the other flavors rise during the first few 100 ms while the shock stalls, matter is accreted, and the initially bloated outer region of the protoneutron star contracts. The temperature of the region near the edge of the lepton number step rises substantially during this epoch (Fig. 11.2). During the first 0.5 s somewhere between 10% and 25% of the total binding energy is radiated away; the remainder follows during the Kelvin-Helmholtz cooling phase of the settled star.

Detailed parametric studies of the Kelvin-Helmholtz cooling phase were performed by Burrows (1988), and more recently by Keil and Janka (1995). In these works the expected SN 1987A detector signal was studied as a function of the assumed nuclear equation of state (EOS), the mass of the collapsed core at bounce, the amount of postbounce accretion, and the temperature profile of the core after collapse.

As expected, a soft EOS leads to a large amount of binding energy and thus to large integrated neutrino luminosities; a large core mass or large postbounce accretion rate has a similar effect. A soft EOS leads to relatively high temperatures during deleptonization, causing large neutrino opacities and thus long emission time scales. If the EOS is too soft, or the core mass too large, the final configuration is not stable and collapses, presumably to a black hole. This must not occur too early to avoid conflict with the duration of the observed SN 1987A

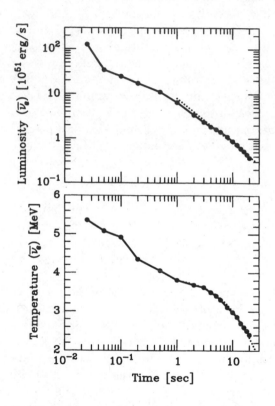

Fig. 11.7. Luminosity and temperature of the $\bar{\nu}_e$ flux from the protoneutron star model 55 of Burrows (1988) which is based on a "stiff equation of state," an initial baryonic core mass of $1.3\,\mathcal{M}_\odot$, and an accretion of $0.2\,\mathcal{M}_\odot$ within the first 0.5 s. The dotted line in the upper panel indicates a t^{-1} behavior, in the lower panel it indicates $e^{-t/4\tau}$ with $\tau \approx 10$ s. The neutrino spectral distribution was taken to be thermal.

signal (Sect. 11.3). However, in some cases studied by Keil and Janka (1995) with an EOS including hyperons this final collapse occurs so late (at 8 s after bounce in one example) that black-hole formation is difficult to exclude on the basis of the SN 1987A observations, notably as no pulsar has yet been found there.

Needless to say, with so many parameters to play it is not difficult to find combinations of EOS, core mass, accretion rate, and initial temperatures which fit the observed SN 1987A signal well within the statistical uncertainties of the observations. An example is model 55 of Burrows (1988) which is based on a "stiff EOS," an initial baryonic core mass of $1.3\,\mathcal{M}_\odot$, and an accretion of $0.2\,\mathcal{M}_\odot$ within the first 0.5 s. The evolution of the effective $\bar{\nu}_e$ luminosity and temperature is shown in Fig. 11.7. After about 1 s the decay of the temperature is fit well by an exponential $e^{-t/4\tau}$ with $\tau \approx 10$ s while the decay of the luminosity is poorly fit by an exponential; it decays approximately as t^{-1} after 1 s. For later reference, the time-integrated flux (fluence) of this model is shown in Fig. 11.8.

It must be stressed that the "cooling behavior" (decrease of the average $\bar{\nu}_e$ energy) shown in Fig. 11.7 may not be generic at early times. Initially the star is quite bloated, and relatively cold. Therefore, the

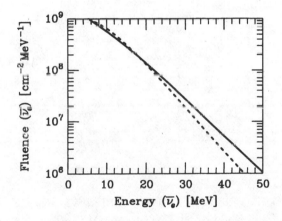

Fig. 11.8. Expected $\bar{\nu}_e$ fluence (time-integrated flux) from SN 1987A, assuming a distance of 50 kpc, and taking Burrows' (1988) model 55 neutrino luminosity shown in Fig. 11.7. The solid line is for an assumed Maxwell-Boltzmann energy spectrum for a given average neutrino energy, the dashed line for a Fermi-Dirac spectrum with a degeneracy parameter $\eta = 2$ as in Fig. 11.6.

initial neutrino flux may be characterized by a decreasing radius (decreasing flux), yet increasing temperature. For an overview of various model calculations see Burrows (1990b). One should be careful not to take details of the time evolution of the temperature and luminosity of any specific calculation too seriously.

11.3   SN 1987A Neutrino Observations

11.3.1   Supernova 1987A

Shelton's (1987) sighting of a supernova (SN 1987A) in the Large Magellanic Cloud (LMC), a small satellite galaxy of the Milky Way at a distance from us of about 50 kpc (165,000 ly), marked the discovery of the closest visual SN since Kepler's of 1604. It was close enough that several underground detectors which were operational at the time were able to measure the neutrino flux from the core collapse of the progenitor star, the blue supergiant Sanduleak $-69\,202$. The observed neutrinos were registered within a few seconds of 7:35:40 UT (universal

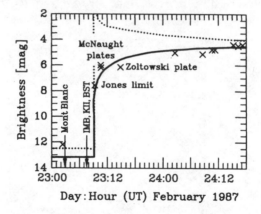

Fig. 11.9. Early optical observations of SN 1987A according to the IAU Circulars, notably No. 4316 of February 24, 1987. The times of the IMB, Kamiokande II (KII) and Baksan (BST) neutrino observations (23:07:35) and of the Mont Blanc events (23:02:53) are also indicated. The solid line is the expected visual brightness, the dotted line the bolometric brightness according to model calculations. (Adapted, with permission, from Arnett et al. 1989, Annual Review of Astronomy and Astrophysics, Volume 27, © 1989, by Annual Reviews Inc.)

time) on 23 February 1987 while the first evidence for optical brightening was found at 10:38 UT on plates taken by McNaught (1987)—see Fig. 11.9.

The main neutrino observations come from the Irvine-Michigan-Brookhaven (IMB) and the Kamiokande II water Cherenkov detectors, facilities originally built to search for proton decay, while a less significant measurement is from the Baksan Scintillator Telescope (BST). A likely spurious observation is from the Mont Blanc Liquid Scintillator Detector (LSD). It preceded the other observations by about 5 h, with no contemporaneous signal at Mont Blanc with the other signals, and no contemporaneous signal at the other detectors with the Mont Blanc event. The Mont Blanc detector was built to search for neutrinos from core collapse supernovae, except that it was optimized for galactic events within a distance of about 10 kpc. The neutrino output of a normal SN in the LMC could not have caused an observable signal at Mont Blanc; the reported events probably represent a background fluctuation.

Koshiba (1992) has given a lively account of the exciting and initially somewhat confusing story of the neutrino measurements and their interpretation. Early summaries of the implications for astrophysics and particle physics of the neutrino and electromagnetic observations were written, for example, by Schramm (1987), Arnett et al. (1989), and Schramm and Truran (1990). A more recent review of SN 1987A is McCray (1993). A nontechnical overview was provided in a book by Murdin (1990).

For the present purposes the bottom line is that SN 1987A broadly confirmed our understanding of SN physics as outlined in Sect. 11.1. A remaining sore point is the lack of a pulsar observation in the SN remnant so that one may continue to speculate that a black hole has formed in the collapse.

11.3.2 Neutrino Observations

In the IMB and Kamiokande water Cherenkov detectors neutrinos are measured by the Cherenkov light emitted by secondary charged particles, $e^{\pm}$ for the relatively low-energy (anti)neutrinos emitted from a stellar collapse. The IMB detector is now defunct while Kamiokande continues to measure, for example, solar neutrinos until the much larger Superkamiokande detector will take up operation in 1996. In the Baksan Scintillator Telescope (BST) one measures the scintillation light produced by charged secondary particles.

The relevant neutrino interaction processes in water are elastic scattering on electrons (Sect. 10.3.4), and the charged-current reactions $\bar{\nu}_e\, p \to n\, e^+$ and $\nu_e\, {}^{16}O \to {}^{16}F\, e^-$ (Arafune and Fukugita 1987; Haxton 1987). The cross section for the $\bar{\nu}_e p$ reaction is given by

$$\sigma = \frac{G_F^2}{\pi}\cos^2\theta_C\left(C_V^2 + 3C_A^2\right)p_e E_e\,(1+\delta)$$

$$= 9.4\times10^{-44}\,\mathrm{cm}^2\,(1+\delta)\,p_e E_e/\mathrm{MeV}^2, \qquad (11.6)$$

where G_F is the Fermi constant and $\cos^2\theta_C \approx 0.95$ refers to the Cabibbo angle. The charged-current vector and axial-vector weak coupling constants are $C_V = 1$ and $C_A = 1.26$, and δ incorporates small corrections from recoil, Coulomb, radiative and weak magnetism corrections (Vogel 1984). Further, p_e and E_e refers to the positron momentum and energy. Ignoring recoil effects, the latter is $E_e = E_\nu - m_n + m_p \approx E_\nu - 1.3\,\mathrm{MeV}$; the threshold is $1.8\,\mathrm{MeV}$ because the minimum E_e is m_e.

A general expression for the $\nu_e{}^{16}O$ cross section is much more complicated. A simple approximation, taking only the 2^- state of the ${}^{16}F$ nucleus, is

$$\sigma \approx 1.1\times10^{-44}\,\mathrm{cm}^2\,(E_\nu/\mathrm{MeV} - 13)^2 \qquad (11.7)$$

(Arafune and Fukugita 1987).

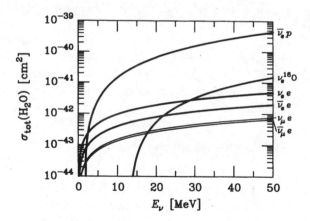

Fig. 11.10. Total cross sections for the measurement of neutrinos in a water Cherenkov detector according to Eqs. (10.17), (11.6), and (11.7). The curves refer to the total cross section per water molecule so that a factor of 2 for protons and 10 for electrons is already included.

All of the relevant cross sections per water molecule are shown in Fig. 11.10 as a function of E_ν. The curves incorporate a factor of 2 for proton targets (two per H_2O), and a factor of 10 for electrons (ten per H_2O). Above its threshold, the $\nu_e {}^{16}O$ cross section rises very fast; it is then the dominant detection process for ν_e's. Still, the $\bar{\nu}_e p$ reaction is the absolutely dominant mode of observing SN neutrinos.

The BST detector is filled with an organic scintillator based on "white spirit" $C_n H_{2n+2}$ with $n \approx 9$. The dominant detection reaction is also $\bar{\nu}_e p \to n e^+$. In addition, elastic scattering on electrons is possible, and the process $\nu_e {}^{12}C \to {}^{12}N\, e^-$ occurs for $E_{\nu_e} \gtrsim 30\,\mathrm{MeV}$.

The trigger efficiencies relevant for the three detectors are shown as a function of the $e^\pm$ energy in Fig. 11.11. Analytic fit formulae to these curves were given by Burrows (1988) for IMB and Kamiokande. IMB reports a dead time of 13% during the SN burst (Bratton et al. 1988); the IMB curve includes a factor 0.87 to account for this effect. The fiducial volume of Kamiokande II relevant for the SN 1987A observations was 2,140 tons, for IMB 6,800 tons, and for BST 200 tons. It corresponds to a target of 1.43×10^{32} protons at Kamiokande, 4.6×10^{32} at IMB, and 1.88×10^{31} at BST.

With the $\bar{\nu}_e p$ cross section of Eq. (11.6) and the efficiency curves of Fig. 11.11 one may compute a prediction for the number of events

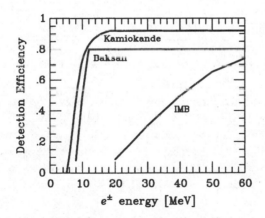

Fig. 11.11. Trigger efficiency for electron (positron) detection at the Kamiokande (Hirata et al. 1988) and IMB (Bratton et al. 1988) water Cherenkov detectors, and the Baksan scintillator telescope (Alexeyev et al. 1988), relevant for the SN 1987A neutrino observations. In the IMB curve a factor 0.87 is included to account for their reported dead time of 13%.

per energy interval due to the dominant $\bar{\nu}_e p$ reaction in the detectors. For Kamiokande and IMB an example is shown in Fig. 11.12, based on Burrows' (1988) model 55 flux calculation which was tuned to fit the data. The predicted fluence at Earth per unit energy was shown in Fig. 11.8 where a distance of 50 kpc was adopted. The solid lines are for the case when the instantaneous $\bar{\nu}_e$ spectra are assumed to be Maxwell-Boltzmann. Of course, the time-integrated flux is then no longer thermal as it is a superposition of Maxwell-Boltzmann spectra at different temperatures. The solid line leads to a total expectation of 13.1 events at Kamiokande and 6.3 at IMB. The dashed lines correspond to spectra which are instantaneously pinched (Fig. 11.6) with a degeneracy parameter $\eta = 2$. Again, the time-integrated spectra are not necessarily pinched. This case leads to 11.6 events at Kamiokande and 3.9 at IMB.

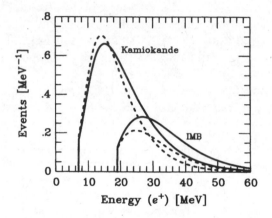

Fig. 11.12. Expected number of events per energy interval at Kamiokande and IMB from the SN 1987A $\bar{\nu}_e$ flux on the basis of the $\bar{\nu}_e p \to n e^+$ reaction. The flux prediction is based on Burrows' (1988) model 55 which was tuned to fit the SN 1987A data. The distance is taken to be 50 kpc, and the detector efficiency curves of Fig. 11.8 are used. The dashed lines refer to neutrino spectra which are instantaneously "pinched" as in Fig. 11.6.

Water Cherenkov detectors, as opposed to scintillation ones, are imaging devices in that they can resolve the direction of the electron (positron) because of the directionality of the emitted Cherenkov light. The main limitation is multiple Coulomb scattering of low-energy $e^\pm$ in the medium. A typical path length in water is only a few cm so that hard collisions are relatively unlikely. The rms angular deviation due to multiple scattering varies from about $34°$ at $E_e = 5\,\mathrm{MeV}$ to $22°$ at $20\,\mathrm{MeV}$ (Hirata 1991).

The direction of motion of the charged lepton relative to the incident neutrino is well preserved in νe collisions (Sect. 10.3.4) so that the main limitation to a reconstruction of the primary neutrino direction is multiple Coulomb scattering. For the $\bar{\nu}_e p$ reaction, the angular distribution is isotropic, apart from a small (about 10%) backward asymmetry (e.g. Boehm and Vogel 1987). The $\nu_e{}^{16}\mathrm{O}$ reaction yields a distribution approximately proportional to $1 - \frac{1}{3}\cos\Theta$, i.e. it is also nearly isotropic with a backward bias (Haxton 1987). As a SN is expected to produce all (anti)neutrino flavors in about equal numbers, the signal is dominated by the isotropic $\bar{\nu}_e p$ reaction.

Table 11.1. Neutrino burst at the Kamiokande detector (Hirata et al. 1988). The time is relative to the first event at 7:35:35 ± 0:01:00 UT, 23 Feb. 1987. The energy refers to the detected $e^{\pm}$, not to the primary neutrino.

Event	Time [s]	Angle [degree]	Energy [MeV]
1	0.000	18 ± 18	20.0 ± 2.9
2	0.107	40 ± 27	13.5 ± 3.2
3	0.303	108 ± 32	7.5 ± 2.0
4	0.324	70 ± 30	9.2 ± 2.7
5	0.507	135 ± 23	12.8 ± 2.9
6[a]	0.686	68 ± 77	6.3 ± 1.7
7	1.541	32 ± 16	35.4 ± 8.0
8	1.728	30 ± 18	21.0 ± 4.2
9	1.915	38 ± 22	19.8 ± 3.2
10	9.219	122 ± 30	8.6 ± 2.7
11	10.433	49 ± 26	13.0 ± 2.6
12	12.439	91 ± 39	8.9 ± 1.9
13[a,b]	17.641	...	6.5 ± 1.6
14[a,b]	20.257	...	5.4 ± 1.4
15[a,b]	21.355	...	4.6 ± 1.3
16[a,b]	23.814	...	6.5 ± 1.6

[a]Usually attributed to background.
[b]Quoted after Loredo and Lamb (1995).

Table 11.2. Neutrino burst at the IMB detector (Bratton et al. 1988). The time is relative to the first event at 7:35:41.374 ± 0:00:00.050 UT, 23 Feb. 1987. The energy refers to the detected $e^{\pm}$, not to the primary neutrino.

Event	Time [s]	Angle [degree]	Energy [MeV]
1	0.000	80 ± 10	38 ± 7
2	0.412	44 ± 15	37 ± 7
3	0.650	56 ± 20	28 ± 6
4	1.141	65 ± 20	39 ± 7
5	1.562	33 ± 15	36 ± 9
6	2.684	52 ± 10	36 ± 6
7	5.010	42 ± 20	19 ± 5
8	5.582	104 ± 20	22 ± 5

Table 11.3. Neutrino burst at the Baksan detector (Alexeyev et al. 1987, 1988). The time is relative to the first event at $7:36:06.571^{+02.000}_{-54.000}$ UT, 23 Feb. 1987. The energy refers to the detected $e^{\pm}$, not to the primary neutrino.

Event	Time [s]	Energy [MeV]
0[a]	0.000	17.5 ± 3.5
1	5.247	12.0 ± 2.4
2	5.682	18.0 ± 3.6
3	6.957	23.3 ± 4.7
4	12.934	17.0 ± 3.0
5	14.346	20.1 ± 4.0

[a]Usually attributed to background.

The energy of the electron (positron) can be reconstructed from the total amount of Cherenkov or scintillation light emitted. For small energies it is roughly proportional to the number of photomultipliers hit in a given event. Because of the reaction threshold and recoil effects, the energy of the primary neutrino in the $\bar{\nu}_e p \to n e^+$ reaction is about 2 MeV larger than the measured e^+ energy. For the rare νe collisions,

the final-state electron energy distribution is broad so that one can infer
only a lower limit to the ν energy.

The measured events at the Kamiokande (Hirata et al. 1987, 1988),
IMB (Bionta et al. 1987; Bratton et al. 1988), and BST (Alexeyev et al.
1987, 1988) detectors are summarized in Tabs. 11.1, 11.2, and 11.3.
The absolute timing at IMB is accurate to within ± 50 ms while at
Kamiokande only to within ± 1 min. At BST, the clock exhibited an
erratic behavior which led to an uncertainty of $+2/-54$ s. Within the
timing uncertainties the three bursts are contemporaneous and may

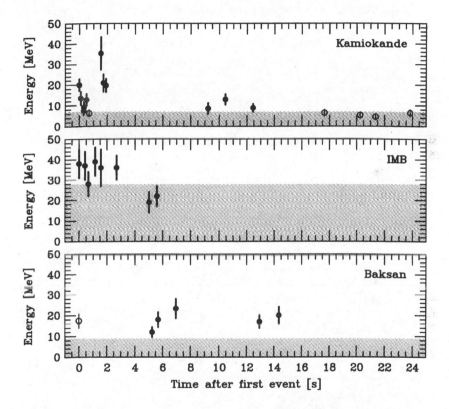

Fig. 11.13. SN 1987A neutrinos at Kamiokande, IMB, and Baksan. The
energies refer to the secondary positrons, not the pimary neutrinos. In the
shaded area the trigger efficiency is less than 30%. The detector clocks have
unknown relative offsets; in each case the first event was shifted to $t = 0$. In
Kamiokande and Baksan, the events marked with open circles are usually
attributed to background.

thus be simultaneously attributed to SN 1987A. The events are shown in the *t-E*-plane in Fig. 11.13 and for the Cherenkov detectors in the $\cos\Theta$-*E*-plane in Fig. 11.14 where Θ is the angle relative to the opposite direction of the SN, i.e. relative to the direction of the neutrino flux.

In principle, the bursts of events observed in the detectors could be due to rare background fluctuations rather than due to SN 1987A. The Kamiokande group has performed a detailed analysis of this possibility by analyzing the multiplicity of events in 10 s time intervals, i.e. the number of chance events in an arbitrarily chosen 10 s time interval. They found a probability of about 0.6×10^{-7} that the observed burst is a random fluctuation of a constant background.

However, there are time-correlated backgrounds, notably the spallation of oxygen induced by primary muons which can cause clusters of events with large multiplicities; the Kamiokande group found one clus-

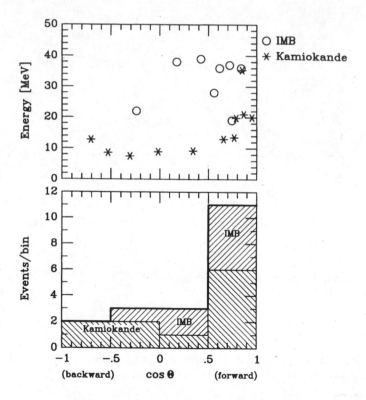

Fig. 11.14. SN 1987A neutrinos at Kamiokande and IMB, excluding the ones which likely are due to background.

ter of 53 events! Therefore, any cluster following a high-energy muon would be very suspicious. Performing a cut on the data for this background leaves no burst with multiplicity 3 or larger in several data sets of several hundred days each (Hirata 1991). Therefore, it is extremely unlikely that the event cluster 10–12 in the Kamiokande data has been caused by background.

The BST detector has a relatively large background rate. Event clusters of multiplicity 5 or more within 9 s occur about once per day. Thus the probability for such a background cluster to fall within a minute of the IMB and Kamiokande events is about 5×10^{-4}.

11.3.3 Analysis of the Pulse

Many authors have studied the distribution of energies and arrival times of the reported events. Probably the most significant work is that of Loredo and Lamb (1989, 1995) who performed a maximum-likelihood analysis, carefully including the detector backgrounds and trigger efficiencies. Loredo and Lamb also gave detailed references to previous works, and in some cases offered a critique of the statistical methodology employed there. Their more extensive 1995 analysis supersedes certain aspects of the earlier methodology and results.

Because of the small number of neutrinos observed, a relatively crude parametrization of the time-varying source is enough. Among a variety of simple single-component emission parametrizations, Loredo and Lamb (1989, 1995) found that an exponential cooling model was preferred. It is characterized by a constant radius of the neutrino sphere, R, and a time-varying effective temperature

$$T(t) = T_0 \, e^{-t/4\tau}, \tag{11.8}$$

so that τ is the decay time scale of the luminosity which varies with the fourth power of the temperature according to the Stefan-Boltzmann law. It should be noted, however, that numerical cooling calculations do not yield exponential lightcurves. For example, the model shown in Fig. 11.7 displays an exponential decline of the effective temperature, but a power-law decline of the neutrino luminosity. Other calculations even yield early heating and a constant temperature for some time (see Burrows 1990b for an overview).

Of course, for the time-integrated spectrum the exponential cooling law is just another assumption concerning the overall spectral shape. For example, one easily finds that the average $\bar{\nu}_e$ energy of the time-integrated spectrum is $\langle E_{\bar{\nu}_e} \rangle = 2.36 \, T_0$ if Fermi-Dirac distributions with

$\eta = 0$ are taken for the instantaneous spectra. The time-integrated spectrum of the exponential cooling model looks quite similar to the time-integrated spectrum shown in Fig. 11.8.

Loredo and Lamb also used $\alpha \equiv (R/10\,\mathrm{km})\,(50\,\mathrm{kpc}/D)\,g^{1/2}$ as a fit parameter where D is the distance to SN 1987A and g a statistical weight factor which is unity if only left-handed, massless or low-mass neutrinos of the three sequential flavors are emitted. The registration time of the first neutrino in each detector is taken as a free parameter relative to the arrival time of the first neutrinos. In the 1995 analysis, Loredo and Lamb included the Baksan signal without "event 0" which is attributed to background because it precedes the main bunch by 5 s.

The following six parameters are then allowed to float freely in order to achieve a maximum-likelihood result: T_0, τ, α, $t_{\mathrm{off}}(\mathrm{IMB})$, $t_{\mathrm{off}}(\mathrm{KII})$, and $t_{\mathrm{off}}(\mathrm{BST})$. All best-fit offset times are found to be zero. The other best-fit values are $\alpha = 4.02$, $\tau = 4.37\,\mathrm{s}$, and $T_0 = 3.81\,\mathrm{MeV}$. This initial temperature of the exponential cooling model corresponds to an average neutrino energy of the time-integrated flux of $\langle E_{\bar{\nu}_e}\rangle = 9.0\,\mathrm{MeV}$. In Fig. 11.15, the 68% and 95% credible regions are shown in the T_0-τ-plane where T_0 has been translated into $\langle E_{\bar{\nu}_e}\rangle$ which is of greater direct relevance.

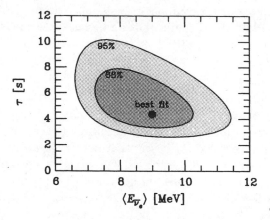

Fig. 11.15. Two-dimensional marginal distribution for the parameters τ and $\langle E_{\bar{\nu}_e}\rangle = 2.36\,T_0$ of the exponential cooling model. (Curves courtesy of Tom Loredo, taken from Loredo and Lamb 1995.)

Given these parameters one infers that the number of expected Kamiokande events is 16.9 plus 5.6 background, 4.0 events at IMB, and 1.8 plus 1.0 background at Baksan. The inferred best-fit neutrino-sphere radius is 40.2 km, and the inferred total emitted $\bar{\nu}_e$ energy is 0.84×10^{53} erg which corresponds to a total binding energy of the neutron star of 5.02×10^{53} erg if exact equipartition of the energy among the neutrino flavors is assumed.

The inferred average neutrino energy, the luminosity-decay time scale, radius of the source, and total energy emitted all agree reasonably well with what one expects from a core collapse SN, even though $\langle E_{\bar{\nu}_e} \rangle$ is somewhat low, the radius and inferred binding energy somewhat large. Of course, all of the inferred quantities carry large uncertainties because of the sparse data.

Loredo and Lamb (1995) have also considered two-component cooling schemes where the neutrino signal is modelled to consist of Kelvin-Helmholtz cooling, plus a low-energy component which mimics the neutrinos emitted by the accreting matter during the stalled-shock phase in the delayed-explosion scenario. With more parameters they naturally find a better fit to the data. More interestingly, the inferred

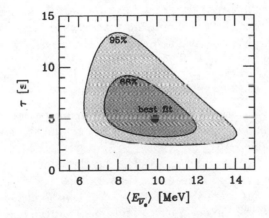

Fig. 11.16. Two-dimensional marginal distribution for the parameters τ and $\langle E_{\bar{\nu}_e} \rangle = 2.13\,T_0$ of the Kelvin-Helmholtz component of the Loredo and Lamb (1995) best-fit two-component cooling model. The parameters τ and T_0 are those of a displaced power law as described in the text. (Curves courtesy of Tom Loredo, taken from Loredo and Lamb 1995.)

neutrino-sphere radius of 18 km and the inferred total binding energy of 3.08×10^{53} erg correspond much better to theoretical expectations. Perhaps this finding can be taken as a hint that the delayed-explosion scenario with a significant matter accretion phase is favored by the SN 1987A data over a prompt-explosion picture.

In Loredo and Lamb's best-fit two-component model the Kelvin-Helmholtz signal is described by a "displaced power law" cooling model with a neutrino sphere of fixed radius R and thermal neutrino emission with $T(t) = T_0/(1 + t/3\tau)$. It turns out that the luminosity, which is proportional to T^4, follows a surprisingly similar curve to the exponential $e^{-t/\tau}$ so that the parameter τ has practically the same meaning as before. $\langle E_{\bar{\nu}_e} \rangle$ for the time-integrated flux is given by $2.13\,T_0$, very similar to $2.36\,T_0$ for the exponential. In Fig. 11.16 the 68% and 95% credible regions are shown in the T_0-τ-plane. While the best-fit value is not too different from the single-component exponential model of Fig. 11.15, the 95% credible region is much larger, including the lowest values of the typical theoretical $\langle E_{\bar{\nu}_e} \rangle$ predictions quoted in Eq. (11.4).

11.3.4 Neutrino Mass and Pulse Duration

Zatsepin (1968) was the first to point out that the $\bar{\nu}_e$ burst expected from stellar collapse offers a possibility to measure or constrain small neutrino masses. Because a neutrino with mass m_ν travels slower than the speed of light its arrival at Earth will be delayed by

$$\Delta t = 2.57\,\text{s} \left(\frac{D}{50\,\text{kpc}} \right) \left(\frac{10\,\text{MeV}}{E_\nu} \right)^2 \left(\frac{m_\nu}{10\,\text{eV}} \right)^2 . \qquad (11.9)$$

Because the measured $\bar{\nu}_e$'s from SN 1987A were registered within a few seconds and had energies in the 10 MeV range, the m_{ν_e} mass is limited to less than about 10 eV.

A detailed study must proceed along the lines of the maximum-likelihood analysis of Loredo and Lamb (1989, 1995) quoted in the previous section where the detector background is included, and such parameters as the unknown offset times between the detectors are left unconstrained. Including the possibility that some of the registered events are due to background is particularly important because the neutrino mass limit is very sensitive to the early low-energy events at Kamiokande which have a relatively high chance of being due to background. Loredo and Lamb (1989) found a vanishing best-fit neutrino mass and a 95% CL upper limit of $m_{\nu_e} < 23\,\text{eV}$. This bound is less restrictive than limits found by previous authors on the basis of less thorough statistical analyses. In their 1995 paper, Loredo and Lamb have not studied neutrino mass limits which likely would change somewhat because of corrections to their previous approach.

An analysis by Kernan and Krauss (1995) on the basis of a similar method yields a limit 19.6 eV at 95% CL. Apparently, the reduction of the limit is due to their inclusion of the 13% dead-time effect in the IMB detector. The difference to the Loredo and Lamb (1989) limit illustrates that changing a relatively fine point of the analysis procedure can significantly change a so-called 95% CL limit. Therefore, instead of quoting a specific confidence limit it is at present more realistic to state qualitatively that a violation of the mass limit

$$m_{\nu_e} \lesssim 20 \, \text{eV} \qquad\qquad (11.10)$$

would have caused a significant and perhaps intolerable modification of the SN 1987A signal.

This limit is weaker than the current bounds from the tritium β decay endpoint spectrum (Sect. 7.1.3). Therefore, the above analysis can be turned around in the sense that the observed neutrino signal duration is probably representative of the duration of neutrino emission at the source. The observed long time scale of Kelvin-Helmholtz cooling which is indicated by the late IMB and Kamiokande events cannot be blamed on neutrino dispersion effects. In this context it is interesting to observe that Loredo and Lamb (1989) also performed a maximum-likelihood analysis with m_{ν_e} held fixed at their 95% CL upper limit 23 eV. The best-fit time scale in an exponential cooling model changed from 4.15 to 2.96 s. Therefore, even assuming a large value for m_{ν_e} did not allow one to contemplate a significantly shorter Kelvin-Helmholtz cooling phase than implied by massless neutrinos.

11.3.5 Anomalies in the Signal?

The distribution of the Kamiokande and IMB events shows a number of puzzling features. The least worrisome of them is a certain discrepancy between the neutrino energies observed in the two detectors which point to a harder spectrum at IMB. The maximum-likelihood analysis in the exponential cooling model of Loredo and Lamb (1989) was also performed for the two detectors separately. The 95% confidence volumes projected on the T_0-E_b-plane are shown in Koshiba (1992); a similar result is found in Janka and Hillebrandt (1989b). There is enough overlap between the confidence contours to allow for a joint analysis. Still, the best-fit value for Kamiokande lies outside the 95% CL volume of IMB.

Including the pinching effect discussed in Sect. 11.2.2 would enhance the discrepancy between the signals in the two detectors. Either way, the IMB detector with its high energy threshold is mostly sensitive to the high-energy tail of the neutrino spectrum. Therefore, the IMB-inferred $\langle E_{\bar{\nu}_e} \rangle$ depends sensitively on the assumed spectral shape and is thus a poor indicator of the true average energies.

A more conspicuous anomaly is the 7.3 s gap between the first 9 and last 3 events at Kamiokande. Ideas proposed to explain the alleged pulsed structure of the signal range from the occurrence of a phase transition in the nuclear medium (pions, quarks) to a secondary collapse to a black hole. It should be noted, however, that the gap is partially filled in by the IMB and Baksan data, thus arguing against a physical cause at the source. The random occurrence of a gap exceeding 7 s with three or more subsequent events can be as high as several percent, but naturally it is sensitive to the expected late-time signal (Lattimer and Yahil 1989).

The most significant and thus the most troubling anomaly is the remarkable deviation from isotropy of the events in both detectors, in conflict with the expected signature from $\bar{\nu}_e p \rightarrow n e^+$ which actually predicts a slight (about 10%) backward bias. LoSecco (1989) found a probability of about 1.5% that the combined Kamiokande and IMB data set was drawn from an isotropic distribution. Kiełczewska (1990) analyzed the expected signal from standard SN cooling calculations and found agreement only at the 0.8% CL with the measured angular distribution. The combined set of IMB plus those Kamiokande data which are above the IMB threshold, i.e. the combined set of "high-energy" events is consistent with isotropy only at the 0.07% level (van der Velde 1989); the four relevant events at Kamiokande are all very forward.

The IMB collaboration claims that their reconstruction of the event direction was not seriously impeded by the outage of about a quarter of their phototubes due to the failure of a high-voltage supply. They conducted a detailed calibration of their detector to investigate this point (Bratton et al. 1988). Therefore, one must accept that the forward-peaked angular distribution shown in Fig. 11.14 is not a problem of the detectors or the event reconstruction.

A forward-peaked distribution is expected from νe elastic scattering which has a much lower cross section than the $\bar{\nu}_e p$ process (Fig. 11.8). One expects far less than one event due to νe scattering from the cooling phase, although the first Kamiokande event has sometimes been interpreted as being a scattering event due to the prompt ν_e burst.

The forward events in both detectors have relatively large energies compared with the isotropic ones at Kamiokande,[66] contrary to what would be expected from $\nu e \to e\nu$ where some of the energy is carried away by the secondary neutrino. Assuming a larger-than-standard flux of ν_e's with larger-than-standard energies (LoSecco 1989) does not solve the problem because above $30-35\,\mathrm{MeV}$ the process $\nu_e\,^{16}\mathrm{O} \to\,^{16}\mathrm{F}\,e^-$ takes over (Fig. 11.8) which has a backward bias.

Anomalously large fluxes of $\nu_{\mu,\tau}$ or $\bar{\nu}_{\mu,\tau}$ are difficult to arrange on energetic grounds—the binding energy of the neutron star is limited. Even allowing for extreme values of E_b and extreme temperature differences between (anti)electron neutrinos and the other flavors improves the agreement only marginally; one can achieve an agreement at the 5% CL with the observed angular distribution (Kiełczewska 1990). The simple problem with elastic νe scattering to explain the data is that this process is too strongly forward peaked, especially for high-energy neutrinos, hence it does not fit the data very well either. This is especially true for the IMB events which are selected for high energies by the detector threshold, and yet are very broadly distributed around the forward direction.

A very speculative idea was put forth by van der Velde (1989) who proposed the existence of a new neutral boson $X°$ which could produce photons when interacting with nucleons. These MeV photons would look very similar to charged particles in the detectors. In the forward direction, the $X°$ cross section on $^{16}\mathrm{O}$ would be coherently enhanced, causing the observed forward bias for high-energy events while low-energy ones would naturally follow a more isotropic distribution. However, the opposite process $\gamma + {}^4\mathrm{He} \to {}^4\mathrm{He} + X°$ would then contribute to the energy-loss of horizontal-branch stars (Raffelt 1988b). The resulting bound on the interaction cross section (Tab. 2.5), valid at an energy of about $10\,\mathrm{keV}$, excludes van der Velde's scenario unless σ increases with energy at least as E^2, a scaling which could bring it up to the requisite level for E in the $10\,\mathrm{MeV}$ range, relevant for the SN detection. The Primakoff conversion of axions or similar particles on oxygen is much too inefficient in view of the restrictive limits on the axion-photon interaction strength.

In summary, the angular and energy distributions of the IMB and Kamiokande events appear to indicate a low-energy isotropic and a

---

[66]The Kamiokande events have an obvious correlation between energy and direction. The application of Spearman's rank-ordering test (e.g. Press et al. 1986) gives a confidence level of 0.06% where event No. 6 was excluded as background. Therefore, the Kamiokande data alone show a fairly significant "angular anomaly."

high-energy forward component; they are not fit well by the assumed dominant detection process $\bar{\nu}_e p \to n e^+$ which is supposed to yield an isotropic positron distribution with no directional correlation with energy. Elastic νe scattering, however, does not fit the data well either because it is too forward peaked, and anyhow it is disfavored by a small cross section unless the flux of $\nu_{\mu,\tau}$ or $\bar{\nu}_{\mu,\tau}$ was extremely high.

However, because no plausible and/or viable nonstandard cause for the observed events has been proposed one has settled for the interpretation of a statistical fluctuation for the apparent anomalies. After all, it is difficult to imagine a small sample drawn from any distribution without some "anomalies" which are easy to overinterpret. Still, if a reasonable alternative to the standard interpretation of the signal were to come forth this topic would have to be reconsidered.

11.4   Neutrino Oscillations

11.4.1   Overview

The expected neutrino signature from a stellar collapse and conversely the inferred protoneutron star properties from the SN 1987A neutrino signal both depend on the assumption that "nothing happens" to the neutrinos on their way to us. One simple modification of the expected signal is a dispersion of the $\bar{\nu}_e$ burst caused by a nonvanishing m_{ν_e} in the 10 eV range (Sect. 11.3.4). Dispersion effects could also be caused by novel interactions with the galactic magnetic field, dark matter, the neutrino background, or simply by decays. All of these scenarios require relatively exotic particle-physics assumptions which can be constrained by the SN 1987A signal (Chapter 13). The assumption of small neutrino masses and mixings, however, fits into the standard model with minimal extensions, and may already be implied by the solar neutrino observations (Sect. 10.6). Therefore, it is prudent not to ignore the possible impact of oscillations on SN neutrinos.

The most obvious consequence is that the prompt ν_e burst could oscillate into another flavor which then would be much harder to observe because of the reduced ν-e cross section for non-ν_e flavors (Fig. 11.10). Notably, if the solar ν_e flux is depleted by resonant oscillations one may expect the same in the SN mantle and envelope where a large range of densities and density gradients is available. It will turn out, however, that the small-angle MSW solution to the solar neutrino problem leaves an observable ν_e burst (Sect. 11.4.2).

Another interesting possibility is a partial swap $\nu_e \leftrightarrow \nu_{\mu,\tau}$ and $\bar{\nu}_e \leftrightarrow \bar{\nu}_{\mu,\tau}$ by oscillations. Because the energy spectrum of the non-ν_e flavors is much harder than that of ν_e or $\bar{\nu}_e$, a number of interesting consequences obtain. First, the detected $\bar{\nu}_e$'s could have larger average energies than expected. Conversely, the SN 1987A-implied emission temperature and neutron-star binding energy could be an overestimate of the true values. It will turn out that one seriously needs to worry about these effects, for example, if the large-angle MSW solution or the vacuum solution to the solar neutrino problem obtain, or if the atmospheric neutrino anomaly is caused by oscillations (Sect. 11.4.3).

Even more importantly, a swap $\nu_e \leftrightarrow \nu_\mu$ or $\nu_e \leftrightarrow \nu_\tau$ would cause a more efficient energy transfer from the neutrino flux to the matter behind the stalled shock after core bounce but before the final explosion. As enhanced neutrino heating actually appears to be required to obtain successful and sufficiently energetic explosions, neutrino oscillations may help to explode supernovae! This scenario works only if the spectral swap occurs inside of the stalled shock wave. In view of the relevant medium densities, resonant transitions obtain for neutrino masses in the cosmologically interesting range of $10-100\,$eV. A mixing angle as small as $\sin^2 2\theta \gtrsim 3\times 10^{-8}$ would be enough (Sect. 11.4.4).

Hardening the ν_e spectrum by a swap with ν_μ or ν_τ, however, can suppress r-process nucleosynthesis just outside of the nascent neutron star a few seconds after core bounce. Normally the $\bar{\nu}_e$ spectrum is harder than the ν_e spectrum, driving β equilibrium in the hot bubble to the required neutron-rich phase. The oscillation scenario can cause the reverse. For this effect the oscillations would need to oc cur close to the protoneutron star surface and so again a relatively large neutrino mass-square difference is required which falls into the cosmologically interesting range. However, the required mixing angle is larger ($\sin^2 2\theta \gtrsim 10^{-5}$), leaving ample room for, say, small ν_e-ν_τ mixing angles where ν_τ's with cosmologically relevant masses could help explode supernovae without disturbing r-process nucleosynthesis (Sect. 11.4.5).

Finally, neutrino oscillations could allow the non-ν_e flavors to participate in β equilibrium in the inner core and thus build up their own degenerate Fermi seas. This possibility has been studied in Sect. 9.5 where it turned out that a significant flavor conversion obtains only for large neutrino masses (keV range and above). Such large masses are cosmologically forbidden unless neutrinos decay fast into invisible channels, a hypothesis that would require novel neutrino interactions beyond masses and mixings (Sect. 12.5.2).

11.4.2   Prompt ν_e Burst

The prompt ν_e burst from a core collapse SN can be detected, in principle, by the forward-peaked signal from the elastic $\nu_e e \to e\nu_e$ scattering
in a water Cherenkov detector. In the Kamiokande SN 1987A observations, the first event could have been caused by the prompt ν_e burst,
but naturally one event contains no statistically significant information.
It could have been caused by $\bar{\nu}_e p \to ne^+$ and simply happen to point
in the forward direction. Therefore, the main interest in the prompt ν_e
burst is the possibility that it could be observed from a future galactic
SN by the Superkamiokande or SNO detectors which would yield statistically significant signatures. A possible oscillation of the ν_e burst
into other flavors would reduce the number of forward events because
of the reduced ν-e scattering cross section of non-ν_e flavors (Fig. 11.10).

If the neutrino mass hierarchy is normal where the lightest mass
eigenstate is the dominant ν_e admixture, the medium-induced neutrino
refractive index in the stellar mantle and envelope can cause a "mass
inversion" and thus level crossing between, say, ν_e and ν_μ in analogy
to the solar MSW effect. Therefore, one may expect resonant flavor
conversion of the prompt ν_e burst in a collapsing star as shown by a
number of authors;[67] I follow the analysis of Nötzold (1987).

When the shock wave breaks through the neutrino sphere and liberates the prompt ν_e burst, the overlaying part of the progenitor star
has not yet noticed the collapse of its core so that the density profile is
given by that of the progenitor star. The electron density is reasonably
well approximated by a simple power law for which Nötzold (1987) used

$$n_e \approx 10^{34}\,\mathrm{cm}^{-3}\,r_7^{-3}, \tag{11.11}$$

where $r_7 \equiv r/10^7\,\mathrm{cm}$. Note that $10^7\,\mathrm{cm} = 100\,\mathrm{km}$ is the approximate
radius of the shell from where the ν_e's originate. According to the
discussion in Sect. 6.7.1 the electron density causes an energy shift
between ν_e and ν_μ or ν_τ of $\Delta V = \sqrt{2}G_F n_e = 1.3\times 10^{-3}\,\mathrm{eV}\,r_7^{-3}$. Comparing this with the energy shift $\Delta m_\nu^2/2p_\nu$ of neutrinos with momentum
p_ν one finds an effective medium-induced effect of $\Delta m_{\rm eff}^2 = 2p_\nu\,\Delta V = 3\times 10^4\,\mathrm{eV}^2\,p_{10}\,r_7^{-3}$ where $p_{10} = p_\nu/10\,\mathrm{MeV}$.

However, because the prompt ν_e burst itself constitutes a large local
ν_e density, the neutrino-induced refractive index may be more important than the standard electron-induced contribution. The total num-

[67]Mikheev and Smirnov (1986), Arafune et al. (1987a,b), Lagage et al. (1987),
Minakata et al. (1987), Nötzold (1987), Walker and Schramm (1987), Kuo and
Pantaleone (1988), Minakata and Nunokawa (1988), and Rosen (1988).

ber of ν_e's in the burst is of order 10^{56}, its duration of order 50 ms. Thus, while it passes it represents a ν_e density of about $10^{32}\,\mathrm{cm}^{-3}\,r_7^{-2}$ which exceeds the local electron density for $r \gtrsim 10^9\,\mathrm{cm}$. However, the phase-space distribution of the neutrinos is locally far from isotropic and so the energy shift involves a factor $\langle 1 - \cos\Theta \rangle$ where Θ is the angle between the "test neutrino" and a "background neutrino;" the average is to be taken over all background neutrinos (Sect. 9.3.2). A typical angle between two neutrinos moving within the burst at the same location is given by the angle subtended by the neutrino sphere as viewed from the relevant radial position, i.e. $\Theta \approx R/r$ with R the radius of the neutrino sphere. For a large r one thus finds $\langle 1 - \cos\Theta \rangle \approx \Theta^2 \approx (R/r)^2$. Then, with $R \approx 10^7\,\mathrm{cm}$ the effective neutrino density is approximately $10^{32}\,\mathrm{cm}^{-3}\,r_7^{-4}$, a value which is always smaller than the electron density Eq. (11.11). Therefore, in the present context one may ignore neutrino-neutrino interactions. This will not be the case for the issue of r-process nucleosynthesis (Sect. 11.4.5).

One may proceed to determine the MSW triangle as in Fig. 8.9 for solar neutrinos, except that there an exponential electron density profile was used while now Eq. (11.11) pertains. Nötzold (1987) found a conversion probability in excess of 50% if

$$\Delta m_\nu^2 \sin^3 2\theta \gtrsim 4\times 10^{-9}\,\mathrm{eV}^2\,E_\nu/10\,\mathrm{MeV}, \tag{11.12}$$

assuming that $\Delta m_\nu^2 \lesssim 3\times 10^4\,\mathrm{eV}^2\,E_\nu/10\,\mathrm{MeV}$ so that a resonance can occur outside of the neutrino sphere. The region in the Δm_ν^2-$\sin^2 2\theta$-plane (mixing angle θ) with a conversion probability exceeding 50% for $E_\nu = 20\,\mathrm{MeV}$ is shown as a shaded area in Fig. 11.17, together with the MSW solutions to the solar neutrino problem. For orientation, the Kamiokande-allowed range for solar neutrinos is also indicated.

The solar small-angle MSW solution would seem to have a small impact on the prompt ν_e burst from a collapsing star. Thus, the first Kamiokande SN 1987A event may still be interpreted as a prompt ν_e, and one may well observe nearly the full ν_e burst from a future SN.

It is interesting that the MSW triangle for the prompt ν_e burst reaches to relatively large neutrino masses. In the 3−30 eV regime neutrino masses would play an important cosmological role as dark matter and for the formation of structure in the universe. Such massive neutrinos likely would mix with ν_e. Unless the mixing angle is very small the appearance of an unoscillated prompt ν_e burst from a stellar collapse would be in conflict with a cosmological role of massive neutrinos (Arafune et al. 1987b).

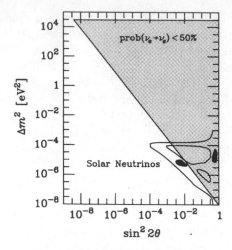

Fig. 11.17. MSW triangle for the prompt ν_e burst from a stellar collapse. In the shaded area the conversion probability exceeds 50% for $E_\nu = 20\,\mathrm{MeV}$, assuming the electron density profile of Eq. (11.11). The MSW solutions to the solar neutrino problem and the Kamiokande allowed range for solar neutrinos are indicated (see Fig. 10.19).

11.4.3   Cooling-Phase $\bar\nu_e$'s

Neutrino oscillations would cause a partial swap $\nu_e \leftrightarrow \nu_{\mu,\tau}$ and $\bar\nu_e \leftrightarrow \bar\nu_{\mu,\tau}$ so that the *measured* $\bar\nu_e$ flux at Earth could be a mixture of the original $\bar\nu_e$ and $\bar\nu_\mu$ or $\bar\nu_\tau$ source spectra (Wolfenstein 1987). The energy spectra of the neutrinos emitted during the Kelvin-Helmholtz cooling phase are flavor dependent (Eq. 11.4); typically, one finds $\langle E_{\bar\nu_\mu} \rangle = (1.3 - 1.7) \times \langle E_{\bar\nu_e} \rangle$. From the SN 1987A measurements one infers a low value of roughly $\langle E_{\bar\nu_e} \rangle \approx 10\,\mathrm{MeV}$. While the lowest typical predictions are about 14 MeV, this discrepancy is not a serious problem. However, it is probably not tolerable that a significant fraction of the observed events were due to oscillated $\bar\nu_\mu$'s.

One may contemplate an "inverted" mass hierarchy where the predominant mass component of ν_e is larger than that of, say, ν_τ. In this case one would obtain resonant oscillations and thus a complete spectral swap in the shaded triangle of mixing parameters shown in Fig. 11.17. Even if $\langle E_{\bar\nu_\tau} \rangle$ is only $1.3 \times \langle E_{\bar\nu_e} \rangle$ this would be in contradiction with the soft $\bar\nu_e$ energies observed from SN 1987A. Therefore, an inverted mass scheme looks excluded for a large range of masses and mixings.

A "normal" mass hierarchy prevents a level crossing among antineutrinos. Still, if the large-angle or the vacuum solution to the solar neutrino problem obtain, the mixing angle would be so large that the spectral swapping could still be uncomfortably large. Smirnov, Spergel, and Bahcall (1994) have considered in detail the probability for swapping the $\bar{\nu}_e$ with the $\bar{\nu}_\mu$ or $\bar{\nu}_\tau$ spectrum. For vacuum oscillations, relevant for small Δm_ν^2, the fractional exchange of the spectra is $p = \frac{1}{2}\sin^2 2\theta$ (vacuum mixing angle θ) with a maximum of $p = 0.5$, i.e. the observed spectrum could be as much as an equal mixture of the primary ones. In general, medium refractive effects must be included, although for a large range of Δm^2 one may still take p to be independent of energy. Contours for p in the $\sin^2 2\theta$-Δm_ν^2-plane are shown in Fig. 11.18. In the shaded area (A) the Earth effect is important so that the amount of conversion is energy dependent and differs between the

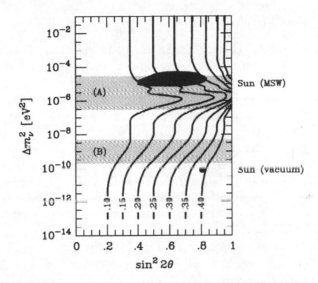

Fig. 11.18. Contours for the "swap fraction" p between the $\bar{\nu}_e$ and the $\bar{\nu}_\mu$ or $\bar{\nu}_\tau$ fluxes from the protoneutron star cooling phase. The matter effect of the stellar envelope and of the Earth are included. In (A) the Earth effect is important; in this area p is an average over neutrino energies while otherwise it does not depend on the energy. In (B) the exact contours depend on the detailed matter distribution of the stellar envelope. Black areas indicate the approximate mixing parameters which would explain the solar neutrino problem. (Adapted from Smirnov, Spergel, and Bahcall 1994.)

two detectors. In this case the contours refer to an average value $\langle p \rangle$. In the shaded area (B) the contours are not independent of the detailed matter distribution in the stellar envelope.

In Fig. 11.18 the large-angle MSW and the vacuum oscillation solutions of the solar neutrino problem are indicated. If either one of them is correct the measurable $\bar{\nu}_e$ spectrum in a detector is a substantial mixture of different-flavor source spectra. Smirnov, Spergel, and Bahcall (1994) argued on the basis of a joint analysis between the SN 1987A signals at the Kamiokande and IMB detectors that $p < 0.17–0.27$ at the 95% CL, depending on the assumed primary neutrino spectra. If p were any larger, the expected spectra would be much harder than has been observed. This analysis excludes the vacuum oscillation solution to the solar neutrino problem. Kernan and Krauss (1995) arrive at the opposite conclusion that all mixing angles are permitted by the SN 1987A signal, and that $\sin^2 2\theta = 0.45$ is actually a favored value.

11.4.4 Shock Revival

The "swap fraction" of the $\bar{\nu}_e$ with the more energetic $\bar{\nu}_\mu$ or $\bar{\nu}_\tau$ spectrum discussed in the previous section is always very small unless the mixing angle is very large because of the assumed normal mass hierarchy which prevents resonant conversions. By the same token a swap of the ν_e with the ν_μ or ν_τ spectrum will be resonant for certain mixing parameters and so it can be almost complete even for very small mixing angles. If the resonance is located between the neutrino sphere and the stalling shock wave after bounce, but before the final explosion, the shock would be helped to rejuvenate because the higher-energy $\nu_{\mu,\tau}$'s are more efficient at transferring energy once they have converted into ν_e's (Fuller et al. 1992). Of course, approximately equal luminosities in all flavors have been assumed.

A typical density profile for a SN model 0.15 s after bounce is shown in Fig. 11.19; the step at a radius of about 400 km is due to the shock front. A resonance occurs if $\Delta m_\nu^2 / 2E_\nu = \sqrt{2} G_F n_e$. On the right scale of the plot the quantity $m_{\text{res}} \equiv (\sqrt{2} G_F n_e \, 2 E_\nu)^{1/2}$ is shown for $E_\nu = 10\,\text{MeV}$ and $Y_e = 0.5$. A resonance occurs inside of the shock wave only for neutrino masses in the cosmologically interesting regime of order 10 eV and above. Fuller et al. (1992) have performed a detailed numerical calculation of the additional heating effect for $(\Delta m_\nu^2)^{1/2} = 40\,\text{eV}$; they found a 60% increase of the energy of the shock wave.

The conversion probability between neutrinos is large if the adiabaticity parameter γ defined in Eq. (8.39) far exceeds unity. Accord-

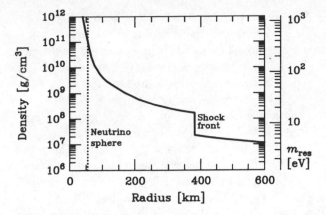

Fig. 11.19. Typical density profile for a SN model at 0.15 s after the core bounce (Fuller et al. 1992). The right-hand scale is $m_{res} = (\sqrt{2}G_F n_e 2E_\nu)^{1/2}$ which indicates the resonance value for $(\Delta m_\nu^2)^{1/2}$ for $E_\nu = 10\,\text{MeV}$, assuming an electron number fraction of $Y_e = 0.5$.

ing to the Landau-Zener formula Eq. (8.41) the swap probability is $1 - e^{-\pi\gamma/2}$. It exceeds 86% if one requires $\gamma > 4/\pi$ or

$$\sin^2 2\theta > \frac{8}{\pi} \frac{E_\nu}{\Delta m_\nu^2} |\nabla \ln n_e|_{res}, \tag{11.13}$$

where the vacuum mixing angle θ was assumed to be small. For $(\Delta m_\nu^2)^{1/2} - 40\,\text{eV}$ the density scale height at the resonance region is $|\nabla \ln n_e|_{res}^{-1} \approx 50\,\text{km}$ so that

$$\sin^2 2\theta \gtrsim 10^{-8} E_\nu/10\,\text{MeV}. \tag{11.14}$$

Therefore, unless the mixing angle with ν_e is very small a cosmologically interesting neutrino mass for, say, the ν_τ may help to explode supernovae! According to the discussion in Sect. 11.4.2 this would imply that the prompt ν_e burst would also oscillate. Then the first Kamiokande event could not be associated with the prompt ν_e burst.

11.4.5 R-Process Nucleosynthesis

a) Basic Picture

A partial swap of the ν_e cooling flux with the more energetic ν_μ or ν_τ flux can prevent the synthesis of heavy nuclei by the r-process neutron capture. As discussed in Sect. 11.1.4, the hot bubble between the settled protoneutron star and the escaping shock wave at a few seconds after core bounce might be an ideal high-entropy environment for this process for which no other site is currently known that could reproduce the observed galactic heavy element abundance and isotope distribution. Naturally, the r-process can only occur in a neutron-rich medium ($Y_e < \frac{1}{2}$). The p/n ratio in the hot bubble is governed by the β reactions $\nu_e n \leftrightarrow pe^-$ and $\bar{\nu}_e p \leftrightarrow ne^+$. Because the neutrino number density is much larger than the ambient $e^+ e^-$ population the proton/neutron fraction is governed by the neutrino spectra and fluxes. The system is driven to a neutron-rich phase because normally the $\bar{\nu}_e$'s are more energetic than the ν_e's. They emerge from deeper and hotter regions of the star because their opacity is governed by the same reactions, and because the core is neutron rich, yielding a larger opacity for ν_e.

If an exchange $\nu_e \leftrightarrow \nu_{\mu,\tau}$ occurs outside of the neutrino sphere the subsequent ν_e flux is more energetic than the $\bar{\nu}_e$ flux which did not undergo a swap. (A normal mass hierarchy has been assumed.) Even a partial swap of a few 10% is enough to shift the medium to a proton-rich state, i.e. to $Y_e > \frac{1}{2}$, to be compared with the standard values of 0.35–0.46. Therefore, the occurrence of such oscillations would be in conflict with r-process nucleosynthesis in supernovae (Qian et al. 1993).

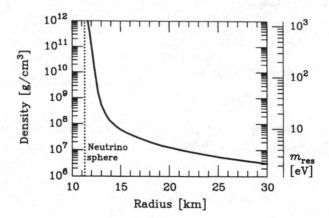

Fig. 11.20. Density profile for a SN model at 6 s after the core bounce, typical for the "hot bubble phase" (Qian et al. 1993). The right-hand scale is $m_{res} = (\sqrt{2} G_F n_e 2 E_\nu)^{1/2}$ which indicates the resonance value for $(\Delta m_\nu^2)^{1/2}$ for $E_\nu = 10\,\text{MeV}$, assuming $Y_e = 0.5$. (Note that out to a few km above the neutrino sphere $Y_e \ll 0.5$.)

The approximate parameter range for which this effect is important can be estimated from the density profile of a typical SN core a few seconds after bounce (Fig. 11.20). Again, one expects resonant conversions for the cosmologically interesting neutrino mass range as in the above shock revival scenario. However, the present effect reaches down to m_ν of about 3 eV; for smaller masses the oscillations would occur at radii too large to have an impact on nucleosynthesis.

With regard to the required mixing angle there is an important difference to the previous case because the neutron star has already settled so that the neutrino sphere is now at a radius of about 11 km rather than at 50 km. Therefore, the relevant length scales in Fig. 11.20 are reduced relative to Fig. 11.19 which corresponds to a postbounce but preexplosion configuration. For example, at $m_{\mathrm{res}} \approx 40$ eV the density scale height is now $|\nabla \ln n_e|_{\mathrm{res}}^{-1} \approx 0.3$ km, about two orders of magnitude smaller than before so that the lower limit on $\sin^2 2\theta$ is about two orders of magnitude larger than it was for the shock revival scenario. From a more detailed analysis Qian et al. (1993) found the hatched area in Fig. 11.21 where Y_e would be driven beyond 0.5 and so this area would be in conflict with r process nucleosynthesis in super novae.

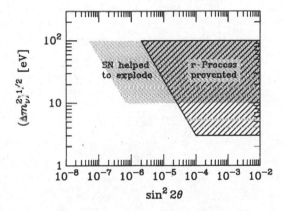

Fig. 11.21. Mass difference and mixing angle of ν_e with ν_μ or ν_τ where a spectral swap would be efficient enough to help explode supernovae (schematically after Fuller et al. 1992), and where it would prevent r-process nucleosynthesis (schematically after Qian et al. 1993; Qian and Fuller 1994).

b) Impact of Neutrino-Neutrino Interactions

The discussion so far has been relatively simplistic because the role of neutrino-neutrino interactions has been ignored. During the hot-bubble phase the neutrino refractive index caused by other neutrinos is not necessarily negligible (Pantaleone 1995; Qian and Fuller 1995). In terms of the neutrino luminosity the number flux and thus the density of neutrinos of a given species at a radius r is given by $n_\nu = L_\nu \langle E_\nu \rangle^{-1} (4\pi r^2)^{-1}$. Moreover, as in the discussion in Sect. 11.4.2 one must include an average of the factor $(1 - \cos\Theta)$ to account for the anisotropy of the neutrino phase space distribution (angle Θ between test and background neutrino). At a given distance r the neutron star (neutrino sphere radius R) subtends an angle given by $\sin\Theta_R = R/r$. For a radially moving test neutrino one finds (for a more rigorous treatment see e.g. Qian and Fuller 1995)

$$\langle 1 - \cos\Theta \rangle = \int_{\cos\Theta_R}^{1} (1 - \cos\Theta)\, d\cos\Theta \Big/ \int_{\cos\Theta_R}^{1} d\cos\Theta$$

$$= \tfrac{1}{2}\left[1 - \sqrt{1 - (R/r)^2} \right], \tag{11.15}$$

which for large r approaches $\tfrac{1}{4}(R/r)^2$. Therefore, the effective neutrino density $n_\nu \langle 1 - \cos\Theta \rangle$ varies as r^{-4} at large distances.

The electrons and positrons cause a refractive energy shift between ν_e and, say, ν_μ of $\Delta V = \sqrt{2} G_F (n_{e^-} - n_{e^+})$ while the effect of the neutrinos is

$$\Delta V \approx \sqrt{2} G_F \left(n_{\nu_e} - n_{\bar\nu_e} - n_{\nu_\mu} + n_{\bar\nu_\mu} \right)\langle 1 - \cos\Theta \rangle$$

$$\approx \frac{\sqrt{2} G_F \langle 1 - \cos\Theta \rangle}{4\pi r^2} \left(\frac{L_{\nu_e}}{\langle E_{\nu_e} \rangle} - \frac{L_{\bar\nu_e}}{\langle E_{\bar\nu_e} \rangle} - \frac{L_{\nu_\mu}}{\langle E_{\nu_\mu} \rangle} + \frac{L_{\bar\nu_\mu}}{\langle E_{\bar\nu_\mu} \rangle} \right). \tag{11.16}$$

The tau-flavored neutrino contribution has not been included because it cancels exactly between ν_τ and $\bar\nu_\tau$. The same is true for the mu-flavored terms before oscillations have taken place. For a test-neutrino of momentum p_ν one finds numerically

$$2 p_\nu \Delta V \approx 420\,\mathrm{eV}^2\,(10\,\mathrm{km}/r)^2\,\langle 1 - \cos\Theta \rangle \times$$

$$\times \frac{p_\nu}{10^{51}\,\mathrm{erg\,s^{-1}}} \left(\frac{L_{\nu_e}}{\langle E_{\nu_e} \rangle} - \frac{L_{\bar\nu_e}}{\langle E_{\bar\nu_e} \rangle} - \frac{L_{\nu_\mu}}{\langle E_{\nu_\mu} \rangle} + \frac{L_{\bar\nu_\mu}}{\langle E_{\bar\nu_\mu} \rangle} \right). \tag{11.17}$$

A few seconds after bounce typical values might be $\langle E_{\nu_e} \rangle = 11\,\mathrm{MeV}$, $\langle E_{\bar\nu_e} \rangle = 16\,\mathrm{MeV}$, and $\langle E_{\nu_\mu} \rangle = \langle E_{\bar\nu_\mu} \rangle = 25\,\mathrm{MeV}$, the luminosities can be taken to be the same at $3\times10^{51}\,\mathrm{erg\,s^{-1}}$ each, and $R = 11\,\mathrm{km}$. For these conditions the estimated contributions to ΔV from electrons and neutrinos are shown in Fig. 11.22 as a function of radius.

As long as no swap has occurred the neutrinos do not play a major role because the contribution of ν_μ cancels exactly against $\overline{\nu}_\mu$. Further, the difference between ν_e and $\overline{\nu}_e$ is smaller than the electron contribution, and it has the same sign whence it simply causes a slightly larger effective matter density.

However, on resonance a relatively large number of ν_e's exchange flavor with ν_μ's so that the neutrino contribution changes sign. Moreover, on resonance the vacuum Δm_ν^2 by definition cancels against the medium-induced contribution. Therefore, "switching on" the neutrino term shifts the resonance position for given vacuum mixing parameters. Also, in a self-consistent treatment one needs to consider the full non-

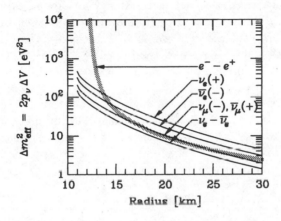

Fig. 11.22. "Mass splitting" between ν_e and ν_μ (equivalently ν_τ) of momentum $p_\nu = 10\,\mathrm{MeV}$ caused by the regular medium (the electrons), and by different neutrino flavors according to Eq. (11.17). For the electrons the density profile of Fig. 11.20 was used with $Y_e = 0.5$; in a real SN core Y_e is much lower near the neutrino sphere so that the thick line would increase less steeply toward the neutrino sphere than shown here. For the neutrinos, a luminosity in each degree of freedom of $3\times10^{51}\,\mathrm{erg\,s^{-1}}$ was assumed, the average energies were taken to be $\langle E_{\nu_e}\rangle = 11\,\mathrm{MeV}$, $\langle E_{\overline{\nu}_e}\rangle = 16\,\mathrm{MeV}$, and $\langle E_{\nu_\mu}\rangle = \langle E_{\overline{\nu}_\mu}\rangle = 25\,\mathrm{MeV}$, and the neutrino-sphere radius is 11 km. The signs of the contributions relative to electrons are: $+$ for ν_e and $\overline{\nu}_\mu$, $-$ for $\overline{\nu}_e$ and ν_μ. The contribution of ν_μ cancels exactly against $\overline{\nu}_\mu$ unless a swap $\nu_e \leftrightarrow \nu_\mu$ has taken place.

linear equations of motion for the neutrino density matrix which causes an "off-diagonal refractive index" as discussed in Sect. 9.3.2. Qian and Fuller (1995) have performed an approximately self-consistent analysis of this problem. All told, they found that the neutrino-neutrino interactions have a relatively small impact on the parameter space where flavor conversion disturbs the r-process. Within the overall precision of these arguments and calculations, their final exclusion plot is nearly identical with the schematic picture shown in Fig. 11.21. Qian and Fuller's findings are corroborated by a study performed by Sigl (1995a).

c) Summary

In summary, there remains a large range of mixing angles where neutrino oscillations between ν_e and ν_μ or ν_τ with a cosmologically interesting mass could help to explode supernovae, and yet not disturb r-process nucleosynthesis. If one assumes a mass hierarchy with ν_e dominated by the lightest, ν_τ by the heaviest mass eigenstate, the cosmologically relevant neutrino would be identified with ν_τ. It is interesting that the relevant range of mixing angles, $3\times10^{-4} \lesssim \theta \lesssim 3\times10^{-3}$, overlaps with the mixing angle among the first and third family quarks which is in the range $0.002-0.005$ (Eq. 7.6). Therefore, a scenario where a massive ν_τ plays a cosmologically important role, helps to explode supernovae, and leaves r-process nucleosynthesis unscathed does not appear to be entirely far-fetched. This scenario leaves the possibility open that the MSW effect solves the solar neutrino problem by ν_e-ν_μ oscillations.

11.4.6 A Caveat

All existing discussions of neutrino flavor oscillations in SNe were based on spherically symmetric, smooth density profiles. However, there can be significant density variations, convection, turbulence, and so forth. Therefore, it is clear that many of my statements about medium-induced oscillation effects are provisional. Further studies will be required to develop a more complete picture of SNe and their neutrino oscillations as 3-dimensional events.

A first study of SN neutrino oscillations with an inhomogeneous density profile was recently performed by Loreti et al. (1995). They added a random density field to the standard smooth profile and studied the impact on neutrino oscillations. They found that the shock-revival scenario involving MSW oscillations can be significantly affected in the sense that less additional energy is transferred because the stochastic density field can prevent a complete swap of the neutrino spectra. On the other hand, for the r-process prevention a complete swap is not necessary. Therefore, the mixing parameters for which r-process nucleosynthesis is prevented by oscillations is not significantly changed, even if the amplitude of the stochastic density component is as large as 1%.

11.5 Neutrino Propulsion of Neutron Stars

Shortly after the discovery of pulsars (neutron stars) it became clear that they tend to have the largest peculiar velocities of all stellar populations. Recent determinations of pulsar proper motions by radio-interferometric methods (Bailes et al. 1990; Fomalont et al. 1992; Harrison, Lyne, and Anderson 1993) and by interstellar scintillation observations (Cordes 1986) reveal typical speeds of a few $100\,\mathrm{km\,s^{-1}}$. Taking into account selection effects against high-velocity pulsars, Lyne and Lorimer (1994) argued that the mean pulsar velocity at birth was $450 \pm 90\,\mathrm{km\,s^{-1}}$. Associating certain pulsars and SN remnants would indicate velocities of up to $2000\,\mathrm{km\,s^{-1}}$ (Frail and Kulkarni 1991; Caraveo 1993; Stewart et al. 1993) while the interaction of PSR 2224+65 with its local environment produces a nebula which reveals a transverse velocity of at least $800\,\mathrm{km\,s^{-1}}$ (Cordes, Romani, and Lundgren 1993).

Therefore, the distribution of pulsar peculiar speeds appears to have a mean of $400-500\,\mathrm{km\,s^{-1}}$ with the largest measured values of $1000-2000\,\mathrm{km\,s^{-1}}$. Recall that the galactic rotation velocity is about $200\,\mathrm{km\,s^{-1}}$, the escape velocity about $500\,\mathrm{km\,s^{-1}}$. Therefore, the fastest pulsars will eventually escape from the galaxy. In most cases the migration is away from the galactic disk in agreement with the picture that pulsars are born in the disk where massive stars can form from the interstellar gas. However, there seem to be a few puzzling cases of pulsars which move toward the disk, apparently having formed in the galactic halo.

Massive stars probably do not form with much larger velocities than other stars and so pulsars are likely accelerated in conjunction with the SN collapse that produced them or during their early evolution. One suggestion for an acceleration mechanism holds that large neutron-star "kick velocities" are related to the breakup of close binaries, notably during the SN explosion of the second binary member (Gott, Gunn, and Ostriker 1970; Dewey and Cordes 1987; Bailes 1989; see also the review by Bhattacharya and van den Heuvel 1991).

Another possibility is that the SN explosion itself is not spherically symmetric and thus imparts a kick velocity on the neutron star (Shklovskiĭ 1970). Indeed, as discussed in Sect. 11.1.3 SN explosions likely involve large-scale convective overturns below and above the neutrino sphere which could lead to an explosion asymmetry of a few percent, enough to accelerate the compact core to a speed of order $100 \, \mathrm{km \, s^{-1}}$, but not enough to account for the typically observed pulsar velocities (Janka and Müller 1994).

An interesting acceleration mechanism was proposed by Harrison and Tademaru (1975) who considered the rotation of an oblique magnetic dipole which is off-center with regard to the rotating neutron star. The radiation of electromagnetic power is then asymmetric relative to the rotation axis and so a substantial accelerating force obtains, enough to cause velocities of several $100 \, \mathrm{km \, s^{-1}}$. The velocity reached should not depend on the magnitude of the magnetic dipole moment while its direction should correlate with the pulsar rotation axis. These predictions do not seem to be borne out by the data sample of Anderson and Lyne (1983) although the more recent observations may be less disfavorable to the "electromagnetic rocket engine." The correlation between peculiar velocity and pulsar magnetic moment may now be less convincing (Harrison, Lyne, and Anderson 1993; Itoh and Hiraki 1994).

Another intriguing mechanism first proposed by Chugaĭ (1984) relies on the asymmetric emission of neutrinos ("neutrino rocket engine"). Recall that the total amount of binding energy released in neutrinos is about $3 \times 10^{53} \, \mathrm{erg}$; because neutrinos are relativistic they carry the same amount of momentum. If the neutron-star mass is taken to be $1 \, \mathcal{M}_\odot$, and if all neutrinos were emitted in one direction, a recoil velocity of $0.17 \, c = 5 \times 10^4 \, \mathrm{km \, s^{-1}}$ would obtain. Thus an asymmetric emission of 1.5% would be enough to impart a kick velocity of $800 \, \mathrm{km \, s^{-1}}$.

Neutrino emission deviates naturally from spherical symmetry if large-scale convection obtains in the region of the neutrino sphere. Janka and Müller (1994) believe that $500 \, \mathrm{km \, s^{-1}}$ is a generous upper limit on the kick velocity that can be achieved by this method. For a reliable estimate one needs to know the typical size of the convective cells as well the duration of the convective phase in the protoneutron star. To this end one needs to perform a fully 3-dimensional calculation. No such results are available at the present time.

An asymmetric neutrino emission would also obtain in strong magnetic fields because the opacity is directional for processes involving initial- or final-state charged leptons, i.e. URCA processes of the type $e^- + p \rightarrow n + \nu_e$ or $e^+ + n \rightarrow p + \bar{\nu}_e$. The rates for such processes in the presence of magnetic fields were discussed by a number of authors.[68] Because of the left-handedness of the weak interaction both neutrinos and antineutrinos would be emitted preferentially in the same direction singled out by the magnetic field. Its effects become substantial only for field strengths near and above the critical strength $m_e^2/e = 4.4 \times 10^{13}$ G. In order to obtain neutron-star kick velocities of several $100 \, \mathrm{km \, s^{-1}}$ it appears that magnetic fields several orders of magnitude larger are required which, however, may possibly exist in some SN cores after collapse. Also, the asymmetric emission of neutrinos may be aided by the formation of a pion condensate and perhaps other processes (Parfenov 1988, 1989). Certain special field configurations seem to allow for significantly anisotropic neutrino emission (Bisnovatyi-Kogan and Janka 1995).

It remains to be seen if sufficiently anisotropic neutrino emission can be established as a generic property of a protoneutron star's Kelvin-Helmholtz cooling phase. Meanwhile, the neutrino rocket engine remains a fascinating speculation for accelerating neutron stars.

11.6 Future Supernovae

The neutrino observations from SN 1987A gave us a wealth of information in the sense that they confirmed the broad picture of neutrino cooling of the compact object formed after collapse. The data were much too sparse, however, to distinguish between, say, different equations of state or different assumptions concerning neutrino transport, or to detect or significantly constrain neutrino masses and mixing parameters. Some worry is caused by the apparent anomalies of the SN 1987A data, notably the angular distribution of the secondary charged particles. No doubt it would be extremely important to observe a SN neutrino signal with greater statistical significance, or from a greater distance. What is the prospect for such an observation?

SN neutrinos can be observed in a number of underground detectors which are operational now or in the near future, or which have only been proposed. An extensive overview was given by Burrows, Klein, and Gandhi (1992). Of the experiments which will become operational within the foreseeable future, the upcoming Superkamiokande water Cherenkov detector (Sect. 10.9) would yield by far the largest num-

---

[68] O'Connell and Matese (1969a,b); Matese and O'Connell (1969); Ivanov and Shul'man (1980, 1981); Dorofeev, Rodionov, and Ternov (1984); Loskutov (1984a,b); Cheng, Schramm, and Truran (1993).

ber of events. Its fiducial mass for the detection of SN neutrinos is about 32,000 t, to be compared with 2,140 t for Kamiokande, i.e. it has a target mass about 15 times larger. Thus one may be able to recognize a neutrino signal from a SN perhaps as much as 4 times farther away than the Large Magellanic Cloud, which is out to about 200 kpc. However, the closest large galaxy is M31 (Andromeda) at a distance of about 700 kpc, allowing Superkamiokande (and all other near-future detectors) to observe SNe only in our own galaxy and in the Large and Small Magellanic Clouds.

The rate at which SNe occur in our Galaxy as well as in the LMC is rather uncertain. From observations in other galaxies the rate of core-collapse SNe in the Milky Way is estimated to be about $7.3 \, h^2$ per century with h the Hubble parameter in units of $100 \, \mathrm{km \, s^{-1} \, Mpc^{-1}}$ (van den Bergh and Tammann 1991). Thus, for a low h of order 0.5 one may expect only about 2 such events per century. Roughly the same number was found in a more recent study by Tammann, Löffler, and Schröder (1994). The record of historical SNe, on the other hand, suggests a significantly larger number. Thus it is optimistic, but not entirely implausible, to hope for an observation within a decade of Superkamiokande running time. The rate for the LMC is thought to be about 0.5 per century (Tammann, Löffler, and Schröder 1994)—one cannot reasonably expect another SN there within our lifetime.

To reach beyond the limits of our own galaxy and the LMC one would need much more sensitive (much bigger) detectors. It would not be enough to go as far as Andromeda because this galaxy appears to have an anomalously low SN rate (van den Bergh and Tammann 1991). In order to achieve a SN rate of at least 1 per year one may need to use the Virgo cluster of galaxies at about 15 Mpc (300 times the distance to the LMC) although it may be enough to reach to the nearby starburst galaxies M82 and NGC 253 within about 4 Mpc which have a very high SN rate because of their high rate of star formation (Becklin 1990). However, a recent estimate of the SN rate for each of these galaxies is only about 1 per 10 years (van Buren and Greenhouse 1994).

A novel detection scheme (Cline et al. 1990) that may allow one to build big enough detectors is based on the neutral-current reaction $\nu + (Z, N) \rightarrow (Z, N - 1) + n + \nu$ which can have a much enhanced cross section in some nuclei due to collective effects; one would detect the final-state neutron. The necessary detector volume can be achieved by using natural deposits of minerals which contain the relevant target nuclei. Naturally, the main concern would be to reduce sources of background in order to isolate the feeble signal from a distant SN.

An even more ambitious goal would be to measure the cosmic $\bar{\nu}_e$ background flux from all past SNe in the universe. For energies below around 10 MeV this flux is swamped by many orders of magnitude by that from the nuclear power plants on Earth. Above a few 10 MeV the atmospheric neutrino flux would dominate and so there is only a small window where the cosmic SN flux might be detectable. However, even a moderately sized (200 tons) scintillation detector located on the moon would have a chance of measuring this flux because these backgrounds do not exist there (Mann and Zhang 1990).

Returning to Earth one may speculate about what could be learned if a galactic SN were indeed observed at Superkamiokande. For the purpose of argument a distance of 10 kpc (5 times closer than the LMC) is assumed. (Recall that the solar system is at a distance of about 8 kpc from the galactic center.) Then one expects at Superkamiokande about 4000 events from the reaction $\bar{\nu}_e p \rightarrow ne^+$, compared with 270 at Kamiokande, and with 12 measured there from SN 1987A (Totsuka 1990). This would be enough to determine a statistically very significant and very detailed "neutrino lightcurve."

Interestingly, one expects about 13 events within the first few ms from the prompt ν_e burst. Its presence would indicate that the ν_e's have not oscillated, say, into ν_τ's as would be expected for a cosmologically interesting ν_τ mass (Sect. 11.4.2), thus excluding a large range of masses and mixing angles (Fig. 11.17). Moreover, the prompt burst could not have been dispersed by a neutrino mass and so a m_{ν_e} bound of order 1 eV could be derived. A number of other conclusions tentatively reached for SN 1987A could be affirmed (Sect. 13.2). The nonobserva tion of the prompt burst, on the other hand, would be more difficult to interpret as its absence could have a variety of causes ranging from neutrino oscillations to some flaw in the standard picture of SN collapse.

The $\bar{\nu}_e$ lightcurve which would last, say, 10 s would not allow one to extract interesting bounds on m_{ν_e} relative to the ones already obtained from SN 1987A and from laboratory experiments. From Eq. (11.9) one concludes that for a distance $D \approx 10$ kpc one is sensitive to masses in the 100 eV range.

In Superkamiokande one would expect to see about 40 events for each $\nu_\mu + \bar{\nu}_\mu$ and $\nu_\tau + \bar{\nu}_\tau$ from the elastic ν-e scattering process. Because the final-state electrons are strongly forward peaked one can separate them from the isotropic $\bar{\nu}_e p \rightarrow ne^+$ signal. Therefore, a ν_μ or ν_τ mass would manifest itself through late forward events. Seckel, Steigman, and Walker (1991) found that from the signal in water Cherenkov detectors one could be sensitive to a mass down to about 75 eV. Of course, masses so large and larger are excluded from cosmology—such neutrinos would have to be unstable. In order to obey the cosmological limits neutrinos emitted at a distance of 10 kpc would decay before reaching Earth if their mass exceeds a few 10 keV. Thus, the observation of a galactic SN would allow one, at best, to probe the mass window $100\,\text{eV} \lesssim m_\nu \lesssim 30\,\text{keV}$. A similar conclusion was reached by Acker, Pakvasa, and Raghavan (1990) who considered the signature in the proposed BOREX detector.

It may be possible, however, to probe a somewhat smaller mass for, say, the ν_τ down to the cosmologically interesting range of 30 eV if one takes advantage of all aspects of the observed neutrino signal (Krauss et al. 1992). Because the late part of the neutrino lightcurve is expected to be similar for $\bar\nu_e$ and the other flavors one can hope to extract the behavior of the source from the $\bar\nu_e$ signal. Then one would be more sensitive to modifications of the ν_τ lightcurve caused by dispersion effects. However, in order to identify the ν_τ's one would have to use energy cuts (ν_e's and ν_τ's have different spectra!) in addition to angular cuts. In the analysis of Krauss et al. (1992) the possibility of $\nu_e \leftrightarrow \nu_\tau$ oscillations was not included which would weaken the range of accessible ν_τ masses because of the modified energy spectra. Of course, the r-process nucleosynthesis argument of Sect. 11.4.5 would indicate that MSW oscillations did not take place late even if they took place early and rendered the prompt ν_e burst unobservable.

Still, a ν_τ mass relevant for the dark matter content of the universe as well as for scenarios of structure formation may be much lower than 30 eV. Therefore, on the basis of current analyses the prospect of being able to recognize a cosmologically relevant ν_μ or ν_τ in the neutrino signal of a galactic SN appears relatively dim.

A far more positive view was taken by Cline et al. (1994) who argued that masses down to 15 eV may be accessible by the simultaneous operation of Superkamiokande and their previously proposed (Cline et al. 1990) Supernova Burst Observatory (SNBO) which is based on neutral-current reactions alone. This would obviate the need to separate the charged-current $\bar\nu_e n \rightarrow p e^+$ detection at Superkamiokande from the neutral-current reaction $\nu e \rightarrow e \nu$ by angular and energy cuts. One could use the Superkamiokande $\bar\nu_e$ signal to monitor the SN neutrino lightcurve, notably its sharp onset, and relate it to the onset of the neutral-current events at SNBO which would be "washed out" for a cosmologically interesting mass of, say, the ν_τ. One must hope that SNBO will become a real project in the near future.

Chapter 12

Radiative Particle Decays from Distant Sources

If neutrinos, axions, or other low-mass particles had radiative decay channels, the decay photons would appear as x- or γ-ray fluxes from stellar sources where these particles can be produced by nuclear or plasma processes. This chapter is devoted to limits on such decays, including decays into charged leptons, that are based on observational limits on photon or positron fluxes from stellar sources, notably the Sun and supernova 1987A. For comparison, laboratory and cosmological limits are also reviewed.

12.1 Preliminaries

This book is largely about the properties of electrically neutral particles whose electromagnetic interactions are correspondingly weak. However, because they can virtually dissociate into charged states, they will still interact with photons through higher-order amplitudes. The focus of the present chapter is the possibility of radiative decays of the form $\nu \to \nu'\gamma$ (neutrinos) or $a \to \gamma\gamma$ (axions), but also $\nu \to \nu'e^+e^-$ and $\nu \to \nu'e^+e^-\gamma$. Cowsik (1977) was the first to recognize that the huge path lengths available in the astrophysical environment allow one to obtain much more restrictive limits on such decays than from laboratory experiments. For example, the absence of single-photon counts in a detector near a fission reactor indicates a bound[69] $\tau_\gamma/m_{\nu_e} > 22\,\text{s/eV}$

---

[69]In this chapter τ_γ will always denote the partial neutrino decay time into radiation while τ_{tot} is the total decay time if hypothetical invisible channels are included. Then $\tau_\gamma^{-1} = B_\gamma\,\tau_{\text{tot}}^{-1}$ with the branching ratio B_γ.

on the process $\nu_e \to \nu'\gamma$. The corresponding limit based on a comparison between the measured solar neutrino flux and the measured limit on x- or γ-rays from the quiet Sun is $7 \times 10^9\,\mathrm{s/eV}$, almost 9 orders of magnitude more restrictive. Moreover, this method is fully analogous to a laboratory experiment as it is based on a *measured* neutrino flux and a *measured* upper limit photon flux. To a lesser degree this remark also applies to the even better SN 1987A constraints which are applicable to all neutrino flavors.

In addition, less directly established particle fluxes can be used such as those from the stars in the galactic bulge or from all hydrogen-burning stars or supernovae in the universe. Even more indirectly, one may study the impact of the radiative decay of the cosmic background sea of neutrinos or axions which are predicted to exist in the framework of the big-bang theory of the early universe.

Usually, the bounds on $\nu \to \nu'\gamma$ thus obtained are presented as limits on the radiative decay time τ_γ. Even if one does not aim at an immediate theoretical interpretation, however, the significance of τ_γ is limited because it represents a combination of the final-state phase-space volume and the matrix element. The latter can be expressed in terms of an effective transition moment μ_{eff} as in Eq. (7.12) of Sect. 7.2.2. This moment characterizes the interaction strength independently of phase-space effects, providing a much more direct link between the experimental results and an underlying theory.

A "heavy" neutrino ν_h with $m_h > 2m_e \approx 1\,\mathrm{MeV}$ can decay into $\nu_e e^+ e^-$. This channel is often included in the notion of "radiative" decays because relativistic charged leptons cause experimental signatures similar to γ rays. If this decay proceeds by virtue of a mixing amplitude U_{eh} between ν_h and ν_e the rate is given by Eq. (7.9). Therefore, it is characterized by $|U_{eh}|$ in a phase-space independent way.

If ν_e is a mixture of different mass eigenstates, any ν_e source such as a power reactor or the Sun produces all components. If their mass differences are small one needs to consider in detail the phenomenon of neutrino oscillations as in Chapter 8. However, if the oscillation length is much smaller than the distance between the detector and the source, the neutrino flux can be considered an incoherent mixture of all mass eigenstates; at a reactor this is the case for $\Delta m^2 \gtrsim 1\mathrm{eV}^2$. The flux of "heavy" neutrinos from a ν_e source is then given by

$$F_{\nu_h}(E_\nu) = |U_{eh}|^2 \, \beta(E_\nu) \, F_{\nu_e}(E_\nu) \,. \tag{12.1}$$

The velocity $\beta(E_\nu) = (1 - m_h^2/E_\nu^2)^{1/2}$ enters from the phase space of nonrelativistic neutrinos in the production process. From ν_e sources one can then derive limits on U_{eh} for any ν_h, even a hypothetical sterile one, if $m_h \gtrsim 1\,\mathrm{MeV}$ because it is the same mixing amplitude that allows for its production in the source and for its $\nu_h \to \nu_e e^- e^+$ decay.

In the following I will discuss radiative lifetime limits from different sources approximately in the order of available decay paths, from laboratory experiments (a few meters) to the radius of the visible universe (about 10^{10} light years).

12.2 Laboratory Experiments

12.2.1 Spectrum of Decay Photons

Perhaps the simplest neutrino source to use for laboratory experiments is a nuclear power reactor which produces a strong $\bar\nu_e$ flux from the weak decays of the uranium and plutonium fission products. If a γ detector is placed at a certain distance from the reactor core one may assume that the local neutrino flux $F_\nu(E_\nu)$ is known (units $\mathrm{cm^{-2}\,s^{-1}\,MeV^{-1}}$). As a first step one then needs to compute the expected flux $F_\gamma(E_\gamma)$ of photons from the decay $\nu \to \nu'\gamma$. The result will also apply to stationary stellar sources such as the Sun while for the short neutrino burst from SN 1987A one needs to derive a separate expression (Sect. 12.4).

Because the decay is a dipole transition, the general form of the photon angular distribution in the rest frame of the parent neutrino is

$$dN_\gamma/d\cos\vartheta = \tfrac{1}{2}(1 - \alpha\cos\vartheta), \tag{12.2}$$

where ϑ is the angle between the ν spin polarization vector and the photon momentum. For Majorana neutrinos the decay is isotropic, independently of their polarization, and thus $\alpha = 0$. Similarly for axion decays, $a \to \gamma\gamma$, as these particles have no spin so that in their rest frame no spatial direction is favored. For polarized Dirac neutrinos the possible parameter range is $-1 \le \alpha \le 1$.

The decay photon has an energy $\omega = \delta_m m_\nu/2$ in the rest frame of the parent neutrino—see Eq. (7.12). If it is emitted in a direction θ with regard to the laboratory direction of motion, the energy in the laboratory frame is $E_\gamma = \omega\,(E_\nu + p_\nu\cos\theta)/m_\nu$. Hence,

$$E_\gamma = \tfrac{1}{2}\delta_m\,E_\nu\,(1 + \beta\cos\theta), \tag{12.3}$$

where $\beta = p_\nu/E_\nu = (1 - m_\nu^2/E_\nu^2)^{1/2}$ is the neutrino velocity. For a left-handed parent neutrino the spin is polarized opposite to its momentum so that $\cos\theta = -\cos\vartheta$ and $dN_\gamma/d\cos\theta = \tfrac{1}{2}(1 + \alpha\cos\theta)$. For a very relativistic parent neutrino $\beta = 1$ and so the normalized photon energy distribution corresponding to Eq. (12.2) is

$$\frac{dN_\gamma}{dE_\gamma} = \frac{1}{\delta_m E_\nu} \left(1 - \alpha + 2\alpha \frac{E_\gamma}{\delta_m E_\nu} \right), \qquad 0 < E_\gamma < \delta_m E_\nu. \quad (12.4)$$

The decays of Majorana neutrinos or axions ($\alpha = 0$) produce a box-shaped spectrum while for $\alpha = \pm 1$ it is triangle shaped (Fig. 12.1).

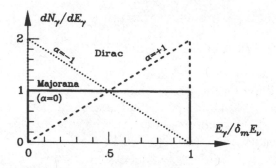

Fig. 12.1. Photon spectrum from the decay of a relativistic neutrino (energy E_ν) according to Eq. (12.4).

A fraction $(m_\nu/E_\nu)(d_\gamma/\tau_\gamma)$ of neutrinos decay before reaching the detector if the laboratory lifetime is large compared with the decay path d_γ. Here, τ_γ is the rest-frame radiative decay time and E_ν/m_ν the time dilation factor. Integrating over the neutrino source spectrum then yields

$$F_\gamma(E_\gamma) = \frac{m_\nu}{\tau_\gamma} d_\gamma \int_{E_\gamma/\delta_m}^{\infty} \frac{dE_\nu}{\delta_m} \left(1 - \alpha + 2\alpha \frac{E_\gamma}{\delta_m E_\nu} \right) \frac{F_\nu(E_\nu)}{E_\nu^2}. \quad (12.5)$$

Most sources emit either antineutrinos (for example $\bar{\nu}_e$'s from a fission reactor) or neutrinos (for example ν_e's from the Sun). In these cases Eq. (12.5) gives us directly the expected γ flux as a function of the assumed value for α. However, there are some examples for simultaneous ν and $\bar{\nu}$ sources. In those cases one needs to know which value of α to use for $\bar{\nu}$ if a certain value for ν has been assumed.

Under very general assumptions the laws of particle physics are invariant under a simultaneous transformation which takes particles into antiparticles (charge conjugation C), reflects all spatial coordinates (parity transformation P), and inverts motions (time reversal T). In this case the CPT theorem states that the masses and *total* decay times of corresponding particles and antiparticles are the same. However, the *partial* decay rates into specific channels need not be identical, and indeed, K° and $\overline{K}^\circ$ show such CP-violating decays: A transformation under CP leads to a "mirror world" which is different from the one we live in. Therefore, in the most general case one may only assume that ν and $\overline{\nu}$ have the same total lifetimes while their radiative decay rates may be different.

However, it is common to analyze the available data under the assumption of CP conservation for all neutrino interactions. For the radiative decay of a polarized ν in its rest frame, the CP-mirrored decay is one where a polarized $\overline{\nu}$ with the same spin decays into $\overline{\nu}'$ and γ of reversed momenta.[70] Therefore, ν and $\overline{\nu}$ of the same polarization are characterized by opposite values for α in Eq. (12.2). In the source, however, the ν's and $\overline{\nu}$'s are produced by weak interactions which violate parity maximally. If they are relativistic they both have negative (left-handed) chiralities which means that the ν's have negative and the $\overline{\nu}$'s positive helicities. Thus, relative to their momentum left-handed ν's and $\overline{\nu}$'s show the same distribution of decay photons so that in Eq. (12.5) one must use the same α for both.

12.2.2 Electron Neutrinos from Reactors

Fission reactors are superb neutrino sources. At a thermal power of 2800 MW, for example, one expects about $5 \times 10^{20}\,\overline{\nu}_e/\mathrm{s}$. At a distance of 30 m this corresponds to a flux of about $4 \times 10^{12}\,\mathrm{cm}^{-2}\,\mathrm{s}^{-1}$, almost a hundred times larger than the solar ν_e flux of about $6.6 \times 10^{10}\,\mathrm{cm}^{-2}\,\mathrm{s}^{-1}$. Reactor $\overline{\nu}_e$'s emerge from many weak decays of the products of the neutron induced fission of ^{235}U and ^{239}Pu while the fission of ^{238}U and ^{241}Pu contributes less than 10% to the total rate. The spectral distribution can be inferred from a measurement of the corresponding β spectra together with the reasonably well-known distribution of the end point energies of the fission products (von Feilitzsch et al. 1982; Schreckenbach et al. 1985). The distribution of $\overline{\nu}_e$ energies per fission of ^{235}U and ^{239}Pu is shown in Fig. 12.2 with a total of about $6\,\overline{\nu}_e$'s per fission.

An early, relatively crude, but often-quoted (Particle Data Group 1994) limit on the decay $\overline{\nu}_e \to \overline{\nu}'\gamma$ was derived by Reines, Sobel, and Gurr (1974) on the basis of the upper limit γ flux in a scintillation detector near the Savannah River reactor (U.S.A.). A weaker but more reli-

---

[70] A parity transformation P inverts all polar vectors, e.g. momenta, currents, or electric fields, while it leaves axial vectors unchanged, e.g. angular momenta, magnetic moments, or magnetic fields.

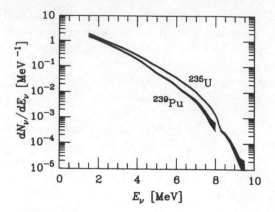

Fig. 12.2. Spectrum of reactor $\bar{\nu}_e$'s per fission of ^{239}Pu and ^{235}U (von Feil-itzsch et al. 1982; Schreckenbach et al. 1985). The width of the lines gives the total error of the spectra.

able bound was inferred by Vogel (1984) from data taken at the Gösgen reactor (Switzerland). The most recent analysis is, again, based on data taken with a scintillation counter at Gösgen (Oberauer, von Feil-itzsch, and Mössbauer 1987). From a comparison of the "reactor on" with the "reactor off" photon counts for several energy channels in the MeV range and using the experimentally established neutrino spectrum (Fig. 12.2), these authors found the 68% CL lower limits on the $\bar{\nu}_e$ ra-diative decay times of $\tau_\gamma/m_{\nu_e} > 22\,\text{s/eV}$ for $\alpha = -1$, $38\,\text{s/eV}$ for $\alpha = 0$, and $59\,\text{s/eV}$ for $\alpha = +1$. It was assumed that ν' is massless so that $\delta_m = 1$ in Eq. (12.5). With Eq. (7.12) the $\alpha = -1$ constraint translates into a bound on the effective electromagnetic transition moment of

$$\mu_{\text{eff}} < 0.092\,\mu_{\text{B}}\,m_{\text{eV}}^{-2}\,, \qquad\qquad (12.6)$$

not a very restrictive limit even if ν_e saturates its upper mass bound of about $5\,\text{eV}$.

Even this weak limit would cease to apply if ν_e and ν' became nearly degenerate. Therefore, Bouchez et al. (1988) performed an ex-periment where they searched for *optical* decay photons at the Bugey reactor (France). They excluded a certain region in the parameter plane spanned by τ_γ/m_{ν_e} and δ_m. Their greatest sensitivity was ap-proximately at $\delta_m = 2{\times}10^{-5}$ where they found $\tau_\gamma/m_{\nu_e} \gtrsim 0.04\,\text{s/eV}$. With Eq. (7.12) this is $\mu_{\text{eff}} \lesssim 10^7\,\mu_{\text{B}}/m_{\text{eV}}^2$ which, unfortunately, is ir-relevant as a constraint. Hence, for nearly degenerate neutrinos there is no meaningful limit on μ_{eff} from decay experiments. This example highlights the importance of separating the intrinsic coupling strength, or the magnitude of the matrix element, from phase-space effects in the interpretation of such results.

12.2.3 Heavy Neutrinos from Reactors

Considering the decay of ν_e into a different neutrino species implies entertaining the notion of the nonconservation of the electron lepton number, forcing one to contemplate the possibility of neutrino flavor mixing as well. Therefore, reactors will be sources for other flavors and notably of "heavy neutrinos" according to Eq. (12.1). The photon flux from radiative ν_h decays can now be calculated as before, except that the dwelling time for nonrelativistic ν_h's in the decay volume is increased by β^{-1} so that the velocity cancels between this factor and Eq. (12.1). However, the photon spectrum shown in Fig. 12.1 must be modified for the decays of nonrelativistic neutrinos because $\beta < 1$ in Eq. (12.3). Therefore, the overall expression for $F_\gamma(E_\gamma)$ becomes somewhat more involved.

As long as $m_h \lesssim 1\,\text{MeV}$ one may treat the ν_h flux as relativistic. Then the radiative decay limits are the same as for $\bar{\nu}_e$, except that they are diminished by the reduced neutrino flux. Thus, the bound Eq. (12.6) translates into

$$\mu_{\text{eff}} < 0.92 \times 10^{-13}\,\mu_{\text{B}}\,|U_{eh}|^{-1} m_{\text{MeV}}^{-2}\,. \tag{12.7}$$

With m_h up to an MeV this bound has a lot more teeth than the one on electron neutrinos.

For $m_h > 2m_e \approx 1\,\text{MeV}$ the decays $\nu_h \to \nu_e e^+ e^-$ will become kinematically possible and probably dominate. The scintillation counters that were used to search for decay photons near a power reactor are equally sensitive to electrons and positrons—for many purposes relativistic charged particles may be treated almost on the same footing as γ rays. Therefore, the same Gösgen data have been analyzed to constrain the mixing amplitude U_{eh} (Oberauer, von Feilitzsch, and Mössbauer 1987; Oberauer 1992). Even more restrictive limits were obtained from data taken at the Rovno reactor (Fayons, Kopeykin, and Mikaelyan 1991), and most recently at the Bugey reactor (Hagner et al. 1995). Note that in this method the same mixing probability $|U_{eh}|^2$ appears in Eq. (12.1) to obtain the ν_h flux from a ν_e source, and in Eq. (7.9) to obtain the decay probability. Hence, the expected e^+e^- flux is proportional to $|U_{eh}|^4$.

In Fig. 12.3 the excluded range of masses and mixing angles[71] is shown together with similar constraints from other neutrino sources. The reactor bounds are weaker than those from the Sun and SN 1987A, but they remain important because of the *short* decay path involved! If one accepts the big-bang nucleosynthesis bounds (Sect. 7.1.5) the ν_τ total lifetime must be so short that it would not escape from the mantle of a SN before decaying, and perhaps not even from the Sun.

12.2.4 Neutrinos from a Beam Stop

Another powerful laboratory source for both ν_e's and ν_μ's is a beam stop where neutrinos are produced from the decay of stopped pions, $\pi^+ \to \mu^+ \nu_\mu$ and the subsequent decay of stopped muons, $\mu^+ \to e^+ \nu_e \bar{\nu}_\mu$. In a recent experiment of this sort (Krakauer et al. 1991), the neutrino intensity was $4.3 \times 10^{13}\, \nu_\mu$/s with a total of $8.52 \times 10^{19}\, \nu_\mu$. In these decays the ν_μ has a fixed energy of $(m_\pi^2 - m_\mu^2)/2m_\pi = 29.8\,\mathrm{MeV}$ while the other normalized spectra are $(3/Y^4)(3Y - 2E_\nu)E_\nu^2$ for $\bar{\nu}_\mu$ and $(12/Y^4)(Y - E_\nu)E_\nu^2$ for ν_e with $Y \equiv \frac{1}{2}m_\mu = 52.8\,\mathrm{MeV}$. For ν_e, the 90% CL radiative lifetime limit as a function of the "anisotropy parameter" α is $\tau_\gamma/m_{\nu_e} > (15.9 + 9.8\alpha + 0.3\alpha^2)\,\mathrm{s/eV}$, somewhat less restrictive than the reactor results.

For ν_μ and $\bar{\nu}_\mu$ one obtains slightly different limits because of the different source spectra. Under the assumption of CP invariance the radiative lifetimes for ν_μ and $\bar{\nu}_\mu$ are the same. In this case the combined limit is $\tau_\gamma/m_{\nu_\mu} > (36.3 + 21.65\alpha + 0.75\alpha^2)\,\mathrm{s/eV}$ while the individual limits are about half this value. In terms of an effective transition moment the most conservative case ($\alpha = -1$) yields

$$\mu_{\text{eff}} < 0.11\, \mu_\mathrm{B}\, m_{\text{eV}}^{-2}. \tag{12.8}$$

For the admixture of other mass eigenstates this result may be translated in a fashion analogous to the discussion of reactor neutrinos.

Beam-stop neutrinos from meson decays may also be used to constrain the $e^+ e^-$ decays of heavy admixtures. Because of the larger amount of available energy one may probe higher masses for ν_h while the reactor bounds drop out above a few MeV because of the relatively soft spectrum. In Fig. 12.3 the most restrictive such constraints are summarized.

---

[71]It is customary to display $|U_{eh}|^2$ when constraining the mixing parameters of heavy, decaying neutrinos while one shows $\sin^2 2\theta_{eh}$ for light, oscillating neutrinos as in Chapter 8. For easier comparison I always use the mixing angle. Recall that $|U_{eh}| = \sin\theta_{eh}$ so that for small mixing angles $\sin^2 2\theta_{eh} = 4|U_{eh}|^2$.

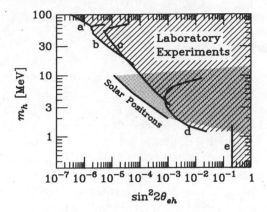

Fig. 12.3. Bounds on ν_e-ν_h mixing from the absence of $\nu_h \to \nu_e e^+ e^-$ decays. *Meson decays:* (a,b) Leener-Rosier et al. (1986) and (c) Bryman et al. (1983). *Reactor neutrinos:* (d) Hagner et al. (1995). *Absence of reactor neutrino oscillations:* (e) Zacek et al. (1986). *Absence of solar positrons:* Toussaint and Wilczek (1981); see also Sect. 12.3.2.

These results are based on ν_e sources which produce ν_h by their mixing which also leads to the subsequent $\nu_h \to \nu_e e^+ e^-$ decay. Of the known sequential neutrinos ν_h can be identified only with ν_3, the dominant mass component of ν_τ with an allowed mass of up to 24 MeV. If one makes this identification, stronger limits are obtained from direct $\nu_\tau \approx \nu_3$ sources. One example is the beam stop at the Big European Bubble Chamber (BEBC) where a strong flux of charmed strange mesons[72] D_s was produced which subsequently can decay as $D_s \to \tau \nu_\tau$ besides the dominant hadronic modes (WA66 Collaboration 1985). According to Babu, Gould, and Rothstein (1994) who quote a private communication from the WA66 collaboration, a model-independent constraint from the BEBC experiment is $\tau_\gamma/m_3 > 0.15\,\mathrm{s/MeV}$ or

$$\mu_{\mathrm{eff}} < 1.1 \times 10^3\, \mu_{\mathrm{B}}\, m_{\mathrm{eV}}^{-2} \,. \tag{12.9}$$

Another constraint is $\tau_{e^+ e^-}/m_3 > 0.18\,\mathrm{s/MeV}$. As a constraint on the mixing amplitude, $|U_{e3}|^2 < 1.6 \times 10^5\, m_{\mathrm{MeV}}^{-6}$, it is weaker than those shown in Fig. 12.3. However, as it is based on a direct ν_τ flux it is valid even if the decays are not induced by mixing but by exotic intermediate states (Babu, Gould, and Rothstein 1993).

---

[72]The D_s used to be called F as in the quoted reference.

12.3   Particles from the Sun

12.3.1   Electron Neutrinos

Like a terrestrial power reactor, the Sun is a prolific neutrino source except that it emits ν_e's rather than $\bar{\nu}_e$'s. The expected spectrum as well as the relevant measurements were discussed in Chapter 10. Suffice it to recall that the solar neutrino flux is now experimentally established without a shred of doubt. There remain significant discrepancies between the predicted and measured spectral shape of the spectrum which may be explained by neutrino oscillations. However, the solar neutrino problem is a fine point in the context of the present discussion because the following results depend mostly on the low-energy *pp* flux.

One may proceed exactly as in the previous section in order to translate the solar neutrino spectrum shown in Fig. 10.1 into an expected flux of x- and γ-rays from the Sun. In Fig. 12.4 I show this flux at Earth for $\tau_\gamma/m_{\nu_e} = 10\,\mathrm{s/eV}$ (about the laboratory lifetime limit) and for the values ± 1 for the "anisotropy parameter" α. The spectrum as shown is based on the *calculated* neutrino flux. The shoulders corresponding to other than the *pp* neutrinos likely would have to be reduced somewhat. The magnitude of the photon flux is enormous because the decay path d_γ is the entire distance to the Sun of $1.5 \times 10^{13}\,\mathrm{cm} = 500\,\mathrm{s}$.

The quiet Sun is a significant source of soft x-rays from the quasi-thermal emission of the hot corona at $T \approx 4.5 \times 10^6\,\mathrm{K}$. The flux measurements of Chodil et al. (1965) are marked as open diamonds in Fig. 12.4. The flux of decay photons in Fig. 12.4 would outshine the solar corona by some 4 orders of magnitude! Moreover, the corona spectrum falls off sharply at larger energies. In the hard x- and soft γ-ray band very restrictive upper limits exist on the emission of the quiet Sun that are shown in Fig. 12.4. They are based on balloon-borne detectors flown many years ago (Frost et al. 1966; Peterson et al. 1966). These upper limits are still far above the estimated albedo (radiation from cosmic rays hitting the surface of the Sun), leaving much room for improvement. Alas, the quiet Sun is not an object of great interest to γ-ray astronomers and so more recent measurements do not seem to exist.

In order to respect these measured upper limit photon fluxes one must shift the decay spectrum in Fig. 12.4 down by about 8 orders of magnitude (thin solid line in Fig. 12.4). This yields a lower radiative liftime limit for ν_e of $\tau_\gamma/m_{\nu_e} \gtrsim 7 \times 10^9\,\mathrm{s/eV}$ (Cowsik 1977; Raffelt 1985).

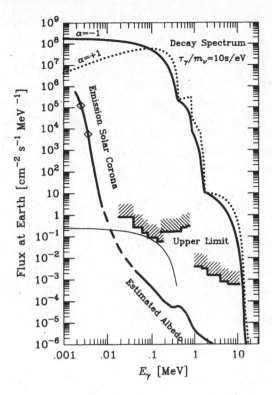

Fig. 12.4. Spectrum of photons from the solar neutrino decay $\nu_e \rightarrow \nu'\gamma$ for the indicated values of the anisotropy parameter α. Measurements of the x-ray emission of the solar corona (open diamonds) according to Chodil et al. (1965). Upper limit x- and γ-ray fluxes according to Frost et al. (1966) and Peterson et al. (1966). Estimated albedo according to Peterson et al. (1966). Thin solid line: Maximally allowed photon spectrum from neutrino decay. (Figure adapted from Raffelt 1985.)

With Eq. (7.12) this translates into

$$\mu_{\text{eff}} < 5 \times 10^{-6} \, \mu_{\text{B}} \, m_{\text{eV}}^{-2} \,. \tag{12.10}$$

It must be stressed, again, that the recent progress in solar neutrino astronomy has placed this result on the same footing as a terrestrial experiment—the magnitude of the solar neutrino flux and its main spectral features are now *experimentally* established!

This bound is diminished if ν_e and ν' are nearly degenerate so that in Eq. (7.12) $\delta_m \ll 1$. The structure of the photon flux as a function of δ_m in Eq. (12.5) is such that for $\delta_m < 1$ the spectrum can be obtained, in a doubly logarithmic representation such as Fig. 12.4, by shifting it "to the left" and "upward" by the amount $|\log \delta_m|$ each, the shape itself remaining unchanged (Raffelt 1985). The excluded regime in the plane of τ_γ / m_{ν_e} and δ_m is shown in Fig. 12.5 (left panel). Again, it is more appropriate to express these limits in terms of μ_{eff} and δ_m by virtue of Eq. (7.12), leading to Fig. 12.5 (right panel). As expected, for fixed m_{ν_e} the limits on μ_{eff} quickly degrade with small δ_m.

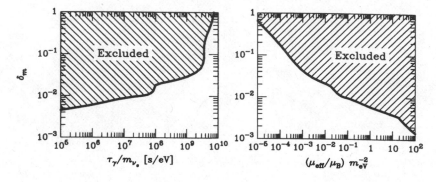

Fig. 12.5. Excluded parameters for $\nu_e \to \nu'\gamma$ from the Sun for nearly degenerate neutrino masses (adapted from Raffelt 1985).

A limit on τ_γ / m_{ν_e} similar to the solar one can be obtained from the central bulge of the galaxy. Within 2.5 kpc it contains a luminosity of about $2 \times 10^{10} L_\odot$ and thus a neutrino luminosity similarly enhanced. With its distance of (8.7 ± 0.6) kpc it is about 2×10^9 times farther away from us, leading to a much smaller local neutrino flux than that from the Sun. However, the neutrino decay path is also 2×10^9 times larger. Even though the neutrino flux scales with the inverse of the distance squared from the source, the flux of decay photons scales only with the inverse distance! Therefore, the local flux of decay photons would be larger than the solar one by perhaps a factor of ten. The measured hard x- and soft γ-ray flux from the central region of the galaxy is similar in magnitude to the upper limit solar flux so that one obtains a similar constraint on radiative decays. Because no dramatic improvement is expected a detailed analysis is not warranted. However, a substantial improvement is achieved by considering all hydrogen-burning stars in the universe as a source (Sect. 12.6).

12.3.2 Heavy Neutrino Admixtures

In full analogy to the case of reactor experiments the solar bound Eq. (12.10) can be reinterpreted as a limit on radiative decays of heavy ν_e admixtures. To this end one interprets $m_{eV} = m_h/eV$ and introduces the factor $|U_{eh}|^{-1}$ on the r.h.s. of Eq. (12.10). It must be stressed, however, that this simple procedure is only applicable to $m_h \lesssim 30\,keV$ because the main part of the solar neutrino spectrum is relatively soft. In the remaining range of interest, $30\,keV \lesssim m_h \lesssim 1\,MeV$, the ν_h flux is partly suppressed. Moreover, a large fraction of it will be nonrelativistic or only moderately relativistic so that the flux of decay photons will have a nonnegligible angular divergence. The detectors used to derive the constraints shown in Fig. 12.4 had a limited forward aperture of about 0.15 sr, relevant in the energy range $18-185\,keV$ (Peterson et al. 1966), and 1 sr, relevant in the energy range $163-774\,keV$ (Frost et al. 1966). Therefore, a certain part of the photon flux that would have come from angles relatively far away from the Sun would have been cut out, weakening the bounds on radiative decays of ν_h.

For $m_h > 2m_e$ one would, again, expect the decay $\nu_h \to \nu_e e^+ e^-$ to dominate. For m_h up to about $14\,MeV$ one may use the ^8B neutrinos from the Sun as a source spectrum. The flux of interplanetary positrons from cosmic ray secondaries is measured to be approximately $10^{-4}\,cm^{-2}\,s^{-1}\,sr^{-1}\,MeV^{-1}$ at kinetic energies of about $5\,MeV$. Interpreting this flux as an upper limit to possible decay positrons from solar neutrinos, Toussaint and Wilczek (1981) derived

$$|U_{eh}|^2 \lesssim \begin{cases} 2\times10^{-4} & (m_h = 2\,MeV), \\ 2\times10^{-5} & (m_h = 5\,MeV), \\ 3\times10^{-6} & (m_h = 10\,MeV). \end{cases} \tag{12.11}$$

Because they used the theoretically expected rather than the experimentally measured solar ^8B neutrino flux I have discounted their original numbers by a factor of 3. These bounds are included in Fig. 12.3; in the applicable mass range they are more restrictive than those from laboratory experiments. They are valid only if the ν_h flux is not diminished by invisible decay channels on its way between Sun and Earth, i.e. the total ν_h (laboratory) lifetime must exceed about 500 s. Because in the mass range of a few MeV the time dilation factor for about $10\,MeV$ neutrinos is not large, the bounds apply for total ν_h lifetimes exceeding about 100 s. Such "long-lived" MeV-mass neutrinos are in conflict with the big-bang nucleosynthesis constraints shown in Fig. 7.2.

12.3.3 New Particles

Besides neutrinos, stars can also produce other weakly interacting particles by both plasma and nuclear processes. With a temperature of about 1.3 keV in the solar center the former reactions would produce a relatively soft spectrum and so I focus on nuclear reactions where MeV energies are available. If the new particle is a boson and if it couples to nucleons, it will substitute for a photon with certain relative rates r in reactions with final-state γ-rays. A short glance at the nuclear reaction chains shown in Fig. 10.2 reveals that a particularly useful case is $p + d \to {}^3\text{He} + \gamma$ with $E_\gamma = 5.5$ MeV. This reaction occurs about 1.87 times for every ${}^4\text{He}$ nucleus produced by fusion in the Sun and so it must occur about $1.7 \times 10^{38}\,\text{s}^{-1}$. A certain fraction of the particles produced will be reabsorbed or decay within the Sun. If their probability for escaping is p the Sun emits $r\,p\,1.7 \times 10^{38}\,\text{s}^{-1}$ of the new objects.

If the new particle is a scalar boson a it will have a decay channel $a \to 2\gamma$. Because this decay is isotropic in a's rest frame the spectrum of decay photons is box-shaped (Fig. 12.1) with an upper endpoint of 5.5 MeV. If a fraction q of the particles decays between the Sun and Earth the local γ flux is $r\,p\,q\,2.2 \times 10^{10}\,\text{cm}^{-2}\,\text{s}^{-1}\,\text{MeV}^{-1}$. The upper limit photon flux shown in Fig. 12.4 at 5.5 MeV is $0.8 \times 10^{-3}\,\text{cm}^{-2}\,\text{s}^{-1}\,\text{MeV}^{-1}$ (Peterson et al. 1966). From there, Raffelt and Stodolsky (1982) found the general upper bound $r\,p\,q < 4 \times 10^{-14}$. They also calculated r, p, and q for the specific case of "standard axions" and were able to derive a strong limit on the properties of this hypothetical particle. Together with many laboratory constraints (Particle Data Group 1994) standard axions are now entirely excluded, the main motivation to consider "invisible axions" instead (Chapter 14).

12.4 Supernova 1987A

12.4.1 Decay Photons from Low-Mass Neutrinos

The most significant constraints on radiative particle decays from stellar sources can be derived on the basis of the neutrino burst from supernova (SN) 1987A. The neutrino observations and their interpretation were discussed in Chapter 11. For the present purpose it is enough to know that the neutrinos arrived in a short burst lasting a few seconds with a total emitted energy per flavor of about $1 \times 10^{53}\,\text{erg} = 6.2 \times 10^{58}\,\text{MeV}$. For $\bar{\nu}_e$ the measured spectral distribution is consistent with a thermal emission at $T_{\nu_e} \approx 4\,\mathrm{MeV}$ so that the fluence[73] of ν's plus $\bar{\nu}$'s per flavor was about

$$\mathcal{F}_\nu = 1.4{\times}10^{10}\,\mathrm{cm}^{-2}\,(4\,\mathrm{MeV}/T_\nu)\,, \tag{12.12}$$

taking $\langle E_\nu \rangle = 3T_\nu$. Approximately the same result is thought to apply to ν_μ and ν_τ with about 1.3–1.7 times the temperature, although for those flavors there is no direct measurement.

Because SN 1987A occurred in the Large Magellanic Cloud (LMC) at an approximate distance of $d_{\mathrm{LMC}} = 50\,\mathrm{kpc} = 1.5{\times}10^{23}\,\mathrm{cm}$ an enormous decay path was available for the neutrinos from this *measured* source. A very restrictive upper limit photon flux was provided by the gamma ray spectrometer on the solar maximum mission (SMM) satellite which was operational at the time of the neutrino signal and did not register any excess γ counts above the normal background. In order to use this result one needs to compute the expected γ signal from neutrino decay. Because one is dealing with a short neutrino burst the previous results for stationary sources do not apply directly: the time structure of the expected photon burst must be taken into account.

This is easy when the mass of the parent neutrino is below about 40 eV; the pulse dispersion is then not much larger than the duration of the observed $\bar{\nu}_e$ burst. Because for low-mass neutrinos the decay photons have essentially the same time structure as the neutrino burst one considers the γ fluence for a time interval of about 10 s around the first neutrino arrival. The non-ν_e flavors could be heavier if they violate the cosmological mass limit of a few 10 eV. Then the photon pulse will be correspondingly stretched, a case to be studied in Sect. 12.4.4 below.

If the neutrino masses are not degenerate so that in Eq. (12.5) $\delta_m = 1$ the expected differential fluence is

$$\mathcal{F}'_\gamma(E_\gamma) = \mathcal{F}_\nu \frac{m_\nu}{\tau_\gamma} d_{\mathrm{LMC}} \int_{E_\gamma}^\infty dE_\nu \left(1 - \alpha + 2\alpha \frac{E_\gamma}{E_\nu}\right) \frac{\Phi_\nu(E_\nu)}{E_\nu^2},$$

$$\tag{12.13}$$

where $\Phi_\nu(E_\nu) \equiv \mathcal{F}'_\nu(E_\nu)/\mathcal{F}_\nu$ is a normalized spectrum (units MeV^{-1}). For this expression CP conservation was assumed so that ν's and $\bar{\nu}$'s are characterized by the same values of τ_γ and α as discussed in Sect. 12.2.1.

---

[73]With fluence one means the time-integrated flux. In this book I use the symbol F for a differential particle flux $(\mathrm{cm}^{-2}\,\mathrm{s}^{-1}\,\mathrm{MeV}^{-1})$, the symbol $\mathcal{F}$ for a fluence (cm^{-2}), and $\mathcal{F}' = d\mathcal{F}/dE$ for a "differential fluence" which includes spectral information $(\mathrm{cm}^{-2}\,\mathrm{MeV}^{-1})$.

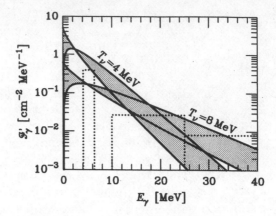

Fig. 12.6. Expected fluence of photons from the decay $\nu \to \nu'\gamma$ of low-mass SN neutrinos according to Eq. (12.15) with $m_\nu/\tau_\gamma = 10^{-15}\,\text{eV/s}$ and the indicated neutrino temperatures. The shaded bands for each temperature are for the range $-1 \le \alpha \le +1$ with the harder edge corresponding to $\alpha = +1$. Also shown are the upper limits from the GRS channels taken from the 10 s column of Tab. 12.1.

Because details of the spectral form are not known it is easiest to use a Boltzmann distribution as a generic case,

$$\Phi_\nu(E_\nu) = \frac{E_\nu^2 \, e^{-E_\nu/T_\nu}}{2T_\nu^3}. \tag{12.14}$$

Then one finds explicitly[74]

$$\mathcal{F}_\gamma'(E_\gamma) = \mathcal{F}_\nu \, \frac{m_\nu}{\tau_\gamma} \, \frac{d_{\text{LMC}}}{2T_\nu^2} \left[(1-\alpha)\,e^{-\varepsilon} + 2\alpha\,\varepsilon\,E_1(\varepsilon) \right], \tag{12.15}$$

where $\varepsilon \equiv E_\gamma/T_\nu$. In Fig. 12.6 this spectrum is shown for $T_\nu = 4$ and $8\,\text{MeV}$ with $m_\nu/\tau_\gamma = 10^{-15}\,\text{eV/s}$. The envelopes of the shaded bands in Fig. 12.6 correspond to $\alpha = \pm 1$ where for each temperature the "harder" edge corresponds to $\alpha = +1$.

The observational constraints give a limiting γ fluence for certain energy bands. Thus one needs the expected fluence for a given energy

---

[74]The exponential integral function is defined as $E_n(x) = \int_1^\infty dt\, e^{-xt}/t^n$. Note that $E_n(\infty) = 0$ while for $n > 1$ $E_n(0) = (n-1)^{-1}$.

range $(E_{\gamma,1}, E_{\gamma,2})$ for which one finds by integration of Eq. (12.15)

$$\mathcal{F}_{\gamma,1,2} = \mathcal{F}_\nu \, \frac{m_\nu}{\tau_\gamma} \, \frac{d_{\mathrm{LMC}}}{2T_\nu} \left\{ (1-\alpha) \, e^{-\varepsilon} + 2\alpha \left[\varepsilon \, E_2(\varepsilon) + E_3(\varepsilon) \right] \right\} \bigg|_{\varepsilon_2}^{\varepsilon_1},$$

$$(12.16)$$

where the expression in braces is meant to be taken as a difference between the two limits for ε. For $\varepsilon_1 = 0$ and $\varepsilon_2 = \infty$ it is equal to 1, independently of α, as it must because the angular distribution of photon emission leaves the total number of decay photons unchanged.

12.4.2 SMM Observations

The gamma ray spectrometer (GRS) on the SMM satellite consists of seven NaI detectors surrounded on the sides by a CsI annulus and at the back by a CsI detector plate (Forrest et al. 1980). The three energy bands shown in Tab. 12.1 have been analyzed for γ-ray emission from SN 1987A (Chupp, Vestrand, and Reppin 1989; Oberauer et al. 1993). At the detection time of the first neutrino event at IMB (7:35:41.37 UT) the GRS was observing the Sun. A time interval of 223.232 s until it went into calibration mode was used to search for a photon signal above background in each energy band. The background was determined by analyzing the rates measured during an interval of 151.6 s before the first neutrino event. In Fig. 12.7 the recorded number of events per 2.048 s is shown in each band as a function of time. No excess counts were found in any of them and the distributions of the rates are in good agreement with a Gaussian shape.

Because the GRS was observing the Sun, γ rays associated with the neutrino burst would have hit the instrument almost exactly from the side and so they had to traverse about 2.5 g cm^2 of spacecraft aluminum before being recognized in one of the detectors. This effect has been included to calculate the effective detector areas which allow one to convert "counts" into a γ-ray fluence. In Tab. 12.1 the corresponding 3σ limits are shown for the time until 223.232 s after the first neutrino arrival (Oberauer et al. 1993). In order to constrain low-mass neutrinos, only a time interval of 10 s around the burst is of interest; the corresponding limits are also given (Chupp, Vestrand, and Reppin 1989). The fluence limits (cm^{-2}) can be expressed as limits on an average differential fluence (cm^{-2} MeV^{-1}) by dividing with the width ΔE_γ of a given channel. For the 10 s column they are shown as dotted histograms in Fig. 12.6 while for 223.2 s they are shown in Fig. 12.13.

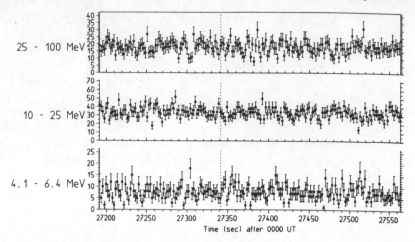

Fig. 12.7. Event rates measured in the Gamma Ray Spectrometer (GRS) of the Solar Maximum Mission (SMM) satellite encompassing the observed neutrino burst of SN 1987A (the dashed line is for the first neutrinos observed in the IMB detector). The time interval for each bin is 2.048 s. The rates to the left of the dashed line are used to determine the background while the ones to the right would include photons from neutrino decay. (Figure from Oberauer et al. 1993 with permission.)

Table 12.1. GRS 3σ upper fluence limits.

Channel	Energy Band [MeV]	γ Fluence Limit [cm^{-2}] $(10\,\text{s})$[a]	$(223.2\,\text{s})$[b]
1	4.1−6.4	0.9	6.11
2	10−25	0.4	1.48
3	25−100	0.6	1.84

[a]Chupp, Vestrand, and Reppin (1989)
[b]Oberauer et al. (1993)

For the 10 s and 223.2 s time intervals one can compute an average flux limit (cm^{-2} s^{-1}) for each channel. Then one expects that for the longer time interval it is more restrictive by the ratio of $(\Delta t)^{1/2}$, i.e. by $(10\,\text{s}/223.2\,\text{s})^{1/2} = 0.21$. This expectation is approximately borne out by the data in Tab. 12.1, confirming their consistency.

12.4.3   Radiative Decay Limit: Low-Mass Neutrinos

We are now armed to derive a radiative lifetime limit for low-mass neutrinos ($m_\nu \lesssim 40\,\text{eV}$) by comparing the expected fluence according to Eqs. (12.12) and (12.16) for channels 1, 2, and 3 with the observational upper limits given in Tab. 12.1 for a 10 s time interval surrounding the observed SN 1987A $\bar{\nu}_e$ burst. Depending on the assumed neutrino spectral distribution which was parametrized by T_ν and the anisotropy parameter α, different channels give the most restrictive limits. These are shown in Fig. 12.8 as a function of T_ν for $\alpha = 0, \pm 1$.

For normal neutrinos the relevant temperature range is between 4 and 8 MeV. In Fig. 12.8 a much larger range is shown because one may also consider the emission of sterile neutrinos or axions from the deep interior of a SN core where temperatures of several 10 MeV and Fermi energies of several 100 MeV are available (Sect. 12.4.6).

For Dirac neutrinos the most conservative case is $\alpha = -1$ where for $4\,\text{MeV} \lesssim T_\nu \lesssim 8\,\text{MeV}$ the bound is approximately constant at $\tau_\gamma/m_\nu > 0.8 \times 10^{15}\,\text{s/eV}$. Therefore, it applies equally to ν_e and low-mass ν_μ and ν_τ. For Majorana neutrinos ($\alpha = 0$) the limit is more restrictive by a factor of 2–3, depending on the assumed T_ν. With Eq. (7.12) the most conservative overall limit[75] ($\alpha = -1$) translates into

$$\mu_{\text{eff}} < 1.5 \times 10^{-8} \mu_{\text{B}}\, m_{\text{eV}}^{-2}. \tag{12.17}$$

It applies if the total laboratory lifetime exceeds the time of flight of $5.7 \times 10^{12}\,\text{s}$ from the LMC to us. With a typical $E_\nu = 20\,\text{MeV}$ one must require $\tau_{\text{tot}}/m_\nu \gtrsim 3 \times 10^5\,\text{s/eV}$ in the neutrino rest frame.

If the total lifetime is shorter than this limit all neutrinos decay before they reach the Earth. Therefore, one can only derive a limit on the branching ratio B_γ of the radiative channel. One easily finds that Eq. (12.13) is to be replaced by

$$\mathcal{F}'_\gamma(E_\gamma) = \mathcal{F}_\nu\, B_\gamma \int_{E_\gamma}^\infty dE_\nu\, (1 - \alpha + 2\alpha\, E_\gamma/E_\nu)\, \Phi_\nu(E_\nu)/E_\nu\,. \tag{12.18}$$

Therefore, using the same Boltzmann source spectrum the photon spectrum is slightly harder. Going through the same steps as before one finds the upper limits on B_γ as a function of the assumed T_ν and α

---

[75]Somewhat stronger constraints found in the literature were based on a less detailed analysis, notably with regard to the spectral dependence and the dependence on α. Published results are $\tau_\gamma/m_\nu > 0.83 \times 10^{15}\,\text{s/eV}$ (von Feilitzsch and Oberauer 1988), 1.7×10^{15} (Kolb and Turner 1989), 6.3×10^{15} (Chupp, Vestrand, and Reppin 1989), and 2.8×10^{15} (Bludman 1992).

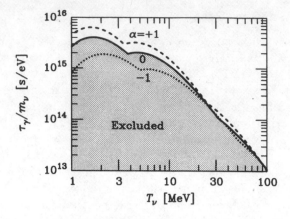

Fig. 12.8. Lower limit on τ_γ/m_ν for $m_\nu \lesssim 40\,\text{eV}$ and $\tau_{\text{tot}}/m_\nu \gtrsim 5\times10^5\,\text{s/eV}$ (most neutrinos pass the Earth before decaying).

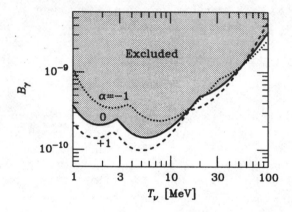

Fig. 12.9. Upper limit on the radiative branching ratio B_γ for $m_\nu \lesssim 40\,\text{eV}$ and $\tau_{\text{tot}}/m_\nu \lesssim 5\times10^5\,\text{s/eV}$ (most neutrinos decay between SN 1987A and Earth).

shown in Fig. 12.9. The most conservative case yields approximately $B_\gamma < 3\times10^{-10}$. As a bound on an effective transition moment this is

$$\mu_{\text{eff}} < 0.7\times10^{-5}\mu_{\text{B}}\, m_{\text{eV}}^{-2}\,(m_{\text{eV}}/\tau_{\text{s}})^{1/2}, \tag{12.19}$$

where $\tau_{\text{s}} = \tau_{\text{tot}}/\text{s}$. Of course, if τ_{tot} became so short that the neutrinos would decay while still within the envelope of the progenitor even this weak limit would not apply. This is the case for $\tau_{\text{lab}} \lesssim R_{\text{env}} \approx 100\,\text{s}$ (envelope radius R_{env} of the progenitor star) and so with $E_\nu \approx 20\,\text{MeV}$ one needs to require $\tau_{\text{tot}}/m_\nu \gtrsim 10^{-5}\,\text{s/eV}$.

12.4.4 Decay Photons from High-Mass Neutrinos

For "high-mass" neutrinos with $m_\nu \gtrsim 40\,\text{eV}$ a calculation of the photon fluence is more involved because of the dispersion of the neutrino burst and the corresponding delay of the decay photons. The parent neutrino travels with a velocity $\beta = (1 - m_\nu^2/E_\nu^2)^{1/2} \approx 1 - m_\nu^2/2E_\nu^2$ so that it arrives at Earth with a time delay of about $(m_\nu^2/2E_\nu^2)\,d_{\text{LMC}}$ relative to massless ones; here, $d_{\text{LMC}} = 50\,\text{kpc} = 5.1{\times}10^{12}\,\text{s}$ is our distance to the LMC where SN 1987A had occurred. With $T_\nu \approx 6\,\text{MeV}$ for ν_μ or ν_τ an average neutrino energy is $3T_\nu \approx 20\,\text{MeV}$, spreading out the arrival times of massive neutrinos over an approximate interval of $10^{-2}\,\text{s}\,(m_\nu/\text{eV})^2$. As the radiative decay may occur anywhere between the LMC and here, the arrival times of the decay photons will be spread out by a similar amount even though the photons themselves travel with the speed of light. Thus for $m_\nu \lesssim 40\,\text{eV}$ all decay photons fall within about a 10 s time window around the arrival time of the first $\bar{\nu}_e$; the results of Sect. 12.4.3 apply to this case. For $m_\nu \lesssim 200\,\text{eV}$ they fall within the 223.2 s interval for which GRS fluence limits exist. Therefore, one may easily scale the previous limits to this case by using the 223.2 s fluence limits in Tab. 12.1 instead of the 10 s ones. Of course, for a given neutrino mass there would be an optimum time window for which fluence limits could be derived on the basis of the original data.

For larger masses only a certain portion of the photon pulse falls into the 223.2 s window. In order to calculate this fraction, I follow Oberauer et al. (1993) and begin with the simple case where all neutrinos are emitted at the same time with a fixed energy E_ν. The radiative decay occurs at a time t_{D} after emission and thus at a distance $d_{\text{D}} = \beta t_{\text{D}}$ from the source (neutrino velocity β), and the photon is emitted at an angle θ_{lab} relative to the neutrino momentum (Fig. 12.10). It has to travel a distance d_γ until it arrives here; elementary geometry yields $d_\gamma = [d_{\text{LMC}}^2 - d_{\text{D}}^2(1 - \cos^2\theta_{\text{lab}})]^{1/2} - d_{\text{D}}\cos\theta_{\text{lab}}$. Therefore, relative to the first (massless) neutrinos the photons are delayed by $t = t_{\text{D}} + d_\gamma - d_{\text{LMC}}$. This delay has two sources: The parent moves with a speed less than that of light, and the photon is emitted at an angle so that a detour is taken from the LMC to us. For ultrarelativistic parents both effects disappear as the relativistic transformations squeeze all laboratory emission angles into the forward direction.

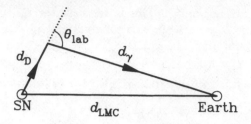

Fig. 12.10. Geometry of the radiative neutrino decay.

The photons which arrive first are the ones from decays near the source. Then the limit $d_D \ll d_{LMC}$ leads to $d_\gamma = d_{LMC} - d_D \cos\theta_{lab}$ and $t = t_D(1 - \beta\cos\theta_{lab})$. With $\cos\theta_{lab} = (\beta + \cos\theta)/(1 + \beta\cos\theta)$ where θ is the angle of photon emission in the parent frame one finds

$$t = \frac{t_D}{\gamma^2(1 + \beta x)}, \tag{12.20}$$

where $\gamma = E_\nu/m_\nu$ is the neutrino Lorentz factor and $x \equiv \cos\theta$. Therefore, photons detected between t and $t + dt$ result from decays at $t_D = \gamma^2(1 + \beta x)\,t$ during an interval $dt_D = \gamma^2(1 + \beta x)\,dt$. (Recall that t is measured after the first massless neutrinos arrived while t_D is measured after emission at the source.) The number of parent neutrinos diminishes in time as $e^{-t_D/\gamma\tau_{tot}}$ with τ_{tot} the *total* decay time. Therefore, the number of photons traversing a spherical shell of radius d_{LMC} per unit time is

$$\dot{N}_\gamma(t) = \frac{\gamma(1 + \beta x)}{\tau_\gamma}\, e^{-\gamma(1+\beta x)\,t/\tau_{tot}}, \tag{12.21}$$

where as before τ_γ is the radiative decay time.

Photons produced before the parent has left the envelope of the progenitor star (radius R_{env}) cannot be detected at Earth. If this absorption effect is to be included, Eq. (12.21) will involve a step function[76] $\Theta(d_D - R_{env})$ with $d_D = \beta t_D = \beta\gamma^2(1 + \beta x)\,t$. Put another way, photons emitted at an angle θ in the rest frame will first arrive at a time $t_0 = R_{env}[\beta\gamma^2(1 + \beta x)]^{-1}$. The progenitor of SN 1987A has been unambiguously identified as the blue supergiant Sanduleak $-69\,202$ (Schramm and Truran 1990). From its surface temperature $(15,000\,\mathrm{K})$, its luminosity $(5\times10^{38}\,\mathrm{erg/s})$ and the distance to the LMC one can infer its radius to be $R_{env} \approx 3\times10^{12}\,\mathrm{cm} = 100\,\mathrm{s}$. Using this

---

[76]The step function is defined by $\Theta(z) = 0$ for $z < 0$ and $\Theta(z) = 1$ for $z > 0$.

value, t_0 is shown in Fig. 12.11 as a function of the neutrino velocity β for several values of $x = \cos\theta$. The absorption effect is negligible except for nonrelativistic neutrinos or for backward emission which, in the laboratory frame, corresponds to very soft photon energies.

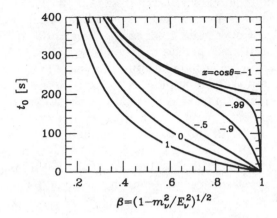

Fig. 12.11. Arrival time t_0 of first decay photons from a parent neutrino with velocity β, taking the envelope radius of the source to be $R_{\text{env}} = 100$ s. The curves are marked with the respective values of $x = \cos\theta$, the direction of photon emission in the neutrino rest frame.

In the parent frame the photons (energy ω) follow a normalized distribution $f(\omega, x)$ which yields $d^3 N_\gamma(t, \omega, x) = \dot{N}_\gamma(t) \, f(\omega, x) \, dt \, d\omega \, dx$ with $\dot{N}_\gamma(t)$ from Eq. (12.21). The usual relativistic transformations lead to a laboratory photon energy of $E_\gamma = \gamma(1 + \beta x) \omega$. Transforming from $d\omega dx$ to $d\omega dE_\gamma$ and integrating over the unobserved rest-frame energy ω yields

$$\frac{d^2 N_\gamma}{dE_\gamma dt} = \frac{1}{\tau_\gamma} \frac{E_\gamma}{\beta\gamma} \int_{\omega_+}^{\omega_-} d\omega \, \frac{f(\omega, x)}{\omega^2} \, e^{-E_\gamma t/\omega\tau_{\text{tot}}} \, \Theta\left(\frac{\beta\gamma E_\gamma t}{R_{\text{env}}} - \omega\right),$$

$$(12.22)$$

where $\omega_\pm = E_\gamma \left[\gamma(1 \pm \beta)\right]^{-1}$ and $x = (E_\gamma/\gamma\omega - 1)/\beta$. The photon flux at Earth is obtained by multiplication with the neutrino fluence $\mathcal{F}_\nu$ of Eq. (12.12) and integration over a suitable spectrum $\Phi_\nu(E_\nu)$ of neutrino energies.

If the neutrinos are sufficiently long-lived (the exact meaning of this is quantified below) the exponential can be ignored. If one also ignores the absorption effect by the progenitor star ($R_{\text{env}} = 0$), the time structure of Eq. (12.22) reduces to $\Theta(t)$, i.e. its spectral form is time independent except that it begins at $t = 0$. This is somewhat surprising because the energy of a photon in the laboratory frame is related to the angle of emission in the neutrino rest frame which in turn determines the "detour" taken from the source to us (Fig. 12.10). However, the first photons come from decays immediately at the source and so any angle of emission leads to the same initial arrival time.

It must be stressed that the expression Eq. (12.22) depends on the assumption of decays not too far from the source ($d_D \ll d_{LMC}$) and so only the head of the photon pulse is correctly described while its tail would require including decays even close to the Earth. Strictly speaking, the photon burst never ends because even if the parent neutrinos have passed the Earth, *some* photons will be received from backward emission. However, because one is interested in neutrino masses so large ($m_\nu \gtrsim 200\,\text{eV}$) that the photon burst is much longer than the GRS measurement window, it is enough to account for the head of the photon pulse.

12.4.5 Radiative Decay Limits: High-Mass Neutrinos

As a first explicit case for the distribution of photon energies and emission angles I take the two-body decay $\nu \to \nu'\gamma$ with a massless daughter neutrino and with the dipole angular distribution of Eq. (12.2) with $x = \cos\theta = -\cos\vartheta$ for a left-handed parent. This amounts to

$$f(\omega, x) = \tfrac{1}{2}(1 + \alpha x)\,\delta(\omega - \tfrac{1}{2}m_\nu) \tag{12.23}$$

in Eq. (12.22). Because of the δ function it is trivial to integrate,

$$\frac{d^2 N_\gamma}{dE_\gamma dt} = \frac{1}{m_\nu \tau_\gamma} \frac{2E_\gamma}{p_\nu} \left[1 + \alpha\,\frac{2E_\gamma - E_\nu}{p_\nu}\right] e^{-t/\tau_*}\,\Theta(t - t_{env}), \tag{12.24}$$

where

$$\tau_* \equiv \frac{m_\nu}{2E_\gamma}\,\tau_{tot}\,, \qquad t_{env} \equiv \frac{m_\nu^2}{2E_\gamma p_\nu}\,R_{env}\,. \tag{12.25}$$

Moreover, the flux vanishes if the chosen value for E_γ does not fall between $\tfrac{1}{2}(E_\nu \pm p_\nu)$, or equivalently, unless

$$E_\nu > E_\gamma + m_\nu^2/4E_\gamma\,, \tag{12.26}$$

a condition on the minimum required neutrino energy.

For the simplest case when the neutrinos are relativistic ($p_\nu = E_\nu$), long-lived ($e^{-t/\tau_*} = 1$), and absorption effects by the progenitor can be ignored ($t_{\text{env}} = 0$) the spectrum is shown in Fig. 12.12 for the anisotropy parameters $\alpha = 0, \pm 1$. Note the difference to the triangular shape of Fig. 12.1 for a stationary source. High-energy photons are now enhanced because lower-energy ones correspond to larger emission angles in the parent frame and so they take a larger "detour" from the source to us (Fig. 12.10). Hence, their flux is spread out over a larger time interval even though photons of all energies begin to arrive at the same time if $t_{\text{env}} = 0$.

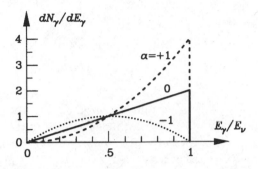

Fig. 12.12. Photon spectrum from the decay of a short burst of relativistic neutrinos, energy E_ν, according to Eq. (12.24) taking $p_\nu = E_\nu$ and $e^{-t/\tau_*} = 1$ (relativistic and long-lived parent), and ignoring absorption effects by the progenitor ($t_{\text{env}} - 0$).

In order to compare with the GRS fluence limits one needs to integrate the expected flux between $t - 0$ and $t - t_{\text{GRS}} - 223.2\,\text{s}$. The Θ function in Eq. (12.24) is accounted for by using t_{env} as a lower limit of integration. Integrating also over the neutrino source spectrum $\mathcal{F}_\nu \Phi_\nu(E_\nu)$ yields

$$\mathcal{F}'_\gamma = \mathcal{F}_\nu \frac{t_{\text{GRS}}}{m_\nu \tau_\gamma} \int_{E_{\text{min}}}^{\infty} dE_\nu \, \Phi_\nu \frac{2E_\gamma}{p_\nu} \left[1 + \alpha \frac{2E_\gamma - E_\nu}{p_\nu} \right] I, \qquad (12.27)$$

where

$$I \equiv \frac{e^{-t_{\text{env}}/\tau_*} - e^{-t_{\text{GRS}}/\tau_*}}{t_{\text{GRS}}/\tau_*}. \qquad (12.28)$$

For sufficiently long-lived parents ($\tau_* \gg t_{\text{GRS}}$) the exponentials can be expanded and $I = (t_{\text{GRS}} - t_{\text{env}})/t_{\text{GRS}}$. The lower limit of integration is set by the condition Eq. (12.26) and by the requirement that $t_{\text{env}} < t_{\text{GRS}}$, i.e. that $I > 0$. This condition may be expressed as $p_\nu > (m_\nu^2/2E_\gamma)(R_{\text{env}}/t_{\text{GRS}})$.

The two-body decay is mostly interesting for $m_\nu \lesssim 2m_e$, a limit in which one may safely ignore all nonrelativistic corrections, including the progenitor absorption effect. In this case

$$I = \frac{1 - e^{-t_{\rm GRS}/\tau_*}}{t_{\rm GRS}/\tau_*}, \tag{12.29}$$

which is unity for $\tau_* \gtrsim t_{\rm GRS}$. With typical photon energies of $3T_\nu \approx$ 20 MeV and $t_{\rm GRS} = 223.2$ s this requirement translates into $m_\nu \tau_{\rm tot} \gtrsim 10^{10}$ eV s.

In the relativistic limit and with $I = 1$ one can easily integrate Eq. (12.27) with the Boltzmann spectrum Eq. (12.14) and finds

$$\mathcal{F}'_\gamma = \mathcal{F}_\nu \frac{t_{\rm GRS}}{m_\nu \tau_\gamma} \left[(1 - \alpha)\,\varepsilon + (1 + \alpha)\,\varepsilon^2 \right] e^{-\varepsilon}, \tag{12.30}$$

where $\varepsilon = E_\gamma / T_\nu$. This spectrum is shown in Fig. 12.13 for $T_\nu = 4$ and 8 MeV with $m_\nu \tau_\gamma = 10^{18}$ eV s. The envelopes of the shaded bands in Fig. 12.13 correspond to $\alpha = \pm 1$ where for each temperature the "harder" edge corresponds to $\alpha = +1$. The expected fluence for each GRS channel of Tab. 12.1 is found by integration. The GRS fluence limits then yield the lower bounds on $m_\nu \tau_\gamma$ shown in Fig. 12.14.

For Dirac neutrinos the most conservative case is $\alpha = -1$ and the temperature range relevant for ν_μ and ν_τ is between 6 and 8 MeV. This yields an approximately temperature independent bound of $m_\nu \tau_\gamma > 7 \times 10^{18}$ eV s. For Majorana neutrinos ($\alpha = 0$) the limit is about[77] $m_\nu \tau_\gamma > 12 \times 10^{18}$ eV s. The most conservative case ($\alpha = -1$) translates with Eq. (7.12) into

$$\mu_{\rm eff} < 1.6 \times 10^{-10} \mu_{\rm B}\, m_{\rm eV}^{-1}, \tag{12.31}$$

assuming $m_\nu \tau_{\rm tot} \gtrsim 10^{10}$ eV s. Note the different dependence on $m_{\rm eV}$ relative to the low-mass result Eq. (12.17).

If $m_\nu \tau_{\rm tot}$ violates this condition because it is below 10^{10} eV s means that the neutrinos decay so fast that the photon burst effectively ends before the GRS integration time $t_{\rm GRS}$ is over. Then the photon burst is again "short" even though the neutrino mass is large. In this case one can state a limit on B_γ as in Sect. 12.4.3. If $\tau_{\rm tot}$ is only slightly shorter so that $m_\nu \tau_{\rm tot} \lesssim 4 \times 10^8$ eV s, all photons arrive within $t_{\rm GRS} = 10$ s and one may directly apply Eq. (12.19), originally derived for small

[77]The result here is after Oberauer et al. (1993) which is similar to 6×10^{18} eV s of Bludman (1992) but substantially more restrictive than 0.84×10^{18} eV s of Kolb and Turner (1989). These works all refer to the isotropic case ($\alpha = 0$).

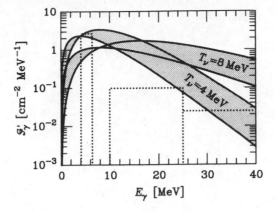

Fig. 12.13. Expected fluence of photons from the decay $\nu \to \nu'\gamma$ of "high-mass" but relativistic SN neutrinos, $200\,\text{eV} \lesssim m_\nu \lesssim 1\,\text{MeV}$, according to Eq. (12.30) with $m_\nu \tau_\gamma = 10^{18}\,\text{eV s}$ and the indicated neutrino temperatures. The shaded bands for each temperature are for the range $-1 \le \alpha \le +1$ with the harder edge corresponding to $\alpha = +1$. Also shown are the upper limits from the GRS channels taken from the 223.2 s column of Tab. 12.1.

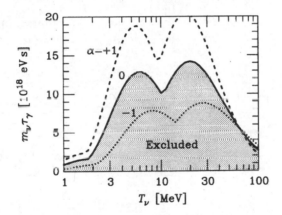

Fig. 12.14. Lower limit on $m_\nu \tau_\gamma$ for $200\,\text{eV} \lesssim m_\nu \lesssim 1\,\text{MeV}$, assuming $m_\nu \tau_{\text{tot}} \gtrsim 10^{10}\,\text{eV s}$.

neutrino masses. For the narrow region where the photon burst ends between 10 and 223.2 s this bound must be discounted by about a factor of 2 because of the less restrictive fluence limits for the larger integration time.

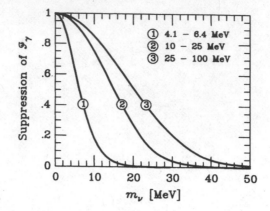

Fig. 12.15. Nonrelativistic correction of the expected photon fluence for the GRS channels of Tab. 12.1 according to Eq. (12.32) with $R_{\text{env}} = 100 \, \text{s}$, $t_{\text{GRS}} = 223.2 \, \text{s}$, and $T_\nu = 6 \, \text{MeV}$.

So far the nonrelativistic corrections for a neutrino mass in the 10 MeV mass range have been ignored. As an example I consider the photon fluence of Eq. (12.27) in the limit of a large τ_* where the exponentials in Eq. (12.28) can be expanded. The Boltzmann spectrum for nonrelativistic neutrinos should include an extra factor $\beta = p_\nu/E_\nu$. Then for $\alpha = 0$ the fluence is[78]

$$\mathcal{F}'_\gamma(E_\gamma) = \mathcal{F}_\nu \, \frac{t_{\text{GRS}}}{m_\nu \tau_\gamma} \int_{E_{\text{min}}}^{\infty} dE_\nu \, \frac{E_\nu \, e^{-E_\nu/T_\nu}}{T_\nu^3} \left(E_\gamma - \frac{m_\nu^2}{2 p_\nu} \frac{R_{\text{env}}}{t_{\text{GRS}}} \right), \, (12.32)$$

with

$$E_{\text{min}} = \max \left\{ m_\nu \left[1 + \left(\frac{m_\nu}{2 E_\gamma} \frac{R_{\text{env}}}{t_{\text{GRS}}} \right)^2 \right]^{1/2}, \, \left(E_\gamma + \frac{m_\nu^2}{4 E_\gamma} \right) \right\}. \, (12.33)$$

Relative to the massless case, the integral expression is suppressed if $m_\nu \gtrsim T_\nu$. A straightforward numerical integration then yields the suppression of the expected fluence for each GRS channel as shown in

_____

[78]Strictly speaking, the fluence of massive neutrinos must be calculated by determining their neutrino sphere which is different from the massless case. This problem was recently tackled by Sigl and Turner (1995) by solving the Boltzmann collision equation by means of an approximation method known from calculations of particle freeze-out in the early universe. However, because only masses of up to 24 MeV are presently considered, a precise treatment of the neutrino spectrum changes the resulting limits only by a small amount.

Fig. 12.15. (The overall factor m_ν^{-1} of Eq. 12.32 is not included, of course.) Up to neutrino masses of about 10 MeV one may essentially ignore the nonrelativistic corrections while for larger masses one has to worry about them. However, because this discussion applies to standard neutrinos, the largest relevant mass is about 24 MeV and so the nonrelativistic corrections never overwhelm the result.

12.4.6 Summary of $\nu \to \nu'\gamma$ Limits

In order to summarize the decay limits I begin in Fig. 12.16 with the relevant regimes of m_ν and $\tau_{\rm tot}$. Above the upper dotted line the neutrinos live long enough so that most of them pass the Earth before decaying while below the lower dotted line they decay within the envelope of the progenitor star. In the areas 1 and 5 the photon burst is "short" ($\Delta t_\gamma \lesssim 10\,{\rm s}$), in 2 and 4 it is "intermediate" ($10\,{\rm s} \lesssim \Delta t_\gamma \lesssim 223.2\,{\rm s}$), and in 3 it is "long" ($223.2\,{\rm s} \lesssim \Delta t_\gamma$). The exact boundaries as well as the relevant constraints are summarized in Tab. 12.2. In Fig. 12.17 the limits on $\mu_{\rm eff}$ are summarized as a contour plot.

If one restricts possible neutrino decays to the radiative channel one has $\tau_{\rm tot} = \tau_\gamma$ which depends only on $\mu_{\rm eff}$ and m_ν. Then one may use directly the upper limits on $\mu_{\rm eff}$ given in Tab. 12.2 for the areas 1–3, depending on the assumed mass. Put another way, the conditions on $\tau_{\rm tot}$ are then automatically satisfied.

These limits certainly apply to ν_e as the $\overline{\nu}_e$ burst from SN 1987A has been measured. The fluxes of the other flavors were only theoretically implied. If they have only standard weak interactions they must have been emitted approximately with the same efficiency as ν_e. Large dipole moments, however, imply large nonstandard interactions: The same electromagnetic interaction vertex that allows for radiative decays also allows for scattering on charged particles by photon exchange! For MeV energies, for example, the scattering cross section on electrons by regular weak interactions and that by photon exchange are the same for $\mu_{\rm eff}$ of order $10^{-10}\mu_{\rm B}$. Hence in the lower left corner of Fig. 12.17 the neutrinos would interact much more strongly by photon exchange than by ordinary weak interactions, causing them to emerge from higher layers of the SN core than normally assumed. Their fluence and effective temperature is then much smaller than standard. Put another way, for $\mu_{\rm eff} \gtrsim 10^{-10}\,\mu_{\rm B}$ the above constraints are not self-consistent (Hatsuda, Lim, and Yoshimura 1988). However, because large dipole moments can be constrained by other methods (Sect. 7.5.1) a detailed investigation of their impact on SN physics is not warranted.

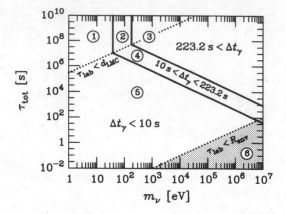

Fig. 12.16. Different regimes of neutrino masses and total lifetimes referred to in the text. Δt_γ is the duration of the burst of decay photons. The radiative lifetime limits in the areas 1−6 are summarized in Tab. 12.2.

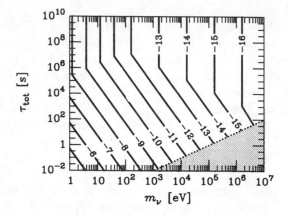

Fig. 12.17. Upper limits on effective neutrino transition moments from SN 1987A as given in Tab. 12.2. The contours are marked with $\log(\mu_{\rm eff}/\mu_{\rm B})$. In the shaded lower right region the neutrinos decay within the progenitor. Toward the lower left side the bounds are not self-consistent because large dipole moments induce large nonstandard scattering cross sections which enhance neutrino trapping.

The discussion of the previous sections focussed on standard neutrinos which are emitted approximately with the same efficiency and similar energies as ν_e's and $\bar{\nu}_e$'s. It is possible, however, that these neu-

Table 12.2. Neutrino radiative lifetime limits[a] from SN 1987A.

Area[b]	Boundaries[c]	Radiative lifetime limit[c]	Transition moment[d] $\mu_{\mathrm{eff}}/\mu_{\mathrm{B}}$
1	$m_\nu < 40$ $3\times10^5 < \tau_{\mathrm{tot}}m_\nu^{-1}$	$\tau_\gamma m_\nu^{-1} > 0.8\times10^{15}$	$1.5\times10^{-8}\,m_\nu^{-2}$
2	$40 < m_\nu < 200$ $3\times10^5 < \tau_{\mathrm{tot}}m_\nu^{-1}$	$\tau_\gamma m_\nu^{-1} > 0.2\times10^{15}$	$0.8\times10^{-8}\,m_\nu^{-2}$
3	$200 < m_\nu < 10^7$ $9\times10^9 < \tau_{\mathrm{tot}}m_\nu$	$\tau_\gamma m_\nu > 7\times10^{18}$	$1.6\times10^{-10}\,m_\nu^{-1}$
4	$m_\nu < 10^7$ $4\times10^8 < \tau_{\mathrm{tot}}m_\nu < 9\times10^9$ $10^{-5} < \tau_{\mathrm{tot}}m_\nu^{-1} < 3\times10^5$	$B_\gamma < 1.2\times10^{-9}$	$1.4\times10^{-5}\,m_\nu^{-3/2}\tau_{\mathrm{tot}}^{-1/2}$
5	$m_\nu < 10^7$ $\tau_{\mathrm{tot}}m_\nu < 4\times10^8$ $10^{-5} < \tau_{\mathrm{tot}}m_\nu^{-1} < 3\times10^5$	$B_\gamma < 3\times10^{-10}$	$0.7\times10^{-5}\,m_\nu^{-3/2}\tau_{\mathrm{tot}}^{-1/2}$
6	$\tau_{\mathrm{tot}}m_\nu^{-1} < 10^{-5}$	$B_\gamma < 0.01$	—

For the anisotropy parameter $\alpha = -1$.
Numbered as in Fig. 12.16.
Neutrino masses in eV, lifetimes in s.
Upper limit.

trinos have Dirac masses and thus right-handed partners which could be emitted from the inner core of the SN by helicity-flipping processes (Sect. 13.8). Moreover, entirely new particles could be produced and escape from there.

The present bounds can be scaled to such cases if one calculates the total energy $E_{x,\mathrm{tot}} = f_x\,10^{53}$ erg emitted in the new x particles, where 1×10^{53} erg is the total energy that was used for a standard ν plus $\bar\nu$. Self-consistency requires $f_x < 1$, of course. In addition, one needs the average energy $\langle E_x\rangle$ of the new objects which allows one to define an approximate equivalent temperature $T_x = \frac{1}{3}\langle E_x\rangle$. Depending on the x mass and total lifetime one can then read the radiative lifetime limits directly from Figs. 12.8, 12.9, and 12.14, except that they must be relaxed by a factor f_x for the reduced fluence.

12.4.7 Limit on $\nu_\tau \to \nu_e e^+ e^-$

Neutrinos with a mass exceeding $2m_e$ can decay into $\nu_e e^+ e^-$, a channel which probably dominates over $\nu'\gamma$. Among the standard neutrinos, the role of the parent can be played only by ν_τ (or rather ν_3) with its upper experimental mass limit of about 24 MeV. Within the standard model where the decay is due to flavor mixing the rate is given by Eq. (7.9).

In order to derive bounds on the $e^+ e^-$ channel from the GRS observations, photons need to be produced. At first one may think that the positrons would quickly annihilate so that a strong prompt γ flux can be expected (Takahara and Sato 1987; Cowsik, Schramm, and Höflich 1989). Following Mohapatra, Nussinov, and Zhang (1994), however, the gas density outside of the progenitor is too low, in spite of a substantial stellar wind during the progenitor's supergiant evolution. Also, the annihilation of the charged leptons from the decay among each other is moderately efficient only if the decays occur close to the source. Typical galactic magnetic fields have a strength of about $3\,\mu$G; they may well be larger in the Large Magellanic Cloud, and the circumstellar field of the SN 1987A progenitor may have been larger still. The gyromagnetic radii for 5 MeV positrons is then less than 10^{10} cm $\ll R_{\rm env}$ so that one may think that the charged leptons were locally trapped (Cowsik, Schramm, and Höflich 1989). However, the momentum carried by the flux of the charged decay products is so large that those fields would have been swept away (Mohapatra, Nussinov, and Zhang 1994).

Altogether it appears that the decay positrons will linger in interstellar space for a long time before meeting annihilation partners unless most decays occur immediately outside of the progenitor. Therefore, prompt photons are mostly produced by bremsstrahlung $\nu_\tau \to \nu_e e^+ e^- \gamma$ which is suppressed relative to the decay rate only by a factor of about $\alpha/\pi \approx 10^{-3}$ (Dar and Dado 1987).

Neutrinos with masses in the MeV range which are emitted at MeV temperatures are nearly nonrelativistic. Their rest frame is then approximately equal to the laboratory frame, but they still move essentially with the speed of light. To escape from the progenitor before decaying their rest-frame lifetime $\tau_{\rm tot}$ must exceed a few 100 s. This also guarantees that the pulse of decay photons will outlast the GRS integration time: one is automatically in the region of a "long" photon burst which was "case 3" in Fig. 12.16. In fact, if one assumes that ν_τ decays are induced by mixing, the decay rate Eq. (7.9) together with the laboratory bounds on U_{eh} shown in Fig. 12.3 easily guarantees that they fulfill this requirement.

In order to estimate the expected photon flux from Eq. (12.22) one needs to know the distribution of photon energies and emission angles in the frame of the parent neutrino. In the absence of a detailed calculation I follow Oberauer et al. (1993) and assume approximate isotropy for the photon emission. The soft part of the spectrum $dN_\gamma/d\omega$ from a bremsstrahlung process is given by the rate of the primary process times $(\alpha/\pi)\,\omega^{-1}$ (Jackson 1975). Extending this behavior up to photon energies of $\frac{1}{2}m_\nu$ I use

$$\frac{1}{\tau_\gamma}\,f(\omega,x) = \frac{\alpha/\pi}{\tau_{e^+e^-}}\,\frac{1}{2\omega}\,\Theta(\tfrac{1}{2}m_\nu - \omega)\,, \tag{12.34}$$

an expression which does not depend on x because of the assumed isotropy. ($\alpha = 1/137$ is the fine-structure constant, not the previous anisotropy parameter.)

After dropping the exponential in Eq. (12.22) because the neutrinos are long-lived, one integrates over a Boltzmann source spectrum for nonrelativistic neutrinos, integrates over the GRS energy channels, and compares the expected fluence with the measured upper limits of Tab. 12.1. The resulting bound on $\tau_{e^+e^-}$ corresponds with Eq. (7.9) directly to a limit on $|U_{e3}|^2$. In Fig. 12.18 I show these bounds (transformed into bounds on the mixing angle) as a function of the assumed neutrino mass for $T_\nu = 4$ and $6\,\mathrm{MeV}$. There remains a strong limit even for masses far exceeding the temperature because the exponential

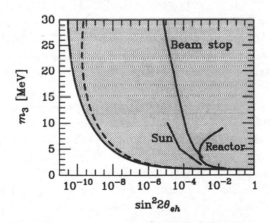

Fig. 12.18. SN 1987A limits on $\sin^2 2\theta_{e3} = 4|U_{e3}|^2$ with $\nu_3 \approx \nu_\tau$ for $T_\nu = 6\,\mathrm{MeV}$ (solid line) and $4\,\mathrm{MeV}$ (dashed line). Also shown are the corresponding laboratory and solar limits from Fig. 12.3.

suppression of the flux is compensated by the steep phase-space factor m_ν^5 in the expression for the decay rate.

These bounds are far more restrictive than those from laboratory experiments (Fig. 12.3). The reason is that a SN explosion is a strong ν_τ source; such a source is difficult to make in the laboratory. This is the reason why the ν_τ has never been directly measured by its charged-current conversion into τ.

The strong SN 1987A bounds on $|U_{e3}|^2$ imply that a heavy ν_τ with only standard-model interactions must be rather long-lived, in fact too long to be compatible with cosmological limits derived from big-bang nucleosynthesis (Sect. 7.1.5). Those bounds together with the present results imply that a heavy ν_τ cannot exist unless it has fast, invisible decays induced by interactions beyond the standard model. If this were the case it could well decay before leaving the SN progenitor. Therefore, the laboratory experiments with their short distance between source and decay volume remain important for anomalously short-lived neutrinos.

12.4.8 Heavy, Sterile Neutrinos

The bounds on U_{e3} from reactors, beam stops, or the Sun were based on a ν_e flux which partially converts into ν_3's which subsequently decay. Because the SN emits about equal numbers of all ordinary neutrino flavors, this approach is obsolete with regard to ν_3. However, one may still consider hypothetical sterile neutrinos which interact only by virtue of their mixing with ν_e. By assumption these states would not interact through ordinary weak interactions and so they would not be trapped in the SN core. Hence the expected ν_h flux would emerge from the deep interior rather than the surface of the core.

In this case, however, the sterile neutrinos would carry away energy much more efficiently than the ordinary ones and so the requirement that enough energy was left for the observed $\bar{\nu}_e$'s from SN 1987A already gives one the approximate limit $|U_{eh}|^2 \lesssim 10^{-10}$ (Sect. 9.6). If this limit is approximately saturated one expects that about as much energy is carried away by ν_h as by the ordinary flavors. Taking account of the harder energies of neutrinos emitted from the SN core one still obtains about the same limit on $|U_{eh}|^2$ as on $|U_{e3}|^2$ before. Put another way, the GRS observations do not dramatically improve on the cooling argument of Sect. 9.6, although they range in the same general magnitude.

12.4.9   Axions

Another hypothetical particle that could have been emitted abundantly from SN 1987A is the axion. The impact of the axionic energy loss is discussed in Sect. 13.5. If axions are more strongly interacting than a certain limit, implying that their mass is larger than a few eV, they are emitted from the surface of the SN core with a luminosity similar to that of neutrinos. Kolb and Turner (1989) found that the axion fluence from SN 1987A would have been $\mathcal{F}_a \approx 6\times10^{10}\,\mathrm{cm}^{-2}\,m_{\mathrm{eV}}^{-12/11}$ with $m_{\mathrm{eV}} \equiv m_a/\mathrm{eV}$ at a temperature of $T_a \approx 15\,\mathrm{MeV}\,m_{\mathrm{eV}}^{-4/11}$. From the GRS fluence limits, Kolb and Turner found that m_a must be less than a few 10 eV, a bound which is less restrictive than, for example, the limit from globular cluster stars (Sect. 5.2.5).

12.4.10   Supernova Energetics

To derive the various GRS limits one had to assume that the radiative decays occurred outside of the progenitor's envelope and so neutrinos falling into the shaded area in Fig. 12.16 were not accessible to these arguments. However, in this case the stellar envelope itself serves as a "detector" as discussed by Falk and Schramm (1978) many years ago; see also Takahara and Sato (1986). Supernova observations in general, and those of SN 1987A in particular, indicate that of the approximately 3×10^{53} erg of released gravitational binding energy only a small fraction on the order of one percent becomes directly visible in the form of the optical explosion as well as the kinetic energy of the ejecta. In contrast, even if only one of the neutrino species decayed radiatively within the progenitor, about 30% of the binding energy would light up!

If the lifetime were so short that the parent neutrinos would never get far from the SN core one would not have to worry. Therefore, the critical range of decay times is between the core dimensions of about $30\,\mathrm{km} = 10^{-4}\,\mathrm{s}$ and the envelope radius of about 100 s. If the neutrinos had nonradiative decay modes, and if their total laboratory lifetime fell into this range, one could only conclude that $B_\gamma \lesssim 10^{-2}$.

If they decayed only into radiation, and taking E_ν to be 10 MeV, the quantity τ_γ/m_ν cannot lie between about 10^{-11} and 10^{-5} s/eV (for heavy neutrinos τ_γ includes the e^+e^- channel). For an effective transition moment μ_{eff}, an interval between about 10^2 and $10^5\,\mu_B\,m_{\mathrm{eV}}^{-2}$ is excluded, moderately interesting only for large masses. Then, however, the cosmological limits strongly suggest the presence of nonradiative decay channels.

12.5 Galactic Supernovae and $\nu_\tau \to \nu_e e^+ e^-$

12.5.1 Bounds on the Positron Flux

A core-collapse SN produces about 3×10^{57} ν_τ's which may subsequently decay into $\nu_e e^+ e^-$. What is the long-term fate of all these positrons? If the ν_τ's decay mostly outside of the galaxy it is well possible that the positrons will linger in intergalactic space "forever." Those positrons produced within the galaxy, however, will be trapped by the magnetic fields (typical strength a few μG) which render the galactic disk a magnetic "bottle" for charged particles. The interstellar electron density is on the order $1 \, \mathrm{cm}^{-3}$, leading to a positron lifetime against annihilation of order 10^5 years. Moreover, elastic $e^+ e^-$ scattering (Bhabha scattering) is very efficient at slowing down relativistic positrons because of the perfect mass match which allows for an efficient energy exchange in collisions. Therefore, most annihilations occur at rest, producing a sharp γ-ray feature at 511 keV.

The galactic SN rate is a few per century while the decay positrons annihilate on a much longer time scale. Therefore, the galactic disk should contain a stationary positron population with a density determined by the galactic SN rate and the ν_τ lifetime. A comparison with the measured photon flux at $E_\gamma = 511 \, \mathrm{keV}$ of about $5 \times 10^{-3} \, \mathrm{cm}^{-2} \, \mathrm{s}^{-1}$ then leads to a very restrictive limit on the $e^+ e^-$ decay channel (Dar, Goodman, and Nussinov 1987).

In detail these authors used a galactic rate of two core-collapse SN per century, an e^+ lifetime against annihilation of 10^5 yr, and a typical distance of the decays from Earth of 10 kpc. If most neutrinos decay within the galactic disk, these assumptions lead to an expected photon flux of $400 \, \mathrm{cm}^{-2} \, \mathrm{s}^{-1}$. Thus, it is enough that one in 10^5 neutrinos decays within the galactic disk to outshine the measured flux.

This estimate is corroborated by the more recent work of Skibo, Ramaty, and Leventhal (1992) who devised detailed models of the positron distribution in the galaxy in order to account for the 511 keV diffuse galactic line feature measured in the direction away from the galactic center. They found a total stationary positron annihilation rate in the galaxy of $0.6 - 3 \times 10^{43} \, \mathrm{s}^{-1}$, where the precise coefficient depends on model assumptions. With two core-collapse SN per century the average galactic ν_τ production rate is $2 \times 10^{48} \, \mathrm{s}^{-1}$. Again, it is enough if one ν_τ in 10^5 injects an e^+ into the galaxy to account for the observations.

If all positrons produced within about $1 \, \mathrm{kpc} = 3 \times 10^{21} \, \mathrm{cm}$ from the source (the scale height of the galactic disk) were magnetically trapped, while those produced further away escaped into intergalactic space, a lifetime below $10^5 \times 1\,\text{kpc} \approx 10^{16}\,\text{s}$ in the laboratory frame is excluded. Because SN neutrinos with MeV masses are nearly nonrelativistic the rest-frame lifetime is identical with the laboratory lifetime to within a factor of a few, excluding $\tau_{e^+e^-} \lesssim 10^{15}\,\text{s}$.

If the decays occur too close to the source the annihilation with electrons from the neutrino decay is of some importance (Mohapatra, Nussinov, and Zhang 1994). Therefore, decay times below about $10^4\,\text{s}$ cannot be excluded by the present argument (Dar, Goodman, and Nussinov 1987).

Actually, the galactic positron flux is thought to be associated with supernovae, albeit not from ν_τ decay but rather from the β^+ decays of certain nuclei which are synthesized in a SN explosion (Chan and Lingenfelter 1993). These authors performed a detailed analysis of the probability for positrons to escape without annihilation from the SN environment into the galaxy.

12.5.2 Can the Tau Neutrino Be Heavy?

Armed with this result we can return to the question raised in Sect. 7.2.2 if a ν_τ with a mass exceeding $2m_e$ is compatible with the cosmological requirement that such particles and their decay products do not "overclose" the universe. It turns out that the SN constraints on $\nu_\tau \to \nu_e e^+ e^-$ presented in this chapter exclude this possibility so that either ν_τ respects the cosmological mass limit of a few $10\,\text{eV}$ or else it must have fast invisible decay channels which inevitably require interactions beyond the standard model.

In Fig. 12.19 the available constraints on $\tau_{e^+e^-}$ are summarized. The SN 1987A bound from the absence of a prompt γ burst, the cosmological requirement, and the above limit from galactic positron annihilation together exclude the entire range of possible masses and lifetimes. The margins of overlap are so enormous that each of the arguments has several orders of magnitude to spare for unaccounted uncertainties.

A heavy standard ν_τ is also excluded on the basis of arguments involving big-bang nucleosynthesis (BBN). The usual limit on the number of effective neutrino degrees of freedom at nucleosynthesis alone is enough to reach this conclusion (Sect. 7.1.5). Moreover, charged leptons and secondary photons from the e^+e^- decay channel would destroy some of the synthesized nuclei (Lindley 1979, 1985; Krauss 1984; Kawasaki, Terasawa, and Sato 1986). The main virtue of the SN limits is, therefore, that no reference to BBN is required to exclude a heavy ν_τ.

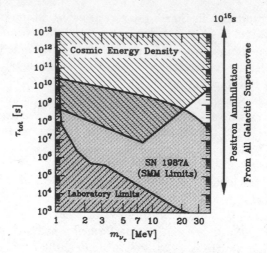

Fig. 12.19. Excluded areas of the ν_τ mass and lifetime if the standard-model decay $\nu_\tau \to \nu_e e^+ e^-$ is the only available channel. The laboratory results refer to the bounds on $\sin^2 2\theta_{e3}$ of Fig. 12.3, translated into a limit on $\tau_{e^+e^-}$ by virtue of Eq. (7.9). The SN 1987A bound is that from Fig. 12.18 while the cosmological one is from Fig. 7.2. The excluded range indicated by the vertical arrow refers to the argument of Sect. 12.5.1.

12.6   Neutrinos from All Stars

All stars in the universe contribute to a diffuse cosmic background flux of MeV neutrinos. If they decayed radiatively they would produce a cosmic x- and γ-ray background which must not exceed the measured levels. Because the entire radius of the visible universe is available as a decay path, one can derive rather restrictive limits on τ_γ (Cowsik 1977).

In order to derive such limits I assume that neutrinos of energy E_ν are produced with a constant rate $\dot{N}_\nu$ (cm^{-3} s^{-1}). Assuming a zero-curvature model of the universe, Kolb and Turner (1989) found for the resulting isotropic flux of decay photons

$$\frac{d^2 F_\gamma}{dE_\gamma \, d\Omega} = \frac{m_\nu}{\tau_\gamma} \frac{1}{4\pi} \frac{9}{2^{1/2} \, 5} \frac{\dot{N}_\nu t_U^2}{E_\nu^{3/2} E_\gamma^{1/2}} \, , \tag{12.35}$$

where t_U is the age of the universe. Moreover, it was assumed that in $\nu \to \nu' \gamma$ the daughter neutrino is massless, and that the decays are isotropic in the parent frame (anisotropy parameter $\alpha = 0$). In a flat universe one has $t_U = \frac{2}{3} H_0^{-1} = h^{-1} 2.05 \times 10^{17}$ s where $H_0 =$

$h\,100\,\mathrm{km\,s^{-1}\,Mpc^{-1}}$ is the present-day Hubble expansion parameter and where observationally $0.4 \lesssim h \lesssim 1$.

There exist numerous measurements of the diffuse cosmic x- and γ-radiation (for example Schönfelder, Graml, and Penningfeld 1980). Between a few 100 keV and a few 10 MeV the isotropic flux is reasonably well approximated by

$$\frac{d^2 F_\gamma}{dE_\gamma \, d\Omega} = 2\times10^{-2}\,\mathrm{cm^{-2}\,s^{-1}\,sr^{-1}\,MeV^{-1}} \left(\frac{\mathrm{MeV}}{E_\gamma}\right)^2. \tag{12.36}$$

Comparing this with Eq. (12.35), the most restrictive limit on τ_γ is obtained for the highest possible photon energy, $E_\gamma = E_\nu$. The requirement that the decay flux does not exceed the measurements at this energy leads to the upper limit

$$\frac{m_\nu}{\tau_\gamma} \lesssim 1.6\times10^{-40}\,\frac{\mathrm{eV}}{\mathrm{s}}\,\frac{\mathrm{cm^{-3}\,s^{-1}}}{\dot{N}_\nu}\,h^2, \tag{12.37}$$

which does not depend on the assumed value for E_ν because Eq. (12.35) and (12.36) both scale with E^{-2}. This limit applies if the total neutrino lifetime τ_{tot} exceeds t_U; otherwise only a limit on the branching ratio B_γ can be found.

The most prolific stellar neutrino source in the universe are hydrogen-burning stars which produce two ν_e's with MeV energies for every synthesized ^{4}He nucleus. Because most of the binding energy that can be liberated by nuclear fusion is set free when single nucleons are combined to form ^{4}He, most of the energy emitted by stars can be attributed to hydrogen burning. Therefore, it is easy to translate the optical luminosity density of the universe into an average rate of neutrino production.

The average luminosity density of the universe in the blue (B) spectral band is about $h\,2.4\times10^8\,L_{\odot,B}\,\mathrm{Mpc^{-3}}$ where $L_{\odot,B}$ is the solar B luminosity. The Sun produces about $1\times10^{38}\,\nu_e/\mathrm{s}$ and so one arrives at $\dot{N}_{\nu_e} \approx h\,2.4\times10^{46}\,\mathrm{Mpc^{-3}\,s^{-1}} = 0.8\times10^{-27}\,\mathrm{cm^{-3}\,s^{-1}}$. (Of course, this estimate is relatively crude in that the neutrino luminosity scales directly with the average bolometric luminosity of a stellar population, but not precisely with L_B.) With Eq. (12.37) this leads to a constraint for ν_e of $\tau_\gamma/m_{\nu_e} \gtrsim 5\times10^{12}\,\mathrm{s/eV}$, or $\mu_{\mathrm{eff}} \lesssim 2\times10^{-7}\,\mu_B\,m_{\mathrm{eV}}^{-2}$.

The (core-collapse) supernovae in the universe are also very prominent neutrino sources, and, more importantly, they are thought to produce MeV neutrinos of all flavors. Such SNe do not occur in elliptical galaxies, and their present-day rate in spirals depends sensitively on the Hubble type, varying from $0.2\,h^2$ SNu for Sa spirals to about $5\,h^2$ SNu for Sd (van den Bergh and Tammann 1991) where the supernova unit is defined by $1\,\mathrm{SNu} \equiv 1$ SN per century per $10^{10}\,L_{\odot,B}$. Adopting $1\,h^2$ SNu as a representative value and about 5×10^{57} neutrinos plus antineutrinos of a given flavor per SN yields for each flavor $\dot{N}_\nu \approx h^3\,1.3\times10^{-27}\,\mathrm{cm}^{-3}\,\mathrm{s}^{-1}$, a rate almost identical to that from hydrogen-burning stars.[79] The radiative lifetime limit is then also identical, except that it applies to neutrinos of all flavors.

All of these bounds are weaker than those from SN 1987A. Therefore, decaying stellar neutrinos cannot actually contribute to the observed x- and γ-ray background.

12.7 Cosmological Bounds

12.7.1 Neutrinos

Within the big-bang scenario a cosmic background sea of neutrinos is an inevitable consequence of the hot early universe. Its contribution to the cosmic energy density was already used in Sect. 7.1.5 to derive extremely restrictive neutrino mass limits. If neutrinos decay radiatively, further constraints can be obtained. For one, the decay photons can show up directly as a diffuse, isotropic cosmic background radiation. If the decays occur before recombination, i.e. before the universe became transparent to radiation, but so late that the photons could not be thermalized entirely, they contribute to a spectral distortion of the cosmic microwave background radiation (CMBR). The resulting limits were discussed, for example, by Kolb and Turner (1990) who found that those areas of masses and lifetimes are excluded that are hatched in Fig. 12.20. It was assumed that neutrinos decay only radiatively.

The contribution of the neutrinos and their decay products to the mass density of the universe leads to the constraints shown in Fig. 7.2 which are based on the present-day value of Ωh^2 and on the expansion

---

[79]Multiplying this rate with the age of the universe of about 3×10^{17} s and the speed of light of 3×10^{10} cm/s, and using $h = 0.5$ one finds an estimated present-day flux at Earth of about $1\,\mathrm{cm}^{-2}\,\mathrm{s}^{-1}$. In a recent detailed study, Totani and Sato (1995) find a flux which is larger than this crude estimate by as much as a factor of 30. The Kamiokande II detector has set an upper limit on the cosmic background flux of $\bar{\nu}_e$ of about $10^3\,\mathrm{cm}^{-2}\,\mathrm{s}^{-1}$ for effective temperatures in the $3-4\,\mathrm{MeV}$ range (Zhang et al. 1988). It is conceivable that this background will be measured by the Superkamiokande detector. Note that at the Kamiokande site the $\bar{\nu}_e$ flux from power reactors is roughly 1000 times larger than the background flux, except that it falls off sharply beyond about $10\,\mathrm{MeV}$.

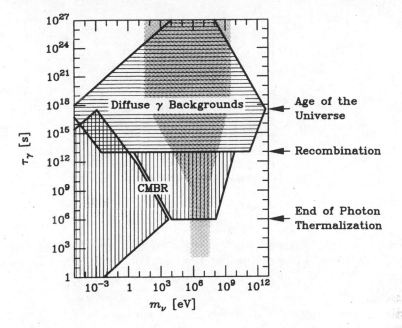

Fig. 12.20. Cosmological limits on neutrino radiative lifetimes according to Kolb and Turner (1990). The radiative mode is assumed to be the only decay channel. The shaded area is excluded according to Sect. 7.1.5 (Fig. 7.2).

rate at nucleosynthesis. These limits (shaded area in Fig. 12.20) are more general because they do not depend on the nature of the final states in the decay.

Kolb and Turner's (1990) exclusion plot is somewhat schematic. Ressell and Turner (1990) performed a much more detailed analysis on the basis of the diffuse photon backgrounds in all wavebands. Probably the most interesting region is that of small neutrino masses and large lifetimes (the upper left corner of Fig. 12.20). The excluded range of effective electromagnetic transition moments for $m_\nu < 30\,\mathrm{eV}$ is shown in Fig. 12.21. It may be useful to approximate the excluded range analytically by

$$\mu_{\mathrm{eff}} \lesssim 3{\times}10^{-11}\,\mu_{\mathrm{B}}\,(\mathrm{eV}/m_\nu)^{2.3} \tag{12.38}$$

which is shown as a dashed line in Fig. 12.21.

Using favored cosmological parameters ($\Omega h^2 \approx 0.3$) neutrinos with $m_\nu \approx 30\,\mathrm{eV}$ would be the dark matter of the universe. With smaller masses there would have to be another component, but neutrinos could

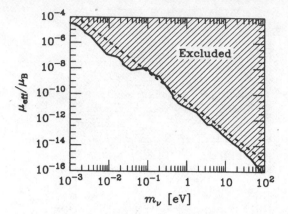

Fig. 12.21. Limits on μ_{eff} according to Ressell and Turner's (1990) bounds on the radiative lifetime of long-lived neutrinos from the diffuse cosmic background radiations. The dashed line corresponds to Eq. (12.38).

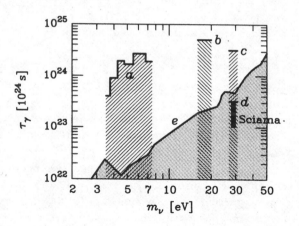

Fig. 12.22. Limits on decaying neutrinos in clusters of galaxies. (a) A1413, A2218, and A2256 (Bershady, Ressell, and Turner 1991). (b) Coma and Virgo (Henry and Feldman 1981). (c) A665 (Davidsen et al. 1991). (d) Extragalactic background light (Overduin, Wesson, and Bowyer 1993). (e) Background light from decay of unclustered neutrinos (Ressell and Turner 1990).

still play a significant dynamical role. Recently, such mixed dark matter scenarios have received much attention where $m_\nu = 5\,\text{eV}$ is a favored value. Such low-mass particles cannot cluster on galactic scales, but likely they would reside in clusters of galaxies. With radiative decays $\nu \to \nu'\gamma$ and a total lifetime exceeding the age of the universe one then expects clusters of galaxies to be strong sources of optical or ultraviolet photons. Several limits are summarized in Fig. 12.22.

A case has been made that radiatively decaying neutrino dark matter is actually required to solve certain problems, notably the ionization of galactic hydrogen clouds (e.g. Melott and Sciama 1981; Melott, McKay, and Ralston 1988; Sciama 1990a,b; Sciama 1993a,b, 1995). The predictions are very specific: An energy of decay photons of $E_\gamma = (14.4 \pm 0.5)\,\text{eV}$ and thus a neutrino mass of $m_\nu = (28.9 \pm 1.1)\,\text{eV}$ with a radiative lifetime of $\tau_\gamma = (2 \pm 1) \times 10^{23}\,\text{s}$ which translates into $\mu_{\text{eff}} = (6.3 \pm 2) \times 10^{-15}\,\mu_{\text{B}}$. Such a large transition moment would require particle physics beyond the standard model.

Nominally, this possibility is already excluded by the absence of a uv line from the cluster A665 (Davidsen et al. 1991). However, a bound from a single source is always subject to the uncertainty of unrecognized absorbing material in the line of sight or internal absorption. Moreover, the dark matter in the core of this cluster may be mainly baryonic (Sciama, Persic, and Salucci 1993; Melott et al. 1994). Bounds from the diffuse extragalactic background light are more reliable in this regard. While they marginally exclude Sciama's neutrino (Overduin, Wesson, and Bowyer 1993) it is perhaps too early to pronounce it entirely dead. A decisive test will be performed with a future satellite experiment where the uv line from neutrinos decaying in the solar neighborhood definitely would have to show up if neutrinos were the bulk of the galactic dark matter (e.g. Sciama 1993b).

12.7.2 Axions

The axion lifetime from $a \to 2\gamma$ is $\tau = 6.3 \times 10^{24}\,\text{s}\,(m_a/\text{eV})^5/\xi^2$ where ξ is a model-dependent number of order unity (Sect. 14.3.2). Therefore, axions with eV masses have radiative lifetimes in the neighborhood of the above neutrino limits whence they can be constrained by similar methods (Kephart and Weiler 1987). Moreover, such axions would contribute substantially to the mass density of the universe because they would have been in thermal equilibrium until relatively late.[80] Their

---

[80]In the early universe, axions are also produced by the relaxation of the coherent initial field configuration at the onset of the QCD phase transition. This process

contribution to the cosmic mass density would be $\Omega_a h^2 = 0.082\, m_a/\mathrm{eV}$ (Turner 1987; Ressell 1991) and so one can expect that clusters of galaxies contain substantial amounts of axions even if they are not the main dark matter component.

Observations of the diffuse extragalactic background radiation limit the axion mass to values below about 8 eV unless ξ is very small (Ressell 1991). Overduin and Wesson (1993) found $\xi < 0.43$, 0.07, and 0.02 for $m_a/\mathrm{eV} = 5.3$, 8.6, and 13, respectively (Fig. 12.23). Moreover, some axions would reside in the halo of our own galaxy so that their decays would light up the night sky. Its brightness yields a conservative bound of $\xi < (6\,\mathrm{eV}/m_a)^5$ (Ressell 1991).

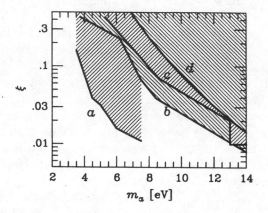

Fig. 12.23. Constraints on axion decays in galaxies and galaxy clusters; ξ parametrizes the coupling to photons with $\xi = 1$ corresponding to common axion models. (a) Line emission from clusters A2256 and A2218 (Bershady, Ressell, and Turner 1991; Ressell 1991). (b) Diffuse extragalactic background radiation according to Ressell (1991) and (c) Overduin and Wesson (1993). (d) Our galaxy (Ressell 1991).

The most interesting limits arise from a search for axion decay lines from the intergalactic space in the clusters of galaxies A2256 and A2218. Bershady, Ressell, and Turner (1991) and Ressell (1991) found $\xi < 0.16$ 0.078, 0.039, 0.032, 0.016, and 0.011 for $m_a/\mathrm{eV} = 3.5$, 4.0, 4.5, 5.0, 6.0, and 7.5 respectively (Fig. 12.23). These limits are placed into the context of other constraints in Fig. 5.9.

---

yields $\Omega_a h^2 \approx (10^{-5}\mathrm{eV}/m_a)^{1.175}$—see Eq. (14.5). While the overall coefficient of this expression is very uncertain it is clear that $\Omega_a = 1$ saturates for m_a somewhere between $1\,\mu\mathrm{eV}$ and $1\,\mathrm{meV}$. In this range axions never achieved thermal equilibrium.

Chapter 13

What Have We Learned from SN 1987A?

The lessons for particle physics from the SN 1987A neutrino burst are studied. First, neutrinos could have decayed or oscillated into other states on their way out of the SN core and to us. Second, propagation effects could have caused a time delay between photons and neutrinos or between $\bar{\nu}_e$'s of different energy. Third, nonstandard cooling agents could have shortened the neutrino burst below its observed duration. These arguments are applied to a variety of specific cases.

13.1 Introduction

In Chapter 11 the neutrino observations from SN 1987A were discussed and it was shown that they agree well with standard theoretical expectations from the core collapse and subsequent explosion of an evolved massive star. The signal displays several anomalies (time gap at Kamiokande, anisotropy in both detectors) which render it a less beautiful specimen of the expected signal characteristics than is sometimes stated in the literature. Still, in the absence of plausible alternatives one must accept that the Kamiokande II, IMB, and Baksan event clusters observed at 7:35 UT on 27 February 1987 represent the $\bar{\nu}_e$ component of the neutrino burst from the core collapse of the SN 1987A progenitor star rather than some other particle flux, or some other reaction than the expected dominant $\bar{\nu}_e p \to n e^+$ process.

Accepting this, there is a host of consequences concerning a variety of fundamental physics issues. The first and simplest set of arguments is based on the fact that the $\bar{\nu}_e$ pulse and perhaps the prompt ν_e burst

493

were observed, constraining various mechanisms that could have removed neutrinos from the beam such as decays. Equally important, the nonobservation of a γ-ray burst in coincidence with the neutrino burst constrains radiative decays of neutrinos and other particles (Sect. 12.4).

More intricate arguments involve signal dispersion, either between photons and neutrinos, between $\bar{\nu}_e$'s and ν_e's, or the intrinsic dispersion of the $\bar{\nu}_e$ burst, constraining various effects that could cause signal dispersion such as a nonzero neutrino mass or charge.

Most importantly, the inferred cooling time scale of a few seconds of the newborn neutron star precludes an efficient operation of a nonstandard cooling agent and thus yields constraints on the emission of new particles from the SN core, notably of right-handed (r.h.) neutrinos or axions. This line of reasoning is analogous to the "energy-loss argument" which for normal stars has been advanced in Chapter 2.

13.2 Basic Characteristics of the Neutrino Burst

13.2.1 Fluence

The neutrinos from a SN are expected to consist of two major components: the prompt ν_e burst, and quasi-thermal emission of about equal total amounts of energy in (anti)neutrinos of all flavors. The water Cherenkov detectors would register the ν_e burst by virtue of the reaction $\nu_e + e^- \rightarrow e^- + \nu_e$ where the scattered electron is strongly forward peaked, while the cooling signal is registered by $\bar{\nu}_e + p \rightarrow n + e^+$ with an essentially isotropic e^+ signal. Even though both detectors observed a forward peaked overall signal, it cannot be associated with ν_e-e collisions (Sect. 11.3.5). Most or all of the events are interpreted as $\bar{\nu}_e$'s.

The observation of a $\bar{\nu}_e$ fluence (time-integrated flux) roughly in agreement with what is expected from a stellar collapse precludes that these particles have decayed on their way from the SN to us, yielding a constraint on their lifetime of (Frieman, Haber, and Freese 1988)

$$\tau_{\nu_e}/m_{\nu_e} \gtrsim 6 \times 10^5 \, \text{s/eV}. \tag{13.1}$$

However, this simple result must be interpreted with care because massive neutrinos are expected to mix. The heavy ν_e admixtures could decay and may violate this bound.

The $\bar{\nu}_e$'s were not removed by excessive scattering on cosmic background neutrinos, majorons, dark-matter particles etc., leading to constraints on "secret interactions" (Kolb and Turner 1987). Take the scattering on cosmic background neutrinos as an example. The present-day density of primordial neutrinos is about $100 \, \mathrm{cm}^{-3}$ in each neutrino and antineutrino flavor. With a distance to the Large Magellanic Cloud of about $50 \, \mathrm{kpc} = 1.5 \times 10^{23} \, \mathrm{cm}$ one has a column density between SN 1987A and Earth of about $10^{25} \, \mathrm{cm}^{-2}$ so that the $\bar{\nu}_e$-ν cross section must be less than about $10^{-25} \, \mathrm{cm}^2$. If the cosmic background neutrinos are massless they have a temperature of about $1.8 \, \mathrm{K}$ and so $\langle E_\nu \rangle \approx 3T_\nu \approx 5 \times 10^{-4} \, \mathrm{eV}$. Because the measured SN neutrinos have a characteristic energy of $30 \, \mathrm{MeV}$ the center of mass energy is $\sqrt{s} \approx 200 \, \mathrm{eV}$.

The cross-section bound is not particularly impressive compared with a standard weak cross section of order $G_F^2 s \approx 10^{-51} \, \mathrm{cm}^2$. However, $\nu\nu$ cross sections have never been directly measured and so the SN 1987A limit provides nontrivial information. As an example, neutrinos could scatter by majoron exchange, or they could scatter directly on a background of primordial majorons. Kolb and Turner then found a certain constraint on the neutrino-majoron Yukawa coupling (Sect. 15.7.2). As another example, the proposition that the solar neutrino flux could be substantially depleted by scatterings on cosmic background particles (Slad' 1983) is excluded.

Other particles besides neutrinos may have been emitted from the SN and could have caused detectable events. Engel, Seckel, and Hayes (1990) have discussed the case of axions; they can be absorbed in water by oxygen nuclei, $a^{16}\mathrm{O} \to {}^{16}\mathrm{O}^*$, which subsequently produce γ rays by decays of the sort ${}^{16}\mathrm{O}^* \to {}^{16}\mathrm{O} \, \gamma$, ${}^{16}\mathrm{O}^* \to {}^{15}\mathrm{O} \, n \, \gamma$, and ${}^{16}\mathrm{O}^* \to {}^{15}\mathrm{N} \, p \, \gamma$. The γ rays would cause electromagnetic cascades and so they are detectable about as efficiently as $e^\pm$. The axion emission was estimated by identifying their unit optical depth for a given interaction strength in a simplified model of the SN temperature and density profile. More than 10 extra events would be expected at Kamiokande for an axion-nucleon Yukawa coupling in the range

$$1 \times 10^{-6} \lesssim g_{aN} \lesssim 1 \times 10^{-3} \tag{13.2}$$

which is thus excluded. In the middle of this interval, up to 300 additional events would have been expected. However, axions with couplings in this interval are also excluded by other methods (Sect. 14.4).

13.2.2 Energy Distribution

The energy distribution of the events at the IMB and Kamiokande detectors broadly confirms the expected quasi-thermal emission with a temperature of around 4 MeV. This- precludes that a major swap by oscillations with the higher-energetic $\bar{\nu}_\mu$ or $\bar{\nu}_\tau$ flux has taken place. The impact of neutrino oscillations on the observable signal has been discussed in Sect. 11.4.

Also, $\nu_{\mu,\tau}$ or $\bar{\nu}_{\mu,\tau}$ decays with final-state $\bar{\nu}_e$'s would produce additional higher-energy events. While the SN 1987A data are probably too sparse to extract significant information on the presence or absence of this effect, a future galactic SN would certainly allow one to exclude a certain range of masses and decay times or to detect this effect (Soares and Wolfenstein 1989).

The trapping of neutrinos in a SN core together with the condition of β equilibrium inevitably implies that there is a large ν_e chemical potential, leading to typical ν_e energies of order 200 MeV. Moreover, the inner temperature during deleptonization reaches values of up to $40-70$ MeV so that typical thermal (anti)neutrino energies of up to $100-200$ MeV are available. Therefore, if neutrinos could escape directly from the inner core they would cause high-energy events in the detectors which have not been observed.

A mechanism to tap the inner-core heat bath directly is the production of r.h. neutrinos by a variety of possible effects such as spin-flip scattering by a Dirac mass term or a magnetic dipole moment (Sect. 13.8). R.h. states could not be detected directly because they are sterile with regard to standard l.h. weak interactions, a property which allows them to avoid the SN trapping. However, they could produce detectable l.h. states by decays (Dodelson, Scott, and Turner 1992) or by magnetic oscillations (Nötzold 1988; Barbieri and Mohapatra 1988).

13.2.3 Prompt ν_e Burst

The prompt ν_e burst can be seen in a water Cherenkov detector by the reaction $\nu_e e \to e\nu_e$ where the final-state electron essentially preserves the direction of the incident neutrino. At Kamiokande, the directionality of the first event[81] is consistent with the interpretation that it was caused by this reaction. However, the expected fluence corresponds only to a fraction of an event and so the first event may also be due to the $\bar{\nu}_e p \to ne^+$ reaction and point coincidentally in the forward direction. A random direction has about a 5% chance of being forward within 25° which is approximately the uncertainty of the Kamiokande directional event reconstruction.

---

[81] In the first publication of the Kamiokande group (Hirata et al. 1987) the second event was also reported forward; its most probable direction was later revised.

Still, the observation of the prompt ν_e' burst from a future galactic SN would allow for a number of interesting conclusions. For example, one could exclude or find evidence for neutrino oscillations (Sect. 11.4). Signal dispersion caused by a neutrino mass or other effects which are discussed below for the cooling-phase signal would be even more significant for the prompt burst because of its short duration. For example, the cooling signal with a duration of about 10 s is sensitive to ν_e masses in the 10 eV regime. As the prompt burst is at least 100 times shorter one is sensitive to a factor of 10 smaller masses, i.e. to m_{ν_e} in the eV range.

The (anti)neutrino signal during the prompt burst phase allows one to decide if the SN consisted of antimatter rather than matter. In that case one would expect a prompt $\overline{\nu}_e$ burst with a scattering cross section on electrons which is about a factor of 2.4 smaller (Eq. 10.17). Moreover, $\overline{\nu}_e$'s are dominantly absorbed by the isotropic $\overline{\nu}_e p \rightarrow n e^+$ reaction and so the prompt burst would cause a substantial isotropic signal within the first 50 ms. In a matter SN the cooling $\overline{\nu}_e$'s have larger energies than the ν_e's; the reverse for antimatter. Therefore, the observable $\overline{\nu}_e$ signal from the cooling phase would be reduced. In a detector like Kamiokande one would then expect $6-20\%$ of the total $\overline{\nu}_e$ signal from the prompt burst, in contrast with at most 1% for a regular matter SN (Barnes, Weiler, and Pakvasa 1987). Of course, in the foreseeable future one can hope to acquire the relevant data only from a galactic SN which, no doubt, consists of matter.

13.2.4 Nonobservation of a γ-Ray Burst

No γ rays in conjunction with the SN 1987A neutrino burst were observed by the solar maximum mission (SMM) satellite which was operational at the relevant time. Therefore, one can derive some of the most restrictive limits on neutrino radiative decays as detailed in Sect. 12.4.

13.3 Dispersion Effects

13.3.1 Photons vs. Antineutrinos

The optical sighting of SN 1987A followed the detection of the $\overline{\nu}_e$ burst by only a few hours (Fig. 11.7), a delay which is expected on the basis of the simple reasoning that some time must pass before the mantle of a SN "notices" the collapse of the inner core. Hence the two signals must have propagated through space with an almost identical velocity so that the speed of light and that of neutrinos are equal to within (Longo 1987; Stodolsky 1988)

$$\left| \frac{c_\nu - c_\gamma}{c_\gamma} \right| \lesssim 2\times 10^{-9}, \tag{13.3}$$

assuming an uncertainty of $\pm 3\,\mathrm{h}$ in the relative duration of the transit times from the LMC to us. This was interpreted as the most stringent test of special relativity to date in the sense that it proves with high precision the universality of a relativistic limiting velocity.[82]

This result can also be interpreted as testing the weak equivalence principle of general relativity (Krauss and Tremaine 1988). In the post-Newtonian approximation one predicts that a gravitational potential $V(\mathbf{r})$ delays a light signal (Shapiro time delay) by an amount

$$\Delta t = -2 \int_{\mathrm{E}}^{\mathrm{A}} V[\mathbf{r}(t)]\, dt, \tag{13.4}$$

where the integral is taken along the trajectory $\mathbf{r}(t)$ of the beam between the points of emission (E) and absorption (A). This delay is the same for neutrinos and photons to within

$$\left| \frac{\Delta t_\nu - \Delta t_\gamma}{\Delta t_\gamma} \right| < 0.7 - 4\times 10^{-3}, \tag{13.5}$$

where the uncertainty reflects the uncertain modelling of the gravitational potential between Earth and SN 1987A.[83] This result has been used to constrain the parameters of a specific model of C- and P-violating gravitational forces (Almeida, Matsas, and Natale 1989), and to constrain the parameters of a class of nonmetric theories of gravity (Coley and Tremaine 1988).

13.3.2   Neutrinos vs. Antineutrinos

Assuming that the first event at Kamiokande represents the prompt ν_e burst one may also constrain the difference in transit time between ν_e and $\bar{\nu}_e$ and thus confirm the equivalence principle between matter and antimatter (LoSecco 1988; Pakvasa, Simmons, and Weiler 1989). Of course, in order to make such results reliable one would need to observe the prompt burst from a future SN with greater statistical significance.

---

[82]For a recent laboratory experiment which addresses the Lorentz limiting velocity, see Greene et al. (1991), and references there to earlier works. See also the book by Will (1993).

[83]See Will (1993) for a review of many other empirical tests of general relativity.

13.3.3   Intrinsic Dispersion of the $\bar{\nu}_e$-Pulse

a) Neutrino Mass

So far the transit time of different particle species was compared under the assumption of a fixed velocity each. However, the most likely effect of signal propagation over large distances is dispersion due to an energy-dependent speed of propagation. The most widely discussed[84] case is that of a nonzero neutrino mass (Zatsepin 1968). The main problem at extracting information about the signal dispersion is the unknown behavior of the source which must be modelled according to some theoretical assumptions. A particularly detailed discussion is that of Loredo and Lamb (1989) who found a mass limit of $m_{\nu_e} < 23\,\mathrm{eV}$ (Sect. 11.3.4). In a similar analysis which included the 13% dead-time effect at IMB, Kernan and Krauss (1995) found $m_{\nu_e} < 20\,\mathrm{eV}$.

b) Neutrino Charge

The absence of an energy-dependent dispersion of the neutrino pulse can be used to constrain other neutrino properties. A small electric charge e_ν would bend the neutrino path in the galactic magnetic field, leading to a time delay of

$$\frac{\Delta t}{t} = \frac{e_\nu^2 \, (B_{\mathrm{T}} d_B)^2}{6 E_\nu^2}, \tag{13.6}$$

where B_{T} is the transverse magnetic field and d_B the path length within the field. This leads to a constraint of

$$\frac{e_\nu}{e} \lesssim 3 \times 10^{-17} \left(\frac{1\,\mu\mathrm{G}}{B_{\mathrm{T}}} \right) \left(\frac{1\,\mathrm{kpc}}{d_B} \right) \tag{13.7}$$

(Barbiellini and Cocconi 1987; Bahcall 1989). Note that a typical field strength for the ordered magnetic field in the galactic spiral arms is $2-3\,\mu\mathrm{G}$ and that the path length of the neutrinos within the galactic disk is only of order 1 kpc because the LMC lies high above the disk (galactic latitude about 33°).

---

[84]Limits on m_{ν_e} from the SN 1987A data were derived, among others, by Abbott, de Rújula, and Walker (1988), Adams (1988), Arnett and Rosner (1987), Bahcall and Glashow (1987), Burrows and Lattimer (1987), Burrows (1988a), Chiu, Chan, and Kondo (1988), Cowsik (1988), Kolb, Stebbins, and Turner (1987a,b), Midorikawa, Terazawa, and Akama (1987), Sato and Suzuki (1987a,b), Spergel and Bahcall (1988), Loredo and Lamb (1989), and Kernan and Krauss (1995).

c) Long-Range Forces

Speculating further one may imagine some sort of neutrino "fifth-force charge." If electrons, protons, or dark-matter particles also carry such a charge the bending of the neutrino trajectory in the fifth-force field of the galaxy would lead to an energy-dependent time delay. This and related arguments were advanced by a number of authors (Pakvasa, Simmons, and Weiler 1989; Grifols, Massó, and Peris 1988, 1994; Fiorentini and Mezzorani 1989; Malaney, Starkman, and Tremaine 1995).

The most plausible form for such a long-range interaction is one mediated by a massless vector boson, i.e. a new gauge interaction, perhaps related to a novel leptonic charge (Sect. 3.6.4). In this case neutrinos and antineutrinos would carry opposite charges so that the cosmic neutrino background would be essentially a neutral plasma with regard to the new interaction. The resulting screening effects then invalidate the SN 1987A argument (Dolgov and Raffelt 1995).

Screening effects would not operate if the force were due to a spin-0 or spin-2 boson which always cause attractive forces. However, any force mediated by a massless spin-2 boson must couple to the energy-momentum tensor and thus is identical with gravity. The force mediated by a scalar boson between a static source and a relativistic neutrino is suppressed by a Lorentz factor. Therefore, even if scalar-mediated forces existed between macroscopic bodies, their effect would be weakened for relativistic neutrinos.

In summary, the SN 1987A signal does not seem to carry any simple information concerning putative nongravitational long-range forces.

d) Fundamental Length Scale

Fujiwara has proposed a quantum field theory where the velocity of particles increases with energy, leading to an energy-dependent advance of the arrival times by $\Delta t/t = -\frac{1}{2}(\ell_0 E_\nu)^2$. Here, ℓ_0 is a fundamental length scale. Whatever the merits of this theory, a value $\ell_0 \lesssim 10^{-18}$ cm would not be in conflict with the SN 1987A neutrino signal (Fujiwara 1989).

e) Lorentz Addition of Velocities

If relativistic particles (photons, massless neutrinos) are emitted by a moving source (velocity v_S) their velocity c' in the laboratory frame should be equal to c (velocity in the frame of the source). The Galilean addition of velocities, on the other hand, would give $c' = c + v_S$. In general one may assume that velocities add according to $c' = c + K v_S$

with $K = 0$ representing the Lorentzian, $K = 1$ the Galilean law of adding velocities. For photons, the most stringent laboratory bound is $K_\gamma \lesssim 10^{-4}$ derived from the time of flight of decay photons $\pi^\circ \to 2\gamma$ from a pulsed π° source (Alväger et al. 1964). A much more stringent constraint $(K_\gamma \lesssim 2 \times 10^{-9})$ obtains from an analysis of the photon signal from a pulsed x-ray source (Brecher 1977).

The absence of dispersion of the $\bar{\nu}_e$ pulse can be used to derive constraints on K_ν (Atzmon and Nussinov 1994). The observed SN 1987A neutrinos were produced by microscopic processes involving nearly relativistic nucleons, and subsequently scattered several times on such nucleons before leaving the star. If the last nucleon on which they scatter is considered the source with $v_S \approx 0.2\,c$ their laboratory speed c'_ν will be represented by a distribution of approximate width $0.2\,K_\nu$ around c because of the random orientation and distribution in magnitude of v_S. Thus one derives a bound $K_\nu \lesssim 10^{-11}$ from the absence of a spread in arrival times exceeding about 10 s.

Atzmon and Nussinov (1994) warn, however, that this simple argument may be too naive as the motion through the progenitor's envelope may cause the particles of the envelope to be the true source of the "neutrino waves" as there is a substantial amount of refraction between the neutrino sphere and the stellar surface. If one follows Atzmon and Nussinov's reasoning, there remains only a much weaker bound of $K_\nu \lesssim 10^{-5}$ from the absence of an anomalous time delay between the neutrino signal and the optical sighting of the SN.

13.4 Duration of Neutrino Emission

13.4.1 General Argument

The most intricate way to use SN 1987A as a laboratory arises from the observed duration of neutrino cooling. While the neutrino luminosity during the first few 100 ms until the shock has been revived is largely powered by accretion and by the contraction and settling of the bloated outer core, the long tail is associated with cooling, i.e. emission from the neutrino sphere which is powered by energy originally stored deep in the inner core. If a direct cooling channel existed for that region, such as the emission of r.h. neutrinos or axions, the late cooling phase would be deprived of energy. Put another way, a novel cooling channel from the inner core would leave the schematic neutrino light curves of Fig. 11.3 more or less unchanged before about 1 s while the long Kelvin-Helmholtz cooling phase would be curtailed.

Of course, this reasoning is identical with the energy-loss argument previously studied for normal stars in Chapters 1 and 2. The main difference is that neutrinos are trapped so that particles which interact more weakly can dominate the thermal evolution by volume emission. In normal stars, photons are trapped and neutrinos can dominate the energy loss by volume emission as, for example, in the early cooling of a white dwarf.

This general argument is best illustrated with axion emission. These particles are pseudoscalars which for the purpose of this argument are taken to interact with neutrons and protons with a common Yukawa coupling strength g_a which is the only free parameter in the problem. For very small values of g_a axions will play no role, but with an increasing coupling strength their emission from the inner core by bremsstrahlung processes, $NN \to NNa$, will begin to compete with neutrino cooling. Of course, if g_a exceeds some critical value axions will be trapped and emitted from an "axion sphere" at about unit optical depth. Beyond some large coupling they will be trapped so effectively that their contribution to the cooling of the SN core is, again, negligible and the neutrino signal assumes its standard duration. This general behavior is shown in Fig. 13.1 on the basis of the numerical cooling calculations[85] of Burrows, Turner, and Brinkmann (1989) and Burrows, Ressell, and Turner (1990). These authors used the quantity $\Delta t_{90\%}$ as a measure of the cooling time; it represents the time at which 90% of the expected number of events have arrived at a detector. $\Delta t_{90\%}$ was calculated separately for Kamiokande II and IMB; in Fig. 13.1 an average relative signal duration is shown, normalized to the value when axions are not important. It is apparent that a large range of g_a values can be excluded on the basis of the observed duration of the neutrino signal.

One is here considering the time scale of neutrino emission at the source while the detectors register a pulse which conceivably could have been lengthened by dispersion effects. However, in view of the recent laboratory limits of $m_{\nu_e} \lesssim 5\,\mathrm{eV}$ this is not a serious concern.

If one contemplates nonstandard neutrinos, a relatively short emission time scale at the source is compatible with the observations if ν_μ's

---

[85]In the free-streaming regime these calculations were based on axion emission rates which do not take the high-density multiple-scattering effects into account that were discussed in Sect. 4.6.7. Therefore, the free-streaming part of Fig. 13.1 probably overestimates the import of axion emission. For the present purpose of discussing the general aspects of a novel cooling channel, however, this problem is of no concern. The axion case is the only one where numerical cooling calculations are available for both the volume-emission (free-streaming) and the trapping limit.

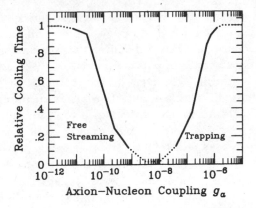

Fig. 13.1. Relative duration of neutrino cooling of a SN core as a function of the axion-nucleon Yukawa coupling g_a. In the free-streaming limit axions are emitted from the entire volume of the protoneutron star, in the trapping limit from the "axion sphere" at about unit optical depth. The solid line is according to the numerical cooling calculations (case B) of Burrows, Turner, and Brinkmann (1989) and Burrows, Ressell, and Turner (1990); the dotted line is an arbitrary completion of the curve to guide the eye. The signal duration is measured by the quantity $\Delta t_{90\%}$ discussed in the text; an average for the IMB and Kamiokande detectors was taken.

or ν_τ's decay. The final states could include $\bar{\nu}_e$'s which are detectable at IMB and Kamiokande so that one can obtain late-time events by a suitable combination of mass and lifetime. For a 17 keV Majorana neutrino a lifetime around 10^4 s would allow one to explain the signal duration even with a short emission time scale at the source (Simpson 1991; see also Cline 1992). In the following it will be assumed that this is not the explanation of the observed signal duration.

Another loophole is that some or all of the late-time events at Kamiokande, which are separated from the main bunch by a 7 s gap, were caused by effects other than core cooling. Recall that a similar problem exists with the x-ray observations of old neutrons stars (Sect. 2.3) where it is not always clear that one is observing blackbody surface emission from thermal cooling rather than magnetospherically produced x-rays. In the present case, one possibility is the fall-back of material onto the core, i.e. late-time accretion which could cause significant neutrino emission. However, on the basis of an analytic estimate Janka (1995b) has argued that even with extreme assumptions this is not a likely explanation of the late events.

While it is clear that a large range of g_a values can be excluded, the quantity $\Delta t_{90\%}$ is a relatively crude measure of the length of the cooling phase. In principle, one should perform a maximum likelihood analysis for a given range of particle properties. Moreover, one would need to consider a variety of models for the protoneutron star where the equation of state (EOS), mass, accretion rate, neutrino opacities, and perhaps other parameters should be varied to optimize the agreement with the observed signal when a novel cooling mechanism operates.

Another caveat applies to the "trapping regime" of the new particles. One may expect that they play a significant role during the infall phase and shock formation of a SN collapse, an issue that was addressed only by a small number of authors in the context of majoron bounds (Fuller, Mayle, and Wilson 1988) and bounds on neutrino dipole moments (Nötzold 1988). Hence, in general it is not obvious that parameters allowed by the cooling argument on the trapping side would remain allowed if one took account of these effects. Moreover, on the trapping side the novel particles interact about as strongly as neutrinos and so they could also cause a signal in the detectors. For axions, this argument rules out the values of g_a given in Eq. (13.2).

13.4.2 Analytic Criterion in the Free-Streaming Limit

In order to estimate the impact of a novel cooling channel on the neutrino signal it is obviously useful to evolve a protoneutron star numerically with the new physics included, and to calculate the expected neutrino signal for a varying strength of the new effect. Considering the many uncertainties involved in this procedure one may well ask if it is not just as reliable to perform a simple analytic estimate.

At about 1 s after core bounce the neutrino luminosity in all six (anti)neutrino degrees of freedom together is about $3\times10^{52}\,\mathrm{erg\,s^{-1}}$. The mass of the object is around $1.5\,\mathcal{M}_\odot = 3\times10^{33}\,\mathrm{g}$ so that its average energy-loss rate is $L_\nu/\mathcal{M} \approx 1\times10^{19}\,\mathrm{erg\,g^{-1}\,s^{-1}}$. A novel cooling agent would have to compete with this energy-loss rate in order to affect the total cooling time scale significantly. Therefore, the observed signal duration indicates that a novel energy-loss rate is bounded by

$$\epsilon_x \lesssim 10^{19}\,\mathrm{erg\,g^{-1}\,s^{-1}}. \tag{13.8}$$

It is to be evaluated at typical core conditions, i.e. at a temperature of around 30 MeV and a density of around $3\times10^{14}\,\mathrm{g\,cm^{-3}}$. The nuclear medium is then at the borderline between degeneracy and nondegeneracy while the electrons are highly degenerate.

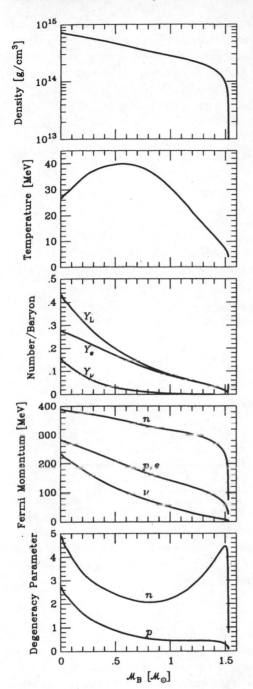

Fig. 13.2. Profile of various parameters for the protoneutron star model S2BH_0 of Keil, Janka, and Raffelt (1995), 1 s after core bounce. The degeneracy parameters were approximated by $\eta_N = (E_F - m_N)/T$ with $E_F^2 = p_F^2 + m_N^2$ and with the effective nucleon mass.

The profile of various parameters as a function of the mass coordinate is shown in Fig. 13.2 for model S2BH_0 of the cooling calculations of Keil, Janka, and Raffelt (1995); it illustrates typical physical conditions encountered in the core of a protoneutron star during the Kelvin-Helmholtz phase. For this model, the average value of $(\rho/\rho_0)^n$ with the nuclear density $\rho_0 = 3\times10^{14}\,\mathrm{g\,cm^{-3}}$ and of $(T/30\,\mathrm{MeV})^n$ is shown in Fig. 13.3 as a function of n.

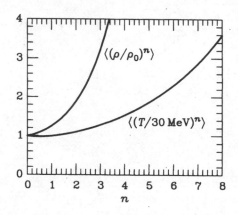

Fig. 13.3. Average values for $(\rho/\rho_0)^n$ with the nuclear density $\rho_0 = 3\times10^{14}\,\mathrm{g\,cm^{-3}}$ and of $(T/30\,\mathrm{MeV})^n$ for the protoneutron star model of Fig. 13.2.

As an example one may apply this criterion to the bremsstrahlung energy-loss rate $NN \to NNa$ for the emission of some pseudoscalar boson a (axion) with a Yukawa coupling g_a. The nondegenerate energy-loss rate Eq. (4.8) is $\epsilon_a = g_a^2\, 2\times10^{39}\,\mathrm{erg\,g^{-1}\,s^{-1}}\,\rho_{15}\,T_{30}^{3.5}$ where $T_{30} = T/30\,\mathrm{MeV}$ and $\rho_{15} = \rho/10^{15}\,\mathrm{g\,cm^{-3}}$. From Fig. 13.3 one finds that $\langle\rho_{15}\rangle \approx 0.4$ and $\langle T_{30}^{3.5}\rangle \approx 1.4$. The criterion Eq. (13.8) then yields $g_a \lesssim 10^{-10}$, similar to what one would conclude from Fig. 13.1. Using the degenerate emission rate Eq. (4.10) yields an almost identical result. Therefore, a simple criterion like Eq. (13.1) is not a bad first estimate for the import of a novel energy-loss rate.

13.4.3 Trapping Limit

When the new particles (for example, axions) interact strongly enough, they will be emitted from a spherical shell where their optical depth is about unity rather than by volume emission. Again, one is concerned

mostly with a time later than $0.5-1\,\text{s}$ where the outer core has settled and the shock has begun to escape. The density of the protoneutron star falls within a thin shell from supranuclear levels to nearly
zero, causing the "photosphere" radius r_x of the new particles to be
essentially the radius $R \approx 10\,\text{km}$ of the settled compact star. With a
"photosphere" temperature T_x of the new objects their luminosity is
$4\pi r^2 \sigma T_x^4$ with the Stefan-Boltzmann constant σ which is $g\pi^2/120$ in
natural units with g the effective number of degrees of freedom (2 for
photons). Therefore, one must demand that

$$T_x \lesssim 8\,\text{MeV}\,g^{-1/4}, \tag{13.9}$$

in order to stay below the total neutrino luminosity of $3\times10^{52}\,\text{erg}\,\text{s}^{-1}$.

It is nontrivial, however, to determine the temperature T_x which
corresponds to about unit optical depth. Following the approach of
Turner (1988) who carried this analysis through for axions one may
assume a simple model for the run of temperature and density above
the settled inner core. A simple power-law ansatz is $\rho(r) = \rho_R\,(R/r)^n$
with the density $\rho_R = 10^{14}\,\text{g}\,\text{cm}^{-3}$ at a radius $R \approx 10\,\text{km}$. A plausible
ansatz for the temperature profile is $T(r) - T_R\,[\rho(r)/\rho_R]^{1/3}$ with T_R
(temperature at radius R) of around $10\,\text{MeV}$. From the opacity κ as a
function of density and temperature one may then calculate the optical
depth as $\tau(r_x) = \int_{r_x}^{\infty} \kappa\rho\,dr$. From the condition $\tau(r_x) \approx \frac{2}{3}$ one can
determine the "photosphere" radius r_x and thus its temperature T_x.

The opacity of axions for a medium of nondegenerate nucleons was
given in Eq. (4.28). One may define $\tau_R \equiv \kappa_R \rho_R R$ so that $\kappa\rho R = \tau_R\,(\rho/\rho_R)^2(T_R/T)^{1/2}$ where Eq. (4.28) yields $\tau_R - g_a^2\,3.4\times10^{16}$. Then
one finds for Turner's model an optical depth τ_a at the axion-sphere
temperature T_a

$$\tau_a = \tau_R\left(\tfrac{11}{6}\,n - 1\right)(T_a/T_R)^{11/2-3/n}. \tag{13.10}$$

Because n is a relatively large number such as $3-7$ the criterion $\tau_a \lesssim \frac{2}{3}$
yields $\tau_R \gtrsim n\,(T_R/T_a)^6$. With the requirement $T_a \lesssim 8\,\text{MeV}$ and with
$T_R \approx 20\,\text{MeV}$ as taken by Turner one finds $g_a \gtrsim 2\times10^{-7}$, not in bad
agreement with what one would conclude from the numerical results
shown in Fig. 13.1.

Still, this argument is rather sensitive to the detailed model assumptions concerning the protoneutron star structure. Also, as axions
contribute to the transfer of energy within the star, a self-consistent
model must take this effect into account. Moreover, for novel fermions
such as r.h. neutrinos one must distinguish carefully between their neutrino sphere (from where they can escape almost freely) and the deeper

region where their energy flux is set. Put another way, for fermions the concept of blackbody emission from a neutrino sphere is not adequate, making it impossible to apply the Stefan-Boltzmann in a simplistic way. The transport of r.h. neutrinos in the trapping limit is an equally complicated problem as that of l.h. ones! Therefore, a proper treatment of the trapping limit is generally a tricky subject; axions are the only case where it has been studied in some detail.

Occasionally one may wish to construct a particle-physics model that avoids the SN limit. It would be incorrect to believe that this is achieved when the interaction strength has been tuned such that the mean free path is of order the neutron star radius. On the contrary, when this condition obtains the impact on the cooling rate is maximized. This is analogous to the impact of novel particles on the structure and evolution of the Sun as depicted in Fig. 1.2; the cooling rate is maximized when the mfp corresponds to a typical geometric dimension of the object. In the trapping regime a new particle is harmless only if it interacts about as strongly as the particles which provide the standard mode of energy transfer.

13.5 Axions

13.5.1 Numerical Studies

The most-studied application of the SN cooling-time argument is that of invisible axions as these particles are well motivated (Chapter 14). Moreover, they have attracted much interest because they are one of the few particle-physics motivated candidates for the cosmic dark matter. Early analytic studies in the free-streaming limit are Ellis and Olive (1987), Raffelt and Seckel (1988), and Turner (1988) who also discussed the trapping regime; his line of reasoning was presented in Sect. 13.4.3 above. Numerical studies in the free-streaming limit were performed by Mayle et al. (1988, 1989) and by Burrows, Turner, and Brinkmann (1989) while the trapping regime was numerically studied by Burrows, Ressell, and Turner (1990). The numerical studies by different workers in the free-streaming limit used different assumptions concerning the axion couplings, emission rates, and other aspects. In my previous review (Raffelt 1990d) I have attempted to reduce the results of these works to a common and consistent set of assumptions; apart from relatively minor differences which could be blamed on different input physics (e.g. softer equation of state and thus higher temperatures in the Mayle et al. papers) the results seemed reasonably consistent. A recent numerical study by Keil (1994) who used the same axion emission rates as Burrows, Turner, and Brinkmann (1988) confirmed their results.

Here, I present the numerical studies of Burrows and his collaborators where axions were assumed to couple with equal strength to protons and neutrons. The axial-vector coupling to nucleons is written in the form $(C/2f_a)\,\overline{\psi}\gamma_\mu\gamma_5\psi\,\partial^\mu a$ with a model-dependent numerical factor C, the Peccei-Quinn energy scale f_a, the nucleon Dirac field ψ, and the axion field a. Under certain assumptions detailed in Sect. 14.2.3 it can be written in the pseudoscalar form $-i\,g_a\,\overline{\psi}\gamma_5\psi$ where $g_a = Cm_N/f_a$ is a dimensionless Yukawa coupling (nucleon mass m_N); Burrows et al. used $C = \frac{1}{2}$. All results will be discussed in terms of g_a and as such they apply to any pseudoscalar particle which couples to nucleons accordingly. In Sect. 14.4 the available constraints on axions will be expressed in terms of the axion mass m_a.

In the free-streaming limit the energy loss by axions was implemented according to the numerical rates of Brinkmann and Turner (1988); limiting cases of these rates were discussed in Sect. 4.2. In the trapping regime, the transfer of energy by axions as well as axion cooling from an "axion sphere" was implemented by means of an effective radiative opacity as discussed in Sect. 4.4. The protoneutron star models are those of Burrows and Lattimer (1986) and of Burrows (1988b). In the latter study, cooling sequences were presented for different equations of state (EOS), and different assumptions concerning the mass and early accretion rate of the stars. A fiducial case in these studies is model 55 with a "stiff EOS," an initial baryon mass of $1.3\,\mathcal{M}_\odot$, and an initial accretion of $0.2\,\mathcal{M}_\odot$.

The compatibility of a given model with the SN 1987A observations should be tested by a maximum-likelihood analysis of the time and energy distributions of the events in both the IMB and Kamiokande II detectors. In practice, it is easier to consider a few simple observables. Burrows and his collaborators chose the total number of events N_{KII} and N_{IMB} in the two detectors as well as the signal duration defined by the expected times t_{KII} and t_{IMB} it takes to accrue 90% of the expected total number of events. As both detectors measured approximately 10 events each, the time of the last event probably is a reasonable estimate of t_{KII} and t_{IMB}. Finally, Burrows et al. calculated the total energy carried away by neutrinos and axions.

The run of these quantities with g_a is shown in Fig. 13.4. Recall from Sect. 11.3.2 that the observed SN 1987A numbers of events are $N_{\mathrm{IMB}} = 8$ and $N_{\mathrm{KII}} = 10\text{--}12$, depending on whether event No. 6

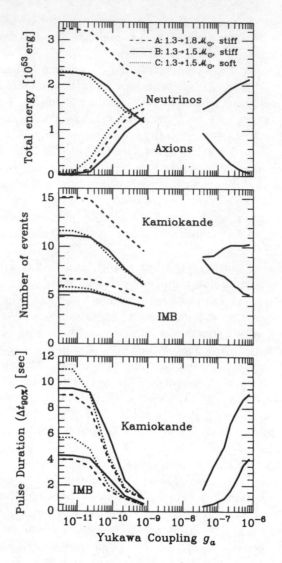

Fig. 13.4. Results from protoneutron star cooling sequences with axions. The free-streaming regime (small g_a) is according to Burrows, Turner, and Brinkmann (1989), the trapping regime (large g_a) according to Burrows, Ressell, and Turner (1990). For models A, B, and C (corresponding to models 57, 55, and 62 of Burrows 1988b) the amount of early accretion and the type of EOS ("stiff" or "soft") is indicated. The models were calculated until 20 s after collapse.

was actually due to background (quite possible) and whether event No. 1 was due to the prompt ν_e burst (possible but not necessary). The last events were registered at 5.6 s after the first (IMB) and 12.4 s (Kamiokande II). Recall also that the absolute timing between the two detectors is uncertain to within a minute although it seems plausible that in both cases the first event essentially marks the arrival of the first neutrinos. Finally, recall that the signal at Kamiokande II exhibits a peculiar 7.3 s time gap before the last three events; event No. 9 was registered at 1.9 s after the first. Naturally, it is worrisome that the large Kamiokande time scale rests on the last three events, i.e. in order to take the Kamiokande pulse duration seriously one needs to appeal to a rare statistical fluctuation.

The total number of events observed is not very sensitive to the amount of axion cooling which has an impact mostly on the late-time neutrino signal. Interestingly, in the trapping regime (large g_a) the number of events at IMB actually increases because the axionic energy transfer heats the neutrino sphere to higher temperatures. N_{IMB} responds sensitively to the neutrino spectrum because of the high threshold at IMB. However, for this reason it is a bad indicator for the actual neutrino flux because the high-energy tail of the spectrum is relatively uncertain. For example, if it is described by a Fermi-Dirac function with a degeneracy parameter $\eta = 2-3$ rather than $\eta = 0$ reduces N_{IMB} by about a factor of two (Fig. 11.11).

No numerical results are available in the intermediate regime between free streaming and trapping where the axion mean free path is of order the neutron star radius. In this range of coupling constants the impact of axions on the star is maximized. Moreover, a substantial modification of the initial collapse phase obtains.

As emphasized before, the most sensitive observable is the duration of the neutrino signal at the detectors. Therefore, nominally a range of coupling constants $1 \times 10^{-10} \lesssim g_a \lesssim 3 \times 10^{-7}$ is excluded. Within this range the observed neutrino signal likely would be shortened too much to be compatible with the observations.

13.5.2   Impact of Multiple-Scattering Effects

The results presented in the previous section were based on a naive perturbative calculation of the axion emission rate without taking the modification of the spin-density structure function into account that must occur at high density as outlined in Sect. 4.6.7. The density dependence of the axion emission rate is encapsuled in the spin-fluctuation

rate Γ_σ which, in the nondegenerate limit, was given in Eq. (4.7) on the
basis of a perturbative one-pion exchange (OPE) calculation. For the
protoneutron star model displayed in Fig. 13.2 the profile of this Γ_σ/T
is shown in Fig. 13.5. In Fig. 4.8 the axion emission rate was shown as a
function of Γ_σ, revealing that for the conditions of interest one is in the
neighborhood of the maximum of the solid curve. In a realistic nuclear
medium, the true spin fluctuation rate may be smaller than the OPE
calculated value, taking one perhaps somewhat to the left of the maximum. Therefore, the true axion emission rate corresponds to the naive
one (dashed line in Fig. 4.8) at $\Gamma_\sigma/T \approx 3-5$ which at temperatures
around 30 MeV corresponds to around 20% nuclear density.

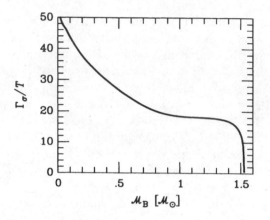

Fig. 13.5. Profile for the nondegenerate spin-fluctuation rate Γ_σ of Eq. (4.7)
in the protoneutron star model S2BH_0 of Keil, Janka, and Raffelt (1995)
shown in Fig. 13.2.

Given the overall uncertainties involved in this discussion it is best
to derive a plausible limit on g_a by the analytic criterion Eq. (13.8). The
relevant average temperature is about 30 MeV for which the maximum
emission rate corresponds to the naive one at about $5 \times 10^{13}\,\mathrm{g\,cm^{-3}}$.
From Eq. (4.8) one finds an approximate axion energy-loss rate of
$g_a^2\,1 \times 10^{38}\,\mathrm{erg\,g^{-1}\,s^{-1}}$. Then Eq. (13.8) indicates that one needs to require $g_a \lesssim 3 \times 10^{-10}$, about a factor of 3 less restrictive than the nominal bound from the numerical calculations above. Altogether one may
adopt

$$3 \times 10^{-10} \lesssim g_a \lesssim 3 \times 10^{-7} \tag{13.11}$$

as a range excluded by the SN 1987A cooling-time argument.

13.6   How Many Neutrino Flavors?

One may ask how the neutrino signal from a SN would be modified if there existed additional light sequential neutrino flavors beyond ν_e, ν_μ, and ν_τ. Of course, the Z° decay width measured at CERN already reveals that there are exactly three sequential neutrino flavors (Particle Data Group 1994); the same conclusion is reached from studies of big bang nucleosynthesis (e.g. Kolb and Turner 1990).

Burrows, Ressell, and Turner (1990) calculated several protoneutron star cooling sequences, varying the number of flavors from 3, the standard value, to 11. This increases the efficiency of energy transfer within the SN core and also allows for a more efficient radiation from the neutrino sphere as there are more degrees of freedom. Thus one expects a shortened signal in the Kamiokande II and IMB detectors, as well as a reduced number of events because the available energy is shared between more neutrino degrees of freedom of which mostly the

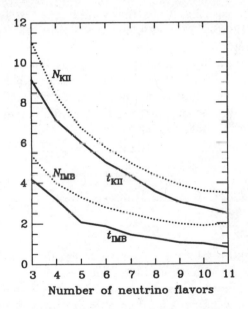

Fig. 13.6. Number of events N_{KII} and N_{IMB} in the Kamiokande and IMB detectors as well as the signal duration t_{KII} and t_{IMB} (in sec) as a function of the assumed number of neutrino flavors (Burrows, Ressell, and Turner 1990). The signal duration is defined as the time it takes to accrue 90% of the total expected number of events.

$\bar{\nu}_e$'s are detected. These expectations are borne out by the numerical results shown in Fig. 13.6. A doubling of the number of flavors is probably excluded by the observed signal duration.

13.7 Neutrino Opacity

The cooling time scale of a young SN core is determined by the neutrino opacities which in turn are dominated by the neutral-current scattering $\nu + N \to N + \nu$ of neutrinos on nucleons. Apart from final-state Pauli blocking effects these opacities are given in terms of the scattering cross section $\sigma = (G_F^2/\pi)\,(C_V^2 + 3C_A^2)\,E_\nu^2$ where the neutral-current nucleon weak-coupling constants $C_{V,A}$ were given in Appendix B. Therefore, the neutrino opacities are dominated by the axial-vector, i.e. the nucleon spin-dependent interaction. In Sect. 4.6.7 it was discussed that a naive application of perturbation theory in a nuclear medium likely is not appropriate because of the large spin fluctuation rate implied by this method. It would indicate that the spin of a given nucleon fluctuates so fast in a SN core that a neutrino would "see" on average a nearly vanishing contribution. This would lead to a decrease of a typical axial-vector scattering rate as estimated in Fig. 4.9.

To test if such a suppression effect is compatible with the SN 1987A neutrino signal, Keil, Janka, and Raffelt (1995) calculated a series of protoneutron star cooling sequences with modified neutrino opacities. To this end they substituted $C_A^2 \to F C_A^2$ in the numerical subroutine which evaluates the opacities where

$$F = (1 - a) + \frac{a}{1 + b} \tag{13.12}$$

with

$$b = \frac{1}{12}\left(\frac{\Gamma_\sigma}{T}\right)^2 \approx \left(\frac{\rho}{3\times 10^{13}\,\mathrm{g\,cm^{-3}}}\right)^2 \frac{10\,\mathrm{MeV}}{T}. \tag{13.13}$$

Here, $a = 1$ represents full suppression while smaller values of a allow one to dial a lesser reduction of the opacities.

The predicted neutrino signal at IMB and Kamiokande II was, again, characterized by the total number of expected events N_{IMB} and N_{KII} as well as the signal durations t_{IMB} and t_{KII} which represent the time at which 90% of the total number of expected events have been accrued. In Fig. 13.7 these quantities are shown as a function of a where the suppression effect was implemented for both, neutral- and charged-current axial-vector interactions. The results marked with open circles

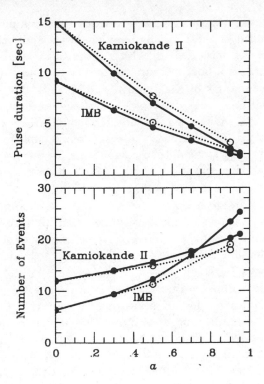

Fig. 13.7. Number of events in the Kamiokande and IMB detectors as well as the signal durations as a function of the assumed "opacity suppression parameter" defined by Eq. (13.12). Filled circles refer to a suppression of both neutral- and charged-current axial-vector interactions while open circles refer to a suppression of neutral-current interactions only. (Adapted from Keil, Janka, and Raffelt 1995.)

refer to a suppression of the neutral-current reactions alone. The modification of the results between those cases is relatively minor, indicating that the neutral-current interactions represent the dominant opacity source for the overall cooling time scale.

The increase of the counting rates at the two detectors with decreasing opacities is explained by the neutrino sphere moving to deeper and hotter layers, yielding larger neutrino energies. The detectors register mostly $\bar{\nu}_e$'s so that the number of events is relatively sensitive to the charged-current opacity which affects only the electron flavor. N_{IMB} is particularly sensitive to the $\bar{\nu}_e$ spectrum because of its high threshold.

By the same token, N_{IMB} is quite sensitive to spectral pinching, an effect not included in these calculations where an equilibrium neutrino transport scheme was used. Therefore, the total number of events, notably at IMB, is a poor measure to characterize the neutrino signal.

In the calculations of Keil, Janka, and Raffelt (1995) the mass of the initial neutron-star model as well as its temperature profile and the equation of state were varied. While such modifications cause changes in the predicted signal durations and event counts, none of these parameters appears likely to be able to compensate for an extreme suppression of the neutrino opacities. The SN 1987A neutrino signal excludes a suppression effect stronger than, say, $a \gtrsim 0.5$.

These findings are in agreement with those of the previous section where the number of neutrino flavors had been increased. Essentially that procedure amounted to increasing the efficiency of neutral-current energy transfer and so it is not very different from the decreased opacities used here.

Therefore, it appears that the "standard" opacities which ignore fast spin fluctuations provide a reasonable representation of what is observed. This result appears to imply that the spin-fluctuation rate Γ_σ does not exceed $\mathcal{O}(T)$ in a SN core—see Sect. 4.6.7 for a discussion of these matters. However, it is surprising that the best fit is achieved by the naive opacities because the spin-fluctuation effect is only one reason to expect reduced axial-vector opacities. Other reasons include reduced effective values for C_A in a nuclear medium, and spin-spin correlations which tend to "pair" the spins and thus tend to reduce the opacities. The question of the appropriate neutrino opacities in a SN medium remains worrisome.

However, with regard to particle bounds the effect of reduced opacities goes in the direction of making those constraints more conservative as a reduction of the opacities, like an anomalous energy loss, shortens the neutrino signal.

13.8   Right-Handed Neutrinos

13.8.1   Dirac Mass

Right-handed neutrinos (helicity-minus neutrinos, helicity-plus antineutrinos) do not interact by the standard weak interactions and so they would not be trapped in the interior of a SN core. Therefore, the SN 1987A neutrino signal allows one to constrain any mechanism that could produce these "wrong-helicity" states.[86] The main possibilities are the existence of novel r.h. interactions which couple directly to r.h. neutrinos, the existence of neutrino magnetic or electric dipole moments which allow for left-right scatterings or magnetic oscillations, and the existence of neutrino Dirac masses. Of course, these possibilities are not necessarily distinct as the existence of r.h. currents or a Dirac mass would usually also induce magnetic dipole moments.

Beginning with the assumption that neutrinos have a Dirac mass, the mismatch between chirality and helicity for massive fermions implies that in purely l.h. interactions a final-state neutrino or antineutrino sometimes has the "wrong" helicity and thus nearly r.h. chirality. Then it is essentially noninteracting and thus may escape almost freely. In principle, there are two production channels, the spin-flip scattering of trapped l.h. states, and the production of pairs $\nu_L\bar{\nu}_R$ or $\nu_R\bar{\nu}_L$ by the medium. In Sect. 4.10 it was shown that the neutrino phase space favors the spin-flip process by a large margin. Moreover, in a nonrelativistic medium the spin-flip scattering rate was found to be simply the nonflip scattering rate times the factor $(m_\nu/2E_\nu)^2$. The energy-loss rate was then given by Q_{scat} in Eq. (4.94) in terms of a dynamic structure function of the medium.[87]

In a dilute medium consisting of only one species of nucleons one has $S(\omega) = (C_V^2 + 3C_A^2)\, 2\pi\delta(\omega)$, leading to an energy-loss rate of

$$\epsilon_R = \frac{3(C_V^2 + 3C_A^2)\, G_F^2 m_\nu^2 T^4}{2\pi^3\, m_N}$$

$$\approx 0.7\times 10^{19}\,\text{erg g}^{-1}\text{s}^{-1} \left(\frac{m_\nu}{30\,\text{keV}}\right)^2 \left(\frac{T}{30\,\text{MeV}}\right)^4, \tag{13.14}$$

where $C_V^2 + 3C_A^2 \approx 1$ was used (see Appendix B). Even though the axial-vector structure function is not a δ function, Eq. (13.14) is probably a reasonable estimate because the results of the previous section indicate that the neutrino scattering rate in a dense medium is probably not

---

[86]Such bounds naturally can be avoided if one assumes that the r.h. neutrinos have other novel interactions which are strong enough to trap them efficiently in a SN core. Explicit models were constructed, for example, by Babu, Mohapatra, and Rothstein (1992) or Rajpoot (1993). These authors aimed at avoiding the SN 1987A bound on Dirac neutrino masses.

[87]Besides the neutrino spin-flip rate due to the neutrino weak interactions with nucleons or other particles, there is also a spin-flip scattering term in the gravitational field of the entire neutron star (Choudhury, Hari Dass, and Murthy 1989). However, the resulting energy loss was found to be small except for low-energy neutrinos.

too different from that found in the dilute-medium limit. If one applies
the analytic criterion Eq. (13.8) one finds

$$m_\nu \lesssim 30\,\mathrm{keV} \tag{13.15}$$

as a limit on a possible Dirac neutrino mass. This agrees with the
bounds originally estimated by Raffelt and Seckel (1988) and Gaemers,
Gandhi, and Lattimer (1989) while Grifols and Massó (1990a) estimated a slightly more restrictive limit (14 keV). This sort of bound
only applies if the mass is not so large that the "wrong-helicity" states
interact strongly enough to be trapped themselves. This would occur
for a mass beyond a few MeV. Therefore, a Dirac-mass ν_τ with, say,
$m_{\nu_\tau} = \mathcal{O}(10\,\mathrm{MeV})$ is not excluded by this argument.

In a numerical study Gandhi and Burrows (1990) implemented the
spin-flip energy-loss rate and calculated the expected event counts and
signal durations at the IMB and Kamiokande II detectors. They found
a bound almost identical with Eq. (13.15). A similar numerical study
by Burrows, Gandhi, and Turner (1992) corroborated this result. Another numerical study was performed by Mayle et al. (1993) who found
a somewhat more restrictive limit of about 10 keV, essentially because
their equation of state allows the core to heat up to much higher temperatures than are found in Burrows' implementation with a stiff EOS.

In the Mayle et al. (1993) study, a more restrictive bound of around
3 keV was claimed if the pion-induced pair emission process $\pi + N \rightarrow
N + \nu_R + \bar{\nu}_L$ was included. This result is incorrect if one accepts the pre-
dominance of the spin-flip scattering over the pair-emission processes
that was discussed in Sect. 4.10. It does not seem believable that the
presence of pions would enhance the scattering cross section on nucleons. In the form implemented by Mayle et al. (1993), the pair emission
rate and their neutrino opacities were not based on a common and
consistent axial-vector dynamical structure function.

A massive ν_μ or ν_τ likely would mix with ν_e. In this case the degenerate ν_e sea initially present in a SN core would partially convert into a
degenerate ν_μ or ν_τ sea (Maalampi and Peltoniemi 1991; Turner 1992;
Pantaleone 1992a). In Sect. 9.5 it was shown that the flavor conversion
would be very fast even for rather small mixing angles. In this case the
spin-flip scattering energy-loss rate involves initial-state neutrinos with
much larger average energies than those of a nondegenerate distribution that was used above. However, even though the initial energy-loss
rate in r.h. neutrino is much larger than before, the degeneracy effect
disappears after the core has been deleptonized and so the late-time

neutrino signal is not affected as significantly as one might have expected. The Dirac mass limit becomes only slightly more restrictive if mixing is assumed (Burrows, Gandhi, and Turner 1992).

Neutrinos with masses in the keV range must decay sufficiently fast in order to avoid "overclosing" the universe. If r.h. Dirac-mass neutrinos escape directly from the inner core of a SN they have energies far in excess of l.h. neutrinos emitted from the neutrino sphere. If their decay products involve sequential l.h. neutrinos or antineutrinos, these daughter states would have caused high-energy events at IMB or Kamiokande II, contrary to the observations. Dodelson, Frieman, and Turner (1992) found that this argument excludes the lifetime range

$$10^{-9}\,\mathrm{s/keV} \lesssim \tau/m_\nu \lesssim 5{\times}10^7\,\mathrm{s/keV}, \tag{13.16}$$

for Dirac masses in the range $1\,\mathrm{keV} \lesssim m_\nu \lesssim 300\,\mathrm{keV}$, assuming that the "visible" channel dominates.

13.8.2 Right-Handed Currents

On some level r.h. weak gauge interactions may exist as, e.g. in left-right symmetric models where the gauge bosons which couple to r.h. currents would differ from the standard ones only in their mass. In the low-energy limit relevant for processes in stars one may account for the novel couplings by a "r.h. Fermi constant" which is given as ϵG_F with ϵ some small dimensionless number which may be different for charged- and neutral-current processes. In left-right symmetric models one finds explicitly for charged-current reactions (Barbieri and Mohapatra 1989)

$$\epsilon_{CC}^2 - \zeta^2 + (m_{W_L}/m_{W_R})^4, \tag{13.17}$$

where $m_{W_{R,L}}$ are the r.h. and l.h. charged gauge boson masses while ζ is the left-right mixing parameter.

In order to constrain ϵ_{CC} one assumes the existence of r.h. ν_e's so that the dominant energy-loss mechanism of a SN core is $e + p \to n + \nu_{e,R}$ where the final-state r.h. neutrino escapes freely. Initially, a substantial fraction of the thermal energy of a SN core is stored in the degenerate electron sea. Therefore, the time scale of cooling is estimated by the inverse scattering rate for $e+p \to n+\nu_{e,R}$. The usual charged-current weak scattering cross section involving nonrelativistic nucleons is $G_F^2(C_V^2+3C_A^2)E_e^2/\pi$ with $C_V^2+3C_A^2 \approx 4$ in a nuclear medium (Appendix B). Using a proton density corresponding to nuclear matter at $10^{15}\,\mathrm{g\,cm^{-3}}$ and using $100\,\mathrm{MeV}$ for a typical electron energy one finds a charged-current scattering rate of $0.6 \times 10^{10}\,\text{s}^{-1}$ or an approximate cooling time-scale by r.h. neutrinos of $\epsilon_{CC}^{-2}\, 2 \times 10^{-10}\,\text{s}$. The requirement that this timescale exceeds a few seconds leads to the constraint

$$\epsilon_{CC} \lesssim 10^{-5}, \tag{13.18}$$

in agreement with a result of Barbieri and Mohapatra (1989) while Raffelt and Seckel (1988) found a somewhat less restrictive limit of $\epsilon_{CC} \lesssim 3 \times 10^{-5}$. Laboratory experiments yield a limit of order $\epsilon_{CC} \lesssim 3 \times 10^{-2}$ (e.g. Jodidio et al. 1986) which is much weaker but does not depend on the assumed existence of r.h. neutrinos. Mohapatra and Nussinov (1989) extended the SN 1987A bound to the case of r.h. Majorana neutrinos which mix with ν_e.

In order to constrain r.h. neutral currents, equivalent to constraining the mass of putative r.h. Z° gauge bosons, one considers the emission of r.h. neutrino pairs $\nu_R \bar{\nu}_R$. The dominant emission process is by the nucleons of the medium; in a dilute medium it can be represented as the bremsstrahlung process $NN \to NN\nu_R \bar{\nu}_R$. Apart from a global scaling factor ϵ_{NC}^2, the bremsstrahlung energy-loss rate for a nondegenerate medium was given in Eq. (4.23). However, in a dense medium this rate probably saturates at around 10% nuclear density as in the case of axion emission (Sect. 4.6.7). Evaluating Eq. (4.23) at 10% nuclear density ($\rho_{15} = 0.03$) and at $T = 30\,\text{MeV}$, and applying the analytic criterion Eq. (13.8) one finds

$$\epsilon_{NC} \lesssim 3 \times 10^{-3}. \tag{13.19}$$

This is less restrictive from what was found by Raffelt and Seckel (1988) or Barbieri and Mohapatra (1989).

The translation of a limit on ϵ_{NC} into one on a r.h. gauge boson mass depends on details of the couplings to quarks and leptons, and notably on the mixing angle between the new and the standard Z bosons. Detailed analyses were presented by Grifols and Massó (1990b), Grifols, Massó, and Rizzo (1990), and Rizzo (1991). Because these authors did not consider multiple-scattering effects and the resulting saturation of the bremsstrahlung process, their bounds on the Z' mass are somewhat too restrictive, perhaps by a factor of 2 or 3. Still, $m_{Z'}$ has to exceed at least 1 TeV, except for special choices of the mixing angle.

The SN 1987A limits on r.h. neutral currents are weaker than those from big bang nucleosynthesis ($\epsilon_{NC} \lesssim 10^{-3}$) which are based on the requirement that r.h. neutrinos must not have come to thermal equilibrium after the QCD phase transition (at $T \lesssim 200\,\text{MeV}$) in the early universe (e.g. Olive, Schramm, and Steigman 1981; Ellis et al. 1986).

13.8.3 Magnetic Dipole Moments

Turning to neutrino magnetic dipole moments, the main production process of r.h. states would be spin-flip scattering on charged particles, notably on protons. The scattering cross section was discussed in Sect. 7.4. It involves the usual Coulomb divergence which in a medium is cut off by screening effects. In a SN core, the electrons are initially very degenerate while the protons are essentially nondegenerate and so the main contribution to screening is from the protons. According to Eq. (D.17) they are electrically weakly coupled at the prevailing temperatures and densities so that Debye screening should be an approximately adequate prescription; the Debye scale for the protons is found to be around 30 MeV. As typical neutrino energies are somewhat larger but of the same order, the Coulomb logarithm is approximately unity. Thus the cross section is approximately $\alpha\mu_\nu^2$ with μ_ν the magnetic dipole or transition moment.

This is to be compared with the spin-flip cross section from a Dirac mass which is approximately $G_F^2 m_\nu^2/4\pi$. In Sect. 13.8.1 a Dirac mass bound of about 30 keV was derived which translates into

$$\mu_\nu \lesssim 4\times10^{-12}\,\mu_B \qquad\qquad (13.20)$$

with $\mu_B - e/2m_e$ the Bohr magneton. This bound is similar to that derived by Barbieri and Mohapatra (1988), but less restrictive by about an order of magnitude than that claimed by Lattimer and Cooperstein (1988). See also Nussinov and Rephaeli (1987), Goldman et al. (1988), and Goyal, Dutta, and Choudhury (1995).

Eq. (13.20) applies to all magnetic, electric, and transition moments of Dirac neutrinos. Numerically, it is similar to the bound Eq. (6.97) derived from the absence of excessive plasmon decay in globular cluster stars. However, because this latter result applies also to Majorana transition moments it is more general than the SN limit. The SN limit, on the other hand, applies to masses up to a few MeV while the globular cluster bound only for $m_\nu \lesssim 5$ keV.

The simple cooling argument that led to Eq. (13.20) is not necessarily the end of the story of neutrino magnetic dipole moments in SNe. If r.h. neutrinos were indeed produced in the inner core and escaped freely, they could rotate back into l.h. ones in the magnetic field around the SN and in the galaxy. For a galactic magnetic field of order 10^{-6} Gauss, extended over, say, 1 kpc (the field is confined to the disk, the LMC lies far above the disk) this effect would be important for $\mu_\nu \gtrsim 10^{-12}\,\mu_B$.

As the r.h. neutrinos escape from the inner core with much larger energies than those from the neutrino sphere one would expect high-energy events in the Kamiokande and IMB detectors, contrary to the observations. Therefore, one probably needs to require $\mu_\nu \lesssim 10^{-12} \mu_B$ for the diagonal dipole moments; spin-flavor oscillations could be suppressed by the neutrino mass differences. As the spin precession is the same for all neutrino energies, this limit would not apply if the Earth happened to be in a node of the oscillation pattern between SN 1987A and us.

Neutrino magnetic moments of order $10^{-12} \mu_B$ could also affect the infall phase of SNe. The spin-flip scattering on nuclei would be coherently enhanced relative to protons. Therefore, neutrinos could escape in the r.h. channel for much longer so that effectively trapping would set in much later than in the standard picture (Nötzold 1988).

In and near the SN core there probably exist strong magnetic fields of order 10^{12} Gauss or more which would induce spin-precessions between r.h. and l.h. neutrinos. Therefore, the sterile states produced in the deep interior by spin-flip scattering could back-convert into active ones near the neutrino sphere. Depending on details of the matter-induced neutrino energy shifts, the vacuum mass differences, and the magnetic field strengths and configurations this conversion could take place inside or outside of the neutrino sphere. The observable neutrino signal could be affected, but also the energy transfer within the SN core and outside of the neutrino sphere. Perhaps, a more efficient transfer of energy to the stalled shock wave could help to explode SNe in the delayed explosion scenario. Various aspects of these scenarios have been studied by Dar (1987), Nussinov and Rephaeli (1987), Goldman et al. (1988), Voloshin (1988), Okun (1988), Blinnikov and Okun (1988), and Athar, Peltoniemi, and Smirnov (1995).

Clearly, Dirac magnetic or transition moments in the $10^{-12} \mu_B$ range and below would affect SN dynamics and the observable neutrino signal in interesting ways. However, because there are so many parameters and possible field configurations, it is hard to develop a clear view of the excluded or desired neutrino properties. If compelling evidence for nonstandard neutrino electromagnetic properties in this range were to emerge, SN dynamics likely would have to be rethought from scratch.

13.8.4   Millicharges

Within the particle physics standard model it is not entirely impossible that neutrinos have small electric charges (Sect. 15.8). In this case neutrinos would have to be Dirac fermions and so the r.h. states can be produced in pairs by their coupling to the electromagnetic field. Obvious production processes are the plasmon decay $\gamma_{\rm pl} \to \bar{\nu}_R \nu_R$ and pair annihilation $e^+ e^- \to \bar{\nu}_R \nu_R$ with an intermediate photon. An intermediate photon can also be coupled to hadronic components of the medium. Mohapatra and Rothstein (1990) considered nucleon-nucleon bremsstrahlung, and notably an amplitude where the electromagnetic field is coupled to an intermediate charged pion. One may well wonder, however, if this sort of naive perturbative bremsstrahlung calculation is adequate in a nuclear medium.

A simple estimate of the plasmon decay process begins with the energy-loss rate Eq. (6.94). Because the electrons in a SN core are highly relativistic the plasma frequency is given by Eq. (6.43) as $\omega_{\rm P}^2 = (4\alpha/3\pi)\,(\mu_e^2 + \frac{1}{3}\pi^2 T^2)$ where μ_e is the electron chemical potential. For a SN core $\omega_{\rm P} \approx 10\,{\rm MeV}$ is a reasonable estimate while the relevant temperature is about $T = 30\,{\rm MeV}$. Taking approximately $Q_1 = 1$ in Eq. (6.94) and applying the approximate criterion Eq. (13.8) one finds

$$e_\nu \lesssim 10^{-9} e. \tag{13.21}$$

This is similar to Mohapatra and Rothstein's (1990) result. Including the $e^+ e^-$ annihilation process would slightly improve this limit and extend it to somewhat larger masses.

The bound Eq. (13.21) would apply to any millicharged particle which is not trapped in the SN core. Mohapatra and Rothstein (1990) estimated that for a charge in excess of about $10^{-7} e$ the particles would be sufficiently trapped by scatterings off electrons to leave the SN cooling time scale essentially unaffected. If they are Dirac neutrinos with a mass in excess of a few MeV trapping by spin-flip scattering would become important. Again, it is not obvious how strong the impact of such trapped particles would be during the infall phase of SN collapse, i.e. one should not infer that millicharged particles in the trapping regime would not have a strong impact on SN physics just because their impact on the Kelvin-Helmholtz cooling phase is small.

13.8.5 Charge Radius

If r.h. neutrinos existed and had an effective electromagnetic interaction by virtue of a charge radius, they would be produced in a SN core by the same processes as above where they were assumed to have a charge. On the basis of the $e^+ e^-$ annihilation process Grifols and Massó (1989) found a limit of about $3 \times 10^{-17}\,{\rm cm}$ on a r.h. charge radius.

Chapter 14

Axions

The idea of axions is introduced and their phenomenological properties are reviewed. The constraints on pseudoscalars that have been derived throughout this book are systematically applied to axions.

14.1   The Strong CP-Problem

All fermions, with the possible exception of neutrinos, have magnetic dipole moments—see Tab. 14.1 for several important examples. The minimal electromagnetic coupling of charged spin-$\frac{1}{2}$ fermions (charge q, mass m) automatically yields a "Dirac moment" $q/2m$. Higher-order amplitudes lead to an additional "anomalous" contribution, which is the only one for neutral particles. For example, the magnetic moments of massive neutrinos calculated in the standard model were given in Eq. (7.16). The QED prediction of the electron anomalous magnetic moment is perhaps the most stunning quantitative success of theoretical physics.

On the other hand, no particle electric dipole moment has ever been detected—see Tab. 14.1 for some upper limits. At first this is quite satisfying because an electric dipole moment would allow one to distinguish between matter and antimatter in an absolute sense.[88] The electroweak and strong gauge interactions are CP-conserving, so the observed broad symmetry between particles and antiparticles appears natural.

---

[88]In the nonrelativistic limit the electric dipole operator $\mathbf{d}$ must be proportional to the spin operator $\mathbf{s}$ which is an axial vector. Because the electric field is a polar vector the energy $\mathbf{d}\cdot\mathbf{E}$ reverses sign under P and thus under CP whence its absolute sign yields an absolute distinction between fermions and antifermions.

Table 14.1. Particle magnetic and electric dipole moments.

Fermion	Magnetic Moment[c]	Electric Moment[d] $[10^{-26} e\,\mathrm{cm}]$
Proton[a]	$2.792,847,3(86)\,\mu_N$	4000 ± 6000
Neutron[a]	$-1.913,042,(75)\,\mu_N$	$< 11^e$
Electron[a]	$1.001,159,652,1(93)\,\mu_B$	-0.3 ± 0.8
Neutrino[b]	$\lesssim 3\times10^{-12}\,\mu_B$	$\lesssim 6000$

[a]Particle Data Group (1994).
[b]All flavors with $m_\nu \lesssim 5\,\mathrm{keV}$ (Sect. 6.5.6).
[c]Bohr magneton $\mu_B = e/2m_e$; nuclear magneton $\mu_N = e/2m_p$.
[d]$1\,e\,\mathrm{cm} = 5.18\times10^{10}\,\mu_B = 0.951\times10^{14}\,\mu_N$ (Appendix A).
[e]95% CL.

This symmetry is not respected, however, by the generally complex Yukawa couplings to the Higgs field which are thought to induce the fermion masses (Sects. 7.1 and 7.2). The resulting complex quark mass matrix M_q can be made real and diagonal by suitable transformations of the quark fields. This involves a global chiral phase transformation (angle $\Theta = \arg\det M_q$), leading to a term in the QCD Lagrangian

$$\mathcal{L}_\Theta = \Theta\,\frac{\alpha_s}{8\pi}\,G\tilde{G}\,. \tag{14.1}$$

Here, α_s is the fine-structure constant of strong interactions and $G\tilde{G} \equiv G_b^{\mu\nu} G_{b\mu\nu}$ where $G_b^{\mu\nu}$ is the color field strength tensor, $\tilde{G}_{b\mu\nu} = \frac{1}{2}\epsilon_{\mu\nu\rho\sigma}G_b^{\rho\sigma}$ its dual, and the implied summation over b refers to the color degrees of freedom. Of course, $\det M_q$ and thus Θ would vanish if one of the quarks were exactly massless, but this does not seem to be the case.

Under the combined action of charge conjugation (C) and a parity transformation (P) the Lagrangian Eq. (14.1) changes sign,[89] violating the CP invariance of QCD. It leads to a neutron electric dipole moment $|d_n| \approx |\Theta|\,(0.04 - 2.0)\times10^{-15}\,e\,\mathrm{cm}$ (Baluni 1979; Crewther et al. 1979; see also Cheng 1988). This is $|d_n| \approx |\Theta|\,(0.004 - 0.2)\,\mu_N$ in units of nuclear magnetons. Hence, for $|\Theta|$ of order unity one expects a neutron electric dipole moment almost as large as its magnetic one. The

---

[89]The structure of $G\tilde{G}$ is $\mathbf{E}_{\mathrm{color}} \cdot \mathbf{B}_{\mathrm{color}}$, i.e. the scalar product of a polar with an axial vector and so it is CP-odd.

experimental limit (Tab. 14.1), however, indicates $|\Theta| \lesssim 10^{-9}$, surprisingly small in view of the phase $\delta = 3.3 \times 10^{-3}$ which appears in the Cabbibo-Kobayashi-Maskawa matrix Eq. (7.6) and which explains the observed CP-violating effects in the K°-$\overline{K}^{\circ}$ system.

Even worse, QCD alone produces a term like Eq. (14.1) because of the nontrivial topological structure of its ground state (Callan, Dashen, and Gross 1976; Jackiw and Rebbi 1976; t'Hooft 1976a,b). The coefficient Θ_{QCD} is a parameter characterizing the "Θ-vacuum." It is mapped onto itself by a transformation $\Theta_{\text{QCD}} \rightarrow \Theta_{\text{QCD}} + 2\pi$ so that *different* ground states are characterized by values in the range $0 \leq \Theta_{\text{QCD}} < 2\pi$.

The phase of the quark mass matrix and the QCD-vacuum together yield $\overline{\Theta} \equiv \Theta_{\text{QCD}} + \arg \det M_q$ as a compound coefficient for Eq. (14.1). The experimental bounds then translate into

$$\left| \Theta_{\text{QCD}} + \arg \det M_q \right| \lesssim 10^{-9} . \tag{14.2}$$

The CP-Problem of strong interactions consists of the smallness of $\overline{\Theta}$ which implies that the numbers Θ_{QCD} and $\arg \det M_q$ are either separately very small, or cancel each other with very high accuracy. However, both are expected to be of order unity, or perhaps of order δ in the case of $\arg \det M_q$, and completely unrelated to each other.

14.2   The Peccei-Quinn Mechanism

14.2.1   Generic Features

An attempt to explain the smallness of $\overline{\Theta}$ may be overambitious as long as we do not have an understanding of the origin of the Yukawa couplings that went into M_q and of the other seemingly arbitrary parameters of the standard model. Still, whatever determines these "constants of nature," the strong CP-problem can be elegantly explained by the existence of a new physical field, the axion field, which allows $\overline{\Theta}$ to vanish dynamically (Peccei and Quinn 1977a,b; Weinberg 1978; Wilczek 1978). In this scheme, the CP-violating Lagrangian Eq. (14.1) is literally switched off by its own force.

To this end the new field $a(x)$ must be a pseudoscalar which couples to gluons according to Eq. (14.1) with Θ replaced by $-a/f_a$. The constant[90] f_a with the dimension of an energy is the *Peccei-Quinn scale*

---

[90]In the literature one often finds f_a/N, F_a/N, v_{PQ}/N etc. for what I call f_a. It was stressed, e.g. by Georgi, Kaplan, and Randall (1986) that a discussion of the *generic* properties of all axion models does not require a specification of the model-dependent integer N which can be conveniently absorbed in the definition of f_a.

or *axion decay constant*. Axions must be fundamentally massless so that all observable effects remain unchanged under a global shift $a(x) \to a(x) + a_0$ where a_0 is a constant. (A mass term $\frac{1}{2}m_a^2 a^2$ would spoil this possibility.) This invariance allows one to absorb $\overline{\Theta}$ in the definition of the axion field. Including a kinetic term, Eq. (14.1) is replaced by

$$\mathcal{L}_{\Theta} \to \mathcal{L}_a = \tfrac{1}{2}(\partial_\mu a)^2 - \frac{\alpha_s}{8\pi f_a}\, a\, G\widetilde{G}\,. \tag{14.3}$$

It conserves CP because axions were assumed to be pseudoscalar (odd under CP), similar to neutral pions.

Even though axions were constructed to be massless they acquire an effective mass by their interaction with gluons. It induces transitions to $q\bar{q}$ states and thus to neutral pions (Fig. 14.1) which means physically that a and π° mix with each other. Axions thereby pick up a small mass which is approximately given by (Bardeen and Tye 1978; Kandaswamy, Salomonson, and Schechter 1978)

$$m_a f_a \approx m_\pi f_\pi\,, \tag{14.4}$$

where $m_\pi = 135\,\mathrm{MeV}$ is the pion mass and $f_\pi \approx 93\,\mathrm{MeV}$ its decay constant. This mass term implies that at low energies the axion Lagrangian contains a potential $V(a)$ which expands to lowest order as $\frac{1}{2}m_a^2 a^2$. Because of the invariance of $\mathcal{L}_\Theta$ with respect to $\Theta \to \Theta + 2\pi$ the gluon-induced potential $V(a)$ is a function periodic with $2\pi f_a$.

Fig. 14.1. Axion mixing with $q\bar{q}$ states and thus π°. The curly lines represent gluons, the solid lines quarks.

The ground state of the axion field is at the minimum of its potential at $a = 0$, explaining the absence of a neutron electric dipole moment. If one could produce a static nonvanishing axion field a_0 in some region of space, neutrons there would exhibit an electric dipole moment corresponding to $\overline{\Theta} = -a_0/f_a$.

The Lagrangian Eq. (14.3) is the minimal ingredient for any axion model: the $aG\widetilde{G}$ coupling is their defining feature as opposed to other pseudoscalar particles. Then axions inevitably acquire an effective mass at low energies. Thus the concept of a "massless axion" for some arbitrary pseudoscalar is a contradiction in terms.

Because of the mixing with π°, axions share not only their mass, but also their couplings to photons and nucleons with a strength reduced by about f_π/f_a. Therefore, they generically couple to photons so that the general discussion of Chapter 5 applies directly except for those aspects which required massless pseudoscalars.

The effective axion mass is a low-energy phenomenon below $\Lambda_{QCD} \approx$ 200 MeV. Above this energy pions and other hadrons dissociate in favor of a quark-gluon plasma. Then $a = 0$ is no longer singled out so that any value in the interval $0 \leq a < 2\pi f_a$ is physically equivalent. Because the universe is believed to begin with a hot and dense "big bang," any initial value for a is equally plausible, or different initial conditions in different regions of space. As the universe expands and cools below Λ_{QCD}, however, the axion field must relax to its newly singled-out ground state at $a = 0$. This relaxation process produces a population of cosmic background axions which is, in units of the cosmic critical density (e.g. Kolb and Turner 1990),

$$\Omega_a h^2 \approx (f_a/10^{12}\,\text{GeV})^{1.175}\,. \tag{14.5}$$

The exact value depends on details of the cosmic scenario and of the relaxation process. Modulo this uncertainty, values exceeding $f_a \approx 10^{12}$ GeV are excluded as axions would overdominate the dynamics of the universe. With Eq. (14.4) this corresponds to $m_a \lesssim 10^{-5}$ eV; axions near this bound would be the cosmic dark matter. A search strategy for galactic axions in this mass range was discussed in Sect. 5.3.

14.2.2 Axions as Nambu-Goldstone Bosons

The invariance of $\mathcal{L}_\Theta$ in Eq. (14.1) against transformations of the form $\Theta \to \Theta + 2\pi$, and the corresponding invariance of the axion Lagrangian against transformations $a \to a + 2\pi f_a$, calls for a very simple interpretation of the axion field as the phase of a new scalar field.

A transparent illustration is provided by the KSVZ axion model (Kim 1979; Shifman, Vainshtein, and Zakharov 1980) where one introduces a new complex scalar field Φ which does not participate in the weak interactions, i.e. an $SU(2)\times U(1)$ singlet. There is also a new massless fermion field Ψ and one considers a Lagrangian with the usual kinetic terms, a potential V for the scalar field, and an interaction term,

$$\mathcal{L} = \left(\tfrac{i}{2}\overline{\Psi}\partial_\mu\gamma^\mu\Psi + \text{h.c.}\right) + \partial_\mu\Phi^\dagger\partial^\mu\Phi - V(|\Phi|)$$
$$- h\left(\overline{\Psi}_L\Psi_R\Phi + \text{h.c.}\right)\,. \tag{14.6}$$

The Yukawa coupling h is chosen to be positive, and $\Psi_L \equiv \frac{1}{2}(1-\gamma_5)\Psi$ and $\Psi_R \equiv \frac{1}{2}(1+\gamma_5)\Psi$ are the usual left- and right-handed projections.

This Lagrangian is invariant under a chiral phase transformation of the form

$$\Phi \to e^{i\alpha}\Phi, \quad \Psi_L \to e^{i\alpha/2}\Psi_L, \quad \Psi_R \to e^{-i\alpha/2}\Psi_R, \tag{14.7}$$

where the left- and right-handed fields pick up opposite phases. This chiral symmetry is usually referred to as the Peccei-Quinn (PQ) symmetry $U_{PQ}(1)$.

The potential $V(|\Phi|)$ is chosen to be a "Mexican hat" with an absolute minimum at $|\Phi| = f_{PQ}/\sqrt{2}$ where f_{PQ} is some large energy scale. The ground state is characterized by a nonvanishing vacuum expectation value $\langle\Phi\rangle = (f_{PQ}/\sqrt{2})e^{i\varphi}$ where φ is an arbitrary phase. It spontaneously breaks the PQ symmetry because it is not invariant under a transformation of the type Eq. (14.7). One may then write

$$\Phi = \frac{f_{PQ}+\rho}{\sqrt{2}}\, e^{ia/f_{PQ}} \tag{14.8}$$

in terms of two real fields ρ and a which represent the "radial" and "angular" excitations.

The potential V provides a large mass for ρ, a field which will be of no further interest for these low-energy considerations. Neglecting all terms involving ρ the Lagrangian Eq. (14.6) is

$$\mathcal{L} = \left(\tfrac{i}{2}\overline{\Psi}\partial_\mu\gamma^\mu\Psi + \text{h.c.}\right) + \tfrac{1}{2}(\partial_\mu a)^2 - m\overline{\Psi}e^{i\gamma_5 a/f_{PQ}}\Psi, \tag{14.9}$$

where $m \equiv hf_{PQ}/\sqrt{2}$. The variation of the fermion fields under a PQ transformation is given by Eq. (14.7) while $a \to a + \alpha f_{PQ}$. The invariance of Eq. (14.9) against such shifts is a manifestation of the $U_{PQ}(1)$ symmetry. It implies that a represents a massless particle, the *Nambu-Goldstone boson of the PQ symmetry*.

Expanding the last term in Eq. (14.9) in powers of a/f_{PQ}, the zeroth-order term $m\overline{\Psi}\Psi$ plays the role of an effective fermion mass. Higher orders describe the interaction of a with Ψ,

$$\mathcal{L}_{\text{int}} = -i\frac{m}{f_{PQ}}a\,\overline{\Psi}\gamma_5\Psi + \frac{m}{2f_{PQ}^2}a^2\,\overline{\Psi}\Psi + \ldots. \tag{14.10}$$

The dimensionless Yukawa coupling $g_a \equiv m/f_{PQ}$ is proportional to the fermion mass.

The fermion Ψ is taken to be some exotic heavy quark with the usual strong interactions, i.e. an $SU_C(3)$ triplet. The lowest-order interaction of a with gluons is then given by the triangle graph of Fig. 14.2. With the first term of Eq. (14.10) it yields an effective a-gluon interaction of

$$\mathcal{L}_{aG} = -\frac{g_a}{m}\frac{\alpha_s}{8\pi}\,aG\widetilde{G}\,, \qquad (14.11)$$

where $\alpha_s \equiv g_s^2/4\pi$. All external momenta were taken to be small relative to the mass m of the loop fermion.

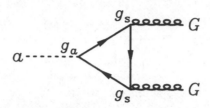

Fig. 14.2. Triangle loop diagram for the interaction of axions with gluons (strong coupling constant g_s, axion-fermion Yukawa coupling g_a). An analogous graph pertains to the coupling of axions with photons if the fermion carries an electric charge which replaces g_s.

In more general models, several conventional or exotic quark fields Ψ^j may participate in this scheme. The transformation of each field under a $U_{PQ}(1)$ transformation is characterized by its PQ charge X_j,

$$\Psi_L^j \to e^{iX_j\alpha/2}\,\Psi_L^j\,. \qquad (14.12)$$

The total $aG\widetilde{G}$ interaction is obtained as a sum over Eq. (14.11) for all Ψ^j. Because $g_{aj} = X_j m_j/f_{PQ}$ the fermion masses drop out. With

$$N \equiv \sum_j X_j \quad \text{and} \quad f_a \equiv f_{PQ}/N \qquad (14.13)$$

one has then found the required coupling Eq. (14.3) which allows one to interpret a as the axion field.

The potential $V(a)$ is periodic with $2\pi f_a = 2\pi f_{PQ}/N$. The interpretation of a as the phase of Φ, on the other hand, implies a periodicity with $2\pi f_{PQ}$ so that N must be a nonzero integer. This requirement restricts the possible assignment of PQ charges to the quark fields. It also implies that there remain N different equivalent ground states for the axion field, each of which satisfies $\overline{\Theta} = 0$ and thus solves the CP problem.

14.2.3   Pseudoscalar vs. Derivative Interaction

There has been considerable confusion in the literature concerning the proper structure for the coupling of axions to fermions. The lowest-order term in the expansion Eq. (14.10) is a pseudoscalar interaction which is frequently used because of its simplicity. However, there is an infinite series of terms which sometimes must be taken into account. For example, the axion scattering on fermions is second order in the first term of Eq. (14.10), but first order in the second whence both must be included.

Such complications can be avoided if one redefines the fermion field in Eq. (14.9) by a local transformation,

$$\psi_L \equiv e^{-ia/2f_{PQ}}\,\Psi_L\,, \quad \psi_R \equiv e^{ia/2f_{PQ}}\,\Psi_R\,. \tag{14.14}$$

The last term in Eq. (14.10) is then a simple mass term $m\overline{\psi}\psi$. The interaction between ψ and a now arises from the kinetic Ψ term in Eq. (14.9),

$$\mathcal{L}_{\text{int}} = \frac{1}{2f_{PQ}}\,\overline{\psi}\gamma_\mu\gamma_5\psi\,\partial^\mu a\,. \tag{14.15}$$

This interaction is of *derivative* nature, and it is linear in a with no higher-order terms. The fermion part has the form of an axial-vector current, bringing out a useful similarity between neutrino and axion interactions.

Pions play the role of Nambu-Goldstone bosons of a spontaneously broken $U(2)_{L-R}$ symmetry of QCD and so their interactions with nucleons should involve similar higher-order terms. The difference between derivative and "naive" pseudoscalar couplings should become apparent in bremsstrahlung processes of the type shown in Fig. 14.3 where two Nambu-Goldstone bosons are attached to one fermion line. A useful template is provided by $p + p \rightarrow p + p + \pi^\circ$ where existing data, indeed, favor the derivative case (Choi, K. Kang, and Kim 1989; Turner, H.-S. Kang, and Steigman 1989).

It should be noted in this context that the pion-nucleon interaction is described by the Lagrangian (Carena and Peccei 1989),

$$\mathcal{L}_{\text{int}} = g_{\pi N}\overline{N}\gamma^\mu\gamma_5\boldsymbol{\tau}N\cdot\partial_\mu\boldsymbol{\pi} + f_{\pi N}^2\overline{N}\gamma^\mu\boldsymbol{\tau}N\cdot\partial_\mu\boldsymbol{\pi}\times\boldsymbol{\pi}\,, \tag{14.16}$$

where $g_{\pi N} \equiv f/m_\pi \approx 1/m_\pi$ and $f_{\pi N} \equiv 1/2f_\pi$ are the relevant coupling constants, $\boldsymbol{\tau}$ is a vector of Pauli isospin matrices, $\boldsymbol{\pi}$ is the isovector of the neutral and charged pion fields, and N is the isodoublet of neutron and proton. Thus there appears an extra dimension-6 term compared

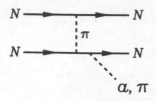

Fig. 14.3. Nucleon-nucleon bremsstrahlung emission of axions or pions.

with the axion example Eq. (14.15) where only one Nambu-Goldstone boson was present as opposed to the pion isotriplet. However, this additional term does not contribute to the bremsstrahlung process in the limit of nonrelativistic nucleons so that the above conclusion regarding the derivative coupling remains valid.

In the process $NN \to NNa$ (Fig. 14.3) axions *and* pions appear so that again two Nambu-Goldstone bosons are attached to one fermion line. It is then necessary to use a derivative coupling for at least one of them (Raffelt and Seckel 1988). For other bremsstrahlung processes such as $e^-p \to pe^-a$, where the particles interact through a virtual photon (a gauge boson) the pseudoscalar coupling causes no trouble. Also for the Compton process $\gamma e^- \to e^- a$ one may use either the pseudoscalar or the derivative axion coupling: both yield the same result. Because it is not always a priori obvious whether the pseudoscalar and derivative couplings yield the same result it is a safe strategy to use the derivative coupling in all calculations.

14.2.4 The Onslaught of Quantum Gravity

A heavy critique was levied against the PQ mechanism by quantum-gravity inspired phenomenological considerations. The main idea is that generally the PQ symmetry, like any other global symmetry, will not be respected by gravity (Georgi, Hall, and Wise 1981). For example, a black hole can "swallow" any amount of PQ charge without a trace, while a swallowed electric charge remains visible by its Coulomb force. At energy scales exceeding the Planck mass $m_{\rm Pl} = 1.2 \times 10^{19}$ GeV quantum gravitational effects are expected and so $m_{\rm Pl}$ is a phenomenological cutoff for any quantum theory which does not fundamentally include gravitation. In the "low-energy" world it should manifest itself by all sorts of effective interactions which are not forbidden by a symmetry and which likely involve inverse powers of the cutoff scale $m_{\rm Pl}$.

Notably, the Higgs field Φ which gives rise to the axion probably exhibits effective interactions of dimension $2m + n$

$$V_{\text{grav}}(\Phi) = g\,e^{i\delta}\,\frac{(\Phi\Phi^\dagger)^m\,\Phi^n}{m_{\text{Pl}}^{2m+n-4}}\,, \tag{14.17}$$

where g and δ are real numbers. Because such interactions violate the PQ symmetry for $n \neq 0$ they induce an effective potential for the axion after spontaneous symmetry breaking. The full potential is then of the form (Kamionkowski and March-Russell 1992)

$$\frac{V(a)}{f_a^2} = m_{\text{QCD}}^2\left[1 - \cos(a/f_a)\right] + m_{\text{grav}}^2\left[1 - \cos(\delta + na/f_a)\right],\tag{14.18}$$

where m_{QCD} is the usual QCD axion mass while gravity induces

$$m_{\text{grav}}^2 = g\,m_{\text{Pl}}^2(f_a/\sqrt{2}\,m_{\text{Pl}})^{2m+n-2}. \tag{14.19}$$

For g of order unity and for low values of m and n one needs a very small f_a for the QCD effect to dominate. Therefore, unless gravity for some reason favors a minimum at the CP-conserving position for a the PQ scheme will be ruined entirely.[91]

Whatever the ultimate quantum theory of gravitation, no doubt it will be very special. Therefore, it is by no means obvious that the above arguments, which do not go far beyond a dimensional analysis, correctly represent the low-energy effects of Planck-scale physics. Even then the PQ mechanism still works if the PQ global symmetry is an "automatic symmetry" of a gauge theory; in this case it is protected from the assault of quantum gravity. Such models can be constructed (Holman et al. 1992) and in fact may be quite generic (Barr 1994). Either way, in order for axions to solve the strong CP problem one must assume that the PQ scheme is not ruined by quantum gravity.

This discussion illustrates an important feature of axion models, or any model involving a broken global symmetry and its Nambu-Goldstone boson. These particles are interlopers in the low-energy world—axions really belong to the high-energy world at the PQ scale. These roots make them susceptible to physics at large energy scales, at the Planck mass, for example. By the same token, if axions were ever detected, for example by the galactic axion search (Sect. 5.3), they would be one of the few messengers that we can ever hope to receive from a high-energy world which is otherwise inaccessible to experimental enquiry.

---

[91] These issues were studied by Barr and Seckel (1992) and by Kamionkowski and March-Russell (1992). See also the earlier papers by Georgi, Hall, and Wise (1981), Lazarides, Panagiotakopoulos, and Shafi (1986), and Dine and Seiberg (1986).

14.3   Fine Points of Axion Properties

14.3.1   The Most Common Axion Models

Axions generically mix with pions so that their mass and their couplings to photons and nucleons are crudely f_π/f_a times those of π°. In detail, however, these properties depend on the specific implementation of the PQ mechanism. Therefore, it is useful to review briefly the most common axion models which may serve as generic examples for an interpretation of the astrophysical evidence.

In the standard model, the would-be Nambu-Goldstone boson from the spontaneous breakdown of $SU(2)\times U(1)$ is interpreted as the third component of the neutral gauge boson Z°, making it impossible for the scalar field Φ of which axions are the phase to be the standard Higgs field. Therefore, one needs to introduce two independent Higgs fields Φ_1 and Φ_2 with vacuum expectation values $f_1/\sqrt{2}$ and $f_2/\sqrt{2}$ which must obey $(f_1^2 + f_2^2)^{1/2} = f_{\text{weak}} \equiv (\sqrt{2}\,G_F)^{-1/2} \approx 250\,\text{GeV}$. In this *standard axion model* (Peccei and Quinn 1977a,b; Weinberg 1978; Wilczek 1978) Φ_1 gives masses to the up- and Φ_2 to the down-quarks and charged leptons. With $x \equiv f_1/f_2$ and 3 families the axion decay constant is $f_a = f_{\text{weak}}\,[3\,(x+1/x)]^{-1} \lesssim 42\,\text{GeV}$. This and related "variant" models (Peccei, Wu, and Yanagida 1986; Krauss and Wilczek 1986), however, are ruled out by overwhelming experimental and astrophysical evidence; for reviews see Kim (1987), Cheng (1988), and Peccei (1989).

Therefore, one is led to introduce an electroweak singlet Higgs field with a vacuum expectation value $f_{\text{PQ}}/\sqrt{2}$ which is not related to the weak scale. Taking $f_{\text{PQ}} \gg f_{\text{weak}}$, the mass of the axion becomes very small, its interactions very weak. Such models are generically referred to as *invisible axion models*. The first of its kind was the KSVZ model (Kim 1979; Shifman, Vainshtein, and Zakharov 1980) discussed in Sect. 14.2.2. It is very simple because the PQ mechanism entirely decouples from the ordinary particles: at low energies, axions interact with matter and radiation only by virtue of their two-gluon coupling which is generic for the PQ scheme. The KSVZ model in its simplest form is determined by only one free parameter, $f_a = f_{\text{PQ}}$, although one may introduce $N > 1$ exotic quarks whence $f_a = f_{\text{PQ}}/N$.

Also widely discussed is the DFSZ model introduced by Zhitnitskiĭ (1980) and by Dine, Fischler, and Srednicki (1981). It is a hybrid between the standard and KSVZ models in that it uses an electroweak singlet scalar field Φ with a vacuum expectation value $f_{\text{PQ}}/\sqrt{2}$ *and* two electroweak doublet fields Φ_1 and Φ_2. There is no need, however, for exotic heavy quarks: only the known fermions carry Peccei-Quinn charges. Therefore, N is the number of standard families. Probably $N = 3$ so that the remaining free parameters of this model are $f_a = f_{PQ}/N$ and $x = f_1/f_2$ which is often parametrized by $x = \cot \beta$ or equivalently by $\cos^2 \beta = x^2/(x^2 + 1)$.

From a practical perspective, the main difference between the KSVZ and DFSZ models is that in the latter axions couple to charged leptons in addition to nucleons and photons. The former is an example for the category of *hadronic axion models*.

Because $f_{PQ} \gg f_{weak}$ in these models, one may attempt to identify f_{PQ} with the grand unification scale $f_{GUT} \approx 10^{16}$ GeV (Wise, Georgi, and Glashow 1981; Nilles and Raby 1982). However, the cosmological bound $f_a \lesssim 10^{12}$ GeV disfavors the GUT assignment. There exist numerous other axion models, and many attempts to connect the PQ scale with other scales—for a review see Kim (1987). In the absence of a compelling model f_a should be viewed as a free phenomenological parameter.

14.3.2 Axion Mass and Coupling to Photons

The axion mass which arises from its mixing with π° can be obtained with the methods of current algebra to be (Bardeen and Tye 1978; Kandaswamy, Salomonson, and Schechter 1978; Srednicki 1985; Georgi, Kaplan, and Randall 1986; Peccei, Bardeen, and Yanagida 1987)

$$
\begin{aligned}
m_a &= \frac{f_\pi m_\pi}{f_a} \left(\frac{z}{(1 + z + w)(1 + z)} \right)^{1/2} \\
&= 0.60 \, \text{eV} \, \frac{10^7 \, \text{GeV}}{f_a},
\end{aligned}
\tag{14.20}
$$

where the quark mass ratios are (Gasser and Leutwyler 1982)

$$
\begin{aligned}
z &\equiv m_u/m_d = 0.568 \pm 0.042, \\
w &\equiv m_u/m_s = 0.0290 \pm 0.0043.
\end{aligned}
\tag{14.21}
$$

Aside from these uncertainties there are higher-order corrections to the current-algebra axion mass which have not been estimated in the literature.

By their generic coupling to gluons, axions necessarily mix with pions and hence couple to photons according to

$$\mathcal{L}_{\text{int}} = -\tfrac{1}{4} g_{a\gamma} F_{\mu\nu} \tilde{F}^{\mu\nu} a = g_{a\gamma} \mathbf{E} \cdot \mathbf{B} \, a, \qquad (14.22)$$

where F is the electromagnetic field strength tensor and $\tilde{F}$ its dual. In models where the quarks and leptons which carry PQ charges also carry electric charges, there is a contribution from a triangle loop diagram as in Fig. 14.2, replacing g_s with the electric charge $Q_j e$ of the lepton. It yields an axion-photon coupling proportional to

$$E \equiv 2 \sum_j X_j Q_j^2 D_j, \qquad (14.23)$$

where $D_j = 3$ for color triplets (quarks) and 1 for color singlets (charged leptons). The total axion-photon coupling strength is then (Kaplan 1985; Srednicki 1985)

$$g_{a\gamma} = -\frac{\alpha}{2\pi f_a} \frac{3}{4} \xi = \frac{m_{\text{eV}}}{0.69 \times 10^{10}\,\text{GeV}} \xi, \qquad (14.24)$$

where

$$\xi \equiv \frac{4}{3} \left(\frac{E}{N} - \frac{2}{3} \frac{4+z+w}{1+z+w} \right) = \frac{4}{3} \left(\frac{E}{N} - 1.92 \pm 0.08 \right) \qquad (14.25)$$

and $m_{\text{eV}} \equiv m_a/\text{eV}$.

In the DFSZ or grand unified models one has for a given family of quarks and leptons $E/N = 8/3$. Neglecting w this yields $\xi \approx (8/3)\, z/(1+z) \approx 1$. However, one may equally consider models where $E/N = 2$ so that $\xi = 0.1 \pm 0.1$, i.e. the axion-photon coupling is strongly suppressed and may actually vanish (Kaplan 1985).

14.3.3   Model-Dependent Axion-Fermion Coupling

The discussion in Sect. 14.2.3 implies that axions interact with a given fermion j (mass m_j) according to a pseudoscalar or a derivative axial-vector interaction,

$$\mathcal{L}_{\text{int}} = -i \frac{C_j m_j}{f_a} \overline{\Psi}_j \gamma_5 \Psi_j \, a \quad \text{or} \quad \frac{C_j}{2 f_a} \overline{\Psi}_j \gamma^\mu \gamma_5 \Psi_j \, \partial_\mu a, \qquad (14.26)$$

where C_j is an effective PQ charge of order unity to be defined below. Evidently $g_{aj} \equiv C_j m_j / f_a$ plays the role of a Yukawa coupling and $\alpha_{aj} = g_{aj}^2/4\pi$ that of an "axionic fine structure constant." Numerically,

$$\begin{aligned} g_{ae} &= C_e m_e/f_a \ = C_e \, 0.85 \times 10^{-10} \, m_{\text{eV}}, \\ g_{aN} &= C_N m_N/f_a = C_N \, 1.56 \times 10^{-7} \ m_{\text{eV}} \end{aligned} \qquad (14.27)$$

for electrons and nucleons.

Various axion models differ in their assignment of PQ charges. However, in all models $N = \sum_{\text{quarks}} X_j$ is a nonzero integer. The assignment at high energies is not maintained in the low-energy sector because the spontaneous breakdown of the weak $SU_L(2) \times U_Y(1)$ symmetry at $f_{\text{weak}} \approx 250 \, \text{GeV}$ mixes the axion with the would-be Nambu-Goldstone boson which becomes the longitudinal component of the Z° gauge boson. Hence the PQ charges must be shifted such that the physical axion does not mix with the Z°; these shifted values are denoted as X_j'. Also, below the QCD scale $\Lambda_{\text{QCD}} \approx 200 \, \text{MeV}$ free quarks do not exist, so one needs to consider the effective coupling to nucleons which arises from the direct axion coupling to quarks and from the mixing with π° and η, leading to PQ charges X_p' and X_n' for protons and neutrons. The above effective PQ charges are obtained by $C_j \equiv X_j'/N$ in order to absorb N in their definition just as it was absorbed in $f_a = f_{\text{PQ}}/N$.

In the KSVZ model $C_e = 0$ at tree level ("hadronic axions") although there are small radiatively induced couplings (Srednicki 1985). In the DFSZ model

$$C_e = \cos^2 \beta / N_{\text{f}}, \tag{14.28}$$

where N_{f} is the number of families, probably 3.

The nucleon interactions in general axion models were investigated by Kaplan (1985) and Srednicki (1985). They were revisited by Mayle et al. (1988, 1989),

$$\begin{aligned}
C_p &= (C_u - \eta)\Delta u + (C_d - \eta z)\Delta d + (C_s - \eta w)\Delta s, \\
C_n &= (C_u - \eta)\Delta d + (C_d - \eta z)\Delta u + (C_s - \eta w)\Delta s,
\end{aligned} \tag{14.29}$$

where $\eta \equiv (1 + z + w)^{-1}$ with z and w were given in Eq. (14.21).

For a given quark flavor, $q = u$, d, or s, the interaction strength with protons depends on the proton spin content carried by this particular quark flavor, $S_\mu \Delta q \equiv \langle p|\bar{q}\gamma_\mu\gamma_5 q|p \rangle$ where S_μ is the proton spin. Similar expressions pertain to the coupling with neutrons; the two sets of expressions are related by isospin invariance. Neutron and hyperon β-decays as well as polarized lepton scattering experiments on nucleons yield a consistent set of Δq's (Ellis and Karliner 1995)

$$\Delta u = +0.85, \quad \Delta d = -0.41, \quad \Delta s = -0.08 \tag{14.30}$$

with an approximate uncertainty of ± 0.03 each.

In the DFSZ model, $C_s = C_d = C_e$, $C_u + C_d = 1/N_{\text{f}}$, and $C_u - C_d = -\cos^2 \beta / N_{\text{f}}$, leading to $C_u = \sin^2\beta/N_{\text{f}}$ and $C_d = C_s = C_e = \cos^2 \beta / N_{\text{f}}$.

With $N_f = 3$ one finds

$$C_p = -0.10 - 0.45 \cos^2 \beta,$$
$$C_n = -0.18 + 0.39 \cos^2 \beta.$$

(14.31)

In the KSVZ model and in other hadronic axion models $C_u = C_d = C_s = 0$ which yields

$$C_p = -0.39,$$
$$C_n = -0.04.$$

(14.32)

In Fig. 14.4 these couplings are shown, for DFSZ axions as a function of $\cos^2 \beta$. They are all uncertain to within about ± 0.05, but even then C_p and C_n never seem to vanish simultaneously.

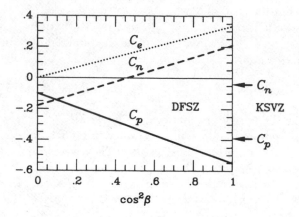

Fig. 14.4. Axion couplings to fermions according to Eqs. (14.28), (14.31), and (14.32).

14.4 Astrophysical Axion Bounds

Because axions couple to nucleons, photons, and electrons it is easy to translate the bounds on such couplings derived in previous chapters of this book into bounds on the Peccei-Quinn scale or equivalently, on the axion mass. A possible modification of the usual axion models by the quantum gravity effects discussed in Sect. 14.2.4 is ignored for the present discussion.[92]

---

[92]Barr and Seckel (1992) studied astrophysical axion bounds when quantum gravity effects are taken seriously.

Bounds on the Yukawa coupling to electrons of various novel particles were derived in Chapter 3; for pseudoscalars a summary was given
in Tab. 3.1. The most restrictive limit was obtained from the delay of
helium ignition in low-mass red giants that would be caused by excessive axion emission; in terms of the axion-electron Yukawa coupling it
is $g_{ae} \lesssim 2.5 \times 10^{-13}$. With Eq. (14.27) this translates into

$$m_a C_e \lesssim 0.003 \, \text{eV} \quad \text{and} \quad f_a/C_e \gtrsim 2 \times 10^9 \, \text{GeV}. \tag{14.33}$$

Axions which interact too strongly to escape freely from the interior
of stars would still contribute to the transfer of energy. For the Sun,
this issue was studied in Sect. 1.3.5. One easily finds that for $C_e = 1$
Fig. 1.2 excludes axion masses below about 50 keV.

In hadronic axion models $C_e = 0$ at tree level and so no interesting
bounds on m_a and f_a obtain. In the DFSZ model, C_e was given in
Eq. (14.28). Taking the number of families to be $N_f = 3$ one finds

$$m_a \cos^2 \beta \lesssim 0.01 \, \text{eV} \quad \text{and} \quad f_a/\cos^2 \beta \gtrsim 0.7 \times 10^9 \, \text{GeV}. \tag{14.34}$$

These limits depend on the parameter $\cos^2 \beta$ which, in principle, can
be equal to 0.

The axion-photon coupling is best constrained by the lifetime of
horizontal-branch (HB) stars as outlined in Sect. 5.2.5. The limit
Eq. (5.23) translates into

$$m_a \xi \lesssim 0.4 \, \text{eV} \quad \text{and} \quad f_a/\xi \gtrsim 1.5 \times 10^7 \, \text{GeV}, \tag{14.35}$$

where ξ was defined in Eq. (14.25). In addition, approximately the
mass range $4-14 \, \text{eV}$ is excluded by the "telescope search" for a line
from the radiative decay of cosmic axions (Fig. 12.23).

The most restrictive limit on the axion-nucleon coupling arises from
the duration of the neutrino signal of SN 1987A. The formally excluded range for the axion-nucleon Yukawa coupling was specified in
Eq. (13.11); as discussed in Sect. 13.5 it is fraught with uncertainties
because no reliable calculation of the axion emission rate from a nuclear
medium is available at the present time. In terms of the axion mass
and axion decay constant the nominally excluded range is

$$0.002 \, \text{eV} \lesssim C_N m_a \lesssim 2 \, \text{eV},$$

$$3 \times 10^6 \, \text{GeV} \lesssim f_a/C_N \lesssim 3 \times 10^9 \, \text{GeV}. \tag{14.36}$$

The case of large m_a (small f_a) is the trapping regime where axions
contribute to the energy transfer in a SN core, and where they are

emitted from an "axion sphere" rather than the entire volume of the protoneutron star.

These bounds were derived assuming equal couplings to protons and neutrons. However, a glance at Fig. 14.4 reveals that KSVZ axions essentially do not couple to neutrons while $C_p \approx -0.36$. For DFSZ axions the couplings vary with $\cos^2 \beta$, although for $\cos^2 \beta \approx 0.5$ about the same values as for KSVZ axions apply which are thus taken as generic. Assuming a proton fraction of about 0.3 for the relevant regions of the SN core I estimate an effective nucleon coupling of $C_N \approx 0.3^{1/2} \, 0.36 \approx 0.2$. Therefore,

$$0.01 \, \text{eV} \lesssim m_a \lesssim 10 \, \text{eV},$$

$$0.6 \times 10^6 \, \text{GeV} \lesssim f_a \lesssim 0.6 \times 10^9 \, \text{GeV} \qquad (14.37)$$

are formally adopted as the SN 1987A excluded axion parameters.

Axions on the "trapping side" of the SN argument can still be excluded because they would have caused additional events in the IMB and Kamiokande water Cherenkov detectors. The excluded range of Eq. (13.2) translates into the approximate mass exclusion range of $20 \, \text{eV} - 20 \, \text{keV}$.

These limits on the axion mass and decay constant are summarized in Fig. 14.5. The slanted end of the SN 1987A exclusion bar is a reminder of the potentially large uncertainty of this limit. Except for very special choices of model-dependent parameters axions with a mass above $0.01 \, \text{eV}$ are excluded.

The high-mass end of the stellar exclusion bars in Fig. 14.5 has not been worked out in detail because of their overlap with laboratory limits. The globular cluster limits apply without modification up to a mass of, say, $30 \, \text{keV}$ because the temperature in the cores of HB stars and red giants are about $10 \, \text{keV}$. However, axions with masses in this range interact much more strongly than those at the low-mass end of the exclusion bar so that the Boltzmann suppression of the emission rate for a large mass is partly balanced by the increased coupling strength.

14.5   Cosmological Limits

The stellar-evolution bounds on axions have received much attention because they push the Peccei-Quinn scale to such large values that axions appear to play a significant cosmological role if they exist at all. Therefore, to place the stellar constraints into context it may be useful to close this chapter with a brief summary of the cosmological

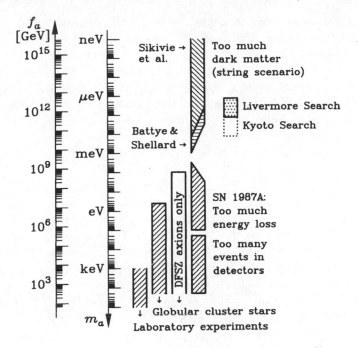

Fig. 14.5. Astrophysical and cosmological bounds on "invisible axions." One globular-cluster limit (white exclusion bar) is based on the axion-electron coupling and thus applies only if axions are of the DFSZ type ($\cos^2 \beta = 1$ was used). For the coupling to photons $\xi - 1$ was assumed. Slanted ends of exclusion bars indicate an estimated uncertainty of the bounds. The anticipated range of sensitivity of the Livermore and Kyoto search experiments are also indicated (Sect. 5.3).

limits which I quote from my contribution to the axion session of the XVth Moriond Workshop *Dark Matter in Cosmology, Clocks, and Tests of Fundamental Laws*, (Villars-sur-Ollon, Switzerland, January 21–28, 1995). The proceedings of this meeting will provide many up-to-date accounts of different aspects of the axion saga.

If axions were sufficiently strongly interacting ($f_a \lesssim 10^8 \, \text{GeV}$) they would have come into thermal equilibrium before the QCD phase transition and so we would have a background sea of invisible axions in analogy to the one expected for neutrinos (Turner 1987). This parameter range is excluded by the astrophysical arguments summarized in Fig. 14.5. Hence axions must be so weakly interacting that they have never come into thermal equilibrium. Still, the well-known mis-

alignment mechanism will excite coherent oscillations of the axion field
(Abbott and Sikivie 1983; Dine and Fischler 1983; Preskill, Wise, and
Wilczek 1983; Turner 1986). When the temperature of the universe
falls below f_a the axion field settles somewhere in the brim of its Mex-
ican hat potential. When the hat tilts at the QCD phase transition,
corresponding to the appearance of a mass term for the axion, the field
begins to move and finally oscillates when the expansion rate of the universe has become smaller than the axion mass. In units of the cosmic
critical density one finds for the axionic mass density

$$\Omega_a h^2 \approx 0.23 \times 10^{\pm 0.6} (f_a/10^{12}\,\mathrm{GeV})^{1.175}\, \Theta_i^2\, F(\Theta_i), \qquad (14.38)$$

where h is the present-day Hubble expansion parameter in units of
$100\,\mathrm{km\,s^{-1}\,Mpc^{-1}}$. The stated range reflects recognized uncertainties of
the cosmic conditions at the QCD phase transition and uncertainties in
the calculations of the temperature-dependent axion mass. The cosmic
axion density thus depends on the initial misalignment angle Θ_i which
could have any value between 0 and π. The function $F(\Theta_i)$ encapsules
anharmonic corrections to the axion potential for $\Theta \gg 0$. (For a recent
analytic determination of F see Strobl and Weiler 1994.)

The age of the universe indicates that $\Omega h^2 \approx 0.3$, causing a problem
with $\Omega = 1$ models if h is around 0.8 as indicated by recent measurements. For the present purpose I take $\Omega_a h^2 = 0.3 \times 2^{\pm 1}$ for axions
which constitute the dark matter where the adopted uncertainty of a
factor of 2 likely covers the whole range of plausible cosmological models. Then, axions with $m_a = \mathcal{O}(1\,\mu\mathrm{eV})$ are the cosmic dark matter if
Θ_i is of order 1.

Because the corresponding Peccei-Quinn scale of $f_a = \mathcal{O}(10^{12}\,\mathrm{GeV})$
is far below the GUT scale one may speculate that cosmic inflation, if
it occurred at all, did not occur after the PQ phase transition. If it
did not occur at all, or if it did occur before the PQ transition with
$T_{\mathrm{reheat}} > f_a$, the axion field will start with a different Θ_i in each region
which is causally connected at $T \approx f_a$ and so one has to average over all
possibilities to obtain the present-day axion density. More importantly,
because axions are the Nambu-Goldstone mode of a complex Higgs field
after the spontaneous breaking of a global U(1) symmetry, cosmic axion
strings will form by the Kibble mechanism (Davis 1986). The motion of
these global strings is damped primarily by the emission of axions rather
than by gravitational waves. At the QCD phase transition, the U(1)
symmetry is explicitly broken (axions acquire a mass) and so domain
walls bounded by strings will form, get sliced up by the interaction
with strings, and the entire string and domain wall system will quickly

decay into axions. This complicated sequence of events leads to the production of the dominant contribution of cosmic axions. Most of them are produced near the QCD transition at $T \approx \Lambda_{\text{QCD}} \approx 200\,\text{MeV}$. After they acquire a mass they are nonrelativistic or mildly relativistic so that they are quickly redshifted to nonrelativistic velocities. Thus, even the string and domain-wall produced axions form a cold dark matter component.

In their recent treatment of axion radiation from global strings, Battye and Shellard (1994a,b) found that the dominant source of axion radiation are string loops rather than long strings, contrary to what was assumed in the previous works by Davis (1986) and Davis and Shellard (1989). At a given cosmic time t the average loop creation size is parametrized as $\langle \ell \rangle = \alpha t$ while the radiation power from loops is $P = \kappa \mu$ with μ the renormalized string tension. The exact values of the parameters α and κ are not known; the cosmic axion density is a function of the combination α/κ. For $\alpha/\kappa < 1$ the dependence of $\Omega_a h^2$ on α/κ is found to be rather weak. Battye and Shellard favor $\alpha/\kappa \approx 0.1$ for which $\Omega_a h^2 = 18 \times 10^{\pm 0.6} (f_a/10^{12}\,\text{GeV})^{1.175}$, about an order of magnitude smaller than originally found by Davis (1986) and Davis and Shellard (1989). The overall uncertainty has the same source as in Eq. (14.38) above. With $\Omega_a h^2 = 0.3 \times 2^{\pm 1}$ the mass of dark-matter axions is found to be $m_a = 30 - 1000\,\mu\text{eV}$; the cosmologically excluded range of axion masses is indicated in Fig. 14.5.

These results are plagued with systematic uncertainties. Battye and Shellard (1994a,b) argue that the largest uncertainty was the impact of the backreaction of axion emission on the string network. (They believe that a current numerical study will allow them to pin down the parameter α/κ to within, say, a factor of 2.) Further, Sikivie and his collaborators (Harari and Sikivie 1987; Hagmann and Sikivie 1991) have consistently argued that the motion of global strings was over-damped, leading to an axion spectrum emitted from strings or loops with a flat frequency spectrum. In Battye and Shellard's treatment, wavelengths corresponding to the loop size are strongly peaked, the motion is not overdamped. In Sikivie et al.'s picture, much more of the string-radiated energy goes into kinetic axion energy which is red-shifted so that ultimately there are fewer axions; it was argued that the cosmic axion density was then of order the misalignment contribution. Therefore, following Sikivie et al. one would estimate the mass of dark-matter axions at about $m_a = 4 - 150\,\mu\text{eV}$ where the range reflects the same overall uncertainties that bedevil the Battye and Shellard estimate, or the misalignment contribution.

While the cosmic axion bounds claimed by both groups of authors still differ significantly, the overall uncertainty within either scenario is larger than the mutual disagreement, i.e. the range of masses where axions could be the dark matter overlaps significantly between the predictions of the two groups (Fig. 14.5). Moreover, there remain difficult to control uncertainties, for example, with the "dilute instanton gas" calculation of the temperature-dependent axion mass near the QCD phase transition. There may be other unaccounted systematic problems which may increase the adopted uncertainty of the cosmic mass prediction which is represented in Fig. 14.5 by the slanted end of the cosmic exclusion bar.

The astrophysical and cosmological limits on axions leave a narrow window of parameters where axions could still exist (Fig. 14.5); they would then be some or all of the dark matter of the universe. While the SN 1987A as well as the cosmological bound are each very uncertain, there is a recent trend toward an allowed range near $m_a = \mathcal{O}(1\,\mathrm{meV})$ where no current or proposed experimental effort appears to be sensitive. Of course, the overall quantitative uncertainty of the predicted cosmic axion density is large, especially if one includes the possibility of late-time inflation after the Peccei-Quinn phase transition or late-time entropy production after the QCD phase transition. Therefore, the microwave cavity experiments (Sect. 5.3) in Livermore and Kyoto no doubt have a fair chance of detecting galactic axions if they are the dark matter. In September 1995, the Livermore search experiment has taken up operation (K. van Bibber, private communication).

Chapter 15

Miscellaneous Exotica

Stellar-evolution constraints on a variety of hypotheses are discussed and compared with limits from other sources. Specifically, a possible time variation of Fermi's and Newton's constant, the validity of the equivalence principle, a photon mass and charge, the existence of free quarks and supersymmetric particles, and the role of majorons and millicharged particles are considered.

15.1 Constancy of Fermi's Constant

One of the basic physical assumptions commonly made in astrophysical research is that the laws of nature are the same at different places in the universe, and at earlier times here and elsewhere. Apparently this assumption has never been challenged seriously by any experiment or observation that would have indicated a spatial or temporal variation of parameters such as particle masses or coupling constants. At the present time it is not known what fixes the values of such "fundamental numbers." Therefore, the possibility that they vary in time or space cannot be a priori rejected. Notably, Dirac (1937, 1938) is often quoted for his speculation that the large value of some dimensionless numbers occurring in physics are related to variations of some physical constants on cosmological time scales. Whatever the merit of Dirac's large numbers hypothesis, it remains an interesting task to isolate simple observables that are sensitive to variations of certain "constants."

One instructive stellar-evolution example was discussed by Scherrer and Spergel (1993) who considered the constancy of Fermi's constant G_F which governs weak-interaction physics. The particle-physics standard model gives $G_F^{-1} = \sqrt{2}\,\Phi_0^2$ in terms of the Higgs-field vacuum expectation value Φ_0. Scherrer and Spergel noted that a nonconstant vacuum expectation value of a physical field was more plausibly subject to variations than dimensionless constants such as gauge and Yukawa couplings.

Type Ia supernovae are thought to represent the nuclear deflagration of a white dwarf pushed beyond its Chandrasekhar limit by accretion (e.g. Woosley and Weaver 1986b). Therefore, their light curves are very reproducible and indeed serve as standard candles in an attempt to improve measurements of the cosmic expansion rate. The shape of the lightcurve is determined by radioactive heating of the SN remnant by the decay of ^{56}Co; its lifetime is proportional to G_F^{-2}. A change of order 10% in the slope of type Ia SN lightcurves in neighboring galaxies (Leibundgut et al. 1991) would be readily observable so that G_F must be constant within at least 5% over 30 Mpc distances.

In addition, Scherrer and Spergel showed that the agreement between the observed primordial light element abundances and big-bang nucleosynthesis calculations imply that G_F lay within -1% and $+9\%$ of its standard value at this early epoch.

A stringent constraint on the local[93] time variation of G_F obtains from the analysis of ores from the Oklo uranium mine where a natural fission reactor is thought to have operated about 2 Gyr ago (e.g. Maurette 1976). A shift of the ground state difference between ^{150}Sm and ^{149}Sm by more than 0.02 eV is inconsistent with the isotope ratios at Oklo (Shlyakhter 1976, 1983). According to this author the contribution of the weak interaction to the nuclear binding energy is about 200 eV, excluding a change of G_F by more than 0.1% over the past 2 Gyr.

15.2   Constancy of Newton's Constant

15.2.1   Present-Day Constraints from Celestial Mechanics

Another "constant of nature" that might vary in time is Newton's constant G_N. Indeed, there exist self-consistent alternative theories to general relativity which actually predict a temporal variation of G_N on cosmological time scales—see Will (1993) for a summary of such theories and detailed references. A typical scale for the rate of change is the cosmic expansion parameter H so that it is natural to write $\dot{G}_N/G_N = \sigma H$ with σ a dimensionless model-dependent number. Some

---

[93] The Earth moves with the galaxy and the local group relative to the cosmic microwave background (CMB). Taking 600 km s^{-1} for this peculiar velocity the Earth moves by about 1.2 Mpc in 2 Gyr relative to a frame defined by the CMB.

Table 15.1. Bounds on the present-day $\dot{G}_N/G_N$. (Adapted from Will 1993.)

Method	$\dot{G}_N/G_N$ $[10^{-12}\,\mathrm{yr}^{-1}]$	References
Laser ranging (Moon)	0 ± 10	Müller et al. (1991)
Radar ranging (Mars)	-2 ± 10	Shapiro (1990)
Binary pulsar 1913+16	11 ± 11	Damour and Taylor (1991)
Spin-down PSR 0655+64	< 55	Goldman (1990)

recent discussions have also addressed the possibility of an oscillating G_N (Hill, Steinhardt, and Turner 1990; Accetta and Steinhardt 1991) with a rate of change much faster than H. The following discussion does not generically address such extreme model assumptions.

The G_N rate of change can be tested by a detailed study of the orbits of celestial bodies. Particularly precise data exist in the solar system from laser ranging of the moon and radar ranging of planets, notably by the Viking landers on Mars. Very precise orbital data also exist beginning 1974 for the binary pulsar PSR 1913+16; among other things its orbital decay reveals the emission of gravitational radiation. A weaker but also less model-dependent bound can be derived from the spin-down rate of the pulsar PSR 0655+64. These present-day limits on $\dot{G}_N/G_N$ are summarized in Tab. 15.1; altogether $|\dot{G}_N/G_N| \lesssim 20 \times 10^{-12}\,\mathrm{yr}^{-1}$ is probably a safe limit. The Hubble expansion parameter today is $H_0 = h\,100\,\mathrm{km\,s^{-1}\,Mpc^{-1}} = h\,1.02 \times 10^{-10}\,\mathrm{yr}^{-1}$ (observationally $0.4 \lesssim h \lesssim 1$). Thus today $|\sigma| \lesssim 0.2\,h^{-1}$.

15.2.2 Big-Bang Nucleosynthesis

An interesting constraint on the value of G_N in the early universe arises from the observed primordial light element abundances as first discussed by Barrow (1978).[94] In a Friedman-Robertson-Walker model of the universe the expansion rate is given by $H^2 = (\dot{R}/R)^2 = \frac{8\pi}{3} G_N \rho$ in terms of the energy density ρ which, during the epoch of nucleosynthesis, is dominated by radiation (photons, neutrinos). It is a standard

---

[94] Apparently there is an earlier discussion of this limit by G. Steigman in an unpublished essay for the 1976 Gravity Research Foundation Awards. Subsequent refinements include Rothman and Matzner (1982), Accetta, Krauss, and Romanelli (1990), Damour and Gundlach (1991), and Casas, García-Bellido, and Quirós (1992).

argument to constrain ρ from the yield of ^{4}He and other light elements, and thus to constrain the effective number of neutrino degrees of freedom at nucleosynthesis (Yang et al. 1979, 1984; Olive et al. 1990; Walker et al. 1991). Because the number of low-mass sequential neutrino families is now known to be 3, such constraints can be translated into constraints on the value of G_N pertaining to the nucleosynthesis epoch.

An extra neutrino species would add around 15% to ρ. It appears reasonably conservative to assume that big-bang nucleosynthesis does not allow for a deviation of the standard number of effective neutrino degrees of freedom by more than 1 so that G_N at that time must have been within about ±15% of its present-day value. It is possible, however, that G_N was considerably smaller if this reduction was compensated by additional exotic degrees of freedom such as right-handed neutrinos which increase ρ. Moreover, it was assumed that Fermi's constant had its present-day value at nucleosynthesis, contrary to the speculations discussed in Sect. 15.1. Still, barring fortuitous compensating effects, nucleosynthesis excludes an $\mathcal{O}(1)$ deviation of G_N from its standard value at nucleosynthesis.

This sort of result can be compared with the present-day bounds of Sect. 15.2.1 only by assuming a specific functional form for $G_N(t)$ which is often taken to be

$$G_N(t) = G_N(t_0)\,(t_0/t)^\beta, \tag{15.1}$$

where t_0 refers to the present epoch. Assuming that G_N at nucleosynthesis was within ±50% of its standard value one finds $|\beta| \lesssim 0.01$ which would imply $|\dot{G}_N/G_N|_{\text{today}} \lesssim 10^{-12}\,\text{yr}^{-1}$, at least a factor of ten below the present-day limits. One should keep in mind, however, that a power-law variation of G_N is a relatively arbitrary assumption. For example, in scalar-tensor extensions of general relativity such as the Brans-Dicke theory G_N varies as a power law during the matter-dominated epoch while it remains constant when radiation dominates. Either way, while the nucleosynthesis bounds are probably somewhat more restrictive than the celestial-mechanics ones it is interesting that the resulting bounds $|\dot{G}_N/G_N|_{\text{today}} \lesssim 1-10 \times 10^{-12}\,\text{yr}^{-1}$ are of the same general order of magnitude. They leave room for a considerable variation of G_N over cosmic time scales.

15.2.3 Properties of the Sun

If G_N did vary in time one would expect a modification of the standard course of stellar evolution as first stressed by Teller (1948). By means of a beautifully simple homology argument, Teller showed that the luminosity of the Sun is approximately proportional to $G_N^7 \mathcal{M}^5$ with $\mathcal{M}$ the solar mass which was also allowed to vary in the spirit of Dirac's (1937, 1938) large numbers hypothesis. Teller then proceeded to estimate the temperature on Earth in the past, taking a modification of its orbit from the G_N variation into account. Depending on whether G_N was larger or smaller in the past the average terrestrial surface temperature would have been larger or smaller. Teller estimated that if G_N fell outside $\pm 10\%$ of its standard value it would be unlikely that life on Earth could be sustained. Therefore, G_N should have remained constant to within this accuracy at least during the past 500 million years or more where life has been known to exist on Earth.[95]

Later, a similar homology argument based on the solar age was presented by Gamow (1967). Detailed models of the Sun with a varying G_N were constructed by Pochoda and Schwarzschild (1964), Ezer and Cameron (1966), Roeder and Demarque (1966), Shaviv and Bahcall (1969), Chin and Stothers (1975, 1976), Demarque et al. (1994), and Guenther et al. (1995). The crux with constraining $\dot{G}_N$ from the Sun is that the presolar helium abundance Y_{initial} and the mixing-length parameter α can and must be tuned to reproduce the Sun's present-day luminosity and radius. In Sect. 1.3.2 it became clear that even extreme anomalous energy-loss rates could be compensated by an adjustment of Y_{initial}; a similar effect pertains to variable G_N solar models. Even though the present-day central temperature, density, and helium abundance could differ vastly from standard predictions, their main impact would be on the neutrino flux which, however, is not a reliable probe of the solar central conditions as it may get modified by neutrino oscillations.

At the present time the most sensitive probe of a variant internal solar structure is afforded by the measured p-mode frequencies which agree well with standard predictions, especially when the gravitational settling of helium is taken into account. Demarque et al. (1994) have constructed solar models with a varying G_N and then analyzed their p-mode spectra in comparison with the observations. They assumed a G_N time variation of the form Eq. (15.1) with $t_0 = 15\,\mathrm{Gyr}$ for the age of

---

[95]Apparently there is more recent evidence for primitive lifeforms on Earth as early as 3.5 Gyr ago (e.g. Gould 1994).

Table 15.2. Characteristics of the Demarque et al. (1994) solar models with a varying G_N according to Eq. (15.1).

β	α	$X_{initial}$ [%]	X_c [%]	T_c [10^6 K]	ρ_c [g/cm^3]	R_{env} [$R_\odot$]	^{37}Cl [SNU]	^{71}Ga [SNU]
−0.4	1.832	69.42	45.6	15.14	125.5	0.739		
−0.2	1.895	70.10	42.1	15.28	134.0	0.731	5.5	117
−0.1	1.936	70.51	40.0	15.37	139.5	0.724		
0.0	1.983	70.98	37.6	15.47	146.2	0.721	6.8	124
0.1	2.036	71.51	34.8	15.58	154.6	0.716		
0.2	2.104	72.11	31.7	15.72	165.1	0.710	8.7	134
0.4	2.291	73.56	23.9	16.07	197.3	0.695		

the universe. Some characterisitics of their solar models as a function of β are summarized in Tab. 15.2 where α is the mixing-length parameter, $X_{initial}$ the presolar hydrogen abundance, the quantitites with index c refer to central conditions of the present-day Sun, and R_{env} to the radius of its convective envelope. The last two columns are the predicted counting rates in the chlorine and gallium solar neutrino experiments under the assumption that there are no neutrino oscillations.

The most important effect of the G_N variation is a shift of the base of the convective envelope which is caused by the required change of α and the initial helium abundance $Y_{initial}$. It is this modification of the convection zone which has the largest impact on the observable p-mode frequencies. However, an identical shift can be produced by other effects such as modified opacities, equation of state, surface boundary conditions, degree of gravitational helium settling, and perhaps by magnetic fields. Therefore, only relatively crude limits can be extracted at the present time. Demarque et al. (1994) believe that $|\beta| \lesssim 0.4$ is a reasonably conservative limit which probably can be improved by a factor of four within the next decade by more precise p-mode observations. With Eq. (15.1) the current limit corresponds to

$$|\dot{G}_N/G_N| \lesssim 30 \times 10^{-12} \, \text{yr}^{-1}, \tag{15.2}$$

similar to the celestial-mechanics bounds of Tab. 15.1. The precise functional form Eq. (15.1) is not crucial for the solar bound as it probes G_N only for the last 4.5 Gyr of the assumed 15 Gyr cosmic age. Therefore, one could have equally assumed a linear form for $G_N(t)$.

Most recently, a similar study was completed by Guenther et al. (1995) who focussed on the predicted g-mode spectrum. At the present time there is no generally accepted observation of solar g-modes (recall that they are evanescent in the convection zone). If one were to take the claimed observations by Hill and Gu (1990) seriously, a bound $|\beta| \lesssim 0.05$ would obtain. Therefore, if an unambiguous identification of g-modes would emerge from a number of forthcoming observational projects, the Sun may yet provide one of the most restrictive limits on the constancy of Newton's constant.

15.2.4 White Dwarfs

A large impact of a time-varying gravitational constant can be expected on the oldest stars which "integrate" $G_N(t)$ into the more distant past than does the evolution of the Sun. One well understood case are white dwarfs, the faintest of which likely formed shortly after the birth of the galactic disk. Therefore, the age of the galactic disk implied by the fast drop of the white-dwarf luminosity function at the faint end (Sect. 2.2.1) depends on the G_N evolution in the past.

In an early study Vila (1976) concluded on the basis of the observations then available that $\dot{G}_N/G_N$ as large as $75 \times 10^{-12} \, \text{yr}^{-1}$ was not excluded. García-Berro et al. (1995) constructed detailed luminosity functions under the assumption of a decreasing G_N. For an assumed age of the galactic disk of 7 Gyr, which probably is a lower plausible limit, the best fit for the faintest data point requires $\dot{G}_N/G_N = -10 \times 10^{-12} \, \text{yr}^{-1}$ while the curves for 0 and $30 \times 10^{-12} \, \text{yr}^{-1}$ lie somewhat outside of the 1σ error bar of this all-important data point. Still, the white-dwarf luminosity function does not seem to yield significant limits relative to the celestial-mechanics ones.

15.2.5 Globular Clusters

The oldest stellar objects in the galaxy are globular-cluster stars which are thus expected to yield the most restrictive stellar-evolution limits on $\dot{G}_N/G_N$. A color-magnitude diagram for an intermediate-aged galactic cluster was constructed by Roeder (1967) while detailed studies of globular clusters were performed by Prather (1976) and Degl'Innocenti et al. (1995). Roeder (1967) and Prather (1976) used a time variation for a specific Brans-Dicke cosmology where G_N decreases approximately as in Eq. (15.1) with $\beta \approx 0.03$ while Degl'Innocenti et al. (1995) considered more generic cases of $G_N(t)$.

The main impact of the assumed G_N variation is a change in the time it takes for a star to burn out hydrogen at its center and thus to leave the main sequence (MS). The subsequent fast evolution (ascending the RGB, HB evolution, etc.) is determined by the present-day value of G_N. The stellar evolutionary tracks in the color-magnitude diagram can look significantly different from the standard ones if the G_N variation was sufficiently severe. However, within the range of possibilities left open by the above $\dot{G}_N/G_N$ bounds, the present-day isochrone cannot be observationally distinguished from the standard case (Degl'Innocenti et al. 1995). Apparently, then, the only significant consequence of a time-varying gravitational constant is that the true age τ of a globular cluster is different from its apparent age τ_* which is inferred from its color-magnitude diagram in the framework of a constant-gravity scenario.

The change of the MS lifetime can be estimated by Teller's (1948) homology relation $L \propto G_N^\gamma$. Based on a specific assumption for the opacity variation with temperature and density Teller found $\gamma = 7$ while a more appropriate value for low-metallicity globular-cluster stars is $\gamma = 5.6$ (Degl'Innocenti et al. 1995). Either way, one can easily show that the true (τ) and apparent age (τ_*) at the MS turnoff are approximately related by

$$\tau_* = \int_{t_0-\tau}^{t_0} dt \left[G_N(t)/G_N(t_0) \right]^\gamma \tag{15.3}$$

(Prather 1976; Degl'Innocenti et al. 1995). For all practical purposes this analytic result can be considered to be exact because it agrees with numerical calculations surprisingly well.

Unless one wishes to probe the very early universe it is fairly generic to assume a linear G_N variation of the form

$$G_N(t) = \left[1 + \Gamma_0(t - t_0) \right] G_N(t_0), \tag{15.4}$$

where $\Gamma_0 = \dot{G}_N(t_0)/G_N(t_0)$ is the present-day rate of change of Newton's constant. Then one finds explicitly

$$\frac{\tau}{\tau_*} = \frac{\gamma_1 \Gamma_0 \tau}{1 - (1 - \Gamma_0 \tau)^{\gamma_1}} = \frac{1 - (1 - \gamma_1 \Gamma_0 \tau_*)^{1/\gamma_1}}{\Gamma_0 \tau_*}, \tag{15.5}$$

where $\gamma_1 \equiv \gamma + 1$. Given a present-day rate of change Γ_0 one can thus determine the modification of the globular-cluster age if a certain apparent age or a certain true age is assumed.

The observed color-magnitude diagrams of globular clusters yield apparent ages τ_* in the range 14 to 18 Gyr. With these values one

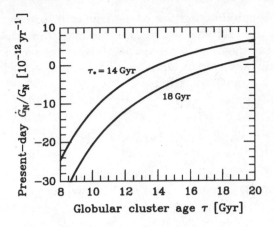

Fig. 15.1. Required present-day $\dot{G}_N/G_N$ in order to achieve a true globular-cluster age τ, given that the apparent age is τ_*. A linear $G_N(t)$ variation as in Eq. (15.4) was assumed.

can relate a desired true age τ to a required value for Γ_0 (Fig. 15.1). Conversely, it is probably safe to assume that the true ages of globular clusters do not exceed 20 Gyr. Then Fig. 15.1 implies that today

$$\dot{G}_N/G_N \lesssim 7 \times 10^{-12}\,\mathrm{yr}^{-1}. \tag{15.6}$$

A lower age limit is less certain. Taking 8 Gyr one finds $\dot{G}_N/G_N \gtrsim -35 \times 10^{-12}\,\mathrm{yr}^{-1}$ which is less certain, and also less interesting relative to the limits discussed in the previous sections.

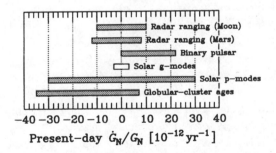

Fig. 15.2. Summary of limits on the present-day $\dot{G}_N/G_N$. The big-bang nucleosynthesis limit is not shown as it depends sensitively on the assumed $G_N(t)$ variation at early cosmic times.

Actually, a certain reduction of the true globular-cluster ages relative to their apparent ones would be a welcome cosmological effect as they are, at best, marginally compatible with other cosmic age indicators. In view of the current limits on $\dot{G}_N/G_N$ summarized in Fig. 15.2 this possibility cannot be excluded at present.

15.3   Test of the Equivalence Principle

The equivalence principle of Einstein's general theory of relativity implies that the space-time trajectories of relativistic particles should be independent of internal degrees of freedom such as spin or flavor, and independent of the type of particle under consideration (photons, neutrinos). A number of astronomical observations allow one to test this prediction. Laboratory tests of various consequences of the equivalence principle are discussed in Will's (1993) book.

Nonsymmetric extensions of general relativity (e.g. Moffat 1991) predict that different polarization components of electromagnetic waves propagate with different phase velocities in gravitational fields. This birefringence effect would lead to the depolarization of the Zeeman components of spectral lines emitted in magnetically active regions of the Sun. The absence of this depolarization effect leads to significant constraints on Moffat's theory and others (Gabriel et al. 1991).

In a similar approach one uses the difference of the Shapiro time delay between different particles or between different polarization states of a given particle which propagate through the same gravitational field. In Sect. 13.3 the absence of an anomalous shift between the SN 1987A photon and neutrino arrival times gave limits on violations of the equivalence principle because both pulses moved through the same galactic gravitational potential.

Also, one may search for differences in the arrival times of left- and right-handed polarized electromagnetic signals from distant pulsars (LoSecco et al. 1989). The best bound was obtained from an analysis of the pulse arrival times from PSR 1937+21 which is about 2.5 kpc away from Earth. One may write the effective gravitational potential in the form $V(\mathbf{r}) = V_0(\mathbf{r})\left[1 + A_1\boldsymbol{\sigma}\cdot\hat{\mathbf{r}} + A_2\boldsymbol{\sigma}\cdot\mathbf{v} + A_3\hat{\mathbf{r}}\cdot(\mathbf{v}\times\boldsymbol{\sigma})\right]$ where $\mathbf{r}$, $\mathbf{v}$, and $\boldsymbol{\sigma}$ represent the location, velocity, and spin of the particles (photons, neutrinos). The PSR 1937+21 data then yield a constraint $|A_1| < 4\times10^{-12}$ and $|A_2| < 1\times10^{-12}$ (Klein and Thorsett 1990), apparently the most restrictive limits of their kind.

A violation of the equivalence principle could also manifest itself by a relative shift of the energies of different neutrino flavors in a gravitational field. For a given momentum p the matrix of energies in flavor space (relativistic limit) is $E = p + M^2/2p + 2p\phi(\mathbf{r})(1 + F)$ where M^2 is the squared matrix of neutrino masses, $\phi(\mathbf{r})$ is the Newtonian gravitational potential, and F is a matrix of dimensionless constants which parametrize the violation of the equivalence principle; in general relativity $F = 0$. A nontrivial matrix F can lead to neutrino oscillations in analogy to the standard vacuum oscillations which are caused by the matrix M^2 (Gasperini 1988, 1989; Halprin and Leung 1991; Pantaleone, Halprin, and Leung 1993; Iida, Minakata, and Yasuda 1993; Minakata and Nunokawa 1995; Bahcall, Krastev, and Leung 1995). Values for F_{ij} in the general 10^{-14}–10^{-17} range could account for the solar neutrino problem and perhaps could be probed with future long-baseline oscillation experiments.

15.4   Photon Mass and Charge

Even though in classical electrodynamics gauge invariance implies that photons must be massless, quantum electrodynamics (QED) can be formulated consistently with the inclusion of a photon mass, and the limit $m_\gamma \to 0$ takes the modified theory smoothly over to massless QED (Stückelberg 1941). Therefore, the possibility of a small photon mass cannot be excluded theoretically; limits must be set by laboratory and astrophysical methods. A still up-to-date review of the laboratory limits was given by Goldhaber and Nieto (1971); the best is $m_\gamma \lesssim 10^{-14}\,\mathrm{eV}$ from a test of Coulomb's law (a photon mass would modify the inverse-square behavior). More recent experiments worked at low temperature (Ryan, Accetta, and Austin 1985; Chernikov et al. 1992); the resulting limits on m_γ are relatively weak, however.

An astrophysical limit may be set by the absence of an anomalous dispersion of photon signals from distant sources, notably the pulsed signal from radiopulsars. This method is limited by the presence of the ionized interstellar medium. It causes a dispersion relation for photons which mimics the effect of a photon mass $m_\gamma = \omega_P$ where the plasma frequency is given by $\omega_P^2 = 4\pi\alpha n_e/m_e$. With a typical electron density n_e of order $0.1\,\mathrm{cm}^{-3}$ the photon plasma mass is of order $10^{-11}\,\mathrm{eV}$ so that a vacuum mass much smaller than this value cannot be probed.

In Sect. 13.3.3 a limit on a hypothetical ν_e charge was derived from the absence of a dispersion of the SN 1987A neutrino pulse. The path of charged particles in the galactic magnetic field would be curved, leading to an energy-dependent time-delay (Barbiellini and Cocconi 1987). Because the same argument can be applied to photons, the signals from radio pulsars also allow one to set a limit on a putative photon electric charge (Cocconi 1988). However, the resulting dispersion effect scales with photon frequency in the same way as the effect caused by a photon mass or by the plasma effect so that this method, again, is limited by the standard dispersion effect (Raffelt 1994). One finds a bound on the photon charge of $Q_\gamma \lesssim 10^{-29}\,e$.

Returning to a hypothetical photon mass, its value can be extracted, in principle, from the spatial distribution of static magnetic fields of celestial bodies. The measured fields can be fitted by an appropriate multipole expansion in which m_γ is kept as a free parameter. The most restricitve limit of this sort was derived from Jupiter's magnetic field on the basis of the Pioneer-10 observations; Davis, Goldhaber, and Nieto (1975) found a limit $m_\gamma \lesssim 0.6 \times 10^{-15}\,\mathrm{eV}$. The same method applied to the Earth's magnetic field yields an almost equivalent bound of $m_\gamma \lesssim 0.8 \times 10^{-15}\,\mathrm{eV}$ (Fischbach et al. 1994).

As detailed in a review by Barrows and Burman (1984) more restrictive limits obtain from detailed considerations of astrophysical objects in which magnetic fields, and hence the Maxwellian form of electrodynamics, play a key role in maintaining equilibrium or creating long-lived stable structures. The most restrictive such limit of $m_\gamma \lesssim 10^{-27}\,\mathrm{eV}$ is based on an argument by Chibisov (1976) concerning the magneto-gravitational equilibrium of the gas in the Small Magellanic Cloud which requires that the range of the interaction exceeds the characteristic field scale of about 3 kpc. This limit, if correct, is surprisingly close to $10^{-33}\,\mathrm{eV}$ where the photon Compton wavelength would exceed the radius of the observable universe and thus would cease to have any observable consequences.

15.5   Free Quarks

It is thought that quarks cannot exist as free particles; they always occur bound in hadrons which are neutral ("white") with regard to the "color charge" of the strong interaction. In order to test this hypothesis of confinement it remains an important task to search for single quarks. The observation of fractional charges in the experiment of LaRue, Phillips, and Fairbanks (1981) has never been confirmed. However, if their observations were caused by unconfined quarks bound to nuclei it would correspond to an abundance of one quarked nucleus (Q-nucleus) in 6×10^{17} normal ones.

A small abundance of Q-nuclei could significantly alter the stellar thermonuclear reaction chains and among other effects change the solar neutrino predictions (Boyd et al. 1983). Detailed nuclear reaction chains involving Q-nuclei were studied by Boyd et al. (1985). For strangelets (lumps of strange quark matter) trapped in stars the nuclear networks were investigated by Takahashi and Boyd (1988); the effect of strangelets is similar to that of Q-nuclei. The Q-nuclear reactions were implemented in a stellar evolution code by Joseph (1984). Predictions for the solar neutrino flux were worked out by Sur and Boyd (1985).

With a Q-nuclear abundance of order 10^{-15} the modified reaction chains would compete with the standard ones. In the Sun one could achieve a reduction of the high-energy solar neutrino flux and thus solve the "old solar neutrino problem" (missing boron neutrinos). However, Sur and Boyd (1985) predict an increase of the low-energy flux, corresponding to a substantially increased counting rate at the gallium solar neutrino experiments. As this contradicts the findings of SAGE and GALLEX (Sect. 10.3) one concludes that Q-nuclear burning is not the answer to the solar neutrino problem. Turning the SAGE/GALLEX observations around one concludes that in the Sun the abundance of Q-nuclei is below about 10^{-15}.

15.6 Supersymmetric Particles

Supersymmetric extensions of the particle-physics standard model are very popular, among other reasons because the lightest supersymmetric particle (LSP) could play the role of the cosmic dark matter. In these models, there is a fermionic partner to all standard bosons, and a bosonic partner to all standard fermions. The supersymmetric partners of the photon, the Z° gauge boson, and the neutral Higgs boson (photino, Zino and Higgsino) would be Majorana fermions. They would be very much like Majorana neutrinos except that their interaction strength is not fixed by the Fermi constant but rather depends on details of the supersymmetric models.

If these "neutralinos" had low enough masses they would be produced in the interior of stars by the same processes that create neutrinos, except that the coupling strength has to be adjusted according to the particular model that one has in mind. Limits to an anomalous energy loss of stars yielded early constraints on supersymmetric models

(Bouquet and Vayonakis 1982; Fukugita and Sakai 1982; Anand et al. 1984). The neutrino burst of SN 1987A yielded more interesting limits for the case of low-mass photinos (Ellis et al. 1988; Grifols, Massó, and Peris 1989; Grifols and Massó 1990b). To avoid that too much energy is carried away by photinos they inferred that squark masses in the approximate range 60 GeV to 2.5 TeV were excluded.

Low-mass neutralinos are disfavored by laboratory limits. Moreover, if the LSP plays the role of cold dark matter its mass likely is above several 10 GeV. In this case the stellar energy-loss arguments would not yield any constraints as all supersymmetric particles would be too heavy to be emitted. Stars would still play an interesting role as they could trap the dark-matter particles. Their annihilation in the Sun or Earth would lead to a high-energy neutrino signal which has been constrained by the Kamiokande detector (Mori et al. 1992). It may well be found at the Cherenkov detectors Superkamiokande, NESTOR, DUMAND, or AMANDA and thus lead to the indirect discovery of particle dark matter in the galaxy. These important issues are discussed at length in the forthcoming review *Supersymmetric Dark Matter* by Jungman, Kamionkowski, and Griest (1995).

15.7   Majorons

15.7.1   Particle-Physics and Cosmological Motivations

Axions (Chapter 14) are one representative of a variety of Nambu-Goldstone bosons of spontaneously broken global symmetries that have appeared in the literature over the years. Another widely discussed example are the majorons first introduced by Chicashige, Mohapatra, and Peccei (1981) as a scheme to generate small neutrino Majorana masses. An important variation by Gelmini and Roncadelli (1981) and Georgi, Glashow, and Nussinov (1981) led to a model where neutrinos had small Majorana masses and coupled to the massless majoron (a pseudoscalar boson like the axion) with a relatively large Yukawa strength. The main phenomenological interest in this sort of conjecture lies in the intriguing possibility that neutrinos could have relatively strong interactions with the majorons and with each other by virtue of majoron exchange. As majorons would not necessarily show up in interactions with ordinary matter one could well speculate that neutrinos might have "secret interactions" which would be of relevance only in a neutrino-dominated environment such as the early universe, perhaps the present-day universe if neutrinos have a cosmologically significant mass, and in supernovae. Another motivation to consider majoron models is the possibility to account for fast neutrino decays in order to avoid cosmological neutrino mass bounds. In the laboratory, majorons could show up in experiments searching for neutrinoless 2β decays.

The main motivation for the introduction of majorons is the puzzling smallness of neutrino masses (if they have nonvanishing masses at all) relative to other fermions. As outlined in Sect. 7.1, in the particle-physics standard model it is thought that all Dirac fermions acquire a mass by their interaction with a background Higgs field which takes on a classical value (vacuum expectation value) Φ_0 everywhere; neutrino masses could well arise in the same fashion, except that the Yukawa couplings to the Higgs field would have to be extremely small. Alternatively, one may speculate that neutrino masses are so small because they arise in a different fashion. Notably, the known sequential neutrinos ν_e, ν_μ, and ν_τ could well be Majorana fermions, i.e. their own antiparticles so that the $\bar\nu_e$ is really equivalent to a helicity-plus ν_e. As long as neutrinos are massless this picture is equivalent to an interpretation where the standard left-handed neutrinos are the two active components of a four-component Dirac spinor while the two remaining sterile components would never have been observed because they do not interact. With a nonvanishing mass these interpretations are vastly different because helicity flips in collisions would allow one to produce the (almost) sterile "wrong-helicity" Dirac components. This possibility was exploited in Sect. 13.8.1 to set bounds on a neutrino Dirac mass from the SN 187A neutrino signal. For Majorana neutrinos, a helicity-flipping collision takes an active ν_e into an active $\bar\nu_e$, thus violating lepton number by two units. Therefore, Majorana masses could not arise from the coupling to the standard Higgs field which is lepton-number conserving.

Majorana masses could arise, however, by interacting with a different Higgs field which would develop a vacuum expectation value by virtue of the usual spontaneous breakdown of a global symmetry. The resulting Nambu-Goldstone boson is the majoron. (Recall that the Nambu-Goldstone boson of the standard Higgs field shows up as the third polarization degree of the massive Z° gauge boson so that there is no massless Nambu-Goldstone degree of freedom in the standard model.) In the original model of Chicashige, Mohapatra, and Peccei (1981), the "singlet majoron model," the new vacuum expectation value was considered to be much larger than the standard $\Phi_0 \approx 250\,\mathrm{GeV}$. Large masses would be given primarily to sterile neutrinos postulated to exist; the standard sequential neutrinos would obtain their small

560

Chapter 15

masses by a see-saw type mixing effect with the heavy states. The majoron coupling to standard neutrinos would be extremely small in this model, leading to no interesting consequences besides small Majorana masses for ν_e, ν_μ, and ν_τ.

Gelmini and Roncadelli (1981) and Georgi, Glashow, and Nussinov (1981) suggested instead to do away with the unobserved heavy sterile neutrinos and give a small Majorana mass directly to the sequential neutrinos by the interaction with the new Higgs field. The intriguing feature of this model is that it requires a very small vacuum expectation value v, perhaps in the keV regime. As all couplings of the new Higgs field to fermions scale with the inverse of v, the majoron would have a rather strong coupling to neutrinos. Among many fascinating phenomenological and astrophysical consequences (e.g. Georgi, Glashow, and Nussinov 1981; Gelmini, Nussinov, and Roncadelli 1982) this model predicted, however, that the new Higgs field should contribute precisely the equivalent of two massless neutrino species to the Z° decay width. The measurements of this width at CERN and SLAC in 1989–1990, however, correspond exactly to the known three neutrino flavors (Particle Data Group 1994), leaving no room for this "triplet majoron model." Other "doublet majoron models" which would contribute one-half of an effective neutrino species to the Z° decay width are also excluded (for references see, e.g. Berezhiani, Smirnov, and Valle 1992). It is possible, however, to construct majoron models for Majorana neutrino masses which retain the original idea of Chicashige, Mohapatra, and Peccei (1981) and yet provide large majoron-neutrino couplings (e.g. Berezhiani, Smirnov, and Valle 1992 and references therein; see also Burgess and Cline 1994a; Kikuchi and Ma 1994, 1995).

The main motivation for going out of one's way to construct such models does not arise from particle theory but rather from experiments and astrophysics. In Sect. 7.1.4 it was outlined that those nuclei which decay predominantly by a double beta channel (emission of $2e^-$ and $2\bar\nu_e$) can also decay in a neutrinoless mode if ν_e has a Majorana mass, allowing an emitted $\bar\nu_e$ to be effectively reabsorbed as a ν_e. In majoron models of Majorana neutrino masses there is a third decay channel where the intermediate ν_e in the 0ν mode radiates a majoron so that effectively $2e^-$ plus one majoron χ are emitted. The expected sum spectrum of the electron energies would be continuous as in the 2ν mode, but with a different spectral shape. Once in a while, experiments which search for the 0ν mode (a sharp endpoint peak of the $2e^-$ sum spectrum) have reported a continuous spectral signature which allegedly could not be ascribed to the dominant 2ν mode or other backgrounds, although such claims have tended to disappear with the collection of more significant data. If interpreted in terms of an upper limit to the majoron-ν_e Yukawa coupling, current experiments give about $g \lesssim 2 \times 10^{-4}$ (Beck et al. 1993 and references therein). At the present time there does not appear to exist a compelling signature for majorons in any of the experiments, although several of them seem to find certain spectral anomalies—see, e.g. Burgess and Cline (1993, 1994b) for an overview and references.

If the spectral anomalies were to represent the first evidence of majoron emission in 2β decays, the Yukawa coupling would have to be near the 10^{-4} level. The neutrino-majoron coupling is given as $g = m_\nu/v$ in terms of the neutrino mass and the symmetry breaking scale v. With $m_{\nu_e} \lesssim 1\,\mathrm{eV}$ for a Majorana mass from measured limits on the 0ν decay mode one finds the requirement $v \lesssim 10\,\mathrm{keV}$ which is an extremely small scale of symmetry breaking. The majoron is the "angular degree of freedom" of a complex scalar field; the "radial degree of freedom," often referred to as the ρ field, has a mass typically of order v. Therefore, a low-mass scalar particle beyond the majoron would appear in the low-energy sector of the theory.

One severe limitation on such models is provided by big-bang nucleosynthesis which has been widely used to set limits on additional low-mass degrees of freedom which are thermally excited during the epoch of nucleosynthesis; an upper limit of about 0.3 is often quoted as the maximum allowed extra contribution in units of effective neutrino degrees of freedom (e.g. Walker et al. 1991). On the face of it, this limit excludes majoron models where the majoron (and possibly the ρ) interact sufficiently strongly with neutrinos to reach thermal equilibrium. In one recent study Chang and Choi (1994) found that one must require $g \lesssim 10^{-5}$ for the largest majoron Yukawa coupling to any neutrino species in order to avoid thermalization of the majorons before nucleosynthesis.

Such constraints rely on the assumption of 3 standard light neutrino species being in thermal equilibrium at the epoch of nucleosynthesis. However, in majoron models the usual cosmological neutrino mass bound does not apply because of the possibility of fast decays of the type $\nu \to \nu' \chi$ (majoron χ) which are induced by flavor off-diagonal Yukawa couplings. Therefore, ν_τ's may have a mass of, say, a few MeV and may have disappeared by decays and annihilations before nucleosynthesis, thus making room for majorons; for detailed numerical studies see Kawasaki et al. (1994) and papers quoted there. Such heavy, short-lived ν_τ's may provide interesting effects on scenarios of galaxy formation and thus could be a novel ingredient for cold dark matter cosmological models (Dodelson, Gyuk, and Turner 1994).

Even if in the long run the 2β experiments do not yield any compelling evidence for majoron emission, one may consider a decaying-neutrino cosmology as a motivation in its own right for majoron models. Such cosmologies may explain the discrepancy between the cosmic density fluctuation spectrum inferred from the cosmic microwave background and from galaxy correlations which persists in a purely cold dark matter cosmology (Bond and Efstathiou 1991; Dodelson, Gyuk, and Turner 1994; White, Gelmini, and Silk 1995). Another neutrino-related explanation of this discrepancy is a hot plus cold dark matter cosmology which involves neutrinos with a mass of a few eV.

In summary, the simplest majoron model which implied large couplings to neutrinos (the Gelmini-Roncadelli model) is experimentally excluded although one can construct more complicated ones which retain sizeable neutrino-majoron couplings and yet are compatible with the Z° decay width. The possibility of such models is entertained because 2β decay experiments may yet turn up compelling evidence for majoron decays, and because certain cosmological models of structure formation may be taken to suggest massive, decaying neutrinos.

15.7.2   Majorons and Stars

Majorons could also have an impact on stellar evolution. Besides interacting with neutrinos, they typically also couple to other fermions, allowing one to apply the astrophysical bounds on pseudoscalars derived throughout this book to majoron models.

Because of the possibility of fast decays $\nu \to \nu'\chi$ the neutrino signal from distant sources, notably from the Sun or from SN 1987A, would be affected. Such decays are not likely to be able to explain the solar neutrino problem as discussed in Sect. 10.8. It remains interesting, however, that the matter-induced ν_e-$\bar{\nu}_e$ energy splitting allows for decays $\nu_e \to \bar{\nu}_e\chi$ (Sect. 6.8). Should a solar $\bar{\nu}_e$ flux show up in future measurements, it could be an indication for such decays.

As for SN neutrinos, the decay of massive ν_μ's or ν_τ's with final-state $\bar{\nu}_e$'s could modify the $\bar{\nu}_e$ signal observed in a detector, notably the energy distribution and duration of the observed pulse (Soares and Wolfenstein 1989; Simpson 1991). The SN 1987A data have not been analyzed in detail with regard to this possibility, although the observed signal can be accounted for without invoking such effects. It is not clear if one could derive significant constraints on majoron models from

SN 1987A on the basis of this decay argument. Aharonov, Avignone, and Nussinov (1988a) predicted for certain parameters a dramatic increase of the number of observable events from the prompt ν_e burst.

One interesting SN 1987A limit is based on the interaction of the pulse of observed $\overline{\nu}_e$'s with the cosmic majoron background that would be expected to exist for majorons which thermalized in the early universe. In order not to deplete the pulse too much by collisions, Kolb and Turner (1987) found a certain upper limit on the majoron-neutrino Yukawa coupling. Unfortunately, their bound was based on an incorrect cross section for the process $\nu\chi \rightarrow \nu\chi$. They used a pseudoscalar coupling of the form $ig\overline{\psi}_\nu\gamma_5\psi_\nu\,\chi$ rather than a derivative coupling $(1/2v)\,\overline{\psi}_\nu\gamma_\mu\gamma_5\psi_\nu\,\partial^\mu\chi$ where $g = m_\nu/v$ and v is the majoron symmetry breaking scale. As discussed in Sect. 14.2.3, in processes which involve two Nambu-Goldstone bosons attached to one fermion it is mandatory to use the derivative coupling in order to obtain the correct interaction rate. The pseudoscalar coupling yields a scattering cross section $(g^4/64\pi)\,s^{-1}$ times an expression of order unity which depends on the neutrino mass and s, the squared CM energy. Put another way, the cross section is $(m_\nu/v)^4\,s^{-1}$ times numerical factors. The derivative coupling, on the other hand, leads to a cross section m_ν^2/v^4 times numerical factors (Choi and Santamaria 1990). Therefore, Kolb and Turner's (1987) bound translates approximately into $g\,(\mathrm{eV}/m_\nu)^{1/2} \lesssim 3\times10^{-4}$. It is not a bound on g alone.

Typical energies of neutrinos and other particles which prevail in the interior of a SN core are in the range of tens to hundreds of MeV. In majoron models with a symmetry breaking scale below this range, the symmetry may be restored in the interior of the star, and both components of the complex Higgs field will be thermally excited. Quick deleptonization may be achieved by reactions involving the majoron field, leading to a high-entropy collapse. Substantial majoron emission may shorten the SN 1987A $\overline{\nu}_e$ signal too much, although decays of heavier neutrinos in flight could, perhaps, provide the late events observed in the detectors. SN scenarios involving majorons were discussed by a number of authors.[96] There is little doubt that majoron models will have an important impact on SN physics for Yukawa couplings somewhere in the range 10^{-6}–10^{-3}. From the available literature, however,

---

[96]Kolb, Tubbs, and Dicus (1982); Dicus, Kolb, and Tubbs (1983); Manohar (1987); Fuller, Mayle, and Wilson (1988); Aharonov, Avignone, and Nussinov (1988b, 1989); Choi et al. (1988); Grifols, Massó, and Peris (1988); Konoplich and Khlopov (1988); Dicus et al. (1989); Berezhiani and Smirnov (1989); Choi and Santamaria (1990).

the present author has not been able to develop a clear view of the precise range of parameters that can be ruled out or ruled in by the SN 1987A neutrino signal.[97]

15.8   Millicharged Particles

It is commonly assumed that all particles have charges in multiples of $\frac{1}{3} e$ (electron charge); notably neutrinos are thought to be electrically neutral. While gauge invariance and anomaly cancellation constraints pose limits on the possible charge assignments in the standard model, electric charge quantization is not entirely assured. Two of the three neutrino species may carry small charges if one gives up the assumption that the three fermion families differ only in the mass of their members (Takasugi and Tanaka 1992; Babu and Volkas 1992; Foot, Lew, and Volkas 1993). Moreover, if one allows for charge nonconservation, all neutrinos could have small charges (Babu and Mohapatra 1990; Maruno, Takasugi, and Tanaka 1991). Finally, the existence of novel particles with small electric charges is possible and actually motivated by certain models involving a "mirror sector" where the mirror symmetry is slightly broken (Holdom 1986; see also Davidson, Campbell, and Bailey 1991). Therefore, it is interesting to study the experimental, astrophysical, and cosmological bounds on the existence of particles with small electric charge.[98]

The most severe constraints obtain for nonstandard charge assignments in the first family of quarks and leptons. The most model-independent charge limit on ν_e was derived from the absence of an anomalous dispersion of the SN 1987A neutrino signal (Sect. 13.3.3)

---

[97] An incomplete list of issues that ought to be considered in a study of majorons in SNe are the following. If the symmetry is broken within the SN core, a derivative majoron coupling to neutrinos should be used instead of a pseudoscalar one (Choi and Santamaria 1990). For light neutrinos, the medium-induced dispersion relation may dominate the cross section result. Besides the medium-induced processes $\nu \to \bar{\nu}\chi$, the process $\chi \to \bar{\nu}\nu$ should be included (Sect. 6.8). The decay of heavy neutrinos outside of the SN could contribute to the measurable signal. The trapping of neutrinos and majorons due to reactions with each other should be properly understood along the lines discussed by Dicus et al. (1989); it is dubious, for example, that a process like $\chi\chi \to \chi\chi$ really contributes to the majoron opacity. The effect of majorons during the infall phase must be understood, especially the possibility of early deleptonization.

[98] Such studies were performed by Dobroliubov and Ignatiev (1990), Davidson, Campbell, and Bailey (1991), Babu and Volkas (1992), Mohapatra and Nussinov (1992), and Davidson and Peskin (1994).

which led to (Barbiellini and Cocconi 1987; Bahcall 1989)

$$e_{\nu_e} \lesssim 3\times 10^{-17}\, e. \tag{15.7}$$

If electric charge conservation is assumed to hold in β processes such as neutron decay, one finds a more restrictive limit of

$$e_{\nu_e} \lesssim 3\times 10^{-21}\, e. \tag{15.8}$$

It is based on a limit for the neutron charge of $e_n = (-0.4\pm 1.1)\times 10^{-21}e$ (Baumann et al. 1988) and on the neutrality of matter which was found to be $e_p + e_e = (0.8\pm 0.8)\times 10^{-21}e$ assuming a vanishing neutron charge (Marinelli and Morpurgo 1984).

The deflection of charged neutrinos in the toroidal magnetic field in the solar convection zone would modify the observable flux at Earth (Ignatiev and Joshi 1994, 1995). However, in view of the above limits an unrealistically large field gradient is required to obtain significant flux modifications.

Babu and Volkas (1992) derived a limit on the ν_μ electric charge from the measured $\nu_\mu e$ cross section which would receive a contribution from photon exchange,

$$e_{\nu_\mu} \lesssim 10^{-9}\, e. \tag{15.9}$$

Similar limits could be derived for ν_e.

The following arguments apply to millicharged neutrinos or novel particles alike. They would appear as virtual states in higher-order amplitudes. For example, they would contribute to the anomalous magnetic moment of electrons and muons, and to the Lamb shift between the $2P_{1/2}$ and $2S_{1/2}$ states of the hydrogen atom. Of these quantitities, the Lamb shift gives the most restrictive limit (Davidson, Campbell, and Bailey 1991),

$$e_x < 0.11 e\, m_x/\text{MeV}, \tag{15.10}$$

where e_x and m_x are the charge and mass of the millicharged particle, respectively. This result applies to $m_x \gtrsim 1\,\text{keV}$.

Davidson, Campbell, and Bailey (1991) have reviewed more restrictive bounds from a host of accelerator experiments (Fig. 15.3).

A simple astrophysical constraint is based on avoiding excessive energy losses of stars which can produce millicharged particles by various reactions, most notably the plasma decay process. To avoid an unacceptable delay of helium ignition in low-mass red giants, and to avoid an

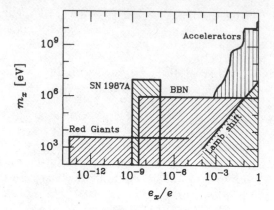

Fig. 15.3. Summary on limits on the electric charge e_x and mass m_x of generic millicharged particles which may be sequential neutrinos or novel particles. (Adapted from Davidson, Bailey, and Campbell 1991.) In order to avoid overclosing the universe, additional model-dependent parameter regions are excluded. The big-bang nucleosynthesis (BBN) excluded region is larger in some models.

undue shortening of the lifetime of horizontal-branch stars, one needs to require (Sect. 6.5.6)

$$e_x \lesssim 2 \times 10^{-14} e. \tag{15.11}$$

It applies for $m_x \lesssim \omega_P/2$ with ω_P the plasma frequency. It is larger for red giants before helium ignition than for HB stars because of the larger average density of $2 \times 10^5 \, \mathrm{g\,cm^{-3}}$ which corresponds to $\omega_P \approx 8.6 \, \mathrm{keV}$. The white-dwarf luminosity function yields about the same limit.

If e_x exceeds around $10^{-8}e$ the mean free path of the millicharged particles will be less than the physical size of a white dwarf or red-giant core. For larger e_x the particles will be trapped and contribute to the transfer of energy. Their impact on stellar evolution will become negligible when e_x is so large that other forms of energy transfer (photon radiation, convection) are more important. Because the new particles act essentially as radiation their mean free path must be less than that of photons, or very crudely, their charge must be of order an electron charge. (The main opacity source is probably Coulomb scattering on charged particles.) Therefore, if their mass is below a few keV, even the properties of the Sun would imply that there is not an allowed range of large e_x on the trapping side of the red-giant argument, except perhaps for e_x so large that it is excluded by experimental arguments.

For a narrow range of charges, these limits can be extended to larger masses by the SN 1987A cooling argument (Sect. 13.8.4). Accordingly,

$$10^{-9}e \lesssim e_x \lesssim 10^{-7}e \qquad (15.12)$$

is excluded for m_x up to several MeV, perhaps up to 10 MeV. On the trapping side of this range (large e_x) these particles surely would have an important impact on SN physics even though they cannot be excluded on the basis of the simple cooling argument.

If millicharged particles reach thermal equilibrium in the early universe before nucleosynthesis they contribute to the energy density and thus to the expansion rate. If they are one of the sequential neutrinos, this means that the right-handed degrees of freedom of that species are excited (they must be Dirac particles!), adding an effective neutrino degree of freedom. If they are nonneutrinos, even more energy is contributed, depending on their spin degrees of freedom. Even one additional effective neutrino degree of freedom is excluded and so for any millicharged particle Davidson, Campbell, and Bailey (1991) found

$$e_x \lesssim 3 \times 10^{-9}e \qquad (15.13)$$

if $m_x \lesssim 1$ MeV. In certain models where the millicharged particles are associated with a shadow sector, more stringent limits apply (Davidson and Peskin 1994).

Additional regions in the mass-charge plane can be excluded by the requirement that the novel objects do not overclose the universe. However, these arguments depend on the annihilation cross section in the early universe so that one needs to know all of their interactions apart from the millicharge. It is hard to imagine novel particles which interact *only* by their small electric charge! For certain specific cases the excluded regime was derived by Davidson, Campbell, and Bailey (1991) and Davidson and Peskin (1994).

All of these constraints leave the possibility open that ν_τ's have a mass in the 1–24 MeV range and a charge in the $10^{-5} - 10^{-3}e$ range. Then they would annihilate sufficiently fast before nucleosynthesis to actually *reduce* their effective contribution to the expansion rate (Foot and Lew 1993). In the standard model with small neutrino charges, however, *two* sequential neutrino species must carry a millicharge of equal but opposite magnitude (Babu and Volkas 1992; Takasugi and Tanaka 1992). Because the large charges required for Foot and Lew's scenario are excluded for ν_e and ν_μ one would need to require that only ν_τ carries a relatively large charge, forcing one to espouse even more exotic particle-physics models.

Chapter 16

Neutrinos: The Bottom Line

Most of the particle-physics arguments discussed in this book are closely related to neutrino physics because these particles play an important role in stellar evolution whether or not they have nonstandard properties. Besides a summary of some recent developments of standard-neutrino astrophysics, a synthesis is attempted of what stars as neutrino laboratories have taught us about these elusive objects, and what one might reasonable hope to learn in the foreseeable future.

16.1 Standard Neutrinos

The main theme of this book has been an attempt to extract information about the properties of neutrinos and other weakly interacting particles from the established properties of stars. However, even standard-model neutrinos (massless, no mixing, no exotic properties) play a significant role in astrophysics. There have been some recent developments in "standard-neutrino astrophysics" which deserve mention in a summary.

It is now thought that neutrinos play an active role in supernovae besides carrying away the binding energy of the newborn neutron star (Chapter 11). In the delayed-explosion scenario they are crucial to revive the stalled shock wave which is supposed to expel the stellar mantle and envelope. Moreover, they have a strong impact on r-process nucleosynthesis which is thought to occur in the high-entropy region above the neutron star a few seconds after collapse. For both purposes it is crucial to calculate the SN "neutrino lightcurve" for the first seconds after collapse. Convection below the neutrino sphere and large-scale convective turnovers in the region between the neutron star and the shock wave are both important and need to be understood better on the basis of 2- and 3-dimensional hydrodynamic calculations which are only beginning to appear in the literature.

In addition, however, the neutrino opacities must be calculated with greater reliability. In Chapter 4 the weaknesses of a naive calculation of the axial-vector opacities have been amply demonstrated. In my opinion, a far better understanding of the interaction rates of neutrinos with a hot nuclear medium is required before one can calculate a SN neutrino lightcurve (notably the duration of Kelvin-Helmholtz cooling) with a reasonable precision.

Neutrinos presumably play a key role for the self-acceleration of neutron stars which are observed to have huge "kick velocities" (Sect. 11.5). The required anisotropic neutrino emission of 1–2% may be caused by temperature fluctuations on the neutrino sphere (convection!), or by magnetic-field induced anisotropies of the neutrino opacities. Of course, other phenomena may be responsible for anisotropic neutrino emission and for the kick velocities.

For the first time ever neutrinos have been observed from a collapsing star (SN 1987A), confirming the expected signal behavior within the large uncertainties caused by the small number of observed events (Chapter 11).

Neutrinos are routinely observed from the Sun—currently in four different detectors with three different spectral response characteristics (Chapter 10). At least three further detectors will soon take up operations. The measured solar neutrino spectrum differs significantly from theoretical predictions. This "solar neutrino problem" has no obvious "astrophysical solution" that would involve plausible variations of input parameters such as nuclear cross sections or photon opacities. Still, the gallium detectors SAGE and GALLEX have for the first time measured the dominant low-energy flux of pp neutrinos, confirming that the Sun cannot be completely different at its center from what had been thought.

There is also some more benign news. The interest of some authors to apply the methods of finite temperature field theory to astrophysical problems has led to a new formulation of the photon dispersion relation in a plasma (Sect. 6.3). It helped to correct some "fossilized errors" in the literature on plasma neutrino emission (see also Appendix C.1). It may well be worthwhile to scrutinize other standard aspects of stellar-evolution input physics that involve subtle dispersion or screening effects.

16.2   Minimally Extended Standard Model

16.2.1   Cosmological Mass Limit for All Flavors

Neutrinos with nonstandard properties would have more radical implications in astrophysics. A minimal and most plausible extension of the standard model is the possibility that they have masses and mixings like the other fermions. Taking this hypothesis in a literal sense means that neutrinos would need to have Dirac masses like the charged fermions. Therefore, one needs to postulate the existence of right-handed neutrinos and Yukawa couplings to the standard Higgs field to generate masses and mixings. What do we know about neutrinos in the context of this Minimally Extended Standard Model?

Perhaps the most dramatic lesson is that the mass of all sequential neutrinos (ν_e, ν_μ, ν_τ) must be less than the cosmological limit of approximately 30 eV (Sect. 7.1.5). This conclusion is not to be taken for granted as massive neutrinos with mixings can decay by virtue of $\nu \to \nu'\gamma$, and by $\nu \to \nu_e e^+ e^-$ if m_ν exceeds about $2m_e$. While the standard-model radiative decays are too slow to avoid the cosmological limit, the $e^+ e^-$ channel can be fast on cosmological time scales if the mixing angle is not too small. This decay channel is an option only for ν_τ which may have a mass of up to 24 MeV while the experimental mass limits on the other flavors are below 0.16 MeV.

Such a "heavy" ν_τ, however, can be excluded by several arguments. The stellar evolution one based on SN arguments was presented in Sect. 12.5.2. The bremsstrahlung emission of photons in $\nu_\tau \to \nu_e e^+ e^- \gamma$ would produce a γ-ray flux in excess of the SMM limits for SN 1987A unless $\sin^2 2\theta_{e3} \lesssim 10^{-9}$. (For a detailed dependence of this limit on the assumed mass see Fig. 12.18.) In addition, the integrated positron flux from all galactic supernovae over the past, say, 100,000 years would exceed the observed value unless the decays are very fast (near the SN) or very slow (outside of the galactic disk). Together, these limits leave no room for a heavy ν_τ (Fig. 12.19). In addition, the mass range $0.5 \, \text{MeV} \lesssim m_\nu \lesssim 35 \, \text{MeV}$ can be excluded on the basis of big bang nucleosynthesis arguments (Fig. 7.2) unless the neutrinos are shorter lived than permitted by the Minimally Extended Standard Model.

Neutrino masses near the cosmological limit would be important for cosmology as they could contribute some or all of the dark matter of the universe. The latter option is disfavored by theories of structure formation. However, a subdominant "hot dark matter" contribution in the form of, say, 5 eV neutrinos might be cosmologically quite welcome.

Neutrino masses in this general range are also accessible to time-of-flight measurements. The neutrino burst of SN 1987A has already provided a limit of $m_{\nu_e} \lesssim 20\,\text{eV}$ (Sect. 11.3.4), a result which remains interesting in view of the confusing situation with the tritium β decay endpoint experiments which seem to be plagued by systematic effects which cause the appearance of a negative neutrino mass-square (Sect. 7.1.3). The observation of the prompt ν_e burst from a future galactic SN would allow one to reduce this limit to a few eV. Moreover, one may well be able to detect or constrain a ν_μ or ν_τ mass in the $10-20\,\text{eV}$ range, assuming the simultaneous operation of a water Cherenkov detector such as Superkamiokande and a neutral-current detector such as the proposed Supernova Burst Observatory (Sect. 11.6).

16.2.2 Oscillations of Solar Neutrinos

Small neutrino masses or rather, small neutrino mass differences can have dramatic consequences if neutrinos also mix; this is expected in the present scenario. Neutrino oscillations then lead to the possibility that a different neutrino flavor is measured in a detector than was produced in the source. The observed characteristics of the solar neutrino flux strongly suggest that neutrino oscillations may in fact be occurring. In terms of the mass difference and mixing angle between ν_e and ν_μ or ν_τ there remain three solutions which account for the presently available data from the chlorine (Homestake), gallium (GALLEX and SAGE), and Cherenkov (Kamiokande) experiments (Tab. 16.1). In view of possible systematic uncertainties concerning such quantities as the solar opacities and the $p\,^7\text{Be}$ cross section the values of the favored mixing angles can be somewhat different from those shown in Tab. 16.1. However, the required mass differences remain rather stable against large nonstandard modifications of the solar model.

Table 16.1. Approximate neutrino parameters which explain all current solar neutrino observations in the framework of standard solar model assumptions.

Solution	Δm^2 [eV^2]	$\sin^2 2\theta$
Large-angle MSW	2×10^{-5}	0.6
Nonadiabatic MSW	0.6×10^{-5}	0.006
Vacuum oscillations	0.8×10^{-10}	$0.8-1$

The main aspect of the current situation is that there does not seem to be a simple "astrophysical solution" to reconcile the solar source spectrum with the measured fluxes in experiments with three different spectral response characteristics. Even allowing for large modifications of the $p\,^7Be$ cross section or the solar central temperature does not yield consistency unless one stretches the experimental uncertainties of the flux measurements beyond reasonable limits. Still, a final verdict on the question of solar neutrino oscillations can be expected only from the near-future experiments Superkamiokande, SNO, and BOREXINO as discussed in Chapter 10.

16.2.3   Oscillation of Supernova Neutrinos

Naturally, neutrino oscillations would also affect the characteristics of SN neutrinos which have been observed only from SN 1987A, although it is not unrealistic to hope for the observation of a galactic supernova at Superkamiokande or SNO within, say, a decade of operation. In water Cherenkov detectors, the main signal of SN neutrinos is thought to be from the $\bar{\nu}_e p \rightarrow n e^+$ reaction, although the angular characteristics of the SN 1987A observations do not square well with this assumption except that there is no convincing alternate interpretation (Sect. 11.3.5). The MSW solution in the Sun requires a "normal" mass hierarchy with ν_e being dominated by the smaller mass eigenstate. In this case resonant oscillations do not occur among the $\bar{\nu}$'s.

Still, for large-angle vacuum oscillations such as those corresponding to the solar vacuum solution, the signal in the IMB and Kamiokande detectors would have been "hardened" by the partial swap of, say, the $\bar{\nu}_e$ with the $\bar{\nu}_\mu$ spectrum. It is not entirely obvious from the current literature if this effect is ruled in or ruled out by the SN 1987A observations (Sect. 11.4.3).

If the mixing angle of ν_e with both ν_μ and ν_τ is small ($\sin^2 2\theta \lesssim 0.1$) there is no impact on the $\bar{\nu}_e$ SN signal. Still, the prompt ν_e burst could be affected even for small mixing angles because of the possibility of resonant oscillations (Sect. 11.4.2) and so the observation of a future galactic supernova could serve to measure this effect.

If one were to contemplate more general neutrino mass matrices with an "inversion" so that ν_e is not dominated by the lowest-mass eigenstate, there could be resonant oscillations in the $\bar{\nu}$ sector which would then lead to dramatic modifications of the SN signal in a large range of masses and mixing angles. This possibility has not been explored much in the literature.

Apart from the detector signal, the oscillation of SN neutrinos can have important implications for SN physics itself because of the swap of, say, the ν_e with the ν_τ spectrum which is much harder. For a mass difference corresponding to the cosmologically interesting range of a few to a few tens of eV, resonant oscillations could occur so close to the neutrino sphere that the "crossover" point is within the stalling shock wave in the delayed-explosion scenario. The effective hardening of the ν_e spectrum would then enhance the neutrino energy transfer to the shock wave, thus helping to explode supernovae (Sect. 11.4.4). Conversely, a few seconds after collapse the same effect would drive the hot wind proton rich which is driven from the surface of the compact remnant. This effect would prevent the occurrence of r-process nucleosynthesis which requires a neutron-rich environment (Sect. 11.4.5). Interestingly, because the remnant is more compact at "late" times (few seconds after collapse), the adiabaticity condition can be met only for relatively large mixing angles. Thus, there is a plausible range of neutrino parameters where oscillations may help to explode supernovae, and still r-process nucleosynthesis may proceed undisturbed (Fig. 11.20). At any rate, it is impossible to ignore neutrino oscillations for SN physics if neutrino masses happen to lie in the cosmologically interesting range.

16.2.4 Electromagnetic Properties

In the Minimally Extended Standard Model neutrinos have magnetic and electric diagonal and transition moments which are proportional to their assumed masses (Sect. 7.2.2). Because of the cosmological mass limit these quantities are so small that they do not seem to be important anywhere. However, one may toy with the idea that neutrinos actually carry small electric charges, a possibility that is not entirely excluded by the structure of the Standard Model if one gives up the notion that the second and third particle families are exact replicas of the first except for the masses (Sect. 15.8).

The possible magnitude of a ν_e charge is limited by $e_{\nu_e} \lesssim 3 \times 10^{-17} e$ from the absence of an anomalous dispersion of the SN 1987A neutrino burst (Sect. 13.3.3). All neutrino charges are limited by $e_\nu \lesssim 2 \times 10^{-14} e$ from the absence of anomalous cooling of globular-cluster stars (Sect. 6.5.6). For ν_μ and ν_τ the cosmological mass limit is crucial for this bound because their emission would be suppressed by threshold effects if their mass exceeded the relevant plasma frequency of a few keV. The assumption of charge conservation in β decay yields a more restrictive limit $e_{\nu_e} \lesssim 3 \times 10^{-21}$ (Sect. 15.8).

Either way, possible charges of all neutrinos must be so small that the quantization of charge is very accurately realized among the fermions of the Standard Model. Therefore, charge quantization is probably exact so that the neutrino electric charges vanish exactly, those of the quarks are exactly $\frac{1}{3}e$ and $\frac{2}{3}e$, respectively.

16.3   New Interactions

16.3.1   Majorana Masses

Because of the cosmological limit all neutrino masses are found to be so small relative to those of the corresponding charged fermions (Fig. 7.1) that it is hard to maintain the pretense that neutrinos are essentially like the other fermions. In this sense the assumption of a Minimally Extended Standard Model is self-defeating. One reaction may be to return to the assumption of massless two-component neutrinos. It remains to be seen for how much longer this option remains viable in view of the expected progress in solar neutrino astronomy and, perhaps, in laboratory experiments (including atmospheric neutrino observations). They may soon yield unrefutable evidence for neutrino oscillations.

Another reaction is to embrace the notion of neutrinos being very different from the charged leptons with a possible wealth of novel and unexpected properties. The most benign assumption is to maintain the notion of only two neutrino components per family which are characterized by a Majorana mass term which may arise by new physics at some large energy scale. In this case all that was said about neutrino masses and oscillations in the previous section remains applicable.

16.3.2   "Heavy" Neutrinos and Fast Decays

If one postulates novel neutrino interactions one may speculate about the possibility of neutrino decays which proceed faster than in the Minimally Extended Standard Model. Besides accelerated radiative decays one may speculate about "fast invisible decays" of the form $\nu \to \nu' \nu'' \bar{\nu}'''$ or $\nu \to \nu' \chi$ where χ is some new boson such as the majoron (Sect. 15.7). In this case one can escape the cosmological mass limit. In fact, such fast-decaying neutrinos can be a welcome feature of theories for the formation of structure in the universe (Sect. 7.1.5).

"Heavy" ν_μ's or ν_τ's are not in obvious conflict with normal stellar evolution even though their emission could now be suppressed by the mass threshold. Thermal production of neutrinos from a stellar plasma is dominated by the plasmon decay process $\gamma \to \bar{\nu}\nu$ for a large range of stellar conditions (Appendix C). This process occurs mostly by the vector-current coupling to electrons of the medium. Because the weak mixing angle has the special value $\sin^2 \Theta_W = 0.23 \approx \frac{1}{4}$ the vector-current coupling of ν_μ and ν_τ to electrons nearly vanishes (Appendix B) and so $\gamma \to \bar{\nu}_\mu\nu_\mu$ and $\bar{\nu}_\tau\nu_\tau$ plays near to no role anyway—in this sense stars are "blind" to the issue of heavy ν_μ or ν_τ masses.

However, assuming that the heavy neutrinos are Dirac particles one can limit their mass by a SN argument: The energy loss by right-handed states produced in spin-flip collisions must not be too large, yielding a limit of $m_\nu \lesssim 30\,\text{keV}$ (Sect. 13.8.1). In addition, these neutrinos have very high energies, typical of SN core temperatures, and so their decays might produce high-energy daughter $\bar{\nu}_e$'s ($100-200\,\text{MeV}$) which were not observed from SN 1987A (Sect. 13.8.1).

Even Majorana neutrinos would not be harmless for SN physics if they would mix sufficiently strongly with ν_e. In this case they would effectively participate in β equilibrium (Sect. 9.5) so that the spin-flip scattering of, say, ν_τ would effectively lead to $\bar{\nu}_\tau$'s and thus to deleptonization without the need of transporting lepton number to the stellar surface.

In addition, even though heavy Majorana neutrinos would be emitted from the neutrino sphere so that their decays would not produce high energy daughter products, the decays could still add to the detectable signal. It is not obvious from the existing literature which (if any) range of masses and decay times is ruled out or ruled in by the SN 1987A observations (Sect. 13.3.2).

If the fast decays were due to some sort of majoron model, large "secret" neutrino-neutrino interactions are conceivable, possibly in conjunction with a small vacuum expectation value of a new Higgs field so that the symmetry may be restored in a SN core. While a substantial body of literature exists on this sort of scenario (Sect. 15.7.2) I believe that in this context the story of SN physics would have to be rewritten more systematically than has been done so far. However, there appears to be little doubt that for neutrino-majoron Yukawa couplings in excess of about 10^{-5} one would expect dramatic modifications of the transport of energy and lepton number.

In summary, "heavy" neutrinos with fast invisible decays are a way to circumvent the cosmological mass limit, and may indeed be desirable in certain scenarios of cosmic structure formation. Depending on their detailed properties they could have a substantial impact on SN physics and the signal observable in a detector. It is difficult, however, to state general constraints as one may easily postulate, for example, that the decays do not involve final-state $\bar{\nu}_e$'s, that even "wrong-helicity" Dirac neutrinos are trapped in a SN core by novel interactions, or that heavy Majorana ν_τ's have only negligible mixings with ν_e. Surely other loopholes could be found.

16.3.3 Electromagnetic Properties

a) Spin and Spin-Flavor Oscillations

If one contemplates neutrino interactions beyond the Minimally Extended Standard Model, neutrino dipole and transition moments no longer need to be small. In particular, they do not need to be proportional to the neutrino masses; one example are left-right symmetric models where even massless neutrinos would have large dipole moments (Sect. 7.3.1). Dipole and transition moments can lead to spin or spin-flavor oscillations in external magnetic fields, they allow for spin-flip scattering on charged particles, for the plasmon decay $\gamma \to \bar{\nu}\nu$ in stars, and for radiative decays $\nu \to \nu'\gamma$.

One motivation for studying neutrino dipole moments is the apparent flux variability of the solar neutrino signal in the Homestake detector. It anticorrelates with solar magnetic activity too closely to blame it comfortably on a statistical fluke (Sect. 10.4.3). The only physical explanation put forth to date is that of Voloshin, Vysotskiĭ, and Okun of a partial depletion of left-handed (measurable) neutrinos by spin or spin-flavor oscillations. Unfortunately, the required value for $\mu_\nu B$ (neutrino dipole moment μ_ν, magnetic field B in the solar convection zone) exceeds by about two orders of magnitude what is allowed by typical models of the solar magnetic field and by limits on μ_ν. Therefore, this scenario appears to be in big trouble. Still, if one ignores the μ_ν limits or speculates about large convection-zone magnetic fields one may fit all currently available solar neutrino data (Sect. 10.7).

Spin or spin-flavor oscillations can be very important in and near the cores of supernovae where fields of order 10^{12} G exist, and perhaps pockets with much larger fields. If neutrinos are Dirac particles so that their spin-flipped (right-handed) states are sterile, the combination of spin-flip scattering on charged particles and the magnetic spin oscillation in large-scale magnetic fields can lead to nonlocal modes of energy transfer where energy can be deposited in one region that was depleted from a distant other region. Thus, energy transfer could no longer be treated with simple differential equations which involve local gradients of temperature and lepton number. (Of course, a nonlocal energy transfer mechanism is already thought to be important for reviving the shock wave in the delayed-explosion scenario.) As far as I know, nothing more quantitative than back-of-the-envelope estimates of this scenario exist in the literature. Therefore, alleged bounds of order $10^{-12}\,\mu_B$ (Bohr magneton $\mu_B = e/2m_e$) on Dirac-neutrino dipole moments probably have to be used with some reservation (Sect. 13.8.3). Conversely, such dipole moments may actually help to explode supernovae.

With regard to the observable neutrino signal, spin and spin-flavor oscillations both in the SN and in the galactic magnetic field may cause vast modifications of the $\bar{\nu}_e$ fluxes and spectra observable in water Cherenkov detectors if neutrinos have dipole moments in the ballpark of 10^{-12}–$10^{-14}\,\mu_B$.

Spin and spin-flavor oscillations can be very important in the early universe where strong magnetic fields may exist, and where a population of the r.h. degrees of freedom would accelerate the expansion rate of the universe. These issues are being investigated in the current literature; final conclusions do not seem to be available at the present time. Still, it appears that this effect may well be the most significant impact of small Dirac neutrino dipole or transition moments anywhere in nature.

b) Laboratory Limits

Less problematic bounds on neutrino dipole moments arise from laboratory experiments where one studies the recoil spectrum of electrons in the reaction $\nu + e \rightarrow e + \nu'$ where ν' can be the same or a different flavor (Sect. 7.5.1). A sensitivity down to, perhaps, as low as $10^{-11}\,\mu_B$ can be expected from a current effort involving reactor neutrinos as a source (MUNU experiment). Current limits on dipole or transition moments are about $2\times10^{-10}\mu_B$ if ν_e is involved, and about $7\times10^{-10}\mu_B$ if ν_μ is involved. For transition moments, these limits are subject to the assumption that there is no cancellation between a magnetic and an electric dipole scattering amplitude.

c) Huge ν_τ Dipole Moments or Millicharges

In principle, the possibility of a large diagonal moment for ν_τ remains open as it has not been possible to produce a strong ν_τ source in the laboratory so that only extremely crude limits exist on the ν_τ-e-scattering cross section. The globular-cluster μ_ν bounds discussed below do not apply to a "heavy" ν_τ so that one is confronted with a nontrivial allowed region in μ_{ν_τ}-m_{ν_τ} space where even MeV masses become cosmologically allowed because of the dipole-induced annihilation process $\bar{\nu}_\tau \nu_\tau \rightarrow e^+ e^-$ (μ_{ν_τ} in the ballpark of $10^{-7}\,\mu_B$). Of course, such large dipole moments must be caused by a fairly nontrivial arrangement of intermediate charged states and so one may wonder if a large ν_τ magnetic moment could be realistically the only manifestation of these new particles and/or interactions.

Still, a large ν_τ dipole moment and the correspondingly large annihilation cross section in the early universe is one possibility to tolerate a large ν_τ mass without the need for fast decays—such a particle could be entirely stable. Another similar possibility is that ν_τ has a "huge millicharge" in the neighborhood of 10^{-5}–$10^{-3}e$. Such a scheme would require the violation of charge conservation as the possibility of neutrino charges within a simple extension of the Standard Model discussed above always gives charges to two neutrino flavors; for ν_e or ν_μ the required value is not tolerable (Sect. 15.8).

For the issues of stellar evolution, the only conceivable consequence of such large ν_τ electromagnetic interaction cross sections would be a reduced contribution to the energy transfer in SNe because of the reduced mean free path. In the study discussed in Sect. 13.6 one should have included the possibility of only two effective flavors! Still, there is little doubt that large ν_τ cross sections could be accommodated in what one knows about SNe today.

d) Astrophysical Bounds on Dipole and Transition Moments

For all neutrinos with a mass below a few keV a very restrictive limit on dipole or transition magnetic or electric moments arises from the absence of anomalous neutrino emission from the cores of evolved globular-cluster stars, notably of red-giant cores just before helium ignition (Sect. 6.5.6). One finds a limit $\mu_\nu \lesssim 3 \times 10^{-12}\,\mu_B$ which applies to Dirac and Majorana neutrinos, and which does not allow for a destructive interference between electric and magnetic amplitudes.

Neutrino transition moments would reveal themselves by radiative decays. Because the decay rate involves a phase-space factor m_ν^3 this method is suitable only for large masses. In the cosmologically allowed range with $m_\nu \lesssim 30$ eV, the only radiative limit which can compete with the globular-cluster bound is from the cosmic diffuse background radiations (Fig. 12.21). In fact, Sciama has proposed a scheme where a 28.9 eV neutrino with a radiative decay time corresponding to a transition moment of $0.6 \times 10^{-14}\,\mu_B$ plays a significant cosmological role (Sect. 12.7.1). While this scenario is probably excluded it highlights the possibility of interesting cosmological effects for radiatively decaying neutrinos in the range allowed by the globular-cluster bound.

Radiative decay limits which are based on astronomical decay paths suffer from the uncertainty of other invisible decay channels which may compete with the radiative mode. Limits based on the cosmic background radiations (Fig. 12.20) imply that in a large range of cosmologically allowed neutrino masses and lifetimes the dominant decay channel must be nonradiative. Therefore, one should use the cosmic background radiations to derive limits on the branching ratio. One could then construct a contour plot of the excluded transition moments in the m_ν-τ_ν plane.

From the SN 1987A radiative lifetime limits I have constructed such a plot in Fig. 12.17. For large dipole moments in excess of, say, $10^{-10}\,\mu_B$ these bounds are not self-consistent because neutrinos would be trapped too strongly by electromagnetic scatterings. However, the laboratory limits exclude large transition moments. Moreover, the globular-cluster bound yields more restrictive limits if m_ν is less than a few keV. However, for relatively large neutrino masses, and for lifetimes not so short that the decays would have occurred within the progenitor star, SN 1987A yields the most restrictive limits on transition moments. Naturally, one must assume that ν_μ or ν_τ were actually emitted with about the standard fluxes. Trapping effects by additional new interactions could circumvent this assumption.

16.3.4 Summary

Neutrinos with nonstandard interactions may well saturate the experimental mass limits, and may have a variety of novel properties. However, it is nearly impossible to derive *generic* constraints on quantities like magnetic transition moments without specifying an underlying particle physics model. Many constraints, notably those related to SN 1987A or to cosmology, can be circumvented by postulating sufficiently bizarre neutrino properties. Therefore, it is probably more important to know the arguments that can serve to learn something about neutrinos in astrophysics than it is to know a list of alleged limits. If a concrete conjecture turns up, or a specific theoretical model needs to be constrained, one can easily go through the list of arguments and check if they apply or not. Perhaps this book can be of help at this task.

Appendix A

Units and Dimensions

In the astrophysical context, frequently occurring units of length are centimeters, (light) seconds, light years, and parsecs. Conversion factors are given in Tab. A.1. For example, $1 \, \text{pc} = 3.26 \, \text{ly}$.

Using both centimeters and (light) seconds as units of length implies a system of units where the speed of light c is dimensionless and equal to unity. In this book I always use natural units where Planck's constant $\hbar$ and Boltzmann's constant k_B are also dimensionless and equal to unity. This implies that $(\text{length})^{-1}$, $(\text{time})^{-1}$, mass, energy, and temperature can all be measured in the same unit by virtue of $x = ct$, $E = mc^2$, $E = \hbar\omega$, $\omega = 2\pi/t$, and $E = k_B T$. In Tab. A.2 conversion factors are given. For example, $1 \, \text{K} = 0.862 \times 10^{-4} \, \text{eV}$ or $1 \, \text{erg} = 0.948 \times 10^{27} \, \text{s}^{-1}$.

The most confusing aspect of natural units is that of an electromagnetic field strength. The square of a field strength is an energy density (erg/cm^3) which, in natural units, is $(\text{energy})^4$ or $(\text{length})^{-4}$. Thus, an electric or magnetic field may be measured, for example, in eV^2 or cm^{-2}. In natural units, electric charges are dimensionless numbers.

However, there is a general ambiguity in the definition of charges and field strengths because only their product (a force on a charged particle) is operationally defined. All physical quantities stay the same if the charges are multiplied with an arbitrary number and the field strengths are divided by it. However, the fine-structure constant $\alpha \approx 1/137$ is dimensionless in all systems of units, and its value does not depend on this arbitrary choice. If e is the charge of the electron one has $\alpha = e^2/4\pi$ in the rationalized system of (natural) units which is always used in modern works on field theory, and is used throughout this book. The energy density of an electromagnetic field is then $\frac{1}{2}(E^2 + B^2)$. In the older literature and some texts on electromagnetism, unrationalized units are used where $\alpha = e^2$ and the energy density is $(E^2 + B^2)/8\pi$.

580

In the astrophysical literature the cgs system of units is very popular where magnetic fields are measured in Gauss (G). Confusingly, this system happens to be an unrationalized one. Field strengths given in Gauss can be translated into our rationalized natural units by virtue of

$$1\,G \rightarrow \sqrt{\frac{1\,\mathrm{erg/cm^3}}{4\pi}} = 1.953 \times 10^{-2}\,\mathrm{eV^2} = 0.502 \times 10^8\,\mathrm{cm^{-2}}, \quad (A.1)$$

where I have converted erg and $\mathrm{cm^{-1}}$ into eV according to Tab. A.2. The energy density of a magnetic field of strength $1\,G$ is, therefore, $\frac{1}{2}(1.953 \times 10^{-2}\,\mathrm{eV^2})^2 = 1.908 \times 10^{-4}\,\mathrm{eV^4} = 3.979 \times 10^{-2}\,\mathrm{erg\,cm^{-3}} = (1/8\pi)\,\mathrm{erg\,cm^{-3}}$. For a further discussion of electromagnetic units see Jackson (1975).

It is sometimes useful to measure very strong magnetic fields in terms of a critical field strength B_{crit} which is defined by the condition that the quantum energy corresponding to the classical cyclotron frequency $\hbar\,(eB/m_ec)$ of an electron equals its rest energy m_ec^2 so that in natural units

$$B_{\mathrm{crit}} = m_e^2/e. \quad (A.2)$$

Note that the Lorentz force on an electron in this field is proportional to eB_{crit} so that the electron charge cancels. Hence, Eq. (A.2) is the same in a rationalized or unrationalized system of units. In our rationalized units $e = \sqrt{4\pi\alpha} = 0.303$ so that $B_{\mathrm{crit}} = (0.511\,\mathrm{MeV})^2/0.303 = 0.862 \times 10^{12}\,\mathrm{eV^2}$ which, with Eq. (A.1), corresponds to $4.413 \times 10^{13}\,G$, in accordance to what is found in the literature (Mészáros 1992).

Magnetic dipole moments of electrons and neutrinos are usually discussed in terms of Bohr magnetons $\mu_{\mathrm{B}} = e/2m_e$. For particle electric dipole moments, on the other hand, one commonly uses $1\,e\,\mathrm{cm}$ as a unit. The conversion is achieved by $1\,e\,\mathrm{cm} = (2m_e\,\mathrm{cm})\,(e/2m_e) = 5.18 \times 10^{10}\,\mu_{\mathrm{B}}$.

Tab. A.1. Conversion factors between different units of length.

	cm	s	ly	pc
cm	1	0.334×10^{-10}	1.06×10^{-18}	0.325×10^{-18}
s	2.998×10^{10}	1	0.317×10^{-7}	0.973×10^{-8}
ly	0.946×10^{18}	3.156×10^{7}	1	0.307
pc	3.08×10^{18}	1.028×10^{8}	3.26	1

Tab. A.2. Conversion factors in the system of natural units.

	s^{-1}	cm^{-1}	K	eV	amu[a]	erg	g
s^{-1}	1	0.334×10^{-10}	0.764×10^{-11}	0.658×10^{-15}	0.707×10^{-24}	1.055×10^{-27}	1.173×10^{-48}
cm^{-1}	2.998×10^{10}	1	0.2289	1.973×10^{-5}	2.118×10^{-14}	3.161×10^{-17}	0.352×10^{-37}
K	1.310×10^{11}	4.369	1	0.862×10^{-4}	0.926×10^{-13}	1.381×10^{-16}	1.537×10^{-37}
eV	1.519×10^{15}	0.507×10^{5}	1.160×10^{4}	1	1.074×10^{-9}	1.602×10^{-12}	1.783×10^{-33}
amu	1.415×10^{24}	0.472×10^{14}	1.081×10^{13}	0.931×10^{9}	1	1.492×10^{-3}	1.661×10^{-24}
erg	0.948×10^{27}	0.316×10^{17}	0.724×10^{16}	0.624×10^{12}	0.670×10^{3}	1	1.113×10^{-21}
g	0.852×10^{48}	2.843×10^{37}	0.651×10^{37}	0.561×10^{33}	0.602×10^{24}	0.899×10^{21}	1

[a]Atomic mass unit.

Appendix B

Neutrino Coupling Constants

Neutrinos can interact with other fermions and with each other by the exchange of W or Z bosons. Because the astrophysical phenomena relevant for this book take place at very low energies compared with the W or Z mass, one may always use an effective four-fermion coupling which is parametrized in terms of the Fermi constant and the weak mixing angle

$$G_{\mathrm{F}} = 1.166 \times 10^{-5} \,\mathrm{GeV}^{-2},$$
$$\sin^2 \Theta_{\mathrm{W}} = 0.2325 \pm 0.0008. \tag{B.1}$$

The tree-level relationship of these quantities with the gauge-boson masses is

$$\sqrt{2}\,G_{\mathrm{F}} = \frac{\pi \alpha}{m_W^2 \sin^2 \Theta_{\mathrm{W}}} = \frac{\pi \alpha}{m_Z^2 \sin^2 \Theta_{\mathrm{W}} \cos^2 \Theta_{\mathrm{W}}}, \tag{B.2}$$

where $m_Z = 91.2\,\mathrm{GeV}$ and $m_W = 80.2\,\mathrm{GeV}$.

The effective charged-current interaction between nucleons and leptons is written in the form

$$\mathcal{H}_{\mathrm{int}} = \frac{G_{\mathrm{F}}}{\sqrt{2}} \overline{\psi}_p \gamma_\mu (C_V - C_A \gamma_5) \psi_n \, \overline{\psi}_\ell \gamma^\mu (1 - \gamma_5) \psi_{\nu_\ell}, \tag{B.3}$$

where the ψ_j are the proton, neutron, charged-lepton, and the corresponding neutrino field. The vector-current coupling constant is $C_V = 1$ while the axial-vector coupling for free nucleons is $C_A = 1.26$. However, in large nuclei this value is suppressed somewhat, and the commonly used value for nuclear matter is $C_A = 1.0$ (e.g. Castle and Towner 1990).

This quantity would be relevant, for example, for β reactions in supernova cores and neutron stars.

The effective charged-current interaction between charged leptons and their own neutrinos, e.g. between e^- and ν_e, is written in the same form with $C_V = C_A = 1$. By virtue of a Fierz transformation it is brought into the form of a neutral current (see below).

Neutral-current interactions between a neutrino ν and a fermion f are written in the form

$$\mathcal{H}_{\text{int}} = \frac{G_F}{\sqrt{2}} \bar{\psi}_f \gamma_\mu (C_V - C_A \gamma_5) \psi_f \, \bar{\psi}_\nu \gamma^\mu (1 - \gamma_5) \psi_\nu. \tag{B.4}$$

If f is the charged lepton corresponding to ν, there is a contribution with $C_V = C_A = 1$ from a Fierz-transformed charged current. The compound effective C_V's and C_A's for various combinations of f and ν are given in Tab. B.1. (Note that the $|C_{V,A}|$ for neutral currents are typically $\frac{1}{2}$, a factor which is sometimes pulled out front so that the global coefficient is $G_F/2\sqrt{2}$ while the couplings are then twice those of Tab. B.1.) For neutrinos interacting with neutrinos of the same flavor a factor 2 for an exchange amplitude for identical fermions was included.

The C_A's for nucleons were thought to be given by isospin invariance to be $\pm 1.26/2$. However, because of the strange-quark contribution to the nucleon spin there is an isoscalar piece as well giving rise to the values shown in Tab. B.1—for a discussion and references to the original literature see Raffelt and Seckel (1995). Moreover, in a nuclear medium a certain suppression is expected to occur. In analogy to the charged-current couplings Raffelt and Seckel (1995) suggested the values $C_A^p \approx 1.09/2$ and $C_A^n \approx -0.91/2$.

Table B.1. Neutral-current couplings for the effective Hamiltonian Eq. (B.4) in vacuum.

Fermion f	Neutrino	C_V	C_A	C_V^2	C_A^2
Electron	ν_e	$+\frac{1}{2} + 2\sin^2\Theta_W$	$+\frac{1}{2}$	0.9312	0.25
	$\nu_{\mu,\tau}$	$-\frac{1}{2} + 2\sin^2\Theta_W$	$-\frac{1}{2}$	0.0012	0.25
Proton	$\nu_{e,\mu,\tau}$	$+\frac{1}{2} - 2\sin^2\Theta_W$	$+1.37/2$	0.0012	0.47
Neutron	$\nu_{e,\mu,\tau}$	$-\frac{1}{2}$	$-1.15/2$	0.25	0.33
Neutrino (ν_a)	ν_a	$+1$	$+1$	1	1
	$\nu_{b\neq a}$	$+\frac{1}{2}$	$+\frac{1}{2}$	0.25	0.25

Appendix C

Numerical Neutrino Energy-Loss Rates

In normal stars with densities below nuclear there are four main reactions that contribute to the energy loss by neutrino emission:

$$
\begin{array}{rcll}
\gamma_{\rm pl} & \to & \nu\bar{\nu} & \text{Plasma process,} \\
\gamma\, e^- & \to & e^-\, \nu\bar{\nu} & \text{Photoneutrino process,} \\
e^+ e^- & \to & \nu\bar{\nu} & \text{Pair annihilation,} \\
e^-\,(Z, A) & \to & (Z, A)\, e^-\, \nu\bar{\nu} & \text{Bremsstrahlung.}
\end{array}
\tag{C.1}
$$

Individual processes dominate in the regions of density and temperature indicated in Fig. C.1. In the following, various analytic fit formulae for these neutrino emission rates are reviewed.

C.1 Plasma Process

Widely used formulae for the plasma process are those of Beaudet, Petrosian, and Salpeter (1967), Munakata, Kohyama, and Itoh (1985), and Schinder et al. (1987), which all agree with each other to better than 1% if the same effective coupling constants are used. All of these rates are poor approximations for $T \lesssim 10^8\,K$ which is relevant for low-mass stars because they were optimized for higher temperatures. Itoh et al. (1989) have attempted to improve the accuracy at low temperatures, and Blinnikov and Dunina-Barkovskaya (1994) gave rates which were optimized for low-mass stars but fail for temperatures above about $10^8\,K$. At high temperatures and densities, a poor approximation to the photon dispersion relation was used in all of these works (Braaten 1991) whence none of these rates are satisfactory. A new fit by Itoh et al. (1992) still contains islands in the ρ-T-plane with errors of several 10%.

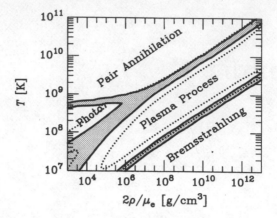

Fig. C.1. Regions of density and temperature where the indicated neutrino emission processes contribute more than 90% of the total. μ_e is the electron "mean molecular weight," i.e. roughly the number of baryons per electron. The bremsstrahlung contribution depends on the chemical composition. The solid lines are for helium, the dotted ones for iron which yields a larger bremsstrahlung rate.

A detailed comparison of these formulae with the exact rates was performed by Haft, Raffelt, and Weiss (1994). They provided a new fitting formula which approximates the analytic emission rate to within 5% in the entire regime where the plasma process dominates.

C.2   Photoneutrino and Pair-Annihilation Process

Beaudet, Petrosian, and Salpeter (1967) provided analytic approximations for the photoneutrino and pair-annihilation processes. Dicus (1972) gave global correction factors to these rates to include neutral-current effects. Schinder et al. (1987) numerically recalculated the emission rates in the standard model and found good agreement with the BPS formulae together with the Dicus correction factors. They supplemented the BPS rates for the temperature range 10^{10}–10^{11} K.

An alternate set of approximation formulae was provided by Itoh et al. (1989) who improved on their previous work (Munakata, Koh-yama, and Itoh 1985). In Fig. C.2 I show the relative deviation between the Itoh et al. (1989) with the Schinder et al. (1987) rates. The total

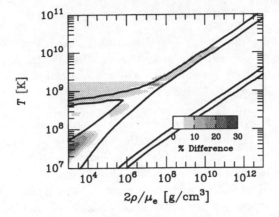

Fig. C.2. Deviation between the Schinder et al. (1987) and the Itoh et al. (1989) rates for the photoneutrino and pair-annihilation processes. Compared are the total emission rates where the plasma rate of Haft, Raffelt and Weiss (1994) and the bremsstrahlung rate (helium) of Itoh and Kohyama (1983) were used. The contours indicate were the individual processes dominate (Fig. C.1).

energy-loss rates for the photo and pair process was calculated according to these authors, while in each case the plasma rate of Haft, Raffelt, and Weiss (1994) and the bremsstrahlung rate for helium of Itoh and Kohyama (1983) were taken. Therefore, deviations occur only in the range of temperatures and densities where the photo or pair process dominates. The largest deviations in the lower left corner of Fig. C.2 are around 25%. However, there the absolute magnitude of neutrino emission is very small (see below) so that the difference between the rates in this regime does not appear to be of much practical significance.

Another analytic approximation formula for the pair process was derived by Blinnikov and Rudzskiĭ (1989).

C.3   Bremsstrahlung

Bremsstrahlung dominates for low temperatures and high densities where electrons are degenerate and the nuclei are strongly correlated. In a series of papers the emission rate was calculated by Itoh and Kohyama (1983), Itoh et al. (1984a,b), and Munakata, Kohyama, and Itoh (1987).

For simple estimates one may use the approximate rate given in Eq. (11.40). In Fig. C.3 I display the error of this approximation for iron relative to the results of Itoh and Kohyama (1983). For orientation the contours of Fig. C.1 for iron are also shown, but only the bremsstrahlung rates are compared. The simple approximation is not a bad fit in the regions where bremsstrahlung could be of interest. For carbon the fit is almost as good, but it is substantially worse for helium.

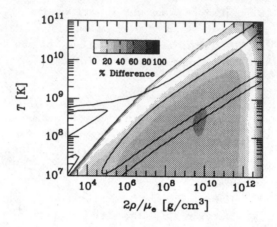

Fig. C.3. Relative deviation between the bremsstrahlung rates of Itoh and Kohyama (1983) for iron and the simple approximation formula Eq. (11.40). The contours where different processes dominate are for iron.

C.4   Total Emission Rate

The total neutrino energy-loss rate for helium is shown in Figs. C.4–C.6 where the photoneutrino and pair-annihilation rates are from Schinder et al. (1987), the plasma process from Haft, Raffelt, and Weiss (1994), and bremsstrahlung (helium) from Itoh and Kohyama (1983)

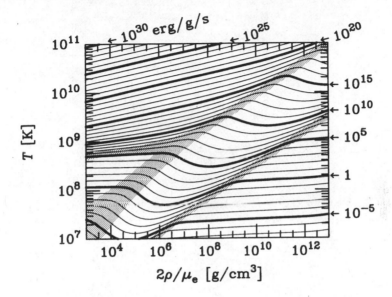

Fig. C.4. Contour plot for the total neutrino energy-loss rate ϵ_ν per unit mass for helium. The thin contours are at intervals of a factor of 10 for ϵ_ν. The regions where the individual processes dominate are also indicated.

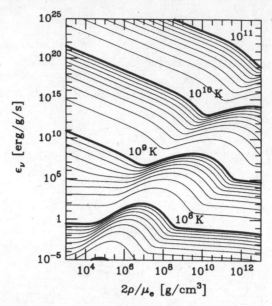

Fig. C.5. Neutrino energy-loss rate as a function of density. The thin lines are for temperatures 2, 3, 4, etc. times the value indicated on the corresponding thick line.

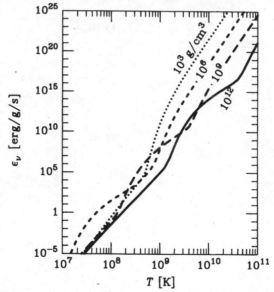

Fig. C.6. Neutrino energy-loss rate as a function of temperature for the indicated values of $2\rho/\mu_e$.

Appendix D

Characteristics of Stellar Plasmas

D.1 Normal Matter

D.1.1 Temperatures and Densities

The material encountered in stars is usually in a state of thermal equilibrium. In the absence of strong magnetic fields, the plasma is entirely characterized by its temperature T, mass density ρ, and a set of chemical composition parameters X, Y, X_{12}, etc. which determine the mass fractions of the elements ^{1}H, ^{4}He, ^{12}C, and so forth. The mass fraction of all elements heavier than helium ("metals") is denoted by Z.

The number density of a species with mass fraction X_j, atomic weight A_j, and charge $Z_j e$ is given by

$$n_j = (\rho/m_u)\, X_j/A_j, \tag{D.1}$$

where $m_u = 1.66\times10^{-24}\,\text{g} = 0.932\,\text{GeV}$ is the atomic mass unit.[99] The number density of electrons is

$$n_e = \sum_j Z_j n_j = \frac{\rho}{m_u} \sum_j \frac{X_j Z_j}{A_j} = \frac{\rho}{\mu_e m_u}, \tag{D.2}$$

where μ_e is the "mean molecular weight" per electron, not to be confused with the electron chemical potential. (Strictly speaking $n_e = n_{e^-} - n_{e^+}$, the number density of electrons minus that of positrons.)

---

[99] The proton and neutron mass are 0.9383 and 0.9396 GeV, respectively. An exact translation between mass and number density thus requires taking nuclear binding energies into account whence the A_j are not exact integers. For the purposes of this book these differences are negligible.

For all elements except hydrogen $Z_j/A_j \approx \frac{1}{2}$. Notably, this applies to helium and compounds of α-particles such as ^{12}C and ^{16}O. Therefore,

$$Y_e \approx \mu_e^{-1} \approx X + \tfrac{1}{2}(Y + Z), \qquad (D.3)$$

where Y_e is the mean number of electrons per baryon. Here, the mass fraction Z of metals must not be confused with a nuclear charge.

Some examples for typical conditions encountered in stars are shown in Fig. D.1, ignoring neutron stars (density around nuclear). Aside from the example of an evolved massive star, all conditions refer to the centers of stars. Except for the hydrogen main squence, the abscissa is essentially the physical density because for most chemical compositions $Y_e \approx \frac{1}{2}$. Horizontal-branch (HB) stars correspond essentially to the helium main sequence at $0.5\,\mathcal{M}_\odot$.

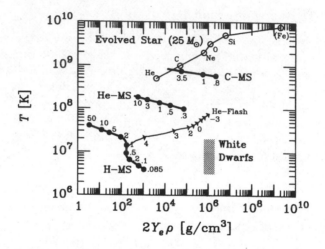

Fig. D.1. Typical temperatures and densities encountered in stars, ignoring neutron stars. The hydrogen, helium, and carbon main sequence (H-MS, He-MS and C-MS) represent the conditions at the center of zero-age models; for selected cases their $\mathcal{M}/\mathcal{M}_\odot$ is indicated (adapted from Kippenhahn and Weigert 1990). The highly evolved $25\,\mathcal{M}_\odot$ star is according to Woosely and Weaver (1986b) where the open circles are marked with the energy source of the different burning shells (He-burning etc.). There is no nuclear burning at the center (Fe). Also shown is the evolution of the central conditions of a $0.8\,\mathcal{M}_\odot$ star from the hydrogen main sequence to the helium flash (Haft, Raffelt, and Weiss 1994). The rear ends of the arrows mark the indicated values of the absolute surface brightness in magnitudes.

D.1.2 Relativistic Conditions for Electrons

The nuclei in normal stellar matter are always nonrelativistic; relativistic corrections begin to be important only in neutron stars. The electrons, on the other hand, tend to be at least partially relativistic. Even at the center of the Sun at a temperature of 1.3 keV, a typical thermal electron velocity is about 9% of the speed of light. In Fig. D.2 contours for the thermal average $\langle v^2 \rangle^{1/2}$ are shown in the ρ-T-plane. The loci of the stellar models of Fig. D.1 are also indicated. For low-mass stars, the electrons are mildly relativistic, although a nonrelativistic treatment is often enough as a first approximation.

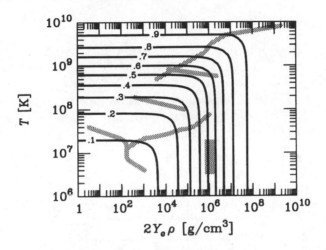

Fig. D.2. Contours for $\langle v^2 \rangle^{1/2}$, the average thermal velocity of electrons. The loci of the stellar models of Fig. D.1 are also indicated.

D.1.3 Electron Degeneracy

The phase-space occupation numbers of fermions in thermal equilibrium are characterized by a Fermi-Dirac distribution

$$f_{\mathbf{p}} = \frac{1}{e^{(E_{\mathbf{p}} - \mu)/T} + 1}, \tag{D.4}$$

where $E_{\mathbf{p}}$ is the energy of the momentum mode $\mathbf{p}$. If dispersion effects can be ignored, $E_{\mathbf{p}}^2 = m^2 + \mathbf{p}^2$ with the fermion vacuum mass m. The (relativistic) chemical potential is denoted by μ; for electrons it should not be confused with the mean molecular weight μ_e. The distribution of antifermions is given by the same expression with $\mu \to -\mu$. Then μ is implicitly given by the phase-space integral

$$n_f = \int \frac{2\,d^3\mathbf{p}}{(2\pi)^3} \left(\frac{1}{e^{(E_\mathbf{p}-\mu)/T}+1} - \frac{1}{e^{(E_\mathbf{p}+\mu)/T}+1} \right), \qquad (D.5)$$

where the second term represents antifermions and the factor 2 is for the two spin degrees of freedom. Again, the fermion density is understood to mean the density of fermions minus that of antifermions.

At vanishing temperature, $f_\mathbf{p}$ becomes a step function $\Theta(\mu - E_\mathbf{p})$. If $\mu > 0$ so that $n_f > 0$, i.e. an excess of fermions over antifermions, there are no antifermions at all at $T = 0$. The fermion integral yields

$$n_f = p_F^3/3\pi^2, \qquad (D.6)$$

where the Fermi momentum is defined by $\mu_0^2 = p_F^2 + m^2$ with μ_0 the zero-temperature chemical potential. The Fermi energy is defined by $E_F^2 = p_F^2 + m_e^2$, i.e. $E_F = \mu_0$.

Equation D.6 is taken as the definition of the Fermi momentum even at $T > 0$; it is a useful parameter to characterize the fermion density, whether or not they are degenerate. Numerically it is

$$p_F = 5.15\,\text{keV}\,(Y_e\rho)^{1/3} \qquad (D.7)$$

for electrons with the mass density ρ in units of g cm^{-3}.

In general, Eq. (D.5) cannot be made explicit for μ; it has to be solved numerically or by an approximation method. In the ρ-T-plane, contours for the electron chemical potential are shown in Fig. D.3. Above the main plot, the electron density is characterized by p_F. On the right side, the temperature is shown in units of keV. Recall that $10^7\,\text{K} = 0.8621\,\text{keV}$ (Appendix A).

For nonrelativistic electrons the contours in Fig. D.3 are very sensitive to the exact value of μ. Therefore, in this regime the nonrelativistic chemical potential

$$\hat{\mu} \equiv \mu - m \qquad (D.8)$$

is a more appropriate parameter. Often $\hat{\mu}$ is referred to as *the* chemical potential. This can be very confusing when relativistic effects are important. In terms of $\hat{\mu}$, the relativistic Fermi-Dirac distribution is

$$f_\mathbf{p} = \frac{1}{e^{(E_\text{kin}-\hat{\mu})/T}+1}, \qquad (D.9)$$

with the kinetic energy $E_\text{kin} = E_\mathbf{p} - m \to \mathbf{p}^2/2m$ (nonrelativistic limit).

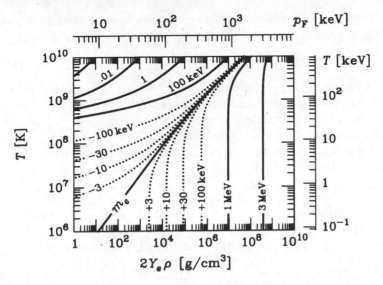

Fig. D.3. Contours for the electron chemical potential μ. The solid lines are marked with the relevant value for μ, the dotted lines with $\mu - m_e$.

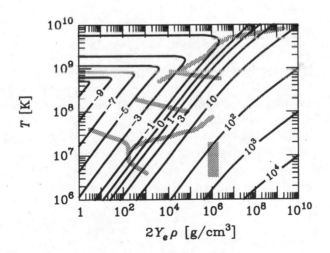

Fig. D.4. Contours for the electron degeneracy parameter $\eta = (\mu - m_e)/T$. Also shown are the loci of the stellar models of Fig. D.1.

A Fermi gas becomes degenerate when a typical thermal (kinetic) energy is on the order of $\hat{\mu}$. Therefore, the degeneracy parameter

$$\eta \equiv \hat{\mu}/T = (\mu - m)/T \tag{D.10}$$

is frequently used to characterize the fermions. They are degenerate for η larger than a few, and nondegenerate for $\eta < 0$. For electrons, contours of η in the ρ-T-plane are shown in Fig. D.4. A comparison with the loci of the stellar models of Fig. D.1 reveals that electrons in stellar plasmas are often at least partially degenerate.

D.1.4 Plasma Frequency

Other characteristic properties of a plasma refer to the behavior of electromagnetic waves. The photon dispersion relation (Sect. 7.4) is characterized by the plasma frequency which at $T = 0$ is

$$\omega_0^2 = 4\pi\alpha\, n_e/E_F = (4\alpha/3\pi)\, p_F^3/E_F. \tag{D.11}$$

Nonrelativistically, it takes on the familiar form $\omega_0^2 = 4\pi\alpha\, n_e/m_e$. Numerically it is

$$\omega_0 = 28.7\,\mathrm{eV}\, \frac{(Y_e\rho)^{1/2}}{[1 + (1.019{\times}10^{-6}\, Y_e\rho)^{2/3}]^{1/4}}, \tag{D.12}$$

where the mass density ρ is in units of $\mathrm{g\,cm^{-3}}$.

D.1.5 Screening Scale

An electric test charge will be screened by the polarization of the plasma. If the plasma is weakly coupled (see below) the screened Coulomb potential takes the form of a Yukawa potential $r^{-1}\, e^{-k_S\, r}$ where k_S is the screening scale; its inverse is the screening radius. The plasma is polarized by the test charge because the positive constituents of the plasma are repelled while the negative ones are attracted, or the reverse. Therefore, both electrons and ions contribute to screening. If both are nondegenerate the total contribution is $k_S^2 = k_D^2 + k_i^2$ where the electron contribution is known as the Debye scale,

$$k_D^2 = 4\pi\alpha n_e/T = (4\alpha/3\pi)\, p_F^3/T. \tag{D.13}$$

The ions (charge $Z_j e$, atomic weight A_j) contribute

$$k_i^2 = \frac{4\pi\alpha}{T} \sum_j Z_j^2 n_j = \frac{4\pi\alpha}{T} \frac{\rho}{m_u} \sum_j \frac{X_j Z_j^2}{A_j}. \tag{D.14}$$

For only one species of ions with charge Ze one has $k_i^2 = Z k_D^2$. Numerically,

$$k_D = 222 \,\text{eV}\, (Y_e\rho/T_8)^{1/2}, \qquad\qquad (D.15)$$

where ρ is in units of g cm^{-3} and $T_8 = T/10^8 \text{K}$. Contours in the ρ-T-plane are shown in Fig. D.5.

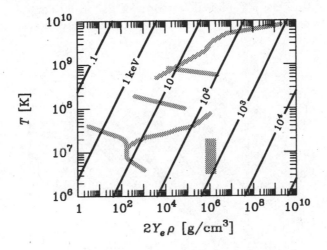

Fig. D.5. Contours for the Debye scale k_D in keV.

When the electrons are degenerate they cannot form a Debye-Hückel cloud around a test charge. Rather, their distribution is characterized by a Thomas-Fermi model which results in the screening scale $k_{TF}^2 = 4\alpha p_F E_F/\pi$. Because $k_{TF} \ll k_D$ the electron screening can be neglected relative to the ions whence $k_S \approx k_i$.

Degenerate electrons form an essentially inert background of negative charge in which the ions move, subject to their mutual Coulomb interaction. They can be treated as a weakly coupled Boltzmann gas as long as a typical thermal energy exceeds a typical Coulomb interaction energy. As a quantitative measure one uses the plasma parameter

$$\Gamma = Z^2\alpha/a_i T, \qquad\qquad (D.16)$$

where Ze is the nuclear charge and a_i the ion-sphere radius defined by $n_i^{-1} = 4\pi a_i^3/3$ with the ion density n_i. Numerically this is

$$\Gamma = 1.806\times10^{-3}\, T_8^{-1}\, (Z^5 2Y_e\rho)^{1/3}, \qquad\qquad (D.17)$$

with the mass density in units of $\mathrm{g\,cm^{-3}}$ and $T_8 = T/10^8\mathrm{K}$. Recall that for a single nuclear species $Y_e = Z/A$ (atomic weight A).

The plasma is weakly coupled for $\Gamma \lesssim 1$, it is in the liquid metal phase for $1 \lesssim \Gamma < 178$, and forms a body-centered cubic lattice for $\Gamma > 178$ (Slattery, Doolen, and DeWitt 1980, 1982). Debye screening by the ions is appropriate for a weakly coupled plasma; otherwise the Debye approximation for the ion-ion correlations is misleading. Contours for Γ in the ρ-T-plane are shown in Fig. D.6. For the purposes of this book, a strongly coupled plasma occurs only in the interior of white dwarfs.

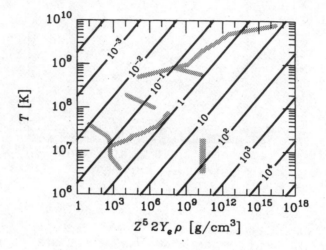

Fig. D.6. Contours for the plasma coupling parameter Γ. Also shown are the loci of the stellar models of Fig. D.1 which had to be shifted relative to each other according to the nuclear charge Z relevant for each chemical composition.

D.1.6   Summary

The characteristic plasma properties for a number of typical astrophysical sites that are important in the main body of the book are summarized in Tab. D.1.

Table D.1. Plasma characteristics for some typical astrophysical sites.

Characteristic	Center of standard solar model	Core of HB stars	Red-giant core just before helium ignition	White dwarf
	nondegenerate nonrelativistic	nondegenerate nonrelativistic	degenerate weakly coupled	degenerate strongly coupled
Temperature	1.55×10^7 K $= 1.3$ keV	$\approx 10^8$ K $= 8.6$ keV	$\approx 10^8$ K $= 8.6$ keV	$3\times10^6-2\times10^7$ K $= 0.3-1.7$ keV [a]
Density	156 g cm^{-3}	$\approx 10^4$ g cm^{-3}	$\approx 10^6$ g cm^{-3}	1.8×10^6 g cm^{-3}
Composition	$X = 0.35$	^{4}He, ^{12}C, ^{16}O	^{4}He	^{12}C, ^{16}O
Electron density	6.3×10^{25} cm^{-3}	3.0×10^{27} cm^{-3}	3.0×10^{29} cm^{-3}	5.3×10^{29} cm^{-3}
Fermi momentum	24.3 keV	88 keV	409 keV	495 keV
Fermi energy[b]	0.58 keV	7.6 keV	144 keV	200 keV
Plasma frequency	0.3 keV	2.0 keV	18 keV	23 keV
Plasma coupling	$\Gamma = 0.07$	0.12	0.57	144–22
Debye screening	electrons + ions $k_S = (k_D^2 + k_i^2)^{1/2}$ $= 9.1$ keV	electrons + ions $k_S = (k_D^2 + k_i^2)^{1/2}$ $= 27$ keV	ions $k_S = k_i$ $= 222$ keV	— (strong screening)

[a]Center of WD with $\mathcal{M} = 0.66\,\mathcal{M}_\odot$.   [b]Nonrelativstic Fermi energy $E_F - m_e$.

D.2   Nuclear Matter

D.2.1   The Ideal $p\,n\,e\,\nu_e$ Gas

For the topics discussed in this book, the properties of hot nuclear matter in a young supernova core are of great interest. The relevant range of densities and temperatures is about $3 \times 10^{12} - 3 \times 10^{15}\,\mathrm{g\,cm^{-3}}$ and $3 - 100\,\mathrm{MeV}$, respectively. The properties of matter at such conditions is determined by its equation of state which takes the nuclear interaction fully into account. However, in order to gain a rough understanding of the behavior of the main constituents of the medium (protons, neutrons, electrons, and electron neutrinos) it is worthwhile to study a simple toy model where these particles are treated as ideal Fermi gases.

To this end, neutrinos and electrons are treated as massless. Their dispersion relation is dominated by the interaction with the medium. Because they interact only by electroweak forces their "effective mass" is always much smaller than their energies.

D.2.2   Kinetic and Chemical Equilibrium

The reaction $e\,p \leftrightarrow n\,\nu_e$ which establishes β equilibrium is fast compared to other relevant time scales. Therefore, the relative abundances of n, p, e, and ν_e are determined by the conditions of kinetic and chemical equilibrium. The physical condition of the medium is then determined by the baryon density n_B, the temperature T, and the condition of electric charge neutrality

$$
\begin{aligned}
n_B &= n_n + n_p &&\text{Baryon density,} \\
n_p &= n_e &&\text{Charge neutrality,} \\
\mu_e + \mu_p &= \mu_n + \mu_{\nu_e} &&\beta\ \text{equilibrium.}
\end{aligned}
\qquad \text{(D.18)}
$$

Here, the μ_j are the relativistic chemical potentials of the fermions which determine their number densities n_j according to the Fermi-Dirac distribution Eq. (D.5). Recall that n_j is the difference between fermions and antifermions of a given species.

In addition, one of two extreme assumptions is made. In a young SN core the neutrinos are trapped so that the local lepton number is conserved. In this case the lepton fraction Y_L is the fourth required input parameter,

$$
Y_L\,n_B = n_e + n_{\nu_e} \qquad\text{Lepton conservation.} \qquad \text{(D.19)}
$$

As a neutron star cools it becomes transparent to neutrinos. In this case their chemical potential vanishes which yields

$$\mu_{\nu_e} = 0 \qquad\qquad \text{Free neutrino escape} \qquad\qquad \text{(D.20)}$$

as the other extreme additional condition.

D.2.3 Cold Nuclear Matter

The limit $T \to 0$ relevant for old neutron stars is particularly simple because it allows one to express the Fermi-Dirac distributions as step-functions. One may express all Fermi momenta in units of the effective nucleon mass, i.e., $x_j \equiv p_F^j/m_N^*$. Then baryon conservation is

$$x_B^3 = x_p^3 + x_n^3, \qquad\qquad \text{(D.21)}$$

where $n_j = (p_F^j)^3/3\pi^3$ was used, and

$$x_B \equiv (3\pi^2 n_B)^{1/3}/m_N^* = 0.255\, \rho_{14}^{1/3}\,(m_N/m_N^*), \qquad\qquad \text{(D.22)}$$

with ρ_{14} the baryonic mass density in units of $10^{14}\,\text{g}\,\text{cm}^{-3}$.

Because the star is transparent to neutrinos one may use $\mu_{\nu_e} = 0$. Then the equation of β-equilibrium becomes $\mu_e + \mu_p - \mu_n = 0$ or

$$x_p + (1 + x_p^2)^{1/2} - (1 + x_n^2)^{1/2} = 0, \qquad\qquad \text{(D.23)}$$

where $x_e = x_p$ was used from the condition of charge neutrality. This is easily solved to yield (Shapiro and Teukolsky 1983)

$$x_p^2 = \frac{x_n^4}{4\,(1 + x_n^2)}. \qquad\qquad \text{(D.24)}$$

This result may be expressed in terms of the usual composition parameters Y_p and Y_n which give the number of protons and neutrons per baryon; $Y_p + Y_n = 1$. Then $Y_{n,p} = (x_{n,p}/x_B)^3$ so that

$$Y_p = \left(\frac{x_B}{2}\right)^3 \frac{(1 - Y_p)^2}{[1 + x_B^2(1 - Y_p)^{2/3}]^{3/2}}. \qquad\qquad \text{(D.25)}$$

When $x_B \ll 1$ this is $Y_p = (x_B/2)^3 = 2.1 \times 10^{-3}\, \rho_{14}\,(m_N/m_N^*)^3$ so that the proton fraction is small—hence the term "neutron star"—although the exact Y_p for a given density depends sensitively on the nucleon dispersion relation. For infinite density ($x_B \to \infty$) a maximum of $Y_p = \frac{1}{9}$ is reached.

D.2.4   Hot Nuclear Matter

In a supernova core right after collapse the temperature is so high (several tens of MeV) that the nucleons are nearly nondegenerate. Moreover, the neutrinos are trapped so that locally a fixed value for Y_L determined by initial conditions is assumed. Most of the lepton number will reside in electrons, causing the proton concentration Y_p to be approximately equal to Y_L. A more accurate determination requires a numerical solution of Eqs. (D.18) and (D.19).

In Figs. D.7 (a)–(c) the results of such an exercise are presented for $Y_L = 0.3$, which is a typical value for the material in a SN core just after collapse. The proton concentration Y_p, the difference between the neutron and proton chemical potentials, and the degeneracy parameters for neutrons and protons are shown. In each case, a solid line refers to the assumption of an effective nucleon mass as in Fig. 4.10 while a dotted line refers to the vacuum mass. As expected, the reduced effective mass increases somewhat the medium's degeneracy.

Because $\mu_n - \mu_p = \mu_e - \mu_{\nu_e}$, the contours of Fig. D.7 (b) also give the difference between the surfaces of the electron and neutrino Fermi seas. Note that the leptons are much more degenerate than the nucleons because they are essentially massless. This remark does not apply to the upper-left corner of the plots where actually an excess of *anti*neutrinos is enforced—there are *more* protons than leptons $(Y_p > Y_L)$!

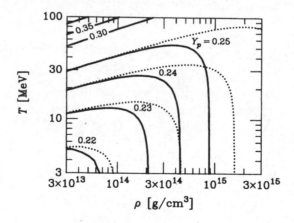

Fig. D.7. (a) Contours for Y_p in hot neutron-star matter with $Y_L = 0.3$. Solid lines for the effective nucleon mass as in Fig. D.7, dotted lines for the vacuum mass.

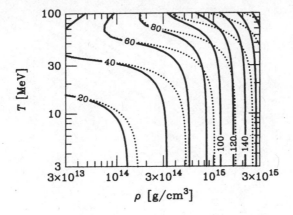

Fig. D.7. (b) $\mu_n - \mu_p = \mu_e - \mu_{\nu_e}$ (in MeV).

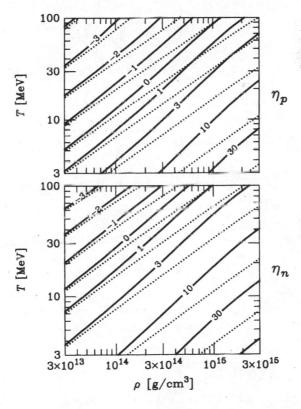

Fig. D.7. (c) Degeneracy parameter for protons and neutrons.

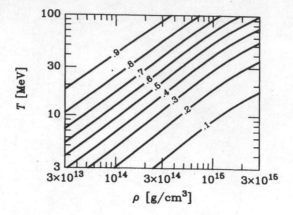

Fig. D.8. Degeneracy suppression of neutrino scattering on "heavy" nucleons for $Y_L = 0.3$ and an effective nucleon mass as in Fig. D.7.

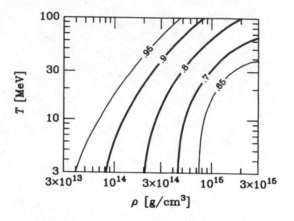

Fig. D.9. Ratio of the suppression factor of Fig. D.8 between the case with an effective nucleon mass and with the vacuum one.

When neutrinos scatter on a "heavy" nucleon recoil effects can be neglected. The nucleon does not change its momentum so that the degeneracy suppression is given by a factor

$$\xi = \frac{2}{n_N} \int \frac{d^3\mathbf{p}}{(2\pi)^3} \, f_{\mathbf{p}}(1 - f_{\mathbf{p}}), \qquad (D.26)$$

where $f_{\mathbf{p}}$ is a Fermi-Dirac occupation number. In Fig. D.8 contours for $Y_n\xi_n + Y_p\xi_p$ are shown for the same parameters as in Fig. D.7, i.e.

$Y_L = 0.3$ and the effective nucleon mass of Fig. 4.10. In Fig. D.9 the ratio of this factor between the case of an effective nucleon mass and the vacuum one are shown, i.e. the additional Pauli suppression of the neutrino scattering rate from using an effective nucleon mass.

References

Prefixes to authors' names have been used as a full part of the name so that, for example, van den Bergh, van Bibber, von Feilitzsch *and others are found under the letter V.*

Abbott, L. F., de Rújula, A., and Walker, T. P. 1988, *Nucl. Phys. B*, 299, 734.

Abbott, L. F., and Sikivie, P. 1983, *Phys. Lett. B*, 120, 133.

Abdurashitov, J. N., et al. 1994, *Phys. Lett. B*, 328, 234.

Accetta, F. S., Krauss, L. M., and Romanelli, P. 1990, *Phys. Lett. B*, 248, 146.

Accetta, F. S., and Steinhardt, P. J. 1991, *Phys. Rev. Lett.*, 67, 298.

Achkar, B., et al. 1995, *Nucl. Phys. B*, 434, 503.

Acker, A., and Pakvasa, S. 1994, *Phys. Lett. B*, 320, 320.

Acker, A., Pakvasa, S., and Raghavan, R. S. 1990, *Phys. Lett. B*, 238, 117.

Adams, E. N. 1988, *Phys. Rev. D*, 37, 2047.

Adams, J. B., Ruderman, M. A., and Woo, C.-H. 1963, *Phys. Rev.*, 129, 1383.

Adler, S. L. 1971, *Ann. Phys. (N.Y.)*, 67, 599.

Aharonov, Y., Avignone III, F. T., and Nussinov, S. 1988a, *Phys. Lett. B*, 200, 122.

Aharonov, Y., Avignone III, F. T., and Nussinov, S. 1988b, *Phys. Rev. D*, 37, 1360.

Aharonov, Y., Avignone III, F. T., and Nussinov, S. 1989, *Phys. Rev. D*, 39, 985.

Ahrens, L. A., et al. 1985, *Phys. Rev. D*, 31, 2732.

Ahrens, L. A., et al. 1990, *Phys. Rev. D*, 41, 3301.

Akhmedov, E. Kh. 1988a, *Yad. Fiz.*, 48, 599 (*Sov. J. Nucl. Phys.*, 48, 382).

Akhmedov, E. Kh. 1988b, *Phys. Lett. B*, 213, 64.

Akhmedov, E. Kh., and Berezin, V. V. 1992, *Z. Phys. C*, 54, 661.

Akhmedov, E. Kh., and Berezhiani, Z. G. 1992, *Nucl. Phys. B*, 373, 479.

Akhmedov, E. Kh., Lanza, A., and Petcov, S. T. 1993, *Phys. Lett. B*, 303, 85.

Akhmedov, E. Kh., Lanza, A., and Petcov, S. T. 1995, *Phys. Lett. B*, 348, 124.

Akhmedov, E. Kh., Lipari, P., and Lusignoli, M. 1993, *Phys. Lett. B*, 300, 128.

Akhmedov, E. Kh., Petcov, S. T., and Smirnov, A. Yu. 1993a, *Phys. Lett. B*, 309, 95.

Akhmedov, E. Kh., Petcov, S. T., and Smirnov, A. Yu. 1993b, *Phys. Rev. D*, 48, 2167.

Alcock, C., and Olinto, A. V. 1988, *Ann. Rev. Nucl. Part. Sci.*, 38, 161.

Alcock, C., et al. 1993, *Nature*, 365, 621.

Alcock, C., et al. 1995, *Phys. Rev. Lett.*, 74, 2867.

ALEPH Collaboration 1995, *Phys. Lett. B*, 349, 585.

Alexeyev, E. N., et al. 1987, *Pis'ma Zh. Eksp. Teor. Fiz.*, 45, 461 (*JETP Lett.*, 45, 589).

Alexeyev, E. N., et al. 1988, *Phys. Lett. B*, 205, 209.

Allen, C. W. 1963, *Astrophysical Quantities* (University of London Press, London).

Almeida, L. D., Matsas, G. E. A., and Natale, A. A. 1989, *Phys. Rev. D*, 39, 677.

Altherr, T. 1990, *Z. Phys. C*, 47, 559.

Altherr, T. 1991, *Ann. Phys. (N.Y.)*, 207, 374.

Altherr, T., and Kraemmer, U. 1992, *Astropart. Phys.*, 1, 133.

Altherr, T., Petitgirard, E., and del Río Gaztelurrutia, T. 1993, *Astropart. Phys.*, 1, 289.

Altherr, T., Petitgirard, E., and del Río Gaztelurrutia, T. 1994, *Astropart. Phys.*, 2, 175.

Altherr, T., and Salati, P. 1994, *Nucl. Phys. B*, 421, 662.

Alväger, T., et al. 1964, *Phys. Lett.*, 12, 260.

Alvarez, L. 1949, University of California Radiation Laboratory Report UCRL-328 (quoted after Davis, Mann, and Wolfenstein 1989).

Anand, J. D., Goyal, A., and Iha, R. N. 1990, *Phys. Rev. D*, 42, 996.

Anand, J. D., et al. 1984, *Phys. Rev. D*, 29, 1270.

Anderhub, H. B., et al. 1982, *Phys. Lett. B*, 114, 76.

Anderson, B., and Lyne, A. G. 1983, *Nature*, 303, 597.

Anderson, S. B., et al. 1993, *Ap. J.*, 414, 867.

Aneziris, C., and Schechter, J. 1991, *Int. J. Mod. Phys. A*, 6, 2375.

Anselm, A. A. 1982, *Pis'ma Zh. Eksp. Teor. Fiz.*, 36, 46 (*JETP Lett.*, 36, 55).

Anselm, A. A. 1985, *Yad. Fiz.*, 42, 1480 (*Sov. J. Nucl. Phys.*, 42, 936).

Anselm, A. A. 1988, *Phys. Rev. D*, 37, 2001.

Anselm, A. A., and Uraltsev, N. G. 1982a, *Phys. Lett. B*, 114, 39.

Anselm, A. A., and Uraltsev, N. G. 1982b, *Phys. Lett. B*, 116, 161.

Arafune, J., and Fukugita, M. 1987, *Phys. Rev. Lett.*, 59, 367.

Arafune, J., et al. 1987a, *Phys. Rev. Lett.*, 59, 1864.

Arafune, J., et al. 1987b, *Phys. Lett. B*, 194, 477.

ARGUS Collaboration 1988, *Phys. Lett. B*, 202, 149.

ARGUS Collaboration 1992, *Phys. Lett. B*, 292, 221.

Arnett, W. D., and Rosner, J. L. 1987, *Phys. Rev. Lett.*, 58, 1906.

Arnett, W. D., et al. 1989, *Ann. Rev. Astron. Astrophys.*, 27, 629.

Ashkin, A. and Dziedzic, J. M. 1973, *Phys. Rev. Lett.*, 30, 139.

Assamagan, K., et al. 1994, *Phys. Lett. B*, 335, 231.

Athanassopoulos, C., et al. 1995, *Phys. Rev. Lett.*, 75, 2650.

Athar, H., Peltoniemi, J. T., and Smirnov, A. Yu. 1995, *Phys. Rev. D*, 51, 6647.

Atzmon, E., and Nussinov, S. 1994, *Phys. Lett. B*, 328, 103.

Aubourg, E., et al. 1993, *Nature*, 365, 623.

Aubourg, E., et al. 1995, *Astron. Astrophys.*, 301, 1.

Auriemma, G., Srivastava, Y., and Widom, A. 1987, *Phys. Lett. B*, 195, 254.

Avignone, F. T., et al. 1987, *Phys. Rev. D*, 35, 2752.

Babu, K. S., Gould, T. M., and Rothstein, I. Z. 1994, *Phys. Lett. B*, 321, 140.

Babu, K. S., and Mohapatra, R. N. 1990, *Phys. Rev. D*, 42, 3866.

Babu, K. S., Mohapatra, R. N., and Rothstein, I. Z. 1992, *Phys. Rev. D*, 45, R3312.

Babu, K. S., and Volkas, R. R. 1992, *Phys. Rev. D*, 46, R2764.

Bahcall, J. N. 1989, *Neutrino Astrophysics* (Cambridge University Press).

Bahcall, J. N. 1990, *Phys. Rev. D*, 41, 2964.

Bahcall, J. N. 1991, *Phys. Rev. D*, 44, 1644.

Bahcall, J. N. 1994a, *Phys. Rev. D*, 49, 3923.

Bahcall, J. N. 1994b, *Phys. Lett. B*, 338, 276.

Bahcall, J. N. 1995, Neutrino 94, *Nucl. Phys. B (Proc. Suppl.)*, 38, 98.

Bahcall, J. N., and Bethe, H. A. 1993, *Phys. Rev. D*, 47, 1298.

Bahcall, J. N., and Frautschi, S. C. 1969, *Phys. Lett. B*, 29, 623.

Bahcall, J. N., and Glashow, S. L. 1987, *Nature*, 326, 476.

Bahcall, J. N., and Holstein, B. R. 1986, *Phys. Rev. C*, 33, 2121.

Bahcall, J. N., Krastev, P. I., and Leung, C. N. 1995, *Phys. Rev. D*, 52, 1770.

Bahcall, J. N., and Pinsonneault, M. H. 1992, *Rev. Mod. Phys.*, 64, 885.

Bahcall, J. N., and Pinsonneault, M. H. 1995, to be published in *Rev. Mod. Phys.*

Bahcall, J. N., and Press, W. H. 1991, *Ap. J.*, 370, 730.

Bahcall, J. N., and Ulrich, R. K. 1988, *Rev. Mod. Phys.*, 60, 297.

Bahcall, J. N., and Wolf, R. A. 1965a, *Phys. Rev. Lett.*, 14, 343.

Bahcall, J. N., and Wolf, R. A. 1965b, *Phys. Rev.*, 5B, 1452.

Bahcall, J. N., et al. 1995, *Nature*, 375, 29.

Bailes, M. 1989, *Ap. J.*, 342, 917.

Bailes, M., et al. 1990, *Mon. Not. R. astr. Soc.*, 247, 322.

Bakalov, D., et al. 1994, *Nucl. Phys. B (Proc. Suppl.)*, 35, 180.

Balantekin, A. B., and Loreti, F. 1992, *Phys. Rev. D*, 45, 1059.

Baluni, V. 1979, *Phys. Rev. D*, 19, 2227.

Balysh, A., et al. 1995, *Phys. Lett. B*, 356, 450.

Barbiellini, G., and Cocconi, G. 1987, *Nature*, 329, 21.

Barbieri, R., and Dolgov, A. 1991, *Nucl. Phys. B*, 349, 743.

Barbieri, R., and Fiorentini, G. 1988, *Nucl. Phys. B*, 304, 909.

Barbieri, R., and Mohapatra, R. N. 1988, *Phys. Rev. Lett.*, 61, 27.

Barbieri, R., and Mohapatra, R. N. 1989, *Phys. Rev. D*, 39, 1229.

Barbieri, R., et al. 1991, *Phys. Lett. B*, 259, 119.

Bardeen, J., Bond, J., and Efstathiou, G. 1987, *Ap. J.*, 321, 28.

Bardeen, W. A., Peccei, R. D., and Yanagida, T. 1987, *Nucl. Phys. B*, 279, 401.

Bardeen, W. A., and Tye, S. H. H. 1978, *Phys. Lett. B*, 74, 580.

Barger, V., Phillips, R. J. N., and Sarkar, S. 1995, *Phys. Lett. B*, 352, 365; (E) *ibid.*, 356, 617.

Barger, V., Phillips, R. J. N., and Whisnant, K. 1991, *Phys. Rev. D*, 44, 1629.

Barger, V., Phillips, R. J. N., and Whisnant, K. 1992, *Phys. Rev. Lett.*, 69, 3135.

Baring, M. G. 1991, *Astron. Astrophys.*, 249, 581.

Barnes, A. V., Weiler, T. J., and Pakvasa, S. 1987, *Ap. J.*, 323, L31.

Barr, S. M., and Seckel, D. 1992, *Phys. Rev. D*, 46, 539.

Barroso, A., and Branco, G. C. 1982, *Phys. Lett. B*, 116, 247.

Barrow, J. D. 1978, *Mon. Not. R. astr. Soc.*, 184, 677.

Barrow, J. D., and Burman, R. R. 1984, *Nature*, 307, 14.

Barstow, M. A. 1993, ed., *White Dwarfs: Advances in Observation and Theory* (Kluwer, Dordrecht).

Battye, R. A., and Shellard, E. P. S. 1994a, *Phys. Rev. Lett.*, 73, 2954.

Battye, R. A., and Shellard, E. P. S. 1994b, *Nucl. Phys. B*, 423, 260.

Baumann, J., et al. 1988, *Phys. Rev. D*, 37, 3107.

Baym, G. 1973, *Phys. Rev. Lett.*, 30, 1340.

Bazilevskaya, G. A., Stozhkov, Yu. I., and Charakhch'yan, T. N. 1982, *Pis'ma Zh. Eksp. Teor. Fiz.*, 35, 273 (*JETP Lett.*, 35, 341).

Beaudet, G., Petrosian, V., and Salpeter, E. E. 1967, *Ap. J.*, 150, 979.

Beck, M., et al. 1993, *Phys. Rev. Lett.*, 70, 2853.

Becker, W., and Aschenbach, B. 1995, in: M. A. Alpar et al. (eds.), *The Lives of the Neutron Stars* (Kluwer Academic Publishers), pg. 47.

Becker, W., et al. 1992, Poster presented at the conference *Physics of Isolated Pulsars* (Taos, New Mexico).

Becker-Szendy, R., et al. 1992, *Phys. Rev. Lett.*, 69, 1010.

Becklin, E. 1990, Talk presented at the *Supernova Watch Workshop* (Santa Monica).

Bég, M. A. B., Marciano, W. J., and Ruderman, M. 1978, *Phys. Rev. D*, 17, 1395.

Belesev, A. I., et al. 1994, Report INR-862/94 (Institute for Nuclear Research, Russia).

Berezhiani, Z. G., Moretti, M., and Rossi, A. 1993, *Z. Phys. C*, 58, 423.

Berezhiani, Z. G., and Rossi, A. 1994, *Phys. Lett. B*, 336, 439.

Berezhiani, Z. G., and Rossi, A. 1995, *Phys. Rev. D*, 51, 5229.

Berezhiani, Z. G., and Smirnov, A. Yu. 1989, *Phys. Lett. B*, 220, 279.

Berezhiani, Z. G., Smirnov, A. Yu., and Valle, J. W. F. 1992, *Phys. Lett. B*, 291, 99.

Berezhiani, Z. G., and Vysotsky, M. I. 1987, *Phys. Lett. B*, 199, 281.

Berezhiani, Z. G., et al. 1992, *Z. Phys. C*, 54, 581.

Berezinsky, V. 1994, *Comm. Nucl. Part. Phys.*, 21, 249.

Berezinsky, V., Fiorentini, G., and Lissia, M. 1994, *Phys. Lett. B*, 341, 38.

Bernabéu, J., et al. 1994, *Nucl. Phys. B*, 426, 434.

Bernstein, J., Ruderman, M. A., and Feinberg, G. 1963, *Phys. Rev.*, 132, 1227.

Bershady, M. A., Ressel, M. T., Turner, M. S. 1991, *Phys. Rev. Lett.*, 66, 1398.

Bethe, H. A. 1935, *Proc. Camb. Phil. Soc.*, 31, 108.

Bethe, H. A. 1939, *Phys. Rev.*, 55, 434.

Bethe, H. A. 1986, *Phys. Rev. Lett.*, 56, 1305.

Bethe, H. A., and Wilson, J. R. 1985, *Ap. J.*, 295, 14.

Bhattacharya, D., and van den Heuvel, E. 1991, *Phys. Rep.*, 203, 1

Bica, E., et al. 1991, *Ap. J.*, 381, L51.

Bieber, J. W., et al. 1990, *Nature*, 348, 407.

Bilenky, S. M., and Giunti, C. 1993, *Phys. Lett. B*, 311, 179.

Bilenky, S. M., and Giunti, C. 1994, *Phys. Lett. B*, 320, 323.

Bionta, R. M., et al. 1987, *Phys. Rev. Lett.*, 58, 1494.

Bionta, R. M., et al. 1988, *Phys. Rev. D*, 38, 768.

Bisnovatyi-Kogan, G. V., and Janka, H.-T. 1995, work in progress.

Bjorken, J. D., and Drell, S. D. 1964, *Relativistic Quantum Mechanics* (McGraw-Hill, New York).

Blinnikov, S. I., and Dunina-Barkovskaya, N. V. 1994, *Mon. Not. R. astr. Soc.*, 266, 289.

Blinnikov, S. I., and Okun, L. B. 1988, *Pis'ma Astron. Zh.*, 14, 867 (*Sov. Astron. Lett.*, 14, 368).

Blinnikov, S. I., and Rudzskiĭ, M. A. (1989), *Astron. Zh.*, 66, 730 (*Sov. Astron.*, 33, 377).

Blinnikov, S. I., et al. 1995, Report ITEP-31-95 and hep-ph/9505444.

Bludman, S. A. 1992, *Phys. Rev. D*, 45, 4720.

Bludman, S., et al. 1993, *Phys. Rev. D*, 47, 2220.

Bludman, S., Kennedy, D., and Langacker, P. 1992a, *Phys. Rev. D*, 45, 1810.

Bludman, S., Kennedy, D., and Langacker, P. 1992b, *Nucl. Phys. B*, 374, 373.

Boehm, F., and Vogel, P. 1987, *Physics of Massive Neutrinos* (Cambridge University Press).

Boguta, J. 1981, *Phys. Lett. B*, 106, 255.

Bond, J. R., and Efstathiou, G., *Phys. Lett. B*, 265, 245.

Bouquet, A., and Vayonakis, C. E. 1982, *Phys. Lett. B*, 116, 219.

Boris, S., et al. 1987, *Phys. Rev. Lett.*, 58, 2019.

Born, M., and Wolf, E. 1959, *Principles of Optics* (Pergamon Press, London).

Börner, G. 1992, *The Early Universe*, 2nd edition (Springer, Berlin).

Borodovsky, L., et al. 1992, *Phys. Rev. Lett.*, 68, 274.

Botella, F. J., Lim, C.-S., and Marciano, W. J. 1987, *Phys. Rev. D*, 35, 896.

Bouchez, J., et al. 1988, *Phys. Lett. B*, 207, 217.

Boyd, R. N., et al. 1983, *Phys. Rev. Lett.*, 51, 609.

Boyd, R. N., et al. 1985, *Ap. J.*, 289, 155.

Braaten, E. 1991, *Phys. Rev. Lett.*, 66, 1655.

Braaten, E. 1992, *Ap. J.*, 392, 70.

Braaten, E., and Segel, D. 1993, *Phys. Rev. D*, 48, 1478.

Bratton, C. B., et al. 1988, *Phys. Rev. D*, 37, 3361.

Brecher, K. 1977, *Phys. Rev. Lett.*, 39, 1051.

Brinkmann, R. P., and Turner, M. S. 1988, *Phys. Rev. D*, 38, 2338.

Brinkmann, W., and Ögelman, H. 1987, *Astron. Astrophys.*, 182, 71.

Brodsky, S. J., et al. 1986, *Phys. Rev. Lett.*, 56, 1763.

Broggini, C., et al. 1990, Experimental Proposal (Univ. Neuchâtel).

Brown, B. A., Csótó, A., and Sherr, R. 1995, Report nucl-th/9506004.

Brown, G. E. 1988, ed., *Phys. Rep.*, 163, 1.

Brown, G. E., Bethe, H. A., and Baym, G. 1982, *Nucl. Phys. A*, 375, 481.

Bruenn, S. W. 1987, *Phys. Rev. Lett.*, 59, 938.

Bruenn, S. W., and Mezzacappa, A. 1994, *Ap. J.*, 433, L45.

Bryman, D. A., et al., 1983, *Phys. Rev. Lett.*, 50, 1546.

Buonanno, R., et al. 1986, *Mem. Soc. Astron. Ital.*, 57, 391.

Buonanno, R., Corsi, C. E., and Fusi Pecci, F. 1989, *Astron. Astrophys.*, 216, 80.

Burbidge, E. M., et al. 1957, *Rev. Mod. Phys.*, 29, 547.

Burgess, C. P., and Cline, J. M. 1993, *Phys. Lett. B*, 298, 141.

Burgess, C. P., and Cline, J. M. 1994a, *Phys. Rev. D*, 5925.

Burgess, C. P., and Cline, J. M. 1994b, Report hep-ph/9401334, *Int. Conf. Nonaccelerator Particle Physics*, Bangalore, India, Jan. 1994.

Burrows, A. 1979, *Phys. Rev. D*, 20, 1816.

Burrows, A. 1988a, *Ap. J.*, 328, L51.

Burrows, A. 1988b, *Ap. J.*, 334, 891.

Burrows, A. 1990a, *Ann. Rev. Nucl. Part. Sci.*, 40, 181.

Burrows, A. 1990b, in: Petschek 1990.

Burrows, A., and Fryxell, B. A. 1992, *Science*, 258, 430.

Burrows, A., and Fryxell, B. A. 1993, *Ap. J.*, 418, L33.

Burrows, A., Gandhi, R., and Turner, M. S. 1992, *Phys. Rev. Lett.*, 68, 3834.

Burrows, A., Hayes, J., and Fryxell, B. A. 1995, *Ap. J.*, 450, 830.

Burrows, A., Klein, D., and Gandhi, R. 1992, *Phys. Rev. D*, 45, 3361.

Burrows, A., and Lattimer, J. M. 1986, *Ap. J.*, 307, 178.

Burrows, A., and Lattimer, J. M. 1987, *Ap. J.*, 318, L63.

Burrows, A., and Lattimer, J. M. 1988, *Phys. Rep.*, 163, 51.

Burrows, A., Ressell, T., and Turner, M. S. 1990, *Phys. Rev. D*, 42, 3297.

Burrows, A., Turner, M. S., and Brinkmann, R. P. 1989, *Phys. Rev. D*, 39, 1020.

Buzzoni, A., et al. 1983, *Astron. Astrophys.*, 128, 94.

Callan, C. G., Dashen, R. F., and Gross, D. J. 1976, *Phys. Lett. B*, 63, 334.

Cameron, A. G. W. 1957, *Publ. Astr. Soc. Pacific*, 69, 201.

Cameron, R., et al. 1993, *Phys. Rev. D*, 47, 3707.

Cannon, R. D. 1970, *Mon. Not. R. astr. Soc.*, 150, 111.

Cantatore, G., et al. 1991, *Phys. Lett. B*, 265, 418.

Caraveo, P. A. 1993, *Ap. J.*, 415, L111.

Carena, M., and Peccei, R. D. 1989, *Phys. Rev. D*, 40, 652.

Carlson, E. D. 1995, *Phys. Lett. B*, 344, 245.

Carlson, E. D., and Garretson, W. D. 1994, *Phys. Lett. B*, 336, 431.

Carlson, E. D., and Salati, P. 1989, *Phys. Lett. B*, 218, 79.

Carlson, E. D., and Tseng, L.-S. 1995, Report HUTP-95/A025 and hep-ph/9507345, to be published in *Phys. Lett. B*.

Casas, J. A., García-Bellido, J., and Quirós, M. 1992, *Phys. Lett. B*, 278, 94.

Castellani, M., and Castellani, V. 1993, *Ap. J.*, 407, 649.

Castellani, M., and Degl'Innocenti, S. 1993, *Ap. J.*, 402, 574.

Castellani, V., Degl'Innocenti, S., and Fiorentini, G. 1993, *Phys. Lett. B*, 303, 68.

Castellani, V., Degl'Innocenti, S., and Romaniello, M. 1994, *Ap. J.*, 423, 266.

Castellani, V., et al. 1994a, *Phys. Lett. B*, 324, 425.

Castellani, V., et al. 1994b, *Phys. Rev. D*, 50, 4749.

Castle, B., and Towner, I. 1990, *Modern Theories of Nuclear Moments* (Clarendon Press, Oxford).

Catelan, M., de Freitas Pacheco, J. A., and Horvath, J. E. 1995, Report astro-ph/9509062, to be published in *Ap. J.*

Cazzola, P., de Zotti, G., and Saggion, A. 1971, *Phys. Rev. D*, 3, 1722.

CDF Collaboration 1995, *Phys. Rev. Lett.*, 74, 2627.

Chaboyer, B., et al. 1992, *Ap. J.*, 388, 372.

Chan, K.-W., and Lingenfelter, R. E. 1993, *Ap. J.*, 405, 614.

Chanda, R., Nieves, J. F., and Pal, P. B. 1988, *Phys. Rev. D*, 37, 2714.

Chandrasekhar, S. 1939, *An Introduction to the Study of Stellar Structure* (University of Chicago Press, Chicago).

Chang, L. N., and Zia, R. K. P. 1988, *Phys. Rev. D*, 38, 1669.

Chang, S., and Choi, K. 1994, *Phys. Rev. D*, 49, R12.

CHARM II Collaboration 1993, *Phys. Lett. B*, 309, 463.

Chen, P. 1995, *Phys. Rev. Lett.*, 74, 634; (E) *ibid.*, 3091.

Cheng, B., Schramm, D. N., and Truran, J. W. 1993, *Phys. Lett. B*, 316, 521.

Cheng, H.-Y. 1988, *Phys. Rep.*, 158, 1.

Chernikov, M. A., et al. 1992, *Phys. Rev. Lett.*, 68, 3383; (E) *ibid.*, 69, 2999.

Chibisov, G. V. 1976, *Sov. Phys. Usp.*, 19, 624.

Chicashige, Y., Mohapatra, R. N., and Peccei, R. D. 1981, *Phys. Lett. B*, 98, 265.

Chieffi, A., Straniero, O., and Salaris, M. 1991, in: K. Janes (ed.), *The Formation and Evolution of Star Clusters* (ASP Conference Series, 13), pg. 219.

Chin, C.-W., and Stothers, R. 1975, *Nature*, 254, 206.

Chin, C.-W., and Stothers, R. 1976, *Phys. Rev. Lett.*, 36, 833.

Chiu, H.-Y., Chan, K. L., and Kondo, Y. 1988, *Ap. J.*, 329, 326.

Chiu, H.-Y., and Salpeter, E. E. 1964, *Phys. Rev. Lett.*, 12, 413.

Chiu, H.-Y., and Stabler, R. C. 1961, *Phys. Rev.*, 122, 1317.

Chodil, G., et al. 1965, *Phys. Rev. Lett.*, 15, 605.

Choi, K., Kang, K., and Kim, J. E. 1989, *Phys. Rev. Lett.*, 62, 849.

Choi, K., and Santamaria, A. 1990, *Phys. Rev. D*, 42, 293.

Choi, K., et al. 1988, *Phys. Rev. D*, 37, 3225.

Choudhury, D., Hari Dass, N. D., and Murthy, M. V. N. 1989, *Class. Quantum Grav.*, 6, L167.

Christensen-Dalsgaard, J. 1992, *Ap. J.*, 385, 354.

Christensen-Dalsgaard, J., Proffitt, C. R., and Thompson, M. J. 1993, *Ap. J.*, 403, L75.

Chugaĭ, N. N. 1984, *Pis'ma Astron. Zh.*, 10, 210 (*Sov. Astron. Lett.*, 10, 87).

Chupp, E. L., Vestrand, W. T., and Reppin, C. 1989, *Phys. Rev. Lett.*, 62, 505.

Çiftçi, A. K., Sultansoi, S., and Türköz, S. 1994, Ankara University Preprint AU/94-03/HEP (quoted after Blinnikov et al. 1995).

Cisneros, A. 1971, *Astrophys. Space Sci.*, 10, 87.

Clayton, D. D. 1968, *Principles of Stellar Evolution and Nucleosynthesis* (University of Chicago Press, Chicago).

CLEO Collaboration 1993, *Phys. Rev. Lett.*, 70, 3700.

Cline, D., et al. 1990, *Astro. Lett. and Communications*, 27, 403.

Cline, D., et al. 1994, *Phys. Rev. D*, 50, 720.

Cline, J. M. 1992, *Phys. Rev. D*, 45, 1628.

Cocconi, G. 1988, *Phys. Lett. B*, 206, 705.

Cohen, J. G., and Frogel, J. A. 1982, *Ap. J.*, 255, L39.

Cohen, J. G., Frogel, J. A., and Persson, S. E. 1978, *Ap. J.*, 222, 165.

Coley, A. A., and Tremaine, S. 1988, *Phys. Rev. D*, 38, 2927.

Colgate, S. A., and Johnson, M. H. 1960, *Phys. Rev. Lett.*, 5, 235.

Colgate, S. A., and White, R. H. 1966, *Ap. J.*, 143, 626.

Commins, E. D., and Bucksbaum, P. 1983, *Weak Interactions of Leptons and Quarks* (Cambridge University Press, Cambridge).

Cooper, L., and Stedman, G. E. 1995, *Phys. Lett. B*, 357, 464.

Cooper-Sarkar, A. M., et al. 1992, *Phys. Lett. B*, 280, 153.

Cooperstein, J. 1988, *Phys. Rep.*, 163, 95.

Cooperstein, J., van den Horn, L. J., and Baron, E. A. 1987, *Ap. J.*, 321, L129.

Cordes, J. M. 1986, *Ap. J.*, 311, 183.

Cordes, J. M., Romani, R. W., and Lundgren, S. C. 1993, *Nature*, 362, 133.

Cowsik, R. 1977, *Phys. Rev. Lett.*, 39, 784.

Cowsik, R. 1988, *Phys. Rev. D*, 37, 1685.

Cowsik, R., Schramm, D., and Höflich, P. 1989, *Phys. Lett. B*, 218, 91.

Cox, J. P. 1980, *Theory of Stellar Pulsation* (Princeton University Press, Princeton).

Crewther, R., et al. 1979, *Phys. Lett. B*, 88, 123; (E) 1980, *ibid.*, 91, 487.

Cung, V. K., and Yoshimura, M. 1975, *Nuovo Cim.*, 29 A, 557.

D0 Collaboration 1995, *Phys. Rev. Lett.*, 74, 2633.

Da Costa, G. S., and Armandroff, T. E. 1990, *Astron. J.*, 100, 162.

Da Costa, G. S., Frogel, J. A., and Cohen, J. G. 1981, *Ap. J.*, 248, 612.

Damour, T., and Gundlach, C. 1991, *Phys. Rev. D*, 43, 3873.

Damour, T., and Taylor, J. H. 1991, *Ap. J.*, 366, 501.

D'Antona, F., and Mazzitelli, I. 1990, *Ann. Rev. Astron. Astrophys.*, 28, 139.

Dar, A. 1987, Report, Institute for Advanced Study (unpublished).

Dar, A., and Dado, S. 1987, *Phys. Rev. Lett.*, 59, 2368.

Dar, A., Goodman, J., and Nussinov, S. 1987, *Phys. Rev. Lett.*, 58, 2146.

Daum, K. 1994, Report WUB 94-9, Contributed paper to Neutrino 94 (unpublished).

Daum, M., et al. 1991, *Phys. Lett. B*, 265, 425.

Davidsen, A. F., et al. 1991, *Nature*, 351, 128.

Davidson, S., Campbell, B., and Bailey, D. 1991, *Phys. Rev. D*, 43, 2314.

Davidson, S., and Peskin, M. 1994, *Phys. Rev. D*, 49, 2114.

Davis, R. L. 1986, *Phys. Lett. B*, 180, 225.

Davis, R. L., and Shellard, E. P. S. 1989, *Nucl. Phys. B*, 324, 167.

Davis Jr., L., Goldhaber, A. S., and Nieto, M. M. 1975, *Phys. Rev. Lett.*, 35, 1402.

Davis Jr., R. 1964, *Phys. Rev. Lett.*, 12, 303.

Davis Jr., R., Harmer, D. S., and Hoffman, K. C. 1968, *Phys. Rev. Lett.*, 20, 1205.

Davis Jr., R., Mann, A. K., and Wolfenstein, L. 1989, *Ann. Rev. Nucl. Part. Sci.*, 39, 467.

Dearborn, D. S. P., Schramm, D. N., and Steigman, G. 1986, *Phys. Rev. Lett.*, 56, 26.

Dearborn, D. S. P., et al. 1990, *Ap. J.*, 354, 568.

Debye, P., and Hückel, E. 1923, *Phys. Z.*, 24, 185.

Degl'Innocenti, S., Fiorentini, G., and Lissia, M. 1995, *Nucl. Phys. B (Proc. Suppl.)*, 43, 66.

Degl'Innocenti, S., et al. 1995, Report MPI-PTh/95-78 and astro-ph/ 9509090, submitted to *Astron. Astrophys.*

Degrassi, G., Sirlin, A., and Marciano, W. J. 1989, *Phys. Rev. D*, 39, 287.

del Campo, S., and Ford, L. H. 1988, *Phys. Rev. D*, 38, 3657.

Demarque, P., et al. 1994, *Ap. J.*, 437, 870.

Dewey, R. J., and Cordes, J. M. 1987, *Ap. J.*, 321, 780.

Dicus, D. A. 1972, *Phys. Rev. D*, 6, 961.

Dicus, D. A., Kolb, E. W., and Teplitz, V. L. 1977, *Phys. Rev. Lett.*, 39, 169.

Dicus, D. A., Kolb, E. W., and Tubbs, D. L. 1983, *Nucl. Phys. B*, 223, 532.

Dicus, D. A., and Repko, W. W. 1993, *Phys. Rev. D*, 48, 5106.

Dicus, D. A., et al. 1976, *Ap. J.*, 210, 481.

Dicus, D. A., et al. 1978, *Phys. Rev. D*, 18, 1829.

Dicus, D. A., et al. 1980, *Phys. Rev. D*, 22, 839.

Dicus, D. A., et al. 1989, *Phys. Lett. B*, 218, 84.

DiLella, L. 1993, Neutrino 92, *Nucl. Phys. B (Proc. Suppl.)*, 31, 319.

Dimopoulos, S., Starkman, G. D., and Lynn, B. W. 1986a, *Phys. Lett. B*, 167, 145.

Dimopoulos, S., Starkman, G. D., and Lynn, B. W. 1986b, *Mod. Phys. Lett. A*, 8, 491.

Dimopoulos, S. et al. 1986, *Phys. Lett. B*, 179, 223.

Dine, M., and Fischler, W. 1983, *Phys. Lett. B*, 120, 137.

Dine, M., Fischler, W., and Srednicki, M. 1981, *Phys. Lett. B*, 104, 199.

Dine, M., and Seiberg, N. 1986, *Nucl. Phys. B*, 273, 109.

Dirac, P. A. M. 1937, *Nature*, 139, 323.

Dirac, P. A. M. 1938, *Proc. R. Soc.*, A165, 199.

Dobroliubov, M. I., and Ignatiev, A. Yu. 1990, *Phys. Rev. Lett.*, 65, 679.

Dodd, A. C., Papageorgiu, E., and Ranfone, S. 1991, *Phys. Lett. B*, 266, 434.

Dodelson, S., and Feinberg, G. 1991, *Phys. Rev. D*, 43, 913.

Dodelson, S., Frieman, J. A., and Turner, M. S. 1992, *Phys. Rev. Lett.*, 68, 2572.

Dodelson, S., Gyuk, G., and Turner, M. S. 1994, *Phys. Rev. Lett.*, 72, 3754.

Dolgov, A. D. 1981, *Yad. Fiz.*, 33, 1309 (*Sov. J. Nucl. Phys.*, 33, 700).

Dolgov, A. D., Kainulainen, K., and Rothstein, I. Z. 1995, *Phys. Rev. D*, 51, 4129.

Dolgov, A. D., and Rothstein, I. Z. 1993, *Phys. Rev. Lett.*, 71, 476.

Dolgov, A. D., and Raffelt, G. G. 1995, *Phys. Rev. D*, 52, 2581.

D'Olivo, J. C., Nieves, J. F., and Pal, P. B. 1989, *Phys. Rev. D*, 40, 3679.

D'Olivo, J. C., Nieves, J. F., and Pal, P. B. 1990, *Phys. Rev. Lett.*, 64, 1088.

Domokos, G., and Kovesi-Domokos, S. 1995, *Phys. Lett. B*, 346, 317.

Donelly, T. W., et al. 1978, *Phys. Rev. D*, 18, 1607.

Dorofeev, O. F., Rodionov, V. N., and Ternov, I. M. 1985, *Pis'ma Astron. Zh.*, 11, 302 (*Sov. Astron. Lett.*, 11, 123).

Dydak, F., et al. 1984, *Phys. Lett. B*, 134, 281.

Dziembowski, W. A., et al. 1994, *Ap. J.*, 432, 417.

Eggleton, P. P., and Cannon, R. C. 1991, *Ap. J.*, 383, 757.

Eggleton, P. P., and Faulkner, J. 1981, in: Iben and Renzini (1981).

Ellis, J., and Karliner, M. 1995, *Phys. Lett. B*, 341, 397.

Ellis, J., and Olive, K. A. 1983, *Nucl. Phys. B*, 233, 252.

Ellis, J., and Olive, K. A. 1987, *Phys. Lett. B*, 193, 525.

Ellis, J., and Salati, P. 1990, *Nucl. Phys. B*, 342, 317.

Ellis, J., et al. 1986, *Phys. Lett. B*, 167, 457.

Ellis, J., et al. 1988, *Phys. Lett. B*, 215, 404.

Ellis, J., et al. 1989, *Phys. Lett. B*, 228, 264.

Engel, J., Krastev, P. I., and Lande, K. 1995, Report hep-ph/9501219, submitted to *Phys. Rev. D*.

Engel, J., Seckel, D., and Hayes, A. C. 1990, *Phys. Rev. Lett.*, 65, 960.

Enqvist, K., Rez, A. I., and Semikoz, V. B. 1995, *Nucl. Phys. B*, 436, 49.

Enqvist, K., and Uibo, H. 1993, *Phys. Lett. B*, 301, 376.

Ezer, D., and Cameron, A. G. W. 1966, *Can. J. Phys.*, 44, 593.

Falk, S. W., and Schramm, D. N. 1978, *Phys. Lett. B*, 79, 511.

Faulkner, J., and Swenson, F. J. 1988, *Ap. J.*, 329, L47.

Fayons, S. A., Kopeykin, V. I., and Mikaelyan, L. A. 1991, quoted after L. Moscoso, Neutrino 90, *Nucl. Phys. B (Proc. Suppl.)*, 19, 147 (1991).

Feinberg, E. L., and Pomeranchuk, I. 1956, *Nuovo Cim. Suppl.*, 3, 652.

Festa G. G., and Ruderman, M. A., *Phys. Rev.*, 180, 1227.

Filippone, B. W., and Vogel, P. 1990, *Phys. Lett. B*, 246, 546.

Filippone, B. W., et al. 1983, *Phys. Rev. C*, 28, 2222.

Finley, J. P., Ögelman, H., Kiziloğlu, Ü. 1992, *Ap. J.*, 394, L21.

Fiorentini, G., and Mezzorani, G. 1989, *Phys. Lett. B*, 221, 353.

Fiorentini, G., et al. 1994, *Phys. Rev. D*, 49, 6298.

Fischbach, E., and Talmadge, C. 1992, *Nature*, 356, 207.

Fischbach, E., et al. 1976, *Phys. Rev. D*, 13, 1523.

Fischbach, E., et al. 1977, *Phys. Rev. D*, 16, 2377.

Fischbach, E., et al. 1986, *Phys. Rev. Lett.*, 56, 3.

Fischbach, E., et al. 1994, *Phys. Rev. Lett.*, 73, 514.

Fleming, T. A., Liebert, J., and Green, R. F. 1986, *Ap. J.*, 308, 176.

Flowers, E. 1973, *Ap. J.*, 180, 911.

Flowers, E. 1974, *Ap. J.*, 190, 381.

Flynn, J. M., and Randall, L. 1988, Report LBL-25115, unpublished.

Fogli, G. L., and Lisi, E. 1995, *Phys. Rev. D*, 52, 2775.

Foldy, L. L. 1945, *Phys. Rev.*, 67, 107.

Fomalont, E. B., et al. 1992, *Mon. Not. R. astr. Soc.*, 258, 497.

Foot, R. 1994, *Phys. Rev. D*, 49, 3617.

Foot, R., and Lew, H. 1993, *Mod. Phys. Lett. A*, 8, 3767.

Foot, R., Lew, H., and Volkas, R. R. 1993, *J. Phys. G: Nucl. Part. Phys.*, 19, 361; (E) *ibid.*, 1067.

Forrest, D. J., et al. 1980, *Sol. Phys*, 65, 15.

Frail, D. A., and Kulkarni, S. R. 1991, *Nature*, 352, 785.

Freedman, D. Z. 1974, *Phys. Rev. D*, 9, 1389.

Fréjus Collaboration 1990, *Phys. Lett. B*, 245, 305.

Fréjus Collaboration 1995, *Z. Phys. C*, 66, 417.

Frieman, J. A., Dimopoulos, S., and Turner, M. S. 1987, *Phys. Rev. D*, 36, 2201.

Frieman, J. A., Haber, H. E., and Freese, K. 1988, *Phys. Lett. B*, 200, 115.

Friman, B. L., and Maxwell, O. V. 1979, *Ap. J.*, 232, 541.

Fritzsch, H., and Plankl, J. 1987, *Phys. Rev. D*, 35, 1732.

Frogel, J. A., Cohen, J. G., and Persson, S. E., 1983, *Ap. J.*, 275, 773.

Frogel, J. A., Persson, S. E., and Cohen, J. G. 1981, *Ap. J.*, 246, 842.

Frogel, J. A., Persson, S. E., and Cohen, J. G. 1983, *Ap. J. Suppl.*, 53, 713.

Frost, K. J., Rothe, E. D., and Peterson, L. E. 1966, *J. Geophys. Res.*, 71, 4079.

Fujikawa, K., and Shrock, E. 1980, *Phys. Rev. Lett.*, 45, 963.

Fujiwara, K. 1989, *Phys. Rev. D*, 39, 1764.

Fukuda, Y., et al. 1994, *Phys. Lett. B*, 335, 237.

Fukugita, M., and Sakai, N. 1982, *Phys. Lett. B*, 114, 23.

Fukugita, M., Watamura, S., and Yoshimura, M. 1982a, *Phys. Rev. Lett.*, 48, 1522.

Fukugita, M., Watamura, S., and Yoshimura, M. 1982b, *Phys. Rev. D*, 26, 1840.

Fukugita, M., and Yanagida, T. 1988, *Phys. Lett. B*, 206, 93.

Fukugita, M., and Yazaki, S. 1987, *Phys. Rev. D*, 36, 3817.

Fuller, G. M., and Malaney, R. A. 1991, *Phys. Rev. D*, 43, 3136.

Fuller, G. M., Mayle, R., and Wilson, J. R. 1988, *Ap. J.*, 332, 826.

Fuller, G. M., et al. 1992, *Ap. J.*, 389, 517.

Fusi Pecci, F., et al. 1990, *Astron. Astrophys.*, 238, 95.

Gabriel, M. D., et al. 1991, *Phys. Rev. Lett.*, 67, 2123.

Gaemers, K. J. F., Gandhi, R., and Lattimer, J. M. 1989, *Phys. Rev. D*, 40, 309.

Gai, M. 1995, Neutrino 94, *Nucl. Phys. B (Proc. Suppl.)*, 38, 77.

Gai, M., and Bertulani, C. A. 1995, *Phys. Rev. C*, 52, 1706.

Galam, S., and Hansen, J.-P. 1976, *Phys. Rev. A*, 14, 816.

GALLEX Collaboration 1994, *Phys. Lett. B*, 327, 377.

GALLEX Collaboration 1995a, *Phys. Lett. B*, 342, 440.

GALLEX Collaboration 1995b, Report GX 75-1995, submitted to *Phys. Lett. B*.

Gamow, G. 1967, *Proc. Natl. Acad. Sci.*, 57, 187.

Gamow, G., and Schoenberg, M. 1940, *Phys. Rev.*, 58, 1117.

Gamow, G., and Schoenberg, M. 1941, *Phys. Rev.*, 59, 539.

Gandel'man, G. M., and Pinaev, V. S. 1959, *Zh. Eksp. Teor. Fiz.*, 37, 1072 (*Sov. Phys. JETP*, 10, 764).

Gandhi, R., and Burrows, A. 1990, *Phys. Lett. B*, 246, 149; (E) 1991, *ibid.*, 261, 519.

Garcia A., et al. 1991, *Phys. Rev. Lett.*, 67, 3654.

García-Berro, E., et al. 1995, in: D. Koester and K. Werner (eds.), *White Dwarfs*, Proc. 9th European Workshop on White Dwarfs, Kiel, Germany, 29 August–1 Sept. 1994 (Springer, Berlin).

Gasperini, M. 1987, *Phys. Rev. Lett.*, 59, 396.

Gasperini, M. 1988, *Phys. Rev. D*, 38, 2635.

Gasperini, M. 1989, *Phys. Rev. D*, 39, 3606.

Gasser, J., and Leutwyler, H. 1982, *Phys. Rep.*, 87, 77.

Gates, E., Krauss, L. M., and White, M. 1995, *Phys. Rev. D*, 51, 2631.

Gavrin, V. 1995, Neutrino 94, *Nucl. Phys. B (Proc. Suppl.)*, 38, 60.

Gell-Mann, M. 1961, *Phys. Rev. Lett.*, 6, 70.

Gelmini, G. B., Nussinov, S., and Roncadelli, M. 1982, *Nucl. Phys. B*, 209, 157.

Gelmini, G. B., and Roncadelli, M. 1981, *Phys. Lett. B*, 99, 411.

Georgi, H., Glashow, S. L., and Nussinov, S. 1981, *Nucl. Phys. B*, 193, 297.

Georgi, H., Hall, L., and Wise, M. 1981, *Nucl. Phys. B*, 192, 409.

Georgi, H., Kaplan, D. B., and Randall, L. 1986, *Phys. Lett. B*, 169, 73.

Ghosh, R. K. 1984, *Phys. Rev. D*, 29, 493.

Ghosh, S. K., Phatak, S. C., and Sahu, P. K. 1994, *Mod. Phys. Lett. A*, 9, 1717.

Gilliland, R. L., and Däppen, W. 1987, *Ap. J.*, 313, 429.

Giudice, G. F. 1990, *Phys. Lett. B*, 251, 460.

Giunti, C., Kim, C. W., and Lam, W. P. 1991, *Phys. Rev. D*, 43, 164.

Giunti, C., Kim, C. W., and Lee, U. W. 1991, *Phys. Rev. D*, 44, 3635.

Giunti, C., Kim, C. W., and Lee, U. W. 1992, *Phys. Rev. D*, 45, 2414.

Giunti, C., et al. 1992, *Phys. Rev. D*, 45, 1557.

Giunti, C., et al. 1993, *Phys. Rev. D*, 48, 4310.

Glashow, S. L., Iliopoulos, J., and Maiani, L. 1970, *Phys. Rev. D*, 2, 1285.

Glass, E. N., and Szamosi, G. 1987, *Phys. Rev. D*, 35, 1205.

Glass, E. N., and Szamosi, G. 1989, *Phys. Rev. D*, 39, 1054.

Gnedin, Yu. N., and Krasnikov, S. V. 1992, *Zh. Eksp. Teor. Fiz.*, 102, 1729 (*Sov. Phys. JETP*, 75, 933).

Goldhaber, A. S., and Nieto, M. M. 1971, *Rev. Mod. Phys.*, 43, 277.

Goldman, I. 1990, *Mon. Not. R. astr. Soc.*, 244, 184.

Goldman, I., et al. 1988, *Phys. Rev. Lett.*, 60, 1789.

Goldman, V. M., Zisman, G. A., and Shaulov, R. Ya. 1972, Tematichenskiĭ Sbornik LGPI (quoted after Blinnikov et al. 1995).

Goldreich, P., and Julian, W. H. 1969, *Ap. J.*, 157, 869.

Goldreich, P., and Weber, S. 1980, *Ap. J.*, 238, 991.

Góngora-T., A., and Stuart, R. G. 1992, *Z. Phys. C*, 55, 101.

Goodman, M. C. 1995, Neutrino 94, *Nucl. Phys. B (Proc. Suppl.)*, 38, 337.

Goodman, J., Dar, A., and Nussinov, S. 1987, *Ap. J.*, 314, L7.

Gordon, J. P. 1973, *Phys. Rev. A*, 8, 14.

Gott, J. R., Gunn, J. E., and Ostriker, J. P. 1970, *Ap. J.*, 160, L91.

Gould, R. J. 1985, *Ap. J.*, 288, 789.

Gould, S. J. 1994, *Scientific American*, 271:4, 85.

Goyal, A., and Anand, J. D. 1990, *Phys. Rev. D*, 42, 992.

Goyal, A., Dutta, S., and Choudhury, S. R. 1995, *Phys. Lett. B*, 346, 312.

Greene, G. L., et al. 1991, *Phys. Rev. D*, 44, R2216.

Gregores, E. M., et al. 1995, *Phys. Rev. D*, 51, 4587.

Gribov, N. V., and Pontecorvo, B. M. 1969, *Phys. Lett. B*, 28, 493.

Grifols, J. A., and Massó, E. 1986, *Phys. Lett. B*, 173, 237.

Grifols, J. A., and Massó, E. 1989, *Phys. Rev. D*, 40, 3819.

Grifols, J. A., and Massó, E. 1990a, *Phys. Lett. B*, 242, 77.

Grifols, J. A., and Massó, E. 1990b, *Nucl. Phys. B*, 331, 244.

Grifols, J. A., Massó, E., and Peris, S. 1988a, *Phys. Lett. B*, 207, 493.

Grifols, J. A., Massó, E., and Peris, S. 1988b, *Phys. Lett. B*, 215, 593.

Grifols, J. A., Massó, E., and Peris, S. 1989a, *Phys. Lett. B*, 220, 591.

Grifols, J. A., Massó, E., and Peris, S. 1989b, *Mod. Phys. Lett. A*, 4, 311.

Grifols, J. A., Massó, E., and Peris, S. 1994, *Astropart. Phys.*, 2, 161.

Grifols, J. A., Massó, E., and Rizzo, T. G. 1990, *Phys. Rev. D*, 42, 3293.

Grifols, J. A., and Tortosa, S. 1994, *Phys. Lett. B*, 328, 98.

Grimus, W., and Neufeld, H. 1993, *Phys. Lett. B*, 315, 129.

Guenther, D. B., et al. 1995, *Ap. J.*, 445, 148.

Guzzo, M. M., Bellandi, J., and Aquino, V. M. 1994, *Phys. Rev. D*, 49, 1404.

Guzzo, M. M., Masiero, A., and Petcov, S. T. 1991, *Phys. Lett. B*, 260, 154.

Guzzo, M. M., and Petcov, S. T. 1991, *Phys. Lett. B*, 271, 172.

Guzzo, M. M., and Pulido, J. 1993, *Phys. Lett. B*, 317, 125.

Gvozdev, A. A., Mikheev, N. V., and Vassilevskaya, L. A. 1992a, *Phys. Lett. B*, 289, 103.

Gvozdev, A. A., Mikheev, N V., and Vassilevskaya, L. A. 1992b, *Phys. Lett. B*, 292, 176.

Gvozdev, A. A., Mikheev, N. V., and Vassilevskaya, L. A. 1993, *Phys. Lett. B*, 313, 161.

Gvozdev, A. A., Mikheev, N. V., and Vassilevskaya, L. A. 1994a, *Phys. Lett. B*, 321, 108.

Gvozdev, A. A., Mikheev, N. V., and Vassilevskaya, L. A. 1994b, *Phys. Lett. B*, 323, 179.

Haensel, P., and Jerzak, A. J. 1987, *Astron. Astrophys.*, 179, 127.

Haft, M. 1993, Master's Thesis, University of Munich, unpublished.

Haft, M., Raffelt, G., and Weiss, A. 1994, *Ap. J.*, 425, 222; (E) 1995, *ibid.*, 438, 1017.

Hagmann, C., et al. 1990, *Phys. Rev. D*, 42, 1297.

Hagmann, C., and Sikivie, P. 1991, *Nucl. Phys. B*, 363, 247.

Hagner, C., et al. 1995, *Phys. Rev. D*, 52, 1343.

Halpern, J. P., and Ruderman, M. 1993, *Ap. J.*, 415, 286.

Halprin, A. 1975, *Phys. Rev. D*, 11, 147.

Halprin, A., and Leung, C. N. 1991, *Phys. Rev. Lett.*, 67, 1833.

Hansen, J.-P. 1973, *Phys. Rev. A*, 8, 3096.

Harari, D., and Sikivie, P. 1987, *Phys. Lett. B*, 195, 361.

Harari, D., and Sikivie, P. 1992, *Phys. Lett. B*, 289, 67.

Harrison, P. A., Lyne, A. G., and Anderson, B. 1993, *Mon. Not. R. astr. Soc.*, 261, 113.

Harrison, E. R., and Tademaru, E. 1975, *Ap. J.*, 201, 447.

Hata, N., Bludman, S., and Langacker, P. 1994, *Phys. Rev. D*, 49, 3622.

Hata, N., and Haxton, W. 1995, *Phys. Lett. B*, 353, 422.

Hata, N., and Langacker, P. 1994, *Phys. Rev. D*, 50, 632.

Hatsuda, T., Lim, C. S., and Yoshimura, M. 1988, *Phys. Lett. B*, 203, 462.

Haxton, W. C. 1987, *Phys. Rev. D*, 36, 2283.

Haxton, W. C. 1995, *Ann. Rev. Astron. Astrophys.*, 33, 459.

Haxton, W. C., and Zhang, W.-M. 1991, *Phys. Rev. D*, 43, 2484.

Heisenberg, W., and Euler, H. 1936, *Z. Phys.*, 98, 714.

Hellings, R. W., et al. 1983, *Phys. Rev. Lett.*, 51, 1609.

Henry, R. C., and Feldmann, P. D. 1981, *Phys. Rev. Lett.*, 47, 618.

Herant, M., Benz, W., and Colgate, S. 1992, *Ap. J.*, 395, 642.

Herant, M., et al. 1994, *Ap. J.*, 435, 339.

Hernanz, M., et al. 1994, *Ap. J.*, 434, 652.

Hill, C. T., Steinhardt, P. J., and Turner, M. S. 1990, *Phys. Lett. B*, 252, 343.

Hill, H. A., and Gu, Ye-ming 1990, *Science in China (Series A)*, 37:7, 854.

Hill, J. E. 1995, *Phys. Rev. Lett.*, 75, 2654.

Hirata, K. S. 1991, Ph. D. Thesis (Univ. of Tokyo); Report ICRR-239-91-8.

Hirata, K. S., et al. 1987, *Phys. Rev. Lett.*, 58, 1490.

Hirata, K. S., et al. 1988, *Phys. Rev. D*, 38, 448.

Hirata, K. S., et al. 1991a, *Phys. Rev. Lett.*, 66, 9.

Hirata, K. S., et al. 1991b, *Phys. Rev. D*, 44, 2241.

Hirata, K. S., et al. 1992, *Phys. Lett. B*, 280, 146.

Hoffmann, S. 1987, *Phys. Lett. B*, 193, 117.

Holdom, B. 1986, *Phys. Lett. B*, 166, 196.

Holman, R., et al. 1992, *Phys. Lett. B*, 132.

Holzschuh, E., et al. 1992, *Phys. Lett. B*, 287, 381.

Hoogeveen, F. 1990, *Phys. Lett. B*, 243, 455.

Hoogeveen, F., and Stuart, R. G. 1992, *Phys. Lett. B*, 286, 165.

Hoogeveen, F., and Ziegenhagen, T. 1991, *Nucl. Phys. B*, 358, 3.

Horowitz, C. J., and Serot, B. D. 1987, *Nucl. Phys. A*, 464, 613.

Horowitz, C. J., and Wehrberger, K. 1991a, *Phys. Rev. Lett.*, 66, 272.

Horowitz, C. J., and Wehrberger, K. 1991b, *Phys. Lett. B*, 266, 236.

Horvat, R. 1993, *Phys. Rev. D*, 48, 2345.

Hubbard, W. B. 1978, *Fund. Cosmic Phys.*, 3, 167.

Hulse, R. A., and Taylor, J. H. 1975, *Ap. J.*, 195, L51.

Iacopini, E., and Zavattini, E. 1979, *Phys. Lett. B*, 85, 151.

Iben Jr., I., and Laughlin, G. 1989, *Ap. J.*, 341, 312.

Iben Jr., I., and Renzini, A. 1981, eds., *Physical Processes in Red Giants* (Reidel, Dordrecht).

Iben Jr., I., and Renzini, A. 1984, *Phys. Rep.*, 105, 329.

Iben Jr., I., and Tutukov, A. V. 1984, *Ap. J.*, 282, 615.

Iglesias, C. A., and Rogers, F. J. 1991a, *Ap. J.*, 371, 408.

Iglesias, C. A., and Rogers, F. J. 1991b, *Ap. J.*, 371, L73.

Iglesias, C. A., Rogers, F. J., and Wilson, B. G. 1990, *Ap. J.*, 360, 221.

Ignatiev, A. Yu., and Joshi, G. C. 1994, *Mod. Phys. Lett. A*, 9, 1479.

Ignatiev, A. Yu., and Joshi, G. C. 1995, *Phys. Rev. D*, 51, 2411.

Iida, K., Minakata, H., and Yasuda, O. 1993, *Mod. Phys. Lett. A*, 8, 1037.

Isern, J., Hernanz, M., and García-Berro, E. 1992, *Ap. J.*, 392, L23.

Ishizuka, N., and Yoshimura, M. 1990, *Prog. Theor. Phys.*, 84, 233.

Itoh, N., and Hiraki, K. 1994, *Ap. J.*, 435, 784.

Itoh, N., and Kohyama, Y. 1983, *Ap. J.*, 275, 858.

Itoh, N., et al. 1984a, *Ap. J.*, 279, 413.

Itoh, N., et al. 1984b, *Ap. J.*, 280, 787; (E) 1987, *ibid.*, 322, 584.

Itoh, N., et al. 1989, *Ap. J.*, 339, 354; (E) 1990, *ibid.*, 360, 741.

Itoh, N., et al. 1992, *Ap. J.*, 395, 622; (E) 1993, *ibid.*, 404, 418.

Itzykson, C., and Zuber, J.-B. 1980, *Quantum Field Theory* (McGraw-Hill, New York).

Ivanov, M. A., and Shul'man, G. A. 1980, *Astron. Zh.*, 57, 537 (*Sov. Astron.*, 24, 311).

Ivanov, M. A., and Shul'man, G. A. 1981, *Astron. Zh.*, 58, 138 (*Sov. Astron.*, 25, 76).

Iwamoto, N. 1980, *Phys. Rev. Lett.*, 44, 1537.

Iwamoto, N. 1982, *Ann. Phys. (Leipzig)*, 141, 1.

Iwamoto, N. 1984, *Phys. Rev. Lett.*, 53, 1198.

Iwamoto, N., and Pethick, C. J. 1982, *Phys. Rev. D*, 25, 313.

Iwamoto, N., et al. 1995, *Phys. Rev. D*, 51, 348.

Jackiw, R., and Rebbi, C. 1976, *Phys. Rev. Lett.*, 37, 177.

Jackson, J. D. 1975, *Classical Electrodynamics*, 2nd ed. (John Wiley, New York).

Jancovici, B. 1962, *Nuovo Cim.*, 25, 428.

Janka, H.-T. 1991, *Astron. Astrophys.*, 244, 378.

Janka, H.-T. 1993, in: F. Giovannelli and G. Mannocchi (eds.), Proc. Vulcano Workshop 1992 *Frontier Objects in Astrophysics and Particle Physics*, Conf. Proc. Vol. 40 (Soc. Ital. Fis.).

Janka, H.-T. 1995a, *Astropart. Phys.*, 3, 377.

Janka, H.-T. 1995b, Report astro-ph/9505034, Proc. Ringberg Workshop *Dark Matter*.

Janka, H.-T., and Hillebrandt, W., 1989a, *Astron. Astrophys. Suppl.*, 78, 375.

Janka, H.-T., and Hillebrandt, W., 1989b, *Astron. Astrophys.*, 224, 49.

Janka, H.-T., and Müller, E. 1993a, in: R. McCray and W. Zhenru (eds.), Proc. of the IAU Coll. No. 145, Xian, China, May 24–29 (Cambridge University Press, Cambridge).

Janka, H.-T., and Müller, E. 1993b, in: Y. Suzuki and K. Nakamura (eds.), *Frontiers of Neutrino Astrophysics* (Universal Academy Press, Tokyo).

Janka, H.-T., and Müller, E. 1994, *Astron. Astrophys.*, 290, 496.

Janka, H.-T., and Müller, E. 1995a, *Phys. Rep.*, 256, 135.

Janka, H.-T., and Müller, E. 1995b, MPA-Report 863, *Astron. Astrophys.*, in press.

Janka, H.-T., et al. 1995, Report astro-ph/9507023, submitted to *Phys. Rev. Lett.*

Jeckelmann, B., Goudsmit, P. F. A., and Leisi, H. J. 1994, *Phys. Lett. B*, 335, 326.

Jodidio, A., et al. 1986, *Phys. Rev. D*, 34, 1967; (E) 1988, *ibid.*, 37, 237.

Johnson, C. W., et al. 1992, *Ap. J.*, 392, 320.

Joseph, C. L. 1984, *Nature*, 312, 254.

Jungman, G., Kamionkowski, M., and Griest, K. 1995, *Phys. Rep.*, in press.

Kainulainen, K., Maalampi, J., and Peltoniemi, J. T. 1991, *Nucl. Phys. B*, 358, 435.

Kamionkowski, M., and March-Russell, J. 1992, *Phys. Lett. B*, 282, 137.

Kandaswamy, J., Salomonson, P., and Schechter, J. 1978, *Phys. Rev. D*, 17, 3051.

Kapetanakis, D., Mayr, P., and Nilles, H. P. 1992, *Phys. Lett. B*, 282, 95.

Kaplan, D. B. 1985, *Nucl. Phys. B*, 260, 215.

KARMEN Collaboration 1995, *Phys. Lett. B*, 348, 19.

Kavanagh, R. W. 1960, *Nucl. Phys.*, 15, 411.

Kavanagh, R. W., et al. 1969, *Bull. Am. Phys. Soc.*, 14, 1209.

Kawakami, H., et al. 1991, *Phys. Lett. B*, 256, 105.

Kawano, L. H. 1992, *Phys. Lett. B*, 275, 487.

Kawasaki, M., Terasawa, N., and Sato, K. 1986, *Phys. Lett. B*, 178, 71.

Kawasaki, M., et al. 1994, *Nucl. Phys. B*, 419, 105.

Kayser, B. 1982, *Phys. Rev. D*, 26, 1662.

Keil, W. 1994, Master's Thesis, University of Munich (unpublished).

Keil, W., and Janka, H.-T. 1995, *Astron. Astrophys.*, 296, 145.

Keil, W., Janka, H.-T., and Raffelt, G. 1995, *Phys. Rev. D*, 51, 6635.

Kephart, T. W., and Weiler, T. J. 1987, *Phys. Rev. Lett.*, 58, 171.

Kepler, S. O., et al. 1991, *Ap. J.*, 378, L45.

Kernan, P. J., and Krauss, L. M. 1995, *Nucl. Phys. B*, 437, 243.

Kiełczewska, D. 1990, *Phys. Rev. D*, 41, 2967.

Kikuchi, H., and Ma, E. 1994, *Phys. Lett. B*, 335, 444.

Kikuchi, H., and Ma, E. 1995, *Phys. Rev. D*, 51, R296.

Kim, C. W., Kim, J., and Sze, W. K. 1988, *Phys. Rev. D*, 37, 1072.

Kim, J. E. 1976, *Phys. Rev. D*, 14, 3000.

Kim, J. E. 1976, 1979, *Phys. Rev. Lett.*, 43, 103.

Kim, J. E. 1976, 1987, *Phys. Rep.*, 150, 1.

Kippenhahn, R., and Weigert, A. 1990, *Stellar Structure and Evolution* (Springer, Berlin).

Kirsten, T. 1995, Neutrino 94, *Nucl. Phys. B (Proc. Suppl.)*, 38, 68.

Kirzhnits, D. A. 1987, *Usp. Fiz. Nauk*, 152, 399 (*Sov. Phys. Usp.*, 30, 575).

Kirzhnits, D. A., Losyakov, V. V., and Chechin, V. A. 1990, *Zh. Eksp. Teor. Fiz.*, 97, 1089 (*Sov. Phys. JETP*, 70, 609).

Klein, J. R., and Thorsett, S. E. 1990, *Phys. Lett. A*, 145, 79.

Klimov, V. V. 1981, *Sov. J. Nucl. Phys.*, 33, 934.

Klimov, V. V. 1982, *Zh. Eksp. Teor. Fiz.*, 82, 336 (*Sov. Phys. JETP*, 55, 199).

Knoll, J., and Voskresensky, D. N. 1995, *Phys. Lett. B*, 351, 43.

Koester, D., and Schönberner, D. 1986, *Astron. Astrophys.*, 154, 125.

Kolb, E. W., Stebbins, A. J., and Turner, M. S. 1987a, *Phys. Rev. D*, 35, 3598.

Kolb, E. W., Stebbins, A. J., and Turner, M. S. 1987b, *Phys. Rev. D*, 36, 3820.

Kolb, E. W., Tubbs, D. L., and Dicus, D. A. 1982, *Ap. J.*, 255, L57.

Kolb, E. W., and Turner, M. S. 1987, *Phys. Rev. D*, 36, 2895.

Kolb, E. W., and Turner, M. S. 1989, *Phys. Rev. Lett.*, 62, 509.

Kolb, E. W., and Turner, M. S. 1990, *The Early Universe* (Addison-Wesley, Reading, Mass.).

Kolb, E. W., et al. 1991, *Phys. Rev. Lett.*, 67, 533.

Kopysov, Yu. S., and Kuzmin, V. A. 1968, Can. J. Phys., 46, S488.

Kohyama, Y., Itoh, N., and Munakata, H. 1986, *Ap. J.*, 310, 815.

Kohyama, Y., et al. 1993, *Ap. J.*, 415, 267.

Konoplich, R. V., and Khlopov, M. Yu. 1988, *Yad. Fiz.*, 47, 891 (*Sov. J. Nucl. Phys.*, 47, 565).

Koshiba, M. 1992, *Phys. Rep.*, 220, 229.

Kostelecký, V. A., Pantaleone, J., and Samuel, S. 1993, *Phys. Lett. B*, 315, 46.

Kovetz, A., and Shaviv, G. 1994, *Ap. J.*, 426, 787.

Krakauer, D. A., et al. 1990, *Phys. Lett. B*, 252, 177.

Krakauer, D. A., et al. 1991, *Phys. Rev. D*, 44, R6.

Krastev, P. I. 1993, *Phys. Lett. B*, 303, 75.

Krastev, P. I., and Petcov, S. T. 1994, *Phys. Rev. Lett.*, 72, 1960.

Krastev, P. I., and Smirnov, A. Yu. 1994, *Phys. Lett. B*, 338, 282.

Krauss, L. M. 1984, *Phys. Rev. Lett.*, 53, 1976.

Krauss, L. M. 1991, *Phys. Lett. B*, 263, 441.

Krauss, L. M., Moody, J. E., and Wilczek, F. 1984, *Phys. Lett. B*, 144, 391.

Krauss, L. M., and Tremaine, S. 1988, *Phys. Rev. Lett.*, 60, 176.

Krauss, L. M., and Wilczek, F. 1986, *Phys. Lett. B*, 173, 189.

Krauss, L. M., et al. 1985, *Phys. Rev. Lett.*, 55, 1797.

Krauss, L. M., et al. 1992, *Nucl. Phys. B*, 380, 507.

Kuhn, J. R. 1988, in: J. Christensen-Dalsgaard and S. Frandsen (eds.), *Advances in Helio- and Asteroseismology* (Reidel, Dordrecht).

Kunihiro, T., et al. 1993, *Prog. Theor. Phys. Suppl.*, No. 112.

Kuo, T. K., and Pantaleone, J. 1988, *Phys. Rev. D*, 37, 298.

Kuo, T. K., and Pantaleone, J. 1989, *Rev. Mod. Phys.*, 61, 937.

Kuo, T. K., and Pantaleone, J. 1990, *Phys. Lett. B*, 246, 144.

Kuznetsov, A. V., and Mikheev, N. V. 1993, *Phys. Lett. B*, 299, 367.

Kwong, W., and Rosen, S. P. 1994, *Phys. Rev. Lett.*, 73, 369.

Kyuldjiev, A. V. 1984, *Nucl. Phys. B*, 243.

Lagage, P. O., et al. 1987, *Phys. Lett. B*, 193, 127.

Lam, W. P., and Ng, K.-W. 1991, *Phys. Rev. D*, 44, 3345.

Lam, W. P., and Ng, K.-W. 1992, *Phys. Lett. B*, 284, 331.

Lamb, D. Q., and van Horn, H. M. 1975, *Ap. J.*, 200 306.

Landau, L. D. 1946, *Sov. Phys. JETP*, 16, 574.

Landau, L. D., and Lifshitz, E. M. 1958, *Statistical Physics* (Addison-Wesley, Reading, Mass.).

Landau, L. D., and Pomeranchuk, I. 1953a, *Dokl. Akad. Nauk. SSSR*, 92, 535.

Landau, L. D., and Pomeranchuk, I. 1953b, *Dokl. Akad. Nauk. SSSR*, 92, 735.

Lande, K. 1995, Neutrino 94, *Nucl. Phys. B (Proc. Suppl.)*, 38, 47, and unpublished viewgraphs at the conference.

Langacker, P., Leveille, J. P., and Sheiman, J. 1983, *Phys. Rev. D*, 27, 1228.

Langacker, P., and Liu, J. 1992, *Phys. Rev. D*, 46, 4140.

Langanke, K., and Shoppa, T. D. 1994, *Phys. Rev. C*, 49, R1771; (E) 1995, *ibid.*, 51, 2844.

Langanke, K., and Shoppa, T. D. 1995, *Phys. Rev. C*, 52, 1709.

Langmuir, I. 1926, *Proc. Natl. Acad. Sci.*, 14, 627.

LaRue, G. S., Phillips, J. D., and Fairbank, W. M. 1981, *Phys. Rev. Lett.*, 46, 967.

Lattimer, J. M., and Cooperstein, J. 1988, *Phys. Rev. Lett.*, 61, 23.

Lattimer, J. M., and Yahil, A. 1989, *Ap. J.*, 340, 426.

Lattimer, J. M., et al. 1991, *Phys. Rev. Lett.*, 66, 2701.

Lattimer, J. M., et al. 1994, *Ap. J.*, 425, 802.

Lazarides, G., Panagiotakopoulos, C., and Shafi, Q. 1986, *Phys. Rev. Lett.*, 56, 432.

Lazarus, D. M., et al. 1992, *Phys. Rev. Lett.*, 69, 2333.

Learned, J. G., and Pakvasa, S. 1995, *Astropart. Phys.*, 3, 267.

Lee, T. D., and Yang, C. N. 1955, *Phys. Rev.*, 98, 1501.

Lee, Y.-W. 1990, *Ap. J.*, 363, 159.

Lee, Y.-W., Demarque, P., and Zinn, R. 1990, *Ap. J.*, 350, 155.

Lee, Y.-W., Demarque, P., and Zinn, R. 1994, *Ap. J.*, 423, 248.

Leener-Rosier, N. de, et al. 1986, *Phys. Lett. B*, 177, 228.

Leibundgut, B., et al. 1991, *Astron. Astrophys. Suppl.*, 89, 537.

Leinson, L. B. 1993, *Ap. J.*, 415, 759.

Lenard, A. 1953, *Phys. Rev.*, 90, 968.

Lesko, K. T., et al. 1993, *Rev. Mex. Fís.*, 39, Supl. 2, 162.

Levine, M. J. 1967, *Nuovo Cim.*, 48, 67.

Li, L. F., and Wilczek, F. 1982, *Phys. Rev. D*, 25, 143.

Liebert, J. 1980, *Ann. Rev. Astron. Astrophys.*, 18, 363.

Liebert, J., Dahn, C. C., and Monet, D. G. 1988, *Ap. J.*, 332, 891.

Lim, C.-S., and Marciano, W. J. 1988, *Phys. Rev. D*, 37, 1368.

Lindley, D. 1979, *Mon. Not. R. astr. Soc.*, 188, 15P.

Lindley, D. 1985, *Ap. J.*, 294, 1.

Lipunov, V. M. 1992, *Astrophysics of Neutron Stars* (Springer, Berlin).

Liu, J. 1991, *Phys. Rev. D*, 44, 2879.

Longo, M. J. 1987, *Phys. Rev. D*, 36, 3276.

Loredo, T. J., and Lamb, D. Q. 1989, in: E. J. Fenyves (ed.), Fourteenth Texas Symposium on Relativistic Astrophysics, *Ann. N.Y. Acad. Sci.*, 571, 601.

Loredo, T. J., and Lamb, D. Q. 1995, Report, submitted to *Phys. Rev. D*.

Loreti, F. N., et al. 1995, Report astro-ph/9508106, submitted to *Phys. Rev. D*.

LoSecco, J. M. 1988, *Phys. Rev. D*, 38, 3313.

LoSecco, J. M. 1989, *Phys. Rev. D*, 39, 1013.

LoSecco, J. M., et al. 1989, *Phys. Lett. A*, 138, 5.

Loskutov, Yu. M. 1984a, *Pis'ma Zh. Eksp. Teor. Fiz.*, 39, 438 (*JETP Lett.*, 39, 531).

Loskutov, Yu. M. 1984b, *Dokl. Akad. Nauk. SSSR*, 275, 1396 (*Sov. Phys. Dokl.*, 29, 322).

Lucio, J. L., Rosado, A., and Zepeda, A. 1985, *Phys. Rev. D*, 31, 1091.

Lyne, A. G., and Lorimer, D. R. 1994, *Nature*, 369, 127.

Lynn, B. W. 1981, *Phys. Rev. D*, 23, 2151.

Maalampi, J., and Peltoniemi, J. T. 1991, *Phys. Lett. B*, 269, 357.

Madsen, J., and Haensel, P. 1992, *Strange Quark Matter in Physics and Astrophysics*, *Nucl. Phys. B (Proc. Suppl.)*, 24B.

Maiani, L., Petronzio, R., and Zavattini, E. 1986, *Phys. Lett. B*, 175, 359.

Maki, Z., Nakagawa, M., and Sakata, S. 1962, *Prog. Theor. Phys.*, 28, 870.

Malaney, R. A., Starkman, G. D., and Butler, M. N. 1994, *Phys. Rev. D*, 49, 6232.

Malaney, R. A., Starkman, G. D., and Tremaine, S. 1995, *Phys. Rev. D*, 51, 324.

Mann, A. K., and Zhang, W. 1990, *Comm. Nucl. Part. Phys.*, 19, 295.

Manohar, A. 1987, *Phys. Lett. B*, 192, 217.

Marciano, W. J., and Sanda, A. I. 1977, *Phys. Lett. B*, 67, 303.

Marinelli, M., and Morpurgo, G. 1984, *Phys. Lett. B*, 137, 439.

Maruno, M., Takasugi, E., and Tanaka, M. 1991, *Prog. Theor. Phys.*, 86, 907.

Matese, J. J., and O'Connell, R. F. 1969, *Phys. Rev.*, 180, 1289.

Matsuki, S., et al. 1995, in: Proc. XVth Moriond Workshop *Dark Matter in Cosmology, Clocks, and Tests of Fundamental Laws*, Villars-sur-Ollon, Switzerland, January 21–28, 1995.

Maurette, M. 1976, *Ann. Rev. Nucl. Sci.*, 26, 319.

Maxwell, O., et al. 1977, *Ap. J.*, 216, 77.

Mayle, R. W., and Wilson, J. R. 1993, *Phys. Rep.*, 227, 97.

Mayle, R. W., Wilson, J. R., and Schramm, D. N. 1987, *Ap. J.*, 318, 288.

Mayle, R., et al. 1988, *Phys. Lett. B*, 203, 188.

Mayle, R., et al. 1989, *Phys. Lett. B*, 219, 515.

Mayle, R., et al. 1993, *Phys. Lett. B*, 317, 119.

Mazurek, T. J. 1974, *Nature*, 252, 287.

Mazurek, T. J. 1975, *Astrophys. Space Sci.*, 35, 117.

Mazurek, T. J. 1976, *Ap. J.*, 207, L87.

Mazzitelli, I. 1989, *Ap. J.*, 340, 249.

Mazzitelli, I., and D'Antona, F. 1986, *Ap. J.*, 308, 706.

McCray, R. 1993, *Ann. Rev. Astron. Astrophys.*, 31, 175.

McNaught, R. H. 1987, IAU Circular No. 4316.

Melott, A. L., McKay, D. W., and Ralston, J. P. 1988, *Ap. J.*, 324, L43.

Melott, A. L., and Sciama, D. W. 1981, *Phys. Rev. Lett.*, 46, 1369.

Melott, A. L., et al. 1994, *Ap. J.*, 421, 16.

Mestel, L. 1952, *Mon. Not. R. astr. Soc.*, 112, 583.

Mészáros, P. 1992, *High-Energy Radiation from Magnetized Neutron Stars* (University of Chicago Press, Chicago).

Meyer, B. S. 1994, *Ann. Rev. Astron. Astrophys.*, 32, 153.

Meyer, B. S. 1995, *Ap. J.*, 449, L55.

Meyer, B. S., et al. 1992, *Ap. J.*, 399, 656.

Midorikawa, S., Terazawa, H., and Akama, K. 1987, *Mod. Phys. Lett. A*, 2, 561.

Migdal, A. B. 1956, *Phys. Rev.*, 103, 1811.

Migdal, A. B., et al. 1990, *Phys. Rep.*, 192, 179.

Mikaelian, K. O. 1978, *Phys. Rev. D*, 18, 3605.

Mikheyev, S. P., and Smirnov, A. Yu. 1985, *Yad. Fiz.*, 42, 1441 (*Sov. J. Nucl. Phys.*, 42, 913).

Mikheyev, S. P., and Smirnov, A. Yu. 1986, *Zh. Eksp. Teor. Fiz.*, 91, 7 (*Sov. Phys. JETP*, 64, 4).

Miller, D. S., Wilson, J. R., and Mayle, R. W. 1993, *Ap. J.*, 415, 278.

Minakata, H., and Nunokawa, H. 1988, *Phys. Rev. D*, 38, 3605.

Minakata, H., and Nunokawa, H. 1995, *Phys. Rev. D*, 51, 6625.

Minakata, H., et al. 1987, *Mod. Phys. Lett. A*, 2, 827.

Moe, M. K. 1995, Neutrino 94, *Nucl. Phys. B (Proc. Suppl.)*, 38, 36.

Moffat, J. W. 1991, in: R. B. Mann and P. Wesson (eds.), *Proc. Banff Summer Institute on Gravitation, Banff, Alberta, 1990* (World Scientific, Singapore).

Mohanty, S., and Nayak, S. N. 1993, *Phys. Rev. Lett.*, 70, 4038; (E) *ibid.*, 71, 1117.

Mohanty, S., and Panda, P. K. 1994, Report hep-ph/9403205 (revised version of 29 Oct. 1994).

Mohapatra, R. N., and Nussinov, S. 1989, *Phys. Rev. D*, 39, 1378.

Mohapatra, R. N., and Nussinov, S. 1992, *Int. J. Mod. Phys. A*, 7, 3817.

Mohapatra, R. N., Nussinov, S., and Zhang, X. 1994, *Phys. Rev. D*, 49, 3434.

Mohapatra, R. N., and Pal, P. 1991, *Massive Neutrinos in Physics and Astrophysics* (World Scientific, Singapore).

Mohapatra, R. N., and Rothstein, I. Z. 1990, *Phys. Lett. B*, 593.

Morgan, J. 1981a, *Phys. Lett. B*, 102, 247.

Morgan, J. 1981b, *Mon. Not. R. astr. Soc.*, 195, 173.

Mori, M., et al. 1992, *Phys. Lett. B*, 289, 463.

Morris, D. E. 1986, *Phys. Rev. D*, 34, 843.

Motobayashi, T., et al. 1994, *Phys. Rev. Lett.*, 73, 2680.

Mourão, A. M., Bento, L., and Kerimov, B. K. 1990, *Phys. Lett. B*, 237, 469.

Mukhopadhyaya, B., and Gandhi, R. 1992, *Phys. Rev. D*, 3682.

Müller, J., et al. 1991, *Ap. J.*, 382, L101.

Munakata, H., Kohyama, Y., and Itoh, N. 1985, *Ap. J.*, 296, 197; (E) 1986, *ibid.*, 304, 580.

Munakata, H., Kohyama, Y., and Itoh, N. 1987, *Ap. J.*, 316, 708.

Murdin, P. 1990, *End in Fire—The Supernova in the Large Magellanic Cloud* (Cambridge University Press, Cambridge).

Musolf, M. J., and Holstein, B. R. 1991, *Phys. Rev. D*, 43, 2956.

Muto, T., and Tatsumi, T. 1988, *Prog. Theor. Phys.*, 80, 28.

Muto, T., Tatsumi, T., and Iwamoto, N. 1994, *Phys. Rev. D*, 50, 6089.

Myra, E. S., and Bludman, S. A. 1989, *Ap. J.*, 340, 384.

Myra, E. S., and Burrows, A. 1990, *Ap. J.*, 364, 222.

Nahmias, M. E. 1935, *Proc. Camb. Phil. Soc.*, 31, 99.

Nakagawa, M., Kohyama, Y., and Itoh, N. 1987, *Ap. J.*, 322, 291.

Nakagawa, M., et al. 1988, *Ap. J.*, 326, 241.

Natale, A. A. 1991, *Phys. Lett. B*, 258, 227.

Natale, A. A., Pleitez, V., and Tacla, A. 1987, *Phys. Rev. D*, 36, 3278.

Nieves, J. F. 1982, *Phys. Rev. D*, 26, 3152.

Nieves, J. F. 1983, *Phys. Rev. D*, 28, 1664.

Nieves, J. F. 1987, Report, Univ. Puerto Rico (unpublished).

Nieves, J. F. 1989, *Phys. Rev. D*, 40, 866.

Nieves, J. F., and Pal, P. B. 1989a, *Phys. Rev. D*, 39, 652; (E) *ibid.*, 40, 2148.

Nieves, J. F., and Pal, P. B. 1989b, *Phys. Rev. D*, 40, 1350.

Nieves, J. F., and Pal, P. B. 1989c, *Phys. Rev. D*, 40, 1693.

Nieves, J. F., and Pal, P. B. 1994, *Phys. Rev. D*, 49, 1398.

Nieves, J. F., Pal, P., and Unger, D. G. 1983, *Phys. Rev. D*, 28, 908.

Nilles, H. P., and Raby, S. 1982, *Nucl. Phys. B*, 198, 102.

Nomoto, K., and Tsuruta, S. 1986, *Ap. J.*, 305, L19.

Nomoto, K., and Tsuruta, S. 1987, *Ap. J.*, 312, 711.

Nötzold, D. 1987, *Phys. Lett. B*, 196, 315.

Nötzold, D. 1988, *Phys. Rev. D*, 38, 1658.

Nötzold, D., and Raffelt, G. 1988, *Nucl. Phys. B*, 307, 924.

Nunokawa, H., and Minakata, H. 1993, *Phys. Lett. B*, 314, 371.

Nussinov, S., and Rephaeli, Y. 1987, *Phys. Rev. D*, 36, 2278.

Oakley, D. S., et al. 1994, *Ap. J.*, 437, L63.

Oberauer, L. 1992 *Nucl. Phys. B (Proc. Suppl.)*, 28A, 165.

Oberauer, L., von Feilitzsch, F., and Mössbauer, R. L. 1987, *Phys. Lett. B*, 198, 113.

Oberauer, L., et al. 1993, *Astropart. Phys.*, 1, 377.

O'Connell, R. F., and Matese, J. J. 1969a, *Phys. Lett. A*, 29, 533.

O'Connell, R. F., and Matese, J. J. 1969b, *Nature*, 222, 649.

Ögelman, H., and Finley, J. 1993, *Ap. J.*, 413, L31.

Ögelman, H., Finley, J., and Zimmermann, H. 1993, *Nature*, 361, 136.

Okun, L. B. 1969, *Yad. Fiz.*, 10, 358 (*Sov. J. Nucl. Phys.*, 10, 206).

Okun, L. B. 1986, *Yad. Fiz.*, 44, 847 (*Sov. J. Nucl. Phys.*, 44, 546).

Okun, L. B. 1988, *Yad. Fiz.*, 48, 1519 (*Sov. J. Nucl. Phys.*, 48, 967).

Olive, K. A., Schramm, D., and Steigman, G. 1981, *Nucl. Phys. B*, 180, 497.

Olive, K. A., et al. 1990, *Phys. Lett. B*, 236, 454.

Oraevskiĭ, V. N., and Semikoz, V. B. 1984, *Zh. Eksp. Teor. Fiz.*, 86, 796 (*Sov. Phys. JETP*, 59, 465).

Oraevskiĭ, V. N., and Semikoz, V. B. 1985, *Yad. Fiz.*, 42, 702 (*Sov. J. Nucl. Phys.*, 42, 446).

Oraevskiĭ, V. N., and Semikoz, V. B. 1987, *Physica*, 142A, 135.

Oraevskiĭ, V. N., and Semikoz, V. B. 1991, *Phys. Lett. B*, 263, 455.

Oraevskiĭ, V. N., and Semikoz, V. B., and Smorodinskiĭ, Ya. A. 1986, *Pis'ma Zh. Eksp. Teor. Fiz.*, 43, 549 (*JETP Lett.*, 43, 709).

Otten, E. W. 1995, Neutrino 94, *Nucl. Phys. B (Proc. Suppl.)*, 38, 26.

Overduin, J. M., and Wesson, P. S. 1993, *Ap. J.*, 414, 449.

Overduin, J. M., Wesson, P. S., and Bowyer, S. 1993, *Ap. J.*, 404, 460.

Page, D., and Applegate, J. H. 1992, *Ap. J.*, 394, L17.

Pakvasa, S., Simmons, W. A., and Weiler, T. J. 1989, *Phys. Rev. D*, 39, 1761.

Pal, P. B., and Pham, T. N. 1989, *Phys. Rev. D*, 40, 259.

Pal, P. B., and Wolfenstein, L. 1982, *Phys. Rev. D*, 25, 766.

Pantaleone, J. 1991, *Phys. Lett. B*, 268, 227.

Pantaleone, J. 1992a, *Phys. Rev. D*, 46, 510.

Pantaleone, J. 1992b, *Phys. Lett. B*, 287, 128.

Pantaleone, J. 1995, *Phys. Lett. B*, 342, 250.

Pantaleone, J., Halprin, A., and Leung, C. N. 1993, *Phys. Rev. D*, 47, R4199.

Pantziris, A., and Kang, K. 1986, *Phys. Rev. D*, 33, 3509.

Papini, G., and Valluri, S. R. 1977, *Phys. Rep.*, 33, 51.

Parfenov, K. V. 1989a, *Yad. Fiz.*, 48, 1023 (*Sov. J. Nucl. Phys.*, 48, 651).

Parfenov, K. V. 1989b, *Yad. Fiz.*, 49, 1820 (*Sov. J. Nucl. Phys.*, 49, 1127).

Parke, S. 1995, *Phys. Rev. Lett.*, 74, 839.

Parker, P. D. 1966, *Phys. Rev.*, 150, 851.

Parker, P. D. 1968, *Ap. J.*, 153, L85.

Particle Data Group 1994, *Phys. Rev. D*, 50, 1173.

Paschos, E. A., and Zioutas, K. 1994, *Phys. Lett. B*, 323, 367.

Peccei, R. D. 1981, in: M. Konuma and T. Maskawa (eds.), *Proc. Fourth Kyoto Summer Institute on Grand Unified Theories and Related Topics* (World Science, Singapore).

Peccei, R. D. 1989, in: C. Jarlskog (ed.), *CP Violation* (World Scientific, Singapore).

Peccei, R. D., and Quinn, H. R. 1977a, *Phys. Rev. Lett.*, 38, 1440.

Peccei, R. D., and Quinn, H. R. 1977b, *Phys. Rev. D*, 16, 1791.

Peccei, R. D., Wu, T. T., and Yanagida, T. 1986, *Phys. Lett. B*, 172, 435.

Peebles, P. J. E. 1993, *Principles of Physical Cosmology* (Princeton University Press, Princeton, N.J.).

Peierls, Sir R. 1976, *Proc. R. Soc. London*, A 347, 475.

Pérez, A., and Gandhi, R. 1990, *Phys. Rev. D*, 41, 2374.

Peterson, L. E., et al. 1966, *J. Geophys. Res.*, 71, 5778.

Pethick, C. J., and Thorsson, V. 1994, *Phys. Rev. Lett.*, 72, 1964.

Petschek, A. G. 1990, ed., *Supernovae* (Springer, New York).

Pinaev, V. S. 1963, *Zh. Eksp. Teor. Fiz.*, 45, 548 (1964, *Sov. Phys. JETP*, 18, 377).

Pines, D., Tamagaki, R., and Tsuruta, S. 1992, *The Structure and Evolution of Neutron Stars*, Conf. Proc. (Addison-Wesley, Redwood City).

Pines, D., and Nozières, P. 1966, *The Theory of Quantum Liquids Vol. I: Normal Fermi Liquids* (Benjamin, New York).

Pisarski, R. D. 1989, *Nucl. Phys. A*, 498, 423.

Pochoda, P., and Schwarzschild, M. 1964, *Ap. J.*, 139, 587.

Pogosyan, D., and Starobinsky, A. 1995, *Ap. J.*, 447, 465.

Poincaré, H. 1892, *Théorie Mathématique de la Lumière*, Vol. 2 (Georges Carré, Paris).

Pontecorvo, B. 1948, Chalk River Laboratory Report PD-205 (quoted after Davis, Mann, and Wolfenstein 1989).

Pontecorvo, B. 1957, *Zh. Eksp. Teor. Fiz.*, 33, 549 (1958, *Sov. Phys. JETP*, 6, 429).

Pontecorvo, B. 1958, *Zh. Eksp. Teor. Fiz.*, 34, 247 (1958, *Sov. Phys. JETP*, 7, 172).

Pontecorvo, B. 1959, *Zh. Eksp. Teor. Fiz.*, 36, 1615 (1961, *Sov. Phys. JETP*, 9, 1148).

Pontecorvo, B. 1967, *Zh. Eksp. Teor. Fiz.*, 53, 1717 (1968, *Sov. Phys. JETP*, 26, 984).

Pontecorvo, B. 1983, *Usp. Fiz. Nauk*, 141, 675 (1983, *Sov. Phys. Usp.*, 26, 1087).

Prather, M. J. 1976, *The Effect of a Brans-Dicke Cosmology upon Stellar Evolution and the Evolution of Galaxies* (Ph.D. thesis, Yale University).

Preskill, J., Wise, M., and Wilczek, F. 1983, *Phys. Lett. B*, 120, 127.

Press, W. H., et al. 1986, *Numerical Recipes—The Art of Scientific Computing* (Cambridge University Press, Cambridge).

Primakoff, H. 1951, *Phys. Rev.*, 81, 899.

Proffitt, C. R. 1994, *Ap. J.*, 425, 849.

Proffitt, C. R., and Michaud, G. 1991, *Ap. J.*, 371, 584.

Pulido, J. 1992, *Phys. Rep.*, 211, 167.

Pulido, J. 1993, *Phys. Rev. D*, 48, 1492.

Pulido, J. 1994, *Phys. Lett. B*, 323, 36.

Qian, Y.-Z., and Fuller, G. M. 1995, *Phys. Rev. D*, 51, 1479.

Qian, Y.-Z., et al. 1993, *Phys. Rev. Lett.*, 71, 1965.

Raffelt, G. 1985, *Phys. Rev. D*, 31, 3002.

Raffelt, G. 1986a, *Phys. Rev. D*, 33, 897.

Raffelt, G. 1986b, *Phys. Lett. B*, 166, 402.

Raffelt, G. 1988a, *Phys. Rev. D*, 37, 1356.

Raffelt, G. 1988b, *Phys. Rev. D*, 38, 3811.

Raffelt, G. 1989, *Phys. Rev. D*, 39, 3378.

Raffelt, G. 1990a, *Phys. Rev. D*, 41, 1324.

Raffelt, G. 1990b, *Ap. J.*, 365, 559.

Raffelt, G. 1990c, *Mod. Phys. Lett. A*, 5, 2581.

Raffelt, G. 1990d, *Phys. Rep.*, 198, 1.

Raffelt, G. 1994, *Phys. Rev. D*, 50, 7729.

Raffelt, G., and Dearborn, D. 1987, *Phys. Rev. D*, 36, 2211.

Raffelt, G., and Dearborn, D. 1988, *Phys. Rev. D*, 37, 549.

Raffelt, G., Dearborn, D., and Silk, J. 1989, *Ap. J.*, 336, 64.

Raffelt, G., and Seckel, D. 1988, *Phys. Rev. Lett.*, 60, 1793.

Raffelt, G., and Seckel, D. 1991, *Phys. Rev. Lett.*, 67, 2605.

Raffelt, G., and Seckel, D. 1995, *Phys. Rev. D*, 52, 1780.

Raffelt, G., and Sigl, G. 1993, *Astropart. Phys.*, 1, 165.

Raffelt, G., Sigl, G., and Stodolsky, L. 1993, *Phys. Rev. Lett.*, 70, 2363.

Raffelt, G., and Starkman, G. 1989, *Phys. Rev. D*, 40, 942.

Raffelt, G., and Stodolsky, L. 1982, *Phys. Lett. B*, 119, 323.

Raffelt, G., and Stodolsky, L. 1988, *Phys. Rev. D*, 37, 1237.

Raffelt, G., and Weiss, A. 1992, *Astron. Astrophys.*, 264, 536.

Raffelt, G., and Weiss, A. 1995, *Phys. Rev. D*, 51, 1495.

Raghavan, R. S. 1991, in: K. K. Phua and Y. Yamaguchi (eds.), *Proc. 25th Int. Conf. High-Energy Physics*, 2–8 August 1990, Singapore (South East Asia Theoret. Phys. Assoc. and Phys. Soc. of Japan).

Raghavan, R. S., He, X.-G., and Pakvasa, S. 1988, *Phys. Rev. D*, 38, 1317.

Raghavan, R. S., Pakvasa, S., and Brown, B. A. 1986, *Phys. Rev. Lett.*, 57, 1801.

Rajpoot, S. 1993, *Mod. Phys. Lett. A*, 8, 1179.

Raychaudhuri, P. 1971, *Astrophys. Space Sci.*, 13, 231.

Raychaudhuri, P. 1986, *Solar Phys.*, 106, 421.

Raychaudhuri, P. 1991, *Mod. Phys. Lett. A*, 6, 2003.

Reines, F., Gurr, H., and Sobel, H. 1976, *Phys. Rev. Lett.*, 37, 315.

Reines, F., Sobel, H., and Gurr, H. 1974, *Phys. Rev. Lett.*, 32, 180.

Renzini, A., and Fusi Pecci, F. 1988, *Ann. Rev. Astron. Astrophys.*, 26, 199.

Renzini, A., et al. 1992, *Ap. J.*, 400, 280.

Ressell, M. T. 1991, *Phys. Rev. D*, 44, 3001.

Ressell, M. T., and Turner, M. S. 1990, *Comm. Astrophys.*, 14, 323.

Rich, J. 1993, *Phys. Rev. D*, 48, 4318.

Rich, J., Lloyd Owen, D., and Spiro, M. 1987, *Phys. Rep.*, 151, 239.

Riisager, K., and Jensen, A. S. 1993, *Phys. Lett. B*, 301, 6.

Ritus, V. I. 1961, *Zh. Eksp. Teor. Fiz.*, 41, 1285 (1962, *Sov. Phys. JETP*, 14, 915).

Rizzo, T. G. 1991, *Phys. Rev. D*, 44, 202.

Robertson, R. G. H., et al. 1991, *Phys. Rev. Lett.*, 67, 957.

Roeder, R. C. 1967, *Ap. J.*, 149, 131.

Roeder, R. C., and Demarque, P. R. 1966, *Ap. J.*, 144, 1016.

Rosen, S. P. 1988, *Phys. Rev. D*, 37, 1682.

Ross, J. E., and Aller, L. H. 1976, *Science*, 191, 1223.

Rothman, T., and Matzner, R. 1982, *Ap. J.*, 257, 450.

Rothstein, I. Z., Babu, K. S., and Seckel, D. 1993, *Nucl. Phys. B*, 403, 725.

Rudzsky, M. A. 1990, *Astrophys. Space Sci.*, 165, 65.

Ruoso, G., et al. 1992, *Z. Phys. C*, 56, 505.

Ryan, J. J., Accetta, F., and Austin, R. H. 1985, *Phys. Rev. D*, 32, 802.

Sakuda, M. 1994, *Phys. Rev. Lett.*, 72, 804.

Sakurai, J. J. 1967, *Advanced Quantum Mechanics* (Addison-Wesley, Reading, Mass.).

Salati, P. 1994, *Astropart. Phys.*, 2, 269.

Salpeter, E. E. 1960, *Phys. Rev.*, 120, 1528.

Saltzberg, D. 1995, *Phys. Lett. B*, 355, 499.

Samuel, S. 1993, *Phys. Rev. D*, 48, 1462.

Sandage, A. 1986, *Ann. Rev. Astron. Astrophys.*, 24, 421.

Sandage, A. 1990a, *Ap. J.*, 350, 603.

Sandage, A. 1990b, *Ap. J.*, 350, 631.

Sandage, A., and Cacciari, C. 1990, *Ap. J.*, 350, 645.

Sarajedini, A., and Demarque, P. 1990, *Ap. J.*, 365, 219.

Sato, K. 1975, *Prog. Theor. Phys.*, 54, 1352.

Sato, K, and Sato, H. 1975, *Prog. Theor. Phys.*, 54, 1564.

Sato, K., Shimizu, T., and Yamada, S. 1993, in: Y. Suzuki and K. Nakamura (eds.), *Frontiers of Neutrino Astrophysics* (Universal Academy Press, Tokyo).

Sato, K., and Suzuki, H. 1987a, *Phys. Rev. Lett.*, 58, 2722.

Sato, K., and Suzuki, H. 1987b, *Phys. Lett. B*, 196, 267.

Sawyer, R. F. 1988, *Phys. Rev. Lett.*, 61, 2171.

Sawyer, R. F. 1989, *Phys. Rev. C*, 40, 865.

Sawyer, R. F. 1995, *Phys. Rev. Lett.*, 75, 2260.

Schäfer, G., and Dehnen, H. 1983, *Phys. Rev. D*, 27, 2864.

Schechter, J., and Valle, J. W. F. 1981, *Phys. Rev. D*, 24, 1883.

Scherrer, R. J., and Spergel, D. N. 1993, *Phys. Rev. D*, 47, 4774.

Schinder, P. J., et al. 1987, *Ap. J.*, 313, 531.

Schmidt, G. 1989, ed., *The Use of Pulsating Stars in Fundamental Problems of Astronomy* (Cambridge University Press, Cambridge).

Schneps, J. 1993, Neutrino 92, *Nucl. Phys. B (Proc. Suppl.)*, 31, 307.

Schneps, J. 1995, Neutrino 94, *Nucl. Phys. B (Proc. Suppl.)*, 38, 220.

Schönfelder, V., Graml, F., and Penningfeld, F.-P. 1980, *Ap. J.*, 240, 350.

Schramm, D. N. 1987, *Comm. Nucl. Part. Phys.*, 17, 239.

Schramm, D. N., and Truran, J. W. 1990, *Phys. Rep.*, 189, 89.

Schreckenbach, K., et al. 1985, *Phys. Lett. B*, 160, 325.

Schwarzschild, M. 1958, *Structure and Evolution of the Stars* (Princeton University Press, Princeton, N.J.).

Sciama, D. W. 1990a, *Phys. Rev. Lett.*, 65, 2839.

Sciama, D. W. 1990b, *Ap. J.*, 364, 549.

Sciama, D. W. 1993a, *Ap. J.*, 409, L25.

Sciama, D. W. 1993b, *Modern Cosmology and the Dark Matter Problem* (Cambridge University Press, Cambridge).

Sciama, D. W. 1995, *Ap. J.*, 448, 667.

Sciama, D. W., Persic, M., and Salucci, P. 1993, *Publ. Astr. Soc. Pacific*, 105, 102.

Seckel, D., Steigman, G., and Walker, T. 1991, *Nucl. Phys. B*, 366, 233.

Segretain, L., et al. 1994, *Ap. J.*, 434, 641.

Sehgal, L. M., and Weber, A. 1992, *Phys. Rev. D*, 46, 2252.

Semertzidis, Y., et al. 1990, *Phys. Rev. Lett.*, 64, 2988.

Semikoz, V. B. 1987a, *Yad. Fiz.*, 46, 1592 (*Sov. J. Nucl. Phys.*, 46, 946).

Semikoz, V. B. 1987b, *Physica*, 142A, 157.

Semikoz, V. B. 1992, *Phys. Lett. B*, 284, 337.

Semikoz, V. B., and Smorodinskiĭ, Ya. A. 1988, *Pis'ma Zh. Eksp. Teor. Fiz.*, 48, 361 (*JETP Lett.*, 48, 399).

Semikoz, V. B., and Smorodinskiĭ, Ya. A. 1989, *Zh. Eksp. Teor. Fiz.*, 95, 35 (*Sov. Phys. JETP*, 68, 20).

Senatorov, A. V., and Voskresensky, D. N. 1987, *Phys. Lett. B*, 184, 119.

Shapiro, I. I. 1990, in: N. Ashby, D. F. Bartlett and W. Wyss (eds.), *General Relativity and Gravitation* (Cambridge University Press, Cambridge).

Shapiro, S. L., and Teukolsky, S. A. 1983, *Black Holes, White Dwarfs, and Neutron Stars* (John Wiley, New York).

Shapiro, S. L., and Wasserman, I. 1981, *Nature*, 289, 657.

Shaviv, G., and Bahcall, J. N. 1969, *Ap. J.*, 155, 135.

Shaviv, G., and Kovetz, A. 1976, *Astron. Astrophys.*, 51, 383.

Shelton, I. 1987, IAU Circular No. 4316.

Shi, X., and Schramm, D. N. 1994, *Particle World*, 3, 109.

Shi, X., Schramm, D. N., and Dearborn, D. S. P. 1994, *Phys. Rev. D*, 50, 2414.

Shi, X., and Sigl, G. 1994, *Phys. Lett. B*, 323, 360.

Shi, X., et al. 1993, *Comm. Nucl. Part. Phys.*, 21, 151.

Shifman, M. A., Vainshtein, A. I., and Zakharov, V. I. 1980, *Nucl. Phys. B*, 166, 493.

Shklovskii, I. S. 1970, *Astron. Zh.*, 46, 715.

Shlyakhter, A. I. 1976, *Nature*, 264, 340.

Shlyakhter, A. I. 1983, ATOMPKI Report A/1 (quoted after Scherrer and Spergel 1993).

Shrock, R. E. 1981, *Phys. Rev. D*, 24, 1275.

Shrock, R. E. 1982, *Nucl. Phys. B*, 206, 359.

Shu, F. 1982, *The Physical Universe—An Introduction to Astronomy* (University Science Books, Mill Valley, Calif.).

Sigl, G. 1995a, *Phys. Rev. D*, 51, 4035.

Sigl, G. 1995b, Fermilab-Pub-95/274-A, submitted to *Phys. Rev. Lett.*

Sigl, G., and Raffelt, G. 1993, *Nucl. Phys. B*, 406, 423.

Sigl, G., and Turner, M. S. 1995, *Phys. Rev. D*, 51, 1499.

Sikivie, P. 1983, *Phys. Rev. Lett.*, 51, 1415.

Sikivie, P. 1984, *Phys. Lett. B*, 137, 353.

Sikivie, P. 1985, *Phys. Rev. D*, 32, 2988; (E) 1987, *ibid.*, 36, 974.

Sikivie, P. 1987, in: E. Alvarez et al. (eds.), *Cosmology and Particle Physics* (World Scientific, Singapore).

Sikivie, P. 1988, *Phys. Rev. Lett.*, 61, 783.

Silin, V. P. 1960, *Zh. Eksp. Teor. Fiz.*, 38, 1577 (*Sov. Phys. JETP*, 11, 1136).

Simpson, J. J. 1991, *Phys. Lett. B*, 269, 454.

Sitenko, A. G. 1967, *Electromagnetic Fluctuations in Plasma* (Academic Press, New York).

Skibo, J. G., Ramaty, R., and Leventhal, M. 1992, *Ap. J.*, 397, 135.

Slad', L. M. 1983, *Pis'ma Zh. Eksp. Teor. Fiz.*, 37, 115 (*JETP Lett.*, 37, 143).

Slattery, W. L., Doolen, G. D., and DeWitt, H. E. 1980, *Phys. Rev. A*, 21, 2087.

Slattery, W. L., Doolen, G. D., and DeWitt, H. E. 1982, *Phys. Rev. A*, 26, 2255.

Smirnov, A. Yu. 1987, in: V. A. Kozyarivsky (ed.), *Proc. Twentieth International Cosmic Ray Conference* (Nauka, Moscow).

Smirnov, A. Yu. 1991, *Phys. Lett. B*, 260, 161.

Smirnov, A. Yu., Spergel, D. N., and Bahcall, J. N. 1994, *Phys. Rev. D*, 49, 1389.

Sneden, C., Pilachewski, C. A., and VandenBerg, D. A. 1986, *Ap. J.*, 311, 826.

Soares, J. M., and Wolfenstein, L. 1989, *Phys. Rev. D*, 40, 3666.

Sommerfeld, A. 1958, *Optik* (Akademische Verlagsgesellschaft, Leipzig).

Spergel, D. N., and Bahcall, J. N. 1988, *Phys. Lett. B*, 200, 366.

Spiro, M., and Vignaud, D. 1990, *Phys. Lett. B*, 242, 279.

Srednicki, M. 1985, *Nucl. Phys. B*, 260, 689.

Steigman, G., and Turner, M. S. 1985, *Nucl. Phys. B*, 253, 375.

Stewart, R. T., et al. 1993, *Mon. Not. R. astr. Soc.*, 261, 593.

Stix, M. 1989, *The Sun—An Introduction* (Springer, Berlin).

Stodolsky, L. 1987, *Phys. Rev. D*, 36, 2273.

Stodolsky, L. 1988, *Phys. Lett. B*, 201, 353.

Stoeffl, W., and Decman, D. J. 1994, submitted to *Phys. Rev. Lett.*

Stothers, R. 1970, *Phys. Rev. Lett.*, 24, 538.

Stothers, R. 1972, *Ap. J.*, 175, 717.

Strobl, K., and Weiler, T. J. 1994, *Phys. Rev. D*, 50, 7690.

Stückelberg, E. C. G. 1941, *Helv. Phys. Acta*, 14, 51.

Subramanian, A. 1979, *Current Science*, 48, 705.

Sudbury Neutrino Observatory Collaboration 1987, *Phys. Lett. B*, 194, 321.

Sur, B., and Boyd, R. N. 1985, *Phys. Rev. Lett.*, 54, 485.

Sutherland, P., et al. 1976, *Phys. Rev. D*, 13, 2700.

Suzuki, Y. 1993, in: *Proc. Int. Symposium on Neutrino Astrophysics* (Takayama Kamioka, Japan, Oct. 19–22, 1992).

Suzuki, Y. 1995, Neutrino 94, *Nucl. Phys. B (Proc. Suppl.)*, 38, 54.

Sweigart, A. V. 1994, *Ap. J.*, 426, 612.

Sweigart, A. V., Greggio, L., and Renzini, A. 1990, *Ap. J.*, 364, 527.

Sweigart, A. V., and Gross, P. G. 1976, *Ap. J. Suppl.*, 32, 367.

Sweigart, A. V., and Gross, P. G. 1978, *Ap. J. Suppl.*, 36, 405.

Sweigart, A. V., Renzini, A., and Tornambè, A. 1987, *Ap. J.*, 312, 762.

Takahara, M., and Sato, K. 1986, *Phys. Lett. B*, 174, 373.

Takahara, M., and Sato, K. 1987, *Mod. Phys. Lett. A*, 2, 293.

Takahashi, K., and Boyd, R. N. 1988, *Ap. J.*, 327, 1009.

Takahashi, K., Witti, J., and Janka, H.-Th. 1994, *Astron. Astrophys.*, 286, 857.

Takasugi, E., and Tanaka, M. 1992, *Prog. Theor. Phys.*, 87, 679.

Tammann, G. A., Löffler, W., and Schröder, A. 1994, *Ap. J. Suppl.*, 92, 487.

Taylor, J. H., and Weisberg, J. M. 1989, *Ap. J.*, 345, 434.

Teller, E. 1948, *Phys. Rev.*, 73, 801.

Thompson Jr., R. J., et al. 1991, *Ap. J.*, 366, L83.

't Hooft, G. 1971, *Phys. Lett. B*, 37, 195.

't Hooft, G. 1976a, *Phys. Rev. Lett.*, 37, 8.

't Hooft, G. 1976b, *Phys. Rev. D*, 14, 3432.

Thorsson, V. 1995, Report nucl-th-9502004.

Tinsley, B. M., and Gunn, J. E. 1976, *Ap. J.*, 206, 525.

Totani, T., and Sato, K. 1995, *Astropart. Phys.*, 3, 367.

Totsuka, Y. 1990, Report ICRR-Report-227-90-20, unpublished.

Totsuka, Y. 1993, Neutrino 92, *Nucl. Phys. B (Proc. Suppl.)*, 31, 428.

Toussaint, D., and Wilczek, F. 1981, *Nature*, 289, 777.

Tsai, W., and Erber, T. 1975, *Phys. Rev. D*, 12, 1132.

Tsai, W., and Erber, T. 1976, *Acta Phys. Austr.*, 45, 245.

Tsuruta, S. 1986, *Comm. Astrophys.*, 11, 151.

Tsuruta, S. 1992, in: Pines, Tamagaki, and Tsuruta 1992.

Tsuruta, S., and Nomoto, K. 1987, in: A. Hewitt et al. (eds.), *Obser-vational Cosmology* (IAU Sympsium No. 124).

Tsytovich, V. N. 1961, *Zh. Eksp. Teor. Fiz.*, 40, 1775 (*Sov. Phys. JETP*, 13, 1249).

Tsytovich, V. N., et al. 1995, *Phys. Lett. A*, 205, 199.

Turck-Chièze, S., and Lopes, I. 1993, *Ap. J.*, 408, 347.

Turck-Chièze, S., et al. 1993, *Phys. Rep.*, 230, 57.

Turner, M. S. 1986, *Phys. Rev. D*, 33, 889.

Turner, M. S. 1987, *Phys. Rev. Lett.*, 59, 2489.

Turner, M. S. 1988, *Phys. Rev. Lett.*, 60, 1797.

Turner, M. S. 1992, *Phys. Rev. D*, 45, 1066.

Turner, M. S., Kang, H.-S., and Steigman, G. 1989, *Phys. Rev. D*, 40, 299.

Umeda, H., Nomoto, K., and Tsuruta, S. 1994, *Ap. J.*, 431, 309.

Umeda, H., Tsuruta, S., and Nomoto, K. 1994, *Ap. J.*, 433, 256.

Ushida, N., et al. 1986, *Phys. Rev. Lett.*, 57, 2897.

Valle, J. W. F. 1987, *Phys. Lett. B*, 199, 432.

van Bibber, K., et al. 1987, *Phys. Rev. Lett.*, 59, 759.

van Bibber, K., et al. 1989, *Phys. Rev. D*, 39, 2089

van Bibber, K., et al. 1992, *Search for Pseudoscalar Cold Dark Matter*, Experimental Proposal, unpublished.

van Bibber, K., et al. 1994, *Status of the Large-Scale Dark-Matter Axion Search*, Report UCRL-JC-118357 (Lawrence Livermore National Laboratory).

van Buren, D., and Greenhouse, M. A. 1994, *Ap. J.*, 431, 640.

VandenBerg, D. A., Bolte, M., and Stetson, P. B. 1990, *Astron. J.*, 100, 445.

van den Bergh, S., and Tammann, G. A. 1991, *Ann. Rev. Astron. Astrophys.*, 29, 363.

van der Velde, J. C. 1989, *Phys. Rev. D*, 39, 1492.

van Horn, H. M. 1971, in: W. J. Luyten (ed.), *White Dwarfs*, IAU-Symposium No. 42 (Reidel, Dordrecht).

Vassiliadis, G., et al. 1995, *Proc. Int. Symp. Strangeness and Quark Matter*, Sept. 1–5, 1994, Crete, Greece (World Scientific, Singapore).

Vaughn, F. J., et al. 1970, *Phys. Rev. C*, 2, 1657.

Vidyakin, G. S., et al. 1987, *Zh. Eksp. Teor. Fiz.*, 93, 424 (*Sov. Phys. JETP*, 66, 243).

Vidyakin, G. S., et al. 1990, *Zh. Eksp. Teor. Fiz.*, 98, 764 (*Sov. Phys. JETP*, 71, 424).

Vidyakin, G. S., et al. 1991, *J. Moscow Phys. Soc.*, 1, 85.

Vidyakin, G. S., et al. 1992, *Pis'ma Zh. Eksp. Teor. Fiz.*, 55, 212 (*JETP Lett.*, 55, 206).

Vila, S. C. 1976, *Ap. J.*, 206, 213.

Vogel, P. 1984, *Phys. Rev. D*, 29, 1918.

Vogel, P. 1984, *Phys. Rev. D*, 30, 1505.

Vogel, P., and Engel, J. 1989, *Phys. Rev. D*, 39, 3378.

Voloshin, M. B. 1988, *Phys. Lett. B*, 209, 360.

Voloshin, M. B., and Vysotskiĭ, M. I. 1986, *Yad. Fiz.*, 44, 845 (*Sov. J. Nucl. Phys.*, 44, 544).

Voloshin, M. B., Vysotskiĭ, M. I., and Okun, L. B. 1986a, *Yad. Fiz.*, 44, 677 (*Sov. J. Nucl. Phys.*, 44, 440).

Voloshin, M. B., Vysotskiĭ, M. I., and Okun, L. B. 1986b, *Zh. Eksp. Teor. Fiz.*, 91, 754 (*Sov. Phys. JETP*, 64, 446); (E) 1987, *ibid.*, 92, 368 (*ibid.*, 65, 209).

von Feilitzsch, F., Hahn, A. A., and Schreckenbach, K. 1982, *Phys. Lett. B*, 118, 162.

von Feilitzsch, F., and Oberauer, L. 1988, *Phys. Lett. B*, 200, 580.

Vorobyov, P. V., and Kolokolov, I. V. 1995, Report astro-ph/9501042.

WA66 Collaboration 1985, *Phys. Lett. B*, 160, 207.

Walker, A. R. 1992, *Ap. J.*, 390, L81.

Walker, T. P., and Schramm, D. N. 1987, *Phys. Lett. B*, 195, 331.

Walker, T. P., et al. 1991, *Ap. J.*, 376, 51.

Wang, J. 1992, *Mod. Phys. Lett. A*, 7, 1497.

Weidemann, V. 1990, *Ann. Rev. Astron. Astrophys.*, 28, 103.

Weidemann, V., and Koester, D. 1984, *Astron. Astrophys.*, 132, 195.

Weinberg, S. 1975, *Phys. Rev. D*, 11, 3583.

Weinberg, S. 1978, *Phys. Rev. Lett.*, 40, 223.

Weinheimer, C., et al. 1993, *Phys. Lett. B*, 300, 210.

Weldon, H. A. 1982a, *Phys. Rev. D*, 26, 1394.

Weldon, H. A. 1982b, *Phys. Rev. D*, 26, 2789.

Weldon, H. A. 1989, *Phys. Rev. D*, 40, 2410.

Werntz, C. W. 1970, unpublished (quoted after Cisneros 1971).

Wheeler, J. C., Sneden, C., and Truran, J. W. 1989, *Ann. Rev. Astron. Astrophys.*, 252, 279.

White, M., Gelmini, G., and Silk, J. 1995, *Phys. Rev. D*, 51, 2669.

Wiezorek, C., et al. 1977, *Z. Phys. A*, 282, 121.

Wilczek, F. 1978, *Phys. Rev. Lett.*, 40, 279.

Will, C. M. 1993, *Theory and Experiment in Gravitational Physics—Revised Edition* (Cambridge University Press, Cambridge).

Wilson, J. R. 1983, in: J. Centrella, J. LeBlanc, and R. L. Bowers (eds.), *Numerical Astrophysics* (Jones and Bartlett, Boston).

Wilson, J. R., and Mayle, R. W. 1988, *Phys. Rep.*, 163, 63.

Winget, D. E., et al. 1987, *Ap. J.*, 315, L77.

Winget, D. E., Hansen, C. J., and van Horn, H. M. 1983, *Nature*, 303, 781.

Winter, K. 1991, ed., *Neutrino Physics* (Cambridge University Press).

Winter, K. 1995, Neutrino 94, *Nucl. Phys. B (Proc. Suppl.)*, 38, 211.

Wise, M. B., Georgi, H., and Glashow, S. L. 1981, *Phys. Rev. Lett.*, 47, 402.

Witti, J., Janka, H.-Th., and Takahashi, K. 1994, *Astron. Astrophys.*, 286, 841.

Wolfenstein, L. 1978, *Phys. Rev. D*, 17, 2369.

Wolfenstein, L. 1986, *Ann. Rev. Nucl. Part. Sci.*, 36, 137.

Wolfenstein, L. 1987, *Phys. Lett. B*, 194, 197.

Wood, M. A. 1992, *Ap. J.*, 386, 539.

Woosley, S. E., and Hoffmann, R. D. 1992, *Ap. J.*, 395, 202.

Woosley, S. E., and Weaver, T. A. 1986a, in: J. Audouze and N. Mathieu (eds.), *Nucleosynthesis and its Implications on Nuclear and Particle Physics* (Reidel, Dordrecht).

Woosley, S. E., and Weaver, T. A. 1986b, *Ann. Rev. Astron. Astrophys.*, 24, 205.

Woosley, S. E., et al. 1994, *Ap. J.*, 433, 229.

Wuensch, W. U., et al. 1989, *Phys. Rev. D*, 40, 3153.

Xu, H. M., et al. 1994, *Phys. Rev. Lett.*, 73, 2027.

Yanagida, T., and Yoshimura, M. 1988, *Phys. Lett. B*, 202, 301.

Yang, J., et al. 1979, *Ap. J.*, 227, 697.

Yang, J., et al. 1984, *Ap. J.*, 281, 493.

Yoshimura, M. 1988, *Phys. Rev. D*, 37, 2039.

Zacek, G., et al. 1986, *Phys. Rev. D*, 34, 2621.

Zaidi, M. H. 1965, *Nuovo Cim.*, 40, 502.

Zatsepin, G. I. 1968, *Pis'ma Zh. Eksp. Teor. Fiz.*, 8, 333 (*JETP Lett.*, 8, 205).

Zhang, W., et al. 1988, *Phys. Rev. Lett.*, 61, 385.

Zhitnitskiĭ, A. P. 1980, *Yad. Fiz.*, 31, 497 (*Sov. J. Nucl. Phys.*, 31, 260).

Zisman, G. A. 1971, Uchenye Zapiski LGPI No. 386, pg. 80 (quoted after Blinnikov et al. 1995).

Acronyms

The acronyms listed below do not include names of experiments such as GALLEX or IMB, and also do not include names of laboratories such as CERN or SLAC. These acronyms really play the role of proper names; usually nobody quite remembers what they stand for.

AGB	asymptotic giant branch
AU	astronomical unit (distance to the Sun)
BBN	big-bang nucleosynthesis
BC	bolometric correction
BP	Bahcall and Pinsonneault
BS	blue straggler
CC	charged current
CKM	Cabbibo-Kobayashi-Maskawa
CL	confidence level
CM	center of mass
CMB	cosmic microwave background
CMBR	cosmic microwave background radiation
CNO	carbon-nitrogen-oxygen
CO	carbon-oxygen
D	degenerate
DAV	DA Variable
DFSZ	Dine-Fischler-Srednicki-Zhitniskiĭ
EOS	equation of state
FTD	finite temperature and density
GIM	Glashow-Iliopoulos-Maiani
GUT	grand unified theory
HB	horizontal branch
IAU	International Astronomical Union
IMF	initial mass function
KII	Kamiokande II (detector)
pc	parsec (3.08×10^{18} cm)
KSVZ	Kim-Shifman-Vainshtein-Zakharov

L	longitudinal
l.h.	left-handed
l.h.s.	left-hand side (of an equation)
ly	light year
LMC	Large Magellanic Cloud
LSP	lightest supersymmetric particle
mfp	mean free path
MS	main sequence
MSW	Mikheyev-Smirnov-Wolfenstein
NC	neutral current
ND	nondegenerate
T	transverse
TL	Turck-Chièze and Lopes
TO	main-sequence turnoff
OPE	one pion exchange
PQ	Peccei-Quinn
QED	quantum electrodynamics
QCD	quantum chromodynamics
RG	red giant
RGB	red-giant branch
r.h.	right-handed
r.h.s.	right-hand side (of an equation)
SGB	sub-giant branch
SN	supernova
SNe	supernovae
SNU	solar neutrino unit
SNu	supernova unit
SNBO	Supernova Burst Observatory
UT	universal time
VVO	Voloshyn-Vysotskiĭ-Okun
WD	white dwarf

Symbols

Symbols which have only a local meaning in, say, one paragraph are not listed here. Four-momenta are usually denoted by uppercase italics such as K, three-momenta are boldface lowercase letters such as $\mathbf{k}$, the modulus of a three-momentum is in lowercase italics such as $k = |\mathbf{k}|$. Lorentz indices are usually given in Greek letters (μ, ν, α, β, etc.), three-indices and flavor indices in Latin letters (i, j, k, etc.).

Functions, Operators

Besides the usual functions and operators, the following convention may be noteworthy.

∂_x partial derivative $\partial/\partial x$

ln   natural logarithm

log   logarithm base 10

Latin Symbols

a axion, axion field, annihilation operator, acceleration, radiation constant ($a = \pi^2/15$ in natural units)

a_i ion-sphere radius

A electromagnetic vector potential, atomic mass number

B magnetic field, operator for "background" medium

B_ω specific energy density of a radiation field

c speed of light ($c = 1$ in natural units)

c_j speed of propagation of particle j

c_p heat capacity at constant pressure

$C_{V,A}$ vector and axial-vector weak-coupling constants

C_j effective Peccei-Quinn charge of fermion j

d electric dipole moment

d, D distance

$d\mathbf{p}$	differential in phase space, $d\mathbf{p} = d^3\mathbf{p}/(2\pi)^3$
e	electron, electric charge of electron, photon polarization four-vector
e_j	charge of particle j
E	electric field, energy of a particle
$E_{\rm F}$	Fermi energy
f	occupation numbers, dimensionless function, energy scale
$f_{\rm PQ}$	Peccei-Quinn scale
f_a	axion decay constant
f_π	pion decay constant (93 MeV)
$f_\mathbf{p}$	occupation number of mode $\mathbf{p}$
F	electromagnetic field-strength tensor, number flux of particles, dimensionless function of order unity
F_ω	specific energy flux in a radiation field
$\mathcal{F}$	fluence (time-integrated flux)
g	dimensionless Yukawa coupling, gluon, graviton, number of spin degrees of freedom
$g_{a\gamma}$	axion-photon coupling strength (units GeV^{-1})
g_{10}	$g_{a\gamma}/10^{-10}\mathrm{GeV}^{-1}$
G	color field-strength tensor, matrix of coupling constants, dimensionless function
$G_{\rm N}$	Newton's constant $(G_{\rm N} = m_{\rm Pl}^{-2})$
$G_{\rm F}$	Fermi's constant $(G_{\rm F} = 1.166{\times}10^{-5}\,\mathrm{GeV}^{-2})$
H	Hubble expansion parameter
H_0	Hubble expansion parameter today $(h \times 100\,\mathrm{km\,s^{-1}\,Mpc^{-1}})$
$\mathcal{H}$	Hamiltonian density
k	momentum of a particle, momentum transfer in a reaction
$k_{\rm B}$	Boltzmann's constant $(k_{\rm B} = 1$ in natural units)
$k_{\rm D}$	Debye screening scale of the electrons
$k_{\rm i}$	Debye screening scale of the ions
$k_{\rm S}$	screening scale
$k_{\rm TF}$	Thomas-Fermi screening scale
l	momentum transfer in a reaction
ℓ	mean free path, flavor index
L	luminosity of a star
$L_\odot$	solar luminosity $(3.85{\times}10^{33}\,\mathrm{erg})$
$\mathcal{L}$	Lagrangian density
m	mass
m_a	axion mass
m_e	electron mass (0.511 MeV)
m_π	pion mass (135 MeV)

m_N	nucleon mass (938 MeV)
m_u	atomic mass unit (931 MeV)
m_{Pl}	Planck mass (1.221×10^{19} GeV)
m_W	W-boson mass (80.2 GeV)
m_Z	Z-boson mass (91.2 GeV)
M	absolute magnitude of a star, mass matrix of quarks or neutrinos
M_{bol}	absolute bolometric magnitude of a star
$\mathcal{M}$	mass of a star, matrix element
$\mathcal{M}_\odot$	solar mass (1.99×10^{33} g)
n	particle density
n_B	baryon density
$n_{e,p,n,\nu}$	electron, proton, neutron, neutrino density
n_{refr}	index of refraction
N	nucleon, number of degenerate Θ vacua in QCD
$N_{\mu\nu}$	matrix element of neutrino current
p	proton, momentum of a particle, pressure, dimensionless power index
p_F	Fermi momentum
P	four-momentum of a particle, stellar pulsation period
q	momentum transfer, charge of a particle
Q	energy loss or generation rate per unit volume
r	radial coordinate
R	radius of a star
$R_\odot$	solar radius (6.96×10^{10} cm)
s	entropy density, dimensionless power index, square of CM energy, dimensionless spin-structure function
S	static or dynamical structure function
t	time
T	temperature
T_7	$T/10^7$ K
T_8	$T/10^8$ K
T_{30}	$T/30$ MeV
U	total internal energy of a star, four-velocity
v	velocity, vacuum expectation value of a Higgs-like field
V	potential energy, volume
w	up/strange quark mass ratio
W	W-boson
$W_{K,K'}$	Transition probability from K to K'.
x	dimensionless energy (ω/T), spatial coordinate, dimensionless Fermi momentum (p_F/m)
X	mass fraction of hydrogen

X_j	Peccei-Quinn charge of a fermion j
Y	mass fraction of helium
Y_j	number of fermions j per baryon
Y_e	mass fraction of helium in stellar envelope
Y_L	lepton number per baryon
z	up/down quark mass ratio, z-coordinate
Z	mass fraction of "metals," nuclear charge
$Z, Z°$	Z-boson

Greek Symbols

α	fine-structure constant ($e^2/4\pi = 1/137$), asymmetry parameter in angular distribution of decay photons
α'	fine-structure constant for general bosons, $\alpha' = g^2/4\pi$ with the Yukawa coupling g
$\alpha_{a,\chi}$	fine-structure constant for axions, majorons
α_s	strong fine-structure constant
α_π	pionic fine-structure constant, $\alpha_\pi = (f2m_N/m_\pi)^2/4\pi \approx 15$ with $f \approx 1.0$
β	velocity of a particle, parameter of axion models, parameter in nucleon-nucleon-axion bremsstrahlung rate
γ	photon, Dirac matrix, adiabatic index, dimensionless fluctuation rate $\gamma = \Gamma/T$
Γ	rate for decay or fluctuations, plasma coupling parameter
Γ_σ	spin-fluctuation rate in a nuclear medium
δ	δ function, Kronecker δ, differential quantity
ϵ	energy loss or generation rate per unit mass, electric dipole moment, polarization vector
ϵ_{ij}	neutrino transition electric moment
η	degeneracy parameter, $\eta = (\mu - m)/T$
θ	mixing angle, scattering angle
Θ	Θ parameter of QCD
Θ_W	weak mixing angle ($\sin^2 \Theta_W = 0.2325$)
κ	opacity, dimensionless screening or momentum scale
κ^*	reduced opacity
λ	wave length, mean free path
μ	magnetic moment, chemical potential, muon, Lorentz index, mean molecular weight
$\hat{\mu}$	nonrelativistic chemical potential, $\hat{\mu} = \mu - m$
μ_{ij}	neutrino transition magnetic moment

μ_B	Bohr magneton $(e/2m_e)$
μ_N	nuclear magneton $(e/2m_p)$
μ_e	electron mean molecular weight $(\mu_e \approx Y_e^{-1})$, electron chemical potential
ν	neutrino, Lorentz index, dimensionless power index
ξ	dimensionless correction factor, dimensionless axion-photon coupling constant
π	$\pi = 3.1415\ldots$, pion
ρ	mass density, density matrix
$\bar{\rho}$	density matrix for antineutrinos
ρ_0	nuclear density $(3 \times 10^{14}\,\mathrm{g\,cm^{-3}})$
ρ_{14}	$\rho/10^{14}\,\mathrm{g\,cm^{-3}}$
ρ_{15}	$\rho/10^{15}\,\mathrm{g\,cm^{-3}}$
σ	scattering cross section, Pauli matrix, spin operator
τ	lifetime of a particle, duration of a stellar evolution phase, Pauli matrix, optical depth
Φ	Higgs field
Φ_0	vacuum expectation value of Higgs field
χ	majoron, majoron field
ψ	fermion field
Ψ	column vector of fermion fields (several flavors)
ω	energy of a particle, energy transfer in reactions
ω_P	plasma frequency
ω_0	zero-temperature plasma frequency
Ω	cosmic density parameter, matrix of energies for mixed neutrinos

Subject Index

649

X, Y, Z